W9-AQE-302

THE NEW TREASURY OF SCIENCE

THE NEW
Treasury of Science

EDITED BY Harlow Shapley,
Samuel Rapport and
Helen Wright

HARPER & ROW, PUBLISHERS

NEW YORK

ACKNOWLEDGMENTS

The editors acknowledge with thanks permission to use copyrighted material, as follows:

THE WONDER OF THE WORLD from *Life: Outlines of General Biology* by Sir J. Arthur Thomson and Patrick Geddes. Reprinted by permission of Harper & Row, Publishers.

GAINS FROM SERENDIPITY from *The Way of an Investigator* by Walter B. Cannon, M.D. Copyright 1945 by W. W. Norton & Company, Inc. Reprinted by permission of the publishers.

TURTLE EGGS FOR AGASSIZ by Dallas Lore Sharp. Reprinted by permission of *The Atlantic Monthly.*

SCIENTISTS ARE LONELY MEN by Oliver La Farge. Reprinted by permission of *Harper's Magazine* and the author.

THE STUDY OF MATHEMATICS from *Mysticism and Logic* by Bertrand Russell. Copyright 1929 by W. W. Norton & Company, Inc. Reprinted by permission of George Allen & Unwin Ltd.

ÉVARISTE GALOIS from *The Life of Science* by George Sarton. Copyright 1948 by Abelard-Schuman Ltd. All rights reserved. Reprinted by permission of the publishers.

GIANT BRAINS by Edmund Callis Berkeley. Copyright 1949 by Edmund Callis Berkeley. Reprinted by permission of John Wiley & Sons, Inc.

MATHEMATICS AND MODERN CIVILIZATION from *Introduction to Mathematics* by Cooley, Gans, Kline and Wahlert. Copyright 1937, 1949 by Hollis R. Cooley, David Gans, Morris Kline and David Wahlert. Reprinted by permission of Houghton Mifflin Company.

OUR NEIGHROUR WORLDS from *This Universe of Space* by Peter Millman. Copyright © 1962 by Schenkman Publishing Company, Inc. Reprinted by permission of Schenkman Publishing Company, Inc, and Routledge & Kegan Paul Ltd.

THE SUN from *Stars in the Making* by Cecilia Payne-Gaposchkin. Copyright 1952 by the President and Fellows of Harvard College. Reprinted by permission of the Harvard University Press, Cambridge, Mass.

RADIO WAVES FROM SPACE from *The Exploration of Outer Space* by Sir Bernard Lovell. Copyright © 1962 by A. C. B. Lovell. Reprinted by permission of Harper & Row, Publishers, and Oxford University Press.

THE MILKY WAY AND BEYOND by Sir Arthur Eddington. Reprinted by permission of the author.

THE ORIGIN AND EVOLUTION OF THE UNIVERSE from *The Exploration of Outer Space* by Sir Bernard Lovell. Copyright © 1962 by A. C. B. Lovell. Reprinted by permission of Harper & Row, Publishers, and Oxford University Press.

ROBERT GODDARD from *The Coming Age of Rocket Power* by G. Edward Pendray. Copyright 1945, 1947 by Harper & Brothers. Reprinted by permission of Harper & Row, Publishers.

A SHOT AROUND THE WORLD from *The Making of a Moon* by Arthur C. Clarke. Copyright © 1957, 1958 by Arthur Charles Clarke. Reprinted by permission of Harper & Row, Publishers, and Frederick Muller Ltd.

THE VOYAGE OF MARINER II by J. N. James. Reprinted with permission. Copyright © 1963 by Scientific American, Inc. All rights reserved.

GEOLOGICAL CHANGE by Sir Archibald Geike, Presidential Address before British Association for the Advancement of Science, 1892. Reprinted by permission of the British Association for the Advancement of Science.

GEOLOGICAL TABLE from *Historical Geology* by Carl O. Dunbar. Copyright © 1960 by John Wiley & Sons, Inc. Reprinted by permission of the publishers.

THE RECORD OF LIVING THINGS from *The Story of a Billion Years* by W. O. Hotchkiss. Copyright 1932 by The Williams and Wilkins Company. Reprinted by permission of the publishers.

EARTHQUAKE: DISCOVERIES IN SEISMOLOGY by L. Don Leet and Florence Leet. Copyright © 1964 by L. Don Leet and Florence Leet. Reprinted by permission of Dell Publishing Company, Inc.

(Continued on next page)

LIBRARY OF CONGRESS CATALOG CARD NUMBER: 65–20439

I–P

THE OCEAN FLOOR and CATASTROPHIC WAVES FROM THE SEA from *The Earth Beneath the Sea* by Francis P. Shepard. Copyright © 1959 by The Johns Hopkins University Press. Reprinted by permission of the publishers.

PIN-POINTING THE PAST WITH THE COSMIC CLOCK by Richard Foster Flint. Copyright 1951 by The American Museum of Natural History. Reprinted by permission of *Natural History Magazine.*

THE WEATHER from *The Story of Weather* by David C. Holmes. Copyright © 1963 by Pyramid Publications Inc. Reprinted by permission of the publishers.

EXPERIMENTS AND IDEAS by Benjamin Franklin, from *The Ingenious Dr. Franklin* edited by Nathan Goodman. Reprinted by permission of The University of Pennsylvania Press.

THE DISCOVERY OF RADIUM from *Madame Curie: A Biography* by Eve Curie. Copyright 1937 by Doubleday, Doran and Company, Inc. Reprinted by permission of Doubleday, Doran and Company, Inc., and William Heinemann Ltd.

FROM X-RAYS TO NUCLEAR FISSION by Henry D. Smyth from *Science in Progress: Sixth Series* edited by George A. Baitsell. Copyright 1949 by Yale University Press. Reprinted with their permission.

THE DEATH OF LOUIS SLOTIN from *Atoms and People* by Ralph E. Lapp. Copyright © 1956 by Ralph E. Lapp. Reprinted by permission of Harper & Row, Publishers.

FUNDAMENTAL PARTICLES by O. R. Frisch, from *Science Survey* I, edited by A. W. Haslett and John St. John. Copyright © 1960 by Vista Books. Reprinted by permission of the publishers.

SPACE, TIME AND EINSTEIN by Paul R. Heyl. Reprinted by permission of the author and *Scientific Monthly.*

SOME EARLY PIONEERS IN CHEMISTRY from *Chemistry, Matter and Life* by Stephen Miall and L. M. Miall. Reprinted by permission of Edward Arnold.

THE RISE OF ORGANIC CHEMISTRY from *Through Alchemy to Chemistry* by John Read. Reprinted by permission of G. Bell & Sons Ltd.

WHERE IS CHEMISTRY TODAY? from *Understanding Chemistry* by Lawrence P. Lessing. Copyright © 1959 by Interscience Publishers, Inc. Reprinted by permission of John Wiley & Sons, Inc.

THE FITNESS OF EARTH from *Where There is Life* by Paul B. Sears. © 1962 by Paul B. Sears. Reprinted by permission of Dell Publishing Company, Inc.

THE NAMING OF ORGANISMS AND THE CATALOGUE OF NATURE from *The Nature of Natural History* by Marston Bates. Copyright 1950 by Marston Bates. Reprinted by permission of Charles Scribner's Sons.

THE ORIGIN OF LIFE by George Wald. Reprinted with permission. Copyright 1954 by Scientific American, Inc. All rights reserved.

OF CELLS, LIFE AND DEATH from *Biology* by Karl von Frisch. Translation copyright © 1964 by Bayerischer Schulbuch-Verlag. Reprinted by permission of Harper & Row, Publishers.

DARWIN AND "THE ORIGIN OF SPECIES" by Sir Arthur Keith, an Introduction to the Everyman's Library edition of *The Origin of Species* by Charles Darwin. Reprinted by permission of E. P. Dutton and Company.

GREGOR MENDEL AND HIS WORK by Hugo Iltis. Reprinted by permission of the author and *The Scientific Monthly.*

DNA by John Pfeiffer. Copyright © 1960 by the American Museum of Natural History. Reprinted by permission of *Natural History Magazine.*

DARWIN'S THEORY IN THE LIGHT OF MODERN GENETICS from *Man and the Living World* by Karl von Frisch. Translated by Elsa B. Lowenstein. Copyright © 1949 by Deutscher Verlag, Berlin; English translation © 1962 by Oliver & Boyd. Reprinted by permission of Harcourt, Brace & World, Inc., and Oliver & Boyd, Ltd. A Helen and Kurt Wolff Book.

ON BEING THE RIGHT SIZE from *Possible Worlds* by J. B. S. Haldane. Copyright 1928 by Harper & Brothers; renewed 1956 by J. B. S. Haldane. Reprinted by permission of Harper & Row, Publishers.

SPACE TRACKS by Dwain W. Warner. Copyright 1963 by The American Museum of Natural History. Reprinted by permission of *Natural History Magazine.*

CAN ANIMALS REASON? by Frank A. Beach. Copyright 1948 by The American Museum of Natural History. Reprinted by permission of *Natural History Magazine.*

THE COMPLEAT BOTANIST by A. J. Sharp. Copyright © 1964 by The American Association for the Advancement of Science. Reprinted by permission of the author.

THE WEB OF LIFE by John H. Storer. Copyright 1953 by John Storer. Reprinted by permission of The Devin-Adair Company.

FEEDING IN PLANTS from *Cells and Society* by John Tyler Bonner. Copyright © 1955 by Princeton University Press. Reprinted by permission of the publishers.

THE POLLINATION OF FLOWERS by Verne Grant. Reprinted with permission. Copyright 1951 by Scientific American, Inc. All rights reserved.

STRANGLER TREES by Theodosius Dobzhansky and João Murça-Pires. Reprinted with permission. Copyright 1953 by Scientific American, Inc. All rights reserved.

EVOLUTION REVISED from *Man, Time and Fossils,* Revised Edition, by Ruth Moore. Copyright © 1953, 1961 by Ruth Moore. Reprinted by permission of Alfred A. Knopf, Inc.

THE DISCOVERY OF ZINJANTHROPUS by L. S. B. Leakey, © 1960 by the National Geographic Society. Reprinted with revisions by permission of the National Geographic Society and the author.

THE GREAT PILTDOWN HOAX by William L. Straus, Jr. Copyright 1954 by The American Association for the Advancement of Science. Reprinted by permission of the author.

THE GROWTH OF CULTURE by Ruth Benedict from *Man, Culture and Society,* edited by Harry Shapiro. Copyright © 1956 by the Oxford University Press, Inc. Reprinted by permission of the publishers.

DIGGING UP THE PAST by Sir Leonard Woolley. Reprinted by permission of Ernest Benn Limited.

BIOGRAPHY OF THE UNBORN, a *Reader's Digest* condensation of the book *Biography of the Unborn* by Margaret Shea Gilbert. Reprinted by permission of *The Reader's Digest* and The Williams & Wilkins Company.

(*Continued on next page*)

VARIATIONS ON A THEME BY DARWIN by Julian Huxley, from *Science in a Changing World* edited by Mary Adams. Reprinted by permission of George Allen & Unwin, Ltd., and D. Appleton-Century Company, Inc.

HIPPOCRATES THE GREEK—THE END OF MAGIC from *Behind the Doctor* by Logan Clendening. Copyright 1933 by Alfred A. Knopf, Inc. Reprinted by permission of and special arrangement with Alfred A. Knopf, Inc.

LOUIS PASTEUR AND THE CONQUEST OF RABIES from *The Life of Pasteur* by René Vallery-Radot. Reprinted by permission of Doubleday, Doran & Company, Inc.

CHEMOTHERAPY from *Chemotherapy: Yesterday, Today, Tomorrow,* by Alexander Fleming. Reprinted by permission of Cambridge University Press.

THE BATTLE AGAINST CANCER by Walter Goodman. Copyright © 1964 by The Curtis Publishing Company. Reprinted by permission of the author.

RADIATIONS AND THE GENETIC THREAT by Warren Weaver. Reprinted by permission of the author.

PSYCHOLOGY, SCIENCE AND MAN from *Psychology* by George A. Miller. Copyright © 1962 by George A. Miller. Reprinted by permission of Harper & Row, Publishers.

INTRODUCING THE BRAIN from *The Human Brain* by John Pfeiffer, copyright 1955 by John Edward Pfeiffer. Reprinted by permission of Harper & Row, Publishers.

SICK MINDS, NEW MEDICINES from *Drugs and the Mind* by Robert S. de Ropp. Copyright © 1957 by Robert S. de Ropp. Reprinted by permission of St. Martin's Press.

IMAGINATION CREATRIX from *The Road to Xanadu* by John Livingston Lowes. Copyright 1927 by John Livingston Lowes. Reprinted by permission of Houghton Mifflin Company.

Contents

Part III: THE WORLD OF LIFE

A. *The Study of Life*

B. *The Evolution of Life*

C. *The Spectacle of Animal Life*

D. *The Spectacle of Plant Life*

Preface

The twentieth century is perhaps the most significant era in the history of the human race. Year by year, almost month by month, new vistas open and new sources of knowledge are tapped. Within recent memory our understanding of the world in which we live has been profoundly altered, and a revolution in both our ways of thinking and our daily lives has taken place. This revolution has been brought about in large part by that branch of human activity known as science. The interpretation of science is the aim of *The New Treasury of Science,* a volume designed for the educated layman, which attempts to offer an understanding of how the scientist works, the body of knowledge that has resulted from his activities, and the intrinsic interest and excitement of the search.

In recent years this search has widened with constantly greater speed. Consider these facts: the number of scientists active today is greater than in all previous history. Well over a million individuals in the English-speaking world alone hold technical and scientific degrees. Scientific reports are appearing at the rate of half a million a year. Much of this work is trivial or unsound; some of it, however, lays the groundwork for great advances in the future; while a very small proportion—for example, the work of Seaborg and his associates on the trans-uranium elements and that of Crick and Watson on DNA—is in itself of the first rank. Out of this vast activity has arisen a new age in human history. Segments of this age have been variously labeled. If we refer to the Atomic Age, we conjure up such achievements as the cracking of the atomic nucleus, the creation and the transmutation of elements, the tapping of new sources of energy resulting in unparalleled capacity for production and destruction, and the possibility of solving the problem of the fundamental nature of matter and energy. If our label is the Age of Space, we think of Sputnik and Cape Kennedy, of the probes which are bringing us new information about our near neighbors in the solar system, of the dawn of an era of extraterrestrial exploration, of new methods of foretelling the weather and scanning the heavens. We are also witnessing the labor pains of still another age, the Age of the Biologist, whose potentialities are great. We begin to understand the processes involved in the origin of life. The possibility that we may at some time in the future be able to reshape the characteristics of living things, including man, is raised by new discoveries in genetics. All these are segments of a whole, which may be characterized as the Age of Science and the Scientist and which is the subject of this book.

The New Treasury of Science is based on a collection of scientific writings first published in 1943. Four subsequent editions of this *Treasury* offered a selection of material dealing with new discoveries as they appeared. Several decades later the purpose of the original work continues to have validity, but great discoveries in almost every branch of science and the changing role of science in our civilization require a comparable re-examination of the literature. *The New Treasury* is the result. As indicated by the headings of the four parts—*Science and the Scientist, The Physical World, The World of Life,* and *The World of Man*—it follows the same general plan as its predecessor. But in internal arrangement and in the choice of material used, it takes account of the achievements of scientists of today.

In the preparation of this volume, as in the original *Treasury,* thousands of books and articles were examined. It was recognized that a group of random selections, however interesting in themselves, would suggest little of the unity, the architectural quality of science. A plan based on the organization of science and of the various scientific disciplines was therefore followed. These structural bones were then clothed in the flesh of contemporary scientific writing as well as of the great contributions of the past. In numerous instances outstanding discoveries were described by the men who made them. In others, in which original discussions were unsuitable for nonscientific readers, sound interpretations were substituted. Exciting episodes and biographies of great scientists were included to add spice and interest to the collection.

In *The New Treasury of Science* all sections have been brought up to date. Separate and expanded sections—The Mathematical Universe, Physical Law, Chemistry, and The Study of Life—are included. Other sections, such as The Earth, The Evolution of Life, The Spectacle of Plant Life, From Ape to Civilization, and Man's Mind, have been materially rearranged and enlarged. Outstanding selections, such as "Proof that the Earth Moves" by Galileo, "The Variolae Vaccinae" by Jenner, "The Evidence of the Descent of Man" by Darwin, "The Discovery of Radium" by Eve Curie, have been preserved from previous editions. To them have been added new selections that give evidence of the power of the scientific method, such as "The Man-Made Chemical Elements Beyond Uranium" by Seaborg and "The Voyage of *Mariner II*" by J. N. James of the Jet Propulsion Laboratory,

California Institute of Technology. Of the eighty-four selections that appear in *The New Treasury of Science*, only twenty-seven, in large part classics of the sort mentioned above, appeared in the original 1943 edition —evidence of the extent of the changes that have occurred and the increasing richness of the literature. It is interesting to note that whereas individual disciplines have been revolutionized, the spirit and methods of science remain the same. Hence *Part I: Science and the Scientist* draws largely on material that appeared in previous editions. The book has been completely reset and printed by offset from new plates.

Over 400,000 copies of this book have been sold. It is hoped that *The New Treasury* will also justify itself in interest and instructional value, whether opened at random or read methodically from cover to cover. It should have value as a general reference work, but it does not aim to be encyclopedic in scope. It is the hope of the editors that the reader will go further into the vast store of available writings to get specialized knowledge of any branch of science that may interest him. It should also be pointed out that because of limitations of space many of the selections have been shortened and compressed. And since the volume is designed for the general reader and not for the specialist except when he is also a general reader, references, technical footnotes, and similar apparatus have been omitted.

THE EDITORS

THE NEW TREASURY OF SCIENCE

HARLOW SHAPLEY

Introduction
On Sharing in the Conquests of Science

•

It's a wonder I can stand it! Tramping for hours through the damp woods back of Walden Pond with Henry Thoreau, checking up on the food preferences of the marsh hawk, and the spread of sumach and goldenrod in old abandoned clearings. It requires stamina to match his stride as he plunges through swamps and philosophy, through underbrush, poetry, and natural history; it takes agility of body and mind if one does a full share of the day's measuring and speculation.

And then I am off on another high adventure, higher than the moon this time; I am entering the study of the Frauenburg Cathedral to help Nicholas Copernicus do calculations on the hypothetical motions of the planets. He is, of course, deeply bemused with that rather queer notion that it might be the Sun that stands still—not the earth. Perhaps he can demonstrate that the planets go around the Sun, each in its own course. Fascinated, I peer over his shoulder at the archaic geometry, watch his laborious penning of the great book, and listen to his troubled murmuring about the inaccuracies of the measured coordinates of Saturn. "There are, you know, two other big ones further out," I put in, "and a system of many moons around Jupiter, which makes it all very clear and obvious." It must startle him no end to have me interrupt in such a confident way. But he does nothing about it. More planets? An incredible idea! Difficulties enough in trying to explain the visible, without complicating the complexities further by introducing invisible planets. My assistance ignored, I experience, nevertheless, a carefree exhilaration; for I have, as it were, matched my wits with the wisdom of the greatest of revolutionaries, and come off not too badly!

Now that I am fully launched in this career of working with the great explorers, and of cooperating in their attacks on the mysteries of the universe, I undertake further heroic assignments. I labor in the laboratories of the world; I maintain fatiguing vigils in the mountains and on the sea, try dangerous experiments, and make strenuous expeditions to Arctic shores and to torrid jungles—all without moving from the deep fireside chair.

Benjamin Franklin has a tempting idea, and I am right there to lend him a hand. We are having a lot of trouble in keeping that cantankerous kite in the thundercloud, from which the electric fluid should flow to charge and animate the house key. "Before long, sir, we shall run printing presses with this fluid, and light our houses, and talk around the world"—but he does not put it in the *Autobiography*. I am clearly a century ahead of my time!

Youthful Charles Darwin is in the Galapagos. The good brig *Beagle* stands offshore. He has with him the collecting kit, the notebooks, and his curiosity. He is making records of the slight variations among closely similar species of plants and animals. Hs is pondering the origin of these differences, and the origin of species, and the whole confounding business of the origin of plants and animals. I sit facing him, on the rocks beside the tide pool, admiring the penetration and grasp of this young dreamer. The goal of his prolonged researches is a revolution in man's conception of life; he is assembling the facts and thoughts, and in this work I am a participant! Nothing could be more exciting. Also I have an advantage. I know about Mendel and Mendelian laws, and genes and chromosomes. I know that X-rays (unknown to Darwin), and other agents, can produce mutations and suddenly create living forms that Nature has not attained. This posterior knowledge of mine enhances the pleasure of my collaboration with the great naturalist; and I need have no fear that my information, or my ethereal presence, might bother him.

There is so much scientific work of this sort for me to do before some tormenting duty draws me out of my strategic chair. The possibilities are nearly endless. Like a benign gremlin, I sit on the brim of a test tube in Marie Curie's laboratory and excitedly speculate with her on that radioactive ingredient in the pitchblende; I help name it radium. With Stefansson and the Eskimos I live for months on a scanty menu, and worry with him about the evils of civilization. And when young Évariste Galois, during his beautiful, brief, perturbed life in Paris, sits down to devise sensationally new ideas and techniques in pure mathematics, I am right there with applause and sympathy.

Whenever I pause to appreciate how simple it is for me to take an active part in unraveling the home life of primitive man; how simple for me, in company with the highest authorities, to reason on the theory of relativity or explore with a cyclotron the insides of atoms, it is then that I call for additional blessings on those artisans who invented printing. They have provided me with guidelines to remote wonders—highly conductive threads that lead me, with a velocity faster than that of light itself, into times long past and into minds that biologically are long extinct.

Through the simple process of learning how to interpret symbols, such as those that make this sentence, I can take part in most of the great triumphs of the human intellect. Blessings and praises, laurel wreaths and myrtle, are due those noble spirits who made writing and reading easily accessible, and thus opened to us all the romance of scientific discovery.

Have you ever heard an ox warble? Probably not. Perhaps it goes through its strange life cycle silent to our gross ears. But I have seen ox warbles, and through the medium of the printed page I have followed their gory careers. The ox warbles to which I refer are, of course, not bovine melodies, but certain flies that contribute to the discomfort of cattle, to the impoverishment of man's property, and to the enrichment of his knowledge of the insect world.

It required a declaration of war on this entomological enemy, by some of the great nations of the planet, in order to discover him completely and entrench mankind against his depredations. It took a century of detective work on the part of entomologists to lay bare the ox warble's secret life. Now that I have the story before me, I can go along with the scientists and experience again their campaigns, their misadventures, and their compensating discoveries. I can see how to connect a number of separate phenomena that long were puzzling—those gay pasture flies that look like little bumblebees; those rows of tiny white eggs on the hairs above the hoofs of cattle; the growing larvae, guided mysteriously by ancestral experience to wind their way for months through the flesh of the legs and bodies of their unknowing hosts; the apparently inactive worms in the cattle's throats; the large midwinter mounds, scattered subcutaneously along the spines of the herd; and eventually those ruinous holes in the leather, which have forced governments into aggressive action—into defense-with-pursuit tactics for the protection of their economic frontiers. It is all clear now. During the millennia of recent geological periods a little fly has learned how to fatten its offspring on a fresh beef diet, and prepare its huge grub for that critical moment when it crawls out, through the hole it has made in the ox hide, and drops to the earth for its metamorphosis—the change from a headless, legless, eyeless, dark childhood to a maturity of wings and sunlight.

The curiosity the scientist strives to satisfy is thus sometimes impelled by economics; more often by the pure desire to know. Our black-on-white guiding threads, which you may call printed books, or recorded history, not only transmit the stories of ancient and modern inquisitiveness and the inquiries it has inspired, but they also report, to the discerning recipient, the inevitability of practiced internationalism. They transmit the message that all races of mankind are curious about the universe, and that, when free and not too depressed by hunger, men instinctively question and explore, analyze and catalogue. They have done it in all ages, in all civilized countries. They work singly, in groups, and increasingly in world-wide organizations. Science recognizes no impossible national boundaries, and only temporary barriers of language. It points the way to international cooperation.

To more than the art of printing, however, do we owe the successes and pleasures of our vicarious adventures in science. We are also greatly indebted to those who can write and will write in terms of our limited comprehension. Not all the scientists have the facility. Sometimes the talk is too tough for us, or too curt. They have not the time to be lucid on our level and within our vocabulary, or perhaps their mental intensity has stunted the faculty of sympathetic explanation. When such technical barriers shut us from the scientific workshop, it is then we like to consult with a clear-spoken and understanding interpreter. We sit on the back porch of the laboratory, while he, as middleman, goes inside to the obscurities and mysteries, to return occasionally with comprehensible reports. In listening to him we hear not only his voice, but the overtones of the master he interprets. I like these men of understanding who play Boswell to the specialist. They often have a gift greater than that of the concentrated workers whom they soften up for us. For they have breadth and perspective, which help us to get at the essence of a problem more objectively than we could even if we were fully equipped with the language and knowledge of the fact-bent explorer and analyst. The scientific interpreters frequently enhance our enjoyment in that they give us of themselves, as well as of the discoverers whose exploits they recount. We are always grateful to them, moreover, for having spared us labor and possibly discouragement.

Perhaps the greatest satisfaction in reading of scientific exploits and participating, with active imagination, in the dull chores, the brave syntheses, the hard-won triumphs of scientific work lies in the realization that ours is not an unrepeatable experience. Tomorrow night we can again go out among the distant stars. Again we can drop cautiously below the ocean surface to observe the unbelievable forms that inhabit those salty regions of high pressure and dim illumination. Again we can assemble the myriad molecules into new combinations, weave them into magic carpets that take us into strange lands of beneficent drugs and of new fabrics and utensils destined to enrich the process of everyday living. Again we can be biologist, geographer, astronomer, engineer, or help the philosopher evaluate the nature and meaning of natural laws.

We can return another day to these shores, and once more embark for travels over ancient or modern seas in quest of half-known lands—go forth as dauntless conquistadores, outfitted with the maps and gear provided through the work of centuries of scientific adventures.

But we have done enough for this day. We have much to dream about. Our appetites may have betrayed our ability to assimilate. The fare has been irresistibly palatable. It is time to disconnect the magic threads; time to wind up the spiral galaxies, roll up the Milky Way and lay it aside until tomorrow.

Part I

•

SCIENCE AND THE SCIENTIST

It is common knowledge that the reality of modern science is far more astonishing than the shadow world of science fiction. It is therefore unfortunate that the story has often been told in selections which present it at its most forbidding. Like Agassiz's monumental work on turtles, Contributions to the Natural History of the United States, *described by Dallas Lore Sharp in the following pages, they are "massive, heavy, weathered, as if dug from the rocks." Yet there is fascination in science, excitement, profound satisfaction. It is fitting that our first selection should be an attempt to describe that feeling,* The Wonder of the World *by Sir J. Arthur Thomson and Patrick Geddes.*

Nor is science something esoteric, something mysterious and incomprehensible to the average person. We are all scientists, as T. H. Huxley shows clearly, whether we are concerned with the properties of green apples or with finding the burglar who stole our spoons. And we are led to our conclusions by "the same train of reasoning as that which a man of science pursues when he is endeavouring to discover the origin and laws of the most occult phenomena." Huxley was one of the great scientists of the nineteenth century, as well as its greatest scientific writer. He is well qualified to instruct us.

Not only the aims but also the activities of the scientist are frequently misunderstood. To the average layman he is a coldly unemotional individual who proceeds purely on the basis of logical reasoning. Such an image has little relationship with actual fact. Like other humans, the scientist is often blundering, illogical, swayed by personal prejudice. And as the great physiologist Walter B. Cannon explains in Gains from Serendipity, *some of his most noteworthy discoveries have been the result of accidental happenings. To characterize such discoveries as pure luck, however, is oversimplification. Luck, in Pasteur's inspired phrase, favors only the prepared mind.*

One quality which sets the scientist apart is the persistence of his curiosity about the world. Like Mr. Jenks of Middleboro, in Turtle Eggs for Agassiz *by Dallas Lore Sharp, he may spend countless hours beside a murky pond, waiting for a turtle to lay her eggs. Or like Oliver La Farge in* Scientists Are Lonely Men, *he may journey to a remote outpost, there to spend months or years on some seemingly trivial quest which may prove a clue to the origin of a race.*

SIR J. ARTHUR THOMSON
AND PATRICK GEDDES

The Wonder of the World

•

ARISTOTLE, who was not unaccustomed to resolute thinking, tells us that throughout nature there is always something of the wonderful—*thaumaston*. What precisely is this "wonderful"? It cannot be merely the startling, as when we announce the fact that if we could place in one long row all the hair-like vessels or capillaries of the human body, which connect the ends of the arteries with the beginnings of the veins, they would reach across the Atlantic. It would be all the same to us if they reached only half-way across. Nor can the wonderful be merely the puzzling, as when we are baffled by the "sailing" of an albatross round and round our ship without any perceptible strokes of its wings. For some of these minor riddles are being read every year, without lessening, however, the fundamental wonderfulness of Nature. Indeed, the much-abused word "wonderful" is properly applied to any fact the knowledge of which greatly increases our appreciation of the significance of the system of which we form a part. *The truly wonderful makes all other things deeper and higher.* Science is always dispelling mists—the minor marvels; but it leaves us with intellectual blue sky, sublime mountains, and deep sea. Their wonder appears—and remains.

There seems to be a rational basis for wonder in the abundance of power in the world—the power that keeps our spinning earth together as it revolves round the sun, that keeps our solar system together as it journeys through space at the rate of twelve miles a second towards a point in the sky, close to the bright star Vega, called "the apex of the sun's way." At the other extreme there is the power of a fierce little world within the complex atom, whose imprisoned energies are set free to keep up the radiant energies of sun and star. And between these extremes of the infinitely great and the infinitely little are the powers of life—the power of winding up the clock almost as fast as it runs down, the power of a fish that has better engines than those of a steamship, life's power of multiplying itself, so that in a few hours an invisible microbe may become a fatal million.

Another, also old-fashioned, basis for wonder is to be found in the immensities. It takes light eight minutes to reach us from the sun, though it travels at the maximum velocity—of about 186,000 miles per second. So we see the nearest star by the light that left it four years ago, and Vega as it was twenty-seven years ago, and most of the stars that we see without a telescope as they were when Galileo Galilei studied them in the early years of the seventeenth century. In any case it is plain that we are citizens of no mean city.

A third basis for rational wonder is to be found in the intricacy and manifoldness of things. We get a suggestion of endless resources in the creation of individualities. For "all flesh is not the same flesh, but there is one kind of flesh of men, another flesh of beasts, another of fishes, and another of birds." The blood of a horse is different from that of an ass, and one can often identify a bird from a single feather or a fish from a few scales. One is not perhaps greatly thrilled by the fact that the average man has twenty-five billions of oxygen-capturing red blood corpuscles, which if spread out would occupy a surface of 3,300 square yards; but there is significance in the calculation that he has in the cerebral cortex of his brain, the home of the higher intellectual activities, some nine thousand millions of nerve cells, that is to say, more than five times the present population of the globe—surely more than the said brain as yet makes use of.

So it must be granted that we are fearfully and wonderfully made! Our body is built up of millions of cells, yet there is a simplicity and the multitudinousness, for each cell has the same fundamental structure. Within the colloid cell-substance there floats a kernel or nucleus, which contains forty-seven (or in woman forty-eight) chromosomes, each with a bead-like arrangement of smaller microsomes, and so on, and so on. Similarly, we know that the various elements differ from one another only in the number and distribution of the electrons and protons that make up their microcosmic planetary system. What artistry to weave the gorgeously varied tapestry of the world out of two kinds of physical thread—besides, of course, Mind, which eventually searches into the secret of the loom.

A fourth basis for rational wonder is in the orderliness of Nature, and that is almost the same thing as saying its intelligibility. What implications there are in the fact that man has been able to make a science of Nature! Given three good observations of a comet, the astronomer can predict its return to a night. It is not a phantasmagoria that we live in, it is a rationalisable cosmos. The more science advances the more the fortuitous shrivels, and the more the power of prophecy grows. Two astronomers foretold the discovery of Neptune; the chemists have anticipated the discovery of new elements; the biologist can not only count but portray his chickens before they are hatched.

Corresponding to the intelligibility of Nature is the pervasiveness of beauty—a fifth basis of rational wonder, appealing to the emotional side of our personality. Surely Lotze was right, that it is of high value to look upon beauty not as a stranger in the world, nor as a casual aspect of certain phenomena, but as "the fortunate revelation of that principle which permeates all reality with its living activity."

A sixth basis of rational wonder is to be found in the essential characteristics of living creatures. We need only add the caution that the marvel of life is not to be taken at its face value; as Coleridge wisely said, the first wonder is the child of *ignorance;* we must attend diligently to all that biochemistry and biophysics can discount; we must try to understand all that can be formulated in terms of colloids, and so on. Yet when all that is said, there seem to be large residual phenomena whose emergence in living creatures reveal a new depth in Nature. Life is an enduring, insurgent activity, growing, multiplying, developing, enregistering, varying, and above all else evolving.

For this is the seventh wonder—Evolution. It is not merely that all things flow; it is that life flows uphill. Amid the ceaseless flux there is not only conservation, there is advancement. While we must consider man in the light of evolution, as most intellectual combatants admit, there is the even more difficult task of envisaging evolution in the light of Man. *Finis coronat opus*—a wise philosophical axiom; and yet the scientist must qualify it by asking who can say Finis to Evolution.

T. H. HUXLEY

We Are All Scientists

•

Scientific investigation is not, as many people seem to suppose, some kind of modern black art. You might easily gather this impression from the manner in which many persons speak of scientific inquiry, or talk about inductive and deductive philosophy, or the principles of the "Baconian philosophy." I do protest that, of the vast number of cants in this world, there are none, to my mind, so contemptible as the pseudo-scientific cant which is talked about the "Baconian philosophy."

To hear people talk about the great Chancellor—and a very great man he certainly was—you would think that it was he who had invented science, and that there was no such thing as sound reasoning before the time of Queen Elizabeth! Of course you say, that cannot possibly be true; you perceive, on a moment's reflection, that such an idea is absurdly wrong.

The method of scientific investigation is nothing but the expression of the necessary mode of working of the human mind. It is simply the mode at which all phenomena are reasoned about, rendered precise and exact. There is no more difference, but there is just the same kind of difference, between the mental operations of a man of science and those of an ordinary person, as there is between the operations and methods of a baker or of a butcher weighing out his goods in common scales, and the operations of a chemist in performing a difficult and complex analysis by means of his balance and finely-graduated weights. It is not that the action of the scales in the one case, and the balance in the other, differ in the principles of their construction or manner of working; but the beam of one is set on an infinitely finer axis than the other, and of course turns by the addition of a much smaller weight.

You will understand this better, perhaps, if I give you some familiar example. You have all heard it repeated, I dare say, that men of science work by means of induction and deduction, and that by the help of these operations, they, in a sort of sense, wring from Nature certain other things, which are called natural laws, and causes, and that out of these, by some cunning skill of their own, they build up hypotheses and theories. And it is imagined by many that the operations of the common mind can be by no means compared with these processes, and that they have to be acquired by a sort of special apprenticeship to the craft. To hear all these large words, you would think that the mind of a man of science must be constituted differently from that of his fellow men; but if you will not be frightened by terms, you will discover that you are quite wrong, and that all these terrible apparatus are being used by yourselves every day and every hour of your lives.

There is a well-known incident in one of Molière's plays, when the author makes the hero express unbounded delight on being told that he had been talking prose during the whole of his life. In the same way, I trust that you will take comfort, and be delighted with yourselves, on the discovery that you have been acting on the principles of inductive and deductive philosophy during the same period. Probably there is not one who has not in the course of the day had occasion to set in motion a complex train of reasoning, of the very same kind, though differing of course in degree, as that which a scientific man goes through in tracing the causes of natural phenomena.

A very trivial circumstance will serve to exemplify this. Suppose you go into a fruiterer's shop, wanting an apple—you take up one, and, on biting it, you find it is sour; you look at it, and see that it is hard and green. You take up another one, and that too is hard, green, and sour. The shopman offers you a third; but, before biting it, you examine it, and find that it is hard and green, and you immediately say that you will not have it, as it must be sour, like those that you have already tried.

Nothing can be more simple than that, you think; but if you will take the trouble to analyse and trace out into its logical elements what has been done by the mind, you will be greatly surprised. In the first place, you have performed the operation of induction. You found that, in two experiences, hardness and greenness in apples went together with sourness. It was so in the first case, and it was confirmed by the second. True, it is a very small basis, but still it is enough to make an induction from; you generalise the facts, and you expect to find sourness in apples where you get hardness and greenness. You found upon that a general law, that all hard and green apples are sour; and that, so far as it goes, is a perfect induction. Well, having got your natural law in this way, when you are offered another apple which you find is hard and green, you say, "All hard and green apples are sour; this apple is hard and green, therefore this apple is sour." That train of reasoning is what logicians call a syllogism, and has all its various parts and terms—its major premiss, its minor premiss, and its conclusion. And, by the help of further reasoning, which, if drawn out, would have to be exhibited in two or three other syllogisms, you arrive at your final determination, "I will not have that apple." So that, you see, you have, in the first place, established a law by induction, and upon that you have founded a deduction, and reasoned out the special conclusion of the particular case. Well now, suppose, having got your law, that at some time afterwards, you are discussing the qualities of apples with a friend: you will say to him, "It is a very curious thing—but I find that all hard and green apples are sour!" Your friend says to you, "But how do you know that?" You at once reply, "Oh, because I have tried them over and over again, and have always found them to be so." Well, if we were talking science instead of common sense, we should call that an experimental verification. And, if still opposed, you go further, and say, "I have heard from the people in Somersetshire and Devonshire, where a large number of apples are grown, that they have observed the same thing. It is also found to be the case in Normandy, and in North America. In short, I find it to be the universal experience of mankind wherever attention has been directed to the subject." Whereupon, your friend, unless he is a very unreasonable man, agrees with you, and is convinced that you are quite right in the conclusion you have drawn. He believes, although perhaps he does not know he believes it, that the more extensive verifications are—that the more frequently experiments have been made, and results of the same kind arrived at—that the more varied the conditions under which the same results are attained, the more certain is the ultimate conclusion, and he disputes the question no further. He sees that the experiment has been tried under all sorts of conditions, as to time, place, and people, with the same result; and he says with you, therefore, that the law you have laid down must be a good one, and he must believe it.

In science we do the same thing—the philosopher exercises precisely the same faculties, though in a much more delicate manner. In scientific inquiry it becomes a matter of duty to expose a supposed law to every possible kind of verification, and to take care, moreover, that this is done intentionally, and not left to a mere accident, as in the case of the apples. And in science, as in common life, our confidence in a law is in exact proportion to the absence of variation in the result of our experimental verifications. For instance, if you let go your grasp of an article you may have in your hand, it will immediately fall to the ground. That is a very common verification of one of the best established laws of nature—that of gravitation. The method by which men of science establish the existence of that law is exactly the same as that by which we have established the trivial proposition about the sourness of hard and green apples. But we believe it in such an extensive, thorough, and unhesitating manner because the universal experience of mankind verifies it, and we can verify it ourselves at any time; and that is the strongest possible foundation on which any natural law can rest.

...ch, then, by way of proof that the method of establishing laws in ...e is exactly the same as that pursued in common life. Let us now ... to another matter (though really it is but another phase of the same question), and that is, the method by which, from the relations of certain phenomena, we prove that some stand in the position of causes towards the others.

I want to put the case clearly before you, and I will therefore show you what I mean by another familiar example. I will suppose that one of you, on coming down in the morning to the parlour of your house, finds that a tea-pot and some spoons which had been left in the room on the previous evening are gone—the window is open, and you observe the mark of a dirty hand on the window-frame, and perhaps, in addition to that, you notice the impress of a hob-nailed shoe on the gravel outside. All these phenomena have struck your attention instantly, and before two seconds have passed you say, "Oh, somebody has broken open the window, entered the room, and run off with the spoons and the tea-pot!" That speech is out of your mouth in a moment. And you will probably add, "I know there has; I am quite sure of it!" You mean to say exactly what you know; but in reality you are giving expression to what is, in all essential particulars, an hypothesis. You do not *know* it at all; it is nothing but an hypothesis rapidly framed in your own mind. And it is an hypothesis founded on a long train of inductions and deductions.

What are those inductions and deductions, and how have you got at this hypothesis? You have observed, in the first place, that the window is open; but by a train of reasoning involving many inductions and deductions, you have probably arrived long before at the general law—and a very good one it is—that windows do not open of themselves; and you therefore conclude that something has opened the window. A second general law that you have arrived at in the same way is, that tea-pots and spoons do not go out of a window spontaneously, and you are satisfied that, as they are not now where you left them, they have been removed. In the third place, you look at the marks on the window-sill, and the shoe-marks outside, and you say that in all previous experience the former kind of mark has never been produced by anything else but the hand of a human being; and the same experience shows that no other animal but man at present wears shoes with hob-nails in them such as would produce the marks in the gravel. I do not know, even if we could discover any of those "missing links" that are talked about, that they would help us to any other conclusion! At any rate the law which states our present experience is strong enough for my present purpose. You next reach the conclusion, that as these kinds of marks have not been left by any other animals than men, or are liable to be formed in any other way than by a man's hand and shoe, the marks in question have been formed by a man in that way. You have, further, a general law, founded on observation and experience, and that, too, is, I am sorry to say, a very universal and unimpeachable one—that some men are thieves; and you assume at once from all these premisses—and that is what constitutes your hypothesis—that the man who made the marks outside and on the window-sill, opened the window, got into the room, and stole your tea-pot and spoons. You have now arrived at a *vera causa*—you have assumed a cause which, it is plain, is competent to produce all the phenomena you have observed. You can explain all these phenomena only by the hypothesis of a thief. But that is a hypothetical conclusion, of the justice of which you have no absolute proof at all; it is only rendered highly probable by a series of inductive and deductive reasonings.

I suppose your first action, assuming that you are a man of ordinary common sense, and that you have established this hypothesis to your own satisfaction, will very likely be to go for the police, and set them on the track of the burglar, with the view to the recovery of your property. But just as you are starting with this object, some person comes in, and on learning what you are about, says, "My good friend, you are going on a great deal too fast. How do you know that the man who really made the marks took the spoons? It might have been a monkey that took them, and the man may have merely looked in afterwards." You would probably reply, "Well, that is all very well, but you see it is contrary to all experience of the way tea-pots and spoons are abstracted; so that, at any rate, your hypothesis is less probable than mine." While you are talking the thing over in this way, another friend arrives. And he might say, "Oh, my dear sir, you are certainly going on a great deal too fast. You are most presumptuous. You admit that all these occurrences took place when you were fast asleep, at a time when you could not possibly have known anything about what was taking place. How do you know that the laws of Nature are not suspended during the night? It may be that there has been some kind of supernatural interference in this case." In point of fact, he declares that your hypothesis is one of which you cannot at all demonstrate the truth and that you are by no means sure that the laws of Nature are the same when you are asleep as when you are awake.

Well, now, you cannot at the moment answer that kind of reasoning. You feel that your worthy friend has you somewhat at a disadvantage. You will feel perfectly convinced in your own mind, however, that you are quite right, and you say to him, "My good friend, I can only be guided by the natural probabilities of the case, and if you will be kind enough to stand aside and permit me to pass, I will go and fetch the police." Well, we will suppose that your journey is successful, and that by good luck you meet with a policeman; that eventually the burglar is found with your property on his person, and the marks correspond to his hand and to his boots. Probably any jury would consider those facts a very good experimental verification of your hypothesis, touching the cause of the abnormal phenomena observed in your parlour, and would act accordingly.

Now, in this suppositious case, I have taken phenomena of a very common kind, in order that you might see what are the different steps in an ordinary process of reasoning, if you will only take the trouble to analyse it carefully. All the operations I have described, you will see, are involved in the mind of any man of sense in leading him to a conclusion as to the course he should take in order to make good a robbery and punish the offender. I say that you are led, in that case, to your conclusion by exactly the same train of reasoning as that which a man of science pursues when he is endeavouring to discover the origin and laws of the most occult phenomena. The process is, and always must be, the same; and precisely the same mode of reasoning was employed by Newton and Laplace in their endeavours to discover and define the causes of the movements of the heavenly bodies, as you, with your own common sense, would employ to detect a burglar. The only difference is, that the nature of the inquiry being more abstruse, every step has to be most carefully watched, so that there may not be a single crack or flaw in your hypothesis. A flaw or crack in many of the hypotheses of daily life may be of little or no moment as affecting the general correctness of the conclusions at which we may arrive; but, in a scientific inquiry, a fallacy, great or small, is always of importance, and is sure to be in the long run constantly productive of mischievous, if not fatal results.

Do not allow yourselves to be misled by the common notion that an hypothesis is untrustworthy simply because it is an hypothesis. It is often urged, in respect to some scientific conclusion, that, after all, it is only an hypothesis. But what more have we to guide us in nine-tenths of the most important affairs of daily life than hypotheses, and often very ill-based ones? So that in science, where the evidence of an hypothesis is subjected to the most rigid examination, we may rightly pursue the same course. You may have hypotheses and hypotheses. A man may say, if he likes, that the moon is made of green cheese: that is an hypothesis. But another man, who has devoted a great deal of time and attention to the subject, and availed himself of the most powerful telescopes and the results of the observations of others, declares that in his opinion it is probably composed of materials very similar to those of which our own earth is made up: and that is also only an hypothesis. But I need not tell you that there is an enormous difference in the value of the two hypotheses. That one which is based on sound scientific knowledge is sure to have a corresponding value; and that which is a mere hasty random guess is likely to have but little value. Every great step in our progress in discovering causes has been made in exactly the same way as that which I have detailed to you. A person observing the occurrence of certain facts and phenomena asks, naturally enough, what process, what kind of operation known to occur in Nature applied to the particular case, will unravel and explain the mystery? Hence you have the scientific hypothesis; and its value will be proportionate to the care and completeness with which its basis has been tested and verified. It is in these matters as in the commonest affairs of practical life: the guess of the fool will be folly, while the guess of the wise man will contain wisdom. In all cases, you see that the value of the result depends on the patience and faithfulness with which the investigator applies to his hypothesis every possible kind of verification.

WALTER B. CANNON

Gains from Serendipity

•

In 1754 Horace Walpole, in a chatty letter to his friend Horace Mann, proposed adding a new word to our vocabulary, "serendipity." The word looks as if it might be of Latin origin. It is rarely used. It is not found in the abridged dictionaries. When I mentioned serendipity to one of my acquaintances and asked him if he could guess the meaning, he suggested that it probably designated a mental state combining serenity and stupidity —an ingenious guess, but erroneous.

Walpole's proposal was based upon his reading of a fairy tale entitled *The Three Princes of Serendip.* Serendip, I may interject, was the ancient name of Ceylon. "As their highnesses traveled," so Walpole wrote, "they were always making discoveries, by *accident* or *sagacity,* of things which they were not in quest of." When the word is mentioned in dictionaries, therefore, it is said to designate the happy faculty, or luck, of finding unforeseen evidence of one's ideas or, with surprise, coming upon new objects or relations which were not being sought.

Readers who remember Bible stories will recall that Saul, the son of Kish, was sent forth to find his father's asses, which were lost. In the discouragement of his failures to find them he consulted one Samuel, a seer. And Samuel told him not to set his mind on them for they had been found, but to know that he was chosen to rule over all the tribes of Israel. So it was announced, and the people shouted their approval. Thus modest Saul, who went out to seek lost asses, was rewarded by a kingdom. That is the earliest record of serendipity I am aware of.

Probably the most astounding instance of accidental discovery in either ancient or modern history was the finding of the Western Hemisphere by Columbus. He sailed away from Spain firm in the faith that by going west he would learn a shorter route to the East Indies; quite unexpectedly he encountered a whole new world. It is noteworthy that he was not aware of the significance of what he had found. Indeed, it has been said that he did not know where, in fact, he was going nor where he was when he arrived nor where he had been after his return, but nevertheless he had had the most unique adventure of all time. He realized that he had had a remarkable experience and, by extending the knowledge of what he had done, he laid a course which others might follow. Such consequences have been common when accident has been favorable to one engaged in a search and the enterprise has proved fruitful.

In the records of scientific investigation this sort of happy use of good fortune has been conspicuous. A good example is afforded by the origin and development of our acquaintance with electrical phenomena. It is reported that some frogs' legs were hanging by a copper wire from an iron balustrade in the Galvani home in Bologna; they were seen to twitch when they were swung by the wind and happened to touch the iron. Whether the twitching was first noted by Luigi Galvani, the anatomist and physiologist, or by Lucia Galvani, his talented wife, is not clear. Certainly that fortuitous occurrence late in the eighteenth century was not neglected, for it started many researches which have preserved the Galvani name in the terms "galvanize" and "galvanism." We now use it, for example, to indicate the disordered state of the heart, because every cardiac contraction sends forth through our bodies an electrical wave, a wave that has a different shape according to the damage in the heart muscle. Only recently have we begun to employ animal electricity to give us information about conditions in the brain. That marvelous organ composed of many billions of nerve cells can display rhythmic electrical pulsations and, when extremely delicate instruments are applied to the scalp, they can reveal the different types of pulsations in rest and activity and the modification in some states of disease.

In the biological sciences serendipity has been quite as consequential as in the physical sciences. Claude Bernard, for example, had the idea that the impulses which pass along nerve fibers set up chemical changes producing heat. In an experiment performed about the middle of the last century he measured the temperature of a rabbit's ear and then severed a nerve which delivered impulses to that structure expecting, in accordance with his theory, that the ear deprived of nerve impulses would be cooler than its mate on the other side. To his great surprise it was considerably warmer! Without at first knowing the import of what he had done, he had disconnected the blood vessels of the ear from the nervous influences that normally hold them moderately contracted; thereupon the warm blood from internal organs was flushed through the expanded vessels in a faster flow and the ear temperature rose. Thus by accident appeared the first intimation that the passage of blood into different parts of the body is under the government of nerves—one of the most significant advances in our knowledge of the circulation since Harvey's proof, early in the seventeenth century, that the blood does indeed circulate in the vessels.

Another striking instance of accidental discovery has been described by the French physiologist, Charles Richet, a Nobel laureate. It was concerned with a peculiar sensitiveness toward certain substances—such as white of egg, strawberries, ragweed pollen, and numerous others—that we now speak of as *anaphylaxis* or *allergy*. This may result from an initial exposure to the substance which later becomes poisonous to the victim. The phenomenon had been noticed incidentally before Richet's studies, but because it did not receive attention its characteristics were virtually unknown. In his charming little book *Le Savant*, he has told the story of how quite unexpectedly he happened upon the curious fact. He was testing an extract of the tentacles of a sea anemone on laboratory animals in order to learn the toxic dose. When animals which had readily survived that dose were given after a lapse of some time a much smaller dose (as little as one-tenth), he was astounded to find that it was promptly fatal. Richet declares that at first he had great difficulty in believing the result could be due to anything *he* had done. Indeed, he testified that it was in spite of himself that he discovered induced sensitization. He would never have dreamt that it was possible.

Pasteur was led by chance to his method of immunization. One day an old and forgotten bacterial culture was being used for inoculating fowls. The fowls became ill but did not die. This happening was illuminative. Possibly by first using cultures that had little virulence and then repeating the injections with cultures of greater virulence, the animals could be made to develop resistance to infection gradually. His surmise proved correct. By this procedure, as readers of his dramatic biography will remember, he was able to immunize sheep against anthrax and human beings against rabies.

It was an accidental observation which ultimately resulted in the discovery of insulin and the restoration of effective living to tens of thousands of sufferers from diabetes. In the late eighties of the last century, von Mering and Minkowski were studying the functions of the pancreas in digestion. While attempting to secure more evidence they removed that organ from a number of dogs. By good luck a laboratory assistant noticed that swarms of flies gathered round the urine of these animals, a fact which he mentioned to the investigators. When the urine was analyzed, it was found to be loaded with sugar. Thus for the first time experimental diabetes was produced, and the earliest glimpse was given into a possible cause of that disease. We now know that small islands of cells in the pancreas produce an internal secretion which exerts control over the use of sugar in the organism. And we know that when these islands are removed or damaged, sugar metabolism is deranged. An extract from the island cells provides the diabetic sufferer with the insulin he needs.

An unforeseen contingency may occasion scientific advances because of the serious problem it presents. A striking instance is afforded in the use of polished rice. There was no reason to anticipate that the polishing of rice would be harmful to those who depended upon it as a food. Yet removal of the covering from the kernels produced in myriads of victims the disease, beriberi, resulting in immeasurable sorrow and distress. As has been pointed out, however, the study of beriberi, thus unwittingly induced, disclosed not only the cause of that disorder but also started explorations in the whole realm of deficiency diseases and thus led to the discovery of some of the most intimate secrets of cellular processes.

A recent instance of serendipity was the finding of vitamin K, lack of which deprives the blood of an essential element for its coagulation. The Danish investigator, Dam, and his collaborators were working on chemical changes in a certain fatty substance in chicks. They noted that the animals on a special restricted diet often suffered from extensive internal hemorrhages. When the diet was changed to seeds and salts, the bleeding failed to occur. By critical tests the abnormal condition was proved to be due not to lack of any previously known vitamin but to lack of a specific agent contained in the liver fat of swine as well as in certain vegetables and in many cereals. This agent, vitamin K, has proved to be important in surgery. For example, patients afflicted with jaundice, owing to an obstruction in the bile duct, can be relieved by operation; unfortunately in jaundice, however, blood clots very slowly; an operation, therefore, may be attended by disastrous bleeding. This danger can now be readily obviated by feeding vitamin K (with bile salts), for it restores to an effective concentration the deficient element of the clotting process, a benefaction which has come to human beings from a chance observation on chicks.

In the life of an investigator whose researches range extensively, advantages from happy chance are almost certain to be encountered. During nearly five decades of scientific experimenting instances of serendipity have several times been my good fortune. Two experiences I mention elsewhere, but not in relation to serendipity. One was stoppage of the movements of the stomach and intestines in times of anxiety. The other was the strange faster beating of the heart, after all its governing nerves were severed, if the animal became excited or if sympathetic fibers were stimulated in some remote region of the body. This effect, due to an agent carried to the heart by the circulating blood, led to the discovery of *sympathin*. Both phenomena were quite unexpected. Proof that the stoppage of digestive movements was due to emotion was the beginning of many years of research on the influence of fear and rage on bodily functions. And the unraveling of the mystery of sympathin led ultimately to prolonged studies on the chemical mediator that serves to transmit influences from nerve endings to the organs they control.

There are many other examples of serendipity which I might detail; among them Nobel's invention of dynamite, Perkin's stumbling upon the coal-tar dyes, and Pasteur's finding that a vegetable mold causes the watery solution in which it is nurtured to change the direction of the light rays as they pass through. Dynamite placed gigantic powers in the hands of man; the coal-tar dyes have fundamentally affected such varied activities as warfare, textile industries, and medical diagnosis; and Pasteur's casual observation has developed into an immense range of chemical theory and research.

Three legends of accidental leads to fresh insight serve to introduce the next point, which is quite as important as serendipity itself. I refer to the presence of a prepared mind. It is said that the idea of specific gravity came to Archimedes as he noted by chance the buoyancy of his body in water. We have all heard the tale, illustrative even if not authentic, that the concept of a universal law of gravitational force occurred to Isaac Newton when he saw an apple fall from a tree while he lay musing on the grass in an orchard. Of similar import is the story that the possibility of the steam engine suddenly occurred to James Watt when he beheld the periodic lifting of the lid of a tea kettle by the steam pressure within it. Many a man floated in water before Archimedes; apples fell from trees as long ago as the Garden of Eden (exact date uncertain!); and the outrush of steam against resistance could have been noted at any time since the discovery of fire and its use under a covered pot of water. In all three cases it was eons before the significance of these events was perceived. Obviously a chance discovery involves both the phenomenon to be observed and the appreciative, intelligent observer.

I may now add to these legends and their illustrative significance the history of that marvelously powerful enemy of infection, penicillin. In 1929 the English bacteriologist, Alexander Fleming, reported noticing that a culture of pus-producing bacteria underwent dissolution in the neighborhood of a mold which accidentally contaminated it. This was the pregnant hint. A careless worker might have thrown the culture away because of the contamination. Instead, Fleming let the mold grow in broth and thus learned that there passed into the broth from the mold a substance which was highly efficacious in stopping the growth of a wide range of disease-producing germs and destroying them. Furthermore he learned that, when injected, this substance was not itself harmful to animals. The mold, a variety of Penicillium, suggested the name "penicillin." The long struggle of Howard Florey and his associates at Oxford in purifying and standardizing this highly potent agent and in proving its value in human cases cannot be recounted here. The record, however, reports one of the most striking instances of immense value that can result from a combination of chance and an alert intelligence; and shows how a brilliant discovery is made practical by hard labor.

Long ago Pasteur recognized that when accident favors an investigator it must be met by sharp insight, for he uttered the wise and discerning dictum, "*Dans les champs de l'observation, le hasard ne favorise que les esprits preparés.*" Even before Pasteur, Joseph Henry, the American physicist, enunciated the same truth when he said, "The seeds of great discoveries are constantly floating around us, but they only take root in minds well prepared to receive them."

DALLAS LORE SHARP

Turtle Eggs for Agassiz

•

IT IS ONE of the wonders of the world that so few books are written. With every human being a possible book, and with many a human being capable of becoming more books than the world could contain, is it not amazing that the books of men are so few? And so stupid!

I took down, recently, from the shelves of a great public library, the four volumes of Agassiz's *Contributions to the Natural History of the United States*. I doubt if anybody but the charwoman, with her duster, had touched those volumes for twenty-five years. They are an excessively learned, a monumental, an epoch-making work, the fruit of vast and heroic labors, with colored plates on stone, showing the turtles of the United States, and their embryology. The work was published more than half a century ago (by subscription); but it looked old beyond its years—massive, heavy, weathered, as if dug from the rocks. It was difficult to feel that Agassiz could have written it—could have built it, grown it, for the laminated pile had required for its growth the patience and painstaking care of a process of nature, as if it were a kind of printed coral reef. Agassiz do this? The big, human, magnetic man at work upon these pages of capital letters, Roman figures, brackets, and parentheses in explanation of the pages of diagrams and plates! I turned away with a sigh from the weary learning, to read the preface.

When a great man writes a great book he usually flings a preface after it, and thereby saves it, sometimes, from oblivion. Whether so or not, the best thing in most books are their prefaces. It was not, however, the quality of the preface to these great volumes that interested me, but rather the wicked waste of durable book material that went into its making. Reading down through the catalogue of human names and of thanks for help received, I came to a sentence beginning:

"In New England I have myself collected largely; but I have also received valuable contributions from the late Rev. Zadoc Thompson of Burlington

. . . from Mr. D. Henry Thoreau of Concord . . . and from Mr. J. W. P. Jenks of Middleboro." And then it hastens on with the thanks in order to get to the turtles, as if turtles were the one and only thing of real importance in all the world.

Turtles no doubt are important, extremely important, embryologically, as part of our genealogical tree; but they are away down among the roots of the tree as compared with the late Rev. Zadoc Thompson of Burlington. I happen to know nothing about the Rev. Zadoc, but to me he looks very interesting. Indeed any reverend gentleman of his name and day who would catch turtles for Agassiz must have been interesting. And as for Henry Thoreau, we know he was interesting. The rarest wood turtle in the United States was not so rare a specimen as this gentleman of Walden Woods and Concord. We are glad even for this line in the preface about him; glad to know that he tried, in this untranscendental way, to serve his day and generation. If Agassiz had only put a chapter in his turtle book about it! But this is the material he wasted, this and more of the same human sort, for the Mr. "Jenks of Middleboro" (at the end of the quotation) was, years later, an old college professor of mine, who told me some of the particulars of his turtle contributions, particulars which Agassiz should have found a place for in his big book. The preface says merely that this gentleman sent turtles to Cambridge by the thousands—brief and scanty recognition. For that is not the only thing this gentleman did. On one occasion he sent, not turtles, but turtle *eggs* to Cambridge—*brought* them, I should say; and all there is to show for it, so far as I could discover, is a sectional drawing of a bit of the mesoblastic layer of one of the eggs!

Of course, Agassiz wanted to make that mesoblastic drawing, or some other equally important drawing, and had to have the fresh turtle egg to draw from it. He had to have it, and he got it. A great man, when he wants a certain turtle egg, at a certain time, always gets it, for he gets someone else to get it. I am glad he got it. But what makes me sad and impatient is that he did not think it worth while to tell about the getting of it, and so made merely a learned turtle book of what might have been an exceedingly interesting human book.

It would seem, naturally, that there could be nothing unusual or interesting about the getting of turtle eggs when you want them. Nothing at all, if you should chance to want the eggs as you chance to find them. So with anything else—good copper stock, for instance, if you should chance to want it, and should chance to be along when they chance to be giving it away. But if you want copper stock, say of C & H quality, *when* you want it, and are bound to have it, then you must command more than a college professor's salary. And likewise, precisely, when it is turtle eggs that you are bound to have.

Agassiz wanted those turtle eggs when he wanted them—not a minute over three hours from the minute they were laid. Yet even that does not seem exacting, hardly more difficult than the getting of hen eggs only three hours old. Just so, provided the professor could have had his private turtle coop in Harvard Yard; and provided he could have made his turtles lay. But turtles will not respond, like hens, to meat scraps and the warm mash. The professor's problem was not to get from a mud turtle's nest in the back yard to the table in the laboratory; but to get from the laboratory in Cambridge to some pond when the turtles were laying, and back to the laboratory within the limited time. And this, in the days of Darius Green, might have called for nice and discriminating work—as it did.

Agassiz had been engaged for a long time upon his *Contributions*. He had brought the great work nearly to a finish. It was, indeed, finished but for one small yet very important bit of observation: he had carried the turtle egg through every stage of its development with the single exception of one—the very earliest—that stage of first cleavages, when the cell begins to segment, immediately upon its being laid. That beginning stage had brought the *Contributions* to a halt. To get eggs that were fresh enough to show the incubation at this period had been impossible.

There were several ways that Agassiz might have proceeded: he might have got a leave of absence for the spring term, taken his laboratory to some pond inhabited by turtles, and there camped until he should catch the reptile digging out her nest. But there were difficulties in all of that—as those who are college professors and naturalists know. As this was quite out of the question, he did the easiest thing—asked Mr. "Jenks of Middleboro" to get him the eggs. Mr. Jenks got them. Agassiz knew all about his getting of them; and I say the strange and irritating thing is that Agassiz did not think it worth while to tell us about it, at least in the preface to his monumental work.

It was many years later that Mr. Jenks, then a gray-haired college professor, told me how he got those eggs to Agassiz.

"I was principal of an academy, during my younger years," he began, "and was busy one day with my classes, when a large man suddenly filled the doorway of the room, smiled to the four corners of the room, and called out with a big, quick voice that he was Professor Agassiz.

"Of course he was. I knew it, even before he had had time to shout it to me across the room.

"Would I get him some turtle eggs? he called. Yes, I would. And would I get them to Cambridge within three hours from the time they were laid? Yes, I would. And I did. And it was worth the doing. But I did it only once.

"When I promised Agassiz those eggs I knew where I was going to get them. I had got turtle eggs there before—at a particular patch of sandy shore along a pond, a few miles distant from the academy.

"Three hours was the limit. From the railroad station to Boston was thirty-five miles; from the pond to the station was perhaps three or four miles; from Boston to Cambridge we called about three miles. Forty miles in round numbers! We figured it all out before he returned, and got the trip down to two hours—record time: driving from the pond to the station; from the station by express train to Boston; from Boston by cab to Cambridge. This left an easy hour for accidents and delays.

"Cab and car and carriage we reckoned into our time-table; but what we didn't figure on was the turtle." And he paused abruptly.

"Young man," he went on, his shaggy brows and spectacles hardly hiding the twinkle in the eyes that were bent severely upon me, "young man, when *you* go after turtle eggs, take into account the turtle. No! no! That's bad advice. Youth never reckons on the turtle—and youth seldom ought to. Only old age does that; and old age would never have got those turtle eggs to Agassiz.

"It was in the early spring that Agassiz came to the academy, long before there was any likelihood of the turtles laying. But I was eager for the quest, and so fearful of failure that I started out to watch at the pond fully two weeks ahead of the time that the turtles might be expected to lay. I remember the date clearly: it was May 14.

"A little before dawn—along near three o'clock—I would drive over to the pond, hitch my horse near by, settle myself quietly among some thick cedars close to the sandy shore, and there I would wait, my kettle of sand ready, my eye covering the whole sleeping pond. Here among the cedars I would eat my breakfast, and then get back in good season to open the academy for the morning session.

"And so the watch began.

"I soon came to know individually the dozen or more turtles that kept to my side of the pond. Shortly after the cold mist would lift and melt away they would stick up their heads through the quiet water; and as the sun slanted down over the ragged rim of tree tops the slow things would float into the warm, lighted spots, or crawl out and doze comfortably on the hummocks and snags.

"What fragrant mornings those were! How fresh and new and unbreathed! The pond odors, the woods odors, the odors of the ploughed fields—of water lily, and wild grape, and the dew-laid soil! I can taste them yet, and hear them yet—the still, large sounds of the waking day—the pickerel breaking the quiet with his swirl; the kingfisher dropping anchor; the stir of feet and wings among the trees. And then the thought of the great book being held up for me! Those were rare mornings!

"But there began to be a good many of them, for the turtles showed no desire to lay. They sprawled in the sun, and never one came out upon the sand as if she intended to help on the great professor's book. The embryology of her eggs was of small concern to her; her contribution to the Natural History of the United States could wait.

"And it did wait. I began my watch on the fourteenth of May; June first found me still among the cedars, still waiting, as I had waited every morning, Sundays and rainy days alike. June first saw a perfect morning, but every turtle slid out upon her log, as if egg laying might be a matter strictly of next year.

"I began to grow uneasy—not impatient yet, for a naturalist learns his lesson of patience early, and for all his years; but I began to fear lest, by some subtle sense, my presence might somehow be known to the creatures; that they might have gone to some other place to lay, while I was away at the schoolroom.

"I watched on to the end of the first week, on to the end of the second week in June, seeing the mists rise and vanish every morning, and along with them vanish, more and more, the poetry of my early morning vigil. Poetry and rheumatism cannot long dwell together in the same clump of cedars, and I had began to feel the rheumatism. A month of morning mists wrapping me around had at last soaked through to my bones. But Agassiz was waiting, and the world was waiting, for those turtle eggs; and I would wait. It was all I could do, for there is no use bringing a china nest egg to a turtle; she is not open to any such delicate suggestion.

"Then came a mid-June Sunday morning, with dawn breaking a little after three: a warm, wide-awake dawn, with the level mist lifted from the level surface of the pond a full hour higher than I had seen it any morning before.

"This was the day: I knew it. I have heard persons say that they can hear the grass grow; that they know by some extra sense when danger is nigh. That we have these extra senses I fully believe, and I believe they can be sharpened by cultivation. For a month I had been watching, brooding over this pond, and now I knew. I felt a stirring of the pulse of things that the cold-hearted turtles could no more escape than could the clods and I.

"Leaving my horse unhitched, as if he too understood, I slipped eagerly into my covert for a look at the pond. As I did so, a large pickerel ploughed a furrow out through the spatter-docks, and in his wake rose the head of an enormous turtle. Swinging slowly around, the creature headed straight for the shore, and without a pause scrambled out on the sand.

"She was about the size of a big scoop shovel; but that was not what excited me, so much as her manner, and the gait at which she moved; for there was method in it, and fixed purpose. On she came, shuffling over the sand toward the higher open fields, with a hurried, determined seesaw that was taking her somewhere in particular, and that was bound to get her there on time.

"I held my breath. Had she been a dinosaurian making Mesozoic footprints, I could not have been more fearful. For footprints in the Mesozoic mud, or in the sands of time, were as nothing to me when compared with fresh turtle eggs in the sands of this pond.

"But over the strip of sand, without a stop, she paddled, and up a narrow cow path into the high grass along a fence. Then up the narrow cow path, on all fours, just like another turtle, I paddled, and into the high wet grass along the fence.

"I kept well within sound of her, for she moved recklessly, leaving a trail of flattened grass a foot and a half wide. I wanted to stand up—and I don't believe I could have turned her back with a rail—but I was afraid if she saw me that she might return indefinitely to the pond; so on I went, flat to the ground, squeezing through the lower rails of the fence, as if the field beyond were a melon patch. It was nothing of the kind, only a wild, uncomfortable pasture, full of dewberry vines, and very discouraging. They were excessively wet vines and briery. I pulled my coat sleeves as far over my fists as I could get them, and, with the tin pail of sand swinging from between my teeth to avoid noise, I stumped fiercely, but silently, on after the turtle.

"She was laying her course, I thought, straight down the length of this dreadful pasture, when, not far from the fence, she suddenly hove to, warped herself short about, and came back, barely clearing me, at a clip that was thrilling. I warped about, too, and in her wake bore down across the corner of the pasture, across the powdery public road, and on to a fence along a field of young corn.

"I was somewhat wet by this time, but not so wet as I had been before, wallowing through the deep dry dust of the road. Hurrying up behind a large tree by the fence, I peered down the corn rows and saw the turtle stop, and begin to paw about in the loose soft soil. She was going to lay!

"I held on to the tree and watched, as she tried this place, and that place, and the other place—the eternally feminine! But *the* place, evidently, was hard to find. What could a female turtle do with a whole field of possible nests to choose from? Then at last she found it, and, whirling about, she backed quickly at it, and, tail first, began to bury herself before my staring eyes.

"Those were not the supreme moments of my life; perhaps those moments came later that day; but those certainly were among the slowest, more dreadfully mixed of moments that I ever experienced. They were hours long. There she was, her shell just showing, like some old hulk in the sand alongshore. And how long would she stay there? And how should I know if she had laid an egg?

"I could still wait. And so I waited, when, over the freshly awakened fields, floated four mellow strokes from the distant town clock.

"Four o'clock! Why, there was no train until seven! No train for three hours! The eggs would spoil! Then with a rush it came over me that this was Sunday morning, and there was no regular seven o'clock train—none till after nine.

"I think I should have fainted had not the turtle just then begun crawling off. I was weak and dizzy; but there, there in the sand, were the eggs! And Agassiz! And the great book! And I cleared the fence, and the forty miles that lay between me and Cambridge, at a single jump. He should have them, trains or no. Those eggs should go to Agassiz by seven o'clock, if I had to gallop every mile of the way. Forty miles! Any horse could cover it in three hours, if he had to; and, upsetting the astonished turtle, I scooped out her round white eggs.

"On a bed of sand in the bottom of the pail I laid them, with what care my trembling fingers allowed; filled in between them with more sand; so with another layer to the rim; and, covering all smoothly with more sand, I ran back for my horse.

"That horse knew, as well as I, that the turtle had laid, and that he was to get those eggs to Agassiz. He turned out of that field into the road on two wheels, a thing he had not done for twenty years, doubling me up before the dashboard, the pail of eggs miraculously lodged between my knees.

"I let him out. If only he could keep this pace all the way to Cambridge! Or even halfway there; and I should have time to finish the trip on foot. I shouted him on, holding to the dasher with one hand, the pail of eggs with the other, not daring to get off my knees, though the bang on them, as we pounded down the wood road, was terrific. But nothing must happen to the eggs; they must not be jarred, or even turned over in the sand before they came to Agassiz.

"In order to get out on the pike it was necessary to drive back away from Boston toward the town. We had nearly covered the distance, and were rounding a turn from the woods into the open fields, when, ahead of me, at the station it seemed, I heard the quick sharp whistle of a locomotive.

"What did it mean? Then followed the *puff, puff, puff* of a starting train. But what train? Which way going? And, jumping to my feet for a longer view, I pulled into a side road that paralleled the track, and headed hard for the station.

"We reeled along. The station was still out of sight, but from behind the bushes that shut it from view rose the smoke of a moving engine. It was perhaps a mile away, but we were approaching, head-on, and, topping a little hill, I swept down upon a freight train, the black smoke pouring from the stack, as the mighty creature pulled itself together for its swift run down the rails.

"My horse was on the gallop, going with the track, and straight toward the coming train. The sight of it almost maddened me—the bare thought of it, on the road to Boston! On I went; on it came, a half—a quarter of a mile between us, when suddenly my road shot out along an unfenced field with only a level stretch of sod between me and the engine.

"With a pull that lifted the horse from his feet, I swung him into the field and sent him straight as an arrow for the track. That train should carry me and my eggs to Boston!

"The engineer pulled the rope. He saw me standing up in the rig, saw my hat blow off, saw me wave my arms, saw the tin pail swing in my teeth, and he jerked out a succession of sharp halts! But it was he who should halt, not I; and on we went, the horse with a flounder landing the carriage on top of the track.

"The train was already grinding to a stop; but before it was near a stand-still I had backed off the track, jumped out, and, running down the rails with the astonished engineers gaping at me, had swung aboard the cab.

"They offered no resistance; they hadn't had time. Nor did they have the disposition, for I looked strange, not to say dangerous. Hatless, dew-soaked, smeared with yellow mud, and holding, as if it were a baby or a bomb, a little tin pail of sand.

" 'Crazy,' the fireman muttered, looking to the engineer for his cue.

"I had been crazy, perhaps, but I was not crazy now.

" 'Throw her wide open,' I commanded. 'Wide-open! These are fresh turtle eggs for Professor Agassiz of Cambridge. He must have them before breakfast.'

"Then they knew I was crazy, and, evidently thinking it best to humor me, threw the throttle wide-open, and away we went.

"I kissed my hand to the horse, grazing unconcernedly in the open field, and gave a smile to my crew. That was all I could give them, and hold myself and the eggs together. But the smile was enough. And they smiled through their smut at me, though one of them held fast to his shovel, while the other kept his hand upon a big ugly wrench. Neither of them spoke to me, but above the roar of the swaying engine I caught enough of their broken talk to understand that they were driving under a full head of steam, with the intention of handing me over to the Boston police, as perhaps the easiest way of disposing of me.

"I was only afraid that they would try it at the next station. But that station whizzed past without a bit of slack, and the next, and the next; when it came over me that this was the through freight, which should have passed in the night, and was making up lost time.

"Only the fear of the shovel and the wrench kept me from shaking hands with both men at this discovery. But I beamed at them; and they at me. I was enjoying it. The unwonted jar beneath my feet was wrinkling my diaphragm with spasms of delight. And the fireman beamed at the engineer, with a look that said, 'See the lunatic grin; he likes it!'

"He did like it. How the iron wheels sang to me as they took the rails! How the rushing wind in my ears sang to me! From my stand on the fireman's side of the cab I could catch a glimpse of the track just ahead of the engine, where the ties seemed to leap into the throat of the mile-devouring monster. The joy of it! Of seeing space swallowed by the mile!

"I shifted the eggs from hand to hand and thought of my horse, of Agassiz, of the great book, of my great luck—luck—luck—until the multitudinous tongues of the thundering train were all chiming 'luck! luck! luck!' They knew! They understood! This beast of fire and tireless wheels was doing its very best to get the eggs to Agassiz!

"We swung out past the Blue Hills, and yonder flashed the morning sun from the towering dome of the State House. I might have leaped from the cab and run the rest of the way on foot, had I not caught the eye of the engineer watching me narrowly. I was not in Boston yet, nor in Cambridge either. I was an escaped lunatic, who had held up a train, and forced it to carry me to Boston.

"Perhaps I had overdone my lunacy business. Suppose these two men should take it into their heads to turn me over to the police, whether I would or no? I could never explain the case in time to get the eggs to Agassiz. I looked at my watch. There were still a few minutes left, in which I might explain to these men, who, all at once, had become my captors. But it was too late. Nothing could avail against my actions, my appearance, and my little pail of sand.

"I had not thought of my appearance before. Here I was, face and clothes caked with yellow mud, my hair wild and matted, my hat gone, and in my full-grown hands a tiny tin pail of sand, as if I had been digging all night with a tiny tin shovel on the shore! And thus to appear in the decent streets of Boston of a Sunday morning!

"I began to feel like a hunted criminal. The situation was serious, or might be, and rather desperately funny at its best. I must in some way have shown my new fears, for both men watched me more sharply.

"Suddenly, as we were nearing the outer freight yard, the train slowed down and came to a stop. I was ready to jump, but I had no chance. They had nothing to do, apparently, but to guard me. I looked at my watch again. What time we had made! It was only six o'clock, with a whole hour to get to Cambridge.

"But I didn't like this delay. Five minutes—ten—went by.

" 'Gentlemen,' I began, but was cut short by an express train coming past. We were moving again, on—into a siding; on—on to the main track; and on with a bump and a crash and a succession of crashes, running the length of the train; on at a turtle's pace, but on, when the fireman, quickly jumping for the bell rope, left the way to the step free, and—the chance had come!

"I never touched the step, but landed in the soft sand at the side of the track, and made a line for the yard fence.

"There was no hue or cry. I glanced over my shoulder to see if they were after me. Evidently their hands were full, and they didn't know I had gone.

"But I had gone; and was ready to drop over the high board fence, when it occurred to me that I might drop into a policeman's arms. Hanging my pail in a splint on top of a post, I peered cautiously over—a very wise thing to do before you jump a high board fence. There, crossing the open square toward the station, was a big, burly fellow with a club—looking for me.

"I flattened for a moment, when someone in the yard yelled at me. I preferred the policeman, and, grabbing my pail, I slid over to the street. The policeman moved on past the corner of the station out of sight. The square was free, and yonder stood a cab!

"Time was flying now. Here was the last lap. The cabman saw me coming, and squared away. I waved a paper dollar at him, but he only stared the more. A dollar can cover a good deal, but I was too much for one dollar. I pulled out another, thrust them both at him, and dodged into the cab, calling, 'Cambridge!'

"He would have taken me straight to the police station had I not said, 'Harvard College. Professor Agassiz's house! I've got eggs for Agassiz'; and pushed another dollar up at him through the hole.

"It was nearly half past six.

" 'Let him go!' I ordered. "Here's another dollar if you make Agassiz's house in twenty minutes. Let him out; never mind the police!'

"He evidently knew the police, or there were none around at that time on a Sunday morning. We went down the sleeping streets as I had gone down the wood roads from the pond two hours before, but with the rattle and crash now of a fire brigade. Whirling a corner into Cambridge Street, we took the bridge at a gallop, the driver shouting out something in Hibernian to a pair of waving arms and a belt and brass buttons.

"Across the bridge with a rattle and jolt that put the eggs in jeopardy, and on over the cobblestones, we went. Half standing, to lessen the jar, I held the pail in one hand and held myself in the other, not daring to let go even to look at my watch.

"But I was afraid to look at the watch. I was afraid to see how near to seven o'clock it might be. The sweat was dropping from my nose, so close was I running to the limit of my time.

"Suddenly there was a lurch, and I dived forward, ramming my head into the front of the cab, coming up with a rebound that landed me across the small of my back on the seat, and sent half of my pail of eggs helter-skelter over the floor.

"We had stopped. Here was Agassiz's house; and without taking time to pick up the scattered eggs I tumbled out, and pounded at the door.

"No one was astir in the house. But I would stir them. And I did. Right in the midst of the racket the door opened. It was the maid.

" 'Agassiz,' I gasped, 'I want Professor Agassiz, quick!' And I pushed by her into the hall.

" 'Go 'way, sir. I'll call the police. Professor Agassiz is in bed. Go 'way, sir!'

" 'Call him—Agassiz—instantly, or I'll call him myself.'

"But I didn't; for just then a door overhead was flung open, a great white-robed figure appeared on the dim landing above, and a quick loud voice called excitedly:

" 'Let him in! Let him in! I know him. He has my turtle eggs!'

"And the apparition, slipperless, and clad in anything but an academic gown, came sailing down the stairs.

"The maid fled. The great man, his arms extended, laid hold of me with both hands, and, dragging me and my precious pail into his study, with a swift, clean stroke laid open one of the eggs, as the watch in my trembling hands ticked its way to seven—as if nothing unusual were happening to the history of the world."

"You were in time, then?" I said.

"To the tick. There stands my copy of the great book. I am proud of the humble part I had in it."

OLIVER LA FARGE

Scientists Are Lonely Men

•

SCIENCE has never been well understood. Scientists have never been good at explaining themselves and, frustrated by this, they tend to withdraw into the esoteric, refer to the public as "laymen," and develop incomprehensible vocabularies from which they draw a naïve, secret-society feeling of superiority.

What is the special nature of a scientist as distinguished from a soda-jerker? Not just the externals such as his trick vocabulary, but the human formation within the man? Most of what is written about him is rot; but there is stuff there which a writer can get his teeth into, and it has its vivid, direct relation to all that we are fighting for.

The inner nature of science within the scientist is both emotional and intellectual. The emotional element must not be overlooked, for without it there is no sound research on however odd and dull-seeming a subject. As is true of all of us, an emotion shapes and forms the scientist's life; at the same time an intellectual discipline molds his thinking, stamping him with a character as marked as a seaman's although much less widely understood.

To an outsider who does not know of this emotion, the scientist suggests an ant, putting forth great efforts to lug one insignificant and apparently unimportant grain of sand to be added to a pile, and much of the time his struggle seems as pointless as an ant's. I can try to explain why he does it and what the long-term purpose is behind it through an example from my own work. Remember that in this I am not thinking of the rare, fortunate geniuses like the Curies, Darwin, or Newton, who by their own talents and the apex of accumulated thought at which they stood were knowingly in pursuit of great, major discoveries. This is the average scientist, one among thousands, obscure, unimportant, toilsome.

I have put in a good many months of hard work, which ought by usual standards to have been dull but was not, on an investigation as yet unfinished to prove that Kanhobal, spoken by certain Indians in Guatemala, is not a dialect of Jacalteca, but that, on the contrary, Jacalteca is a dialect of Kanhobal. Ridiculous, isn't it? Yet to me the matter is not only serious but exciting. Why?

There is an item of glory. There are half a dozen or so men now living who will pay me attention and respect if I prove my thesis. A slightly larger number, less interested in the details of my work, will give credit to La Farge for having added to the linguistic map of Central America the name of a hitherto unnoted dialect. But not until I have told a good deal more can I explain—as I shall presently—why the notice of so few individuals can constitute a valid glory.

There's the nature of the initial work. I have spent hours, deadly, difficult hours, extracting lists of words, paradigms of verbs, constructions, idioms, and the rest from native informants, often at night in over-ventilated huts while my hands turned blue with cold. (Those mountains are far from tropical.) An illiterate Indian tires quickly when giving linguistic information. He is not accustomed to thinking of words in terms of other words; his command of Spanish is so poor that again and again you labor over misunderstandings; he does not think in our categories of words. Take any schoolchild and ask him how you say, "I go." Then ask him in turn, "Thou goest, he goes, we go." Even the most elementary schooling has taught him, if only from the force of staring resentfully at the printed page, to think in terms of the present tense of a single verb—that is, to conjugate. He will give you, in Spanish for instance, "*Me voy, te vas, se va, nos vamos,*" all in order. Try this on an illiterate Indian. He gives you his equivalent of "I go," follows it perhaps with "thou goest," but the next question reminds him of his son's departure that morning for Ixtatán, so he answers "he sets out," and from that by another mental leap produces "we are traveling." This presents the investigator with a magnificently irregular verb. He starts checking back, and the Indian's mind being set in the new channel, he now gets "I travel" instead of "I go."

There follows an exhausting process of inserting an alien concept into the mind of a man with whom you are communicating tenuously in a language which you speak only pretty well and he quite badly.

Then of course you come to a verb which really is irregular and you mistrust it. Both of you become tired, frustrated, upset. At the end of an hour or so the Indian is worn out, his friendship for you has materially decreased, and you yourself are glad to quit.

Hours and days of this, and it's not enough. I have put my finger upon the village of Santa Eulalia and said, "Here is the true, the classic Kanhobal from which the other dialects diverge." Then I must sample the others; there are at least eight villages which must yield me up fairly complete word-lists and two from which my material should be as complete as from Santa Eulalia. More hours and more days, long horseback trips across the mountains to enter strange, suspicious settlements, sleep on the dirt floor of the schoolhouse, and persuade the astonished yokelry that it is a good idea, a delightful idea, that you should put "The Tongue" into writing. Bad food, a bout of malaria, and the early-morning horror of seeing your beloved horse's neck running blood from vampire bats ("Oh, but, yes, señor, everyone knows that here are very troublesome the vampire bats"), to get the raw material for proving that Jacalteca is a dialect of Kanhobal instead of . . .

You bring your hard-won data back to the States and you follow up with a sort of detective-quest for obscure publications and old manuscripts which may show a couple of words of the language as it was spoken a few centuries ago, so that you can get a line on its evolution. With great labor you unearth and read the very little that has been written bearing upon this particular problem.

By now the sheer force of effort expended gives your enterprise value in your own eyes. And you still have a year's work to put all your data in shape, test your conclusions, and demonstrate your proof.

Yet the real emotional drive goes beyond all this. Suppose I complete my work and prove, in fact, that Kanhobal as spoken in Santa Eulalia is a language in its own right and the classic tongue from which Jacalteca has diverged under alien influences, and that, further, I show just where the gradations of speech in the intervening villages fit in. Dear God, what a small, dull grain of sand!

But follow the matter a little farther. Jacalteca being relatively well-known (I can, offhand, name four men who have given it some study), from it it has been deduced that this whole group of dialects is most closely related to the languages spoken south and east of these mountains. If my theory is correct, the reverse is true—the group belongs to the Northern Division of the Mayan Family. This fact, taken along with others regarding physical appearance, ancient remains, and present culture, leads to a new conclusion about the direction from which these tribes came into the mountains: a fragment of the ancient history of what was once a great, civilized people comes into view. So now my tiny contribution begins to be of help to men working in other branches of anthropology than my own, particularly to the archaeologists; it begins to help toward an eventual understanding of the whole picture in this area: the important question of, not what these people are today, but how they got that way and what we can learn from that about all human behavior including our own.

Even carrying the line of research as far as this assumes that my results have been exploited by men of greater attainments than I. Sticking to the linguistic line, an error has been cleared away, an advance has been made in our understanding of the layout and interrelationship of the many languages making up the Mayan Family. With this we come a step nearer to working out the processes by which these languages became different from one another and hence to determining the archaic, ancestral roots of the whole group.

So far as we know at present, there are not less than eight completely unrelated language families in America north of Panama. This is unreasonable: there are hardly that many families among all the peoples of the Old World. Twenty years ago we recognized not eight, but forty. Some day perhaps we shall cut the total to four. The understanding of the Mayan process is a step toward that day; it is unlikely that Mayan will remain an isolated way of speech unconnected with any other. We know now that certain tribes in Wyoming speak languages akin to those of others in Panama; we have charted the big masses and islands of that group of tongues and from the chart begin to see the outlines of great movements and crashing historical events in the dim past. If we should similarly develop a relationship between Mayan and, let's say, the languages of the Mississippi Valley, again we should offer something provocative to the archaeologist, the historian, the student of mankind. Some day we shall show an unquestionable kinship between some of these families and certain languages of the Old World and with it cast a new light on the dim subject of the peopling of the Americas, something to guide our minds back past the Arctic to dark tribes moving blindly from the high plateaus of Asia.

My petty detail has its place in a long project carried out by many men which will serve not only the history of language but the broad scope of history itself. It goes farther than that. The humble Pah-Utes of Nevada speak a tongue related to that which the subtle Montezuma used, the one narrow in scope, evolved only to meet the needs of a primitive people, the other sophisticated, a capable instrument for poetry, for an advanced governmental system, and for philosophical speculation. Men's thoughts make language and their languages make thought. When the matter of the speech of mankind is fully known and laid side by side with all the other knowledges, the philosophers, the men who stand at the gathering-together point of science, will have the means to make man understand himself at last.

Of course no scientist can be continuously aware of such remote possible consequences of his labors; in fact the long goal is so remote that if he kept his eyes on it he would become hopelessly discouraged over the half inch of progress his own life's work will represent. But it was the vision of this which first made him choose his curious career, and it is an emotional sense of the great structure of scientific knowledge to which his little grain will be added which drives him along.

II

I spoke of the item of glory, the half dozen colleagues who will appreciate one's work. To understand that one must first understand the *isolation* of research, a factor which has profound effects upon the scientist's psyche.

The most obvious statement of this is in the public attitude and folk-literature about "professors." The titles and subjects of Ph.D. theses have long been sources of exasperated humor among us; we are all familiar with the writer's device which ascribes to a professorial character an intense interest in some such matter as the development of the molars in pre-Aurignacian man or the religious sanctions of the Levirate in north-eastern Australia, the writer's intention being that the reader shall say "Oh God!," smile slightly, and pigeonhole the character. But what do you suppose is the effect of the quite natural public attitude behind these devices upon the man who is excitedly interested in pre-Aurignacian molars and who knows that this is a study of key value in tracing the evolution of *Homo sapiens*?

Occasionally some line of research is taken up and made clear, even fascinating, to the general public, as in Zinsser's *Rats, Lice and History,* or de Kruif's rather Sunday-supplement writings. Usually, as in these cases, they deal with medicine or some other line of work directly resulting in findings of vital interest to the public. Then the ordinary man will consent to understand, if not the steps of the research itself, at least its importance, will grant the excitement, and honor the researcher. When we read Eve Curie's great biography of her parents our approach to it is colored by our knowledge, forty years later, of the importance of their discovery to every one of us. It would have been quite possible at the time for a malicious or merely ignorant writer to have presented that couple as archetypes of the "professor," performing incomprehensible acts of self-immolation in pursuit of an astronomically unimportant what's-it.

Diving to my own experience like a Stuka with a broken wing, I continue to take my examples from my rather shallow linguistic studies because, in its very nature, the kind of thing a linguist studies is so beautifully calculated to arouse the "Oh God!" emotion.

It happened that at the suggestion of my betters I embarked upon an ambitious, general comparative study of the whole Mayan Family. The farther in I got the farther there was to go and the more absorbed I became. Puzzle piled upon puzzle to be worked out and the solution used for getting after the next one, the beginning of order in chaos, the glimpse of understanding at the far end. Memory, reasoning faculties, realism, and imagination were all on the stretch; I was discovering the full reach of whatever mental powers I had. When I say that I became absorbed I mean absorbed; the only way to do such research is to roll in it, become soaked in it, live it, breathe it, have your system so thoroughly permeated with it that at the half glimpse of a fugitive possibility everything you have learned so far and everything you have been holding in suspension is in order and ready to prove or disprove that point. You do not only think about your subject while the documents are spread before you; everyone knows that some of our best reasoning is done when the surface of the mind is occupied with something else and the deep machinery of the brain is free to work unhampered.

One day I was getting aboard a trolley car in New Orleans on my way to Tulane University. As I stepped up I saw that if it were possible to prove that a prefixed *s-* could change into a prefixed *y-* a whole series of troublesome phenomena would fall into order. The transition must come through *u-* and, thought I with a sudden lift of excitement, there may be a breathing associated with *u-* and that may make the whole thing possible. As I paid the conductor I thought that the evidence I needed might exist in Totonac and Tarascan, non-Mayan languages with which I was not familiar. The possibilities were so tremendous that my heart pounded and I was so preoccupied that I nearly went to sit in the Jim Crow section. Speculation was useless until I could reach the University and dig out the books, so after a while I calmed myself and settled to my morning ration of Popeye, who was then a new discovery too. As a matter of fact, the idea was no good, but the incident is a perfect example of the "professor mind."

Of course, if as I stepped on to the car it had dawned upon me that the reason my girl's behavior last evening had seemed odd was that she had fallen for the Englishman we had met, the incident would not have seemed so funny, although the nature of the absorption, subconscious thinking, and realization would have been the same in both cases.

I lived for a month with the letter *k*. If we have three words in Quiché, one of the major Mayan languages, beginning with *k*, in Kanhobal we are likely to find that one of these begins with *ch*. Moving farther west and north, in Tzeltal one is likely to begin with *k*, one with *ch*, and the one which began with *ch* in Kanhobal to begin with *ts*. In Hausteca, at the extreme northwest, they begin with *k*, *ts*, and plain *s* respectively. Why don't they all change alike? Which is the original form? Which way do these changes run, or from which point do they run both ways? Until those questions can be answered we cannot even guess at the form of the mother tongue from which these languages diverged, and at that point all investigation halts. Are these *k*'s in Quiché pronounced even faintly unlike? I noticed no difference between the two in Kanhobal, but then I wasn't listening for it. I wished someone properly equipped would go and listen to the Quiché Indians, and wondered if I could talk the University into giving me money enough to do so.

This is enough to give some idea of the nature of my work, and its uselessness for general conversation. My colleagues at Tulane were archaeologists. Shortly after I got up steam they warned me frankly that I had to stop trying to tell them about the variability of *k*, the history of Puctun t^y, or any similar matter. If I produced any results that they could apply, I could tell them about it; but apart from that I could keep my damned sound-shifts and intransitive infixes to myself; I was driving them nuts. My other friends on the faculty were a philosopher and two English professors; I was pursuing two girls at the time but had not been drawn to either because of intellectual interests in common; my closest friends were two painters and a sculptor. The only person I could talk to was myself.

The cumulative effect of this non-communication was terrific. A strange, mute work, a thing crying aloud for discussion, emotional expression, the check and reassurance of another's point of view, turned in upon myself to boil and fume, throwing upon me the responsibility of being my own sole check, my own impersonal, external critic. When finally I came to New York on vacation I went to see my Uncle John. He doesn't know Indian languages but he is a student of linguistics, and I shall never forget the relief, the reveling pleasure, of pouring my work out to him.

Thus at the vital point of his life-work the scientist is cut off from communication with his fellow-men. Instead, he has the society of two, six, or twenty men and women who are working in his specialty, with whom he corresponds, whose letters he receives like a lover, with whom when he meets them he wallows in an orgy of talk, in the keen pleasure of conclusions and findings compared, matched, checked against one another—the pure joy of being really understood.

The praise and understanding of those two or six become for him the equivalent of public recognition. Around these few close colleagues is the larger group of workers in the same general field. They do not share with one in the steps of one's research, but they can read the results, tell in a general way if they have been soundly reached, and profit by them. To them McGarnigle "has shown" that there are traces of an ancient, dolichocephalic strain among the skeletal remains from Pusilhá, which is something they can use. Largely on the strength of his close colleagues' judgment of him, the word gets round that McGarnigle is a sound man. You can trust his work. He's the fellow you want to have analyze the material if you turn up an interesting bunch of skulls. All told, including men in allied fields who use his findings, some fifty scientists praise him; before them he has achieved international reputation. He will receive honors. It is even remotely possible that he might get a raise in salary.

McGarnigle disinters himself from a sort of fortress made of boxes full of skeletons in the cellar of Podunk University's Hall of Science, and emerges into the light of day to attend a Congress. At the Congress he delivers a paper entitled *Additional Evidence of Dolichocephaly among the Eighth Cycle Maya* before the Section on Physical Anthropology. In the audience are six archaeologists specializing in the Maya field, to whom these findings have a special importance, and twelve physical anthropologists including Gruenwald of Eastern California, who is the only other man working on Maya remains.

After McGarnigle's paper comes Gruenwald's turn. Three other physical anthropologists, engaged in the study of the Greenland Eskimo, the Coastal Chinese, and the Pleistocene Man of Lake Mojave respectively, come in. They slipped out for a quick one while McGarnigle was speaking because his Maya work is not particularly useful to them and they can read the paper later; what is coming next, with its important bearing on method and theory, they would hate to miss.

Gruenwald is presenting a perfectly horrible algebraic formula and a diagram beyond Rube Goldberg's wildest dream, showing *A Formula for Approximating the Original Indices of Artificially Deformed Crania.* (These titles are not mere parodies; they are entirely possible.) The archaeologists depart hastily to hear a paper in their own section on *Indications of an Early Quinary System at Uaxactún.* The formula is intensely exciting to McGarnigle because it was the custom of the ancient Mayas to remodel the heads of their children into shapes which they (erroneously) deemed handsomer than nature's. He and Gruenwald have been corresponding about this; at one point Gruenwald will speak of his colleague's experience in testing the formula; he has been looking forward to this moment for months.

After the day's sessions are over will come something else he has been looking forward to. He and Gruenwald, who have not seen each other in two years, go out and get drunk together. It is not that they never get drunk at home, but that now when in their cups they can be uninhibited, they can talk their own, private, treble-esoteric shop. It is an orgy of release.

III

In the course of their drinking it is likely—if an archaeologist or two from the area joins them it is certain—that the talk will veer from femoral pilasters and alveolar prognathism to personal experiences in remote sections of the Petén jungle. For in my science and a number of others there is yet another frustration.

We go into the field and there we have interesting experiences. The word "adventure" is taboo and "explore" is used very gingerly. But the public mind has been so poisoned by the outpourings of bogus explorers that it is laden with claptrap about big expeditions, dangers, hardships, hostile tribes, the lighting of red flares around the camp to keep the savages at bay, and God knows what rot. (I can speak freely about this because my own expeditions have been so unambitious and in such easy country that I don't come into the subject.) As a matter of fact it is generally true that *for a scientist on an expedition to have an adventure is evidence of a fault in his technique.* He is sent out to gather information, and he has no business getting into "a brush with the natives."

The red-flare, into-the-unknown, hardship-and-danger boys, who manage to find a tribe of pink-and-green Indians, a lost city, or the original, hand-painted descendants of the royal Incas every time they go out, usually succeed in so riling the natives and local whites upon whom scientists must depend if they are to live in the country as to make work in the zones they contaminate difficult for years afterward. The business of their adventures and discoveries is sickening.

These men by training express themselves in factual, "extensional" terms, which don't make for good adventure stories. They understandably lean over backward to avoid sounding even remotely like the frauds, the "explorers." And then what they have seen and done lacks validity to them if it cannot be told in relation to the purpose and dominant emotion which sent them there. McGarnigle went among the independent Indians of Icaiché because he had heard of a skull kept in one of their temples which, from a crude description, seemed to have certain important characteristics. All his risks and his maneuverings with those tough, explosive Indians centered around the problem of gaining access to that skull. When he tries to tell an attractive girl about his experiences he not only understates, but can't keep from stressing the significance of a skull with a healed, clover-leaf trepan. The girl gladly leaves him for the nearest broker.

It is too bad both for the scientists and the public that they are so cut off from each other. The world needs now not the mere knowledges of science, but the way of thought and the discipline. It is more than skepticism, the weighing of evidence, more even than the love of truth. It is the devotion of oneself to an end which is far more important than the individual, the certainty that the end is absolutely good, not only for oneself but for all mankind, and the character to set personal advantage, comfort, and glory aside in the devoted effort to make even a little progress toward it.

Part II

•

THE PHYSICAL WORLD

A. *The Mathematical Universe*

The development of mathematics has gone hand in hand with the growth of science. From the days of Galileo and Newton the physicist has thought in mathematical terms. The modern biologist measures the exactitude of his research by his ability to describe it mathematically. Even such a formerly "nonexact" science as psychology depends increasingly on the techniques of mathematics. And in practical terms the growth of automation is a tribute to its importance.

What is the nature of this discipline which is held in awe by the layman and in total misunderstanding by many of us? In The Study of Mathematics *a philosopher, essayist, political thinker and outstanding mathematician, Bertrand Russell, discusses the essence of the subject.*

The advance of mathematics has been paralleled not only by the advance of science but also by that of all civilization. In remote ages man learned first to count, then to measure, finally to calculate. From the long and fascinating history of mathematics, we select one brief and tragic episode in Évariste Galois *by George Sarton, the famous historian of science. It is the biography of a great mathematician, done to death at the age of twenty.*

The growth of automation has been mentioned above. It has resulted in large part from the development of computers, of "Giant Brains" which perform many of the functions of human brains with far greater speed and accuracy. Edmund Callis Berkeley, who has participated in the development of large-scale computers, writes amusingly and illuminatingly about these modern Frankensteins. Our final selection, Mathematics and Modern Civilization, *written by members of the staff of the Department of Mathematics, Washington Square College, New York University, examines the cultural aspects of mathematics, the contributions it can make to the individual's appreciation of knowledge in general.*

BERTRAND RUSSELL

The Study of Mathematics

•

IN REGARD to every form of human activity it is necessary that the question should be asked from time to time, What is its purpose and ideal? In what way does it contribute to the beauty of human existence? As respects those pursuits which contribute only remotely, by providing the mechanism of life, it is well to be reminded that not the mere fact of living is to be desired, but the art of living in the contemplation of great things. Still more in regard to those avocations which have no end outside themselves, which are to be justified, if at all, as actually adding to the sum of the world's permanent possessions, it is necessary to keep alive a knowledge of their aims, a clear prefiguring vision of the temple in which creative imagination is to be embodied.

Although tradition has decreed that the great bulk of educated men shall know at least the elements of the subject [of mathematics], the reasons for which the tradition arose are forgotten, buried beneath a great rubbish-heap of pedantries and trivialities. To those who inquire as to the purpose of mathematics, the usual answer will be that it facilitates the making of machines, the travelling from place to place, and the victory over foreign nations, whether in war or commerce. If it be objected that these ends—all of which are of doubtful value—are not furthered by the merely elementary study imposed upon those who do not become expert mathematicians, the reply, it is true, will probably be that mathematics trains the reasoning faculties. Yet the very men who make this reply are, for the most part, unwilling to abandon the teaching of definite fallacies, known to be such, and instinctively rejected by the unsophisticated mind of every intelligent learner. And the reasoning faculty itself is generally conceived, by those who urge its cultivation, as merely a means for the avoidance of pitfalls and a help in the discovery of rules for the guidance of practical life. All these are undeniably important achievements to the credit of mathematics; yet it is none of these that entitles mathematics to a place in every liberal education.

Mathematics, rightly viewed, possesses not only truth, but supreme beauty—a beauty cold and austere, like that of sculpture, without appeal to any part of our weaker nature, without the gorgeous trappings of painting or music, yet sublimely pure, and capable of a stern perfection such as only the greatest art can show. The true spirit of delight, the exaltation, the sense of being more than man, which is the touchstone of the highest excellence, is to be found in mathematics as surely as in poetry. What is best in mathematics deserves not merely to be learnt as a task, but to be assimilated as a part of daily thought, and brought again and again before the mind with ever-renewed encouragement. Real life is, to most men, a long second-best, a perpetual compromise between the ideal and the possible; but the world of pure reason knows no compromise, no practical limitations, no barrier to the creative activity embodying in splendid edifices the passionate aspiration after the perfect from which all great work springs. Remote from human passions, remote even from the pitiful facts of nature, the generations have gradually created an ordered cosmos, where pure thought can dwell as in its natural home, and where one, at least, of our nobler impulses can escape from the dreary exile of the actual world.

So little, however, have mathematicians aimed at beauty, that hardly anything in their work has had this conscious purpose. Much, owing to irrepressible instincts, which were better than avowed beliefs, has been moulded by an unconscious taste; but much also has been spoilt by false notions of what was fitting. The characteristic excellence of mathematics is only to be found where the reasoning is rigidly logical: the rules of logic are to mathematics what those of structure are to architecture. In the most beautiful work, a chain of argument is presented in which every link is important on its own account, in which there is an air of ease and lucidity throughout, and the premises achieve more than would have been thought possible, by means which appear natural and inevitable. Literature embodies what is general in particular circumstances whose universal significance shines through their individual dress; but mathematics endeavours to present whatever is most general in its purity, without any irrelevant trappings.

The nineteenth century, which prided itself upon the invention of steam and evolution, might have derived a more legitimate title to fame from the discovery of pure mathematics. This science, like most others, was baptised long before it was born; and thus we find writers before the nineteenth century alluding to what they called pure mathematics. But if they had been asked what this subject was, they would only have been able to say that it consisted of Arithmetic, Algebra, Geometry, and so on. As to what these studies had in common, and as to what distinguished them from applied mathematics, our ancestors were completely in the dark.

Pure mathematics was discovered by Boole, in a work which he called the *Laws of Thought* (1854). This work abounds in asseverations that it is not mathematical, the fact being that Boole was too modest to suppose his book the first ever written on mathematics. He was also mistaken in supposing that he was dealing with the laws of thought: the question how people actually think was quite irrelevant to him, and if his book had really contained the laws of thought, it was curious that no one should ever have thought in such a way before. His book was in fact concerned with formal logic, and this is the same thing as mathematics.

Pure mathematics consists entirely of assertions to the effect that, if such and such a proposition is true of *anything*, then such and such another proposition is true of that thing. It is essential not to discuss whether the first proposition is really true, and not to mention what the anything is, of which it is supposed to be true. Both these points would belong to applied mathematics. We start, in pure mathematics, from certain rules of inference, by which we can infer that *if* one proposition is true, then so is some other proposition. These rules of inference constitute the major part of the principles of formal logic. We then take any hypothesis that seems amusing, and deduce its consequences. *If* our hypothesis is about *anything*, and not about some one or more particular things, then our deductions constitute mathematics. Thus mathematics may be defined as the subject in which we never know what we are talking about, nor whether what we are saying is true. People who have been puzzled by the beginnings of mathematics will, I hope, find comfort in this definition, and will probably agree that it is accurate.

GEORGE SARTON

Évariste Galois

•

NO EPISODE in the history of thought is more moving than the life of Évariste Galois—the young Frenchman who passed like a meteor about 1828, devoted a few feverish years to the most intense meditation, and died in 1832 from a wound received in a duel, at the age of twenty. He was still a mere boy, yet within these short years he had accomplished enough to prove indubitably that he was one of the greatest mathematicians of all time. When one sees how terribly fast this ardent soul, this wretched and tormented heart, were consumed, one can but think of the beautiful meteoric showers of a summer night. But this comparison is misleading, for the soul of Galois will burn on throughout the ages and be a perpetual flame of inspiration. His fame is incorruptible; indeed the apotheosis will become more and more splendid with the gradual increase of human knowledge.

It is safe to predict that Galois' fame can but wax, because of the fundamental nature of his work. While the inventors of important applications, whose practical value is obvious, receive quick recognition and often very substantial rewards, the discoverers of fundamental principles are not generally awarded much recompense. They often die misunderstood and unrewarded. But while the fame of the former is bound to wane as new processes supersede their own, the fame of the latter can but increase. Indeed the importance of each principle grows with the number and the value of its applications; for each new application is a new tribute to its worth. To put it more concretely, when we are very thirsty a juicy orange is more precious to us than an orange tree. Yet when the emergency has passed, we learn to value the tree more than any of its fruits; for each orange is an end in itself, while the tree represents the innumerable oranges of the future. The fame of Galois has a similar foundation; it is based upon the unlimited future. He well knew the pregnancy of his thoughts, yet they were even more far-reaching than he could possibly dream of. His complete works fill only sixty-one small pages: but a French geometer, publishing a large volume some forty years after Galois' death, declared that it was simply a commentary on the latter's discoveries. Since then, many more consequences have been deduced from Galois' fundamental ideas which have influenced the whole of mathematical philosophy. It is likely that when mathematicians of the future contemplate his personality at the distance of a few centuries, it will appear to them to be surrounded by the same halo of wonder as those of Euclid, Archimedes, Descartes, and Newton.

Évariste Galois was born in Bourg-la-Reine, near Paris, on the 25th of October, 1811, in the very house in which his grandfather had lived and had founded a boys' school. This being one of the very few boarding schools not in the hands of the priests, the Revolution had much increased its prosperity. In the course of time, grandfather Galois had given it up to his younger son and soon after, the school had received from the imperial government a sort of official recognition. When Évariste was born, his father was thirty-six years of age. He had remained a real man of the eighteenth century, amiable and witty, clever at rhyming verses and writing playlets, and instinct with philosophy. He was the leader of liberalism in Bourg-la-Reine, and during the Hundred Days had been appointed its mayor. Strangely enough, after Waterloo he was still the mayor of the village. He took his oath to the King, and to be sure he kept it, yet he remained a liberal to the end of his days. One of his friends and neighbours, Thomas François Demante, a lawyer and judge, onetime professor in the Faculty of Law of Paris, was also a typical gentleman of the "ancien régime," but of a different style. He had given a very solid classical education not only to his sons but also to his daughters. None of these had been more deeply imbued with the examples of antiquity than Adelaïde-Marie who was to be Évariste's mother. Roman stoicism had sunk deep into her heart and given to it a virile temper. She was a good Christian, though more concerned with the ethical than with the mystical side of religion. An ardent imagination had colored her every virtue with passion. Many more people have been able to appreciate her character than her son's, for it was to be her sad fortune to survive him forty years. She was said to be generous to a fault and original to the point of queerness. There is no doubt that Évariste owed considerably more to her than to his father. Besides, until the age of eleven the little boy had no teacher but his mother.

In 1823, Évariste was sent to college in Paris. This college—Louis-le-Grand—was then a gloomy house, looking from the outside like a prison, but within aflame with life and passion. For heroic memories of the Revolution and the Empire had remained particularly vivid in this institution, which was indeed, under the clerical and reactionary regime of the Restoration, a hot-bed of liberalism. Love of learning and contempt of the Bourbons divided the hearts of the scholars. Since 1815 the discipline had been jeopardized over and over again by boyish rebellions, and Évariste was certainly a witness of, if not a partner in, those which took place soon after his arrival. The influence of such an impassioned atmosphere upon a lad freshly emancipated from his mother's care cannot be exaggerated. Nothing is more infectious than political passion, nothing more intoxicating than the love of freedom. It was evidently there and then that Évariste received his political initiation. It was the first crisis of his childhood.

At first he was a good student; it was only after a couple of years that his disgust at the regular studies became apparent. He was then in the second class (that is, the highest but one) and the headmaster suggested to his father that he should spend a second year in it, arguing that the boy's weak health and immaturity made it imperative. The child was not strong, but the headmaster had failed to discover the true source of his lassitude. His seeming indifference was due less to immaturity than to his mathematical precocity. He had read his books of geometry as easily as a novel, and the knowledge had remained firmly anchored in his mind. No sooner had he begun to study algebra than he read Lagrange's original memoirs. This extraordinary facility had been at first a revelation to himself, but as he grew more conscious of it, it became more difficult for him to curb his own domineering thought and to sacrifice it to the routine of classwork. The rigid program of the college was to him like a bed of Procrustes, causing him unbearable torture without adequate compensation. But how could the headmaster and the teachers understand this? The double conflict within the child's mind and between the teachers and himself, as the knowledge of his power increased, was intensely dramatic. By 1827 it had reached a critical point. This might be called the second crisis of his childhood: his scientific initiation. His change of mood was observed by the family. Juvenile gaiety was suddenly replaced by concentration; his manners became stranger every day. A mad desire to march forward along the solitary path which he saw so distinctly, possessed him. His whole being, his every faculty was mobilized in this immense endeavor.

I cannot give a more vivid idea of the growing strife between this inspired boy and his uninspired teachers than by quoting a few extracts from the school reports:

1826–1827

This pupil, though a little queer in his manners, is very gentle and seems filled with innocence and good qualities. . . . He never knows a lesson badly: either he has not learned it at all or he knows it well. . . .

A little later:

This pupil, except for the last fortnight during which he has worked a little, has done his classwork only from fear of punishment. . . . His ambition, his originality—often affected—the queerness of his character keep him aloof from his companions.

1827–1828

> Conduct rather good. A few thoughtless acts. Character of which I do not flatter myself I understand every trait; but I see a great deal of self-esteem dominating. I do not think he has any vicious inclination. His ability seems to me to be entirely beyond the average, with regard as much to literary studies as to mathematics. . . . He does not seem to lack religious feeling. His health is good but delicate.

Another professor says:

> His facility, in which one is supposed to believe but of which I have not yet witnessed a single proof, will lead him nowhere; there is no trace in his tasks of anything but of queerness and negligence.

Another still:

> Always busy with things which are not his business. Goes down every day.

Same year, but a little later:

> Very bad conduct. Character rather secretive. Tries to be original. . . .
> Does absolutely nothing for the class. The furor of mathematics possesses him. . . . He is losing his time here and does nothing but torment his masters and get himself harassed with punishments. He does not lack religious feeling; his health seems weak.

Later still:

> Bad conduct, character difficult to define. Aims at originality. His talents are very distinguished; he might have done very well in "Rhétorique" if he had been willing to work, but swayed by his passion for mathematics, he has entirely neglected everything else. Hence he has made no progress whatever. . . . Seems to affect to do something different from what he should do. It is possibly to this purpose that he chatters so much. He protests against silence.

In his last year at the college, 1828–1829, he had at last found a teacher of mathematics who divined his genius and tried to encourage and to help him. This Mr. Richard, to whom one cannot be too grateful, wrote of him: "This student has a marked superiority over all his schoolmates. He works only at the highest parts of mathematics." You see the whole difference. Kind Mr. Richard did not complain that Évariste neglected his regular tasks, and, I imagine, often forgot to do the petty mathematical exercises which are indispensable to drill the average boy; he does not think it fair to insist on what Évariste does not do, but states what he does do: he is only concerned with the highest parts of mathematics. Unfortunately, the other teachers were less indulgent. For physics and chemistry, the note often repeated was: "Very absent-minded, no work whatever."

To show the sort of preoccupations which engrossed his mind: at the age of sixteen he believed that he had found a method of solving general equations of the fifth degree. One knows that before succeeding in proving the impossibility of such resolution, Abel had made the same mistake. Besides, Galois was already trying to realize the great dream of his boyhood: to enter the École Polytechnique. He was bold enough to prepare himself alone for the entrance examination as early as 1828—but failed. This failure was very bitter to him—the more so that he considered it as unfair. It is likely that it was not at all unfair, at least according to the accepted rules. Galois knew at one and the same time far more and far less than was necessary to enter Polytechnique; his extra knowledge could not compensate for his deficiencies, and examiners will never consider originality with favor. The next year he published his first paper, and sent his first communication to the Académie des Sciences. Unfortunately, the latter got lost through Cauchy's negligence. This embittered Galois even more. A second failure to enter Polytechnique seemed to be the climax of his misfortune, but a greater disaster was still in store for him. On July 2 of this same year, 1829, his father had been driven to commit suicide by the vicious attacks directed against him, the liberal mayor, by his political enemies. He took his life in the small apartment which he had in Paris, in the vicinity of Louis-le-Grand. As soon as his father's body reached the territory of Bourg-la-Reine, the inhabitants carried it on their shoulders, and the funeral was the occasion of disturbances in the village. This terrible blow, following many smaller miseries, left a very deep mark on Évariste's soul. His hatred of injustice became the more violent, in that he already believed himself to be a victim of it; his father's death incensed him, and developed his tendency to see injustice and baseness everywhere.

His repeated failures to be admitted to Polytechnique were to Galois a cause of intense disappointment. To appreciate his despair, one must realize that the École Polytechnique was then, not simply the highest mathematical school in France and the place where his genius would be most likely to find the sympathy it craved, it was also a daughter of the Revolution who had remained faithful to her origins in spite of all efforts of the government to curb her spirit of independence. The young Polytechnicians were the natural leaders of every political rebellion; liberalism was for them a matter of traditional duty. This house was thus twice sacred to Galois, and his failure to be accepted was a double misfortune. In 1829 he entered the École Normale, but he entered it as an exile from Polytechnique. It was all the more difficult for him to forget the object of his former ambition, because the École Normale was then passing through the most languid period of its existence. It was not even an independent institution, but rather an extension of Louis-le-Grand. Every precaution had been taken to ensure the loyalty of this school to the new regime. Yet there, too, the main student body inclined toward liberalism, though their convictions were very weak and passive as compared with the mood prevailing at Polytechnique;

because of the discipline and the spying methods to which they were submitted, their aspirations had taken a more subdued and hypocritical form only relieved once in a while by spasmodic upheavals. Évariste suffered doubly, for his political desires were checked and his mathematical ability remained unrecognized. Indeed he was easily embarrassed at the blackboard, and made a poor impression upon his teachers. It is quite possible that he did not try in the least to improve this impression. His French biographer, P. Dupuy, very clearly explains his attitude:

> There was in him a hardly disguised contempt for whosoever did not bow spontaneously and immediately before his superiority, a rebellion against a judgment which his conscience challenged beforehand and a sort of unhealthy pleasure in leading it further astray and in turning it entirely against himself. Indeed, it is frequently observed that those people who believe that they have most to complain of persecution could hardly do without it and, if need be, will provoke it. To pass oneself off for a fool is another way and not the least savory, of making fools of others.

It is clear that Galois' temper was not altogether amiable, yet we should not judge him without making full allowance for the terrible strain to which he was constantly submitted, the violent conflicts which obscured his soul, the frightful solitude to which fate had condemned him.

In the course of the ensuing year, he sent three more papers to mathematical journals and a new memoir to the Académie. The permanent secretary, Fourier, took it home with him, but died before having examined it, and the memoir was not retrieved from among his papers. Thus his second memoir was lost like the former. This was too much indeed and one will easily forgive the wretched boy if in his feverish mood he was inclined to believe that these repeated losses were not due to chance but to systematic persecution. He considered himself a victim of a bad social organization which ever sacrifices genius to mediocrity, and naturally enough he cursed the hated regime of oppression which had precipitated his father's death and against which the storm was gathering. We can well imagine his joy when he heard the first shots of the July Revolution! But alas! While the boys of Polytechnique were the very first in the fray, those of the École Normale were kept under lock and key by their faint-hearted director. It was only when the three glorious days of July were over and the fall of the Bourbons was accomplished that this opportunist let his students out and indeed placed them at the disposal of the provisional government! Never did Galois feel more bitterly that his life had been utterly spoiled by his failure to become an alumnus of his beloved Polytechnique.

In the meanwhile the summer holidays began and we do not know what happened to the boy in the interval. It must have been to him a new period of crisis, more acute than any of the previous ones. But before speaking of it let me say a last word about his scientific efforts, for it is probable that thereafter political passion obsessed his mind almost exclusively. At any rate it is certain that Évariste was in the possession of his general principles by the beginning of 1830, that is, at the age of eighteen, and that he fully knew their importance. The consciousness of his power and of the responsibility resulting from it increased the concentration and the gloominess of his mind to the danger point; the lack of recognition developed in him an excessive pride. By a strange aberration he did not trouble himself to write his memoirs with sufficient clearness to give the explanations which were the more necessary because his thoughts were more novel. What a pity that there was no understanding friend to whisper in his ear Descartes' wise admonition: "When you have to deal with transcendent questions, you must be transcendently clear." Instead of that, Galois enveloped his thought in additional secrecy by his efforts to attain greater conciseness, that coquetry of mathematicians.

It is intensely tragic that this boy already sufficiently harassed by the turmoil of his own thoughts, should have been thrown into the political turmoil of this revolutionary period. Endowed with a stronger constitution, he might have been able to cope with one such; but with the two, how could he—how could anyone do it? During the holidays he was probably pressed by his friend, Chevalier, to join the Saint-Simonists, but he declined, and preferred to join a secret society, less aristocratic and more in keeping with his republican aspirations—the "Société des amis du peuple." It was thus quite another man who re-entered the École Normale in the autumn of 1830. The great events of which he had been a witness had given to his mind a sort of artificial maturity. The revolution had opened to him a fresh source of disillusion, the deeper because the hopes of the first moment had been so sanguine. The government of Louis-Philippe had promptly crushed the more liberal tendencies, and the artisans of the new revolution, who had drawn their inspiration from the great events of 1789, soon discovered to their intense disgust that they had been fooled. Indeed under a more liberal guise, the same oppression, the same favoritism, the same corruption soon took place under Louis-Philippe as under Charles X. Moreover, nothing can be more demoralizing than a successful revolution (whatever it be) for those who, like Galois, were too generous to seek any personal advantage and too ingenuous not to believe implicitly in their party shibboleths. It is such a high fall from one's dearest ideal to the ugliest aspect of reality—and they could not help seeing around them the more practical and cynical revolutionaries eager for the quarry, and more disgusting still, the clever ones, who had kept quiet until they knew which side was gaining, and who now came out of their hiding places to fight over the spoils and make the most of the new regime. Political fermentation did not abate and the more democratic elements, which Galois had joined, became more and more disaffected and restless. The director of the École Normale had been obliged to restrain himself considerably to brook Galois' irregular conduct, his "laziness," his intractable temper; the boy's political attitude, and chiefly his undisguised contempt for the director's pusillanimity now increased the tension between them to the breaking point. The publication in the "Gazette des Écoles" of a letter of Galois' in which he scornfully criticized the director's tergiversations was but the last of many offenses. On December 9, he was invited to leave the school, and his expulsion was ratified by the Royal Council on January 3, 1831.

To support himself Galois announced that he would give a private course on higher algebra in the back shop of a bookseller, Mr. Caillot, 5 rue de la Sorbonne. I do not know whether this course, or how much of it, was actually delivered. A further scientific disappointment was reserved for him: a new copy of his second lost memoir had been communicated by him to the Académie; it was returned to him by Poisson, four months later, as being incomprehensible. There is no doubt that Galois was partly responsible for this, for he had taken no pains to explain himself clearly.

This was the last straw! Galois' academic career was entirely compromised, the bridges were burned, he plunged himself entirely into the political turmoil. He threw himself into it with his habitual fury and the characteristic intransigency of a mathematician; there was nothing left to conciliate him, no means to moderate his passion, and he soon reached the extreme limit of exaltation. He is said to have exclaimed: "If a corpse were needed to stir the people up, I would give mine." Thus on May 9, 1831, at the end of a political banquet, being intoxicated—not with wine but with the ardent conversation of an evening—he proposed a sarcastic toast to the King. He held his glass and an open knife in one hand and said simply: "To Louis-Philippe!" Of course he was soon arrested and sent to Ste. Pélagie. The lawyer persuaded him to maintain that he had actually said: "To Louis-Philippe, *if he betray*," and many witnesses affirmed that they had heard him utter the last words, though they were lost in the uproar. But Galois could not stand this lying and retracted it at the public trial. His attitude before the tribunal was ironical and provoking, yet the jury rendered a verdict of not proven and he was acquitted. He did not remain free very long. On the following Fourteenth of July, the government, fearing manifestations, decided to have him arrested as a preventive measure. He was given six months' imprisonment on the technical charge of carrying arms and wearing a military uniform, but he remained in Ste. Pélagie only until March 19 (or 16?), 1832, when he was sent to a convalescent home in the rue de Lourcine. A dreadful epidemic of cholera was then raging in Paris, and Galois' transfer had been determined by the poor state of his health. However, this proved to be his undoing.

He was now a prisoner on parole and took advantage of it to carry on an intrigue with a woman of whom we know nothing, but who was probably not very reputable ("une coquette de bas étage," says Raspail). Think of it! This was, as far as we know, his first love—and it was but one more tragedy on top of so many others. The poor boy who had declared in prison that he could love only a Cornelia or a Tarpeia[1] (we hear in this an echo of his mother's Roman ideal), gave himself to this new passion with his frenzy, only to find more bitterness at the end of it. His revulsion is lamentably expressed in a letter to Chevalier (May 25, 1832):

> . . . How to console oneself for having exhausted in one month the greatest source of happiness which is in man—of having exhausted it without happiness, without hope, being certain that one has drained it for life?
>
> Oh! come and preach peace after that! Come and ask men who suffer to take pity upon what is! Pity, never! Hatred, that is all. He who does not feel it deeply, this hatred of the present, cannot really have in him the love of the future. . . .

One sees how his particular misery and his political grievances are sadly muddled in his tired head. And a little further in the same letter, in answer to a gentle warning by his friend:

> I like to doubt your cruel prophecy when you say that I shall not work any more. But I admit it is not without likelihood. To be a savant, I should need to be that alone. *My heart has revolted against my head.*[2] I do not add as you do: It is a pity.

Can a more tragic confession be imagined? One realizes that there is no question here of a man possessing genius, but of genius possessing a man. A man? a mere boy, a fragile little body divided within itself by disproportionate forces, an undeveloped mind crushed mercilessly between the exaltation of scientific discovery and the exaltation of sentiment.

Four days later two men challenged him to a duel! The circumstances of this affair are, and will ever remain, very mysterious. According to Évariste's younger brother the duel was not fair. Évariste, weak as he was, had to deal with two ruffians hired to murder him. I find nothing to countenance this theory except that he was challenged by two men at once. At any rate, it is certain that the woman he had loved played a part in this fateful event. On the day preceding the duel, Évariste wrote three letters of which I translate one:

> *May 29, 1832*
>
> LETTER TO ALL REPUBLICANS.
>
> I beg the patriots, my friends, not to reproach me for dying otherwise than for the country.
>
> I die the victim of an infamous coquette. My life is quenched in a miserable piece of gossip.
>
> Oh! why do I have to die for such a little thing, to die for something so contemptible!
>
> I take heaven to witness that it is only under compulsion that I have yielded to a provocation which I had tried to avert by all means.

[1] He must have quoted Tarpeia by mistake.

[2] The italics are mine.

I repent having told a baleful truth to men who were so little able to listen to it coolly. Yet I have told the truth. I take with me to the grave a conscience free from lie, free from patriots' blood.

Good-bye! I had in me a great deal of life for the public good.

Forgiveness for those who killed me; they are of good faith.

E. GALOIS

Any comment could but detract from the pathos of this document. I will only remark that the last line, in which Galois absolves his adversaries, destroys his brother's theory. It is simpler to admit that his impetuosity, aggravated by female intrigue, had placed him in an impossible position from which there was no honorable issue, according to the standards of the time, but a duel. Évariste was too much of a gentleman to try to evade the issue, however trifling its causes might be; he was anxious to pay the full price of his folly. That he well realized the tragedy of his life is quite clear from the laconic post-scriptum of his second letter: *Nitens lux, horrenda procella, tenebris æternis involuta.* The last letter addressed to his friend, Auguste Chevalier, was a sort of scientific testament. Its seven pages, hastily written, dated at both ends, contain a summary of the discoveries which he had been unable to develop. This statement is so concise and so full that its significance could be understood only gradually as the theories outlined by him were unfolded by others. It proves the depth of his insight, for it anticipates discoveries of a much later date. At the end of the letter, after requesting his friend to publish it and to ask Jacobi or Gauss to pronounce upon it, he added: "After that, I hope some people will find it profitable to unravel this mess. *Je t'embrasse avec effusion.*" The first sentence is rather scornful but not untrue and the greatest mathematicians of the century have found it very profitable indeed to clear up Galois' ideas.

The duel took place on the 30th in the early morning, and he was grievously wounded by a shot in the abdomen. He was found by a peasant who transported him at 9:30 to the Hôpital Cochin. His younger brother —the only member of the family to be notified—came and stayed with him, and as he was crying, Évariste tried to console him, saying: "Do not cry. I need all my courage to die at twenty." While still fully conscious, he refused the assistance of a priest. In the evening peritonitis declared itself and he breathed his last at ten o'clock on the following morning.

His funeral, which strangely recalled that of his father, was attended by two to three thousand republicans, including deputations from various schools, and by a large number of police, for trouble was expected. But everything went off very calmly. Of course it was the patriot and the lover of freedom whom all these people meant to honor; little did they know that a day would come when this young political hero would be hailed as one of the greatest mathematicians of all time.

A life as short yet as full as the life of Galois is interesting not simply in itself but even more perhaps because of the light it throws upon the nature of genius. When a great work is the natural culmination of a long existence devoted to one persistent endeavor, it is sometimes difficult to say whether it is the fruit of genius or the fruit of patience. When genius evolves slowly it may be hard to distinguish from talent—but when it explodes suddenly, at the beginning and not at the end of life, or when we are at a loss to explain its intellectual genesis, we can but feel that we are in the sacred presence of something vastly superior to talent. When one is confronted with facts which cannot be explained in the ordinary way, is it not more scientific to admit our ignorance than to hide it behind faked explanations? Of course it is not necessary to introduce any mystical idea, but it is one's duty to acknowledge the mystery. When a work is really the fruit of genius, we cannot conceive that a man of talent might have done it "just as well" by taking the necessary pains. Pains alone will never do; neither is it simply a matter of jumping a little further, for it involves a synthetic process of a higher kind. I do not say that talent and genius are essentially different, but that they are of different orders of magnitude.

Galois' fateful existence helps one to understand Lowell's saying: "Talent is that which is in a man's power, genius is that in whose power man is." If Galois had been simply a mathematician of considerable ability, his life would have been far less tragic, for he could have used his mathematical talent for his own advancement and happiness; instead of which, the furor of mathematics—as one of his teachers said—possessed him and he had no alternative but absolute surrender to his destiny.

Lowell's aphorism is misleading, however, for it suggests that talent can be acquired, while genius cannot. But biological knowledge points to the conclusion that neither is really acquired, though both can be developed and to a certain extent corrected by education. Men of talent as well as men of genius are born, not made. Genius implies a much stronger force, less adaptable to environment, less tractable by education, and also far more exclusive and despotic. Its very intensity explains its frequent precocity. If the necessary opportunities do not arise, ordinary abilities may remain hidden indefinitely; but the stronger the abilities the smaller need the inducement be to awaken them. In the extreme case, the case of genius, the ability is so strong that, if need be, it will force its own outlet.

Thus it is that many of the greatest accomplishments of science, art and letters were conceived by very young men. In the field of mathematics, this precocity is particularly obvious. To speak only of the two men considered in this essay, Abel had barely reached the age of twenty-two and Galois was not yet twenty, perhaps not yet nineteen, when they made two of the most profound discoveries which have ever been made. In many other sciences and arts, technical apprenticeship may be too long to make such early discovery possible. In most cases, however, the judgment of

Alfred de Vigny holds good. "What is a great life? It is a thought of youth wrought out in ripening years." The fundamental conception dawns at an early age—that is, it appears at the surface of one's consciousness as early as this is materially possible—but it is often so great that a long life of toil and abnegation is but too short to work it out. Of course at the beginning it may be very vague, so vague indeed that its host can hardly distinguish it himself from a passing fancy, and later may be unable to explain how it gradually took control of his activities and dominated his whole being. The cases of Abel and Galois are not essentially different from those contemplated by Alfred de Vigny, but the golden thoughts of their youth were wrought out in the ripening years of other people.

It is the precocity of genius which makes it so dramatic. When it takes an explosive form, as in the case of Galois, the frail carcass of a boy may be unable to resist the internal strain and it may be positively wrecked. On the other hand when genius develops more slowly, its host has time to mature, to adapt himself to his environment, to gather strength and experience. He learns to reconcile himself to the conditions which surround him, widely different as they are, from those of his dreams. He learns by and by that the great majority of men are rather unintelligent, uneducated, uninspired, and that one must not take it too much to heart when they behave in defiance of justice or even of common sense. He also learns to dissipate his vexation with a smile or a joke and to protect himself under a heavy cloak of kindness and humor. Poor Évariste had no time to learn all this. While his genius grew in him out of all proportion to his bodily strength, his experience, and his wisdom, he felt more and more ill at ease. His increasing restlessness makes one think of that exhibited by people who are prey to a larvate form of a pernicious disease. There is an internal disharmony in both cases, though it is physiological in the latter, and psychological in the former. Hence the suffering, the distress, and finally the acute disease or the revolt!

A more congenial environment might have saved Galois. Oh! would that he had been granted that minimum of understanding and sympathy which the most concentrated mind needs as much as a plant needs the sun! But it was not to be; and not only had he no one to share his own burden, but he had also to bear the anxieties of a stormy time. I quite realize that this self-centered boy was not attractive—many would say not lovable. Yet I love him; I love him for all those who failed to love him; I love him because of his adversity.

His tragic life teaches us at least one great lesson: one can never be too kind to the young; one can never be too tolerant of their faults, even of their intolerance. The pride and intolerance of youth, however immoderate, are excusable because of youth's ignorance, and also because one may hope that it is only a temporary disorder. Of course there will always be men despicable enough to resort to snubbing, as it were, to protect their own position and to hide their mediocrity, but I am not thinking of them. I am simply thinking of the many men who were unkind to Galois without meaning to be so. To be sure, one could hardly expect them to divine the presence of genius in an awkward boy. But even if they did not believe in him, could they not have shown more forbearance? Even if he had been a conceited dunce, instead of a genius, could kindness have harmed him? It is painful to think that a few rays of generosity from the heart of his elders might have saved this boy or at least might have sweetened his life.

But does it really matter? A few years more or less, a little more or less suffering . . . Life is such a short drive altogether. Galois has accomplished his task and very few men will ever accomplish more. He has conquered the purest kind of immortality. As he wrote to his friends: "I take with me to the grave a conscience free from lie, free from patriots' blood." How many of the conventional heroes of history, how many of the kings, captains and statesmen could say the same?

EDMUND CALLIS BERKELEY

Giant Brains

•

RECENTLY there has been a good deal of news about strange giant machines that can handle information with vast speed and skill. They calculate and they reason. Some of them are cleverer than others—able to do more kinds of problems. Some are extremely fast. Where they apply, they find answers to problems much faster and more accurately than human beings can; and so they can solve problems that a man's life is far too short to permit him to do. That is why they were built.

These machines are similar to what a brain would be if it were made of hardware and wire instead of flesh and nerves. It is therefore natural to call these machines *mechanical brains.* Also, since their powers are like those of a giant, we may call them *giant brains.*

Can we say that these machines really think? What do we mean by thinking, and how does the human brain think?

Human Thinking

We do not know very much about the physical process of thinking in the human brain. If you ask a scientist how flesh and blood in a human brain can think, he will talk to you a little about nerves and about electrical and chemical changes, but he will not be able to tell you very much about how we add 2 and 3 and make 5. What men do not know about the way in which a human brain thinks would fill many libraries.

Injuries to brains have shown some things of importance; for example, they have shown that certain parts of the brain have certain duties. There is a part of the brain, for instance, where sights are recorded and compared. If an accident damages the part of the brain where certain information is stored, the human being has to relearn—haltingly and badly—the information destroyed.

We know also that thinking in the human brain is done essentially by a process of storing information and then referring to it, by a process of learning and remembering. We know that there are no little wheels in the brain so that a wheel standing at 2 can be turned 3 more steps and the result of 5 read. Instead, you and I store the information that 2 and 3 are 5, and store it in such a way that we can give the answer when questioned. But we do not know the register in our brain where this particular piece of information is stored. Nor do we know how, when we are questioned, we are able automatically to pick up the nerve channels that lead into this register, get the answer, and report it.

Since there are many nerves in the brain, about ten billion of them, in fact, we are certain that the network of connecting nerves is a main part of the puzzle. We are therefore much interested in nerves and their properties.

A single nerve, or *nerve cell* consists of a *cell nucleus* and a *fiber*. This fiber may have a length of anything from a small fraction of an inch up to several feet. In the laboratory, successive impulses can be sent along a nerve fiber as often as 1,000 a second. Impulses can travel along a nerve fiber in either direction at a rate from 3 feet to 300 feet a second. Because the speed of the impulse is far less than 186,000 miles a second—the speed of an electric current—the impulse in the nerve is thought by some investigators to be more chemical than electrical.

We know that a nerve cell has what is called an *all-or-none response*, like the trigger of a gun. If you stimulate the nerve up to a certain point, nothing will happen; if you reach that point, or cross it—bang!—the nerve responds and send out an impulse. The strength of the impulse, like the shot of the gun, has no relation whatever to the amount of the stimulation.

The structure between the end of one nerve and the beginning of the next is called a synapse. No one really knows very much about synapses, for they are extremely small and it is not easy to tell where a synapse stops and other stuff begins. Impulses travel through synapses in from ½ to 3 thousandths of a second. An impulse travels through a synapse only in one direction, from the head (or *axon*) of one nerve fiber to the foot (or *dendrite*) of another. It seems clear that the activity in a synapse is chemical. When the head of a nerve fiber brings in an impulse to a synapse, apparently a chemical called *acteylcholine* is released and may affect the foot of another fiber, thus transmitting the impulse; but the process and the conditions for it are still not well understood.

It is thought that nearly all information is handled in the brain by groups of nerves in parallel paths. For example, the eye is estimated to have about 100 million nerves sensitive to light, and the information that they gather is reported by about 1 million nerves to the part of the brain that stores sights.

Not much more is yet known, however, about the operation of handling information in a human brain. We do not yet know how the nerves are connected so that we can do what we do. Probably the greatest obstacle to knowledge is that so far we cannot observe the detailed structure of a living human brain while it performs without hurting or killing it.

Therefore, we cannot yet tell what thinking is by observing precisely how a human brain does it. Instead, we have to define thinking by describing the kind of behavior that we call thinking. Let us consider some examples.

When you and I add 12 and 8 and make 20, we are thinking. We use our minds and our understanding to count 8 places forward from 12, for example, and finish with 20. If we could find a dog or a horse that could add numbers and tell answers, we would certainly say that the animal could think.

With no trouble a machine can do this. An ordinary 10-column adding machine can be given two numbers like 1,378,917,766 and 2,355,799,867 and the instruction to add them. The machine will then give the answer, 3,734,717,633, much faster than a man.

Or, suppose that you are walking along a road and come to a fork. If you stop, read the signpost, and then choose left or right, you are thinking. You know beforehand where you want to go, you compare your destination with what the signpost says, and you decide on your route. This is an operation of logical choice.

A machine can do this. The mechanical brain can examine any number that cames up in the process of a calculation and tell whether it is bigger than 3 (or any stated number) or smaller. If the number is bigger than 3, the machine will choose one process; if the number is smaller than 3, the machine will choose another process.

Now suppose that we consider the basic operation of all thinking: in the human brain it is called learning and remembering, and in a machine it is called storing information and then referring to it. For example, suppose you want to find 305 Main Street in Kalamazoo. You look up a map of Kalamazoo; the map is information kindly stored by other people for your use. When you study the map, notice the streets and the numbering, and then find where the house should be, you are thinking.

A machine can do this. In the mechanical brain the map could be stored as a long list of the blocks of the city and the streets and numbers that apply to each block. The machine will then hunt for the city block that contains 305 Main Street and report it when found.

A machine can handle information; it can calculate, conclude, and choose; it can perform reasonable operations with information. A machine, therefore, can think.

Now when we speak of a machine that thinks, or a mechanical brain, what do we mean? Essentially, a *mechanical brain* is a machine that handles information, transfers information automatically from one part of the machine to another, and has a flexible control over the sequence of its operations. No human being is needed around such a machine to pick up a physical piece of information produced in one part of the machine, personally move it to another part of the machine, and there put it in again. Nor is any human being needed to give the machine instructions from minute to minute. Instead, we can write out the whole program to solve a problem, translate the program into machine language, and put the program into the machine. Then we press the "start" button; the machine starts whirring; and it prints out the answers as it obtains them. Machines that handle information have existed for more than 2,000 years. These two properties are new, however, and make a deep break with the past.

How should we imagine a mechanical brain? One way to think of a mechanical brain is a railroad line with four stations, marked *input, storage, computer,* and *output.* These stations are joined by little gates or switches to the main railroad line. We can imagine that numbers and other information move along this railroad line, loaded in freight cars. *Input* and *output* are stations where numbers or other information go in and come out, respectively. *Storage* is a station where there are many platforms and where information can be stored. The *computer* is a special station somewhat like a factory; when two numbers are loaded on platforms 1 and 2 of this station and an order is loaded on platform 3, then another number is produced on platform 4.

We see also a tower, marked *control.* This tower runs a telegraph line to each of its little watchmen standing by the gates. The tower tells them when to open and when to shut which gates.

Now we can see that, just as soon as the right gates are shut, freight cars of information can move between stations. Actually the freight cars move at the speed of electric current, thousands of miles a second. So, by closing the right gates each fraction of a second, we can flash numbers and information through the system and perform operations of reasoning. Thus we obtain a mechanical brain.

In general, a mechanical brain is made up of:

1. A quantity of registers where information (numbers and instructions) can be stored.
2. Channels along which information can be sent.
3. Mechanisms that can carry out arithmetical and logical operations.
4. A control, which guides the machine to perform a sequence of operations.
5. Input and output devices, whereby information can go into the machine and come out of it.
6. Motors or electricity, which provide energy.

There are many kinds of thinking that mechanical brains can do. Among other things, they can:

1. Learn what you tell them.
2. Apply the instructions when needed.

3. Read and remember numbers.
4. Add, subtract, multiply, divide, and round off.
5. Look up numbers in tables.
6. Look at a result, and make a choice.
7. Do long chains of these operations one after another.
8. Write out an answer.
9. Make sure the answer is right.
10. Know that one problem is finished, and turn to another.
11. Determine *most* of their own instructions.
12. Work unattended.

They do these things much better than you or I. They are fast. They are reliable. Even with hundreds of thousands of parts, the existing giant brains have worked successfully. They have remarkably few mechanical troubles; in fact, for one of the giant brains, a mechanical failure is of the order of once a month. They are powerful. The mechanical brains that have been finished are able to solve problems that have baffled men for many, many years, and they think in ways never open to men before. Mechanical brains have removed the limits on complexity of routine: the machine can carry out a complicated routine as easily as a simple one. Already, processes for solving problems are being worked out so that the mechanical brain will itself determine more than 99 per cent of all the routine orders that it is to carry out.

Most of the thinking so far done by these machines is with numbers. They have already solved problems in airplane design, astronomy, physics, mathematics, engineering, and many other sciences, that previously could not be solved. To find the solutions of these problems, mathematicians would have had to work for years and years, using the best known methods and large staffs of human computers.

These mechanical brains not only calculate, however. They also remember and reason, and thus they promise to solve some very important human problems. For example, one of these problems is the application of what mankind knows. It takes too long to find understandable information on a subject. The libraries are full of books: most of them we can never hope to read in our lifetime. The technical journals are full of condensed scientific information: they can hardly be understood by you and me. There is a big gap between somebody's knowing something and employment of that knowledge by you or me when we need it. But these new mechanical brains handle information very swiftly. In a few years machines will probably be made that will know what is in libraries and that will tell very swiftly where to find certain information. Thus we can see that mechanical brains are one of the great new tools for finding out what we do not know and applying what we do know.

HOLLIS R. COOLEY,
DAVID GANS, MORRIS KLINE
AND HOWARD E. WAHLERT

Mathematics and Modern Civilization

•

THERE ARE two major ways in which mathematics has become so effective in our age. The first is through its relationships with science. Because the distinguishing characteristic of modern civilization is the extent to which the physical sciences have molded it, this relationship of mathematics to science bears directly on the importance of mathematics in modern civilization. The second way in which mathematics has materially influenced our age has been through its connection with human reasoning. Mathematical method is reasoning on the highest level known to man, and every field of investigation, be it law, politics, psychology, medicine, or anthropology, has felt its influence and modeled itself on mathematics to some extent. This chapter will elaborate on these two major contributions of mathematics.

The Relation of Mathematics to the Sciences

In order to gain a more comprehensive view of the relation of mathematics to the sciences, let us analyze the various ways in which mathematics serves scientific investigation.

A. *Mathematics supplies a language for the treatment of the quantitative problems of the physical and social sciences.* Much of this language takes the form of mathematical symbols. Workable rules for carrying out operations are made possible by symbols; without them the simple operations of arithmetic would be extremely clumsy and the solution of even simple equations would be very difficult.

Symbols also permit concise, unambiguous representation of ideas which are sometimes quite complex. Consider, for example, how much is involved in the calculus symbol, Dy. Once the meaning and use of a symbol has been grasped, there is no need to think through the origin and development of the idea symbolized, each time it is used. One of the chief reasons that mathematics has been so effective in problems that have been insoluble by other methods is that it has powerful techniques based upon the use of symbols. Scientists have learned to use mathematical symbols whenever possible.

B. *Mathematics supplies science with numerous methods and conclusions.* Among the important conclusions which mathematics furnishes are its formulas, which scientists accept and use in solving problems. The use of such formulas is so common that the contribution of mathematics in this direction is not always appreciated. A brief survey will show that this contribution is an important one. Consider, for example, the many geometric formulas, such as those for areas and volumes, the formulas for finding the distance and velocity of moving objects, and formulas for compound interest and annuities. The realization of the importance of such formulas would be further strengthened if we could live for a time in the world as it was before many of the problems of science were solved by mathematical means.

Over and above the content of mathematics are methods which mathematics has developed and which are advantageously employed by scientists. Indirect measurement by means of the trigonometric ratios is an example of such a method, as is the representation of a curve by means of an equation for the purpose of studying the curve. The calculus likewise presents the sciences with a method for finding rates of change of varying quantities, as well as other methods. Of course, in a larger sense all of mathematics is a method, but this aspect of mathematics will be discussed in the latter part of the chapter.

Any discussion of the methodological contribution of mathematics to science cannot overlook the fact that mathematics is often an essential part of scientific method. The scientist observes or experiments so as to discover new facts and, more important, so as to secure a basis for a theory or hypothesis which should explain a large class of phenomena. Often this hypothesis either takes the form of a mathematical law, or includes mathematical laws. Mathematical methods are very frequently used to deduce consequences of the laws. Further observation or experimentation checks these deductions and thereby tests the suitability of the hypothesis.

C. *Mathematics enables the sciences to make predictions.* This is perhaps the most valuable contribution of mathematics to the sciences. If a structural engineer wishes to use a beam of given size at the base of a skyscraper, he wants to know in advance whether or not it is able to carry the load to which it will be subjected. It would not do for him to use the beam only to find out that it is too weak when the building collapses. In order to determine whether the beam is strong enough he uses mathematical analysis. If machines are to be built, an industrialist wants to know beforehand what the machines will cost and what savings in the cost of production they will effect. The answers to these questions are obtained largely by the use of mathematical methods. The accurate determination of the standard of living of a country, comprising such factors as wages and living costs, is made by means of statistical analyses. The occurrence of an eclipse is far less astonishing than man's ability to predict the time of its occurrence. The same can be said for predictions in many sciences.

The ability to make predictions by mathematical means was exemplified in a most remarkable way in 1846 by two astronomers, Leverrier and Adams. Each of these men, working independently, decided, as a result of his calculations, that there must exist another planet beyond those known at the time. A search for it in the sky at the predicted place and time revealed the planet which we now call Neptune. When one considers that Neptune is not visible to the naked eye and can be found with a telescope only with exact knowledge of its position, the extraordinary precision of this prediction is apparent. It should be realized that prediction plays a part in every mathematical solution of a quantitative problem arising in the physical and social sciences.

D. *Mathematics supplies science with ideas with which to describe phenomena.* Among such ideas which mathematics has furnished for science may be mentioned the idea of a functional relation; the graphical representation of functional relations by means of coordinate geometry; the notion of a limit, which provides methods of determining instantaneous velocity and acceleration and of calculating areas and volumes; and the notion of infinite classes, which helps us, among other things, to understand motion. Of special importance are the statistical methods and theories which mathematics furnishes to science and which have led to the idea of a statistical law. It may be worthwhile to call attention to the important fact that by the use of the notion of a statistical law, scientists are enabled to reason about situations which appear to be utterly chaotic. The seemingly random motions of people on a crowded thoroughfare, may be found in some cases to obey a law as closely as does the earth in its motion around the sun.

A description of the extent to which mathematical concepts are used by scientists is not complete without mention of the fact that for many physical phenomena no exact concepts exist other than mathematical ones. The physicist uses extensively the concept of voltage or electromotive force which causes electric current to flow in wires. The layman makes use of this voltage when he connects his lamp or radio to a socket in a room of his home. Yet there is no precise physical concept of voltage nor a good physical explanation of what causes current to flow in wires. There are, however, numerous exact mathematical relationships which involve the concept of voltage, as well as others equally vague physically. For example, Ohm's law, which states that the voltage between any two points on a wire is the product of the current flowing and the resistance in that wire, in such an exact relationship involving voltage. Another physical concept which can be represented and discussed only in mathematical terms is the notion of an ether wave which carries light, radio broadcasts, X-rays, and other electromagnetic phenomena. It is very significant to recall in this connection that Clerk Maxwell predicted the existence of a wide class of these waves (that light was a wave motion in ether was accepted long before Maxwell's time) only because he had a mathematical term in his equations which should have some physical significance. No physical understanding of ether waves exists even today. Nevertheless the marvels of radio broadcasting and television are realized because we can discuss and work with this concept mathematically.

The impress of mathematics on our civilization through the sciences is nowhere better exemplified than by the work of Maxwell. One has only to contemplate the number of people who are entertained, educated, and propagandized by radio broadcasts to realize the magnitude of the effect. The effect of television broadcasting staggers the imagination.

E. *Mathematics has been of use to science in preparing men's minds for new ways of thinking.* The concepts of importance in science today, elementary though some of them may seem, came to men with great difficulty. The concepts of a force of gravity, of energy, and of limitless space, took years to develop, and genius was required to express them precisely. Many times in the history of science, progress was possible only because mathematical thinking led the way. For example, by A.D. 1600 algebra had developed to such an extent that an algebraic expression, like $x^2 + 2x + 5$, suggested the idea of a functional relation between the variable x and a variable y, formed by setting $y = x^2 + 2x + 5$. Moreover, the great number of algebraic expressions called attention to the variety of possible relations between variables. Soon after 1600 scientists began to attack their problems by seeking the mathematical relationships between variables, and they used algebraic expressions to represent these relationships. Descartes, Galileo, Huygens, Leibniz, and Newton were prominent among the discoverers of such relationships.

A more recent example of the way in which mathematical thought has prepared men's minds for new developments in science is furnished by the change in the conception of space. This took place to a great extent during the middle of the nineteenth century, and was produced by discoveries of systems of non-Euclidean geometry. Early in the present century Minkowski emphasized the necessity of linking time and space in order to have a proper understanding of the way in which physical events take place. Still later, Einstein recast some of the most fundamental physical notions by utilizing the mathematical ideas of non-Euclidean geometry and of space-time as a single concept. Great as is the genius of Einstein, it is almost certain that he was able to achieve some of his results only because mathematicians of preceding decades had suggested new ways of thinking about space and time.

The purpose of this article has been to indicate broadly several ways in which mathematics is useful to science. To summarize: mathematics supplies a language, methods, and conclusions for science, enables scientists to predict results, furnishes science with ideas for describing phenomena, and prepares the minds of scientists for new ways of thinking.

It would be quite wrong to think that mathematics gives so much to the sciences and receives nothing in return. The physical objects and quantities with which the sciences work and the observed facts concerning those objects and quantities often serve as a source of the elements and postulates of mathematics. It is true that the elements and postulates thus suggested need not be used, and that a mathematical system can be based on elements and postulates which have no apparent application in the physical world. But this does not alter the fact that actually the fundamental concepts of many branches of mathematics are the ones suggested by physical experiences. This statement is especially true of Euclidean geometry. On a more advanced level the result of the Michelson-Morley experiment in physics was adopted as an important basic assumption in the theory of relativity.

A further service rendered by science to mathematics is found in the fact that scientific theories frequently suggest directions for pursuing mathematical investigation and thus furnish a starting point for mathematical discoveries. For example, Copernican astronomy suggested many new problems involving the effects of gravitational attraction between heavenly bodies in motion. These problems stimulated activity in the field of differential equations.

The Mathematization of Science

The point of the preceding section is that mathematics is useful to science in many ways. But that fact in itself fails to describe completely the relation of mathematics to science. A further fact of importance in this connection is that science is becoming more and more mathematical in its concepts and in its methods. Many fundamental physical concepts are being replaced by mathematical ones. Since the time of Newton physicists have believed confidently in the existence of a force called gravity, which causes objects to attract each other. As a consequence of the theory of relativity, the concept of a force of gravity is replaced by an essentially mathematical concept, namely, that the nature of space is such as to cause objects to move as they do, just as the form of a railroad track determines the path of a train running on it.

Science is tending to become more mathematical not only in its concepts, but also in its methods. For example, science now uses abstract concepts far removed from observed physical substances. These concepts are fictions introduced to form a theory. One such fiction is the substance called ether. No physical evidence showed the existence of this substance, but some medium was considered necessary for the transmission of light, much as air serves as a medium in the transmission of sound; hence, the existence of ether with definite properties was assumed. Surely no concept of mathematics is less "real" than ether. Its position in physics is analogous to the position of an undefined term in mathematics, such as a line. We know nothing about a mathematical line except those properties which the postulates imply, and similarly we know nothing about ether except those properties which it is assumed to have.

The assumption of the existence of abstract elements in science is not limited to ether. Gravity, we found, is in the same class. In addition, in the study of atomic structure today, science assumes the existence of entities such as electrons, protons, positrons, neutrons, and others. The only justification for the assumption of their existence is that they help to simplify the explanation of observed facts. They permit the development of a logical theory of atomic structure. But they are not entities which we can observe directly. That physics and chemistry, which avowedly attempt to explain the world of our sense experience, should resort to these fictions or ideal elements is far more surprising than that mathematics, which acknowledges its abstract nature, should do so. Yet, such is the state of affairs in modern science.

Finally, science has become more mathematical in its greater and greater reliance on deductive reasoning as a means of arriving at truth. This tendency is understandable. The certainty of conclusions obtained by deductive reasoning from accepted facts of experience is preferable to the uncertainty of conclusions gained by experiment and generalization therefrom. Moreover, the use of abstract concepts in science requires the use of deductive reasoning, because one cannot experiment with abstractions as with tangible elements.

One may ask what postulates are at the basis of deductive reasoning in physics, for example. Experience suggests to physicists statements which seem to apply to the world about us. These statements are taken as postulates and are used as a basis for reasoning. The so-called law of the conservation of energy is an example of a physical postulate. It is a common observation that when energy is used to do work, other energy appears. If muscular energy is used in sawing wood, energy in the form of heat raises the temperature of the wood and the saw. The energy latent in coal is used to give energy in the form of electricity. From these, and numerous other examples, many physicists have been willing to accept as axiomatic the statement that in a physical or chemical process energy is never lost, but may be converted into a new form. Newton's laws of motion are further postulates for physics. Of course, in so far as the properties of space are involved in their work, physicists use the postulates about space and the consequent deductions from the postulates made by mathematicians.

To sum up the ways in which science has become mathematical, we may state that science has tended to replace physical concepts by concepts which are mathematical in nature, that it has adopted the use of abstract concepts and postulates to explain phenomena, and, finally, that it has become more mathematical by making greater use of the deductive method of reasoning.

The philosopher Kant once remarked that the degree of development of science depends upon the extent to which it has become mathematized. By this criterion the physical sciences have reached a high degree of development today. But even physics is not entirely mathematical, and there are branches of chemistry and biology in which the use of mathematics is of minor importance, as it is in most social sciences. Even in some of those fields in which mathematics is little employed, however, there are many who believe that if mathematical ideas and methods were used more extensively, progress would be more rapid.

The philosopher Descartes devoted himself to the problem of finding a method of obtaining truths which should be applicable to all fields. He finally decided that mathematics supplied the answer to his problem. Applied to any one field of investigation, the method consists essentially of selecting certain basic concepts, securing facts or relations involving these concepts about which one could be certain, and then deducing conclusions from these fundamental facts. To Descartes the existence of mathematics was proof that the method worked, and he proceeded to apply it to philosophy.

Descartes was both right and wrong. He was wrong in supposing that all mathematical conclusions are truths. The creation of non-Euclidean geometry has taught us to be wary of any statements proclaimed as unquestionable truths. But Descartes was right in his choice of mathematical method as an outstanding method of obtaining useful conclusions.

The method Descartes urged should, in the light of our present knowledge of mathematics, be stated thus: Choose concepts that appear to be basic and accept, as postulates, statements about these concepts that appear to be supported by experience. Deduce conclusions by strict reasoning. The conclusions will then be as certain as the postulates. In addition to the above one might well add: Employ symbols and quantitative and geometrical relationships where possible. Thereby numerous specialized mathematical methods and conclusions will become applicable.

Mathematical method is something over and above the mere use of mathematical formulas and conclusions. It is an approach to problems which may be employed in almost all fields. The physical sciences have consciously employed it for centuries, and the biological and social sciences are doing so more and more.

Let us see how the social sciences use mathematical method. We shall consider the school of economic thought known as the single tax system and proposed by Henry George. From several principles which George believed to be basic in our economic system he sought to deduce further economic laws. The desirability of the society to which these conclusions pointed caused George to urge strongly the acceptance of his principles.

Land, says Henry George, is the basis of all wealth. All taxation should be on land itself but not on improvements on land such as buildings. Moreover, the tax on land should be large enough to discourage the ownership of unimproved land. On the other hand, taxes on labor or the products of industry discourage industry.

From these assumptions Henry George and his followers deduced many interesting conclusions. For example, the high tax rate on unimproved land would make it unprofitable to own land without utilizing it. An owner would therefore build or farm on his land. This step would result in considerable industry and in farming, thereby creating a market for labor and preventing unemployment, and producing goods for consumption. Moreover, since labor would be essential to the utilization of land either as a farm or as an industrial site, labor would be valued and properly rewarded. Involuntary poverty would be abolished.

Whether or not this economic system, which still has many supporters, is attractive to the reader, it does illustrate how one may approach economic problems by means of mathematical method. While this school of thought was chosen because its postulates and reasoning are readily stated and understood, it is not unique in its use of mathematical method.

Philosophers, especially, have been aware of mathematical method and of its value in attempting to arrive at truths. The student who would care to sample the entertaining arguments contained in Plato's dialogues will see example after example of deductive reasoning carried out on the basis of statements and definitions initially accepted by the disputants. He would also enjoy the surprising and sometimes startling conclusions to which the reasoning leads.

Some philosophers have gone further in their use of mathematical method. The philosophers Thomas Aquinas and Baruch Spinoza were not satisfied merely to use the essentials of this method. To insure the accuracy of their reasoning they stated explicitly the postulates and theorems of their respective systems of philosophy, and proved each theorem carefully by reference to previous deductions or to the postulates. While the use of this mathematical style of stating theorems and their proofs breeds a stilted literary style, these philosophers were seeking truth and preferred to sacrifice style in order to guarantee the accuracy of their conclusions.

It may be said of almost all philosophers that they seek to employ deductive reasoning from postulates acceptable to them. Sometimes, it must be admitted, assumptions creep into their reasoning which were not intended to be part of their systems.

Mathematical method has impressed the logicians themselves. They, too, have analyzed human reasoning to find the basic principles of reasoning acceptable to all men. From these basic principles or postulates, "laws" (theorems) of reasoning are deduced, to which all acceptable reasoning must conform. The logicians have gone so far as to develop an extensive symbolism which aids them just as symbols aid mathematicians. Every modern textbook on logic now teaches this symbolic logic.

Knowledge of the nature of mathematical method pays dividends to the average man who seeks to understand and cope with political, religious, and economic problems. When carefully analyzed, different political doctrines, as well as religious and economic doctrines, differ essentially in the postulates on which they are based. Statements acceptable to some people as fundamental truths are regarded by others as unacceptable and sometimes unreasonable assumptions. It follows, therefore, that the conclusions correctly deduced from such postulates will not be equally acceptable to all people.

A glance at current economic theories illustrates the remarks of the preceding paragraph. The differences between social and economic systems, such as socialism and capitalism, might well be reduced to differences in fundamental assumptions concerning the acquisition and ownership of wealth. Shall natural resources such as coal and waterpower be the property of a few people or of the whole population? Shall profits be unlimited or should the tax rate be larger for corporations with higher profits so as to prevent excessive profits? Is the contribution of men's labor to a business an investment as is money, or is labor to be paid for as a commodity, on the basis of the supply and demand? Does the government have the obligation to employ people who are not employed by industry and, if so, can it tax to secure money to pay those people? Such fundamental issues are at the heart of economic systems. Once a person commits himself to one or another side of issues like these, the whole body of his economic beliefs follows as a consequence. Much dispute would be avoided if people would recognize the importance of unearthing the fundamental assumptions on which differing economic beliefs are based and concentrate their discussion on these assumptions.

A person's decision to adopt one or another set of basic economic assumptions is entirely analogous to the scientist's decision to adopt one or another system of geometry. This analogy goes further. When scientists found that a non-Euclidean geometry fitted observations and experience better than Euclidean geometry, the latter was rejected and the former installed in its place. A revolution took place in scientific thought. The same happens in economic thought. Individuals and sometimes nations find that an economic system does not meet the test of experience, that is, does not meet the needs of the people. Individuals react by changing their economic beliefs. Nations sometimes react by revolution, for often the economic system is tied to the political system or imposed by a ruling group.

In political systems, too, basic assumptions determine entire theories. Before each presidential-election campaign in this country the politicians dare to be logical. The leading members of each party gather to draw up the party's platform. This platform contains the basic principles to which party members supposedly adhere. These principles are, in a real sense, postulates. From them the party's position on public issues should be deducible. Needless to say the usual party platform uses numerous undefined terms. Needless to say, also, a party's actual position on public issues does not always follow logically from the postulates contained in the platform. And frequently the postulates are surreptitiously changed after the campaign gets under way.

Mathematics and the Culture of a Civilization

The influence of mathematics on our civilization through the medium of the sciences, and the direct influence of mathematics on all fields of thought as a major method of attacking problems, are the larger values of the subject. These values together with the uncountable applications and relationships of mathematics to engineering, art, philosophy, music, logic, religion, and the social sciences, establish mathematics as having unchallengeable importance for our civilization and for the student of our civilization. It will be noted that the importance of mathematics extends beyond the ways in which man earns his daily bread. It includes those higher forms of human activity such as art, philosophy, and music, which are commonly referred to as the cultural fields.

Some students of the history of culture, in particular the late Oswald Spengler, go further in their judgment of the relation of mathematics to the culture of a civilization. For example, Spengler maintains that a study of the mathematics of any civilization will reveal characteristics which are common to other forms of expression of the culture of that civilization. By other forms of expression is meant literature, painting, music, architecture, science, philosophy, and the like.[1] It is not maintained that this correspondence extends to every detail of a culture but that it is typical of the culture. We shall illustrate this thesis, but must warn the reader that the point of view is not necessarily held by all competent writers on the history of civilizations.

Historians such as Spengler emphasize the following characteristics

[1] A discussion of these interrelationships, which indicates agreement with Spengler's views, will be found in *Science and the Human Temperament* by Erwin Schrödinger, the famous atomic physicist.

of Greek mathematics. First, it was mainly static; it dealt with the properties of figures at rest in space. This is illustrated by the geometry of Euclid. Second, their mathematics was for the most part confined to bounded figures, lying in small regions of space, for example, the circle and the triangle. Although the Greeks did regard a straight line as being infinite in extent, and defined parallel lines to be lines which do not meet however far extended, they did not carry far the idea of a geometrical infinity, nor did they study other concepts of infinity. Third, Greek mathematics was confined to objects which could be visualized, that is, to geometry. Although Pythagoras, in his study of length, discovered irrational numbers in the sixth century B.C., the Greeks never developed the abstract idea of irrational numbers (irrational numbers were studied as line segments), nor, for that matter, did they use or develop algebra to any extent. Negative numbers, zero, and imaginary numbers were unknown to them. Fourth, form in mathematics was valued, as the emphasis on deductive reasoning in geometry shows.

Compare these characteristics of Greek mathematics with the following facts about Greek life and thought. Greek architecture, as revealed in their temples, was, to a large extent, static, that is, it did not suggest the idea of motion. To many people their temples present an appearance of repose, of being well balanced and firmly set on the ground. The type of physics studied by the Greeks was the branch of mechanics now known as static, a study of the forces acting on bodies at rest. Just as their geometry was confined to bounded figures, so in their lives the Greeks were relatively unexplorative. They lived in city-states and, as compared with other great nations, stayed near home. They are accused of a lack of perspective, or the representation of depth, in painting. Their music is called two-dimensional by some writers because it consisted only of rhythm and melody. Harmony was to come later. Their preference for mathematical concepts that can be visualized is reflected in their high development of simple forms in architecture and sculpture.

Let us now make a similar, though partial, comparison of the mathematics and culture of the civilization commencing with the Renaissance, roughly the fourteenth century. The development of algebra by the Hindus and others increased the ability of the mathematician to solve quantitative problems. The notion of a variable, implicit even in the elementary formulas of the Egyptians and the Greeks, became significant. With that, came the idea of a function or relation among variables. Next, the question of rate of change of a function was raised. The attempts to answer this question produced the idea of a limit and its application to functions. Mathematics had become dynamic and was concerned with change. Another characteristic of modern mathematics is the freedom it enjoys to develop concepts having no obvious counterparts in the physical world. This tendency became significant with the rise of non-Euclidean geometry. The realization by mathematicians that they were free to develop any system they chose, regardless of whether or not they started with postulates taken from experience, encouraged them to explore new fields. They constructed a logical theory to answer the bothersome questions about infinity, and they began to study spaces which had, at the time of their investigation, no correspondence to the physical world.

Let us compare these characteristics of mathematics with the following developments in other fields. Science began in the sixteenth century to make a quantitative study of the world. Physical laws appeared in functional form. Velocity and acceleration, which are nothing more than rates of change of functions, became basic objects of study. A science of dynamics, that is, the study of motion, arose and has become very important. The earth was explored and the heavens were studied, thus enlarging our knowledge of space. Even Gothic architecture, with its tall buildings and spires, is regarded by some authorities as a sign of explorative tendency. Science has recently adopted logically constructed descriptions of phenomena which appeal to the mind rather than the senses. Philosophy has become more concerned with, and influenced by, the results of science. Modern art likewise reflects appeal to the mind rather than the eye alone.

This topic could be carried much further, but to do so it would be necessary to analyze other forms of our culture. The foregoing paragraphs have for their purpose only to explain the meaning of the statement that the characteristics of mathematics are related to the characteristics of the other forms in which the culture of a civilization expresses itself. It could be shown, further, by means of a detailed analysis, that the history of mathematical thought is interrelated with the history of civilization. This statement does not imply that mathematics caused the changes which produced one civilization from another, but merely that it changed with the civilizations and to a large extent reflected each civilization.

The greatest distinction between our present Western civilization and all others of which we have any knowledge is the growth of mathematics and the natural sciences and the application of these fields to industry, engineering, and commerce.

Without belittling the merits of our historians, economists, philosophers, writers, poets, painters, and statesmen, it is possible to say that other civilizations have produced their equals, not merely in ability but in accomplishments. On the other hand, though Euclid and Archimedes were undoubtedly great and though our mathematicians and scientists were able to reach higher, only because, as Newton put it, they were able to stand on the shoulders of such giants, nevertheless it is in our age that mathematics and science have attained their maturity and extraordinary applicability.

Because mathematics has left its imprint upon so many aspects of present-day civilization, its position in the modern world is a fundamental one, and a knowledge of mathematics is essential for a comprehensive understanding of current life and thought.

B. *The Heavens*

•

1. Astronomy

Astronomy is the oldest of the sciences. Centuries before Christ, the Chinese are thought to have predicted eclipses of the sun. The Egyptians linked the annual flooding of the Nile with the appearance of Sothis (Sirius) in the predawn sky and prepared a calendar of 365 days for guidance in both agriculture and religious rites. The Babylonians and Assyrians observed the movements of the planets, recorded eclipses and attempted to describe their observations with mathematical tables. These studies were intimately connected with the pseudoscience of astrology, which held that the movements of heavenly bodies controlled men's destinies and that in some cases the reverse was true—that unwise actions on earth, for example, might result in eclipses or other signs of heavenly disapproval. Despite every evidence of its worthlessness, astrology still numbers devotees in the millions.

It was the Greeks who first evolved a unified system, a cosmology, by which the complex movements of the heavenly bodies, particularly the planets, could be explained. They also made calculations of the distance of the sun from the earth and of the size of the latter. Their greatest astronomer, Hipparchus, who lived in the second century B.C., *discovered the precession of the equinoxes. In the century before him, Aristarchus of Samos had even suggested that the center of the universe, about which all else revolved, was the sun. His heliocentric theory was of course rejected by the ancients, partly because of the influence of Aristotle, who gave his support to the theory that the stars and planets were embedded in a series of concentric crystal spheres, all revolving at different rates around the earth at the center. A coherent exposition of this hypothesis was presented in* The Almagest *by Ptolemy of Alexandria, who lived in the second century* A.D. *For over a millennium the Ptolemaic system was to dominate scientific and religious thought. Not until the sixteenth century was its death knell sounded with the publication of one of the greatest books in human history.*

On the twenty-third day of May, in the year 1543, Nicholas Copernicus received on his deathbed the first copy of his immortal work, De Revolutionibus Orbium Coelestium (Concerning the Revolutions of the Heavenly Bodies). *A few hours later he closed his eyes on a medieval world which still believed in the geocentric universe. Considering the state of communications at the time, the news of his hypothesis spread with surprising speed. It was received with derision by philosophers and with violent antagonism by the Church. But a few enlightened men, of whom the greatest was Galileo the Pisan, rose to its defense. In 1610, with a telescope he had made, he watched four small bodies as they revolved around the planet Jupiter. Here was a miniature solar system similar to our own. Here was* Proof that the Earth Moves. *The struggle was not won. Galileo was forced by the Church to recant. But as evidence grew, the heoliocentric theory was accepted by all educated men. In the following pages excerpts from the writings of these two great men offer us glimpses of the story. As we read, some of their own excitement and wonder comes to us across the centuries.*

Since that day, the extent of our knowledge has increased enormously and something of what we have learned is described in the selections which follow. In Our Neighbour Worlds *Peter Millman, the Canadian astronomer, describes the planets, the "wanderers," whose movements, far more obvious than those of the "fixed stars," first excited the interest of the ancients. Study of them decreased as astronomers shifted their interest to the stars and nebulae. Now, with the development of space probes which can make actual measurements of compositions, temperatures and other characteristics, interest in our fellow members of the solar system has revived.*

Our knowledge of the stars—their constitutions, life cycles, the forces which govern their activities—is based in large part on the sun; for the sun, as astronomers constantly emphasize, is "a typical star, a common kind of star." Armed with a variety of complex and almost incredibly sensitive instruments, they have assembled a huge amount of observational material about the sun, and in doing so have thrown light on the characteristics of all the stars. Their findings are described in The Sun *by Cecilia Payne-Gaposchkin, professor of astronomy at Harvard.*

Until recently, with the minor exception of the meteorites which have reached the earth's surface, all man's knowledge of outer space has come to him through the medium of light. With great telescopes such as the Hale telescope at Palomar he has gathered this light; with photographic plates he has fixed it for detailed observation; with spectroscopes and attendant instruments he has analyzed it. Now, with space probes of which Sputnik was the prototype, he can collect direct evidence in a way hitherto impossible. But the limits of observation of such space probes are small. At present, and probably for long periods to come, their range is the solar system. A new discovery, on which radioastronomy is founded, has no such limitation. Its origin and uses are described in Radio Waves from Space *by Sir Bernard Lovell, head of the Jodrell Bank observatory, who has been one of the pioneers in the field.*

Radioastronomy joins with older methods of observation to help us penetrate ever farther into the universe. We journey from the sun to near and distant stars, from our own galaxy to comparable systems, in the classic article The Milky Way and Beyond *by one of England's greatest astronomers, Sir Arthur Eddington. There have been extraordinary developments in cosmology since 1933 when Eddington wrote. Some of these developments, including theories of the origin of the universe currently being considered by astronomers, are the subject of Sir Bernard Lovell's* The Origin and Evolution of the Universe.

NICHOLAS COPERNICUS

Concerning the Revolutions of the Heavenly Bodies

•

That the Universe Is Spherical

FIRST OF ALL we assert that the universe is spherical; partly because this form, being a complete whole, needing no joints, is the most perfect of all; partly because it constitutes the most spacious form, which is thus best suited to contain and retain all things; or also because all discrete parts of the world, I mean the Sun, the Moon, and the planets, appear as spheres; or because all things tend to assume the spherical shape, a fact which appears in a drop of water and in other fluid bodies when they seek of their own accord to limit themselves. Therefore no one will doubt that this form is natural for the heavenly bodies.

That the Earth Is Likewise Spherical

That the earth is likewise spherical is beyond doubt, because it presses from all sides to its center. Although a perfect sphere is not immediately recognized because of the great height of the mountains and the depression of the valleys, yet this in no wise invalidates the general spherical form of the earth. This becomes clear in the following manner: To people who travel from any place to the North, the north pole of the daily revolution rises gradually, while the south pole sinks a like amount. Most of the stars in the neighborhood of the Great Bear appear not to set, and in the South some stars appear no longer to rise. Thus Italy does not see Canopus, which is visible to the Egyptians. And Italy sees the outermost star of the River, which is unknown to us of a colder zone. On the other hand, to people who travel toward the South, these stars rise higher in the heavens, while those stars which are higher to us become lower. Therefore, it is plain that the earth is included between the poles and is spherical. Let us add that the inhabitants of the East do not see the solar and lunar eclipses that occur in the evening, and people who live in the West do not see eclipses that occur in the morning, while those living in between see the former later, and the latter earlier.

That even the water has the same shape is observed on ships, in that the land which can not be seen from the ship can be spied from the tip of the mast. And, conversely, when a light is put on the tip of the mast, it appears to observers on land gradually to drop as the ship recedes until the light disappears, seeming to sink in the water. It is clear that the water, too, in accordance with its fluid nature, is drawn downwards, just as is the earth, and its level at the shore is no higher than its convexity allows. The land therefore projects everywhere only as far above the ocean as the land accidentally happens to be higher.

Whether the Earth Has a Circular Motion, and Concerning the Location of the Earth

Since it has already been proved that the earth has the shape of a sphere, I insist that we must investigate whether from its form can be deduced a motion, and what place the earth occupies in the universe. Without this knowledge no certain computation can be made for the phenomena occurring in the heavens. To be sure, the great majority of writers agree that the earth is at rest in the center of the universe, so that they consider it unbelievable and even ridiculous to suppose the contrary. Yet, when one weighs the matter carefully, he will see that this question is not yet disposed of, and for that reason is by no means to be considered unimportant. Every change of position which is observed is due either to the motion of the observed object or of the observer, or to motions, naturally in different directions, of both; for when the observed object and the observer move in the same manner and in the same direction, then no motion is observed. Now the earth is the place from which we observe the revolution of the heavens and where it is displayed to our eyes. Therefore, if the earth should possess any motion, the latter would be noticeable in everything that is situated outside of it, but in the opposite direction, just as if everything were traveling past the earth. And of this nature is, above all, the daily revolution. For this motion seems to embrace the whole world, in fact, everything that is outside of the earth, with the single exception of the earth itself. But if one should admit that the heavens possess none of this motion, but that the earth rotates from west to east; and if one should consider this seriously with respect to the seeming rising and setting of the sun, of the moon and the stars; then one would find that it is actually true. Since the heavens which contain and retain all things are the common home of all things, it is not at once comprehensible why a motion is not rather ascribed to the thing contained than to the containing, to the located rather than to the locating. This opinion was actually held by the Pythagoreans Heraklid and Ekphantus and the Syracusean Nicetas (as told by Cicero), in that they assumed the earth to be rotating in the center of the universe. They were indeed of the opinion that the stars set due to the intervening of the earth, and rose due to its receding.

Refutation of the Arguments, and Their Insufficiency

It is claimed that the earth is at rest in the center of the universe and that this is undoubtedly true. But one who believes that the earth rotates will also certainly be of the opinion that this motion is natural and not violent. Whatever is in accordance with nature produces effects which are the opposite of what happens through violence. Things upon which violence or an external force is exerted must become annihilated and cannot long exist. But whatever happens in the course of nature remains in good condition and in its best arrangement. Without cause, therefore, Ptolemy feared that the earth and all earthly things if set in rotation would be dissolved by the action of nature, for the functioning of nature is something entirely different from artifice, or from that which could be contrived by the human mind. But why did he not fear the same, and indeed in much higher degree, for the universe, whose motion would have to be as much more rapid as the heavens are larger than the earth? Or have the heavens become infinite just because they have been removed from the center by the inexpressible force of the motion; while otherwise, if they were at rest, they would collapse? Certainly if this argument were true the extent of the heavens would become infinite. For the more they were driven aloft by the outward impulse of the motion, the more rapid would the motion become because of the ever increasing circle which it would have to describe in the space of 24 hours; and, conversely, if the motion increased, the immensity of the heavens would also increase. Thus velocity would augment size into infinity, and size, velocity. But according to the physical law that the infinite can neither be traversed, nor can it for any reason have motion, the heavens would, however, of necessity be at rest.

But it is said that outside of the heavens there is no body, nor place, nor empty space, in fact, that nothing at all exists, and that, therefore, there is no space in which the heavens could expand; then it is really strange that something could be enclosed by nothing. If, however, the heavens were infinite and were bounded only by their inner concavity, then we have, perhaps, even better confirmation that there is nothing outside of the heavens, because everything, whatever its size, is within them; but then the heavens would remain motionless. The most important argument, on which depends the proof of the finiteness of the universe, is motion. Now, whether the world is finite or infinite, we will leave to the quarrels of the natural philosophers; for us remains the certainty that the earth, contained between poles, is bounded by a spherical surface. Why should we hesitate to grant it a motion, natural and corresponding to its form; rather than assume that the whole world, whose boundary is not known and cannot be known, moves? And why are we not willing to acknowledge that the *appearance* of a daily revolution belongs to the heavens, its *actuality* to the earth? The relation is similar to that of which Virgil's Æneas says: "We sail out of the harbor, and the countries and cities recede." For when a ship is sailing along quietly, everything which is outside of it will appear to those on board to have a motion corresponding to the movement of the ship, and the voyagers are of the erroneous opinion that they with all that they have with them are at rest. This can without doubt also apply to the motion of the earth, and it may appear as if the whole universe were revolving.

Concerning the Center of the Universe

Since nothing stands in the way of the movability of the earth, I believe we must now investigate whether it also has several motions, so that it can be considered one of the planets. That it is not the center of all the revolutions is proved by the irregular motions of the planets, and their varying distances from the earth, which cannot be explained as concentric circles with the earth at the center. Therefore, since there are several central points, no one will without cause be uncertain whether the center of the universe is the center of gravity of the earth or some other central point. I, at least, am of the opinion that gravity is nothing else than a natural force planted by the divine providence of the Master of the World into its parts, by means of which they, assuming a spherical shape, form a unity and a whole. And it is to be assumed that the impulse is also inherent in the sun and the moon and the other planets, and that by the operation of this force they remain in the spherical shape in which they appear; while they, nevertheless, complete their revolutions in diverse ways. If then the earth, too, possesses other motions besides that around its center, then they must be of such a character as to become apparent in many ways and in appropriate manners; and among such possible effects we recognize the yearly revolution.

GALILEO GALILEI

Proof that the Earth Moves

•

About ten months ago a report reached my ears that a Dutchman had constructed a telescope, by the aid of which visible objects, although at a great distance from the eye of the observer, were seen distinctly as if near; and some proofs of its most wonderful performances were reported, which some gave credence to, but others contradicted. A few days after, I received confirmation of the report in a letter written from Paris by a noble Frenchman, Jaques Badovere, which finally determined me to give myself up first to inquire into the principle of the telescope, and then to consider the means by which I might compass the invention of a similar instrument, which after a little while I succeeded in doing, through deep study of the theory of Refraction; and I prepared a tube, at first of lead, in the ends of which I fitted two glass lenses, both plane on one side, but on the other side one spherically convex, and the other concave. Then bringing my eye to the concave lens I saw objects satisfactorily large and near, for they appeared one-third of the distance off and nine times larger than when they are seen with the natural eye alone. I shortly afterwards constructed another telescope with more nicety, which magnified objects more than sixty times. At length, by sparing neither labour nor expense, I succeeded in constructing for myself an instrument so superior that objects seen through it appear magnified nearly a thousand times, and more than thirty times nearer than if viewed by the natural powers of sight alone.

First Telescopic Observations

It would be altogether a waste of time to enumerate the number and importance of the benefits which this instrument may be expected to confer, when used by land or sea. But without paying attention to its use for terrestrial objects, I betook myself to observations of the heavenly bodies; and first of all, I viewed the Moon as near as if it was scarcely two semidiameters of the Earth distant. After the Moon, I frequently observed other heavenly bodies, both fixed stars and planets, with incredible delight.

Discovery of Jupiter's Satellites

There remains the matter, which seems to me to deserve to be considered the most important in this work, namely, that I should disclose and publish to the world the occasion of discovering and observing four planets, never seen from the very beginning of the world up to our own times, their positions, and the observations made during the last two months about their movements and their changes of magnitude.

On the 7th day of January in the present year, 1610, in the first hour of the following night, when I was viewing the constellations of the heavens through a telescope, the planet Jupiter presented itself to my view, and as I had prepared for myself a very excellent instrument, I noticed a circumstance which I had never been able to notice before, owing to want of power in my other telescope, namely, that three little stars, small but very bright, were near the planet; and although I believed them to belong to the number of the fixed stars, yet they made me somewhat wonder, because they seemed to be arranged exactly in a straight line, parallel to the ecliptic, and to be brighter than the rest of the stars, equal to them in magnitude. The position of them with reference to one another and to Jupiter was as follows:

Ori. * * O * Occ.

On the east side there were two stars, and a single one towards the west. The star which was furthest towards the east, and the western star, appeared rather larger than the third.

I scarcely troubled at all about the distance between them and Jupiter, for, as I have already said, at first I believed them to be fixed stars; but when on January 8th, led by some fatality, I turned again to look at the same part of the heavens, I found a very different state of things, for there were three little stars all west of Jupiter, and nearer together than on the previous night, and they were separated from one another by equal intervals, as the accompanying figure shows.

Ori. O * * * Occ.

At this point, although I had not turned my thoughts at all upon the approximation of the stars to one another, yet my surprise began to be excited, how Jupiter could one day be found to the east of all the aforesaid fixed stars when the day before it had been west of two of them; and forthwith I became afraid lest the planet might have moved differently from the calculation of astronomers, and so had passed those stars by its own proper motion. I, therefore, waited for the next night with the most intense longing, but I was disappointed of my hope, for the sky was covered with clouds in every direction.

But on January 10th the stars appeared in the following position with regard to Jupiter, the third, as I thought, being

Ori. * * O Occ.

hidden by the planet. They were situated just as before, exactly in the same straight line with Jupiter, and along the Zodiac.

When I had seen these phenomena, as I knew that corresponding changes of position could not by any means belong to Jupiter, and as, moreover, I perceived that the stars which I saw had always been the same, for there were no others either in front or behind, within a great distance, along the Zodiac—at length, changing from doubt into surprise, I discovered that the interchange of position which I saw belonged not to Jupiter, but to the stars to which my attention had been drawn, and I thought therefore that they ought to be observed henceforward with more attention and precision.

Accordingly, on January 11th I saw an arrangement of the following kind:

Ori. * * O Occ.

namely, only two stars to the east of Jupiter, the nearer of which was distant from Jupiter three times as far as from the star further to the east; and the star furthest to the east was nearly twice as large as the other one; whereas on the previous night they had appeared nearly of equal magnitude. I, therefore, concluded, and decided unhesitatingly, that there are three stars in the heavens moving about Jupiter, as Venus and Mercury round the Sun; which at length was established as clear as daylight by numerous other subsequent observations. These observations also established that there are not only three, but four, erratic sidereal bodies performing their revolutions round Jupiter.

These are my observations upon the four Medicean planets, recently discovered for the first time by me; and although it is not yet permitted me to deduce by calculation from these observations the orbits of these bodies, yet I may be allowed to make some statements, based upon them, well worthy of attention.

Orbits and Periods of Jupiter's Satellites

And, in the first place, since they are sometimes behind, sometimes before Jupiter, at like distances, and withdraw from this planet towards the east and towards the west only within very narrow limits of divergence, and since they accompany this planet alike when its motion is retrograde and direct, it can be a matter of doubt to no one that they perform their revolutions about this planet while at the same time they all accomplish together orbits of twelve years' length about the centre of the world. Moreover, they revolve in unequal circles, which is evidently the conclusion to be drawn from the fact that I have never been permitted to see two satellites in conjunction when their distance from Jupiter was great, whereas near Jupiter two, three, and sometimes all four, have been found closely packed together. Moreover, it may be detected that the revolutions of the satellites which describe the smallest circles round Jupiter are the most rapid, for the satellites nearest to Jupiter are often to be seen in the east, when the day before they have appeared in the west, and contrariwise. Also, the satellite moving in the greatest orbit seems to me, after carefully weighing the occasions of its returning to positions previously noticed, to have a periodic time of half a month. Besides, we have a notable and splendid argument to remove the scruples of those who can tolerate the revolution of the planets round the Sun in the Copernican system, yet are so disturbed by the motion of one Moon about the Earth, while both accomplish an orbit of a year's length about the Sun, that they consider that this theory of the universe must be upset as impossible; for now we have not one planet only revolving about another, while both traverse a vast orbit about the Sun, but our sense of sight presents to us four satellites circling about Jupiter, like the Moon about the Earth, while the whole system travels over a mighty orbit about the Sun in the space of twelve years.

PETER MILLMAN

Our Neighbour Worlds

•

MOST OF THE MATTER in our universe exists in the form of very hot gas, making up the stars, or as much cooler gas and dust, scattered through space. A very small fraction is in the form of dark, relatively cool, solid bodies I shall call worlds. There is no particular reason why there may not be countless numbers of such worlds out among the stars, but with our present instruments we cannot observe or even detect them. Our discussion is therefore limited to our own family of worlds, the members of the solar system held by the sun's gravitation in a local group.

I have called these our neighbour worlds and the term is apt in spite of their great distances. To illustrate this, suppose we make a scale model where the distance from the Earth to the sun, ninety-three million miles, is just under one-quarter of an inch. Now take a dime out of your purse. On the scale of our model the orbits of the four inner planets Mercury, Venus, Earth, and Mars, fit comfortably on this coin with the orbit of Mars represented by the circumference. The orbits of all four planets are very nearly in the same plane, and it so happens that the thickness of the dime represents to scale the thickness of the disk-shaped volume of space within which these planets move (Figure 1). The orbit of Neptune, the outermost large planet, will be fourteen inches across centred roughly on the dime and also nearly in the same plane. It is only when we come to little Pluto, with an orbit a bit bigger than Neptune's, that we find its orbit tilted by quite an angle to the rest (Figure 2). All the planets move round the sun counter-clockwise as viewed from the north. And on the scale of our model where will the nearest star be? Exactly one mile away from the dime. This is the closest star. The centre of our star system, or galaxy, would be over 6,000 miles from the dime, and the millions of other galaxies very much further away. I think you will agree that the term neighbor worlds is quite appropriate.

These worlds fall naturally into four groups, a division based in general on size and orbital characteristics. In one group are the four inner planets, Mercury, Venus, Earth, Mars, and, at the outer edge of the planet orbits, Pluto. All but distant Pluto have been known from antiquity. They can conveniently be called the Earth planets, since they resemble our own home planet more closely than does anything else we have found in space. Their diameters range from 3,000 to 8,000 miles (Figure 3).

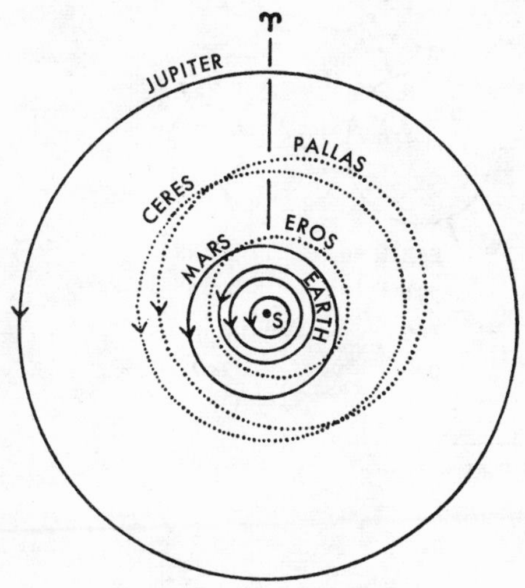

Fig. 1. The orbits about the Sun, S, of the inner planets, three asteroids, and Jupiter. The vertical line indicates zero longitude, or the first point of Aries, whose symbol is the horns of the ram. The scale here is the same as suggested in the text, where the orbit of Mars fits on a dime.

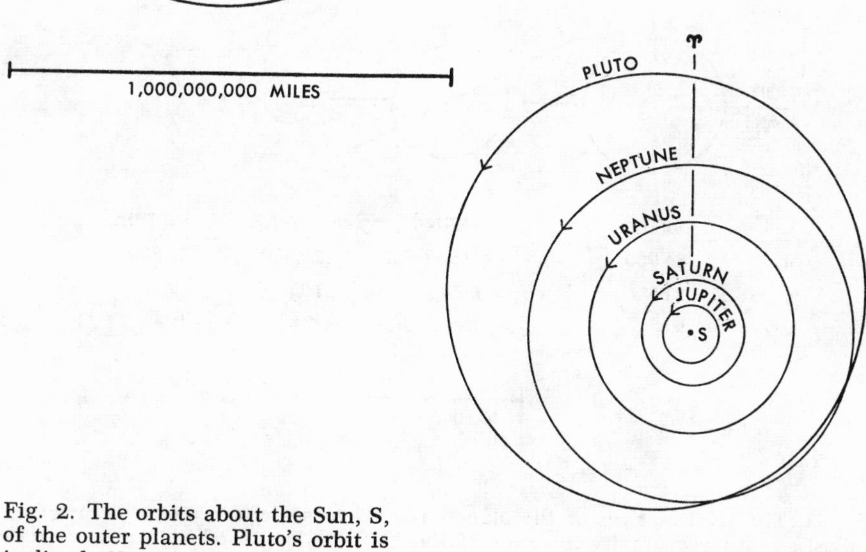

5,000,000,000 MILES

Fig. 2. The orbits about the Sun, S, of the outer planets. Pluto's orbit is inclined 17° to the plane of the paper.

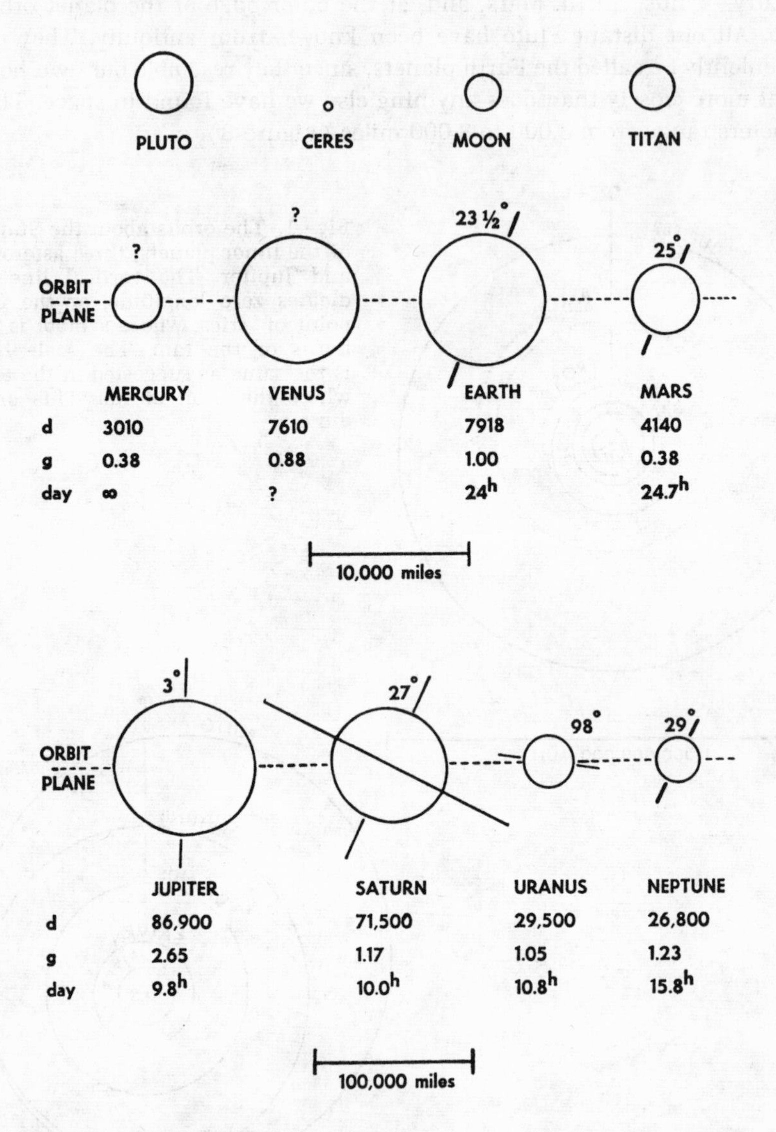

Fig. 3. The relative sizes of the planets plotted to scale; d = mean diameter in miles, g = surface gravity in terms of Earth's. The length of each planet's day is given in terms of hours on the Earth; the angle between a planet's equator and its orbit is given in degrees.

Then we have the four great planets, Jupiter, Saturn, Uranus, and Neptune. These have average diameters ranging from 27,000 to 87,000 miles, roughly ten times those of the Earth planets, and their orbits lie in the outer part of the solar system.

The next group of worlds consists of the moons, or satellites. These are, in general, smaller than the planets and move in orbits about the planets. In all, thirty-one natural satellites of six planets are known. The Earth, in addition to its one satellite of natural origin, has a rapidly growing complex of artificial satellites.

The last of the four groups of worlds is that of the asteroids. These little planets have orbits which form a network, mainly between the orbit of Mars, at the edge of our dime, and the orbit of Jupiter, one and a quarter inches from the dime centre. The orbits of over 1,600 asteroids have already been tabulated. There are probably some 50,000 of these little worlds in all.

Let us visit each of the four groups of worlds in turn, starting with the asteroids. The first asteroid was discovered in 1801, and was called Ceres. It is the largest of the group, having a diameter just under 500 miles. In the period from 1802 to 1807 three more asteroids were discovered. But the fifth was not found until 1845 and it was only after the introduction of photography that the little planets were found in large numbers each year.

The great problem is to keep them all sorted out. In most cases the only sure identification is the position of the orbit around the sun. Even after these orbits have been determined it requires a great deal of work to keep them up to date, since they change slowly as a result of the gravitational pull of the large planets, in particular Jupiter. Asteroids are often lost owing to the impossibility of following them all.

The discoverer of an asteroid has the privilege of naming it, and by tradition most of these objects are given feminine names. When the orbit is unusual a masculine name is given. At first, asteroids were given names taken mainly from classical mythology. But these names were soon used up, and now we find little planets called Jenny, Hilda, Ruth, Marilyn, Susanna, and Alice, to take a few modern examples. One of the most extreme orbits is that of Icarus, aptly named after the youth who singed his wings. This asteroid swings in closer to the sun than the planet Mercury.

The great majority of asteroids are quite small. Less than twenty have diameters over 100 miles, and most of those discovered to date are in the ten- to fifty-mile category. The smallest known are about a mile across. Except for the largest members of the group the asteroids seem to be quite irregular in shape, and show a variation in brightness as they rotate in periods of a few hours. One famous asteroid, Eros, came close enough to the Earth in 1931 for its shape to be observed in the telescope. Eros is a piece of rock, eighteen miles long and five miles wide, and it turns end over end in a period of 5¼ hours. These jagged lumps of space rock seem to be fragments of one or more medium-sized planets that must have broken up when the solar system was much younger.

The surface gravity on these little worlds is extremely weak ranging from one-hundredth to one-thousandth of the Earth's surface gravity, and even less for the smallest asteroids. It would not be difficult to throw a stone right off into space from an asteroid ten or twenty miles across. In fact, you would have to be careful not to jump too hard on a very small asteroid, or you might find yourself unexpectedly launched into space. No asteroid has a force of gravity strong enough to retain an atmosphere of any kind.

The satellites, which make up the next group of worlds, range in size from the two little moons of Mars, five to ten miles across, up to the two largest of the twelve moons of Jupiter. These Jovian satellites are well over 3,000 miles in diameter, and thus bigger than Mercury. The smallest of the satellites, on the other hand, are much like the asteroids, in fact some of them may have been asteroids before being picked up by the parent planet.

As an example of the larger satellites we can take our moon, which is close enough to be studied in some detail. The diameter of the moon is just over 2,000 miles, not far from one-quarter the diameter of the earth. Its mass is roughly one-eightieth of the earth's mass and its distance thirty earth-diameters, or about 240,000 miles. Absence of an atmosphere on our satellite gives a hard, sharp contrast between light and shadow. This is particularly striking when the sun is low in the lunar sky and shadows are very long, greatly exaggerating the rugged character of the surface detail.

The world of the moon may be visited by space travellers within a reasonable period of time. The absence of atmosphere means a dark sky full of stars, even with the sun above the horizon. It is a silent world as no air waves carry the sound of wind or weather, and surface gravity is low, about one-sixth that on Earth. On the moon a 200-pound man would be a light-weight of only thirty-three pounds. The surface is covered with a dusty, grey-brown, ashy substance, that stretches in gentle rolls out to the horizon. And the horizon on the moon seems unusually close, a consequence of the high curvature of the lunar surface.

A day on the moon lasts for 29½ of our days. The sun slowly rises and slowly sets, and all this time the Earth appears to remain nearly stationary in the sky, a great, bluish-white globe, four times as wide across as the sun. The moon keeps the same face always towards the Earth so that from most positions on the lunar surface the Earth will not appear to rise or set, and on the far side of the moon the Earth can never be seen.

There is a very large temperature range between day and night conditions on the moon. When the sun is high the surface temperature rises to over 200° Fahrenheit. At sunset this drops to 50° below zero, and in the middle of the long lunar night it may be colder than 300° below zero.

On the moon's surface we find long mountain ranges with some peaks higher than Mt. Everest, deep valleys, low undulating ridges, and long cracks or crevasses. Steep slopes are uncommon and the dark surface material is fairly smooth on a scale of inches. The most characteristic feature of the entire lunar surface is the presence of thousands of circular mountain forms called craters. These vary all the way from the smallest we can observe, a fraction of a mile across, up to some over 150 miles in diameter. The lunar craters do not resemble volcanic craters, and it is unlikely that enough volcanic activity could exist on the moon to produce all the craters found there. But localized emissions of gas probably still occur from time to time. An event of this type was photographed in the crater Alphonsus on November 3, 1958, by N. A. Kozyrev at the Crimean Astrophysical Observatory. Most astronomers now believe that the majority of the large lunar craters resulted from the impact of great meteorites in the far distant past of the solar system.

The absence of an atmosphere on the moon is not surprising, since the moon's surface gravity is not enough to hold the common atmospheric gases under temperature conditions corresponding to the lunar surface. The largest satellites in the outer part of the solar system, where it is colder, should be just able to retain an atmosphere. In fact methane, or marsh gas, has been detected on Titan, the largest satellite of Saturn. In passing, you may wonder how we know what is in the atmosphere of a distant heavenly body. The light emitted or reflected by atoms and molecules frequently contains certain characteristics peculiar to the elements involved. When this light is analyzed by the spectroscope, the substances in the source can be named with certainty.

We move on now to the four giant worlds, Jupiter, Saturn, Uranus, and Neptune, with equatorial diameters ranging up to 90,000 miles. If we represent the Earth by a large marble, or agate, then Neptune would be the size of a baseball, and Jupiter a basketball. In spite of their size these great planets all spin rapidly on their axes with rotation periods of less than sixteen hours. Jupiter, the largest, has the most rapid rotation, turning once in under ten hours. A point on Jupiter's equator moves eight miles a second, fast enough to throw loose material off into space under the gravity conditions we find on Earth. But the Jovian surface gravity is over two and a half times that found here, so there is no danger of this planet showering space with loose surface material.

The rapid spin does have the effect of flattening these planets at the poles. Jupiter has a polar diameter 6,000 miles less than its equatorial diameter. The chief surface markings visible are great cloud belts drawn out along lines parallel to the direction of rotation. The various cloud belts of Jupiter, and semipermanent features floating in the belts, have rotation periods that differ slightly from each other. For example, the Great Red Spot has been well observed for over one hundred years, and may have been seen as early as A.D. 1664. It has its own rotation period that varies somewhat from the rotation periods of the belts.

One of the more interesting recent discoveries has been the detection of radio noise from Jupiter. This may originate in atmospheric disturbances or storms of some type, and is usually observed at wave-lengths from ten to twenty metres. Much of it seems to originate from localized centres that have been tentatively identified with some long-enduring markings in the cloud belts. Perhaps free-floating storm centres have some way of maintaining themselves or, alternatively, these centres may be tied to a solid surface below the clouds.

The spectroscope shows that the clouds of Jupiter consist of methane—the same gas that was found in Titan—mixed with ammonia. All four planets have methane present in large quantities, but only Jupiter and Saturn show ammonia clearly, while Uranus may have a very faint trace of it. To find an explanation for this fact we take a look at the surface temperatures. Measured values on the Fahrenheit scale range from 210° below zero for Jupiter, down to 330° below zero or lower for Neptune. At these temperatures ammonia freezes while methane would remain a gas over most of the range. Apparently on Jupiter we have frozen crystals of ammonia floating in a methane atmosphere. As we move out to the colder temperatures of the outer planets we find that most of the ammonia has disappeared out of the atmosphere, leaving only the methane which becomes progressively stronger. Heavy layers of methane absorb red light, giving Uranus a greenish colour and making Neptune look blue.

Obviously, in the case of all four of these planets, we see only the outer layers of a very extensive cloud cover. The low mean densities, not much greater than that for water, suggest that a large part of the volume of each planet must consist of the atmosphere, which is probably hydrogen and helium, together with carbon and nitrogen compounds. Under these conditions one would not expect to find free oxygen, for it would combine with the hydrogen or other elements.

One theory about the interiors of the giant planets is that each has a small rocky or metallic core at the centre, surrounded by a deep ice layer. Recently, however, it has been pointed out that hydrogen and other light elements will act like metals under the extreme pressure found in the interior of such planets. Possibly the denser cores are merely hydrogen under high pressure.

All the great planets are well supplied with satellites, and Saturn has, in addition, a unique system of rings that circle the planet well inside the orbits of its nine moons. These make it one of the most beautiful telescopic sights in the heavens. The rings, which are 170,000 miles across but only about ten miles thick, consist of countless small solid particles revolving about the planet. The light reflected from the rings seems to suggest ice particles, mixed probably with stony material. No satellite of any size could exist in the position of the rings; it would be torn apart by tidal forces.

I have left to the last the five small Earth planets: Pluto, Mars, Earth,

Venus, and Mercury. Pluto, on the outer edge of our system, is so far away that we know very little about its physical condition. Even its diameter is not well determined, but is probably about half that of the Earth. This is too small to retain the lightest gases and the others, if present, are probably frozen on the planet's surface. The temperature must be about 350° below zero Fahrenheit.

Mercury, at the other extreme from Pluto, is close to the sun—a scorched little world that hurries around its orbit once every eighty-eight days. Like our moon, which keeps one face to the Earth, Mercury always turns the same side to the sun. The temperature at the centre of the sunward face averages 660° Fahrenheit—hot enough to melt lead. In contrast, the central part of the dark hemisphere on Mercury is never exposed to sunlight and the temperature here is estimated as 410° below zero—colder even than the daylight side of Pluto. Mercury has no atmosphere to moderate these terrible extremes of temperature.

Apart from the Earth only two worlds remain to be discussed: Venus and Mars. These are the two planets closest to the Earth in space and also in size, and here we find physical conditions which most nearly parallel those in which we live.

Venus, 7,600 miles in diameter, is almost the same size as the Earth. It revolves about the sun once every 225 days in an orbit just inside ours. Thanks to the smaller size of its orbit, and a greater speed in this orbit, Venus laps the Earth every 584 days or 19½ months. This is a useful period to remember, since it represents the interval between successive returns to the same relative positions for Earth, Venus, and Sun. For example, if a certain situation is particularly favorable for a space flight to Venus, essentially the same situation will repeat itself just 584 days later. A flight to Venus under minimum power conditions takes 146 days, and a shorter time if extra power is available.[1]

On Mars, in contrast to Venus, we find an atmosphere that is thin and quite transparent so that the solid surface of the planet is visible. Also, Mars revolves about the sun in an orbit outside that of the Earth, which means that when we are closest to this planet we are looking at its daylight hemisphere. Although Mars is just 4,200 miles in diameter, it closely resembles the Earth in a number of ways. The day on Mars is forty minutes longer than ours. The tilt of its equator to the orbit-plane is almost the same as the Earth's, which results in seasonal changes similar to ours. However, it takes Mars 687 of our days to go once about the sun, so its seasons are nearly twice the length of our seasons. We overtake Mars at intervals which average two years and fifty days. Mars is then said to be in opposition, that is opposite to the sun in the sky and closest to the Earth. The last opposition of Mars was on December 30, 1960. The next will be

[1] See "The Voyage of Mariner II" by J. N. James, page 171, for further information about Venus.—Eds.

in the middle of February, 1963. Since the orbits of Mars and the Earth are not circular, we may be as much as sixty-three million miles from Mars at some oppositions, and only thirty-five million miles away at others.

The most evident surface features on Mars are the two white polar caps. During the Martian winter season in either hemisphere, these may extend down to a latitude of sixty degrees, or roughly one-third of the distance from the pole to the equator. They become very small or disappear entirely in late summer. Measurement of the temperatures at the polar caps strongly supports the suggestion that the white material is some form of snow or frost. As spring advances this evaporates into the thin atmosphere of Mars, while at the peak of the evaporation period a narrow band of water forms along the border of the shrinking cap. The maximum thickness of the snow or frost is probably not more than two or three inches.

As the south polar cap shrinks during spring in the southern hemisphere certain areas on the Martian surface start to grow darker. The effect appears first in the south polar region, but spreads down towards the equator and into the opposite hemisphere as spring advances. Towards the end of the southern summer there is a general fading of the dark areas. This is followed by another wave of darkening, proceeding from the north polar regions this time, as the north polar cap melts. The conclusion seems inescapable that these seasonal changes on Mars result from the periodic release of extra moisture into the normally dry atmosphere. The colour of the dark markings has frequently been described as green, but it is believed that much of this impression of a green colour results from contrast with the red-orange shade of the remainder of the Martian surface. The true colour of the dark areas is more likely a bluish-grey.

The so-called "canals" of Mars, long, narrow, dark markings first observed by Schiaparelli in 1877, have been plotted in various forms by most observers of the planet. Over the last eighty years there has been a continued and lively discussion among astronomers about their exact nature. I can sum up the present situation by saying that there is general agreement on the reality of the canals, in other words they are not illusions, but result from something on the Martian surface that produces the effects drawn by the visual observers and recorded on the photographs. The big question is: Are they continuous, essentially uniform in width, and geometrically straight, which would suggest an artificial origin; or are they made up of a maze of irregular fine detail which blends into a line-like image under our difficult conditions of observation? This latter alternative suggests a natural origin. Here we find widely divergent views, but most astronomers feel that the second alternative is more likely.

I can speak more confidently about some of the other aspects of the Martian surface. Gravity is 38 percent, that is about a third, of the value on Earth. The atmospheric pressure at the surface is just one-twelfth of our sea-level value, and corresponds to the pressure about ten miles above the Earth. But at heights near twenty miles the pressure is the same on both Mars and the Earth, about an eightieth sea-level pressure. Above this, Mars actually has a denser atmosphere than the Earth has at corresponding heights.

The Martian atmosphere is predominantly nitrogen with 2 or 3 percent carbon dioxide, possibly a little argon, and very small traces of water vapor and oxygen. The temperature on the night side of Mars drops to 50° or 100° below zero Fahrenheit, but as the sun rises over the equatorial regions the surface rapidly warms up, and by ten o'clock in the morning it is above freezing, reaching a peak of about 70° near midday. Yellow dust clouds are sometimes observed two or three miles above the Martian surface. Above these we have a violet haze-layer, extending up to about ten miles. This haze normally covers most of the planet, but may suddenly clear off in a most unaccountable manner. The nature of the phenomenon is not yet understood. Above the haze, we sometimes see white or blue clouds, probably ice crystals.

What about the possibility of life on these other worlds? Suppose we consider for a moment the physical conditions under which life normally exists. Throughout the universe we have some ninety-two natural elements, with the same properties as found here on Earth. Organic matter, as we know it, is built up of very complicated molecules consisting chiefly of the four elements hydrogen, carbon, nitrogen, and oxygen. In this system carbon is the key element and forms the backbone of the long chains of atoms combining to make up various organic molecules. Only one other element, silicon, seems to be a possibility as a substitute for carbon in the building of life-molecules, but this is a pure supposition and we have no evidence of this hypothetical "silicon-life."

To maintain our terrestrial type of "carbon-life" we must have some readily available form of energy, and temperatures below 150° Fahrenheit —that is, low enough to permit the continued existence of very complex molecules. There should also be carbon dioxide to supply carbon and water or water vapour to provide moisture. This last condition presupposes temperatures high enough to melt ice, but this need only be for a fraction of the time, since presumably life can lie dormant for long periods of extreme cold. Finally there should be an absence of toxic substances such as ammonia, methane, carbon monoxide, and chlorine.

In considering the possibility of organic life on our neighbour worlds we can rule out at once all the little worlds without atmospheres and all the cold giant planets, or large satellites, with extensive but poisonous atmospheres. This leaves only our two immediate neighbours, Venus and Mars, both of which lie in the temperature zone capable of supporting active life. If our present understanding of the surface temperature of Venus is correct, there seems small chance of life forms anywhere near its surface, and it would appear rather doubtful that life could exist only near the top of the cloud layer.

On Mars the situation is different. Here the regular seasonal changes in the dark markings seem best explained by some form of vegetation. Admittedly this does not show the chlorophyl characteristic of the green colouring matter in plants, but the Russian astronomer G. A. Tikhov has pointed out that vegetation on Earth, which grows in cold, high-altitude regions, often lacks this chlorophyl characteristic also. In 1956, W. M. Sinton of the Lowell Observatory, working with the 200-inch telescope on Mt. Palomar, analyzed the light of Mars in the infra-red. With the spectroscope he found that the dark areas on Mars reflected light with features in the infra-red which corresponded to those found in light reflected by certain vegetation on Earth. This gave evidence for the presence on Mars of organic molecules, and in particular, of carbohydrates. However, there was no indication of the presence of these organic features in the light from the orange-red desert regions of Mars.

Although we must be careful not to be too positive, I think most scientists agree that this recent work points strongly to the existence on Mars of some drought-resistant type of vegetation, which revives rapidly from a dormant state when the polar frosts distribute moisture over the Martian surface. And so far as I know, this, together with the other phenomena of the Martian surface, is the only well authenticated and positive evidence we have for the existence of organic life elsewhere than on Earth.

CECILIA PAYNE-GAPOSCHKIN

The Sun

•

FIVE THOUSAND STARS are visible to the unaided eye; a four-inch lens reveals over two million; and over a billion are accessible with the 200-inch mirror. The fainter we go, the more rapidly do the numbers increase. The story is told that Edward C. Pickering, of Harvard, was describing a formula that expressed the number of stars brighter than any given magnitude.[1] One of his hearers remarked that the formula required *two* stars brighter than apparent magnitude—1, whereas there is only one such star—Sirius. "Ah!" said Pickering. *"You've forgotten the sun."* Perhaps familiarity breeds contempt; it is easy to forget that the sun is the nearest of the stars, the most readily studied, the only one that can be kept under continuous surveillance.

The sun is a typical star, a common kind of star. A quarter-million stars have been analyzed in some detail, and 10 percent of them resemble the sun; it merely happens that most of them are far away, and our luminary is near by. A typical specimen of the cosmic population is, so to speak, on our doorstep—giving us a superb opportunity to study the construction and habits of stars in general.

The Light of the Sun

The sun is a gigantic globe of glowing gas, and so is every star that shines, though not one other is near enough to appear as a disk, even to the most powerful telescope. With over a hundred times the diameter of our planet, more than a million times its bulk, three hundred thousand times its mass, the sun is yet a small star and a lightweight. The dazzling surface, intolerable to the eye even at a distance of 93 million miles, many

[1] Magnitude is the astronomer's measure of stellar brightness—a logarithmic scale in which the smaller numbers express the greater brightness. A difference of one magnitude corresponds to a ratio in brightness of 2½.

times brighter than the most powerful artificial light, pales in comparison to those of the hottest stars. Yet, in its degree, the sun displays the same capacities as other stars; close-ups of its face reveal nuances of expression that elude us at greater distances, and provide clues to the behavior of the other members of the cosmic population.

We take the steady dependability of sunlight for granted in everyday life, and not without reason. The most careful measurement has revealed only infinitesimal variations during the past half-century, and most of these have probably been caused by variations in the transparency of our own atmosphere.

The earth's temperature is almost entirely governed by the amount of heat received from the sun, and the very fact that life on earth has existed in unbroken sequence for hundreds of millions of years shows that *the sun has been shining steadily at least as long as that.* The earth receives energy at the rate of 4,690,000 horsepower per square mile from the sun, and has been doing so for hundreds of millions of years, and yet our tiny planet intercepts less than two thousand-millionths of the sun's radiated energy. Such numbers beggar the imagination.

Put in another way, the output of the sun is even more impressive. Modern physics recognizes not only the interconvertibility of various forms of energy, but also the equivalence of energy and mass. The famous Einstein equation states that

$$E = mc^2,$$

where E is the energy in ergs, m the mass in grams, and c the speed of light, 3×10^{10} centimeters, or about 186,000 miles, per second. In the sense of this equation, light has weight just as matter does. The sun pours *four million tons* of radiant energy into space every second, and if (as we believe) this has been going on for at least a hundred million years, more than 10,000,000,000,000,000,000,000 (or 10^{22}) tons of light and heat have issued from our luminary in a steady stream! Large as the figure is, it represents less than one-millionth of the total mass of the sun. As we shall see, the sun is actually converting its own substance into radiant energy. But it draws upon less than one million-millionth of its material capital a year—a very modest expenditure. Many stars are far more prodigal of their resources.

The sun's steady output suggests tranquillity, but its surface is far from quiescent. Dark spots on it are often visible to the naked eye, and a completely unspotted sun is extremely rare. Closer scrutiny reveals a continually changing fine granulation over the entire disk. The sun's face is not a smooth unruffled sea of gas, but a heaving, churning expanse, with whirling tornadoes (sunspots) that break through from below, tongues of gas (spicules) that surge up and subside, clouds of glowing vapor (prominences) that float, swirl, and erupt high above the surface, and sudden localized blazes of intensely brilliant radiation (flares). On the sun these things can be seen; on the stars they can only be surmised or indirectly observed, but we can be sure that they are often even more spectacular than on the sun.

The whole surface of the sun—flares, granulations, even sunspots—glows with intense brilliance. A complete array of color is present, from X-rays through the visible spectrum to radio waves, and perhaps it is no coincidence that the sun shines most brightly in the colors to which our eyes are sensitive. The fact that the sun is brightest in the yellow-green gives a clue to the temperature of its radiating surface. Common experience tells that the hotter a glowing surface, the bluer is the light by which it shines; and the quantitative formulation of this fact (Wien's law) enables us to say that the radiating surface of the sun has a temperature of about 6000° C (11,000° F). And this is true whatever the sun is made of.

The Composition of the Sun

The brightness of the sun is not distributed among all colors with unbroken brilliance. When sunlight is passed through a prism, and spread into the artificial rainbow known as the *spectrum,* some colors are seen to be greatly depleted. The rainbow is broken up into an array of sharply bounded regions of color, separated by others, far less brilliant (the "Fraunhofer lines"). Something has robbed the sunlight of these colors, and the atoms above the solar surface have been convicted as the culprits.

Each atom has its own characteristic array of colors, and it can take up or give out these colors only, absorbing or emitting energy as it does so. The distribution of the colors in sunlight, and our knowledge of the behavior of atoms on earth, make possible a chemical analysis of the sun's surface, actually more delicate than we could perform in the laboratory if we had a chunk of the sun given us to analyze.[2] The results of the analysis show that the sun is made of the familiar chemical elements known on earth.

The matching of characteristic colors led at first to a qualitative analysis, and showed that *all* known atoms with spectrum lines in the accessible region of the sun[3] are represented in its spectrum. A few simple compounds, such as cyanogen, are found, but most of the material is in the form of isolated atoms. In other words, the spectrum shows that *the outer layers of the sun are completely gaseous.* One can go further: the array of colors characteristic of an atom varies with temperature, and so the

[2] This is because of the great depth of the layer of atoms above the sun's surface, and the vast number accordingly available for the analysis. A large number of spectrum lines, only *predictable* on earth, can actually be *observed* in the spectrum of the sun!

[3] Molecules in the atmosphere of the earth, especially those of ozone, oxygen, and water vapor, obscure some parts of the sun's spectrum almost completely.

temperature of the low-lying atmosphere of the sun, which produces the Fraunhofer lines, can be determined; it agrees fairly well with the temperature inferred from the color of sunlight.

Growing knowledge of the physics of spectra has actually made it possible not only to identify the atoms above the sun's surface, but to count them. The sun, we find, is mainly hydrogen; there are more atoms of hydrogen, lightest and simplest of atoms, than of all other kinds put together. Next in order comes the second lightest atom, helium, and, with some notable exceptions, the numbers of heavier, more complex atoms fall off steadily in order of complexity.

This scheme of chemical composition is not peculiar to the sun. It is typical of the composition of the whole cosmos, not only the stars, but also the loose gas and dust that pervade interstellar space. The atomic makeup of all stars is not identical—and the differences, small as they are, may be of great significance—but the general uniformity is amazing, and it would be difficult to point with confidence to any cosmic object that does not consist mainly of hydrogen.

The Sun's Surface

Each atom has its characteristic array of colors, and a photograph of the sun in a single color records the atoms of one kind by themselves. Luckily the sun is so bright that even a very restricted range of color can be photographed, either with the instrument known as a spectroheliograph or by an ingenious arrangement of light filters. Pictures of the sun made by light of calcium or hydrogen show not only increased detail, but revealingly different detail. Whereas a direct photograph, in light of all colors, shows only dark sunspots and vague granulations, calcium light reveals brilliant variegations in the neighborhood of the sunspots, and greatly accentuates the contrast of the granulation. Even before a sunspot swirls through the surface, bright calcium "flocculi" herald its presence; and they remain for some time to mark the place after the spot has died away. The bright areas shown by the calcium atoms reflect the greater disturbances, more violent motions, and probably hotter regions, near the tornadoes that are sunspots.

Hydrogen poses are more difficult to take, because hydrogen does not cut so wide a swath in the spectrum as calcium, and less light is available for the photograph. This may seem surprising, for there is much more hydrogen than calcium above the sun's surface; but it is a consequence of the idiosyncrasies of the two kinds of atoms. At the sun's temperature nearly all the calcium atoms are in the right state to emit light, but the atom of hydrogen is more recalcitrant, and is only about one-millionth as prone to emit as calcium at 6000° C. Hydrogen atoms are more than ten thousand times as common as atoms of calcium, but even so, the lines they produce in the spectrum of the sun are less than one-hundredth as intense.

Photographs in hydrogen light show the disturbed regions near sunspots, but with less brilliance, because of the difficulty of stimulating the atoms. Both the hydrogen and the calcium photographs reveal slowly changing patterns of dark filaments silhouetted against the bright surface of the sun. The filaments are not really dark; they seem so only by contrast. When one of them extends beyond the sun's edge it is seen as a glowing prominence —a great cloud of gas poised above the surface. Some prominences are so brilliant that they show up as bright streaks, even against the face of the sun.

Prominences are protean in form, and have an infinite variety of motions. Some hang poised over the surface. Some spurt upward, in filamentary surges, like geysers, and seem to dissipate into space. Some rise and fall like fountains. But a surprising majority shower downward, not upward, and many give the impression of being sucked into a point at the surface.

We are far from understanding the motions of prominences. Some are associated with sunspots, but many are not. The variegated brilliance of the sun's surface may affect them. Electric forces may be of importance; magnetic forces probably play a crucial role. Whatever be the significant factors in producing the protean variety of solar prominences, they are important also in stars of very different kinds. For, as we shall see, prominence activity is characteristic of many stars, and often on a scale that makes the sun's activities seem puny.

Whether or not the motions of prominences are governed by magnetic forces, very intense magnetic fields are observed on the sun. The spectrum of a sunspot tells the story, through the medium of the spectra of the individual atoms in the tornado. An atom in a magnetic field absorbs and radiates in a special way; its peculiar series of colors subdivides into an intricate pattern, and the stronger the magnetic field, the more is the subdivision accentuated. The sunspot behaves like a tremendous electromagnet, many thousand miles across; no doubt electrically charged particles, whirling around the axis of the tornado, play the part of the current in the electromagnet, and a powerful magnetic field is produced along the axis of the spot. Sunspots, like prominences, are incompletely understood, but that they possess magnetic fields of several thousand gauss[4] is certain. Large as such fields are, even larger magnetic fields are found for certain peculiar stars *as a whole;* and like many sunspots, they reverse their polarity at regular intervals.

The times when sunspots are thickly scattered over the sun's face are marked by striking events nearer home. The Aurorae, the Northern Lights, gleam and shimmer in the sky. Magnetic storms disrupt communications and intrude on radio programs. The disturbances that produce spots on the sun have direct effects on our planet.

Disturbed areas of the sun are showering particles into space at high

[4] The gauss is the unit of magnetic field. The magnetic field of the earth, which affects the compass, is small—a fraction of a gauss.

speeds, and a rain of electrons, protons, and even heavier particles pours down into our atmosphere. An electrically charged shower plays upon the atoms and molecules of the upper air, and excites the auroral glow. Oxygen in the high atmosphere emits its peculiar red and green light; molecules of nitrogen and other substances contribute their characteristic colors. The spectrum of the rain of solar hydrogen has recently been photographed by Meinel. The earth receives showers of particles that left the sun a few hours ago.[5] Even more significant: the sun is continually spraying matter into space. Many stars, as we shall see, are doing the same.

Prominences are not the only features that rise above the sun's bright edge. At the crucial moment of a total eclipse, when the moon's disk cuts off the body of the sun, a brilliant rim of rosy light—the chromosphere—appears around it. The spectrum of the chromosphere shows that it consists of radiating atoms, the same atoms that were revealed by the Fraunhofer lines in a layer nearer the sun's surface, but with a difference. The pattern of colors that they radiate is modified in a way that admits of but one explanation—the temperature of the chromosphere, from five to ten thousand miles above the solar surface, is more than three times the temperature of the atoms that produce the absorption spectrum of the sun! Even helium, which is far more refractory than hydrogen, and requires a far higher temperature to excite it, appears not only in normal form, but even in the "ionized" condition, with one electron torn away—a situation found only at the surface of the hottest stars, at temperatures of over 30,000° C.

The chromosphere consists of a sort of hairy rim of tiny spicules, or jets, which spurt upward and disappear in a few minutes. The spicules may be related to the minute granules that pepper the face of the sun, and seem equally short-lived.

Other stars than the sun possess chromospheres, and with some of them, unlike the sun, the chromosphere is far larger than the star itself, and shines so brilliantly that the glowing atoms in the spectrum produce *bright spectrum lines* on the background of the star's light. Some chromospheres are poised, like the sun's, above the star's surface, with little motion. The shining atoms around other stars are flowing or spurting steadily outward, and some stars occasionally blow great chromosphere bubbles, which thin out gradually and dissipate into space.

Outside the chromosphere of the sun gleam the pearly streamers of the corona, which extend to distances comparable to the size of the sun itself. Like the chromosphere, the corona has a spectrum given by glowing atoms, but for many years its nature was a mystery. No such colors had been produced by any atoms on earth, and they used to be ascribed to a mysterious substance, *coronium*, that was unknown elsewhere. Now we know, from the work of the Swedish physicist Edlén, that the corona consists of

[5] The particles travel at 125 to 625 miles a second, and make the trip from sun to earth in from 200 to 40 hours.

well-known, common elements (such as iron, calcium, and nickel) but under conditions that represent temperatures never attained on earth. The corona—the "iron crown" of the sun—seems to be at a temperature of about a million degrees!

Other stars, too, have coronas, and some of them are intensely brilliant. The spectral colors of the sun's iron crown have been found in the light of certain peculiar stars that have suffered sudden explosion.

Perhaps the most remarkable thing about the outer regions of the sun is the increasing temperature of successive outward layers. The reversing layer and the photosphere[6] have temperatures of about 6000°; the chromosphere is at about 20,000°; and the corona, at 1,000,000°. The sunspots, which look like depressions in the solar surface, are even cooler than the reversing layer; their spectra and colors point to a temperature not far from 4000°.

The remarkable temperature stratification of the sun is not an isolated phenomenon. Other stars show it, and some of them display an even greater span of conditions. If we did not know that the sun is a single star, the variety of its spectra might tempt us to doubt; other stars whose spectra look as though they must be complex may be similarly put together.

The Sun's Rotation

From the human point of view, the most important thing about the sun is the fact that it has planets, and that one of them presents physicochemical conditions favorable to life. But from the standpoint of the sun, all the planets are negligible: even Jupiter, the largest, weighs less than one thousandth as much as our luminary. Within the planetary system, Jupiter is the only really influential member; it is the most potent factor, for example, in governing the motions of the comets and asteroids, the lesser members of the system. In one respect, Jupiter excels even the sun: the giant planet possesses most of the total energy of rotation of the solar system—far more than the sun itself. True, the sun is spinning, but spinning very slowly. It takes nearly a month to make one complete turn. This fact has always been one of the great difficulties in the path of a theory of the origin of the solar system; almost all the theories that have been moderately successful in other ways seem to require that the sun possess the greater part of the energy of rotation of the whole.

The slow spinning of the sun is far from being unusual. Most stars that resemble the sun in size and temperature are also turning slowly on their axes. Some stars, it is true, spin very rapidly, but these are usually

[6] The photosphere, or "sphere of light," is the glowing surface of the sun; the reversing layer is the atmosphere of absorbing atoms that lies above it. Temperatures, here and later, are given on the centigrade scale.

the massive stars of high temperature. Stars such as the sun rotate rapidly only when constrained to do so by being members of twin systems; they raise huge tides in each other and always stay face to face. Any star that spins rapidly is distorted into a spheroid; even the solid or semisolid planets like the earth, Jupiter, and Saturn are more or less flattened at the poles by rotation. Jupiter looks like an orange, even in a small telescope. But the sun is so nearly spherical that no polar flattening has ever been detected.

Slowly as the sun rotates, it still does so in a remarkable manner: it spins faster at the equator than at the poles, so that its surface must be in a state of shear—that is, some parts of the surface must continually be slipping past others. Possibly the differential rotation plays some part in producing sunspot vortexes. And if even the slowly turning sun spins faster at the equator than at the poles, what of the stars that turn on themselves in a few hours, and are highly distorted? What, too, of the internal rotation of the sun? We can see the surface only; and a different internal rate of rotation is not only possible but likely. A star that turns very slowly will probably not churn up its interior and mix its constituents; but one that is spinning fast may be much better mixed. The degree of mixing of the materials within a star may well be a crucial factor in its history.

The Sun's Interior

Which brings us to the problem of the sun's interior. So far we have spoken only of the parts of the sun that can be seen, a mere skin. Conditions within are very different. Without going beyond the elementary laws of physics, it can be shown that the sun, and all other stars, are gaseous not only at the surface, but all the way through. Moreover, both temperature and pressure must necessarily rise toward the center.

In fact, it is well known that the central temperature of a star must depend essentially on its size and mass, and is proportional to the average mass of the individual particles of which it consists. This average mass of the particles, the so-called "mean molecular weight," would be least for a star that was all made of hydrogen, and would increase somewhat —but not very much—with larger proportions of heavier elements. The reason for this rather surprising statement is that *the interiors of the stars are so hot that atoms are stripped of nearly all their electrons*. Each electron counts as one particle in the average mass of the individual particles, and the masses of electrons are negligible, even in comparison with those of the nuclei of hydrogen, lightest of elements. In units of the hydrogen nucleus, a star all of hydrogen would have a "mean molecular weight" of ½; if the star were all helium, whose nucleus weighs four times as much as hydrogen, and which can part with two electrons, the average molecular weight would be 4⁄3, or 1.33; even if the stars were all uranium

(92 electrons, atomic weight 238 times hydrogen), the mean molecular weight would only be 238/93, or 2.55. As most stars consist mainly of hydrogen, the mean molecular weight will usually be between ½ and 1½; and the central temperature, for stars of the same mass and size but different composition, will therefore not differ by a factor of more than 2 or 3.

If, in addition to size and mass, the total energy output (luminosity) of a star is known, the same elementary theory permits the calculation of the mean molecular weight, which can be fitted by a certain number of different combinations of hydrogen, helium, and heavier elements.

The sun is found to have a central temperature near to 18 million degrees; and the temperature almost certainly increases steadily from the surface inward. Thus we have the odd paradox that the sun is actually coolest at the surface, or even a little below it, in the cores of the sunspots, and the temperature goes up again as we pass outward through the chromosphere to the corona.

High temperature and enormous pressure prevail within the sun. The high temperature is responsible for the fact that the sun's substance behaves like an ideal gas, even at the center; it strips the atoms of their attendant electrons, resolves them into fragments far smaller than at the surface, and permits them to pack more closely without violating the laws that govern the behavior of gases.

The Sun's Source of Energy

The hot interior of the sun is the source of its light and heat. At 18 million degrees the atoms are able to interact, to convert some of their substance into energy. No other source is adequate to have produced the tremendous outpouring, steady over millions of years. The source of the sun's energy was long a puzzle. Combustion, chemical reaction, gravitational contraction, the drawing of energy from the environment, all were shown to be hopelessly inadequate. Nuclear energy seemed the only avenue of salvation, long before the details of the actual process were understood. "Does energy issue freely from matter," speculated Eddington a quarter of a century ago, "at 40 million degrees as steam issues from water at 100 degrees?"

Modern nuclear physics provides an affirmative answer. The actual processes have been observed in the laboratory. The interior of the sun liberates energy by a catalytic action similar to those of atomic chemistry; but the reactants are the naked nuclei, not atoms clad in their haze of electrons. Hans Bethe and C. F. von Weizsäcker discovered independently, nearly at the same time, that hydrogen cores combine, by a chain of reactions set off by carbon nuclei; four hydrogens interlock to produce a helium core. The helium is lighter than the sum of the hydrogens by about 0.7 percent, and this deficiency of mass is turned into energy, which passes from the interior to the surface in a steady flow. Only at temperatures between 15 and 20 million degrees can the reaction produce enough energy to supply the sun. The rate of production varies at about the eighteenth power of the temperature, and most of the sun's energy accordingly issues from its substance in the central region where the temperature is highest. At 15 million degrees the "light" given out resembles X-rays (even more "violet" than ultraviolet light); it flows outward, passed from hand to hand, so to speak, by the electrons and atoms of the overlying layers, and is steadily "reddened" in the process so that when it reaches the surface the observable color is primarily yellow-green.[7]

The sun shines by feeding on its own substance, and its diet is exceedingly simple. So far as we know, our luminary subsists entirely upon hydrogen. The same food sustains the other stars; stellar infants may possibly have a somewhat different diet, but their infancy is brief, if only because foods other than hydrogen are in short supply. Digestive processes may differ somewhat from one kind of star to another. If the temperature is well below 15 million degrees, the cycle catalyzed by carbon may be replaced by others, such as the *direct* combination of protons (hydrogen nuclei) to form helium—the "proton-proton" reaction. The food remains the same; hydrogen is consumed and helium is left behind.

Although the sun has been steadily digesting its own interior for tens, and even for thousands, of millions of years it still consists almost entirely of hydrogen, enough to keep things going at the present rate for at least an equal interval. Most other stars are equally rich in the vital substance, and have an equally bright future. In fact, paradoxically enough, the future of stars that are consuming hydrogen is even brighter than their past. For the brightness of a star of given size and mass depends primarily on the mean molecular weight of its substance. As the hydrogen supply slowly falls, the mean molecular weight gradually increases, and the star grows a little brighter, so long as it does not drastically alter its internal arrangements.

The great astronomer Eddington showed, even before the actual process of stellar nutrition was identified, that the more massive a star is, the more energy does it pour out. The energy output of a star of given composition is, in fact, proportional to something between the cube and the fourth power of its mass. The greater majority of the stars whose masses are known conform to this rule (the "mass-luminosity law") fairly closely. Most of the differences are within the limits that would be expected from the possible range of mean molecular weight with composition—a range

[7] Actually the sun's surface gives out a surprising amount of radiation of very short wavelength, far more than would be expected if its light were distributed according to the elementary laws of radiation by a so-called "black body" (a technical term, which sounds rather paradoxical, and denotes a surface that absorbs and radiates ideally according to certain laws that are deducible from quantum theory).

that was shown earlier to involve a factor not greater than two or three. But some stars are nonconformists, and they are important finger posts for theories of stellar evolution.

If a star's luminosity (which can be expressed in terms of tons of radiation per second for all stars, just as for the sun) were simply proportional to its mass, all stars of similar composition would have the same life expectancy. But a star much more massive than the sun is consuming itself much faster. A star of twice the mass is twelves times as prolific; at ten times the mass the factor is over a thousand, and at a hundred times the mass of the sun a star would be well over a million times as prodigal, and its life expectancy a million times shorter. The total life of a star like the sun (in the style to which we are accustomed) is about 5000 million years; if there were stars a hundred times as massive, their active lives would be reckoned in thousands rather than in millions of years, and they must become effectively bankrupt during an interval over which the sun can radiate with virtually unchanging brightness.

Such bankrupt stars actually exist. They can be recognized by the fact that their brightness is far lower than we should expect from the mass-luminosity law. Their light is feeble; they have exhausted their internal nuclear resources, have spent all their available hydrogen, and exist only on their very limited gravitational capital.[8] Such a destiny probably awaits all stars, but for most of them it lies in the far future. Even for the sun we see it as inevitable.

The sun, indeed, holds a mirror to the cosmos. Like all other stars it is a globe of glowing gas, hotter and denser within. Its surface is a seething, surging sea of atoms. Plumes of gas float above it; glowing filaments surge upward; shining fountains cascade downward. Giant tornadoes swirl through the surface. Spicules rise and dissipate like darting flames. Dazzling flares blaze up and vanish. A brilliant chromosphere rings it; and around it gleams the aura of the corona. Powerful magnetic forces play across the surface; atoms and electrons spray into space. As it spins on its axis the equator pulls steadily ahead; and across its face proceeds the slow rhythm of the sunspot cycle, waxing and waning every eleven years. The spectacle is impressive in itself. As the mirror of the cosmos it is stupendous. Other stars are doing the same things, and these stellar habits are the clue to their history.

The sun is made of hydrogen, "with a smell of other substances," and so are most other stars. The steady consumption of hydrogen keeps it shining; most stars are sustained in no other way. Wherever we look in the cosmos we see the play of the same forces, the march of the same phenomena—but often on a scale that makes the sun seem puny.

[8] A star can convert gravitational energy into light and heat by contracting in size.

SIR BERNARD LOVELL

Radio Waves from Space

•

FOR MANY YEARS it seemed that man's only hope of obtaining information about the stars and the galaxies was by the use of big optical telescopes. He has evolved with eyes which are sensitive to that part of the spectrum in the visible region between the ultra-violet and infra-red, and it is over this region of the spectrum that a transparency, or window, exists in the earth's atmosphere. If man had evolved with eyes which were sensitive only in the infra-red or ultra-violet then he would have had very little knowledge indeed of outer space until the present day, when it has become possible to move beyond the obscuring region of the atmosphere with satellites and space probes.

Because the earth's atmosphere almost completely obscures any radiation which lies outside the familiar colors of the rainbow it seemed impossible that any useful knowledge of outer space would ever be accumulated in parts of the spectrum other than in this visible gap. This is in spite of the fact that the early researches with radio waves in the 1920s had led to the realization that there was another transparency or gap in the earth's atmosphere at much longer wavelengths in the radio-wave region. Whereas the wavelengths of visible light are measured in millionths of a centimeter, the radio-wave region, in which there is this other transparency in the atmosphere, extends from a fraction of a centimeter to many meters in wavelength. In the middle of this radio-wave band the ordinary broadcasting and television transmissions are made on earth.

Although the existence of this transparency was known, it seemed unlikely that any use could be made of it for astronomical purposes. After all, the stars and the sun are very hot bodies and the fundamental laws of physics indicate that the maximum output of energy from such hot bodies with surface temperatures of many thousands of degrees is in the visible and near visible regions of the spectrum. It was therefore with some incredulity that astronomers received the news in 1931 and 1932 that an electrical engineer, Karl Jansky, who was working at the Bell Telephone Laboratories in America, had detected some radiations or signals in this part of the spectrum which he was convinced had their origin from regions of space outside the solar system. Jansky's apparatus worked on rather a long wavelength between 14 and 20 meters and the aerial consisted of an array of rods which could be rotated on a brick foundation. Jansky was investigating the static which was interfering with and limiting the usefulness of radio communications around the world. He discovered that even when there was no obvious form of atmospherics such as a thunderstorm, nevertheless there was a residual disturbance in his receiving apparatus and he noticed that this residual noise in his equipment had a diurnal variation; that is, it varied in strength throughout the day. Furthermore he made the classic observation that the maximum

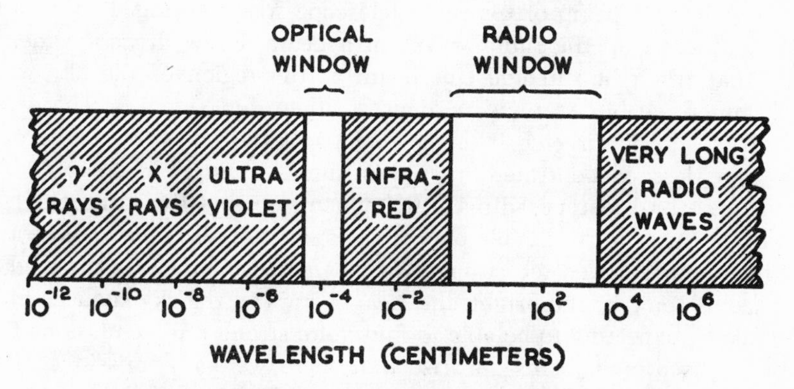

Fig. 1. The complete electromagnetic spectrum showing the regions of transparency of the Earth's atmosphere and ionosphere.

in this signal did not come exactly every day at the same time but that it was four minutes earlier each day. This led Jansky to conclude, quite correctly, that the only possibility of explaining this residual noise must be that it was a result of radio waves generated somewhere in regions of space outside the solar system which were being picked up in his aerial system (the reason being that the period of maximum of 23 hours 56 minutes represents the sidereal day, that is the period of rotation of the earth with respect to the stars, and not the solar day of 24 hours). This perfectly simple and straightforward observation of Jansky led to the correct conclusion that the source of this radio noise had its origin not merely in extraterrestrial space but in extrasolar space, and that it originated either on the stars or in the space between the stars. Astronomers took little notice of Jansky's discovery and the further development of this subject in the years before the second world war was left to an amateur, Grote Reber, who built the prototype of the modern radio telescope in the garden of his home in Illinois. This radio telescope was a bowl, 30 feet in diameter, in the form of a paraboloid, mounted so that it could be directed to any part of the sky. This telescope of Reber's has been re-erected recently as a museum piece at the entrance to the United States National Radio Astronomy Observatory at Greenbank, West Virginia.

The First Radio Map of the Sky

With this instrument Reber was able to explore with a reasonable degree of precision the radio waves from the Milky Way, and, indeed, he made the first real map of the radio sky. He confirmed Jansky's conclusion that the noise in his receiver was the result of emissions in the radio wave part of the spectrum picked up from outer space, and he was able to show that the strength of this radio noise depended on the direction in which the beam of his radio telescope was pointing. Reber's radio telescope picked up the radio waves in a beam a few degrees wide. He found that when he directed this beam at the region of the sky where the common visible stars were densest, then his signal strength was greatest. When he directed it to the regions of the Milky Way where the stars were less concentrated, then his signal strength decreased. This seemed a natural result leading to the conclusion that the stars which we see with our eyes in the visible part of the spectrum also emit radio waves. But this picture proved too simple, for when Reber hopefully directed his telescope at some of the bright and nearer stars in the sky such as Sirius and Capella, expecting to be able to find quite strong radio emissions from them, he discovered to his surprise that there were no emissions at all. Thereby he established a paradox which is still not completely resolved, namely that the Milky Way system with its hundred thousand million stars appears to emit radio waves, as well as the light waves by which we see the stars. Although the radio waves are most intense in the regions where there are the most stars, the individual star members do not seem to be radio emitters.

Apart from one or two doubtful and unusual cases this remains the situation today—that nobody has succeeded in detecting radio waves from the individual or common stars which we see by eye in the Milky Way. The sun is an exception; although, for reasons presumably associated with the long period cycle of sunspots, Reber was unable to find any radio emissions from the sun, which itself appears to be an extremely powerful source of radio emissions. Indeed, under some conditions, the solar radio waves are so intense that they hinder the investigation of the remote parts of the cosmos. This situation is analogous to the optical case where sunlight makes it impossible to see the faint light of the stars by day. Even if all the stars in the Milky Way system were to emit with the same intensity as the sun, they are so distant and the dilution factor is so great, that there would be no significant contribution to the strength of the signal from the regions of space outside the solar system picked up in a radio telescope. The solution of this paradox seemed to Reber to lie in the following argument. The stars were obviously not emitting radio waves which made a significant contribution to his records but, since there was this close relationship between intensity of the visible light and the strength of the radio emission, the sources of the radio waves must be the interstellar hydrogen gas. Since Reber reached that conclusion in the period 1934 to 1938 there have been many changes of opinion about the origin of these radio waves, and now it does seem that Reber was partially correct. First there was a violent swing against this view and it was believed that the interstellar gas did not contribute at all. Now the situation, although extremely complex, involves an explanation in which the radio waves generated in the interstellar hydrogen gas represent a significant component of the extraterrestrial radio emission.

The Development of the Radio Telescope

There was a long hiatus in the development of these discoveries by Jansky and Reber because of the war, but it was the excellence of the techniques and our new ideas about radio and radar which evolved under the stress of miltary requirements which led to a vigorous resurgence of these studies immediately after the war. It soon became obvious in the early stages of this development that the problem of the radio astronomer was analogous to that of the optical astronomer, namely that large radio telescopes were needed for the same kind of reason that the optical astronomer required large optical instruments. The optical instrument needs a large mirror in order to collect as much light as possible to penetrate further into space, and also to obtain good definition. A similar situation exists with the radio telescope, where it is necessary to collect the radio waves over a big area in order to improve the signal strength of the faint emissions which are generated at great distances in the cosmos, and also because the definition or the resolution of the beam of such a telescope depends directly, for a given wavelength, on the size of the telescope.

Reber's radio telescope was a parabolic bowl 30 feet in diameter. The largest contemporary version of this form of radio telescope is the instrument at Jodrell Bank in which a parabolic bowl 250 feet in diameter is mounted so that it can be directed with precision to any part of the sky. Incoming radiation is reflected from the steel bowl on to the primary feed which is mounted on a mast at the focus 62½ feet from the apex of the paraboloid. Then the signals are transformed by electronic apparatus and recorded. It is possible to direct this bowl to any part of the sky, and for a given wavelength the telescope has something like eight times the definition and sixty-four times greater gain than the instrument originally used by Reber. The paraboloidal steel bowl of the radio telescope is formed of 80 tons of steel sheet. This is mounted on a cradle which itself weighs 800 tons and the whole cradle is suspended on trunnion bearings 180 feet above the ground. The elevation drive, or tilt of the bowl, is obtained by electric motors driving through large gun racks, originally part of the *Royal Sovereign* battleship. The 180-foot towers which support the elevation bearings and motors are connected at ground level by a diametral girder, pivoted at its center point. The two towers are each carried on six bogies which move on a double railway track 320 feet in diameter so that the instrument can be given rotation or movement in azimuth. The whole superstructure, which moves on the railway track, weighs over 2,000 tons.

On the reverse side of the steel bowl a small hanging laboratory is suspended so that it always remains in the upright position whatever the tilt of the bowl. This is approached by catwalks when the bowl is in the zenith and contains receiving equipment and, if necessary, the observer. Sometimes in order to avoid losses in cables it is necessary to mount the receiving equipment close to the primary feed at the focus. On the top of the aerial tower which rises 62½ feet from the apex a box about 6-foot cube is fixed and access to this is obtained by a hydraulic platform from the base of the bowl. This box contains some of the receiving equipment and for some work at high frequencies the aerial itself in the form of a horn feed, as distinct from the common rod dipole, emerges directly from this box.

The control room of the telescope is 200 yards from the superstructure and from a main control desk the controller can command the instrument to perform any motion which the experiment requires. It can be driven separately in azimuth and elevation or a sidereal motion can be obtained by driving through a computer in right ascension and declination. Under those conditions the computer works out the position to which the telescope has to be directed and it will then automatically follow a given star from rising to setting, or a planet, or the sun, depending on the needs of the research program.

The Jodrell Bank telescope is still the biggest steerable radio telescope, but several specialized instruments have been built which have a better definition and cover a bigger area of ground. These are not generally in this paraboloidal form and they are usually built so that they work in a restricted wavelength range; whereas the great adaptability of the Jodrell Bank instrument and the fact that it can be used as a transmitter as well as a receiver over a very wide range of wavelengths has turned out to be of considerable importance.

Partly because of their association with space probes, radio telescopes have become rather popular instruments. There are now many completely steerable parabolic radio telescopes of size between 50 feet and 100 feet in operation, but still very few of larger diameter. The nearest competitor to the Jodrell Bank telescope is the 210-foot diameter instrument opened in the autumn of 1961 as part of the equipment of the radio astronomy division of the Council for Scientific and Industrial Research Organization in Australia. The Russians appear to be operating a steerable telescope with an effective aperture of 140 feet in their deep-space tracking network.

The optical telescope produces its records on a photographic plate, either of the stars or nebulae themselves, or in the form of spectrographs. The records of the output of the radio telescope generally appear as an ink trace on a paper chart recording the strength of the signal in relation to the time and position of the beam of the telescope. Even for the investigation of a single object such as the M31 nebula in Andromeda it may be necessary to obtain hundreds of these records with the telescope scanning over different parts of the nebula. So one obtains a mosaic from which can be built up a system of isophotes which give the radio picture of the nebula. Nowadays these final processes of treatment of the individual scans are increasingly the subject for computers and other machines which save hundreds of hours of time in the reduction of the basic records.

Experiments of this type with a radio telescope on the Andromeda nebula reveal a very interesting situation which seems to epitomize many of the investigations with radio telescopes, in that they reveal the same kind of paradox as that found in the early work of Reber. Here the radio contours on the whole show a good concentration in the region of the stars which make up the Andromeda nebula, but they also show a considerable intensity of emission at distances which are far removed from anything which the photographic plate records. In fact, there appears to be nothing but empty space in regions surrounding the nebula of stars where there is still an intense radio output. We know that both in the M31 nebula and in many of the other extragalactic nebulae which are within the range of our radio observations, the visible stars are surrounded by a corona, or halo, of radio emission which stretches for great distances outside the regions where any matter can be seen in the ordinary photographs. The probable explanation of this emission was suggested by Shklovsky, the Russian scientist, many years ago. He suggested that these galaxies are surrounded by magnetic fields and that the radio emission is generated by electrons moving at very high speeds in this magnetic field.

Editors' Note:

Radio astronomy is one of the most rapidly developing fields in science. Since this article was written, new instruments, both in Russia and the West, have been constructed. Others, even larger and more complex, are in the planning stage.

SIR ARTHUR EDDINGTON

The Milky Way and Beyond

•

Editors' Note:

The following article by the great English astronomer, Sir Arthur Eddington, was written in the year 1937. Since that time, and especially since the dedication of the 200-inch Hale telescope on Mount Palomar in California in 1948, our ideas of the size of the universe and the speed of recession of the distant galaxies have changed. While the methods and processes described by Eddington are still considered valid, some of his figures, especially on distances, need revision. Thus, partly from studies of globular clusters in the Magellanic Clouds and in the Andromeda Nebula, it is possible that the Andromeda Nebula may be over two million light years from us and about 2.5 times as large as originally estimated —larger, therefore, than our own galaxy, the Milky Way. Spectroscopic observations with the 200-inch telescope indicate also that the most distant measurable galaxies are receding at a speed of about 75,000 miles per second. It is possible that current research may alter these results still further. It is also possible that research in the field of radio astronomy will aid in the solution of these and other problems.

As Eddington has shown in The Milky Way and Beyond *we see the most distant galaxies as they were a billion years ago when their light left them on its long, long journey to us, and we see the nearest galaxies as they were only a few million years ago—provided, of course, that nothing has happened to alter the nature of that light on its journey. With the Hale telescope we have been able to reach out more than two billion light-years on the new distance scale. With new and greater radio telescopes, we should be able to penetrate even farther. Will observations continue to show, as it now appears, that the population of radio stars increases with distance from the earth? Will they confirm recent tentative observations made with the Hale telescope which indicate that the rate of expan-*

sion of the universe is slowing down? Shall we be able to identify additional nonluminous matter in intergalactic space now invisible with optical telescopes? How will answers to these and other questions help us in our understanding of the origin of the universe? How will they help us to choose between the two presently outstanding theories of that origin—between the one which holds that the universe was created in a single cataclysmic birth from a small but highly concentrated mass of matter (from which the whole has evolved), and the other advocated by a group of English scientists which holds that the universe is the result of a process of continuous creation without beginning or any evident end?

If the rate of expansion is actually slowing down and if therefore the density of matter increases with increasing distance, both these facts would suggest an evolutionary universe and would argue against theories of continuous creation which postulate a more or less uniform distribution of matter in space. We return then to the question of the exact kind of universe with which we are dealing—a subject that is discussed in the following article.

IN ONE OF Jules Verne's stories the astronomer begins his lecture with the words "Gentlemen, you have seen the moon—or at least heard tell of it." I think I may in the same way presume that you are acquainted with the Milky Way, which can be seen on any clear dark night as a faintly luminous band forming an arch from horizon to horizon. The telescopes show that it is composed of multitudes of stars. One is tempted to say "countless multitudes"; but it is part of the business of an astronomer to count them, and the number is not uncountable though it amounts to more than ten thousand millions. The number of stars in the Milky Way is considerably greater than the number of human beings on the earth. Each star, I may remind you, is an immense fiery globe of the same general nature as our sun.

There is no sharp division between the distant stars which form the Milky Way and the brighter stars which we see strewn over the sky. All these stars taken together form one system or galaxy; its extent is enormous but not unlimited. Since we are situated inside it we do not obtain a good view of its form; but we are able to see far away in space other galaxies which also consist of thousands of millions of stars, and presumably if we could see our own galaxy from outside, it would appear like one of them. These other galaxies are known as "spiral nebulae." We believe that our own Milky Way system is more or less like them. If so, the stars form a flat coil—rather like a watch-spring—except that the coil is double.

When we look out in directions perpendicular to the plane of the coil, we soon reach the limit of the system; but in the plane of the coil we see stars behind stars until they become indistinguishable and fade into the hazy light of the Milky Way. It has been ascertained that we are a very long way from the centre of our own galaxy, so that there are many more stars on one side of us than on the other.

Looking at one of these galaxies, it is impossible to resist the impression that it is whirling round—like a Catherine Wheel. It has, in fact, been possible to prove that some of the spiral nebulae are rotating, and to measure the rate of rotation. Also by studying the motions of the stars in our own galaxy, it has been found that it too is rotating about a centre. The centre is situated a long way from us in the constellation Ophiuchus near a particularly bright patch of the Milky Way; the actual centre is, however, hidden from us by a cloud of obscuring matter. My phrase, "whirling round," may possibly give you a wrong impression. With these vast systems we have to think in a different scale of space and time, and the whirling is slow according to our ordinary ideas. It takes about 300 million years for the Milky Way to turn round once. But after all that is not so very long. Geologists tell us that the older rocks in the earth's crust were formed 1300 million years ago,[1] so the sun, carrying with it the earth and planets, has made four or five complete revolutions round the centre of the galaxy within geological times.

The stars which form our Milky Way system show a very wide diversity. Some give out more than 10,000 times as much light and heat as the sun; others less than 1/100th. Some are extremely dense and compact; others are extremely tenuous. Some have a surface temperature as high as 20,000 or 30,000° C.; others not more than 3000° C. Some are believed to be pulsating—swelling up and deflating within a period of a few days or weeks; these undergo great changes of light and heat accompanying the expansion and collapse. It would be awkward for us if our sun behaved that way. A considerable proportion (about 1/3 of the whole number) go about in pairs, forming "double stars"; the majority, however, are bachelors like the sun.

But in spite of this diversity, the stars have one comparatively uniform characteristic, namely their mass, that is, the amount of matter which goes to form them. A range from 1/5 to 5 times the sun's mass would cover all but the most exceptional stars; and the general run of the masses is within an even narrower range. Among a hundred stars picked at random the diversity of mass would not be greater proportionately than among a hundred men, women and children picked at random from a crowd.

Broadly speaking, a big star is big, not because it contains an excessive amount of material, but because it is puffed out like a balloon; and a small star is small because its material is highly compressed. Our sun,

[1] Now estimated to be 4,500 million years.—Eds.

which is intermediate in this, as in most respects, has a density rather greater than that of water. (The sun is in every way a typical middle-class star.) The two extremes—the extremely rarefied and the extremely dense stars—are especially interesting. We find stars whose material is as tenuous as a gas. The well-known star Capella, for example, has an average density about equal to that of air; to be inside Capella would be like being surrounded by air, as we ordinarily are, except that the temperature (which is about 5,000,000° C) is hotter than we are accustomed to. Still more extreme are the red giant stars Betelgeuse in Orion and Antares in Scorpio. To obtain a star like Betelgeuse, we must imagine the sun swelling out until it has swallowed up Mercury, Venus and the Earth, and has a circumference almost equal to the orbit of Mars. The density of this vast globe is that of a gas in a rather highly exhausted vessel. Betelgeuse could be described as "a rather good vacuum."

At the other extreme are the "white dwarf" stars, which have extravagantly high density. I must say a little about the way in which this was discovered.

Between 1916 and 1924 I was very much occupied trying to understand the internal constitution of the stars, for example, finding the temperature in the deep interior, which is usually ten million degrees, and making out what sort of properties matter would have at such high temperatures. Physicists had recently been making great advances in our knowledge of atoms and radiation; and the problem was to apply this new knowledge to the study of what was taking place inside a star. In the end I obtained a formula by which, if you knew the mass of a star, you could calculate how bright it ought to be. An electrical engineer will tell you that to produce a certain amount of illumination you must have a dynamo of a size which he will specify; somewhat analogously I found that for a star to give a certain amount of illumination it must have a definite mass which the formula specified. This formula, however, was not intended to apply to all stars, but only to diffuse stars with densities corresponding to a gas, because the problem became too complicated if the material could not be treated as a perfect gas.

Having obtained the theoretical formula, the next thing was to compare it with observation. That is where the trouble often begins. And there was trouble in this case; only it was not of the usual kind. The observed masses and luminosities agreed with the formulae all right; the trouble was that they would not stop agreeing! The dense stars for which the formula was *not* intended agreed just as well as the diffuse stars for which the formula was intended. This surprising result could only mean that, although their densities were as great as that of water or iron, the stellar material was nevertheless behaving like a gas; in particular, it was compressible like an ordinary gas.

We had been rather blind not to have foreseen this. Why is it that we can compress air, but cannot appreciably compress water? It is because in air the ultimate particles (the molecules) are wide apart, with plenty of empty space between them. When we compress air we merely pack the molecules a bit closer, reducing the amount of vacant space. But in water the molecules are practically in contact and cannot be packed any closer. In all substances the ordinary limit of compression is when the molecules jam in contact; after that we cannot appreciably increase the density. This limit corresponds approximately to the density of the solid or liquid state. We had been supposing that the same limit would apply in the interior of a star. We ought to have remembered that at the temperature of millions of degrees there prevailing the atoms are highly ionized, i.e. broken up. An atom has a heavy central nucleus surrounded by a widely extended but insubstantial structure of electrons—a sort of crinoline. At the high temperature in the stars this crinoline of electrons is broken up. If you are calculating how many dancers can be accommodated in a ball-room, it makes a difference whether the ladies wear crinolines or not. Judging by the crinolined terrestrial atoms we should reach the limit of compression at densities not much greater than water; but the uncrinolined stellar atoms can pack much more densely, and do not jam together until densities far beyond terrestrial experience are reached.

This suggested that there might exist stars of density greater than any material hitherto known, which called to mind a mystery concerning the Companion of Sirius. The dog-star Sirius has a faint companion close to it, visible in telescopes of moderate power. There is a method of finding densities of stars which I must not stop to explain. When it was found to give for the Companion of Sirius a density 50,000 times greater than water, it was naturally assumed that it had gone wrong in its application. But in the light of the foregoing discussion, it now seemed possible that the method had not failed, and that the extravagantly high density might be genuine. So astronomers endeavoured to check the determination of density by another method depending on Einstein's relativity theory. The second method confirmed the high density, and it is now generally accepted. The stuff of the Companion of Sirius is 2000 times as dense as platinum. Imagine a match-box filled with this matter. It would need a crane to lift it—it would weigh a ton.

I am afraid that what I have to say about the stars is largely a matter of facts and figures. There is only one star near enough for us to study its surface, namely our sun. Ordinary photographs of the sun show few features, except the dark spots which appear at times. But much more interesting photographs are obtained by using a spectro-heliograph, which is an instrument blind to all light except that of one particular wave length—coming from one particular kind of atom.

Now let us turn to the rest of the universe which lies beyond the Milky Way. Our galaxy is, as it were, an oasis of matter in the desert of emptiness, an island in the boundless ocean of space. From our own island we see in the far distance other islands—in fact a whole archipelago of islands one beyond another till our vision fails. One of the nearest of them can actually be seen with the naked eye; it is in the constellation Andromeda, and looks like a faint, rather hazy, star. The light which we now see has taken 900,000 years to reach us. When we look at that faint object in Andromeda we are looking back 900,000 years into the past. Some of the telescopic spiral nebulae are much more distant. The most remote that has yet been examined is 300,000,000 light-years away.

These galaxies are very numerous. From sample counts it is found that more than a million of them are visible in our largest telescopes; and there must be many more fainter ones which we do not see. Our sun is just one star in a system of thousands of millions of stars; and that whole system is just one galaxy in a universe of thousands of millions of galaxies.

Let us pause to see where we have now got to in the scale of size. The following comparative table of distances will help to show us where we are:

	Kilometres
Distance of the sun	150,000,000
Limit of the solar system (Orbit of Pluto)	5,800,000,000
Distance of nearest star	40,000,000,000,000
Distance of nearest external galaxy	8,000,000,000,000,000,000
Distance of furthest galaxy yet observed	3,000,000,000,000,000,000,000

Some people complain that they cannot realize these figures. Of course they cannot. But that is the last thing one wants to do with big numbers—to "realize" them. In a few weeks time our finance minister in England will be presenting his annual budget of about £900,000,000. Do you suppose that by way of preparation, he throws himself into a state of trance in which he can visualize the vast pile of coins or notes or commodities that it represents? I am quite sure he cannot "realize" £900,000,000. But he can spend it. It is a fallacious idea that these big numbers create a difficulty in comprehending astronomy; they can only do so if you are seeking the wrong sort of comprehension. They are not meant to be gaped at, but to be manipulated and used. It is as easy to use millions and billions and trillions for our counters as ones and twos and threes. What I want to call attention to in the above table is that since we are going out beyond the Milky Way we have taken a very big step up in the scale of distance.

The remarkable thing that has been discovered about these galaxies is that (except three or four of the nearest of them) they are running away from our own galaxy; and the further they are away, the faster they go. The distant ones have very high speeds. On the average the speed is proportional to the distance.

Why are they all running away from us? If we think a little, we shall see that the aversion is not especially directed against us; they are running away from us, but they are also running away from each other. If this room were to expand 10 per cent in its dimensions, the seats all separating in proportion, you would at first think that everyone was moving away from you; the man 10 metres away has moved 1 metre further off; the man 20 metres away has moved 2 metres further off; and so on. Just as with the galaxies, the recession is proportional to the distance. This law of proportion is characteristic of a uniform expansion, not directed away from any one centre, but causing a general scattering apart. So we conclude that recession of the nebulae is an effect of uniform expansion.

The system of the galaxies is all the universe we know, and indeed we have strong reason to believe that it is the whole physical universe. The expansion of the system, or scattering apart of the galaxies, is therefore commonly referred to as the expansion of the universe; and the problem which it raises is the problem of the "expanding universe."

The expansion is proceeding so fast that, at the present rate, the nebulae will recede to double their present distances in 1300 million years. Astronomers will have to double the apertures of their telescopes every 1300 million years in order to keep pace with the recession. But seriously 1300 million years is not a long period of cosmic history; I have already mentioned it as the age of terrestrial rocks. It comes as a surprise that the universe should have doubled its dimensions within geological times. It means that we cannot go back indefinitely in time; and indeed the enormous time-scale of billions[2] of years, which was fashionable ten years ago, must be drastically cut down. We are becoming reconciled to this speeding up of the time-scale of evolution, for various other lines of evidence have convinced us that it is essential. It seems clear now that we must take an upper limit to the age of the stars not greater than 10,000 million years; previously, an age of a thousand times longer was commonly adopted.

For reasons which I cannot discuss fully we believe that along with the expansion of the material universe there is an expansion of space itself. The idea is that the island galaxies are scattered throughout a "spherical space." Spherical space means that if you keep going straight on in any direction you will ultimately find yourself back at your starting point. This is analogous to what happens when you travel straight ahead on the earth; you reach your starting point again, having gone round the world. But here we apply the analogy to an extra dimension—to *space* instead of to a *surface*. I realize, of course, that this conception of a closed spherical space is very difficult to grasp, but really it is not worst than the older conception of infinite open space which no one can properly imagine. No one can conceive infinity; one just uses the term by habit without trying to grasp it. If I may refer to our English expression, "out of the frying-pan into the fire," I suggest that if you feel that in receiving this modern conception of

[2] The English "billion" is equivalent to the American "trillion."—Eds.

space you are falling into the fire, please remember that you are at least escaping from the frying-pan.

Spherical space has many curious properties. I said that if you go straight ahead in any direction you will return to your starting point. So if you look far enough in any direction and there is nothing in the way, you ought to see—the back of your head. Well, not exactly—because light takes at least 6000 million years to travel round the universe and your head was not there when it started. But you will understand the general idea. However, these curiosities do not concern us much. The main point is that if the galaxies are distributed over the spherical space more or less in the same way that human beings are distributed over the earth, they cannot form an expanding system—they cannot all be receding from one another—unless the space itself expands. So the expansion of the material system involves, and is an aspect of, an expansion of space.

This scattering apart of the galaxies was not unforeseen. As far back as 1917, Professor W. de Sitter showed that there was reason to expect this phenomenon and urged astronomers to look for it. But radial velocities of spiral nebulae have been measured in sufficient numbers to show conclusively that the scattering occurs. It is one of the deductions from relativity theory that there must exist a force, known as "cosmical repulsion," which tends to produce this kind of scattering in which every object recedes from every other object. You know the theory of relativity led to certain astronomical consequences—a bending of light near the sun detectable at eclipses, a motion of the perihelion of Mercury, a red-shift of spectral lines—which have been more or less satisfactorily verified. The existence of cosmical repulsion is an equally definite consequence of the theory, though this is not so widely known—partly because it comes from a more difficult branch of the theory and was not noticed so early, and perhaps partly because it is not so directly associated with the magic name of Einstein.

I can see no reason to doubt that the observed recession of the spiral nebulae is due to cosmical repulsion, and is the effect predicted by relativity theory which we were hoping to find. Many other explanations have been proposed—some of them rather fantastic—and there has been a great deal of discussion which seems to me rather pointless. In this, as in other developments of scientific exploration, we must recognize the limitations of our present knowledge and be prepared to consider revolutionary changes. But when, as in this case, observation agrees with what our existing knowledge had led us to expect, it is reasonable to feel encouraged to pursue the line of thought which has proved successful; and there seems little excuse for an outburst of unsupported speculation.

Now we have been all over the universe. If my survey has been rather inadequate, I might plead that light takes 6000 million years to make the circuit that I have made in an hour. Or rather, that was the original length of the circuit; but the universe is expanding continually, and whilst I have been talking the increase of the circuit amounts to one or two more days' journey for the light. Anyhow, the time has come to leave this nightmare of immensity and find again, among the myriads of orbs, the tiny planet which is our home.

SIR BERNARD LOVELL

The Origin and Evolution of the Universe

•

THE COSMOLOGICAL PROBLEM is epitomized by many typical photographs taken by the large optical telescopes in which it is possible to distinguish almost as many extragalactic nebulae as there are foreground stars in the Milky Way. Wherever one looks in space there seem to be these great numbers of galaxies. Many of these galaxies are similar to the Milky Way and must contain something like 10,000 million to 100,000 million stars of which our sun might be a typical member. On the whole, these galaxies seem to be uniformly distributed throughout the observable universe—at least when the large-scale distribution is considered.

There are indeed indications of a certain structure in the cosmos because few of these galaxies exist on their own; they are nearly all contained in clusters or groups. The Milky Way system belongs to a small group of galaxies which includes the Andromeda nebula, M31, our closest neighbour, at a distance of 2 million light-years, and about a dozen others. This group occupies a volume of space in the shape of a flattened ellipsoid with a major axis of two million light-years, a minor axis of a million light-years, and a thickness of about 500,000 light-years. Our own galaxy is near the extremity of this group, about a million light-years from the centre which lies in the direction of the Andromeda nebula. Compared with some of the clusters which can be photographed this is, indeed, a small group. For example, the cluster of galaxies in Virgo is about 14 million light-years distant from us; it extends for at least 2 or 3 million light-years and contains several hundred galaxies.

In the Coma cluster of galaxies, which is 90 million light-years distant, it is possible to describe an area of sky no bigger than that occupied by the full moon, in which something like 500 galaxies are concentrated. These figures serve to indicate the enormous quantity of material which we have to deal with in the universe and the vastness of the problem. The galaxies contained within the field of view of the 200-inch telescope are effectively uncountable; all that can be said is that down to 22nd and 23rd magnitude, which is near the limit of visibility of the telescope, it has been estimated that there are something of the order of a trillion galaxies.

There is one quite remarkable feature of the observations of these extragalactic nebulae which presents a critical problem in respect of any attempted explanation of the origin of the universe and, indeed, of its future history. This observation is that the spectral lines in the light from the distant galaxies are shifted towards the red end of the spectrum. The first critical measurements were made in 1912 at the Lowell Observatory by Slipher, who found that if the shift were interpreted in the conventional way as a doppler displacement, then the indicated velocities of recession were much greater than for any other known celestial object. The real significance of these measurements was not apparent until Hubble in the period 1922-24 gave conclusive arguments in support of the concept that the nebulae were extragalactic lying far beyond the confines of the Milky Way, and then in 1929 announced that the red-shift seemed to be linearly related to the distance of the galaxy. This implies that the velocity of recession of the extragalactic nebulae increases as their distance from us increases. The most tremendous arguments have taken place about the interpretation of this reddening, but now all astronomers agree to an explanation which implies that this reddening is due to a doppler shift, indicating that the galaxies are moving apart at a great speed.

Some of these early measurements were surprising enough. For example, the well known Virgo cluster of galaxies at a distance of 7 million light-years showed a red-shift which indicated a velocity of 600 miles per second; but with the advent of the 200-inch telescope, the limits of penetration were extended to such a degree that the indicated speeds of recession reached appreciable fractions of the velocity of light. In 1960 the most distant cluster yet identified in Boötes 5,000 million light-years away, showed a recessional velocity of 46 per cent of the velocity of light—86,000 miles per second. It is a fact, difficult to comprehend, that during the time taken to read a few lines of this print we have separated by another million miles from the objects which we can photograph at the extreme limits of our penetration into the universe. The entire universe is expanding at an enormous rate, and there is not yet any reliable evidence which indicates a departure from linear relationship between velocity of recession and distance, at least out to several thousand million light-years. Any theory of the universe has to explain not only the existence of this recession of the nebulae but also the fact that it increases linearly with distance out to the furthest limit of penetration of the big telescopes.

In spite of this dramatic revelation of the content, arrangement, and movement of the galaxies in space, the optical telescopes have failed to show any differences in the large-scale organization of the universe out to these distances. In other words, the telescopes have failed so far to reveal any changes in the past history of the universe. The critical point in this argument is that as we look out into space, so we look back in time, because the light from these distant objects has been thousands of millions of years on its journey to us. We are, therefore, studying the universe as it existed thousands of millions of years ago in time past. Therefore one can say from the observations with the optical telescopes that back to an epoch of a few thousand million years there do not appear to have been any great changes in the organization of the cosmos.

The Impact of Radio Astronomy on Cosmology

The advent of the radio telescopes quite unexpectedly introduced a dramatic new situation into cosmological studies. The radio emission from the Milky Way can be explained in terms of the radiation from hydrogen gas, and from certain discrete sources concentrated in the plane of the Milky Way which are identified with supernova remnants. The radio observations with a high-definition radio telescope also reveal thousands of discrete sources of radio waves having a distribution which is nearly isotropic and only a few of these can be linked up with the individual extragalactic nebulae which are fairly close to us in space, such as M31.

The first suggestion that these radio sources might be of cosmological significance arose when an attempt was made to identify the second strongest source of radio emission in the sky—in the constellation of Cygnus. Nature confused the issue because this source, like the slightly stronger one in Cassiopeia, was close to the galactic plane. The Cassiopeia source was identified with a supernova remnant in the galaxy, but there was no galactic object possessing any features which might have been responsible for the intense source in Cygnus. In fact, the region of the sky in question was rather undistinguished; it contained many faint distant stars in the Milky Way system, but nothing which would indicate that there could be an object responsible for such strong radio emission.

After some years of this dilemma the position of the Cygnus radio source was measured with sufficient accuracy to enable Baade and Minkowski to use the 200-inch telescope on Mount Palomar and make a very long exposure of this region of the sky. In their series of photographs of this region Baade and Minkowski discovered an unusual object which they identified as two spiral galaxies closely interacting. The possibility that galaxies might, under certain conditions, collide with one another had given rise to some speculation before the advent of radio astronomy. It seemed possible that in the very dense clusters the galaxies must be close enough together, and, since they are moving at random under their own gravitational attraction, then some at least must occasionally run into one another. No such cases were discovered until this example of strong radio emission led to the photography of this region of the sky.

There were immediately two important consequences. First, the object was estimated to be at a very great distance—700 million light-years away. Second, although the object was extremely faint optically, the output of energy in the radio wave region was considerable, to such an extent that it was the second strongest source of radio emission known. This interpretation of the phenomenon has turned out to be of the utmost significance to cosmology because it appears that under certain circumstances events take place in the universe which give rise to a relatively large output of energy in the radio wave part of the spectrum, although optically the objects themselves are very faint. Indeed, if these colliding galaxies in Cygnus, which are at 700 million light-years, were ten times further away, that is at 7,000 million light-years, then it would be impossible to photograph them with any of the world's optical telescopes; but the radio telescopes would still record the interacting galaxies as a prominent object. This conception led to the idea that the difficulty of relating many of these radio sources to objects which can be photographed might be that they existed as cases of collisions of galaxies so far away in the cosmos that they were beyond the range of the world's biggest optical telescope. This is indeed what we now believe to be the case. The problem is of great significance because we hope that these radio studies will enable us to penetrate to those regions of space and time which lie beyond the photographic limit, where it may be possible to discover signs of change in the past history of the universe.

It should be mentioned that the interpretation of the Cygnus photograph as an example of two galaxies in collision is not unanimously accepted amongst astronomers. In particular, the Russian astronomer Ambartsumian takes the view that the object is an example of the division of the nucleus of a galaxy to form two separate galaxies. In any case, both in this example and in the others subsequently discovered, the crucial observation for cosmology is that certain processes can take place in distant, faint photographic objects which produce a relatively large output of energy in the radio-wave part of the spectrum.

The Investigation of the Distant Radio Sources

Large-scale investigations of these sources of radio waves at very great distances are in progress, and special telescopes have been developed for this work. One frequently used technique is the employment of a radio telescope which effectively consists of two aerials, separated, perhaps, by many miles. Then instead of the single beam of the parabolic reflector, the reception pattern of the combined aerials is a system of lobes. As the earth sweeps this beam across the sky, a source of radio waves of angular diameter, large compared with the separation of the lobes, gives a smooth pattern. On the other hand, if the radio waves came from a source whose angular extent is small compared with the separation between the lobes, then the received pattern will show maxima and minima as the earth sweeps the lobe pattern across the source.

By using this and similar techniques, Ryle at Cambridge and the radio astronomers in Sydney, Australia, have investigated the distribution of several thousand of these radio sources. If the objects in the universe are distributed uniformly in space and time then, because the radio intensity or optical intensity will decrease as the inverse square of the distance, there will be a definite relation between the number of objects and their brightness (in the radio or optical sense). In fact it is not difficult to calculate that if the logarithm of the number of objects of a given brightness is plotted against the logarithm of the intensity, then the result should be a straight line with a slope of 1.5.

The radio astronomers have plotted their results in this form to show how the number of radio sources increases as the intensity diminishes, and it is to be expected that, if the distribution were uniform in space and time, then the relation would be a straight line with this slope of 1.5. In fact, the Cambridge observations indicate that although there is good agreement with this uniform spatial distribution for the strong radio sources, the agreement disappears as one gets to fainter and fainter sources. There is a departure of the experimental curve from the theoretical straight line, the departure being in the sense which indicates that for very weak radio sources (which are presumably at great distances) the numbers appear to be greater than would be indicated by the concept of a uniform distribution in space and time. If we take the simplest interpretation of this result, the conclusion is inevitable that the content of the universe—the spatial density of the material, the number of galaxies or the number of radio sources per unit volume of space—is greater in those remote parts than in the regions closer to us in space and time. If this result could be substantiated beyond doubt then it would be of the greatest cosmological significance, because for the first time we would have an indication of some change in the organization of the cosmos as we penetrate further into space and further back into time. Of course this result from the counting of the radio sources is exactly what one might expect if the universe were in an evolving state, because as we look back to these regions of time, many thousand million years ago, we would expect to find a universe more densely populated and nearer the point of its origin than it is today. Unfortunately, and perhaps because of the significance of the conclusions, this work has been severely criticized. The derivation of the statistics is a complicated process and, moreover, a similar experiment by the workers in Sydney has yielded a contrary result, indicating that there is no departure from the uniform distribution out to the limits of penetration of the radio telescopes.

An attempt to overcome this dilemma, and to evade the criticism that one could not treat the radio source counts in this statistical manner, has been made at Jodrell Bank. The problem occupied more than a third of the working time of the radio telescope in the first four years of its use, which means that 5,000 or 6,000 hours of work have been spent on this aspect of the cosmological problem. In this system the telescope at Jodrell Bank is used in conjunction with smaller aerials which can be moved to different distances from it. The greatest separation so far achieved is 72 miles. The information from these remote stations is transmitted over a radio link and is correlated with the signals picked up simultaneously in the radio telescope at Jodrell Bank. The aim of this experiment is to measure the strength of the radio waves and the apparent angular diameter of the sources. With this information we would know much more about the distant radio sources because we could calculate the effective temperature of the source and begin to get some scale of distance for these unidentified sources. The experiment is carried out by changing the spacing of the distant aerial until the lobe separation becomes so small that the maximum and minimum in the interferometer fringe pattern of a particular source begins to disappear. The source is then beginning to be resolved, and it is possible to estimate the actual angular diameter of the emitting region.

So far, out of 300 of the most intense unidentified radio sources which we believe to be at distances of cosmical significance, only about 10 per cent of them have angular diameters which would indicate that they were at distances greater than about 2,000 million light-years. Of these, three have proved quite remarkable in that at the greatest spacing of the aerials the intensity of the fringes has remained unchanged, indicating that the radio sources must have angular diameters less than a second of arc. The interpretation to be placed on this result, based on the known distance and characteristics of the Cygnus source, is that these sources must be situated so far away in the universe that the radio waves have been on their journey for probably 7,000 or 8,000 million years.

The import of the radio investigations has been made manifest by the successful optical identification in 1960 of one of these sources of small angular diameter. The actual coordinates of this radio source had been very well determined and its characteristics indicated that it was typical of a Cygnus-type object—that is, where galaxies are in collision or closely interacting. The result of the investigation of the appropriate region of the sky by the 200-inch Palomar telescope was that the radio source was related to a cluster of galaxies in the constellation of Boötes. This cluster of galaxies in Boötes represents at present the greatest distance to which the Palomar telescope has penetrated—5,000 million light-years—with a speed of recession, as measured by the red-shift, of 46 per cent of the velocity of light or 86,000 miles per second. An interesting consequence of this cooperation between the radio telescopes and the optical telescopes has been that within a few years the ultimate range of the 200-inch optical telescope on Palomar has increased by a factor of 3.

The results at the moment are consistent with the belief that many of these unidentified radio sources are cases of dense clusters of galaxies, extremely far away in time and space. For reasons which are not really understood, but which are almost certainly associated with the interaction of the dust and gas in the galaxies during the collisions, they generate relatively intense radio waves although their output of energy in the optical part of the spectrum is rather small. At the moment, in the region of space out to the 2,000 million light-year cluster in Hydra, apart from this new one in Boötes, there are about a dozen of these peculiar nebulae which have been identified. Although there can be little doubt that the radio observations of the unidentified sources have great importance in cosmology, it will probably take many years of work before the results can be used to make firm cosmological predictions.

Theories of the Cosmos

We must now consider what these observations may mean in terms of our ideas about the evolution and origin of the cosmos. Cosmology was given a new impetus with the development of Einstein's theory of general relativity because the solutions of the equations of general relativity appeared to have an important bearing on the arrangement of matter in space. In the solution of these equations there is an arbitrary constant—the lambda term or the cosmic constant. The problem of the value to be attached to this cosmical constant has caused extensive controversies in cosmology over the past few decades because there are a variety of model universes which can be specified by the equations of general relativity depending on this constant. Not all of the possible models have been explored theoretically but they all have this in common: they represent a universe in some kind of evolutionary state; a universe which in time past had some degree of singularity. A universe which has this unique or singular condition in a past epoch presents many associated problems—particularly with regard to the problem of creation at a specific time.

Hoyle, Bondi, Gold, and McCrea have brought forward an entirely new theory of cosmology which uses the equations of general relativity in which the cosmical constant is made zero—the theory of continuous creation or the steady state theory. On this theory the universe has never had any singularity in the past, neither will it in the future. They invoke the perfect cosmological principle in which the universe is the same through all space *and time,* so that however far we go back into the past and however far into the future we will always find the same kind of universe as we find today. In order to maintain this stable situation in face of the obvious expansion of the universe the concept implies that the primeval material of the universe, the hydrogen atoms, are being created now, at this moment, continuously, at such a rate that they form into galaxies to make up for those that are moving out of our field of view.

Today the cosmological situation is one of extreme interest and complexity. The evolutionary and the steady state theories are in sharp contrast, and although there are many variations of the evolutionary theories, they all imply an origin of the universe, or a singular condition, in some finite and predictable past epoch. This contrast between the predictions of the evolutionary and steady state theories in time past should make it possible to devise a specific observational test to distinguish between them. Indeed we think now that the radio records which give us information about the state of the universe at distances of 7,000 or 8,000 million light-years must refer to a region of space and time where the differences between these two theories should be manifest.

The concept of the unchanging nature of the universe in time and space of the steady state theory, carries the implication that however far we recede into time past the large-scale appearance of the universe will be the same as it is today. The primeval hydrogen atoms are being created at the rate of about one atom per cubic mile of space per year, but this means that trillions of tons are being created every second in the observable universe. This creation of hydrogen has two consequences on the steady state theories. First, it provides the necessary material for the galaxies to form at just the right rate to compensate for those which are moving out of our field of view because of the expansion of the universe; and second, it is the pressure arising from the creation of this material which provides the driving force resulting in the expansion of the universe.

In principle there is a simple observational test between the evolutionary and the steady state theories, because if we can penetrate far enough back into past time then on the steady state theory we should find exactly the same number of galaxies per unit volume of space as we do in our local neighbourhood, whereas on the evolutionary theory the predictions are quite different. The spatial density of galaxies will not be the same as in the regions of time and space closer to us. Indeed it is obvious that if the universe did in fact originate from a superdense initial state in an epoch 10,000 million years ago, then in the regions of space some 7,000 million light-years distant to which we believe the radio telescopes are penetrating, the spatial density of galaxies will be much greater than they are today. The optical telescopes have not so far penetrated to regions where any marked differences are apparent, but it is in these regions more than 5,000 million light-years distant where there is reason to hope that the classification of the radio sources may show whether the spatial density of the galaxies is varying as we penetrate into time past.

The Model Universes in Evolutionary Cosmology

If the steady state theory is correct then the universe has never had a beginning and will never have an end, and it is useless to ask when time and space began or when time and space will end because the question is meaningless within the framework of the theory of continuous creation. On the other hand, the situation in the evolutionary theories is completely different because they do accept the possibility of a singular condition of the universe, and the creation of matter in time present has no part in the theories. The essence of the evolutionary ideas can be appreciated if we imagine the recession of the galaxies reversed. We are separating from the cluster of galaxies in Boötes with a speed of nearly half the velocity of light—86,000 miles a second. Therefore a few minutes ago we were millions of miles closer to this cluster than we are now. If we retrace the history of the cosmos in this way we find that about 10,000 million years ago all the material which now forms the galaxies must have been closely packed together. On at least some of the evolutionary models this moment must represent the beginning of the universe, the beginning of time and space. Because of the possible variations in the value of the cosmical constant, the various evolutionary models differ considerably in their description of the cosmos at this epoch.

For example, one model which has had considerable vogue is due to Gamow. In this model it is assumed that 10,000 million years ago the whole material of the universe was indeed contained in a primeval lump of material of fundamental particles—probably neutrons and protons. At the beginning of time this supercondensate exploded, and on Gamow's theory all the elements which we know today were formed within the first few hours of the history of the universe. Actually a situation in which all the elements were formed in this way seems impossible because certain light elements required in the sequence are not stable. On the other hand we believe now that element formation is taking place in the hot central cores of the stars and this criticism of the theory may not therefore be serious. On this theory of Gamow's the origin of the universe occurred 10,000 million years ago, the cosmical constant is zero, and the expansion which we witness today is simply the result of the impetus of this initial explosion.

One of the most thoroughly studied evolutionary models is that of the Abbé Lemaître. His theory also implies that the galaxies began to form at this period of about 10,000 million years ago. But in his view the cosmical constant in Einstein's equation is positive, it has a real meaning, and he does not believe that the expansion of the universe which we see today is the result of the impetus of the initial explosion. In fact, to get to the beginning of the Lemaître universe we have to retrace our steps, not for 10,000 million years but for a period of time which is not exactly defined but is probably between 40,000 and 60,000 million years in the past. At that time, according to Lemaître, the universe was in its original state of the primeval atom. Lemaître's primeval atom must have been a concentrate of neutrons or protons so closely packed that the actual nuclei of the atoms were squashed against one another in a completely degenerate state. This primeval atom must have contained the entire mass of the universe which we see around us today; 10^{21} tons in a volume of space no bigger than the solar system is today. The density in this primeval lump must have been colossal, something like a million million times the density of water.

Such a condensate would obviously be in a most unstable condition and at this moment of time the primeval atom must have suffered some radioactive disintegration and the material began to spread out into space. After a thousand million years of time, space, according to the Lemaître theory, occupied about a thousand million light-years and was full of hydrogen gas uniformly distributed. At this stage it seemed possible that the universe might settle down into a steady condition. Indeed such a condition was the starting point of Eddington's theory of the universe in which there was a uniform distribution of hydrogen atoms spread through a region of space of about a thousand million light-years in diameter.

After this period of a thousand million years the initial impetus of the decay of the primeval atom was exhausted and the gravitational forces had introduced a state of stability. Then condensations in the gas began to occur. The positive cosmical constant in Lemaître's equation is equivalent to a physical force which works in opposition to the Newtonian forces of gravitation, and so on the cosmical scale we have a repulsion which works in opposition to the forces of attraction. Hence when the condensations formed in the gaseous mass the universe began on its career of expansion which we witness today. The time at which the galaxies began to form from the gaseous condensations is not exactly defined by the theory. Neither does Lemaître's equation lead to predictions about the time of appearance of the population II stars in the nuclei of the galaxies, or of the relation between the formation of those stars and the initial formation of the galaxies. The present belief that the main formation of the population II stars occurred 8,000 or 9,000 million years ago, and that population I stars are still in process of formation, is not at variance with the main features of Lemaître's theory. Indeed it is extraordinarily difficult to find any single observational feature, either in terms of the ages of the galaxies or of the formation of stars, which would be completely decisive evidence for or against these various cosmological theories.

In another variant of the evolutionary theory, with a different arrangement of the constants, the universe has a hyperbolic form. After a certain degree of expansion it begins to contract again and the process of expansion and contraction is cyclic, and capable of indefinite repetition.

As far as the future history of the cosmos is concerned, continuous creation leads to the concept of an infinite future existence. The end of the universe has no meaning. Although M31 and the other galaxies will disappear from the region accessible to our telescopes, nevertheless other galaxies will form from the newly created hydrogen atoms, and the universe is ageless. On the evolutionary theories a singularity occurred when all the material must have been created at once and no further creation has been taking place since that time. Therefore, the future outlook for the universe is bleak, because the material is dispersing and the universe will die. This is the essence of the conflicts in cosmology today which we are seeking to clarify by the observations with the radio telescopes. At least we are optimistic enough to believe that the radio telescopes will penetrate to those regions of time and space where the differences between the cosmologies should become apparent.

The bearing of the theories of cosmology on philosophy and theology is often discussed, and indeed the subject is one of extreme interest and importance. It has frequently been said that the steady state theory is entirely a materialistic theory because it does not involve any moment when a unique act of creation could have taken place, and moreover all the tenets of the theory are effectively open to observational test. On the other hand, it has been claimed that the evolutionary theories are entirely in accordance with theological doctrine in that they invoke a moment of singularity which might involve a unique act of creation in time past. My personal attitude is that neither theory necessarily possesses these attributes. There is a fundamental principle in physics which implies that it is impossible to observe any event with precision—the uncertainty principle. For the steady state theory to be proved materialistic, one would have to be able to imagine the development of a perfect scientific instrument which could, in principle, observe the creation of a single atom of the hydrogen which the theory predicts to be taking place continuously. Even if one could devise such an instrument then one would still never be able to obtain any exact knowledge of the creation of the hydrogen atom because in the very act of observing it the uncertainty principle implies that one would disturb it, and therefore one could never obtain any exact information about the basic process of creation.

The evolutionary theories envisage a moment in the remote past when all matter was created. Again for fundamental reasons, in which the finite velocity of light is involved, one can never observe this epoch. At 5,000 million light-years we are already dealing with an object which is receding from us at a speed nearly half the velocity of light, and this velocity of recession is still increasing linearly with distance. Any material which exists near the time of the singularity must be receding with a speed which is closely approaching the velocity of light, and therefore can never be observable. Hence any precise observational knowledge of this critical past epoch is forever forbidden to us. Eddington epitomized the problem when he said that light is like a runner on a track which is ever expanding so that the winning post is always receding from him faster than he can run. This is exactly the situation as far as we are concerned with the observation of the remotest parts of space and time which would contain the singularities on any of the evolutionary theories.

Although with our telescopes we shall no doubt clarify the cosmological problem to a large extent, the ultimate issue of the origin of the cosmos may well be a metaphysical one lying outside the realms which the tools of physics and astronomy can approach for reasons which are inherent in fundamental scientific theory.

B. *The Heavens*

•

2. The Age of Space

On October 4, 1957, with the successful launching into orbit of "Sputnik" by the Russians, a wholly new age, the Age of Space, began. This age has already seen the probing of our near celestial neighbors. It lifts the possibility of interplanetary flight with human cargo from the visionary to the practicable. It poses new military problems and offers new techniques for scientific research. Its instrument is the rocket, and the basic scientific principle on which this instrument is constructed—the law of action and reaction—was first stated in Newton's Third Law of Motion: "To every action there is always an equal and opposite reaction."

The principle had been utilized about the beginning of the Christian era by Hero of Alexandria, who constructed a hollow sphere supported by pivots, with two bent spouts at the ends of an axis perpendicular to the supports. When steam was introduced through one pivot and vented through the spouts, the sphere revolved. The principle was adapted to skyrockets and with indifferent success to weapons of war. It was not until the early nineteenth century that these weapons became valuable. At that time, William Congreve, a colonel in the British Army, constructed rockets which were used effectively in the Battle of Leipzig and against the Americans in the War of 1812. "The rocket's red glare" gave proof of Congreve's achievement, but interest in this type of weapon was halted by the superior fire power and accuracy of newly developed rifled artillery.

The principle was not completely abandoned. Rocket-propelled harpoons, signal rockets and airborne objects were tested. In the 1890's a German inventor, Herman Ganswindt, and a Russian teacher of mathematics, Konstantin Tsiolkovski, suggested rocket-propulsion systems of sufficient power to attain escape velocity. At the beginning of the twentieth century a Swedish inventor, Baron von Unge, developed an aerial torpedo with which the Krupp armament firm experimented. This international research was steadily expanded by experimenters of almost Messianic fervor. In Germany in particular, a group whose members were to include Werner von Braun and Willy Ley was extremely active. In America a young teacher named Robert Goddard was engaged in calculations for perfecting "A Method of Reaching Extreme Altitudes." He it was who conducted "the first actual shot of a liquid-fuel rocket anywhere in the world." His story is told by G. Edward Pendray, who was himself active in pioneering rocket research in the United States.

It is easy to understand why objects like artillery projectiles, under the influence of gravity, return to earth within seconds of firing. It is also easy to comprehend why such projectiles, with sufficient impetus, will escape. The reason why, under carefully calculated conditions, an object may be caused to orbit round the earth is somewhat more difficult to grasp. In A Shot Around the World *Arthur C. Clarke, the well-known writer on scientific subjects, explains the problems and how they are solved.*

Our rocket has been launched successfully into orbit. What are the sensations of a human being encased in such a projectile? Each of the Russians and Americans who have been propelled into space has described his experiences. Their reactions have been startlingly similar and John H. Glenn, Jr.'s account, Pilot's Flight Report, *is typical.*

The rocket which launched Glenn is by contemporary standards already as outdated as a battlewagon. New and far more powerful rockets are probing our near celestial neighbors and giving a new dimension to our knowledge of the solar system. What do they learn and how? The Voyage of Mariner II *by J. N. James, of the Jet Propulsion Laboratory, California Institute of Technology, is an account of one successful probe.*

G. EDWARD PENDRAY

Robert Goddard

•

A NATIVE New Englander, Robert Hutchings Goddard was a slender, quiet young man; early bald, careful of words and precise in thinking. He always disliked publicity, and to the end he was almost unknown as a person, though famous for his work.

Born in Worcester on October 5, 1882, his early schooling and his college work were all obtained at Boston, where he lived until he was sixteen, and at Worcester, where he was graduated from the Worcester Polytechnic Institute in 1908. His academic career was conventional, rising in the usual steps from fellowship to instructor to assistant professor and finally to full professor at Clark University. His only "foreign" adventure was to accept a research fellowship at Princeton, in 1912, the year following completion of his work for the degree of Doctor of Philosophy at Clark.

In his school days he was a serious young man, with an odd streak of scientific speculativeness in his nature. He enjoyed mathematics, was fond of figuring out faster ways to travel, and better ways to do things in general. In his freshman year at college one of his professors assigned the topic "Traveling in 1950" as a theme subject. Goddard produced a bold paper which he read before the class, describing in some detail a railway line in which the cars were supported electromagnetically without any metal-to-metal contact, in a tube from which the air had been exhausted. With such a vacuum railroad he calculated it would be perfectly possible to make enormous velocities safely; for example, a running time of ten minutes from Boston to New York.

It is not known exactly when he made his first experiments with rockets, but he often told friends about carrying on static tests with small rockets in 1908, in the basement of Worcester Tech. He promptly filled the whole place with smoke, and had to talk fast to get out of trouble. While at Princeton in the season of 1912–1913 he made the computations that formed the basis of his Smithsonian paper of 1919. It was in this period, when about thirty, that the great excitement of discovery first began to come to him. The calculations showed that only a little fuel, relatively, would be needed to lift a payload to really great heights by rocket. The theory, in fact, was so promising he could hardly wait to begin transforming his figures into actuality.

Upon returning to Clark in 1914 he began experimenting with ship rockets, which he purchased out of his slender salary as an instructor. Next came tests with steel rockets using smokeless powder, fired both in air and in a vacuum. Connection to a vacuum pump required the installing of extensive piping in the basement workroom at Clark. On one occasion Dr. A. G. Webster, head of the physics department, viewed the pipe system with an admiring eye. "Well, Robert," he exclaimed, "when you go I hope you leave all this tubing here!"

In the course of these experiments, Goddard spent some eight hundred dollars of his own money, and by 1916 had reached the limit of what he thought he could do on his own resources. Being inexperienced in the ways of self-promotion, he could think of no way to obtain a backer except to make out a report of what he had done with rockets, and project what he thought could finally be accomplished. With characteristic thoroughness, he cast the paper into the best scientific form, rewriting it several times. To complete the job he bound it in a special cover with a neat gold border, and sent it away to one foundation after another, hoping for support.

The Smithsonian Institution was almost the last address on his list. After filing away the collection of polite refusals he had received from the others, it was with some hesitation he sent the document forth once again. This time, after an interval of about three weeks, he received a letter from Dr. Charles D. Walcott, then secretary of the Smithsonian, commending him on his report,[1] and asking how much would be needed to continue the work. Goddard debated between asking a safe $2,500, which he felt would be inadequate, and $10,000, which perhaps would be enough but might be refused. Finally he compromised on $5,000—and by return mail received a warm letter granting his request. Folded with the letter was an advance of $1,000: the largest check he had ever seen.

Then began the series of experiments which were to launch modern rocketry and perhaps to change the world's history. Almost nobody except those immediately engaged knew what these experiments were until the first Monday morning in January, 1920, when the Smithsonian Institution issued a news release on the work and simultaneously published the Worcester scientist's first paper on rockets: a modest sixty-nine-page monograph bound in brown paper, entitled "A Method of Reaching Extreme Altitudes." It was issued as a part of the Smithsonian Miscellaneous Col-

[1] Though the report is dated May 26, 1919, it was not actually released to the public until January, 1920.

lections, Volume 71, No. 2. Exactly 1,750 copies were published. This classical treatise, which marked the beginning of an era, is no longer obtainable separately, but has been reprinted in *Rockets,* a book of Goddard material, by The American Rocket Society.

The Smithsonian paper was basically the same report as that submitted to the Institution in 1916. Goddard's experiments simply corroborated his earlier conclusions; only the factual data based on his post-1916 tests needed revision. As to the cost of the work, it had come to something over $11,000. Beginning with the original $5,000, the Smithsonian had put up all of it.

Goddard divided his initial paper into three parts, dealing respectively with the theory of rockets, the experiments he made before 1919 (some of them at Worcester Polytechnic Institute and the others at Mount Wilson Observatory during the first World War) and the calculations derived from theory and experiment.

It is these calculations which are today the most interesting. Among other things he figured out the "initial masses" of a series of theoretical rockets which would be powerful enough to lift a payload of one pound to various heights in the atmosphere. Making some assumptions as to the jet velocities attainable, and the efficiency of his reaction motors, he concluded that to raise one final pound of rocket to a height of 35 miles would require a starting weight of only 3.66 pounds, provided an "effective velocity" of 7,000 feet per second could be obtained.

To show what really great altitudes could be obtained with a relatively small starting weight, continuing to assume an "effective velocity" of 7,000 feet per second, he showed that an altitude of 71 miles would be reached with a total starting weight of 5.14 pounds; 115 miles with a starting weight of 6.40 pounds, and 437 miles with an initial mass of only 12.33 pounds.

As a way to construct rockets with a suitable ratio of propellant to structure to permit of such high altitudes, Goddard went on to give a clear explanation of the theory of the step-rocket, a device upon which he had himself received the basic patent nearly six years earlier, on July 7, 1914. A step-rocket, of course, is a multiple rocket, in which the larger part is consumed, to give high velocity to the smaller part.

He also disclosed that he had put to rest the old fallacy that a rocket thrusts by "pushing against the air." He had shot rocket motors in partial vacuum, and obtained results equal to or even better than at atmospheric pressure. Likewise, he had tackled the problem of constructing dry-fuel motors separate from the powder charge, and had developed successful intermittent motors in which the propellants were inserted by a device working on the general principle of the machine gun.

He ended the report with what was, for 1919, a startling conclusion:

that in theory at least, it should be perfectly possible to shoot a rocket at such velocity that it would not return to the earth. With a total launching weight of only eight or ten tons, he estimated, a rocket could be constructed capable of carrying enough magnesium powder to create a telescopically visible flash against the dark side of the moon.

It was this discussion of a possible moon-rocket, rather than the less spectacular but more practical work reflected by the rest of the book, that most forcefully reached the public. Newspaper readers across the continent were moved to excitement, comment and derision. The rocket came forth once more out of the history books and military museums and began to have the beginnings of a new world prominence—only this time with an ironic twist. During the period when for the first time it was really undergoing something like genuine scientific development, the rocket was to become, to many unthinking people, a symbol of impractical ideas and fantastic schemes. Everyone who had to do with rockets during the next two decades was to be branded as "queer"; and rocketors were to inherit the mantle of ridicule previously worn by airplane pioneers.

After the publication of "A Method of Reaching Extreme Altitudes," which dealt optimistically with dry fuel propellants, Goddard came to the conclusion that despite the convenience of these fuels they could not bring about the results he had in mind. Accordingly he gave them up and turned his attention to the problems of developing liquid-fuel rockets.

From 1920 until 1922 he made what are now known as proving stand tests with liquid-fuel motors, trying liquid oxygen and various liquid hydrocarbons, including gasoline, liquefied propane and ether. He presently decided that liquid oxygen and gasoline made the most practical combination; virtually all of his subsequent liquid-fuel experiments were carried out with these liquids.

By 1923 Goddard had reached the point of trying an actual shot with a liquid-fuel rocket. Still working with funds supplied by the Smithsonian Institution, he constructed a small rocket in which the fuels were fed to the motor through pumps. This contrivance was tried on the proving stand, but was not released for flight.

A year later Goddard finished his third liquid-fuel rocket, and on March 16, 1926, let it fly. This was the first actual shot of a liquid-fuel rocket anywhere in the world. It occurred at Auburn, Massachusetts, on a cold, clear spring day.

The only witnesses to that historic flight other than Dr. Goddard himself were Henry Sachs, machinist and instrument maker of the Clark University shop; Dr. P. M. Roope, assistant professor of physics at Clark; and Mrs. Goddard, who came along to take the pictures which later documented the report.

The rocket was an odd and fragile-looking contrivance. The motor, with its metal nozzle nearly as long as the cylindrical blast chamber, was mounted at the forward part of the rocket in a slender frame consisting of the fuel pipes, crossed by a bracing strut engaging the nozzle. The whole rocket was about ten feet tall; the motor measured over all about two feet, and fuel tanks were about two and a half feet long.

Goddard's 1926 rocket took off with a loud roar, rose in a high trajectory, and flew for two and a half seconds, traveling a distance of 184 feet. Timing it with a stop watch, he calculated later that its average speed was 60 miles an hour.

Following this first shot, other short flights with liquid-fuel rockets were made at Auburn; all very quietly, with such elaborate secrecy that nobody but the experimenter and his immediate circle knew what was going on. But rocket experimentation is hard to keep a secret; the rocket itself is a mighty self-advertiser. On July 17, 1929, Goddard shot a rocket of some size, big enough to carry a small barometer and a camera. It made noise in proportion. Someone who witnessed the flight from a distance mistook the rocket for a burning airplane and notified the police and fire departments.

Goddard was on the front pages of the newspapers again the next day: for the first time, virtually, since the publication of his "A Method of Reaching Extreme Altitudes." This time the uses of publicity were to be proved to him in a most pleasant and exciting way. Colonel Charles A. Lindbergh, then at the height of his popularity, read about Goddard's experiments and became interested. He communicated his interest to Daniel Guggenheim, who promptly made a grant of funds to put Goddard's work on a considerably more adequate financial basis.

Goddard meanwhile had moved his experiments to more secluded quarters at Fort Devens, Massachusetts, on ground placed at his disposal by the United States Army. Here, though he was free from unwelcome visitors, the work was going slowly, principally because of the difficulties of transportation.

When word came of the Guggenheim grant, the first job was to select a suitable site for the experiments. He chose New Mexico, it being a country "of clear air, few storms, moderate winds and level terrain." The final decision fell upon the vicinity of Roswell, in the south, where there were good power and transportation facilities. The actual site was the Mescalero Ranch, where a shop was erected in September, 1930, large enough for himself and four assistants. A small tower 20 feet high was then built near the shop for static tests, equipped with heavy weights to keep the test rockets from rising out of the tower. A 60-foot launching tower previously used at Auburn and Fort Devens was put up about 15 miles away, for flight tests.

The first project was to develop a standard rocket motor which would deliver dependable power. The motor finally produced was 5¾ inches in diameter and weighed 5 pounds. Its maximum thrust was 289 pounds and it could burn 20 seconds or more. Goddard later stated that this motor produced the equivalent of 1,030 horsepower, or 206 horsepower per pound, the horsepower being calculated from the energy of motion of the blast, with the rocket held at rest.

On December 30, 1930, the first flight of a rocket at the New Mexico site took place. The rocket was 11 feet long, and weighed a little over 33 pounds. It reached an altitude of 2,000 feet, and a maximum speed of 500 miles an hour.

It was big, but it set no altitude record, for as Goddard later pointed out, his first objective was to produce a dependable rocket, not to see how high he could shoot. To this end he also began studying the problem of stabilizing the flight, for he had become convinced that it was impossible for a rocket to fly even straight up without some sort of controlling device.

The first flight of a gyroscopically controlled rocket was made on April 19, 1932. In this rocket the steering vanes were forced into the blast of the rocket motor by gas pressure—the pressure, and therefore the amount of steering, being controlled by a small gyroscope. The scheme showed some signs of working, but the test was hardly a complete success. Goddard concluded that the vanes used in the experimental model were too small.

When the original Guggenheim grant was made, it had been agreed to undertake the work in New Mexico for two years; then study the results with a view to a two-year extension. To supervise things, a committee had been named, headed by Dr. John C. Merriam, then president of the Carnegie Institution.

This committee gravely studied Goddard's reports, and recommended the granting of funds for the two additional years. But the great depression was then on. Goddard went back to Clark University to resume his teaching. The Smithsonian Institution, loath to see the research come to an end, made a small grant to permit some laboratory tests that did not require rocket flights. In the following year the Daniel and Florence Guggenheim Foundation came to the rescue, and work was resumed in New Mexico in 1934.

The job, now, as he saw it, was to develop fully stabilized flight. In the beginning he tried a stabilizer operated by a pendulum. Some stabilization effect can be obtained by this means, but as the rocket's acceleration increases, the pendulum becomes less effective, because the direction of the pendulum is inevitably a resultant between the acceleration of gravity and the acceleration of the rocket.

Goddard's pendulum-controlled rocket rose about 1,000 feet, bellied over, flew horizontally for about two miles, and landed 11,000 feet from the launching tower. At one point the speed exceeded 700 miles an hour —or nearly the speed of sound.

Goddard next approached the stabilizer problem by returning to his first idea, a small gyroscope. With his gyro-control, a series of beautiful rocket shots were made, beginning March 8, 1935, when the gyro-rocket reached an altitude of 4,800 feet, flew a horizontal distance of 13,000 feet, and made a maximum speed of 550 miles an hour. The gyroscope was set to correct the flight when it deviated 10 degrees or more from the vertical. Since this permitted a considerable lag, especially at the beginning, the first few hundred feet of flight looked like a fish swimming gracefully upward into the air. As the rocket picked up speed, the oscillations became much smaller, then virtually disappeared.

The equipment was gradually improved. A notable gyro-controlled flight was made on October 14, 1935, when the rocket rose 4,000 feet. On May 31, 1935, a gyro-rocket reached an altitude of 7,500 feet, or nearly a mile and a half. As in the previous experiments, Goddard was not attempting to set altitude marks in these shots, but was still concentrating on the complicated task of developing the apparatus to a state of reliable performance. His gyro-rockets weighed from 58 to 85 pounds at starting, and some were 10 to 15 feet in length.

Goddard concluded his last published report, in 1936, with the remark that the next step in the development of liquid-propellant rockets would be the reduction of weight to a minimum, a natural prelude to really high altitude shots.

"Some progress along this line," he dryly remarked, "has already been made." The exact nature of his progress was not then disclosed, but it is now known to have included the development of high-speed, high-efficiency liquid-fuel pumps, in which, as in so many other matters, he anticipated the much more highly advertised accomplishments of the German rocket engineers.

Every single important development made in rockets during the war, including the basic design of long-range rockets such as the V-2, had been worked out before 1940 by Goddard, and was available to military men in this country.

Goddard's death, on August 10, 1945, unfortunately brought to an untimely end his plans for further experimentation. Nevertheless, he lived to see at least part of his life's dream become reality. Jet propulsion, at least for the uses of war, matured in his lifetime from a fantastic notion into a billion-dollar industry.

ARTHUR C. CLARKE

A Shot Around the World

THE LAWS governing the behavior of satellites are basically simple, but because they involve conditions outside the range of normal human experience many people find it hard to grasp them. Perhaps if we lived on a planet which, like Jupiter, had a dozen natural moons in the sky, the idea of creating a few additional ones would not seem so surprising.

It is "common sense" that it needs force to keep a body in motion, and that if this propulsive force—whatever its origin—is removed the body will sooner or later come to rest. Any kind of vehicle, whether it travels on the sea, the land or in the air, obeys this fundamental law; no one expects an automobile to keep moving indefinitely once the motor is turned off.

Yet common sense—the accumulated experience of all the generations of men—is often a misleading guide. It is so in this case, and it required the genius of Galileo and Sir Isaac Newton to point out the true laws of motion, which here on Earth are disguised and distorted by the confusing effects of friction. The scientific outlook reverses the common sense opinion; it needs *no* force to keep a body moving. The planets circle in their orbits forever, without any help from the angels whom medieval theologians pictured as busily pushing them along.

The development of high-altitude flight has made it common knowledge that great speeds can easily be reached and maintained in the thin air of the upper atmosphere, where frictional drag is reduced to a small fraction of its value at sea level. It requires little effort of imagination to see that where there is no atmosphere at all, the last trace of resistance will also vanish. Once a body in space has been given a certain speed, it can maintain that speed indefinitely.

The only natural force that can affect the motion of a body in space, once it is clear of the atmosphere, is gravity. It is gravity that keeps the

Moon chained in its orbit—and it is gravity that will control the new moons we are building now.

Now one knows what gravity is: we can merely describe its actions, but that is all that we need do for our present purpose. The movement of bodies in gravitational fields has been the subject of intense study ever since the invention of gunpowder gave birth to the applied (or misapplied) science of ballistics. In a sense, a satellite is no more than a missile of infinite range.

Why this is so may be seen from a very simple argument; the fact that it has become almost a cliché in books on space travel will not deter us from using it again. Imagine a gun, pointed horizontally, standing on the edge of a very high cliff, and consider what will happen if the shell is fired at varying speeds. For the moment, assume that there is no such thing as air resistance.

If it merely trickles out of the barrel (*vide* one of early scenes in Chaplin's *The Great Dictator*) the shell will fall vertically and will hit the bottom of the cliff after a time given by the simple equation $s=\frac{1}{2}gt^2$—which is probably all that most people remember of dynamics after they have left school. It would take just one minute, for example, for the shot to reach the base of a cliff eleven miles high, if such a phenomenon can be imagined.

Now this time of fall is quite independent of the velocity with which the shell is projected, as long as it leaves the gun horizontally. In the very hypothetical case we have assumed, the missile will always reach the plain beneath it one minute after firing, however fast it leaves the gun. However, the distance of the point of impact from the base of the cliff will increase steadily with the speed of projection. If the shell leaves the gun at a mile a second—sixty miles a minute—it will hit the ground sixty miles away at the end of its one-minute fall, and so on.

However, we have ignored one fundamental point. The Earth isn't flat,[1] and as we increase the speed of projection of the missile its range will obviously lengthen more rapidly than simple arithmetic would indicate. At two miles a second, it will hit the ground considerably more than 120 miles away; at four miles a second, much more than 240 miles away, because of the curvature of the Earth's surface.

It is obvious from the diagram what is going to happen next. There will be a certain speed at which the projectile never reaches the Earth at all, because the surface drops away at the same rate as the descending shell. When this happens, the missile will be in a state of permanent fall around the world, never getting any closer to the surface.

One can get a rough idea of the speed necessary to achieve this state of affairs by the simple geometry of Figure 1. After its one-minute "fall,"

[1] I am waiting, with considerable interest, for the reactions to the satellite program of the people who still believe that it is.

the projectile must be just as far from the ground as when it started—i.e., it must still be eleven miles up. The problem may therefore be stated in this form: "How far must one travel before the Earth's surface drops away eleven miles below the horizon?"

This is merely the reverse way of looking at a familiar question, "How far away is the visible horizon from a height of eleven miles?" The answer is approximately three hundred miles, which, therefore, gives us the dis-

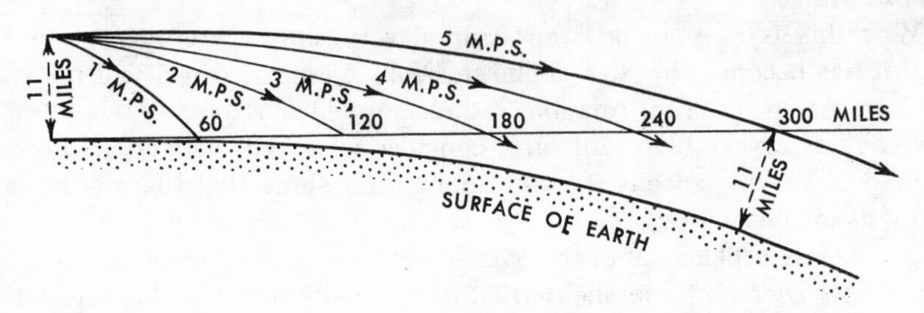

Fig. 1. Distances covered by a projectile over (a) flat and (b) curved Earth.

tance the projectile must cover in one minute. So its launching speed has to be three hundred miles a minute, or five miles a second.

In actual practice, of course, no satellite could be established at such a low altitude, because air resistance would bring it to rest in a few seconds. Even a hundred miles up, a satellite might not be able to complete its 24,000-mile circuit of the globe before being slowed down so seriously that it would fall back to Earth. But from about 150 miles onward, semi-permanent orbits are possible—that is, a satellite would stay up for days or weeks before the faint traces of air resistance brought it spiraling down.

The critical speed of five miles a second—eighteen thousand miles an hour—is known as "circular" or "orbital" velocity, and it is important to realize that it is quite independent of the size or mass of the satellite concerned. If the Earth had a natural moon just outside the atmosphere, this is the speed at which it would have to travel, and since at this speed it takes an hour and a half to go around the world, this would be the duration of the "month."

A permanent satellite can be established at any altitude, as long as it is high enough to avoid atmospheric drag; there is nothing particularly sacrosanct about the eighteen-thousand-miles-per-hour, ninety-minute orbit—except that, being the closest to the Earth, it is the easiest to achieve. At greater distances from the Earth, where the downward pull of gravity begins to weaken, less speed is required to maintain a satellite in its orbit, and the period of revolution is correspondingly increased. For example, 1,075 miles up, the period becomes exactly two hours—one-twelfth of a day—a convenient fraction which may make this particular distance of special importance.

There has been a great deal of popular confusion about the role which the Earth's gravitational field plays in connection with satellites. Indeed, there have been newspaper articles which stated that such satellites maintained themselves in their orbits because they were "beyond the Earth's gravity." This, of course, is utter nonsense; although gravity weakens with increasing distance from the Earth, at the heights where the first satellites will be established it still has 90 per cent of its sea-level value. Without the pull of gravity, in fact, the satellites would at once fly off into space, since there would be no downward force checking their natural tendency to continue moving in a straight line.

At this point it is impossible to avoid another astronautical cliché—the stone-and-string analogy. If you tie a stone on the end of a piece of string, it requires very little effort to keep it whirling around in a circle. The system is in a state of balance because the tension of the string counteracts the tendency of the stone to fly away. One of man's earliest weapons, the sling, depends on this simple application of dynamical laws.

In the case of a satellite, the pull of gravity plays the part of the string. With this analogy in mind, it is obvious that the greater the radius of movement—the longer the string—the more slowly a satellite need move in order to maintain its equilibrium.

Close satellites will revolve around the Earth very much more rapidly than our planet turns on its axis, and as we shall see later this gives rise to some very odd effects. It is necessary to go out to the considerable altitude of 22,000 miles before the orbital period is as much as a day. Figure 2 shows in diagrammatic form the distances which satellites at various altitudes would cover in the same period—in this case, one hour—as compared with the arc through which the Earth itself revolves.

Although we have so far discussed only circular orbits, because these are the simplest and in some ways most useful ones, they really represent a rather exceptional case. In practice, the orbit of a real satellite would not be a perfect circle but an ellipse—perhaps a highly elongated one. To establish an exactly circular orbit would demand an impossible and indeed unnecessary degree of accuracy on the part of the launching rocket.

Circles exist only in geometry books; real planets, comets and satellites (artificial or natural) move round their center of attraction in elliptical paths—a fact which gives them a kind of inherent stability. If a satellite is traveling too slowly at some part of its orbit it will fall inward and thus pick up speed. Conversely, if it were going too fast, it would swing outward.

Thus any error—within limits—in the initial launching speed would not prevent the establishment of a satellite. The orbit would automatically adjust its shape to take up the slack, as it were. If the launching speed were too low, at the other end of its orbit the satellite would drop down toward Earth, but that would not matter as long as it remained clear of the atmosphere.

The actual shape of a satellite's orbit is only one of its characteristics;

the plane in which it lies is another. From the practical point of view, it is easiest to launch a satellite into the Equatorial plane, because when it takes off it can get the maximum boost from the thousand-mile-an-hour spin of the Earth at the Equator. That is, assuming that it is launched in the west-east direction; if fuel economy were no problem, it would be possible to

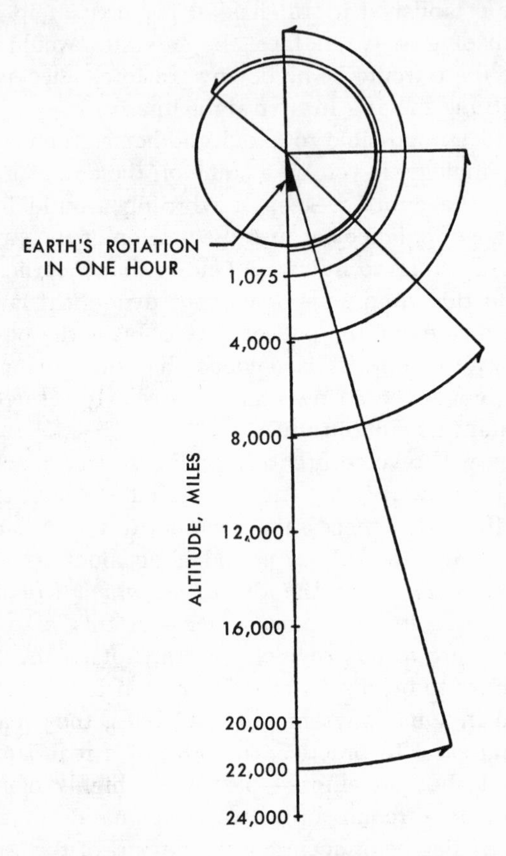

Fig. 2. Arcs covered in one hour by satellites at varying distances from the Earth.

establish a satellite moving from east to west, against the direction of the Earth's spin. Such an orbit would be termed a "retrograde" one, but there would be no particular purpose in trying to achieve it since it would require an extra two thousand miles an hour of rocket impulse over the "direct" orbit.

Of very great scientific interest is the orbit which passes over the Poles. The Polar and Equatorial orbits represent two extreme cases, but orbits at intermediate angles are equally possible. In fact, the plane of a satellite can lie at *any* angle in space; the only restriction is that it must pass through the center of the Earth. A few of the more interesting possible satellite orbits are shown in Figure 3. For simplicity, only circular orbits are shown.

If our planet were not turning on its axis, any satellite would continually retrace the same path above the Earth's surface, and after completing one revolution would be back over the same spot from which it had started. However, the real situation is not as simple as that; only a satellite above

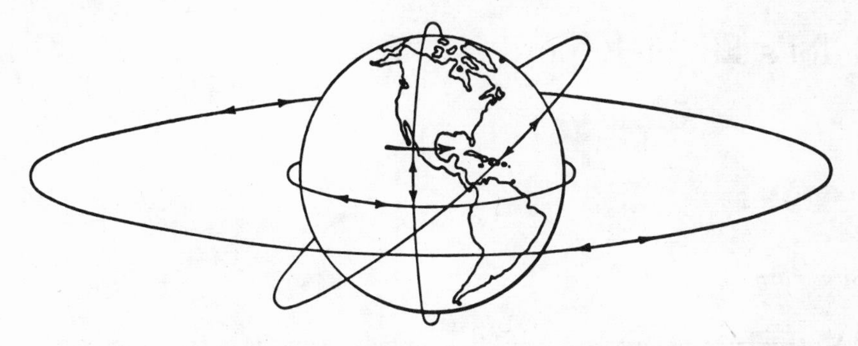

Fig. 3. Some possible satellite orbits.

the Equator would forever trace out the same ground track on the globe below. A satellite in an orbit that passed over the Poles would weave out a pattern embracing the entire Earth as it shuttled from north to south and the planet turned beneath it. This kind of behavior is both an advantage and a disadvantage; it means that a satellite can reconnoiter the whole globe several times during a single day—but conversely it will be visible from any given ground station only for a few minutes in each twenty-four hours.

JOHN H. GLENN, JR.

Pilot's Flight Report

•

Summary

WEIGHTLESS FLIGHT was quickly adapted to, and was found to be pleasant and without discomfort. The chances of mission success are greatly enhanced by the presence of a human crew in the spacecraft. A human crew is vital to future space missions for the purpose of intelligent observation and actions when the spacecraft encounters expected or unexpected occurrences or phenomena.

Introduction

The test objectives for the MA-6 mission of Friendship 7, as quoted from the Mission Directive, were as follows:

(1) Evaluate the performance of a man-spacecraft system in a three-orbit mission

(2) Evaluate the effects of space flight on the astronaut

(3) Obtain the astronaut's opinions on the operational suitability of the spacecraft and supporting systems for manned space flight

My report is concerned mainly with those items in all three objectives where man's observation capabilities provided information not attained by other means. It is in this type of reporting that a manned vehicle provides a great advantage over an unmanned vehicle, which is often deaf and blind to the new and the unexpected. My report, then, will stress what I heard, saw, and felt during the orbital flight.

Preparation and Countdown

Preparation, transfer to the launch pad, and insertion into the spacecraft went as planned. The technicians and I had been through the entry to the spacecraft many times.

As with every countdown, short delays were encountered when problems arose. The support for the microphone in the helmet, an item that had been moved and adjusted literally thousands of times, broke and had to be replaced. While the spacecraft hatch was being secured, a bolt was broken and had to be repaired. During this time I was busy going over my checklist and monitoring the spacecraft instruments.

Many people were concerned about my mental state during this and earlier delays, which are a part of preparation for a manned space flight. People have repeatedly asked whether I was afraid before the mission. Humans always have fear of an unknown situation—this is normal. The important thing is what we do about it. If fear is permitted to become a paralyzing thing that interferes with proper action, then it is harmful. The best antidote to fear is to know all we can about a situation. It is lack of knowledge which often misleads people when they try to imagine the feelings of an astronaut about to launch. During the years of preparation for Project Mercury, the unknown areas have been shrunk, we feel, to an acceptable level. For those who have not had the advantage of this training, the unknowns appear huge and insurmountable, and the level of confidence of the uninformed is lowered by an appropriate amount.

All the members of the Mercury team have been working toward this space flight opportunity for a long time. We have not dreaded it; we have looked forward to it. After three years we cannot be unduly concerned by a few delays. The important consideration is that everything be ready, that nothing be jeopardized by haste which can be preserved by prudent action.

The initial unusual experience of the mission is that of being on top of the Atlas launch vehicle after the gantry has been pulled back. Through the periscope, much of Cape Canaveral can be seen. If you move back and forth in the couch, you can feel the entire vehicle moving very slightly. When the engines are gimbaled, you can feel the vibration. When the tank is filled with liquid oxygen, the spacecraft vibrates and shudders as the metal skin flexes. Through the window and periscope the white plume of the lox (liquid oxygen) venting is visible.

Launch

When the countdown reached zero, I could feel the engines start. The spacecraft shook, not violently but very solidly. There was no doubt when lift-off occurred. When the Atlas was released there was an immediate gentle surge that let you know you were on your way. The roll to the correct azimuth was noticeable after lift-off. I had preset the little window mirror to watch the ground. I glanced up after lift-off and could see the horizon turning. Some vibration occurred immediately after lift-off. It smoothed out after about 10 to 15 seconds of flight but never completely stopped. There was still a noticeable amount of vibration that continued up to the time the spacecraft passed through the maximum aerodynamic pressure or maximum q, at approximately T + 1 minute. The approach of maximum q is signaled by more intense vibrations. Force on the outside of the spacecraft was calculated at 982 pounds per square foot at this time. During this period, I was conscious of a dull muffled roar from the engines. Beyond the high q area the vibration smoothed out noticeably. However, the spacecraft never become completely vibration free during powered flight.

The acceleration buildup was noticeable but not bothersome. Before the flight my backup pilot, Astronaut Scott Carpenter, had said he thought it would feel good to go in a straight-line acceleration rather than just in circles as we had in the centrifuge and he was right. Booster engine cut-off occurred at 2 minutes 9.6 seconds after lift-off. As the two outboard engines shut down and were detached, the acceleration dropped but not as sharply as I had anticipated. Instead, it decayed over approximately ½ second. There is a change in noise level and vibration when these engines are jettisoned. I saw a flash of smoke out the window and thought at first that the escape tower had jettisoned early and so reported. However, this flash was apparently deflected smoke coming up around the spacecraft from the booster engines which had just separated. The tower was jettisoned at 2 minutes, 33.3 seconds, and I corrected my earlier report. I was ready to back up the automatic sequencing system if it did not perform correctly and counted down the seconds to the time for tower jettisoning. I was looking at the nozzles of the tower rockets when they fired. A large cloud of smoke came out but little flame. The tower accelerated rapidly from the spacecraft in a straight line. I watched it to a distance of approximately ½ mile. The spacecraft was programmed to pitch down slowly just prior to jettisoning the tower and this maneuver provided my first real view of the horizon and clouds. I could just see clouds and the horizon behind the tower as it jettisoned.

After the tower fired, the spacecraft pitched slowly up again and I lost sight of the horizon. I remember making a comment at about this time that the sky was very black. The acceleration built up again, but as before, acceleration was not a major problem. I could communicate well, up to the maximum of 7.7g at insertion when the sustainer-engine thrust terminates.

Just before the end of powered flight, there was one experience I was not expecting. At this time the fuel and lox tanks were getting empty and apparently the Atlas becomes considerably more flexible than when filled. I had the sensation of being out on the end of a springboard and could feel oscillating motions as if the nose of the launch vehicle were waving back and forth slightly.

Insertion into Orbit

The noise also increased as the vehicle approached SECO (sustainer engine cutoff). When the sustainer engine cut off at 5 minutes, 1.4 seconds and the acceleration dropped to zero, I had a slight sensation of tumbling forward. The astronauts have often had a similar sensation during training on the centrifuge. The sensation was much less during the flight, and since the spacecraft did pitch down at this point it may have been a result of actual movement rather than an illusion.

There was no doubt when the clamp ring between the Atlas and the Mercury spacecraft fired. There was a loud report and I immediately felt the force of the posigrade rockets which separate the spacecraft from the launch vehicle. Prior to the flight I had imagined that the acceleration from these three small rockets would be insignificant and that we might fail to sense them entirely, but there is no doubt when they fire.

Immediately after separation from the Atlas, the autopilot started to turn the spacecraft around. As the spacecraft came around to its normal aft-viewing attitude, I could see the Atlas through the window. At the time I estimated that it was "a couple of hundred yards away." After the flight an analysis of the trajectory data showed that the distance between the launch vehicle and the spacecraft should, at this point, be 600 feet. Close enough for a rough estimate. I do not claim that I can normally judge distance so close. There was a large-sized luck factor in the estimate; nevertheless, the facts do give an indication that man can make an adequate judgment at least of short distances to a known object in space. This capability will be important in future missions in which man will want to achieve rendezvous, since the pilot will be counted on to perform the final closing maneuver.

I was able to keep the Atlas in sight for 6 or 7 minutes while it traveled across the Atlantic. The last time I reported seeing it the Atlas was approximately 2 miles behind and 1 mile below the spacecraft. It could be seen easily as a bright object against the black background of space and later against the background of earth.

Orbit

The autopilot turned the spacecraft around and put it into the proper attitude. After my initial contact with Bermuda I received the times for firing the retrorockets and started the check of the controls. This is a test of the control systems aboard the spacecraft. I had practiced it many times on the ground in the Mercury procedures trainer and the test went just as it had in the trainer. I was elated by the precision with which the test progressed. It is quite an intricate check. With your right hand you move the control stick, operating the hydrogen peroxide thrusters to move the spacecraft in roll, pitch, and yaw. With your left hand you switch from one control system to another as the spacecraft is manually controlled to a number of precise rates and attitudes.

This experience was the first time I had been in complete manual control, and it was very reassuring to see not only the spacecraft react as expected, but also to see that my own ability to control was as we had hoped.

Following this controls check I went back to autopilot control and the spacecraft operated properly on autopilot throughout the first orbit.

Thruster Problem

Because of a malfunction in a low-torque thruster at the end of the first orbit, it was necessary to control the spacecraft manually for the last two orbits. This requirement introduced no serious problems, and actually provided me with an opportunity to demonstrate what a man can do in controlling a spacecraft. However, it limited the time that could be spent on many of the experiments I had hoped to carry out during the flight.

Flight Plan

The Mercury flight plan during the first orbit was to maintain optimum spacecraft attitude for radar tracking and communication checks. This plan would provide good trajectory information as early as possible and would give me a chance to adapt to these new conditions if such was necessary. Other observations and tasks were to be accomplished mainly on the second and third orbits. Since the thruster problem made it necessary for me to control manually during most of the second and third orbits, several of the planned observations and experiments were not accomplished.

Attitude Reference

A number of questions have been raised over the ability of a man to use the earth's horizon as a reference for controlling the attitude of the space vehicle.

Throughout this flight no trouble in seeing the horizon was encountered. During the day the earth is bright and the background of space is dark. The horizon is vividly marked. At night, before the moon is up, the horizon can still be seen against the background of stars. After the moon rises (during this flight the moon was full), the earth is well enough lighted so that the horizon can be clearly seen.

With this horizon as a reference, the pitch and roll attitudes of the spacecraft can easily be controlled. The window can be positioned where you want it. Yaw, or heading reference, however, is not so good. I believe that there was a learning period during the flight regarding my ability to determine yaw. Use of the view through the window and periscope gradually improved.

To determine yaw in the spacecraft, advantage must be taken of the speed of the spacecraft over the earth which produces an apparent drift of the ground below the spacecraft. When the spacecraft is properly oriented, facing along the plane of the orbit, the ground appears to move parallel to the spacecraft longitudinal axis. During the flight I developed a procedure which seemed to help me use this terrain drift as a yaw reference. I would pitch the small end of the spacecraft down to about −60° from the normal attitude where a fairly good vertical view was available. In this attitude, clouds and land moving out from under me had more apparent motion than when the spacecraft was in its normal orbit attitude and I looked off toward the horizon.

At night with the full moon illuminating the clouds below, I could still determine yaw through the window but not as rapidly as in the daytime. At night I could also use the drift of the stars to determine heading but this took longer and was less accurate.

Throughout the flight I preferred the window to the periscope as an attitude reference system. It seemed to take longer to adjust yaw by using the periscope on the day side. At night, the cloud illumination by the moon is too dim to be seen well through the periscope.

Three times during the flight I turned the spacecraft approximately 180° in yaw and faced forward in the direction of flight. I liked this attitude—seeing where I was going rather than where I had been—much better. As a result of these maneuvers my instrument reference system gave me an inaccurate attitude indication. It was easy to determine the proper attitude, however, from reference to the horizon through the window or the periscope. Maintaining orientation was no problem, but I believe that the pilot automatically relies much more completely on vision in space than he does in an airplane, where gravity cues are available. The success with which I was able to control the spacecraft at all times was, to me, one of the most significant features of the flight.

Weightlessness

Weightlessness was a pleasant experience. I reported I felt fine as soon as the spacecraft separated from the launch vehicle, and throughout the flight this feeling continued to be the same.

Approximately every 30 minutes throughout the flight I went through a series of exercises to determine whether weightlessness was affecting me in any way. To see if head movement in a zero g environment produced any symptoms of nausea or vertigo, I tried first moving, then shaking my head from side to side, up and down, and tilting it from shoulder to shoulder. In other words, moving my head in roll, pitch, and yaw. I began slowly, but as the flight progressed, I moved my head more rapidly and vigorously until at the end of the flight I was moving as rapidly as my pressure suit would allow.

In another test, using only eye motions, I tracked a rapidly moving spot of light generated by my finger-tip lights. I had no problem watching the spot and once again no sensations of dizziness or nausea. A small eye chart was included on the instrument panel, with letters of varying size and with a "spoked wheel" pattern to check both general vision and any tendency toward astigmatism. No change from normal was apparent.

An "oculogyric test" was made in which turning rates of the spacecraft were correlated with sensations and eye movements. Results were normal. Preflight experience in this test and a calibration had been made at the Naval School of Aviation Medicine, Pensacola, Fla., with Dr. Ashton Graybiel, so that I was thoroughly familiar with my reactions to these same movements at 1 g.

To provide medical data on the cardiovascular system, at intervals, I did an exercise which consisted of pulling on a bungee cord once a second for 30 seconds. This exercise provided a known workload to compare with previous similar tests made on the ground. The flight surgeons have reported the effect that this had on my pulse and blood pressure. The effect that it had on me during the flight was the same effect that it had on the ground—it made me tired.

Another experiment related to the possible medical effects of weightlessness was eating in orbit. On the relatively short flight of Friendship 7, eating was not a necessity, but rather an attempt to determine whether there would be any problem in consuming and digesting food in a weightless state. At no time did I have any difficulty eating. I believe that any type of food can be eaten as long as it does not come apart easily or make crumbs. Prior to the flight, we joked about taking along some normal food such as a ham sandwich. I think this would be practical and should be tried.

Sitting in the spacecraft under zero g is more pleasant than under 1 g on the ground, since you are not subject to any pressure points. I felt that I adapted very rapidly to weightlessness. I had no tendency to overreach nor did I experience any other sign of lack of coordination, even on the first movements after separation. I found myself unconsciously taking advantage of the weightless condition, as when I would leave a camera or some other object floating in space while I attended to other matters. This was not done as a preplanned maneuver but as a spur-of-the-moment thing when another system needed my attention. I thought later about how I had done this as naturally as if I were laying the camera on a table in a 1 g field. It pointedly illustrates how rapidly adaptable the human is, even to something as foreign as weightlessness.

We discovered from this flight that some problems are still to be solved in properly determining how to stow and secure equipment that is used in a space vehicle. I had brought along a number of instruments, such as cameras, binoculars, and a photometer, with which to make observations from the spacecraft. All of these were stowed in a ditty bag by my right arm. Each piece of equipment had a 3-foot piece of line attached to it. By the time I had started using items of the equipment, these lines became tangled. Although these lines got in the way, it was still important to have some way of securing the equipment, as I found out when I attempted to change film. The small canisters of film were not tied to the ditty bag by lines. I left one floating in midair while working with the camera, and when I reached for it, I accidentally hit it and it floated out of sight behind the instrument panel.

Color and Light

As I looked back at the earth from space, colors and light intensities were much the same as I had observed when flying at high altitude in an airplane. The colors observed when looking down at the ground appeared similar to those seen from 50,000 feet. When looking toward the horizon, however, the view is completely different, for then the blackness of space contrasts vividly with the brightness of the earth. The horizon itself is a brilliant, brilliant blue and white.

It was surprising how much of the earth's surface was covered by clouds. The clouds can be seen very clearly on the daylight side. The different types of clouds—vertical developments, stratus clouds, and cumulus clouds—are readily distinguished. There is little problem identifying them or in seeing the weather patterns. You can estimate the relative heights of the cloud layers from your knowledge of the types or from the shadows the high clouds cast on those lower down. These observations are representative of information which the scientists of the U.S. Weather Bureau Meteorological Satellite Laboratory had asked Project Mercury to determine. They are interested in improving the optical equipment in their Tiros and Nimbus satellites and would like to know if they could determine the altitude of cloud layers with better optical resolution. From my flight I would say it is quite possible to determine cloud heights from this orbital altitude.

Only a few land areas were visible during the flight because of the cloud cover. Clouds were over much of the Atlantic, but the western (Sahara Desert) part of Africa was clear. In this desert region I could plainly see dust storms. By the time I got to the east coast of Africa where I might have been able to see towns, the land was covered by clouds. The Indian Ocean was the same.

Western Australia was clear, but the eastern half was overcast. Most of the area across Mexico and nearly to New Orleans was covered with high cirrus clouds. As I came across the United States I could see New Orleans, Charleston, and Savannah very clearly. I could also see rivers and lakes. I think the best view I had of any land area during the flight was the clear desert region around El Paso on the second pass across the United States. I could see the colors of the desert and the irrigated area north of El Paso. As I passed off the east coast of the United States I could see across Florida and far back along the Gulf Coast.

Over the Atlantic I saw what I assume was the Gulf Stream. The different colors of the water are clearly visible.

I also observed what was probably the wake of a ship. As I was passing over the recovery area at the end of the second orbit, I looked down at the water and saw a little "V." I checked the map. I was over recovery area G at the time, so I think it was probably the wake from a recovery ship. When I looked again the little "V" was under a cloud. The change in light reflections caused by the wake of a ship are sometimes visible for long distances from an airplane and will linger for miles behind a ship. This wake was probably what was visible.

I believe, however, that most people have an erroneous conception that from orbital altitude, little detail can be seen. In clear desert air, it is common to see a mountain range 100 or so miles away very clearly, and all that vision is through atmosphere. From orbital altitude, atmospheric light attenuation is only through approximately 100,000 feet of atmosphere so it is even more clear. An interesting experiment for future flights can be to determine visibility of objects of different sizes, colors, and shapes.

Obviously, on the night side of the earth, much less was visible. This may have been due not only to the reduced light, but also partly to the fact that I was never fully dark adapted. In the bright light of the full moon, the clouds are visible. I could see vertical development at night. Most of the cloudy areas, however, appeared to be stratoform.

The lights of the city of Perth, in Western Australia, were on and I could see them well. The view was similar to that seen when flying at high altitude at night over a small town. South of Perth there was a small group of lights, but they were much brighter in intensity. Inland there was a series of four or five towns lying in a line running from east to west. Knowing that Perth was on the coast, I was just barely able to see the coastline of Australia. Clouds covered the area of eastern Australia around Woomera, and I saw nothing but clouds from there across the Pacific until I was east of Hawaii. There appeared to be almost solid cloud cover all the way.

Just off the east coast of Africa were two large storm areas. Weather Bureau scientists had wondered whether lightning could be seen on the night side, and it certainly can. A large storm was visible just north of my track over the Indian Ocean and a smaller one to the south. Lightning could be seen flashing back and forth between the clouds but most prominent were lightning flashes within thunderheads illuminating them like light bulbs.

Some of the most spectacular sights during the flight were sunsets. The sunsets always occurred slightly to my left, and I turned the spacecraft to get a better view. The sunlight coming in the window was very brilliant, with an intense clear white light that reminded me of the arc lights while the spacecraft was on the launching pad.

I watched the first sunset through the photometer which had a polarizing filter on the front so that the intensity of the sun could be reduced to a comfortable level for viewing. Later I found that by squinting, I could look directly at the sun with no ill effects, just as I can from the surface of the earth. This accomplished little of value but does give an idea of intensity.

The sun is perfectly round as it approaches the horizon. It retains most of its symmetry until just the last sliver is visible. The horizon on each side of the sun is extremely bright, and when the sun has gone down to the level of this bright band of the horizon, it seems to spread out to each side of the point where it is setting. With the camera I caught the flattening of the sun just before it set. This is a phenomenon of some interest to the astronomers.

As the sun moves toward the horizon, a black shadow of darkness moves across the earth until the whole surface, except for the bright band at the horizon, is dark. This band is extremely bright just as the sun sets, but as time passes the bottom layer becomes a bright orange and fades into reds, then on into the darker colors, and finally off into the blues and blacks. One thing that surprised me was the distance the light extends on the horizon on each side of the point of the sunset. I think that the eye can see a little more of the sunset color band than the camera captures. One point of interest was the length of time during which the orbital twilight persisted. Light was visible along the horizon for 4 to 5 minutes after the sunset, a long time when you consider that sunset occurred 18 times faster than normal.

The period immediately following sunset was of special interest to the astronomers. Because of atmospheric light scattering, it is not possible to study the region close to the sun except at the time of a solar eclipse. It had been hoped that from above the atmosphere the area close to the sun could be observed. However, this would require a period of dark adaptation prior to sunset. An eye patch had been developed for this purpose, which was to be held in place by special tape. This patch was expected to permit one eye to be night adapted prior to sunset. Unfortunately, the tape proved unsatisfactory and I could not use the eyepatch. Observations of the sun's corona and zodiacal light must await future flights when the pilot may have an opportunity to get more fully dark adapted prior to sunset.

Another experiment suggested by our advisors in astronomy was to obtain ultraviolet spectrographs of the stars in the belt and sword of Orion. The ozone layer of the earth's atmosphere will not pass ultraviolet light below 3,000 angstroms. The spacecraft window will pass light down to 2,000 angstroms. It is possible, therefore, to get pictures of the stars from the Mercury spacecraft which cannot be duplicated by the largest telescopes on the ground. Several ultraviolet spectrographs were taken of the stars in the belt of Orion. They are being studied at the present time to see whether useful information was obtained.

The biggest surprise of the flight occurred at dawn. Coming out of the night on the first orbit, at the first glint of sunlight on the spacecraft, I was looking inside the spacecraft checking instruments for perhaps 15 to 20 seconds. When I glanced back through the window my initial reaction was that the spacecraft had tumbled and that I could see nothing but stars through the window. I realized, however, that I was still in the normal attitude. The spacecraft was surrounded by luminous particles.

These particles were a light yellowish green color. It was as if the spacecraft were moving through a field of fireflies. They were about the brightness of a first magnitude star and appeared to vary in size from a pinhead up to possibly ⅜ inch. They were about 8 to 10 feet apart and evenly distributed through the space around the spacecraft. Occasionally, one or two of them would move slowly up around the spacecraft and across the window, drifting very, very slowly, and would then gradually move off, back in the direction I was looking. I observed these luminous objects for approximately 4 minutes each time the sun came up.

During the third sunrise I turned the spacecraft around and faced forward to see if I could determine where the particles were coming from. Facing forward I could see only about 10 per cent as many particles as I had when my back was to the sun. Still, they seemed to be coming toward me from some distance so that they appeared not to be coming from the spacecraft. Just what these particles are is still subject to debate and awaits further clarification.

Other Planned Observations

As mentioned earlier, a number of other observations and measurements during orbit had to be canceled because of the control-system problems. Equipment carried was not highly sophisticated scientific equipment. We believed, however, that it would show the feasibility of making more comprehensive measurements on later missions.

Some of these areas of investigation that we planned but did *not* have an opportunity to check are as follows:

(a) Weather Bureau observations:

(1) Pictures of weather areas and cloud formations to match against map forecasts and Tiros pictures

(2) Filter mosaic pictures of major weather centers

(3) Observation of green air glow from air and weather centers in 5,577-angstrom band with air-glow filter

(4) Albedo intensities—measure reflected light intensities on both day and night side

(b) Astronomical observations:
  (1) Light polarization from area of sun
  (2) Comets close to sun
  (3) Zodiacal light
  (4) Sunlight intensity
  (5) Lunar clouds
  (6) Gegenschein
  (7) Starlight intensity measurements
(c) Test for otolith balance disturbance and autokynesis phenomena
(d) Vision tests:
  (1) Night vision adaptation
  (2) Photometer eye measurements
(e) Drinking

Re-entry

After having turned around on the last orbit to see the particles, I maneuvered into the correct attitude for firing the retrorockets and stowed the equipment in the ditty bag.

This last dawn found my attitude indicators still slightly in error. However, before it was time to fire the retrorockets the horizon-scanner slaving mechanism had brought the gyros back to orbit attitude. I cross-checked repeatedly between the instruments, periscope presentation, and the attitude through the window.

Although there were variations in the instrument presentations during the flight, there was never any difficulty in determining my true attitude by reference to the window or periscope. I received a countdown from the ground and the retrorockets were fired on schedule just off the California coast.

I could hear each rocket fire and could feel the surge as the rockets slowed the spacecraft. Coming out of zero-g condition, the retrorocket firing produced the sensation that I was accelerating back toward Hawaii. This sensation, of course, was an illusion.

Following retrofire the decision was made to have me re-enter with the retro package still on because of the uncertainty as to whether the landing bag had been extended. This decision required me to perform manually a number of the operations which are normally automatically programmed during the re-entry. These maneuvers I accomplished. I brought the spacecraft to the proper attitude for re-entry under manual control. The periscope was retracted by pumping the manual retraction lever.

As deceleration began to increase I could hear a hissing noise that sounded like small particles brushing against the spacecraft.

Due to ionization around the spacecraft, communications were lost. This had occurred on earlier missions and was experienced now on the predicted schedule. As the heat pulse started there was a noise and a bump on the spacecraft. I saw one of the straps that holds the retrorocket package swing in front of the window.

The heat pulse increased until I could see a glowing orange color through the window. Flaming pieces were breaking off and flying past the spacecraft window. At the time, these observations were of some concern to me because I was not sure what they were. I had assumed that the retropack had been jettisoned when I saw the strap in front of the window. I thought these flaming pieces might be parts of the heat shield breaking off. We know now, of course, that the pieces were from the retropack.

There was no doubt when the heat pulse occurred during re-entry but it takes time for the heat to soak into the spacecraft and heat the air. I did not feel particularly hot until we were getting down to about 75,000 to 80,000 feet. From there on down I was uncomfortably warm, and by the time the main parachute was out I was perspiring profusely.

The re-entry deceleration of 7.7g was as expected and was similar to that experienced in centrifuge runs. There had been some question as to whether our ability to tolerate acceleration might be worse because of the 4½ hours of weightlessness, but I could note no difference between my feeling of deceleration on this flight and my training sessions in the centrifuge.

After peak deceleration, the amplitude of the spacecraft oscillations began to build. I kept them under control on the manual and fly-by-wire systems until I ran out of manual fuel. After that point, I was unknowingly left with only the fly-by-wire system and the oscillations increased; so I switched to auxiliary damping, which controlled the spacecraft until the automatic fuel was also expended. I was reaching for the switch to deploy the drogue parachute early in order to reduce these re-entry oscillations, when it was deployed automatically. The drogue parachute stabilized the spacecraft rapidly.

At 10,800 feet the main parachute was deployed. I could see it stream out behind me, fill partially, and then as the reefing line cutters were actuated it filled completely. The opening of the parachute caused a jolt, but perhaps less than I had expected.

The landing deceleration was sharper than I had expected. Prior to impact I had disconnected all the extra leads to my suit, and was ready for rapid egress, but there was no need for this. I had a message that the destroyer *Noa* would pick me up within 20 minutes. I lay quietly in the spacecraft trying to keep as cool as possible. The temperature inside the spacecraft did not seem to diminish. This, combined with the high humidity of the air being drawn into the spacecraft kept me uncomfortably warm and perspiring heavily. Once the *Noa* was alongside the spacecraft, there was little delay in starting the hoisting operation. The spacecraft was pulled part way out of the water to let the water drain from the landing bag.

During the spacecraft pickup, I received one good bump. It was probably the most solid jolt of the whole trip as the spacecraft swung against the side of the ship. Shortly afterwards the spacecraft was on the deck.

I had initially planned egress out through the top, but by this time I had been perspiring heavily for nearly 45 minutes. I decided to come out the side hatch instead.

General Remarks

Many things were learned from the flight of Friendship 7. Concerning spacecraft systems alone, you have heard many reports that have verified previous design concepts or have shown weak spots that need remedial action.

Now, what can be said of man in the system?

Reliability

Of major significance is the probability that much more dependence can be placed on the man as a reliably operating portion of the man-spacecraft combination. In many areas his safe return can be made dependent on his own intelligent actions. Although a design philosophy could not be followed up to this time, Project Mercury never considered the astronaut as merely a passive passenger.

These areas must be assessed carefully, for man is not infallible, as we are all acutely aware. As an inflight example, the face plate on the helmet was open during the re-entry phase. Had cabin pressure started to drop, I could have closed the face plate in sufficient time to prevent decompression, but nevertheless a face-plate-open re-entry was not planned.

On the ground, some things would also be done differently. As an example, I felt it more advisable in the event of suspected malfunctions, such as the heat-shield-retropack difficulties, that require extensive discussion among ground personnel to keep the pilot updated on each bit of information rather than have him wait for a final clearcut recommendation from the ground. This keeps the pilot fully informed if there would happen to be any communication difficulty and it became necessary for him to make all decisions from onboard information.

Many things would be done differently if this flight could be flown over again, but we learn from our mistakes. I never flew a test flight on an airplane that I didn't return wishing I had done some things differently.

Even where automatic systems are still necessary, mission *reliability* is tremendously increased by having the man as a backup. The flight of Friendship 7 is a good example. This mission would almost certainly not have completed its three orbits, and might not have come back at all, if a man had not been aboard.

Adaptability

The flight of the Friendship 7 Mercury spacecraft has proved that man can *adapt* very rapidly to this new environment. His senses and capabilities are little changed in space. At least for the 4.5-hour duration of this mission, weightlessness was no problem.

Man's adaptability is most evident in his powers of observation. He can accomplish many more and varied experiments per mission than can be obtained from an unmanned vehicle. When the unexpected arises, as happened with the luminous particles and layer observations on this flight, he can make observations that will permit more rapid evaluation of these phenomena on future flights. Indeed, on an unmanned flight there likely would have been no such observations.

Future Plans

Most important, however, the future will not always find us as power limited as we are now. We will progress to the point where missions will not be totally preplanned. There will be choices of action in space, and man's intelligence and decision-making capability will be mandatory.

Our recent space efforts can be likened to the first flights at Kitty Hawk. They were first unmanned but were followed by manned flights, completely preplanned and of a few seconds' duration. Their experiments were, again, power limited, but they soon progressed beyond that point.

Space exploration is now at the same stage of development.

J. N. JAMES

The Voyage of *Mariner II*

•

ON DECEMBER 14, 1962, the U.S. spacecraft *Mariner II* completed the first successful interplanetary voyage when it passed within 22,000 miles of the planet Venus. If an observer had been riding in the craft, he would have seen Venus as a brilliant disk 900 times the area of the full moon as viewed from the earth. At its historic rendezvous *Mariner II* was 36 million miles from the earth, having traversed 180.2 million miles of space in 109 days. During that time and for 20 days thereafter the 447-pound vehicle maintained constant communication with the earth, obeying commands and sending back a huge volume of information about itself, interplanetary space and Venus.

The voyage of *Mariner II* was a technological feat of the first magnitude. The craft had in effect been launched three times: first from the surface of the earth, then from a "parking" orbit of the earth and finally from an orbit of the sun, nine days and 1.5 million miles from the earth, where it was put through a maneuver to place it in a new solar orbit. Throughout most of the flight it maintained a rigid orientation in space with respect to the sun and also with respect to the earth. It even recovered its proper orientation after being struck by an object in space. *Mariner II* proved that a spacecraft can be tracked with impressive accuracy on a microwave channel using only three watts of power and can be guided from the earth across tens of millions of miles.

Scientific instruments accounted for less than 10 per cent of the total weight of the craft. All across the void between the two planets, however, they produced unprecedented quantities of data about the magnetic fields of the solar system, cosmic rays and the solar wind—the streams of protons and electrons that issue from the sun. At the rendezvous with Venus the instruments observed the planet with a resolution impossible at this time from the earth. Venus, it can now be said, is covered by cold, dense clouds but has a surface temperature of approximately 800 degrees Fahrenheit on both its dark and its sunlit side. The planet seems to have little or no magnetic field and hence no belts of trapped radiation analogous to the Van Allen belts of the earth, and to be rotating very slowly or not at all. From the tracking of the spacecraft it will also be possible to calculate the astronomical unit (mean distance from the earth to the sun) with greater accuracy than ever before, to figure the mass of Venus and the moon with far more precision than that previously attained and even to locate certain points on the surface of the earth more accurately.

The story of *Mariner II* begins in July, 1960, when the National Aeronautics and Space Administration (NASA) approved the proposal of the Jet Propulsion Laboratory of the California Institute of Technology to send a spacecraft to Venus in the summer of 1962. The choice of Venus as the destination for the first NASA venture in "deep" space was a conservative one: the other planets are even more difficult to reach. Every 19 months Venus and the earth come within 26 million miles of each other; Mars on its closest approaches is between 35 million and 63 million miles away, and these approaches occur at the longer interval of 25 months. To effect an interplanetary journey with a minimum expenditure of energy the spacecraft must be launched, some months before the planet's closest approach to earth, on a trajectory that will bring it into encounter with the planet sometime after the planet's closest approach to earth. A voyage to Venus would take three to four months; a voyage to Mars would normally require seven or eight months. Radio transmission from Venus would require less power and there would be more solar energy available in its vicinity: at a distance of about 67 million miles from the sun the solar-power cells of a spacecraft would receive a great deal more sunlight to convert into electricity than they would at the distance of Mars, some 142 million miles from the sun. A Mars flight would thus have called for a heavier and more expensive spacecraft and a more powerful rocket for launching.

Altogether a rendezvous with Venus presented a much more feasible objective. In the summer of 1960 it was not possible to schedule a flight for the next launching opportunity in January and February, 1961. The choice of the opportunity after that—in July and August, 1962—gave us two years to get ready.

As of July, 1960, it appeared that the Atlas-Centaur would be developed in time to serve as our launching vehicle. This new combination, with a Centaur second-stage rocket fueled by hydrogen and oxygen, made it possible for us to think in terms of a half-ton spacecraft. With this mass available we were able to commit ourselves to the building of an "attitude-stabilized" spacecraft, like the lunar *Ranger*, rather than a "spin-stabilized" craft, like the *Explorers* and *Pioneers*. Because the former could hold Agena. In the three weeks from August 8 to September 1, 1961, using de-

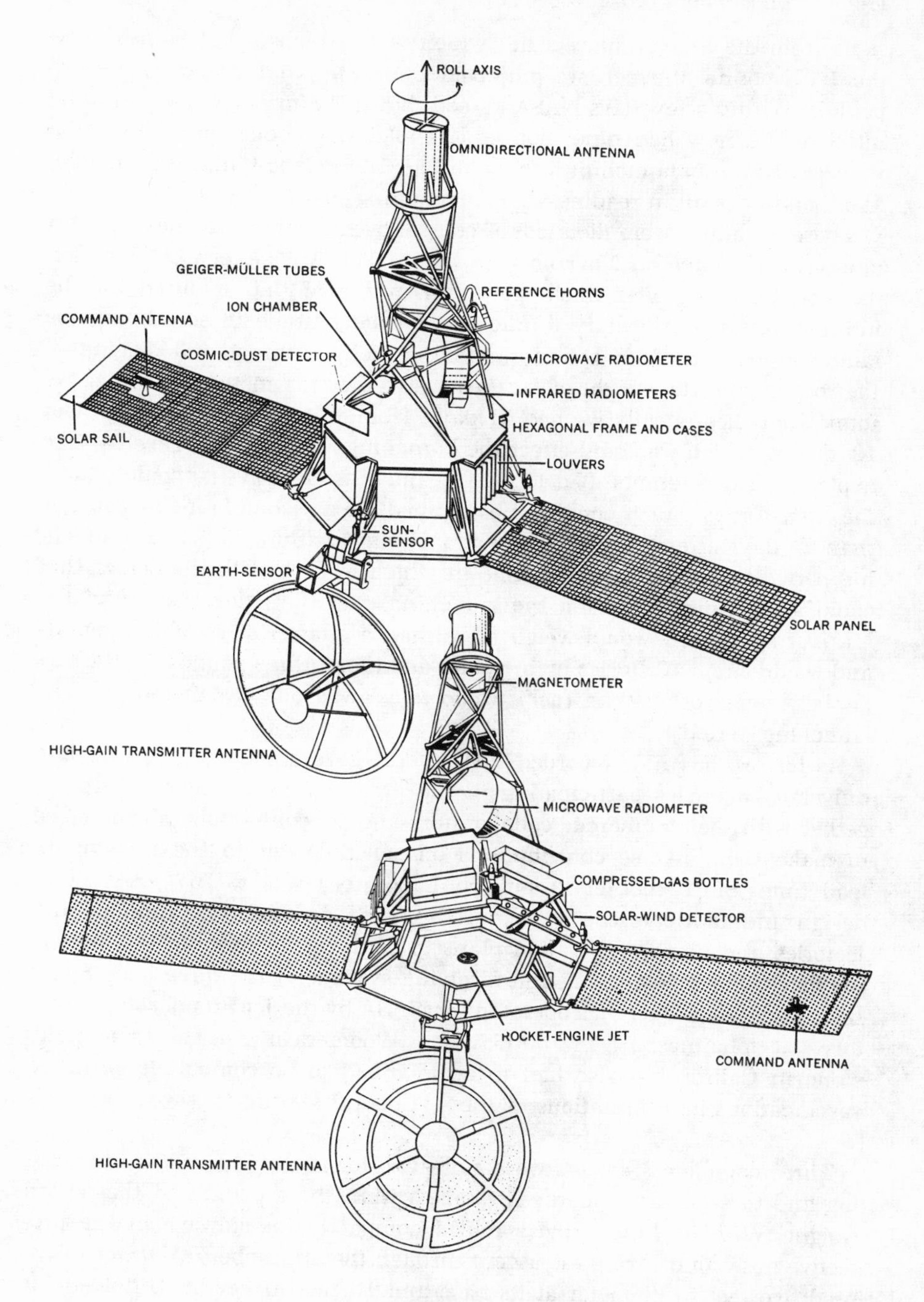

Fig. 1. *Mariner II* is diagramed in two positions to show its major parts. Craft was built around a hexagonal frame. While microwave radiometer scanned Venus, reference horns on it provided calibration by scanning space. Solar cells are seen in top drawing.

signs from the *Ranger* spacecraft (which was launched that month) and the 1,100-pound *Mariner*, we projected a *Mariner* that would weigh 447 pounds. Within a few days NASA agreed that this *Mariner* was equal to its mission. We now had nine months left to design, build and deliver the spacecraft and its launching vehicle to Pad 12 at Cape Canaveral in June, 1962, and have all in readiness for a launching in July or August.

These deadlines were dictated, of course, by the date of the next nearest approach of Venus on November 12, 1962. The Atlas-Agena combination did not, however, offer much power in excess of that required for the minimum-energy trajectory to allow any wide latitude in scheduling the launching. Taking into consideration the weight, thrust and staging of the rockets and the weight of *Mariner*, as well as the motion of Venus, we found that theoretically we had 51 days, from July 22 through September 10, during which we could effect the launching. Happily this gave us time to plan for two attempts; Pad 12 was the only installation that could handle the Atlas-Agena, but it could not be readied for a second launching in less than 24 days after the first. The extra 27 days within the 51-day launching period was barely enough time in which to cope with the delays that could arise. Fully fueled on the launching pad, the Atlas-Agena and the *Mariner* spacecraft would weigh more than a quarter of a million pounds and would stand 10 stories high; yet an additional three pounds on *Mariner* or the Agena (other than fuel) would take away one day of the precious launching period.

As history has now recorded, the several governmental, industrial and university agencies participating in the preparations for the venture, together with their numerous collaborators and subcontractors, all managed to make their diverse contributions on schedule within the nine-month lead time. Since the Jet Propulsion Laboratory was to "fly" as well as design and fabricate the spacecraft, its staff proceeded to compute trajectories and prepare operation plans to conform to the range of launching dates. The transmitting and receiving systems of the three Deep Space Instrumentation Facilities operated for NASA by the Jet Propulsion Laboratory—at Johannesburg in South Africa, Woomera in Australia and Goldstone in California—also had to be modified to be compatible with the specifications and limitations of the 447-pound *Mariner*.

Three complete *Mariners* were built, two for launching and one for testing and to serve as a spare; all came within three pounds of the design weight. With its hinged parts folded compactly for stowage in the protective nose cone during the ascent through the atmosphere, *Mariner* measured five feet in diameter at its base and 10 feet in height. Unfolded for flight in the vacuum of space, the spacecraft gained two feet in height (or length) and an ungainly "wingspread" of 16.5 feet, the "wings" being the two solar-cell panels bearing a total of 9,800 cells.

The term "spacecraft" is no misnomer as applied to *Mariner*. To accomplish its mission something more than a package of scientific instruments was required. *Mariner* itself played an active part in delivering the instruments to their destination and in carrying out the program of observation for which they were designed. In fact, the gear necessary for navigation on command from earth, for transmission of data to earth, for precise stabilization of its attitude in flight and for propulsion during the mid-course maneuver took up a full 406 pounds of the total weight of the craft. The "spaceframe" of *Mariner* was a hexagonal structure that held six cases of electronic and electromechanical apparatus. Two of the cases were occupied by the self-contained electrical power system, consisting of a 33.3-pound rechargeable silver-zinc storage battery and the switchgear to regulate and distribute the power from the solar cells or the battery. A combined radio receiver and three-watt transmitter and the control system for the scientific instruments filled two other segments of the hexagon. The fifth segment contained gear for translating the output of the scientific instruments and the inboard sensing devices into digital code for transmission to earth. The sixth segment of the hexagon bore the three gyroscopes that sensed and controlled the orientation of *Mariner* in space; it also held the heart—more properly, the brain—of the spacecraft: the "central computer and sequencer" that integrated the workings of the entire system around the ticking of a highly accurate electronic clock locked into the timekeeping system of the control stations on earth. In addition this segment contained the memory unit for storing the program of instructions for the expedition and for retaining commands received from earth in the course of the flight.

Mounted in the center of the hexagon and pointing aft along the roll, or long, axis of the spacecraft was the rocket motor that supplied the propulsion for the mid-course maneuver. It had been designed to give enough thrust to correct the predicted inaccuracies in the launching trajectory. Mounted on the outside of the hexagon was the power plant for attitude control: two bottles made of titanium, each holding 4.3 pounds of nitrogen under a pressure of 3,500 pounds per square inch, connected by small pipes and valves to 10 small jet nozzles located about the structure. These "motors" were controlled by the gyroscopes and the earth- and sun-sensors. On command they would jet tiny puffs of gas through orifices aimed along one or another of the three axes of the spacecraft.

Surmounting the open-frame superstructure that rose from the hexagon was an omnidirectional antenna. From this the spacecraft radiated its three-watt signal during periods when it was rolling or tumbling out of its designed orientation in flight. At such times the ground stations on earth had to listen sharply for the much attenuated signal. On the underside of the hexagon—at the bottom, or stern, of the spacecraft—was a four-foot parabolic "high gain" transmitting antenna. During periods of normal flight, when the spacecraft was locked in its proper orientation, this antenna focused the signal toward the earth. An earth-sensor—a photomultiplier tube mounted on the antenna—caused it to swing on its hinge and point steadily earthward as the spacecraft proceeded on its orbit. Signals from the earth, radiated at high power and focused by the great parabolic antenna at each of the three ground stations, were picked up by small antennas carried on the solar panels.

The payload of instrumentation was variously arrayed on the spaceframe. A detector for counting protons in the solar wind and a crystal microphone for registering impact by particles of interplanetary dust were mounted directly on the hexagon. Other instruments, including three Geiger-Müller tubes, an ion chamber and a magnetometer, had to be mounted on the superstructure away from the magnetic fields set up by the electrical and electronic gear. There they were also exposed to a minimum of the secondary radiation generated by the impact of cosmic rays on the metal of the spacecraft. The most elaborate piece of instrumentation was a parabolic antenna 20 inches in diameter and three inches deep, which was to scan the surface of Venus at two microwave frequencies and which, in addition, carried two optical sensors to take parallel readings of the infrared radiation from the planet. This assembly was mounted on the superstructure above the hexagon.

One of the thorniest problems to be anticipated in the design of *Mariner* was temperature control. Some measure of passive control was sought by giving the various surfaces different degrees of reflectivity and insulation against the full power of raw sunlight to which they were to be exposed. Movable louvers on the outside of the case containing the central computer and sequencer provided some active control at this critical point. In the extremes of environmental and mechanical stress to which we subjected a thermal-control model of *Mariner*, we tested particularly for the effects of heating and ways to minimize them. Unfortunately there was no test chamber in the U.S. at the time that could simulate solar radiation adequately. Such a chamber, completed after the launching, yielded some answers immediately, but too late to protect *Mariner II* from what proved to be the most serious jeopardy to its performance at the very moment of encounter. Nor were we able to launch a prototype *Mariner* on a test flight, a routine procedure in the development of launching rockets and some spacecraft and instrument packages. We had no choice but to commit the first U.S. interplanetary vehicle to its full journey into space.

In the first week of June, 1962, three *Mariners*, two Atlases and two Agenas were shipped to Cape Canaveral. The Atlas-Agena-*Mariner* was set up by mid-July. Banking on optimum performance by the rockets and with the aim of stretching the theoretical launching period, we decided to attempt a launching on July 19, three days early. Countdown delays postponed the launching until 4:21 A.M. on July 21. The launching failed, however, owing to a defective signal from the Atlas and the omission some years earlier of a single symbol from the program equations of the computer guiding the flight. The Atlas-Agena began to turn wildly and it was necessary to blow up the rockets. *Mariner I* had been sending signals and continued to do so until it hit the Atlantic Ocean 357 seconds after lift-off.

The schedule now called for launching *Mariner II* 24 days later, on August 14. Problems plaguing the second Atlas caused several postponements and we were not able to bring off the launching until August 27. The huge Atlas fired at 1:53 A.M. and soared to an altitude of 100 miles. The nose cone protecting *Mariner II* was popped off and the Atlas separated from the Agena. The Agena-*Mariner* pitched over almost to the horizontal, pointing in a southeasterly direction. Then the Agena fired for the first time, burning for 147 seconds, just long enough to go into a "parking" orbit at an altitude of 115 miles. There it coasted until 24 minutes after launching, when it reached a latitude of 14.8 degrees south of the Equator and a longitude of 2.1 degrees west—over the middle of the South Atlantic, the optimum location for taking off for Venus on this particular day at this particular time.

The Agena now fired again for 95 seconds, achieving the velocity necessary to escape the earth's gravitational field and accelerate itself and *Mariner II* into an orbit of the sun that would deliver the latter to the vicinity of Venus in mid-December. Explosive charges and springs separated the Agena from *Mariner*. Small nozzles expelled residual gases to turn the spent Agena aside and slow it down, separating it from *Mariner* so that the rocket carcass would not interfere with the optical sensors and placing it on an orbit that would carry it safely wide of Venus. *Mariner II* was now slowly tumbling and was drawing on its battery to send numerous signals, which were arriving at the earth stations right on schedule.

Within an hour after lift-off the central computer and sequencer on board *Mariner II* caused the solar panels to open out; it then instructed the gyros and sun-sensors to activate the nitrogen jets and point the long axis of the spacecraft toward the sun. The stabilization system still allowed the spacecraft to roll slowly about its long axis, but the solar cells were now facing into the glare of the sunlight. Some 195 watts of electricity converted from solar energy promptly relieved the drain on the battery, supplying nearly all the power necessary to operate the spacecraft. The average demand during most of the journey was 151 watts. On closer approach to the sun the output of the solar cells would rise gradually to about 275 watts and the excess energy would continually recharge the battery, which by itself could meet all needs for 9.5 hours.

As *Mariner II* was fired into its parking orbit and then into its interplanetary trajectory by successive thrusts of the Agena, it was moving around the earth more rapidly than the earth turns on its axis. Because the spacecraft was also increasing its distance from the earth, however, it appeared to reverse its direction about 80 minutes after launching. The point on the earth's surface above which it was at zenith had traveled southeastward, had turned northeastward and now swung around to speed westward as the earth rotated. With respect to the earth's revolution about the sun the spacecraft presented a quite different motion. The launching from the night side of the earth at 1.53 A.M. in a southeasterly direction had added a component of the earth's rotational speed to the velocity imparted by the rockets and had hurled the spacecraft out ahead of and around the earth as the latter proceeded on its orbit. By this maneuver the spacecraft had been headed into space in a direction opposite to the motion of the earth about the sun and inward toward the sun. As *Mariner II* fell behind the earth, 8,000 miles per hour had been subtracted from the 66,500-mile-per-hour orbital speed around the sun that it had possessed, along with the earth, at the instant of its launching. The slower speed of revolution would let the gravitational field of the sun pull the spacecraft inward, ultimately to cross the orbit of Venus. Since the plane of the orbit of Venus is tilted at three degrees to the plane of the earth's orbit, the earthbound navigators of *Mariner II* had also canted the orbit of the spacecraft in a direction calculated to intersect the plane of the orbit of Venus at the point of the rendezvous 109 days later.

As soon as *Mariner II* had escaped the earth's gravitational field a computer at the Jet Propulsion Laboratory was calculating its exact course. The angle at which the parabolic antennas on the ground picked up the spacecraft's signal most strongly located its direction from the earth against the background of the stars. Measurement of the Doppler shift (change in frequency) of the signal from the frequency at which it was transmitted yielded a correspondingly precise plot of the spacecraft's radial velocity (speed away from the earth). For this purpose signals were sent out from the earth at a frequency held stable to one part in 100 billion, and they were returned immediately by the spacecraft. The signals, doubly shifted toward longer wavelengths in their round trip to and from the receding spacecraft, produced measurements of its velocity accurate to within a tenth of an inch per second—even at the end of its active life, when it was 54 million miles from the earth and moving away at 49,000 miles per hour. Ten hours after the launching the computer had calculated that *Mariner II* would miss Venus by 233,000 miles, crossing its orbit ahead of it. Happily the necessary correction lay within the capacity of the mid-course propulsion system of the spacecraft.

During the first two days of the trip *Mariner II* transmitted only the readings from the 52 devices that registered temperatures, angles of hinged units, voltages, currents, fuel-tank pressures, transmitter power and other information about its own performance. The computer aboard was programmed to start transmission from the scientific instruments on the seventh day. Because the engineering data showed that the whole craft was working perfectly, the Johannesburg station on the third day commanded it to begin sending data from all the scientific instruments except the Venus radiometers. To conserve power the rate of transmission was slowed from 33 "bits," or units of information, to 8.3 bits: about one message or measurement per second. From then until a few hours before it passed Venus *Mariner II* sent scientific information for 20.16 seconds and then information about "itself" for 16.8 seconds. The latter was essential for evaluating the operation of this first interplanetary craft, for deciding on commands to give it and for designing its successors.

A week after launching, when *Mariner II* was 1.2 million miles from the earth, a command stored in its computer memory caused the spacecraft to point its high-gain parabolic antenna at the earth and assume a stable flight attitude. This could not be attempted earlier because the earth had been too bright for the earth-sensor mounted on the antenna. The spacecraft now ceased rolling; its tendency to pitch and yaw was restrained to within one degree around its long axis, which pointed toward the sun in the orbital plane of the craft. All was supposedly in readiness for the mid-course maneuver. It was apparent, however, that something was wrong just as soon as the transmitter switched over from the omnidirectional antenna to the parabolic one. The signal from the earth-sensor was only half as strong as had been expected—as though the device was "seeing" an object only half as bright as the earth. It might have been "locked" on the moon, but the earth at that distance is not two times but 83 times brighter than the moon. We decided to wait a day and see if the angle of the antenna hinge changed 1.5 degrees, which it would do to stay locked on the earth, or seven degrees, which would show that it was locked on the moon. If it was locked on some other object, such as the Agena carcass, or was not working properly at all, the angle of the hinge would show still another change. In case it was not pointed at the earth, we could command it to unlock and start to roll so that the sensor would see and lock onto the earth before it had rolled all the way around and locked onto the mysterious wrong object again.

Fortunately the signal the next day showed a hinge movement of 1.5 degrees. We still did not know why the sensor was giving such a low signal. Since everything else was working, however, we decided to go ahead with the mid-course maneuver.

On September 5, nine days after launching, punched tapes containing the calculated commands were fed into an encoder at Goldstone. These were transmitted to *Mariner II* for storage in its computer. They told the craft to unlock from earth and sun and roll slightly in order to position the cold-gas jets for the next movement, which was to point the omnidirectional antenna, or "bow," of the spacecraft 140 degrees away from the sun.

Then the hydrazine rocket motor was to fire. The motor was designed to burn for any length of time between .2 second and 57 seconds, and thereby to change the velocity of the spacecraft by as little as .7 foot per second or as much as 200 feet per second (136 miles per hour). Our instructions commanded it to burn for 27.8 seconds. The net effect would be to push *Mariner II* away from the sun and thus make it remain outside the orbit of Venus just long enough to allow Venus to pass by before *Mariner II* crossed its orbit. This maneuver added 45 miles per hour to its previous orbital speed of 60,250 miles per hour with respect to the sun and was chosen from among the several that might have accomplished the course correction for two secondary reasons. First, it would point the omnidirectional antenna toward the earth, ensuring an uninterrupted communication during the anxious moments of the maneuver. Second, and more important, it would fix the time for the encounter with Venus at an hour when the planet would be above the horizon at Goldstone, thereby assuring that the spacecraft could be kept under the surveillance of Goldstone's special equipment.

At 2:49 P.M. Pacific Standard Time, with the spacecraft 1,492,500 miles away, we gave the command for execution of the maneuver. This remote-control "launching in space," as delicate and nerve-racking as any on the earth, required three and three-quarters hours. While it was taking place *Mariner II* drew power from its battery instead of the solar cells and transmitted over the omnidirectional antenna. Afterward it reoriented itself, locking on the sun and then on the earth with the help of the gyroscopes, sun-sensors, earth-sensor and the cold-gas jets.

Within the next few days tracking established that the maneuver was a success. *Mariner II* was traveling, it was true, two miles per hour faster than had been planned. Instead of passing Venus at the hoped-for distance of 10,000 miles it would pass the planet at a distance of 21,648 miles. This was still well within the 40,000-mile target zone.

On the 12th day of the flight *Mariner II* lost its lock on the earth and the sun; it recovered, however, within three minutes. Apparently it had collided with a substantial bit of interplanetary rubble. On the 33rd day the spacecraft momentarily lost its earth and sun lock. When the craft recovered, the earth-sensor was operating at the proper strength! The sensor, incidentally, produced a fascinating record throughout the flight, reporting clear variations in the brightness of the earth as the land masses, clouds and oceans alternated in its field of view.

Meanwhile, as the spacecraft fell toward the sun, it was picking up orbital speed. On the 64th day after launching it overtook and passed the earth. The next day a short circuit occurred in one of the solar-cell panels.

Fortunately the craft was now close enough to the sun for the other panel to supply more than enough power by itself. That panel had more solar cells than the other; we had found in May that the rockets could handle additional weight and we had used the weight to increase the spacecraft's power safety margin. To balance the craft against pressure exerted on it by sunlight, we added a "sail" without power cells to the other panel. It was the panel with the sail that shorted out. The short circuit cured itself and recurred a few days later with no ill effects. We turned off the scientific experiments for eight days, until we were sure the craft had enough power to handle them.

On the 91st day, at 22.5 million miles from the earth, *Mariner II* set a new distance record for communications. It was to exceed this record continuously for the rest of the trip.

By the 100th day of the flight, with Venus nine days away, the temperatures of *Mariner II* had risen alarmingly, averaging 40 degrees Fahrenheit more than expected. Most of the apparatus could function well between 14 and 149 degrees F. The battery, however, had reached the upper limit, the earth-sensor had passed it and other units were approaching it. Moreover, the situation could only grow more taxing as the voyage continued. The solar radiation impinging on *Mariner II*, which amounted to 130 watts per square foot near the earth, would exceed 250 watts at Venus. All this was painful to observe because one week of flight data after launching had told us enough for us to have kept the spacecraft at a safe temperature range throughout the flight by simple adjustment of paint patterns. Now we could only wait and hope. On the 104th day four of the 52 engineering sensory devices failed, undoubtedly because one blew a fuse common to all four. On the 107th day the central computer failed to deliver a calibration signal that normally came every 16.6 hours. We had to assume that heat was to blame and that we could no longer rely on the computer to initiate the operations programmed for the meeting with Venus two days later. As a hedge against this untoward possibility, we had designed the system to accept a "backup" command from earth that would set the scanning program in motion. But there was now some anxiety about our chances for a successful encounter.

We found what comfort we could in the assurance that the flight had already been remarkably successful. The magnetometer, consisting of three magnetic cores each aligned along a different axis, had given a fine picture of magnetic-field strengths and fluctuations in the region of the solar system it had traversed. This instrument was sensitive to half a gamma, a field strength about .00001 that of the earth. (In the laboratory it could detect a nail in the shoe of anyone within 10 feet of it.) At times when the sun was quiet it had found field strengths of two to five gammas, compared with the earth's 30,000 gammas at the Equator and 50,000 gammas at the poles. During magnetic storms it had reported readings as high as 25 gammas. Changes in the flow of the solar wind were associated with such fluctuations, and it also appeared that the solar wind pushes the magnetic fields in interplanetary space around, just as it does the magnetic field of the earth.

The solar-wind detector recorded 40,000 spectra of particle energies throughout the flight and found that the solar plasma—a gas of dissociated charged particles composed in this case chiefly of electrons and protons—constantly pervades inner interplanetary space, a fact not known before. This apparatus was a particle counter ingeniously designed to discriminate and count protons over the energy range from 240 to 8,400 electron volts; it was mounted in the long axis of the spacecraft and hence pointed constantly "upwind." The density of particles, according to its reports, ranges from 10 to 20 per cubic inch and their velocity varies between 200 and 500 miles per second as they stream outward from the sun. Disturbances on the face of the sun caused the solar-wind detector to register increases of 20 to 100 per cent in the velocity of the particles on 20 different occasions during the flight. When these "gusts" in the solar wind subsequently arrived at the earth, they stirred up storms in the terrestrial magnetic field.

The three Geiger-Müller tubes and the ionization chamber carried by *Mariner II* found that the density and energy spectrum of the much-higher-energy cosmic ray particles in interplanetary space remains constant regardless of distance from the sun. This is a clear indication that these instruments were counting true cosmic rays originating outside the solar system. Only one solar disturbance, on October 23 and 24, sent out protons of sufficient energy to penetrate the counters. The total dosage of radiation throughout the flight as measured by the ionization chamber amounted to about three roentgens, approximately what a person on earth receives in 30 years from cosmic and other natural sources. It would seem that, from the standpoint of radiation, interplanetary space travel in 1962 would not have been dangerous for an astronaut.

As for the more macroscopic particles of interplanetary dust, the crystal microphone mounted in the center of a sounding plate counted only two impacts in the course of the flight. This instrument was designed to respond to particles as small as .0000000013 gram. Near the earth the flux of such particles is 10,000 times greater.

All these instruments were to continue in operation throughout the encounter with Venus and to report their readings along with the measurements made by the radiometers. Early in the morning of December 14, the 109th day of the flight, it became apparent that the central computer and sequencer aboard *Mariner II* had failed to start the operations scheduled for the encounter to come a few hours later. We gave the back-up command to start these operations, therefore, shortly after Venus and the approaching spacecraft rose above the horizon at Goldstone. With *Mariner II* 36 million miles away it took 6.5 minutes for the signal, traveling at the speed of light, to reach the craft and make the return trip to the earth, telling us that the small parabolic antenna and all its attachments had gone into operation for the microwave and infrared scans of Venus. Until it reached the scan zone the antenna would move at one degree per second, warming up the stabilization system for the encounter. To compensate for the motion of the antenna, which would otherwise have set the entire spacecraft into equal and opposite motion, the sun- and earth-sensors and the gyroscopes actuated the nitrogen-gas jets so as to maintain a steady platform. *Mariner II* was now so hot that seven of the 18 measurements of temperature had gone off their scales, and the earth-sensor and the communication system were nearing uncertain states. The scientific instruments and the essential working parts of the spacecraft were nevertheless operating properly at 5:30 A.M. Pacific Standard Time, with the encounter with Venus less than six hours away.

The transmitter was accordingly instructed to stop sending engineering data and to switch over to full-time reporting of the readings from the instruments. One of the space-flight operation teams, on the way to breakfast a few minutes late, saw Venus to the southeast shining in the clear California sky in all its brilliance as the morning star. The men were awed to realize that their spacecraft, which had just responded to their signal, was almost within that disk of light, closer to the planet than the moon is to the earth. *Mariner II* was then moving at a speed of 13,000 miles per hour toward Venus, 86,000 miles per hour in its orbit around the sun and 39,000 miles per hour away from the earth.

At 10:54 A.M. the spacecraft reached the scan zone. As the radiometer antenna made its first scan of the planet at 11:03 A.M. it slowed its motion to a tenth of a degree per second. In accordance with instructions, it proceeded to scan the disk of the planet, which was 16 degrees in diameter at this close range. When the radiometer indicated that the antenna had scanned past the limb, or edge, of the planet and was looking into space, the antenna drive reversed its direction and started the next scan. Altogether it made three passes across the disk "up and down," while the orbit of the craft provided the lateral motion. The radiometer sent 18 readings, five from the night side of Venus, eight from the terminator (the shadow line between the dark and light sides) and five from the day side.

Although the antenna was small, its resolution so close to Venus could have been equaled on the earth only by a parabolic antenna thousands of feet in diameter. The two wavelengths on which it operated—13.5 and 19 millimeters—were chosen to settle some long-standing questions and controversies. In the first place, the difference in the readings between the two frequencies would show if there was any water in the planet's atmosphere and how much of it there was, because any water present would strongly absorb and so reduce radiation of the 13.5-millimeter wavelength. in the second place, the 19-millimeter wavelength would provide the surest measurement of the surface temperature of the planet; radiation of this wavelength would come up unhindered from the suface even through the thick cloud cover that gives the planet its high albedo and hence its brightness in the sky.

Radio observations of Venus from the earth had indicated surprisingly high temperatures—in the range of 600 degrees F. Such readings are or course taken from the whole disk of the planet. One school of investigators was therefore able to argue that the radio emissions indicating a high temperature were really coming from an ionosphere, and that these readings did not necessarily mean that Venus had a hot surface. Another group argued for the "greenhouse model," asserting that the radiation does come from a hot planet and that the heat of solar radiation is trapped under the planet's thick clouds.

The radiometer aboard *Mariner II* could settle this question because it could make readings at discrete points across the disk of the planet. If it found "limb-brightening," or high readings near the limb, where the instrument would be measuring the temperature of the planet's upper atmosphere, and relatively lower readings at the center of the disk where the atmosphere would be thinner, then the proponents of a hot ionosphere would have the better of the argument. On the other hand, "limb-darkening," or readings lower near the limb than at the center, would support the greenhouse model.

Mariner II found an unquestionable limb-darkening and found furthermore that there is little difference in temperature on the dark side compared with the sunlit side of the planet. On the basis of the still incomplete analysis of the radiometer scans, the surface of Venus, where the 19-millimeter radiation originates, appears to have a temperature of about 800 degrees F. Thus the surface, which is hot enough to melt lead, could not sustain any form of life that is known on earth. Apparently there is also very little water vapor in the atmosphere—less than a thousandth that of the earth's atmosphere.

The infrared radiometer attached to the microwave antenna essentially confirmed the findings made at the longer wavelengths. It was hoped that this instrument would detect any breaks in the cloud cover and yield a measurement of the carbon dioxide content of the planet's atmosphere. The infrared radiometer was designed to make comparative measurements of the radiation from the atmosphere of Venus at two wavelengths, 8.4 and 10.4 microns respectively. Radiation from the ground and the lower atmosphere would presumably come through breaks in the cloud cover on the 8.4 micron wavelength, whereas a high concentration of carbon dioxide in the atmosphere would absorb and reduce radiation at the 10.4—micron wavelength. The comparative measurements on the two wavelengths by the radiometer gave substantially the same results. Accordingly we can conclude that there were no breaks in the cloud cover over the regions scanned and that there is no significant amount of CO_2 in the upper clouds; the amount of carbon dioxide below the clouds remains unresolved.

The readings on the two wavelengths did, however, indicate the same temperature (about −30 degrees F.) for the top of the clouds on both the day and night side of the planet. These measurements support the greenhouse concept of a thick cloud cover that keeps the heat from radiating away from the surface. One anomaly, a spot 20 degrees cooler than any other point seen at the same angle by the infrared radiometer, showed up in the lower half of the pass along the terminator. It could conceivably have been caused by a surface feature such as a very high mountain.

Meanwhile the other instruments aboard *Mariner II* showed the external atmosphere of Venus to be equally featureless. The magnetometer found no increase in magnetic-field readings near the planet compared with the measurements taken farther out in space, and it recorded fewer fluctuations as it approached. If Venus had a magnetic field similar to that of the earth, the magnetometer would have reported 100 to 200 gammas at the point of closest approach instead of a steady reading of a fraction of this strength. The Geiger-Müller counters detected no increase in the density of particles and found a slight increase rather than a decrease in tthe velocity of the solar wind. In both cases the earth's magnetic field brings about opposite effects near the earth. The low-energy radiation, as measured by the more sensitive Geiger-Müller counter, would have jumped 10,000-fold if Venus had a magnetic field like that of the earth. The magnetic field of the planet, if any, must have a strength less than 10 per cent of the earth's. The conclusion is that Venus rotates very slowly or not at all, confirming studies made from terrestrial observatories. It is the rotation of the earth and the motion of its fluid interior that are believed to generate this planet's magnetism.

The dust detector was not struck by a single particle in the vicinity of Venus; it would have been bombarded by such particles near the earth. The significance of this observation is not fully understood as yet.

The engineering data that shared such important radio time with the observations of the scientific instruments aboard *Mariner II* also produced important findings. So accurate was the tracking and orbital determination that we know to within 10 miles that the craft was 21,648 miles from the surface of Venus, even though at the time it was 36 million miles from the earth. Astronomers have been watching Venus for decades in an effort to calculate its mass, and they have found it to be .8148 times that of the earth to within .05 percent accuracy. Two weeks of tracking *Mariner II* within the gravitational field of Venus show that the planet deflected the spacecraft 24.5 degrees from its approach path and changed its velocity appreciably; these observations have given Venus a mass of .81485 times that of the earth to within .005 per cent. Computer calculations employing the tracking data have also enabled us to arrive at a more accurate measurement of the mass of the moon than we have ever had before and to interlock the cartographic grids of the earth's continents with greater precision.

Most important, the 22,000 Doppler readings from *Mariner II* along with simultaneous tracking of Venus by the Goldstone station will enable us to measure the astronomical unit—the mean orbital distance of the earth from the sun—to several more significant figures. The relative distances of the sun, Venus, the moon and other bodies in the solar system are known from Kepler's laws and centuries of astronomical observation. With the absolute distance to Venus established, it becomes possible to give absolute values to all these other distances, including the astronomical unit itself.

After making its nearest approach to Venus at 32 seconds before noon Pacific Standard Time on December 14, *Mariner II* continued in its now eternal orbit around the sun. The "year" of this new solar satellite has a duration of 345.9 terrestrial days. It made its closest approach to the sun, at 65.5 million miles, on December 27, and on the 129th day, January 2, 1963, ceased to transmit information to the earth. It gave no clue as to why it stopped sending. At that time it was 54 million miles from the earth, 5.7 million miles from Venus and had traveled 223.7 million miles through the solar system in 129 days. Scientists and engineers will spend years studying the 11 million measurements it sent back, all now recorded on magnetic tapes and stored in vaults.

EXPERIMENT	MEASUREMENT RANGES	FINDINGS	EXPERIMENTERS
Microwave radiometer	Wavelengths of 13.5 and 19 millimeters	Venus surface temperature of about 800 degrees F. on both dark and light sides.	A. H. Barrett, Massachusetts Institute of Technology; D. E. Jones, Jet Propulsion Laboratory; J. Copeland, Army Ordnance Missile Command; A. E. Lilley, Harvard College Observatory
Infrared radiometer	Wavelengths of 8.4 and 10.4 microns	Top of Venus' clouds at −30 degrees F. No breaks detected in clouds. Cold spot found, indicating possible high surface feature.	L. D. Kaplan, J.P.L. and University of Nevada; G. Neugebauer, J.P.L.; C. Sagan, University of California at Berkeley
Magnetometer	Up to 64 gammas in interplanetary space, up to 320 gammas near Venus	Venus has little or no magnetic field, is rotating slowly or not at all. Weak, fluctuating solar magnetic fields found in interplanetary space.	P. J. Coleman, Jr., National Aeronautics and Space Administration; L. Davis, California Institute of Technology; E. J. Smith, J.P.L.; C. P. Sonett, NASA
Ion chamber and Geiger-Müller tubes	Protons above 10 million electron volts (Mev) energy, electrons above .5 Mev, alpha particles above 40 Mev. Directional tube counted protons above .5 Mev, electrons above .04 Mev.	Total radiation dosage for whole trip: three roentgens. Cosmic ray flux fairly constant throughout trip, not changing near Venus.	H. R. Anderson, J.P.L.; H. V. Neher, C.I.T.; J. A. Van Allen, State University of Iowa
Crystal microphone	Dust particles as small as .0000000013 gram	Detected only two particles throughout voyage, none near Venus.	W. M. Alexander, NASA Goddard Space Flight Center
Solar-plasma spectrometer	Protons from 240 to 8,400 electron volts	Solar wind "blows" constantly, varies in intensity and temperature with events on sun.	M. M. Neugebauer and C. W. Snyder, J.P.L.
Radio (tracking)	Employed three-watt transmitter, high-gain antenna on Mariner at frequency of 960 megacycles per second.	More precise measurement of astronomical unit, mass of moon and Venus, and location of tracking station on earth	T. W. Hamilton, J. F. Koukol, N. A. Renzetti, D. W. Trask and J. D. Anderson, J.P.L.

Scientific Results of voyage of *Mariner II* are summarized. Radio tracking, while not an experiment, is included because it produced valuable scientific data. Microwave and infrared radiometers scanned Venus only; other instruments studied space also.

C. *The Earth*

•

From outer space to the rocks we observe on any walk in the country is a transition from the new and strange to the old and seemingly commonplace. We are apt to forget that the rocks have a fascinating story to tell. Old rocks hold a key to the age and history of the earth, newer ones to the origin and development of species. In Geological Change *the famous nineteenth-century geologist Sir Archibald Geike tells how the study began and how the fundamental principles of geology were established. He writes of the rhythmic cycles caused by alternating erosion and uplifting of land. He shows how the earth must be far older than previous generations had suspected. One of the major classifications of rocks, the "sedimentary" deposits, tells the story of how living things have evolved. This story, based on the study of fossil remains, is and must continue to be fragmentary, though each year new evidence concerning the past of man, animals and plants helps fill the gaps. In* The Record of Living Things *the American geologist W. O. Hotchkiss leads us into this fascinating branch of geological study. Preceding his article is a* Geological Table *from* Historical Geology *by Dunbar, which gives us a bird's-eye view of the earth's long history.*

We have become accustomed to thinking of earth history in terms of millions and billions of years. (The latest estimate of the age of the earth is about four and one-half billion years.) Most of the processes involved, such as the formation of continents and the building of mountains, must be measured in terms which dwarf all human history. On occasion, however, sudden and cataclysmic happenings occur. Tensions built up over long periods are released with terrifying force, far more powerful than the largest thermonuclear bombs. Earthquakes are the most familiar of such happenings. What are their causes and characteristics? The Harvard seismologist L. Don Leet and his wife Florence Leet answer these questions in Earthquake: Discoveries in Seismology. *Volcanoes, too, are apt to erupt with devastating loss of life and property. Vesuvius, Krakatao and Pelée are well-known examples. The actual formation of a new volcano, one of the most interesting exhibits in geology, took place before the astonished eyes of Mexican peons in 1943 and is described in* The Birth of Parícutin *by Gonzales and Foshag.*

As the surface of the earth has been more thoroughly investigated, scientists have also explored above and beneath it. In The Ocean Floor *Francis P. Shepard discusses one of the new dimensions that has been added. With instruments and by direct observation we have learned surprising new facts about the structure of the ocean floor. The same author was fortunate enough, at least from the viewpoint of scientific study, to be living on the shore when a huge tidal wave struck Hawaii on April 1, 1946. His eyewitness account is contained in* Catastrophic Waves from the Sea.

One of the geologist's most difficult problems is to make sound estimates of geologic time. A number of techniques are available. For the period in which life has existed on our planet, the most accurate is through Carbon 14 dating. Richard Foster Flint, professor of geology at Yale, describes the method used in Pin-pointing the Past with the Cosmic Clock.

Similarly we have at last begun to understand the forces at work in the atmosphere which surrounds us. The answers to the question "What makes the weather?" are becoming clearer and the techniques for predicting it more exact. Something of what we have learned is explained in The Weather *by Captain David C. Holmes of the United States Navy.*

SIR ARCHIBALD GEIKE

Geological Change

•

It was a fundamental doctrine of Hutton[1] and his school that this globe has not always worn the aspect which it bears at present; that on the contrary, proofs may everywhere be culled that the land which we now see has been formed out of the wreck of an older land. Among these proofs, the most obvious are supplied by some of the more familiar kinds of rocks, which teach us that, though they are now portions of the dry land, they were originally sheets of gravel, sand, and mud, which had been worn from the face of long-vanished continents, and after being spread out over the floor of the sea were consolidated into compact stone, and were finally broken up and raised once more to form part of the dry land. This cycle of change involved two great systems of natural processes. On the one hand, men were taught that by the action of running water the materials of the solid land are in a state of continual decay and transport to the ocean. On the other hand, the ocean floor is liable from time to time to be upheaved by some stupendous internal force akin to that which gives rise to the volcano and the earthquake. Hutton further perceived that not only had the consolidated materials been disrupted and elevated, but that masses of molten rock had been thrust upward among them, and had cooled and crystallized in large bodies of granite and other eruptive rocks which form so prominent a feature on the earth's surface.

It was a special characteristic of this philosophical system that it sought in the changes now in progress on the earth's surface an explanation of those which occurred in older times. Its founder refused to invent causes or modes of operation, for those with which he was familiar seemed to him adequate to solve the problems with which he attempted to deal. Nowhere was the profoundness of his insight more astonishing than in the clear, definite way in which he proclaimed and reiterated his doctrine, that every part of the surface of the continents, from mountain top to seashore, is

[1] James Hutton, 1726-1797.—Eds.

continually undergoing decay, and is thus slowly travelling to the sea. He saw that no sooner will the sea floor be elevated into new land than it must necessarily become a prey to this universal and unceasing degradation. He perceived that as the transport of disintegrated material is carried on chiefly by running water, rivers must slowly dig out for themselves the channels in which they flow, and thus that a system of valleys, radiating from the water parting of a country, must necessarily result from the descent of the streams from the mountain crests to the sea. He discerned that this ceaseless and wide-spread decay would eventually lead to the entire demolition of the dry land, but he contended that from time to time this catastrophe is prevented by the operation of the under-ground forces, whereby new continents are upheaved from the bed of the ocean. And thus in his system a due proportion is maintained between land and water, and the condition of the earth as a habitable globe is preserved.

A theory of the earth so simple in outline, so bold in conception, so full of suggestion, and resting on so broad a base of observation and reflection, ought (we think) to have commanded at once the attention of men of science, even if it did not immediately awaken the interest of the outside world; but, as Playfair sorrowfully admitted, it attracted notice only very slowly, and several years elapsed before any one showed himself publicly concerned about it, either as an enemy or a friend. Some of its earliest critics assailed it for what they asserted to be its irreligious tendency,—an accusation which Hutton repudiated with much warmth. The sneer levelled by Cowper a few years earlier at all inquiries into the history of the universe was perfectly natural and intelligible from that poet's point of view. There was then a wide-spread belief that this world came into existence some six thousand years ago, and that any attempt greatly to increase that antiquity was meant as a blow to the authority of Holy Writ. So far, however, from aiming at the overthrow of orthodox beliefs, Hutton evidently regarded his "Theory" as an important contribution in aid of natural religion. He dwelt with unfeigned pleasure on the multitude of proofs which he was able to accumulate of an orderly design in the operations of Nature, decay and renovation being so nicely balanced as to maintain the habitable condition of the planet. But as he refused to admit the predominance of violent action in terrestrial changes, and on the contrary contended for the efficacy of the quiet, continuous processes which we can even now see at work around us, he was constrained to require an unlimited duration of past time for the production of those revolutions of which he perceived such clear and abundant proofs in the crust of the earth. The general public, however, failed to comprehend that the doctrine of the high antiquity of the globe was not inconsistent with the comparatively recent appearance of man,—a distinction which seems so obvious now.

Hutton died in 1797, beloved and regretted by the circle of friends who had learned to appreciate his estimable character and to admire his genius, but with little recognition from the world at large. Men knew not then that a great master had passed away from their midst, who had laid broad and deep the foundations of a new science; that his name would become a household word in after generations, and that pilgrims would come from distant lands to visit the scenes from which he drew his inspiration.

Clear as was the insight and sagacious the inferences of the great masters [of the Edinburgh school] in regard to the history of the globe, their vision was necessarily limited by the comparatively narrow range of ascertained fact which up to their time had been established. They taught men to recognize that the present world is built of the ruins of an earlier one, and they explained with admirable perspicacity the operation of the processes whereby the degradation and renovation of land are brought about. But they never dreamed that a long and orderly series of such successive destructions and renewals had taken place and had left their records in the crust of the earth. They never imagined that from these records it would be possible to establish a determinate chronology that could be read everywhere and applied to the elucidation of the remotest quarter of the globe. It was by the memorable observations and generalizations of William Smith that this vast extension of our knowledge of the past history of the earth became possible. While the Scottish philosophers were building up their theory here, Smith was quietly ascertaining by extended journeys that the stratified rocks of the west of England occur in a definite sequence, and that each well-marked group of them can be discriminated from the others and identified across the country by means of its inclosed organic remains. It is nearly a hundred years since he made known his views, so that by a curious coincidence we may fitly celebrate on this occasion the centenary of William Smith as well as that of James Hutton. No single discovery has ever had a more momentous and far-reaching influence on the progress of a science than that law of organic succession which Smith established. At first it served merely to determine the order of the stratified rocks of England. But it soon proved to possess a world-wide value, for it was found to furnish the key to the structure of the whole stratified crust of the earth. It showed that within that crust lie the chronicles of a long history of plant and animal life upon this planet, it supplied the means of arranging the materials for this history in true chronological sequence, and it thus opened out a magnificent vista through a vast series of ages, each marked by its own distinctive types of organic life, which, in proportion to their antiquity, departed more and more from the aspect of the living world.

Thus a hundred years ago, by the brilliant theory of Hutton and the fruitful generalization of Smith, the study of the earth received in our country the impetus which has given birth to the modern science of geology.

From the earliest times the natural features of the earth's surface have arrested the attention of mankind. The rugged mountain, the cleft ravine, the scarped cliff, the solitary boulder, have stimulated curiosity and prompted many a speculation as to their origin. The shells embedded by millions in the solid rocks of hills far removed from the seas have still further pressed home these "obstinate questionings." But for many long centuries the advance of inquiry into such matters was arrested by the paramount influence of orthodox theology. It was not merely that the church opposed itself to the simple and obvious interpretation of these natural phenomena. So implicit had faith become in the accepted views of the earth's age and of the history of creation, that even laymen of intelligence and learning set themselves unbidden and in perfect good faith to explain away the difficulties which nature so persistently raised up, and to reconcile her teachings with those of the theologians.

It is the special glory of the Edinburgh school of geology to have cast aside all this fanciful trifling. Hutton boldly proclaimed that it was no part of his philosophy to account for the beginning of things. His concern lay only with the evidence furnished by the earth itself as to its origin. With the intuition of true genius he early perceived that the only basis from which to explore what has taken place in bygone time is a knowledge of what is taking place to-day. He thus founded his system upon a careful study of the process whereby geological changes are now brought about.

Fresh life was now breathed into the study of the earth. A new spirit seemed to animate the advance along every pathway of inquiry. Facts that had long been familiar came to possess a wider and deeper meaning when their connection with each other was recognized as parts of one great harmonious system of continuous change. In no department of Nature, for example, was this broader vision more remarkably displayed than in that wherein the circulation of water between land and sea plays the most conspicuous part. From the earliest times men had watched the coming of clouds, the fall of rain, the flow of rivers, and had recognized that on this nicely adjusted machinery the beauty and fertility of the land depend. But they now learned that this beauty and fertility involve a continual decay of the terrestrial surface; that the soil is a measure of this decay, and would cease to afford us maintenance were it not continually removed and renewed, that through the ceaseless transport of soil by rivers to the sea the face of the land is slowly lowered in level and carved into mountain and valley, and that the materials thus borne outwards to the floor of the ocean are not lost, but accumulate there to form rocks, which in the end will be upraised into new lands. Decay and renovation, in well-balanced proportions, were thus shown to be the system on which the existence of the earth as a habitable globe had been established. It was impossible to conceive that the economy of the planet could be maintained on any other basis. Without the circulation of water the life of plants and animals would be impossible, and with the circulation the decay of the surface of the land and the renovation of its disintegrated materials are necessarily involved.

As it is now, so must it have been in past time. Hutton and Playfair pointed to the stratified rocks of the earth's crust as demonstrations that the same processes which are at work to-day have been in operation from a remote antiquity.

Obviously, however, human experience, in the few centuries during which attention has been turned to such subjects, has been too brief to warrant any dogmatic assumption that the various natural processes must have been carried on in the past with the same energy and at the same rate as they are carried on now. It was an error to take for granted that no other kind of process or influence, nor any variation in the rate of activity save those of which man has had actual cognizance, has played a part in the terrestrial economy. The uniformitarian writers laid themselves open to the charge of maintaining a kind of perpetual motion in the machinery of Nature. They could find in the records of the earth's history no evidence of a beginning, no prospect of an end.

The discoveries of William Smith, had they been adequately understood, would have been seen to offer a corrective to this rigidly uniformitarian conception, for they revealed that the crust of the earth contains the long record of an unmistakable order of progression in organic types. They proved that plants and animals have varied widely in successive periods of the earth's history; the present condition of organic life being only the latest phase of a long preceding series, each stage of which recedes further from the existing aspect of things as we trace it backward into the past. And though no relic had yet been found, or indeed was ever likely to be found, of the first living things that appeared upon the earth's surface, the manifest simplification of types in the older formations pointed irresistibly to some beginning from which the long procession has taken its start. If then it could thus be demonstrated that there had been upon the globe an orderly march of living forms from the lowliest grades in early times to man himself to-day, and thus that in one department of her domain, extending through the greater portion of the records of the earth's history, Nature had not been uniform, but had followed a vast and noble plan of evolution, surely it might have been expected that those who discovered and made known this plan would seek to ascertain whether some analogous physical progression from a definite beginning might not be discernible in the framework of the globe itself.

But the early masters of the science labored under two great disadvantages. In the first place, they found the oldest records of the earth's history so broken up and effaced as to be no longer legible. And in the second place, they considered themselves bound to search for facts, not to build up theories; and as in the crust of the earth they could find no facts which threw any light upon the primeval constitution and subsequent development of our planet, they shut their ears to any theoretical interpretations that might be offered from other departments of science.

What the more extreme members of the uniformitarian school failed to perceive was the absence of all evidence that terrestrial catastrophes even on a colossal scale might not be a part of the present economy of this globe. Such occurrences might never seriously affect the whole earth at one time, and might return at such wide intervals that no example of them has yet been chronicled by man. But that they have occurred again and again, and even within comparatively recent geological times, hardly admits of serious doubt.

As the most recent and best known of these great transformations, the Ice Age stands out conspicuously before us. There can not be any doubt that after man had become a denizen of the earth, a great physical change came over the Northern hemisphere. The climate, which had previously been so mild that evergreen trees flourished within ten or twelve degrees of the North Pole, now became so severe that vast sheets of snow and ice covered the north of Europe and crept southward beyond the south coast of Ireland, almost as far as the southern shores of England, and across the Baltic into France and Germany. This Arctic transformation was not an episode that lasted merely a few seasons, and left the land to resume thereafter its ancient aspect. With various successive fluctuations it must have endured for many thousands of years. When it began to disappear it probably faded away as slowly and imperceptibly as it had advanced, and when it finally vanished it left Europe and North America profoundly changed in the character alike of their scenery and of their inhabitants. The rugged rocky contours of earlier times were ground smooth and polished by the march of the ice across them, while the lower grounds were buried under wide and thick sheets of clay, gravel, and sand, left behind by the melting ice. The varied and abundant flora which had spread so far within the Arctic circle was driven away into more southern and less ungenial climes. But most memorable of all was the extirpation of the prominent large animals which, before the advent of the ice, had roamed over Europe. The lions, hyenas, wild horses, hippopotamuses, and other creatures either became entirely extinct or were driven into the Mediterranean basin and into Africa. In their place came northern forms—the reindeer, glutton, musk ox, wooly rhinoceros, and mammoth.

Such a marvellous transformation in climate, in scenery, in vegetation and in inhabitants, within what was after all but a brief portion of geological time, though it may have involved no sudden or violent convulsion, is surely entitled to rank as a catastrophe in the history of the globe. It was probably brought about mainly if not entirely by the operation of forces external to the earth. No similar calamity having befallen the continents within the time during which man has been recording his experience, the Ice Age might be cited as a contradiction to the doctrine of uniformity. And yet it manifestly arrived as part of the established order of Nature. Whether or not we grant that other ice ages preceded the last great one, we must admit that the conditions under which it arose, so far as we know them, might conceivably have occurred before and may occur again. The various agencies called into play by the extensive refrigeration of the Northern hemisphere were not different from those with which we are familiar. Snow fell and glaciers crept as they do to-day. Ice scored and polished rocks exactly as it still does among the Alps and in Norway. There was nothing abnormal in the phenomena, save the scale on which they were manifested. And thus, taking a broad view of the whole subject, we recognize the catastrophe, while at the same time we see in its progress the operation of those same natural processes which we know to be integral parts of the machinery whereby the surface of the earth is continually transformed.

Among the debts which science owes to the Huttonian school, not the least memorable is the promulgation of the first well-founded conceptions of the high antiquity of the globe. Some six thousand years had previously been believed to comprise the whole life of the planet, and indeed of the entire universe. When the curtain was then first raised that had veiled the history of the earth, and men, looking beyond the brief span within which they had supposed that history to have been transacted, beheld the records of a long vista of ages stretching far away into a dim illimitable past, the prospect vividly impressed their imagination. Astronomy had made known the immeasurable fields of space; the new science of geology seemed now to reveal boundless distances of time.

The universal degradation of the land, so notable a characteristic of the earth's surface, has been regarded as an extremely slow process. Though it goes on without ceasing, yet from century to century it seems to leave hardly any perceptible trace on the landscapes of a country. Mountains and plains, hills and valleys appear to wear the same familiar aspect which is indicated in the oldest pages of history. This obvious slowness in one of the most important departments of geological activity doubtless contributed in large measure to form and foster a vague belief in the vastness of the antiquity required for the evolution of the earth.

But, as geologists eventually came to perceive, the rate of degradation of the land is capable of actual measurement. The amount of material worn away from the surface of any drainage basin and carried in the form of mud, sand, or gravel, by the main river into the sea represents the extent to which that surface has been lowered by waste in any given period of time. But denudation and deposition must be equivalent to each other. As much material must be laid down in sedimentary accumulations as has been mechanically removed, so that in measuring the annual bulk of sediment borne into the sea by a river, we obtain a clue not only to the rate of denudation of the land, but also to the rate at which the deposition of new sedimentary formations takes place.

But in actual fact the testimony in favor of the slow accumulation and high antiquity of the geological record is much stronger than might be inferred from the mere thickness of the stratified formations. These sedimentary deposits have not been laid down in one unbroken sequence, but have had their continuity interrupted again and again by upheaval and depression. So fragmentary are they in some regions that we can easily demonstrate the length of time represented there by still existing sedimentary strata to be vastly less than the time indicated by the gaps in the series.

There is yet a further and impressive body of evidence furnished by the successive races of plants and animals which have lived upon the earth and have left their remains sealed up within its rocky crust. No universal destructions of organic life are chronicled in the stratified rocks. It is everywhere admitted that, from the remotest times up to the present day, there has been an onward march of development, type succeeding type in one long continuous progression. As to the rate of this evolution precise data are wanting. There is, however, the important negative argument furnished by the absence of evidence of recognizable specific variations of organic forms since man began to observe and record. We know that within human experience a few species have become extinct, but there is no conclusive proof that a single new species have come into existence, nor are appreciable variations readily apparent in forms that live in a wild state. The seeds and plants found with Egyptian mummies, and the flowers and fruits depicted on Egyptian tombs, are easily identified with the vegetation of modern Egypt. The embalmed bodies of animals found in that country show no sensible divergence from the structure or proportions of the same animals at the present day. The human races of Northern Africa and Western Asia were already as distinct when portrayed by the ancient Egyptian artists as they are now, and they do not seem to have undergone any perceptible change since then. Thus a lapse of four or five thousand years has not been accompanied by any recognizable variation in such forms of plant and animal life as can be tendered in evidence. Absence of sensible change in these instances is, of course, no proof that considerable alteration may not have been accomplished in other forms more exposed to vicissitudes of climate and other external influences. But it furnishes at least a presumption in favor of the extremely tardy progress of organic variation.

If, however, we extend our vision beyond the narrow range of human history, and look at the remains of the plants and animals preserved in those younger formations which, though recent when regarded as parts of the whole geological record, must be many thousands of years older than the very oldest of human monuments, we encounter the most impressive proofs of the persistence of specific forms. Shells which lived in our seas before the coming of the Ice Age present the very same peculiarities of form, structure, and ornament which their descendants still possess. The lapse of so enormous an interval of time has not sufficed seriously to modify them. So too with the plants and the higher animals which still survive.

Some forms have become extinct, but few or none which remain display any transitional gradations into new species. We must admit that such transitions have occurred, that indeed they have been in progress ever since organized existence began upon our planet, and are doubtless taking place now. But we can not detect them on the way, and we feel constrained to believe that their march must be excessively slow.

If the many thousands of years which have elapsed since the Ice Age have produced no appreciable modification of surviving plants and animals, how vast a period must have been required for that marvellous scheme of organic development which is chronicled in the rocks!

I have reserved for final consideration a branch of the history of the earth which, while it has become, within the lifetime of the present generation, one of the most interesting and fascinating departments of geological inquiry, owed its first impulse to the far-seeing intellects of Hutton and Playfair. With the penetration of genius these illustrious teachers perceived that if the broad masses of land and the great chains of mountains owe their origin to stupendous movements which from time to time have convulsed the earth, their details of contour must be mainly due to the eroding power of running water. They recognized that as the surface of the land is continually worn down, it is essentially by a process of sculpture that the physiognomy of every country has been developed, valleys being hollowed out and hills left standing, and that these inequalities in topographical detail are only varying and local accidents in the progress of the one great process of the degradation of the land.

GEOLOGICAL TABLE

from Historical Geology *by Carl O. Dunbar*

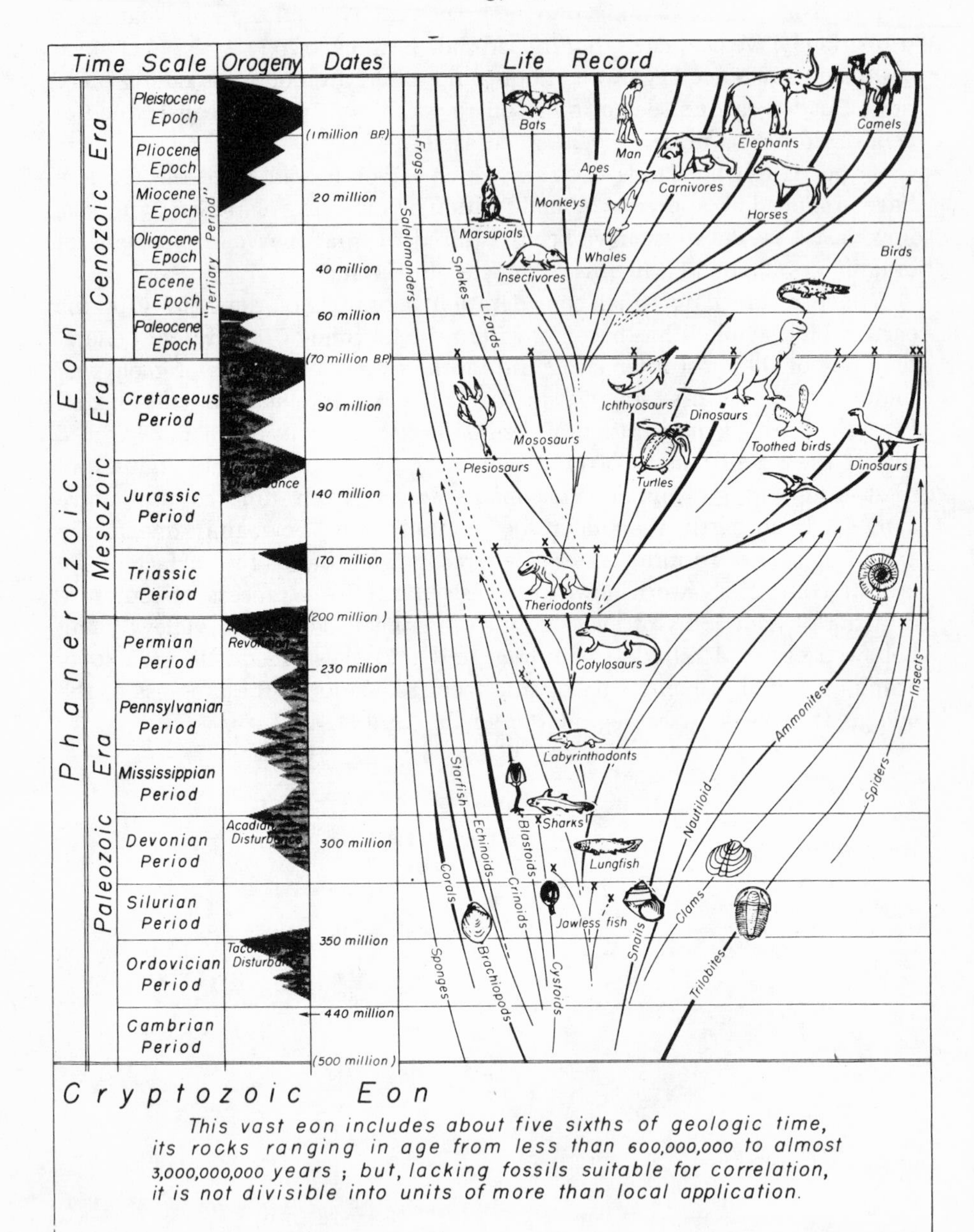

Cryptozoic Eon

This vast eon includes about five sixths of geologic time, its rocks ranging in age from less than 600,000,000 to almost 3,000,000,000 years ; but, lacking fossils suitable for correlation, it is not divisible into units of more than local application.

The geologic time scale in the history of North America. Space allotted to different time units is not to uniform scale. Shading to indicate time of chief orogeny is purely symbolical. A cross at the end of a line indicates extinction; a dart indicates that the group persisted to the present. Dates in parentheses are estimates; other dates are based on radioactive minerals.

W. O. HOTCHKISS

The Record of Living Things

•

Editors' Note:

The reader will observe a number of relatively minor discrepancies between the datings contained in this article and those in the preceding Geological Chart. *The Chart contains more recent estimates. There is, however, complete agreement about the basic thesis of the enormous span of earth history.*

In the story of the discoveries of the Scotch farmer, Hutton, it was indicated that successive beds of sand, lime mud and clay had been deposited over large areas. During the deposition of the sands and clays and lime mud, which were hardened into rock to make the cliffs that Hutton examined, there were likewise deposited the remains of various living things—shells, skeletons of animals, pieces of vegetation, all sorts of things that might leave their imprints when these unconsolidated materials were later hardened into rocks. Such remains of living things have been of great value in helping us to unravel the story of the past from the study of these various rock beds.

As men studied these remains more and more, they noted that individual beds were characterized by certain types of animal and vegetable forms. Other beds above or below were characterized by different types. About one hundred years ago it began to be recognized that these remains, which are called fossils, are so definitely characteristic of the beds in which they occur that they make a most excellent means of identifying particular beds wherever they are found. When a certain group of fossils was found in a bed in eastern New York and the same group of fossils in the same kind of bed could be traced clear across the country from quarry to quarry, from hillside to hillside, it became apparent that this particular bed must have been deposited in a sea of that extent and that conditions in this sea were favorable to the existence of this kind of shells.

As lower, and therefore older, beds were examined it was found that in a general way the living forms were simpler as the age of the beds increased. As the overlying beds were examined, it was found that the forms of life usually became more complex, until in the more recent of these beds remains of higher animals and of man were found. As a result of these studies it was found that the whole series of beds laid down in the past could be divided into groups characterized by the forms of life which they contained. On the basis of this life history of the past the geologic eras have been named.

The oldest rocks found are those of the Proterozoic Era. They contain either no evidence of life or evidence that is very hazy and indefinite. Such fossils as exist are chiefly of microscopic simple forms, single-celled animals and plants like those we find in the waters of the sea and lakes today. Few of these earliest forms of life possessed hard parts that could be readily preserved.

After long ages, toward the end of the Proterozoic Era, and at the beginning of the next era, some living forms began to protect themselves with a hard shell or "exterior skeleton." Still later on larger and more complex organisms appeared, various kinds of shellfish and other small forms of life, much like those we find along our seacoasts and lake beaches today. About the middle of this era the first fishes began to appear, the earliest animals to possess a backbone. Toward the close of the era some of these developed the capacity to breathe and so to live on land as well as in water and thus became the first of what we call amphibians. This period of life development is given the name of the Paleozoic Era, which means "early life era."

The third great era was characterized by the development of enormous land animals and has been called the "age of Reptiles." This era has been given the name of Mesozoic, meaning "middle forms of life." The great dinosaurs, which reached their highest development at this time, were among the largest land animals ever to inhabit the face of the globe. Most of them were animals that laid eggs, just as fish and turtles and alligators do at present. Most of them left their young to hatch out and care for themselves unaided from the day they were born. Their lives were easy, and they prospered and developed many different forms. Some were plant-eating and some were fish-eating. Some of them found it easiest to get their food by swimming in the seas of those days and gradually developed the capacity to live in water. These reversed the experience of their ancestors who had developed from fish that got tired of living in the water and developed lungs so that they could live on the land.

Toward the end of the Mesozoic Era a higher type of animal appeared which brought forth its young alive and cared for them and nursed them through a period of infancy. This great group of animals rules the earth today. They are known as mammals, a term which includes all animals which nurse their young. Just as some of the reptiles of the Mesozoic Era found an easier livelihood by returning to the sea to live, so in this later era some of the mammals found it desirable to live in the water. Thus were developed the whales, dolphins and porpoises, inhabitants of our seas today, which bear living young and nurse them. This latest era of geologic time, which followed the Age of Reptiles, or Mesozoic Era, is given the name of the Age of Mammals, or Cenozoic Era.

You will notice that the grand divisions of geologic time are known as "eras." These are divided into "periods" which are in turn subdivided into "epochs." Epochs are further subdivided into "ages," but that is getting too far into technical detail. The division of the geologic past into epochs and ages has been by no means fully worked out. Much remains to be done before our knowledge is complete. Further study will lead geologists to amend the geologic table as new facts are discovered. Geology, like all other sciences, is a progressing, developing state of knowledge, no field of which will be completely known for long ages.

Each of the divisions of geologic time began with some event that changed the previous conditions, such as the slow submergence by the sea of half a continent, or the elevation of a great chain of mountains. The most important changes marked the close of one era and the beginning of another. Less important changes of this kind marked the ends of periods, epochs and ages.

The close of the Paleozoic Era (early life) was marked by a great earth movement, estimated to have occurred about one hundred and eighty million years ago. This movement compressed and tilted the rocks that now make the Appalachian Mountains and raised them and the whole eastern part of the United States above sea level. How long a time this took we do not know, but we do know enough to feel quite sure that the process was too slow to be noticed by any casual observer, had there been one present. At the rate of one inch per year the highest peak in the Appalachians could have been elevated from sea level in seventy-five thousand years. In some parts of our country there are probably vertical movements of the earth's surface now going on at this rate, or even more rapid rates, which are entirely unnoticed by the people living there. Yet it is probable that the building of the Appalachian Mountains took place at a slower rate and occupied a much longer time.

The close of the Paleozoic and the beginning of the Mesozoic Eras, then, were marked by the mountain-building uplift that resulted in our Appalachian ranges. The end of the Mesozoic and the beginning of the Cenozoic Eras were similarly marked by the elevation and folding of the Rocky

Mountains. This was approximately sixty million years ago. Our Rocky Mountains apparently are mere youths, only a third as old as their feeble grandfathers in the East.

The Sierra and Coast ranges along our west coast are mere infants, only about a third as old as the Rockies or perhaps even less.

The events marking the ends of geologic eras have been mentioned because of their bearing on the story of living things shown by the rocks. The elevation of vast areas above sea level, or depression below it, changed living conditions very greatly. Ocean currents were deflected into new courses. Where there had been warm water there was perhaps now cold, and vice versa. Organisms that had thrived before died under the changed conditions or underwent modifications that adapted them to the change. New Types of organisms immigrated and ate up or drove out the old. So there were vast changes in the kinds of remains of life that were deposited in the sediments of the sea and shore. If you will consult a textbook of geology and will turn first to the illustrations of the fossils found in Paleozoic rocks and then to those found in Mesozoic, you will find that even without knowing anything about their long specific names or attempting to qualify yourself as a paleontologist (one who studies fossils) you can see quite notable differences.

There are a number of epochs in each era and the average length of each of these epochs for the last two hundred and ninety-nine million years is more than twenty million years. In one of the epochs there was time for many changes in living forms to take place. When we consider the wide variety of kinds and sizes of dogs and cattle produced in a relatively few centuries by careful breeding, and then try to think of what might happen in only one geologic epoch of an average length of twenty million years, and then, if imagination is not already stretched to the breaking point, try to still further multiply this by fifty to get the record of a billion years, it is not difficult to see how living things in their struggles for existence have had ample time to develop from the simplest forms of one-celled beings to that exceedingly complex, little understood, but "inordinately proud" being that calls himself man.

One of the great steps toward dominating the earth was the development of life forms that were able to live upon and occupy the land. All the earlier forms were probably water dwellers. In earliest times the land was barren of plants and animals. No living thing grew or crawled beyond the shore. If by accident it was left out of water, it promptly died. There is evidence that this condition prevailed in all the history of the earth up to a half-billion years ago. The landscape then consisted of bare rocks and sand and mud, deposits lying stark and naked as the rains and rivers of those ages left them. The landscape must have looked like a desert; yet in most of the world the rainfall was about as abundant as it is today.

Remains found in the rocks indicate that about five hundred million years ago, in the Cambrian Period, some few of the plants—which previously had all been sea plants—had found a way to live on land. They were the highest plant forms of their time, even though they were only algae and the simplest of mosses. They first found a precarious living along the rocks of the seacoast. Later on, after a hundred million years of progressive adjustment to their "new" surroundings, they began to look somewhat like some of the plants we see today.

About three hundred million years ago plants had developed swamp-living forms of moderate size, and in the Carboniferous Period, two hundred and fifty million years ago, great tree ferns and similar forms grew to a height of eighty feet. These forms and their associates accumulated in their swamp homes to considerable thickness, so that today they make one of our most valuable mineral resources. They used the sunlight of those days to transform water and carbon dioxide from the air into woody cellulose which was altered into the coal which we now use to drive our trains, run our factories and light our homes. In fact, we might truly say that our present civilization is largely based on the sunshine that fell on the earth two hundred and fifty million years ago.

Plants continued to develop newer and better and more complex forms. About two hundred million years ago, in the Permian Epoch, the first cone-bearing trees appeared. About one hundred million years ago, in the Cretaceous Period, seed-bearing plants developed in abundance. Most plants up to that time had been spore bearers like our ferns today. When the first seed bearers appeared, they were so much better fitted to land conditions that they quickly became the dominant type of vegetation, a position which they hold today. They include all our grasses and grains and our fruit- and nut-bearing trees. The spore bearers have literally been relegated to the shade by the seed bearers.

It is a most interesting thing to study this development of plant life as portrayed in the rocks. We see the evident striving for adjustment to prevailing living conditions. We see the plants making experiments, as it were, trying to see whether this kind of change or that would permit them to conquer and occupy territory from which they had hitherto been debarred. We find that some experiments were huge successes, the plants quickly multiplying to cover large areas. If the experiment was not a success or developed a form adapted only to a temporary set of living conditions, the plants soon died out and disappeared from the earth. Thus all through the record in the rocks we can read of gradual improvement and progress toward a perfection still to be attained. It is as though the Creator had implanted a yearning for perfection in the first living things and we were privileged to sit and watch this progress from the crudest beginnings to the present through the whole long procession of a billion years.

The development of animal life has a similar history. The earliest animal forms are concealed in the hazy and indistinct records of the very ancient

—rocks. At first all animals, like plants, were sea dwellers and were long in developing hard parts, such as shells or bones, that would not be easily destroyed and "sunk without trace."

In rocks of the Cambrian Period, which began more than five hundred million years ago, we find our first well-preserved evidence of an abundance of animal life. In older rocks, fossils are few in numbers not because living beings were scarce but because few of them had discovered how to utilize the lime in the water to build themselves stony protective armor. When this discovery was made, the rocks at once began to preserve the remains of an abundance of different forms of life, many of them quite complex in their organization. We find in these Cambrian rocks the remains of many hundreds of species of animals. Some are shellfish somewhat like our modern clams and oysters, others are similar to modern snails in having coiled shells, and still others are like corals in structure.

The largest and most complex animal of those times was a trilobite about twenty inches long. It possessed most of the organs found in the animals of the present day. It had a well-developed digestive tract, feeling organs, a coordinated muscular system, an external protective shell like the crab and lobster of today, eyes, and a well-developed nervous system with a central brain of minute size. It had all of our five senses with the possible exception of hearing and smelling.

The next step in the evolution of animal life, the conquest of the land, required what to us seems a long period of time—perhaps one hundred and fifty million years. It could not be completed until plants had become land dwellers. The earliest land animals found in the rocks are insects, spiders and scorpions, which appeared about three hundred and sixty million years ago in the Silurian Period, and reached a high development in the Mesozoic Era, which covered the period of time from one hundred and eighty million to one hundred million years ago.

The first vertebrate skeleton was owned by an ancestral fish that lived perhaps four hundred million years ago in the Ordovician Period. He had found that he needed something to keep his head from being driven back into his body as he swam about in search of his prey, and so he grew a bony skeleton and discovered that life was easier. His predecessors all had "external skeletons," or shells, that were good armor against his enemies but were cumbersome and cut down the speed with which they could navigate and catch the other organisms upon which they lived. This development of a backbone was an improvement of such great usefulness that all higher types of animals since that time have an internal jointed skeleton, the main feature of which is a flexible, jointed backbone. Nothing has been invented by Nature thus far that is better for its purpose than this great device. It has been the prime factor that has enabled animals to attain to great size.

Since the first animal with a vertebrate skeleton appeared four hundred million years ago, there has been progress in size, until today we now have the largest animal that ever lived, the great blue whale, which is known to have attained a length of one hundred and six feet. Progress was not rapid. One hundred million years had to pass after the invention of the backbone before animals ten feet long developed. It was not until the vertebrates took to living on the land that development to great size occurred. In the Age of Reptiles—the Mesozoic Era—when the dinosaurs and their kin were lords of creation, they so "quickly" added to their size that no more than thirty or forty million years of this era elapsed before they had attained to maximum lengths of seventy feet. This experiment of increasing the bulk of flesh inside one skin—or, to look at it from the inside out, the bulk of flesh surrounding a single backbone—was successful for about seventy-five million years, and the great animals of the period prospered for a time lasting from one hundred and fifty million years ago to seventy-five million years ago.

The development of these animals to greater size was not accompanied by corresponding brain development. The largest brain of those days was less than a quarter the size of yours or mine.

In the meantime Nature was making a different kind of experiment. The great reptiles did not have any marked maternal instincts. Most of them continued the practice which characterized the poor fish and lower animals that had been left behind in the race. They laid eggs and left their young to hatch out and care for themselves. About one hundred and fifty million years ago there appeared some small animals that hatched their eggs inside their bodies and produced living young, which they nursed and cared for through a period of helpless infancy. From this habit of nursing their young they have been given the name of mammals. With them the great quality of mother love first began to be an important factor in the life of the earth. Through at least eight hundred and fifty million long years of the billion-year story living beings got along with little or none of the mother love that is so powerful an influence in the lives of all of the higher animals today.

Mammals also gave up an old practice followed by all other living things, that of being cold-blooded. They found that to elevate the blood temperature gave them advantages over their cold-blooded associates. Warm blood and the habit of nursing their young were the most important new elements in life that distinguished the mammals. They had the same organs, muscles, nervous system and brain as their cold-blooded, egg-laying neighbors, but they had in the two new improvements qualities that were to make them, after a hundred millions of years had elapsed, the dominant type of animals on the face of the globe. For the last sixty million years they have prospered more extensively than any other kind of animal life.

The story told by the rocks relates that, after the success of the mammals as a dominant type of animal life, Nature began another major experiment, the development of a larger brain. The invention of the shell or external skeleton she had bettered by inventing the jointed backbone. This had made possible the reptilian conquest of the land. The reptilian experiment she had largely discarded after the successful development of mammals.

Before this last experiment the largest brain in the world was probably no larger than one-quarter the size of the average human brain, was much less finely organized, and was much less than a quarter as capable.

In early Tertiary time, during the first period of the ascendancy of mammals, there was one kind that began to live in trees. In all animals living on the ground the sense of smell had been highly developed and was most useful. In the tree dwellers this sense lost its importance, and the sense of sight became of greater value. Consequently, sight was developed more highly as the ages rolled on. The conformation of the skull changed so that the two eyes could look straight ahead and both eyes could see the same object. This change was associated with a corresponding development of the parts of the brain which related to vision. This made possible the kind of vision that enables its possessor to estimate distances. If a ground dweller does not estimate distance correctly when leaping for his prey, he only loses a meal. If a tree dweller misjudges the distance of a branch he leaps for, he is in a far worse plight—he is quite likely to lose his life. The quickness of movement of these tree dwellers undoubtedly made increased brain capacity a great asset to them, and so those with better brains prospered and propagated their kind.

Tree dwelling developed the grasping capacity of the extremities and the capacity to balance and walk on the hind legs. All these things made brains more of an asset. The grasping capacity of the forelimbs permitted the use of clubs, which began to come into style as weapons.

When the tailless, hind-foot-walking, manlike apes—anthropoids—appeared in the Oligocene Epoch, about forty million years ago as the story of the rocks reveals, brain capacity increased. Some of the best of these apes had brains one-third the size of man's.

The oldest erect-walking man, whose skull was found in the rocks of Java, had a brain 60 per cent as large as ours.

After the time of the Java man came the great change in climate which produced the Glacial Epoch, when great continental ice sheets spread over the land from the north. South of the margin of the ice the climate became colder. Man had to develop to meet this emergency, or perish. He already had the capacity to use clubs, bones and stones as weapons. At this time he probably learned to use fire, to live in caves, and to build himself shelters in which he could find relief from the cold. The remains found in the rocks tell this story in a sketchy fashion which as yet is far from satisfactory, but they do inform us positively of the facts given above and also of the size of his brain cavity, which was 90 per cent as large as ours.

This brings us to the scientific domain of the anthropologist, where we may leave the development of the human brain and its tremendous significance.

To see the success of this latest experiment of Nature—the development of the brain—you need only look about you, and, with the background of the story of a billion years, consider how completely man dominates the world in which he lives, how he has mastered fire, water, earth and air, how he has tamed the lightning of the heavens and made it his servant, how he has learned to bring wealth from a depth far below the surface and use it for his convenience, how he has multiplied in numbers and learned to control all other forms of life and the great forces of Nature so that they minister to his welfare.

I once heard a friend facetiously remark that it was too bad that Moses was not a better geologist. And continuing, he said, "If he had been, instead of writing a story of the *fall of man* and his decline from previous perfection he could have written a much more inspiring and hopeful tale of the *rise of man* from lowly beginnings in the remote past, with the unlimited possibilities of the future before him." It is truly a most satisfying experience in reverence for the Creator of all things to read in the record not made by human hands this great story of the past and then to turn and look into the future which is developing so rapidly before our eyes.

L. DON LEET
AND FLORENCE LEET

Earthquake: Discoveries in Seismology

•

I

It is generally agreed now that earthquakes originate when rocks of the earth break and snap into new positions. Sometimes the break can be seen, sometimes it cannot. But the instrumentally recorded vibrations are so similar for all earthquakes that we may assume that the vibrations all start in a similar way.

In a way the cause of earthquakes is explained by rocks breaking. But unexplained is the fundamental cause: what makes the rocks distort to the breaking point? One clue is that earthquakes occur around mountains now standing high above the sea, those that geologists call "young" mountains. But they also occur around "old" mountains, and mountains that are completely covered by the sea, like the mid-Atlantic ridge. There is a common factor here, though: earthquakes are associated with mountains and with whatever makes and changes the mountains. We now believe that the fundamental cause of earthquakes is the force that distorts rocks in the making of mountains. So let us look at this geological process which makes mountains.

Mountain chains are ranged around the Pacific Ocean, from east to west across the Mediterranean through the Himalayas to the Pacific, and forty thousand miles of total length across the middle of the ocean floors. Mountains differ widely in appearance, but where they can be studied on land they all have two things in common: (1) they are composed primarily of rocks of sedimentary origin; (2) the rocks have been squeezed mightily until the surface area they once covered has been drastically reduced.

To say that the rocks have sedimentary origin means that they have hardened from accumulations of sediments once under the ocean. As rain, wind, and frost wore away other rocks that once stood above the sea, these sediments resulted. The products of this erosion were carried to the ocean by streams. This natural process of making rock is not unlike the setting of concrete when sand and gravel have been mixed with some cement. In the case of sedimentary rocks, the "cement" varies, so some rocks are harder than others. Also the "setting up" under water is a very, very slow process, taking thousands of years.

An important feature of the sediments that make up mountains is that they accumulated to thicknesses of twenty-five thousand to fifty thousand feet, yet they are composed of sediments that gathered in "shallow," water-filled basins no deeper than a thousand feet. This is shown by the coarseness of sand grains and pebbles, and the types of fossils embedded in the rock. We conclude that as sediment poured into a basin that became a mountain, the basin bent down about as fast as the sediment poured in. These basins are called "geosynclines," which is a Greek-rooted word for "very large troughs." The gradual bending of the basin had to result from the squeezing of the crust from the sides, since the relatively lighter incoming sediments would not depress the heavier rocks on which they came to rest. We do not know what is squeezing the crust—just that it is being squeezed. So we have a simple but majestic picture of a trough slowly deepening in a shrinking crust, with sediments pouring into it. It has been calculated that the troughs deepen from five hundred to one thousand feet per million years; in other words it takes from twenty-five to fifty million years to gather together the sediments for a mountain range.

Sedimentary rocks usually accumulate in layers. Conditions that favor accumulation of sand (later to become sandstone) change to favor accumulation of clay (later to become shale) or lime (later to become limestone). These changes occur back and forth many times until the pile of sediments is layered like a cake, and the layers are nearly or entirely horizontal while accumulating.

Causes of such changes are numerous. For example, heavy rains on rugged hills or mountains might produce coarse sands and gravels. When hills are worn down to low levels, sediments are formed from them more slowly and frequently dominated by clays. Some limestones are formed by the shells of many animals. If there are no animals because of the coolness of ocean waters or muddiness of the seas, there will be none of this kind of limestone.

The sediments harden into rock as the trough deepens and the pile of sediments thickens. But as the rocks at the bottom of the pile are buried more and more deeply, other changes in them take place. They lose the brittleness they had near the surface and become plastic under the loads. Also they are heated as they sink slowly into the hot interior. This heating causes them to expand until they take up as much as 10 percent more space than they did when formed.

The big squeeze continues. But now the sedimentary rocks near the bottom of the pile no longer resist this squeeze as they once did. They flow like stiff tar under pressure and begin to extrude from the viselike sides of the geosyncline like toothpaste from a tube. Slowly, over millions of years, they are pushed up above sea level and slide out over other rocks at the surface. Eventually they reach a climax—the greatest heights they are to attain.

Meanwhile, erosion starts wearing away the rocks from the moment the first ones are shoved up above the sea. For a time—several million years—the squeeze elevates rocks faster than erosion removes them. But every mountain range eventually reaches the point where the rate of erosion exceeds that of development. The force of the squeeze seems to diminish, and the once-towering ranges are cut down to sea level, or near it. Sometimes there is a renewal of uplift, like the final struggles of a dying giant, but it doesn't push the rocks back to their greatest heights. In some of the old mountain structures of the earth, like the Laurentian Highlands of Canada and the Appalachians of eastern United States, there seems to be no push left at all.

During the process of making mountains, some rocks get melted deep down under the folding pile of sediments. The molten rock eats its way upward and feeds volcanoes at the surface. Some doesn't reach the surface, but after millions of years eventually hardens as "fire-made" or igneous rocks, forming a core of the mountain range.

With this story of the making of mountains in mind, we can see how earthquakes would be caused frequently during the process. The tilting, bending, uplifting, and general distortion could hardly help but break rocks from time to time and thereby set up the vibrations that are earthquakes. Where mountain-making forces are at work today, there are earthquakes.

The force that bends, breaks, and pushes up the rocks to make mountains is actually behind the cause of earthquakes. Where it is operating, the force seems to be continuous. It distorts rocks until they are like tightly coiled springs. When they break, the energy that went into distorting them is released all at once, and part of it goes into the movement of the mass of rock and part of it into making the earthquake. The force continues; more distortion is followed by another break and another earthquake. So the process goes on and on. Where earthquakes have happened, they can be expected to happen again, in time. But what is the source of the force, what causes the big squeeze? *We don't know.* It's as simple and as disappointing as that.

II

When rocks of the earth rupture to produce an earthquake, energy is released. The word "energy" is familiar to most people in terms of energy-producing foods, of the energy needed to climb stairs, of energy radiated to us from the sun. But what is energy? Scientists have attempted to define it by saying that *energy is the capacity for producing motion.*

When a baseball leaves the pitcher's hand, it carries to the plate the energy the pitcher imparted to it when he threw it; and when the batter hits it, the ball carries to the field, or over the fence, some of the pitcher's energy and all of the batter's. This moving baseball represents one method by which energy travels from one place to another: an object delivers the energy; it is called "missile transfer." However, there is another method of delivering energy from one place to another: that is by waves. And it is this wave method which is involved in the study of the energy released when an earthquake occurs.

There are many other things in our lives that involve waves. One of the more familiar ones is sound. Sound transfers energy from one place to another. It can be carried by the air about us, but its transfer does not result from a flow of air from the speaker to the hearer. The waves of sound agitate the air, and the agitation is passed along from molecule to molecule. The air does not move bodily from place to place with the sound.

What sort of disturbance does the air undergo as the sound waves pass through it? Particles of air in their path vibrate back and forth as the waves go along. The waves alternately push and pull particles of the air through which they travel. As one particle moves forward, it nudges the particle next to it, this in turn nudges the next, and so on. In this way the disturbance moves along. Each particle in the path, after moving forward a short distance, pulls back in the opposite direction, then vibrates back and forth until all the sound waves have passed.

When particles push together they take up less space, that is, their volume is reduced; when they pull apart, the volume is increased. So sound waves have been called volume-change waves. And they can travel through any material.

There is another way of thinking about the changes in air as sound passes through it: at a point in the wave's path, pressure changes rhythmically. The air, as you know, presses in on us from all sides with a force of about fourteen and a half pounds per square inch. This is called "atmospheric pressure." And as sound passes through the air, it causes increases and decreases in this pressure. These changes are small: about a thousandth to a ten thousandth of a pound per square inch in a loud shout, and three millionths of a pound per square inch in the faintest sound the human ear can hear. Sound is a series of moving pressure changes.

One characteristic of sound is determined by the number of times it causes the air pressure to fluctuate in one second. This is called "frequency." Frequency determines the pitch of a sound. If sound pressures on your ear drums alternate 264 times a second, you hear a pitch called "middle C."

Another characteristic of sound is the speed with which its pressure changes travel. In air, this is about 1130 feet per second, in sea water, about 4900 feet per second, and in granite, about 18,000 feet per second. The speed does not change with the frequency and loudness.

Frequency and speed combine to describe wavelength, another feature of sound waves—in fact, waves of any kind. If the frequency is a hundred, for example, that means that a cluster of one hundred waves passed a point in one second. So if the speed is a thousand feet per second, the first wave of the cluster would be a thousand feet away as the last one passed, and the hundred waves would be spread out evenly over a thousand feet. This would mean that they were ten feet apart, and we call this separation the wavelength. For sound waves, the wavelength is the distance between adjacent places of maximum pressure.

As sound gets farther and farther from its source, it becomes weaker, because its original energy is used up pushing molecules in its path back and forth. Finally it dies out completely.

If sound hits a mountain, wall, or other solid surface some of it bounces back. This bouncing off is called "reflection" or "echoing." And when it takes place back and forth and up and down in a room, we call it "reverberation." The great difference between the sound of things in open air and in a room or auditorium is that in the open, there is no reverberation and inside there is.

Sound travels through anything that resists attempts to change its volume: gas, liquid, or solid. And since sound waves go through the body of these things, they are called body waves. But there are also waves of other kinds that travel only on the surface of a material. Water waves are the most common everyday example of surface waves. Simple water waves are started by tossing a pebble into a quiet pond. Waves spread away from where the pebble hit, carrying with them some of the energy delivered to the water by the pebble. These waves stay on the surface of the water. They travel at a definite speed, however, just as sound waves do. And they have frequency and wavelength. In this case, however, the wavelength can be seen. It is the distance between adjacent crests.

In connection with surface waves, also, there is the question "What is a surface"? Is it a molecule-thick film on the top, or what? It is actually the part of the top of the medium that gets involved in transmitting the waves. You can't wrinkle a surface without disturbing material immediately below, but this has to stop someplace, and it does. Submariners, for example, have learned that when the sea is rough and they submerge, there is some turbulence at shallow depths, but they enter quiet water as they go deeper. Actually, the depth to which water is agitated by the passage of surface waves is about the same number of feet as the wavelength of the waves. So in that sense, the "surface" is a few inches thick for small ripples and tens of feet for large rollers.

We have mentioned the wavelength of water waves as the distance between adjacent crests. Between crests, of course, are troughs, and as waves move along, the originally flat surface of water is shaped into crests and troughs. These may be low and shallow or high and deep. This involves a property called "wave height" or "amplitude," that is, the greatest height to which a particle goes above the level at which it stood when the water was calm.

There are also surface waves from earthquakes. They too, have "crests" and "troughs," frequency, speed, and wavelength. But the crests and troughs are too small to be seen. We know they are there from seismographic records and in addition to frequency, speed, and wavelength, these waves also have amplitude. Their amplitude is the amount the top of the earth's surface moves up or down from where it was before the waves arrived.

Some of the energy released when the rupture of rocks causes an earthquake is carried away by waves. The first of these are just like sound waves: they vibrate particles as they pass by pushing and pulling them. They have frequency, speed, wavelength, amplitude. Amplitude, for them, is the distance a particle is pushed or pulled from where it was when the wave arrived. These waves die out as they get farther and farther from the source; and they echo and reverberate at boundaries where rocks change, just as sound waves in air do when they encounter a change at walls or other solids.

These push-pull waves from earthquakes are called "primary waves" (abbreviated P), because of all earthquake waves they reach distant points first. From earthquake records, their frequencies have been found to be around one per second or less. When less than one wave passes per second, it is convenient to give the length of time the wave takes to pass, rather than to say some fraction of a wave passes per second. The time it takes for a wave to pass is called its "period."

Another part of the energy from an earthquake is carried away by waves called "secondary" (or S) waves. These cause particles in their path to vibrate in directions at right angles to the direction of the waves' movement. It is a shake motion. These waves also have frequency or period, wavelength, and amplitude, and they die out with distance. They echo and reverberate at boundaries where rocks change. Their periods may be a few seconds or up to ten or more. Because of their more complicated motion, their speed is always less than that of P waves in the same material. Unlike P waves or sound waves, however, S waves can't travel in a liquid or a gas, because these materials do not resist the kind of motion S causes.

Both P and S are body waves. They travel into and through all parts of the earth. When they reach the surface and disturb it sufficiently, they generate surface waves. Some earthquake surface waves have periods of twenty or more seconds, and the "surface" for them is the top forty miles of the earth because their wavelengths are forty miles.

Waves in the ground are recorded on instruments called seismographs.

III

The founder of modern seismology was John Milne. In 1875 the Emperor of Japan appointed him professor of geology and mining in the Imperial College of Engineering at Tokyo. Though there were so many small earthquakes that he had them "for breakfast, dinner, and supper, and to sleep on," he didn't become actively curious until he was vigorously shaken by one on February 22, 1880—the strongest felt in Tokyo since his arrival. Then he wanted to know the how, where, and when of these events.

He began to read all the literature on earthquakes he could get. Imagine his surprise to find in this earthquake-ridden land only catalogues listing dates and places where earthquakes occurred, together with much speculation on earthquake causes.

He read about attempts to use mechanical devices that were expected to indicate the direction from which an earthquake came. Milne was not satisfied with these small bits of information. Nobody had determined the direction in which the ground moves: north, south, east, or west; how much it moves: two inches, two feet; whether it moves twenty times a second or twenty times an hour. Milne felt that specific knowledge of the nature of earthquake motion would have specific bearing on the problem of constructing buildings to resist the shaking. He also felt it would aid him in tracking an earthquake to its source and in reasoning how it was produced. So he set about to obtain more exact knowledge of the nature of earthquake motion. Knowing that an earthquake's vibrations could be visibly recorded, Milne set about to develop an instrument that would write them down in a language he could read—a separate line for each principal direction, and spread out in a continuous scrawl rather than jumbled on a fixed piece of paper in a grand tangle.

By November, 1880, we find John Milne and a colleague of his, John Gray, performing experiments to simulate an earthquake and then recording it. He hoped this would aid him in developing better instruments to record natural earthquakes. They arranged for a one-ton ball of iron to drop from heights of ten to thirty-five feet. This compressed the ground and generated earth vibrations similar to those of a natural earthquake. The energy of these vibrations was carried away by waves in the usual way. And Gray and Milne measured vibrations of the ground at different places as these waves passed. They used a simple device Gray had designed, consisting of a free-swinging pendulum suspended on a string. As the ground vibrated, the pendulum's bob remained stationary and a record of the earth vibrations was scratched on a smoked glass plate geared to slide sideways under the pendulum. From their records, they observed that there were to-and-fro motions, "compressional vibrations," along the line from the ball to the observing point (which they had expected). But they also found that there were side-to-side motions, "distortional vibrations," which they had not expected. They could not record any up and down movements because of the instrument's design.

This was a significant start. However, the glass plate on which the artificially generated vibrations were recorded had serious limitations for registering natural earthquakes: it was only large enough to record for a few seconds, while waves from local earthquakes sometimes vibrated the ground for minutes. If Milne had tried to use it to record local earthquakes, it would have given him just the vibrations from the first few waves, and he wanted all of them to give him a complete picture. He also needed to have an instrument capable of recording movements in all directions. The instrument he used to record vibrations caused by the dropped ball was designed only to respond to horizontal motion.

Further research finally resulted in the Gray-Milne seismograph, which was first used in 1883 to record local earthquakes. This instrument had three pendulums, which were fastened in such a way that each responded to motion in a different direction. Two of these directions were horizontal (at right angles to each other). The third direction was vertical. Its recording mechanism was capable of registering earth vibrations continuously for twenty-four hours at a stretch before the paper on which it wrote had to be changed. It wrote in ink, by means of fine siphons, and it drew three lines, one above the other on a continuous strip of paper. Gray and Milne knew that vibration of the ground from some earthquakes is so small it must be magnified to be seen. Their newer seismograph incorporated a system of levers to produce the needed magnification.

The vibrations of the ground showed on the record as wriggles of each line. Their seismograph was oriented so that one line wrote horizontal vibrations in the east-west direction only; one line wrote horizontal vibrations in the north-south direction only; and the third showed vertical vibrations. So if an earthquake vibrated the instrument in any direction, a record of the movement could be seen. There was also an arrangement for marking the hours on each record.

Milne found that this instrument could pick up from nearby earthquakes vibrations that passed at the rate of around sixty per minute, but he felt that a different instrument could be built that would respond to the vibrations from distant earthquakes. He suspected that some of the vibrations from more distant earthquakes would pass at the rate of two to ten per minute. He knew that the pendulums designed for the Gray-Milne seismograph could not respond to such slow vibrations.

Every seismograph has a natural period: the length of time it requires for its pendulum to complete one swing. Earth vibrations also have a period: the length of time required for one complete vibration of a particle of earth. The seismograph's period is determined by the way the instrument is built. Its period must be equal to or greater than the period of earth-particle vibrations the seismograph is to record. Short-period earth vibrations are about one second or less; long periods range up to sixty seconds or even more. If vibrations have a period of one second, sixty of them will pass in a minute; if they have periods of ten, twenty, or thirty seconds, they pass a seismograph at the rate of six, three, or two per minute. Some seismographs are built for recording short-period vibrations; others for long-period vibrations.

Milne looked about for means of constructing a seismograph with a longer period than the Gray-Milne had. While he was doing this, Ernst von Rebeur Paschwitz in Germany was using an instrument to prove the moon caused tides in solid rock similar to the tides of the oceans. The weight that responded to the tides was hung in a special way, so it would swing very, very slowly, that is, with a long period. His recording paper moved about a quarter of an inch an hour through the machine. On April 17, 1889, Rebeur Paschwitz found an unusual set of vibrations lasting a couple of hours and riding as an unwanted blur on the leisurely undulation of his tilt record. He suspected spiders, because these industrious inhabitants of dark vaults occasionally latch onto the movable part of an instrument, tie it to the pier, and then run back and forth to produce remarkable registrations as they construct webs. But he found no spiders, and the mystery was great. Later, he heard and made note of a report that at the time of his strange record there had been an earthquake nine thousand miles away in Japan. The coincidence was remarkable, but Rebeur Paschwitz had to recognize that maybe it was just a coincidence.

Then three months later, on July 28, 1889, there was another earthquake in Japan, and another "coincidence" on Rebeur Paschwitz's records. This opened the door, and as soon as the recording-paper speed was increased to spread out the registration from an unintelligible fuzz to a distinct series of wave groups, the regular recording of distant earthquakes became a reality.

Six years earlier, Milne had written, "It is not unlikely that every large earthquake might, with proper instrumental appliances, be recorded at any point on the land surfaces of the globe." His dream was about to be realized. By 1893, he had built a slow-moving pendulum capable of picking up long-period waves. By easy adjustments, it could be made to respond to almost any long-period earthquake vibrations that Milne expected to find. Now he was regularly recording distant earthquakes.

Milne left Japan in 1894 and settled at Shide, not far from the center of the Isle of Wight off the southern coast of England. There, within three weeks, he built two long-period pendulum seismographs, using as the main support of each an old lamp post. He made six-foot booms of bicycle tubing to support the pendulum bobs, wrapped smoked paper around a revolving drum, and had the records scratched on these by wisps of grass straw grown in his south pasture. Before long, however, he had refined the design, reduced the size, and arranged the recording so it was done by a beam of light directed onto photographic paper.

By this time, seismographic observatories were being set up in many parts of the world. But there was an obvious need for some arrangement to collect and analyze the records from these in some central place. Milne was the first one to do this on a systematic basis, and Shide became the international seismographic capital of the world until the time of Milne's death in 1913.

IV

When Milne and others finally got records of earthquakes spread out before them so that their waves showed clearly, they found that many records could be divided into three well-defined parts. The beginning was small motion, which they called "first preliminary tremors." Then, sometimes a minute or two later, sometimes as long as eight or ten minutes later, the motion suddenly became larger. They called this the "second preliminary tremor." After this continued for a few minutes, a third part began. This part contained the largest motion of all, and Milne called it "large tremors" or "principal part."

At the turn of the century, R. D. Oldham first published the explanation for this three-part character of earthquake records: the "first and second preliminary tremors" are set up by waves that traveled from the quake's focus to the seismograph station through the interior of the earth; the "large tremors" were set up by waves that traveled along the earth's surface. He recognized that the first tremors, which he represented by the symbol "P" (for primary), are vibrations transmitted via waves through the earth by a pushing and pulling motion, just as sound travels through air or a pulse goes along a stretched spring. And the second tremors, which he represented by the symbol "S" (for secondary), traveled via waves through the earth by a shearing or twisting motion. The push-pull motion enables waves to travel through a mass of rock faster than the twist motion, because the twist is a more complicated movement and takes longer to complete. In most rocks, the wave with the push-pull motion travels about 1.7 times as fast as the twist. Waves of push-pull motion and twist motion start together at the focus of an earthquake. The push-pull outruns the twist as the two travel through the rocks of the earth, and the farther they travel, the farther the push-pull gets ahead.

The recognition of push-pull and twist vibrations of the ground is made possible by recording vibrations moving in three directions, such as having one seismograph for picking up north-south movement, one for east-west, and one for vertical. So if the ground moves toward the northeast, the northward movement is recorded by the north-south instrument and the eastward movement by the east-west. Likewise if there is an up-and-down motion, this shows on the records from the vertical seismograph.

The large-motion waves, which are represented by the symbol "L" (for large), travel along the surface of the globe's rocks, just as waves on water move away from a stone thrown into a quiet pond. L waves are started when P and S waves disturb the surface of the earth above the focus of an earthquake. L waves have crests and troughs, like waves on a water surface. And the distance between two consecutive crests is their wavelength. They are called "surface waves" because they travel along the surface of the earth's rocks. But, of course, the surface can't move without

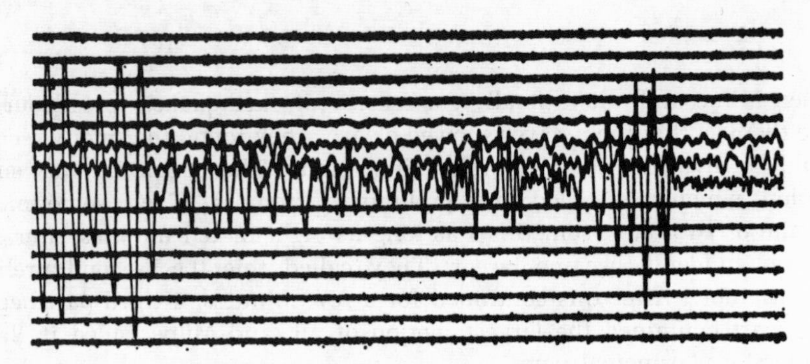

Fig. 1. S waves and the beginning of L waves from a large distant earthquake. Pips along the lines are marks placed there once a minute for time reference. Each line was written half an hour after the one above it.

some associated movement in the material below that supports it and is attached to it. So one question we might ask about surface waves is "How far down is there disturbance as surface waves pass?" The material immediately below the surface moves almost as much as the surface itself. But deeper and deeper down, the movement is less and less, until there is no movement. The total depth of disturbance is about equal to the wavelength of a surface wave. So surface waves that are five thousand feet from crest to crest vibrate a section of crust about five thousand feet thick. And surface waves that are fifty miles between crests vibrate a section of crust about fifty miles thick.

By 1900, Milne was getting enough records from stations around the world to start setting up time schedules for P, S, and L waves. He found that after P waves are recorded, the number of minutes and seconds before S waves arrive depends on the distance of the recording instrument from the earthquake focus; and this time interval *is different for every distance.* This was a major breakthrough in the history of seismology. The time interval between the arrival of P and the arrival of S (called the "S-minus-P interval") turned out to be twenty seconds at a hundred miles from the focus of an earthquake, two minutes and 45 seconds a thousand miles away, eight minutes four thousand miles away, and eleven minutes and 49 seconds seven thousand miles away. Now, a seismograph station's operator would be able to tell how far he was from an earthquake without reading the newspapers: if S arrived eight minutes after P, the quake was four thousand miles away; if S arrived two minutes and forty-five seconds after P, the quake was a thousand miles away. Then if three or more stations recorded the quake and determined their distances from it by using the S-minus-P interval, distance circles drawn around each station would have one common intersection—the place where the quake occurred.

The S-minus-P interval was used to determine a station's distance from an earthquake. It was used from several stations to determine the place where the earthquake occurred. But it could not be used to figure out what time of day the earthquake started. The method for doing this came out of a study of the San Francisco earthquake waves of 1906. By the time this quake sent its waves coursing through and around the earth, there were 96 seismograph stations ready to record them.[1] They ranged in distance from 80 miles in Mount Hamilton, California, to 11,170 miles at the island of Mauritius in the Indian Ocean, nearly the opposite side of the globe. These distances were known because the break that caused the earthquake was evident on the surface. Harry Fielding Reid gathered records from the stations and with them was able to take the next important step: figuring the time of day at which the waves all started out from the focus.

By plotting the arrival time of the P wave at each recording distance on a graph, he found that the points formed a smooth curve. By extending this line backwards to zero distance, he was able to set the exact time of day at which the waves started. Then, with the starting time known, and the arrival times at the recording stations known, he had plotted on the graph between these the actual time of travel for P waves to these stations and to all other distances from the source out to 11,170 miles.

When John Milne died on July 31, 1913, his work of collecting worldwide data, locating earthquakes, and studying the travel times of earthquake waves was taken over by H. H. Turner. Turner was Savilian Professor of Astronomy at the University of Oxford, and had assisted Milne in his work. As years passed and mounting quantities of data accumulated, Turner noticed that every once in a while the regular behavior that characterizes the arrival of P and S waves from most earthquakes was broken. P and S waves from these maverick quakes got through to recording stations too soon. In traveling to the opposite side of the globe, they sometimes arrived more than a minute too soon. He suggested that the *focus* was deeper for these quakes than for most earthquakes, believed to be within the outermost thirty or so miles of the earth's crust. Turner stirred up a hornet's nest with his suggestion, and it was several years before there was general agreement that he was right. The trouble was that seismologists found it difficult to believe that the forces that act on the rocks to cause an earthquake could be effective hundreds of miles below the surface. We have since recorded waves from earthquakes that have originated as far down as 435 miles. These are called "deep-focus" quakes.

[1] Twenty-five of these stations were equipped with Milne seismographs.

Another striking feature of the records from deep-focus earthquakes is the absence of L waves, which are always present on records of shallow quakes. Their absence is explained by the fact that when P and S waves reach the surface from a depth of hundreds of miles, they don't have enough energy left to start surface waves.

Even though P and S waves from deep-focus quakes do not have enough energy to move the earth's surface sufficiently to generate the larger movements of surface waves, they do hit the surface and bounce back into the interior. These waves which bounce back are reflected P and S waves, which travel through the earth as if they had started from a focus on the surface.

More than fifteen hundred miles away, stations record these reflected P and S waves from deep-focus quakes in addition to the P and S waves that travel to their stations directly from the focus. The time interval between the arrival of the direct P and S waves from the focus and the reflected P and S waves from the surface can also be used to determine the depth of focus of an earthquake. P and S waves from a 300-mile-deep focus take roughly three times as long to reach the surface and to be reflected back inward as do the P and S waves from a hundred-mile-deep focus. The reflected P waves of a deep-focus quake are called "pP," and S waves, "sS." The time interval between pP and P for the 435-mile-deep earthquake was two and a half minutes at a distance of 5000 miles.

As years passed, more and more of the details of records from earthquakes were recognized. For example, we learned that the length of time vibrations record at a station depends for a given earthquake upon the distance from the earthquake to the recording station. Within a few miles of the earthquake focus, vibrations are recorded for only a few minutes. At a few hundred miles, vibrations may be recorded for fifteen minutes to half an hour, because P, S, and L become separated by their different speeds of travel. And a few thousand miles away, vibrations may be recorded for several hours.

Historical accounts of large earthquakes have sometimes contained statements that the ground was in continuous motion for hours, days, or months. Seismograph records show that in a sense that is true, but the motion is actually separated into a series of vibrations, with each series of vibrations coming from a separate aftershock.

Vibrations caused by P waves have periods of from one to five seconds; that is, each P wave requires that amount of time to pass a recording station. S waves have periods of from one to ten seconds, and L wave periods range from ten seconds up to a minute or more. The actual amount the ground is displaced by the vibrations from the passing waves is quite small: around .0002 inch for a strong P wave, double that for S, and commonly up to .004 inch for large surface waves (seldom greater than .020 inches for the very largest). P and S waves and some L waves generate such small amounts of movement that seismographs must be made to magnify them thousands of times so they can be seen on a record.

JENARO GONZALEZ
AND WILLIAM FOSHAG

The Birth of Parícutin

•

MANY THOUSANDS of volcanoes, old and young, are scattered over the earth's surface. Some that are very old have been reduced to traces of erosion, others are still perfectly preserved in their essential form, although cold and inactive. About five hundred volcanoes are known to have been active within historical times, although the number of volcanoes in eruption at one time is never large. With very few exceptions the active volcanoes are old and well-established features, antedating the history of man by many years, some having their beginnings a million or more years ago.

In all recorded history there have been but six instances reported of a new volcano being born, that is, originating at a spot with no evidence of previous volcanic outbursts. To these six, we can now add a seventh—Parícutin Volcano—that arose in a corn field in Mexico on February 20, 1943. The only previously recorded instance in which the outbreak of a new volcano was actually observed is that in Chinyero, Tenerife, which opened up about one hundred meters from a farmer and his son.

In the case of Parícutin Volcano, four persons actually saw its beginning at very close hand, and their observations furnish the first adequate account of this rare phenomenon. Others visited the spot after the first outbreak, and scientists were soon on hand to submit it to detailed study. Since there are so many apocryphal accounts of what took place during the first moments of Parícutin Volcano, we will quote, as accurately as possible, the narration of events as recounted to us by the actual eyewitnesses. It is remarkable that, in spite of the tremendous shock and overwhelming fear induced by this sudden apparition, there is so little apparent distortion in their observations.

Before the Volcano

In the lands of Rancho Tepacua there existed for many years a small hole. Both Dionisio Pulido and his brother Dolores mentioned it as having existed all during their tenure of the land. Each year they cast dirt and debris into this cavity, but it showed no appreciable signs of becoming filled. Sra. Severiana Murilla, now an old lady, recalls how as a child, more than fifty years ago, she played about this small pit. She remembered it well for two reasons; first, because her father warned her to avoid the spot, saying that it was the entrance to an old Spanish mine (although no mining activity has been recorded in the area); and second, because one frequently heard subterranean noises, as if made by falling rocks, near the hole. Further, they amused themselves around the hole because it emitted a pleasant warmth.

Early February is the season of Barbecho, the first plowing of the year, in preparation for the season's sowing. At this time the villagers are in their fields busily engaged in their various tasks. On February 5th the first premonition of the impending disaster was noticed—the earth began to tremble. With each succeeding day the tremors increased, both in number and in violence. Subterranean noises, too, could be heard with increasing frequency and intensity. These seemingly unnatural manifestations kept the inhabitants in constant turmoil and fear. The earth tremors became so frequent and so violent that it was feared the great church of Parangaricutiro, with its massive walls of masonry more than a meter thick, would collapse. As a precaution, the sacred image of the church, El Señor de los Milagros, famous throughout the region for its miraculous powers, was placed in the main plaza, near the village cross, and by a strange coincidence faced directly toward the spot where the volcano would appear.

February 20, 1943

February 20th was clear and calm. Dionisio Pulido left his village of Parícutin to prepare his farm "Cuiyútziro" for the coming sowing. With him he took his oxen and his plow. He was accompanied by his wife Paula and his son, who would watch the sheep, and Demetrio Toral (who died a short time ago in Calzontzin) to help with the plowing.

In the afternoon, after midday, I joined my wife and son, who were watching the sheep, and inquired if anything new had occurred, since for two weeks we had felt strong *temblores* in the region. Paula replied, yes, that she had heard noise and thunder underground. Scarcely had she finished speaking when I, myself, heard a noise, like thunder during a rainstorm, but I could not explain it, for the sky above was clear and the day was so peaceful, as it is in February.

At four o'clock I left my wife to set fire to a pile of branches which Demetrio and I and another, whose name I cannot remember, had gathered. I went to burn the branches when I noticed that at a *cueva*,[1] which was situated on one of the knolls of my farm, a fissure had opened, and I noticed that this fissure, as

[1] Variously referred to by Pulido as a *cueva* (cave or grotto), *resumidero* (a hole or crevice into which water disappears during the rainy season) or *agujero* (a hole).

I followed it with my eye, was long and passed from where I stood, through the hole, and continued in the direction of the Cerro de Canijuata, where Canijuata joins the Mesa of Cocojara. Here is something new and strange, thought I, and I searched the ground for marks to see whether or not it had opened in the night but could find none; and I saw that it was a kind of fissure that had only a depth of half a meter. I set about to ignite the branches again, when I felt a thunder, the trees trembled, and I turned to speak to Paula; and it was then I saw how, in the hole, the ground swelled and raised itself—two or two and one-half meters high—and a kind of smoke or fine dust—gray, like ashes—began to rise up in a portion of the crack that I had not previously seen, near the *resumidero.* Immediately more smoke began to rise, with a hiss or whistle, loud and continuous, and there was a smell of sulfur. I then became greatly frightened and tried to help unyoke one of the ox teams. I hardly knew what to do, so stunned was I before this, not knowing what to think or what to do and not able to find my wife or my son or my animals. Finally my wits returned and I recalled the sacred Señor de los Milagros, which was in the church in San Juan (Parangaricutiro) and in a loud voice I cried: "Santo Señor de los Milagros, you brought me into this world—now save me from the dangers in which I am about to die," and I looked toward the fissure from whence rose the smoke, and my fear, for the first time, disappeared. I ran to see if I could save my family and my companions and my oxen, but I did not see them, and thought that they had taken the oxen to the spring for water. When I saw that there was no longer any water in the spring, for it was near the fissure, I thought the water was lost because of the fissure. Then, very frightened, I mounted my mare and galloped to Parícutin, where I found my wife and son and friends awaiting, fearing that I might be dead, and that they would never see me again. On the road to Parícutin I thought of my little animals, the yoke oxen, that were going to die in that flame and smoke but upon arriving at my house I was happy to see that they were there.

At no time did Pulido notice any heat in the ground about the spot.

Aurora de Cuara, wife of Gregorio Cuaro Sota, had been with her family at their farm at San Nicolas, some twenty kilometers from Parangaricutiro. All during the day they felt very strong earth tremors and heard subterranean noises. Aurora and her children were returning afoot to Parangaricutiro along the road that leads directly past Cuiyútziro. At 4:30 P.M. they reached the foot of the Piedra del Sol, precisely at the time when the ground opened up.

As I passed the Piedra del Sol, I felt very heavy earth shocks and saw the earth open up, like a fissure. From this fissure arose a smoke of very fine gray dust to about one-half the height of the nearby pine trees.

Although terribly frightened Aurora clambered to the summit of the rock, in order that she might see what was happening. The fissure was about fifty meters distant. There was no "thunder" but she was able to see that not only smoke and gray dust, but also "sparks" rose from the fissure. She could see Pulido assist his helper unyoke the oxen but could not see Paula because a grove of pines obscured a full view of the farm, and she saw the two men flee in fright toward the village. At the Piedra del Sol, one could hear noises like a roar or like a stone falling down a deep well and striking the sides.

Dolores Pulido, brother of Dionisio, was working in the forest on Cerra de Janánboro. He saw smoke arising from his brother's land and went to see what had taken place. He reached the spot about 6 P.M. and saw, from a distance of eight meters, smoke issuing from a vent in the ground. About this vent were low mounds of fine gray dust. He was unable to approach closer because of falling stones. He then took fright and fled.

In Parangaricutiro, Luis Ortiz Solorio was standing near his house, talking to his neighbor, the shoemaker. It was a quarter past five in the afternoon. Looking toward Quitzocho, he saw a thin column of smoke arising. He went to the plaza, where many people had gathered in front of the church, for news had come that the earth had opened up and smoke was issuing from a crack in the ground. The Cura, José Caballero, with the consent of the Presidente, Felipe Cuara Amezcua, decided to send a group of men to the spot to see what had taken place. Solorio offered to go, also Jesús Anguiano, Jesús Martinez, Antonio Escalera, and Miguel Campoverde. Since the Cura believed this mission would be a dangerous one in which they might lose their lives, and to give them spirit, as well as valor, he gave them his benediction.

They went by horse, riding rapidly, and very soon came to the spot, the first two to arrive being Jesús Anguiano and Jesús Martinez. They found that the earth had opened, forming a kind of fissure, at the extreme southern end of which was a hole about half a meter across, from which issued smoke, and red-hot stones were thrown into the air a short distance. Anguiano, desirous to see what was taking place in the hole, approached the spot, when Solorio cried out to come back, the side was about to collapse. Scarcely had he leapt back when the wall fell in, widening the orifice to two meters across, and the column of smoke increased in size.

According to Anguiano, the orifice was pear-shaped and from this cavity arose a fine gray dust like ashes and "sparks," and stones were thrown out without much force to a height of five meters. A choking odor pervaded the spot. In the vent the sand was "boiling" like the bubbling sand in a rising spring, with a noise like a large jug of water, boiling vigorously, or boulders dragged along a stream bed by a river in flood. About the vent small mounds of fine dust half a meter high had gathered. This fine ash was very hot but Anguiano collected some in his handkerchief as well as two of the hot stones.

The ground shook violently, "jumping up and down, not with the swaying motion they had experienced in Parangaricutiro."

They decided then to return and report what they had seen, and they carried with them the ash and the two stones. The stones were delivered to the Cura, and being still hot, they were placed in a dish, and the Cura exorcised them, that the volcano might cease. The Cura and others then consulted a book on Vesuvius in the library of the church, and it was decided that what they had seen was a volcano, which greatly astonished the gathered people.

Between six and nine o'clock the volcano began to throw out large stones, and at ten o'clock, one could see clearly, from Parangaricutiro, through the pine trees, incandescent rocks hurled out, but without any thunderous noises. Between eleven o'clock and midnight the volcano began to roar, huge incandescent bombs were hurled into the air, and flashes of lightning appeared in the heavy ash column.

On the morning of February 21st, Pulido drove his oxen to the forest to graze and then went to his farm to see what had taken place. At eight o'clock the volcano was about ten meters high. It emitted smoke and hurled out hot rocks with great violence.

With the outbreak of the volcano, the earth tremors ceased, much to the relief of the populace. The Cura and Presidente allayed their fears somewhat, but on the morning of the 21st a strong earthquake threw them into panic and they abandoned their homes, those from Parícutin fleeing to Parangaricutiro, those from Parangaricutiro to Angahuan or Uruapan, and those from Angahuan to the mountains.

During the morning of the 21st the activity of Parícutin Volcano greatly increased in intensity, casting out great quantities of incandescent material to build up its cone. By midday its height was variously estimated at thirty to fifty meters. The amount of ash, however, was relatively small and the eruptive column of less size and vigor than appeared some weeks later.

The first lava began to flow within two days after the initial outburst, perhaps sometime during the day of the 21st. It issued as a viscous mass, spreading slowly over the fields of Cuiyútziro and Quitzocho. It moved slowly, about five meters per hour, forming a rugged sheet of torn and jumbled lava fragments.

Later Growth

Parícutin Volcano continued to grow with startling rapidity. On February 26th it had reached a height of more than one hundred and sixty meters, and its explosive activity had increased to an awesome thunderous bombardment, in which immense quantities of viscous lava were hurled continuously into the air; the noise of these tremendous explosions could be heard in many remote corners of Michoacan, and even in Guanajuto, three hundred and fifty kilometers to the northeast.

In late March the first lava ceased flowing and the eruptive activity changed to a heavy emission of ash, the eruptive column rising to a height of more than twenty thousand feet. This ash covered the countryside for miles around, ruining the fields and destroying the forests.

In time the lavas reached both Parangaricutiro and Parícutin, engulfing and destroying them, and scattering their inhabitants to other areas. Celadonio Gutierrez wrote on February 20, 1946, in his simple diary of the life of the volcano:

> Three years ago my village existed tranquilly, without any warning of the volcano as it exists today. Three years ago all parts of this region were beautiful, with fruit trees in the villages and in the fields, green pastures, beautiful lands that demonstrated the wealth of the region, with cattle and sheep, and droves of horses that grazed in the rich fields. Now there remains for me only a memory and a pride to have known it as it was three years ago.

Parícutin Volcano continued to grow, although most of the later growth was in width rather than height.

FRANCIS P. SHEPARD

The Ocean Floor

•

THE STUDY of the geology of the ocean floor, commonly referred to as submarine or marine geology, has as its purpose the explanation of the continental shelves, the slopes, the canyons, and the deep-sea floor. This study involves a total area of approximately 72 per cent of the earth's surface. The nature and origin of the sediments that only partially cover these vast submerged areas are also involved in the study, as are the rocks that underlie this sediment. The attack of geologists on this territory of the sea has necessarily differed from that on land, since the area studied has been mostly out of the range of vision and too deep to allow the application of the Brunton compass and the hammer, so important in land studies. Instead, dredging, coring, and random photography by means of cameras in protected cylinders have had to act as substitutes. It is only in recent years that a few geologists have started exploring the shallow margins of the ocean by swimming with aqualungs, which consist of one or more cylinders of compressed air strapped to the back and a tube leading to the mouth for breathing. With these they have been able to apply direct methods to the zone where water depths are less than about 180 feet. The possibility of extending these direct studies to much greater depths has come from the development of the bathyscaph. This device, invented by Auguste Piccard in 1948, is literally an underwater balloon having a large envelope filled with gasoline to make it lighter than water for the ascent and, in the place of the gondola, a steel ball with thick windows and enough room inside for two observers. During the descent small steel shot are held in place by a magnet, but these can be released upon reaching the bottom to allow the balloon to rise. There are propellers and a motor, which are capable of giving the bathyscaph only very slow backward and forward motion for a short period.

Despite these pioneer efforts, submarine geology is still largely dependent on observations made from vessels over an area where the scientist can judge the appearance of the bottom that he is exploring from echo-sounding profiles alone.

The difficulties of conducting shipboard operations that involve contact with the sea floor are little appreciated by those who have not been involved. The loss of expensive equipment is commonplace when one is grappling with rock ledges and pinnacles that cannot be seen and working from a rolling and pitching platform that is being carried along by currents of proportions difficult to evaluate. The great pressures of the depths cause apparatus to collapse, and the cold temperatures decrease the effect of lubrication so that parts "freeze" and the salt water and salt air corrode the metal unless very special care is maintained. Finally, many scientists are under a physical and mental strain caused by seasickness (now much less acute because of dramamine and bonamine) and by the dangers of accidents, that result from supporting tons of weight over the side of the vessel on rigging that is none too strong for the job. Thus, in practice seagoing geology lacks much of the glamor that has sometimes been attributed to it by landlubbers. It is hard, trying work, and results are still small.

In some branches of science, such as nuclear physics, many of the great advances have been made by those investigators who have had time or taken time to think. The philosophic approach, however, has not been very successful in many phases of geology, because ideas have been based on too fragmentary a background of facts. Many of the problems of geology long debated in lecture halls have been solved by those who were willing to go out and dig up the contacts between rock formations. In the newest branch of geology, the study of the sea floor, there is still a woeful need for diggers.

Thirty-seven years ago I emerged from the cloistered halls of the University of Chicago with a suspicion that some of the hypotheses that I had been taught with almost religious fervor might not be as well founded as I had been led to suppose. At that time the geology textbooks dealt with the ocean mostly in the light of what geologists had learned from the world-encircling British *Challenger* Expedition of 1872-1876 and from the study of marine sedimentary rocks obtained on land. The *Challenger* Expedition was certainly very productive, but because of the primitive means then existing for deep-sea soundings and for securing samples from the ocean bottom, the achievements of this one expedition were of course quite limited. Similarly the attempt to interpret conditions on the present sea floor from the study of ancient marine sedimentary rocks was obviously the wrong approach, for it was rather like trying to classify existing fauna and flora from a study of the fossil record without collecting any present-day forms.

The first indication I had that all was not the way it was described in the textbooks was in 1923 when I took a series of bottom samples in

Massachusetts Bay, going out from the beach to where the water was a hundred feet or more in depth. I had been taught that sediment on the sea floor grew finer as one went out from the shore because of the decreasing power of the waves and currents. In almost all of my sample lines, however, the sediment after some decrease in grain size grew coarser near the outer end of the line. I began to look for evidence from other localities to see if this finding represented a rare exception to a seemingly well-established rule. I examined the notations concerning the ocean bottom on many navigation charts and again found many examples to correspond with my own findings. No one, apparently, seemed to have consulted the charts before establishing the generalizations. A visit to the National Museum of the Smithsonian Institution, in Washington, produced even more supporting evidence that the textbooks were wrong. Gathering dust in the attic of the National Museum were well-labeled bottles with thousands of samples obtained in the charting operations of the United States Coast and Geodetic Survey—samples that no one had taken the trouble to study. Here was proof that the shelf sediments do not grade outward from coarse to fine in crossing the continental shelf. In many places gravel was sampled from the outer shelf and mud from areas near the shore. More recently oceanographic institutions have provided numerous shelf samples. These also show the lack of outward gradation, but they have provided information that makes it possible to understand this curious enigma.

The taking of long cores in the sediments of the ocean floor has led to revolutionary changes in geological ideas. Pipes up to sixty feet in length[1] have been pushed into the bottom by using great weight on the coring device above the pipe, by utilizing hydrostatic pressure, and by using the principle of the piston to reduce friction. The cores taken from these pipes have shown that the deep-ocean floor has many layers indicative both of climatic changes and of the introduction of coarse sediments from shallow water by powerful currents that move down the continental slopes and out over the plains beyond. Study of the cores is providing the best yardstick for measuring the length of the various glacial stages of the great ice age.

I remember having it impressed upon me in my university work that the ocean bottom was a flat monotonous surface with only a few rather minor irregularities around the margins. Here was a philosophical concept —based, in this case, on the supposition that sediments falling on the sea floor had filled or smoothed the basins and buried most of the hills. The soundings of the *Challenger* Expedition, made at intervals of hundreds of miles, did little to confirm or deny this hypothesis. It was only in 1924, when echo soundings were first taken across an ocean basin on the German ship *Meteor*, that it was found that the irregularities of the ocean bottom

[1] The Soviets claim to have obtained 100-foot cores.

might be as great as those of the continents. This has now been confirmed and amplified by countless fathograms made across the oceans of the world. Hundreds of high mountains have been discovered, especially in the Pacific Ocean, many of them forming parts of great mountain chains. Deep, elongated depressions (some of them known before the days of echo soundings) add to the major relief features that are now unfolding. On the other hand, new, extremely accurate echo-sounding devices now in use show that some of the flattest of all plains exist on the sea floor. The present idea of the origin of these plains is that they are due to deposition from sediment-laden currents that move rapidly down the slopes and out onto the deep-sea floor. This process would never have been suspected from the results of the *Challenger* Expedition.

Among the most perplexing and interesting features of the sea bottom are the submarine canyons. The extension of some of these deep valleys down the great marginal slopes of the continents has been known for over a century through the soundings made by the various marine surveyors, particularly those of the United States Coast and Geodetic Survey and the British Admiralty. At first little attention was paid to these canyons, and most geologists who gave them any thought explained them as having resulted from local sinking of the land margins, bringing river valleys below the level of the ocean. In view of the great crustal movements that have occurred, elevating the sea bottom into high mountain ranges, this did not seem particularly puzzling. As one result of echo-sounding surveys, it has been discovered that the submarine canyons are very widespread, being found, in fact, along virtually all the coasts of the world. It has also been evident that some of the canyons have walls that are even comparable to those of the Grand Canyon of the Colorado. Because of these facts many geologists have become skeptical of the river origin of the sea canyons. Such an origin implies that the coasts of the world must have been depressed thousands of feet.

Maurice Ewing and his group of scientists from Columbia University's Lamont Geological Observatory, located at Palisades, New York, began to trace some of these valleys, such as the submarine canyon off the Hudson River, out to the deep floor of the Altantic. They found a troughlike valley 100 feet or more below its surroundings and extending out to where the ocean has depths of about 15,000 feet. Other shallow valleys have been found as seaward continuations of the submarine canyons of the Pacific, and some of them even extend down the oceanic deeps leading south from Baffin Bay and the Danish Strait, on either side of Greenland. These shallow valleys at great depths are not explained by anyone as having been cut by land rivers, and many geologists are convinced that submarine canyons in general are not submerged river valleys but are the result of some as-yet-unknown process taking place on the ocean bottom. But the fact remains that the well-surveyed inner canyons have a remarkable resemblance to river valleys.

Discoveries concerning the nature of sediments and rocks underlying the sea bottom have also brought great surprises to scientists in the last few years. Carefully computed estimates had been made of the thickness of sediments that mantle the deep-ocean floors. These estimates were based on the calculated amount of sediment now contributed annually to the ocean and on the estimates of the age of the oceans. These figures led the well-known Dutch marine geologist Philip Kuenen to compute in 1949 that there should be an average thickness of about two miles of sediment. However, when seagoing geophysicists started using the technique of the oil geologists known popularly as seismic prospecting and sent sound waves through the ocean-floor sediment, it was found that the thickness was far less than supposed. In fact, in some places there was virtually no sediment. Elsewhere, a few thousand feet of sediment were all that the records indicated. Several explanations of this enigma have now been made, but there is no assurance that any of these are correct.

When the investigations were carried still deeper into the crust below the sediment cover, a different type of rock was found under the oceans from that which exists under the continents. The speed of sound in the rocks underlying the oceans is much faster than in the rocks under the continents at the same crustal depth. As a result of these geophysical investigations, we are making some progress in understanding the reason that the floor of the ocean lies at a much lower level than that of the continents. The rocks with the higher speed of sound are probably heavier, and hence the crust under the oceans has sunk in relation to that under the continents.

Even deep drilling has been used as a means of investigating the history of the ocean. One of the long-controverted subjects in geology has been the origin of the ringlike atoll coral reefs so common in the South Pacific. A hundred years ago Charles Darwin suggested that these atolls represented former volcanic islands that had submerged slowly and upon which coral animals had developed their colonies, growing up to the surface at a rate about equal to that of the sinking of the islands. This was long a hotly debated hypothesis, and it is only in recent years that the drilling conducted by the United States Navy and the United States Geological Survey in preparation for the hydrogen bomb test at Eniwetok helped solve the problem, showing that Darwin's ideas were fundamentally correct.

Waves and currents play an important role in modifying the sea floor, much more than was formerly supposed. Until very recent years ocean currents were thought to die out or become negligible at depths of a few hundred feet. We have found, however, that among the core samples brought up from the deep-sea floor there are clean sand and even gravel layers, both indicative of strong currents at great depths. Currents with velocities comparable at least to those of land rivers are now being considered as developing from time to time on the oceanic marginal slopes and continuing down to the deep-ocean floor. Waves, also supposed in the past to be limited to shallow water in their effect on the bottom, are now thought to exist on surfaces several thousand feet below sea level, although they are of different character than surface waves. Ripple marks of the type caused by waves have been found in flashlight photographs taken on the summits of submarine mountains at these great depths.

The serious erosion problems that have come from the building of jetties and other artificial coastal structures have led to a combined engineering and geological investigation of the marine processes that operate along the shore lines. In recent years there has come to light much new important information, which may lead to the more intelligent development of harbors along such inhospitable coasts as California and Florida, where natural harbors are at a premium.

In some of our early studies of shore processes at Scripps Institution we helped dispel the old myth about the dangers of undertow. Although our measurements show some slight net outward movements along the bottom, the really dangerous currents, which cause so many drownings, move seaward at the surface to an even greater degree than they do along the bottom.

Thus the branch of science that deals with the geology of the sea floor and of the oceanic margins is developing slowly but surely. As we get longer cores penetrating deeper into the bottom sediments, make more deep-water current measurements, take more ocean-floor photographs, and actually see more of the ocean floor with improved, more maneuverable bathyscaphs, the picture will undoubtedly change considerably. Right now we have a fairly good grasp of the problem. In the last few decades we have learned that the deep-ocean floor has tremendous mountains and valleys. The margins have canyons as deep as the deepest on land. The sediments are of an amazing complexity even in deep water, where they were supposed to be monotonously similar. Currents and waves of a special type are operating at the greatest depths. The sedimentary rocks on land are being matched with recent sediments on the sea floor, making it possible to distinguish the environments in which these rocks were deposited. Gradually the ocean bottom is yielding its wealth to man with great promise for the future when land reserves of petroleum, cobalt, nickel, manganese, and no doubt other products become scare.

FRANCIS P. SHEPARD

Catastrophic Waves from the Sea

•

THE TERM *tidal wave* has had an ominous sound in Hawaii since April 1, 1946. My own experience on that day may serve to introduce the discussion of tidal waves, or tsunamis. At that time my wife and I were living in a rented cottage at Kawela Bay on northern Oahu. On the previous day, a Sunday, the beaches and reefs were swarming with people and the cottages alive with activity. Fortunately, almost everybody left to go back to Honolulu that night. Early the next morning we were sleeping peacefully when we were awakened by a loud hissing sound, which sounded for all the world as if dozens of locomotives were blowing off steam directly outside our house. Puzzled, we jumped up and rushed to the front window. Where there had been a beach previously, we saw nothing but boiling water, which was sweeping over the ten-foot top of the beach ridge and coming directly at the house. I rushed and grabbed my camera, forgetting such incidentals as clothes, glasses, watch, and pocketbook. As I opened the door I noticed with some regret that the water was not advancing any further but, instead, was retreating rapidly down the slope.

By that time I was conscious of the fact that we might be experiencing a tsunami. My suspicions became confirmed as the water moved swiftly seaward, and the sea level dropped a score of feet, leaving the coral reefs in front of the house exposed to view. Fish were flapping and jumping up and down where they had been stranded by the retreating waves. Quickly taking a couple of photographs, in my confusion I accidentally made a double exposure of the bare reef. Trying to show my erudition, I said to my wife, "There will be another wave, but it won't be as exciting as the one that awakened us. Too bad I couldn't get a photograph of the first one."

Was I mistaken? In a few minutes as I stood at the edge of the beach ridge in front of the house, I could see the water beginning to rise and swell up around the outer edges of the exposed reef; it built higher and higher and then came racing forward with amazing velocity. "Now," I said, "here is a good chance for a picture." I took one, but my hand was rather unsteady that time. As the water continued to advance I shot another one, fortunately a little better. As it piled up in front of me, I began to wonder whether this wave was really going to be smaller than the preceding one. I called to my wife to run to the back of the house for protection, but she had already started, and I followed her just in time. As I looked back I saw the water surging over the spot where I had been standing a moment before. Suddenly we heard the terrible smashing of glass at the front of the house. The refrigerator passed us on the left side moving upright out into the cane field. On the right came a wall of water sweeping toward us down the road that was our escape route from the area. We were also startled to see that there was nothing but kindling wood left of what had been the nearby house to the east. Finally, the water stopped coming on and we were left on a small island, protected by the undamaged portion of the house, which, thanks to its good construction and to the protecting ironwood trees, still withstood the blows. The water had rushed on into the cane field and spent its fury.

My confidence about the waves getting smaller was rapidly vanishing. Having noted that there was a fair interval before the second invasion (actually fifteen minutes as we found out later), we started running along the emerging beach ridge in the only direction in which we could get to the slightly elevated main road. As we ran, we found some very wet and frightened Hawaiian women standing wringing their hands and wondering what to do. With difficulty we persuaded them to come with us along the ridge to a place where there was a break in the cane field. As we hurried through this break, another huge wave came rolling in over the reef and broke with shuddering force against the small escarpment at the top of the beach. Then, rising as a monstrous wall of water, it swept on after us, flattening the cane field with a terrifying sound. We reached the comparative safety of the elevated road just ahead of the wave.

There, in a motley array of costumes, various other refugees were gathered. One couple had been cooking their breakfast when all of a sudden the first wave came in, lifted their house right off its foundation, and carried it several hundred feet into the cane field where it set it down so gently that their breakfast just kept right on cooking. Needless to say, they did not stay to enjoy the meal. Another couple had escaped with difficulty from their collapsing house.

We walked along the road until we could see nearby Kawela Bay, and from there we watched several more waves roar onto the shore. They came with a steep front like the tidal bore that I had seen move up the Bay of Fundy at Moncton, New Brunswick, and up the channels on the tide flat at Mont-Saint-Michel in Normandy. We could see various ruined houses, some of them completely demolished. One house had been thrown into a pond right on top of another. Another was still floating out in the bay.

Finally, after about six waves had moved in, each one apparently getting progressively weaker, I decided I had better go back and see what I could rescue from what was left of the house where we had been living. After all, we were in scanty attire and required clothes. I had just reached the door when I became conscious that a very powerful mass of water was bearing down on the place. This time there simply was no island in back of the house during the height of the wave. I rushed to a nearby tree and climbed it as fast as possible and then hung on for dear life as I swayed back and forth under the impact of the wave. Like the others, this wave soon subsided, and the series of waves that followed were all minor in comparison.

After the excitement was over, we found half of the house still standing and began picking up our belongings. I chased all over the cane fields trying to find books and notes that had been strewn there by the angry waves. We did, finally, discover our glasses undamaged, buried deep in the sand and debris covering the floor. My waterproof wristwatch was found under the house by the owner a week later.

"Well," I thought, "you're a pretty poor oceanographer not to know that tsunamis increase in size with each new wave." As soon as possible I began to look over the literature, and I felt a little better when I could not find any information to the effect that successive waves increase in size, and yet what could be a more important point to remember? You can be sure that since then those of us who have investigated these waves in the Hawaiian Islands have stressed this danger, and I was most happy to find recently at a local island store a tidal-wave warning that emphasized the crescendo to be anticipated in future disasters. Nowadays, also, there are tidal-wave warning alarms that send out alerts either when reports of earthquakes under the ocean indicate dangerous possibilities, or when early waves arrive at other islands along the general route, or when the tide begins to fluctuate in an abnormal fashion. The importance of these warnings can be seen when it is noted that most of the 159 people who were lost during the 1946 tsunami could have saved their lives by running from the scene to higher ground when the waves first began. The Hawaiians are early risers, and being always attuned to the varying moods of the ocean, almost everyone was conscious of a sudden diminution of the noise of the breakers when the sea withdrew. Most people ran to see the strange sight of the reefs being laid bare, and many went out on the reefs to pick up the stranded fish. The 1957 tsunami was almost as destructive to property in Hawaii as that of 1946, but thanks to the warning system no lives were lost. I was shocked to learn that another house in which I had vacationed was destroyed by the 1957 waves.

Tsunamis and Their Significance

The meaning of the Japanese word *tunami* (pronounced *"tsunami"* and hence written that way) is "large waves in harbors," a good name, as it takes a disturbance of this kind to produce large waves in sheltered bays. The tsunamis certainly do not have anything to do with the tide, although the approach of the waves on an open coast where there are no reefs looks like a rapid rise of the tide, hence tidal wave.

Most tsunamis apparently have their origin in the great sea trenches that surround the margin of the Pacific Ocean. Fortunately for those who live on the west coast of the United States, there are no deep trenches in this section, which is perhaps the reason no appreciable tsunami has been observed along the California coast. Almost all tsunamis are preceded by world-shaking earthquakes, in which all seismograph stations have recorded the earth tremors from the disturbance. It seems likely that the waves are caused by faulting, a sudden dropping or lifting of a segment of the ocean bottom, which results in a displacement of large amounts of water. An alternative explanation is that huge submarine landslides produce the waves, although there is no good confirmation of this idea, and all tsunamis except those caused by volcanic eruptions have followed large earthquakes. If the ocean bottom drops, the surface waters are sucked into the hole, and when the water flowing from either side comes together, the surface of the water rises and waves move out in all directions under the force of gravity. Alternatively, if the bottom rises, the water is lifted and moves outward.

The waves are most violent in their effect in a direction at right angles to the fault. Since the Aleutian Trench, south of the islands of that name, runs east and west, movement along the faults that bound the trench produce waves that are most significant to the north and south, as were those of 1946 and 1957. Almost no one lives along the south exposed side of the Aleutian Islands, so that little damage has resulted in that area, although in 1946 the water rose at Scotch Cap on Unimak Island to over 100 feet, destroying a lighthouse and flowing over a 100-foot terrace. The Hawaiian Island group, more than 2,000 miles to the south, had waves that washed up to a maximum height of 57 feet, as far as we were able to determine. Fortunately, in most places it did not rise nearly as high as that.

Tsunamis move at an enormous speed in the open ocean, averaging about 450 miles an hour. This varies directly with the depth of the water, because these waves, unlike wind waves, have very long periods, commonly fifteen minutes, and have distances of as much as 100 miles between crests. Their height in the open ocean is so small, however, that they may have no erosive effect on the deep-sea floor. It is only along a coast that they become destructive.

The waves took about four hours to reach Hawaiian shores after the Aleutian earthquake of 1946. As the waves came into shallow water, they were greatly slowed down, so that they advanced at a rate of only about 15 miles an hour as they approached the coast. As their energy became confined to shallow water, they grew in height. The exposed coasts on the north of the Hawaiian Islands had large waves, whereas small heights were observed on the protected south side of the islands.

The investigations that followed the tsunami resulted in some conclusions that may prove helpful in ameliorating the effects of future calamities of this sort. The increasing height of the successive waves was perhaps the most important lesson. We found that in some places the second or third waves were the largest, but elsewhere the seventh or eighth reached the greatest height. On the western coast of Hawaii (the big island) some waves actually came in during the following night after an interval of eighteen hours and reached heights greater than those experienced that morning. These surprising reports were confirmed by a considerable number of sources, but are not readily understood. It can only be supposed that the waves represented a reflection from a submarine cliff off Japan and another reflection from an escarpment in Oceania, so that finally the waves, after making what is comparable to a three-cushion shot in billiards, arrived at their destination. In any case the danger of possible late wave arrivals, especially on protected sides of islands, cannot be minimized.

About the most dangerous thing that a person can do during a tsunami is to walk out on the exposed reefs to gather up the fish left by the retreating seas. In 1946 many of the drownings in Hawaii occurred as a result of this activity, a natural reaction of people whose livelihood comes from the sea and who for the most part had never even heard of a tsunami (the last one of any size having occurred in 1877). The building of sea walls in front of a town, as had been done at Hilo, is helpful even if the sea wall is knocked over by the advancing waves. Undoubtedly the friction considerably decreases the power of the waves. Wherever possible the restriction of building to zones that have at least moderate elevation above sea level in danger areas is recommended.

In the tropics the corals have been very helpful in sparing man from even worse trouble from tsunamis by building large protective reefs along many coasts. The widest reef in the Hawaiian Islands, at Kaneohe Bay, is found on the north side of the Island of Oahu and therefore on the side from which the waves approached. Yet, this wide reef seems to have been entirely effective in stopping the progress of the waves. Most people living in its lee were not even aware that a tsunami had occurred. Heights of not more than one or two feet were all that could be found by careful investigations along this shore. Other areas where reefs had smaller widths were less fortunate, as at Kawela Bay, where we were living behind a small reef and where the water rose ten to nineteen feet. However, the height of the raised water level, behind these reefs, was in almost every case less than in adjacent areas, where the water came in unimpeded.

Similarly, the existence of a submarine valley or canyon off a coast definitely has an important effect. Just as ordinary wind waves are small at the heads of submarine valleys, so also tsunamis are greatly reduced by the spreading of the energy as the waves move up the valleys at a faster rate than over the intervening ridges. Conversely, the waves traveling over a submarine ridge are particularly large. For example, there are three ridges extending down the slope on the north side of Kauai Island, and over these the waves attained their greatest heights. So if you live in Hawaii and want to live next to the beach, build your house behind a coral reef or look at a chart to see if there is a submarine valley out in front.

A few destructive tsunamis have occurred in localities where no great trench is known to exist. Among these are the waves that swept in on Lisbon in 1755, moving up the Tagus River and causing a very heavy loss of life. These followed a great earthquake with a center under the Atlantic some distance off the shore.

The most destructive waves of all time have been related to volcanic activity. In 1883 when Krakatoa blew off its head, a sudden engulfment occurred, setting up very unusual waves. These rolled in on the adjacent islands of Java and Sumatra and drowned tens of thousands of natives, rising, it is said, to heights of well over a hundred feet. Curious reports came as a result of these waves. They showed on the tide gauges all the way around the world, even in the English Channel. If these tide-gauge records indicated a tsunami actually coming from the Krakatoa engulfment, the waves must have been reflected from numerous submarine escarpments in order to have reached such a destination. The waves were recorded also in the Hawaiian Islands, although here the time of arrival does not agree with the time that one would predict for a wave traveling from Krakatoa to the Hawaiian Islands. The explanation is still in doubt.

It is disturbing to consider what would happen if a tsunami should come into a shore like Long Island, where some of the beaches have hundreds of thousands of bathers during a warm summer day. We have no records of dangerous waves coming in at these places. However, in the case of the 1929 Grand Banks earthquake, which wrecked a large part of the submarine cables going between our east coast cities and Europe, there had been no previous record of tsunamis in the area. The waves accompanying the Grand Banks earthquake moved in on Burin Peninsula on the south coast of Newfoundland, rising to fifteen feet. Such rises would, of course, sweep over most of the beaches along the exposed portion of the east coast. Let us hope that no new submarine faults come suddenly into being and send waves into this area. The effect would be almost as bad as that of a hydrogen bomb.

The term *tidal wave* has sometimes been used also to describe a rise in sea level that accompanies a hurricane. A *storm tide* seems to be a more acceptable term. In 1900 the sea rose about fifteen feet at Galveston, Texas, and topped the sea wall, sweeping into the city and drowning 6,000 people. A sea wall has now been constructed that will probably prevent any recurrence of this sort, but there are other cities less well protected and subject to dangerous waves. In 1938 a great hurricane moved up the east coast, quite contrary to the predictions that had been made by the meteorologists, and passed inland across Long Island. The sea here rose also about fifteen feet, killing 600 people and causing tremendous amounts of damage to the beach property. It developed numerous new inlets in the beaches and changed the appearance of the coast until it was practically unrecognizable after the waves had stopped. Several other hurricanes, notably those of 1954 on the east coast and that of 1957 in western Louisiana, have produced similar inundations. A far worse catastrophe of this sort occurred at the head of the Bay of Bengal in 1737, when 300,000 people were drowned during a hurricane. The rises accompanying these great storm waves are not unlike tsunamis, except that the waves do not come in rhythmic succession. The rise may be quite as rapid, but the high water usually lasts a longer time, and recurrences are not particularly pronounced.

The damage caused by all of these rises of sea level is related only in part to the high water. In addition, the natural barriers to wind waves that exist along the shore and many of the artificial walls and jetties become less protective, so that the wind waves are superimposed upon the top of the sea-level rise and wreak their havoc at a new high level.

RICHARD FOSTER FLINT

Pin-pointing the Past with the Cosmic Clock

•

IN A Chicago laboratory a radiochemist stood before an oscilloscope connected to a Geiger counter. Across the oscilloscope screen, very much like the screen in a small television set, jumped a never-ending ribbon of green zigzags. The zigzags were made by the impulses coming from disintegrating atoms of radioactive carbon within the counter. The carbon had been extracted from a piece of wood, but the wood was by no means ordinary. In fact, it was a piece of very old wood, part of a spruce tree that grew in a Wisconsin forest so long ago that when the tree was alive the Ice Age still had northern United States in its chill grip. So long ago that mastodons and mammoths still inhabited the country in force. Indeed, it is quite possible that a mastodon, crashing through the spruce forest in the chill glacial air, brushed against this very tree.

The wood had been sent to the laboratory for an exact determination of its age—not the age of the living tree, which can easily be learned by counting its growth rings, but the time elapsed since the tree was alive. When a police surgeon examines a dead body, his medical skill tells him *about* how much time has elapsed since death occurred. But only *about*. He has no means of fixing the time exactly, even though the guilt or innocence of a person accused of murder may depend upon it. The surgeon can only look for certain signs and use his professional judgment.

Like the surgeon, geologists and archaeologists, trying to date the events of prehistoric periods, have only been able to look for signs and on the basis of them to make estimates which, although far better than nothing at all, are certainly not accurate. Some geologists, for example, had estimated the age of this particular spruce forest in Wisconsin at about 25,000 years, although they could not be sure.

The uncertainty, however, was being ended by the green zigzags that continued to jump across the oscilloscope screen. Radiocarbon, the cause of the zigzags, is a recent discovery that promises to make an accurate timetable of prehistoric events. It constitutes a kind of clock that set itself going when the forest tree died and that has been ticking ever since, at a known rate. The piece of wood was the latest of five samples, all collected from the same forest. When tested, all five gave answers very close to each other, and the average of them reads: 11,400 years ago—six times as long ago as the birth of Christ—and the date is probably accurate to within a few per cent.

This date is of tremendous importance to science, for it cuts earlier estimates and guesses more than in half and brings the extinct mastodons and mammoths (among other things) much closer to our own time than had been supposed. Thanks to radiocarbon, we can look forward to the enjoyment of talking about prehistoric dates in figures that are almost precise.

How this has been made possible is a story of research in nuclear physics and in chemistry, with important contributions from archaeology and geology. It is a story of cooperation among scientists, a pooling of knowledge in several fields. The pioneers in this unusual research are Dr. W. F. Libby and Dr. J. R. Arnold, and they did the work at the University of Chicago's Institute for Nuclear Studies.

What Is Radiocarbon?

The story of the research that has led up to this result shows the extent to which science has become integrated, with contributions from nuclear physics, chemistry, archaeology—and a few additions from geology, the field of sanitation, and antarctic exploration!

It all grew out of research on cosmic rays. These are great streams of neutrons that pour in to the Earth from outer space. They bombard the Earth's atmosphere and set up a sort of chain reaction that showers cascades of particles down through the air to the Earth's solid surface, where their arrival is pictured on the oscilloscope screen.

Many of the invading horde of neutrons, penetrating the Earth's atmosphere, collide with atoms of nitrogen, of which the atmosphere is largely composed. When one of these high-speed collisions occurs, a new nucleus is created. This disintegrates, emitting in the process an atom of a newly created element, Carbon 14. In this way great quantities of Carbon 14 are created, miles above the Earth's surface, like sparks struck by an impact. Like sparks, too, the Carbon 14 atoms do not last long, for they are radioactive, and they destroy themselves through spontaneous disintegration. Carbon 14 is "heavy" carbon. It is "heavy" because, whereas ordinary carbon has atomic weight of 12, the weight of the new carbon is 14. The new carbon isotope is familiarly called, by the scientists who study its characteristics, *radiocarbon.*

Although radiocarbon atoms may not last long, still, while they do exist, they get around. The first thing they do is combine with oxygen to form carbon dioxide. This mixes with the ordinary carbon dioxide that contains ordinary carbon and in the course of time becomes evenly mixed throughout the air that surrounds the Earth's solid surface. Wherever we may be, as we breathe we draw into our lungs heavy carbon in this form.

That doesn't mean, though, that radiocarbon is very abundant. Actually, for every trillion atoms of ordinary carbon in the atmosphere there is only one atom of the heavy Carbon 14! So, although we are constantly breathing "hot" carbon, its "hotness" is almost unbelievably faint, and we are none the worse for it.

The proportion of Carbon 14 to Carbon 12 in the atmosphere is believed to be constant. This should be so because it represents a balance between the supposedly steady rate of creation of radiocarbon and the rate of its destruction through its own disintegration.

Both plants and animals absorb carbon dioxide freely from the atmosphere. Probably, therefore, they contain just the same proportion of radiocarbon to ordinary carbon—one to about a trillion—that is present in the air. To be sure, the radiocarbon in plant and animal tissues is constantly disintegrating, but it is being renewed as constantly from the air.

A Calendar of the Past

This continuous renewal of radiocarbon goes on as long as the plant or animal is alive. But when death occurs—in a tree, for instance—the intake of atmospheric carbon ceases abruptly, and so new radiocarbon no longer comes in to replenish what is being continuously lost by radioactive disintegration. Thus from year to year the amount of radiocarbon present in the dead wood becomes gradually less.

The rate of decrease of radiocarbon is the same everywhere, and the rate is known. Because of this it is possible, by measuring the amount of radiocarbon left in the dead wood, to calculate the time that has elapsed since the death of the tree. Actually it isn't the *amount* that is measured; it is the *rate* of disintegration, which constantly diminishes and is always proportional to the amount of radiocarbon remaining. It is the tiny disintegrations—so many per minute per gram of carbon—that produce the green zigzags on the oscilloscope. The measurement is a tricky laboratory operation, one that must be watched carefully at every stage and that must be safeguarded against possible errors of several kinds. But thanks to skillful development, it works.

How the Calendar Was Developed

The radiocarbon calendar is the culmination of a long train of necessary research, without which the calendar would have no value. More than fifteen years ago it occurred to Dr. A. V. Grosse, radiochemist of the Houdry

Process Corporation of Pennsylvania, that bombardment by cosmic rays probably was creating new radioactive elements. Later, this idea led Dr. Libby to believe that "heavy" carbon, created in that way, must be found in living matter. To test this theory, Libby and Grosse obtained from the Department of Public Works of the City of Baltimore a sample of sewage, a pure organic product. Analyzing it, they not only found radiocarbon in it; they also found that radiocarbon was present in just about the proportion they had calculated beforehand.

The next step was to make tests to see whether this same proportion of radiocarbon is present in living matter all over the world. So they obtained samples of wood from Chicago, New Mexico, Panama, South America, a Pacific island, Australia, North Africa, and Sweden and tested them for radiocarbon. This is a pretty good geographic distribution. But seashells (calcium carbonate) from Florida and seal oil specially collected by the Ronne Antarctic Expedition were tested also; and in the whole lot, radiocarbon proved to be present in the same proportion, within the limits of experimental error.

The success of these tests made it seem possible to take any piece of ancient wood or other carbon-bearing organic substance and determine its radiocarbon date. So it was decided to check radiocarbon against historical knowledge. The method was to determine the radiocarbon dates of pieces of wood and to compare these dates with the dates already calculated by archaeologists on the basis of historical evidence. This was done with a variety of samples, and the archaeologists no doubt had an interesting time selecting them. Two of these were pieces of wood from Egyptian tombs known to be about 4600 years old. The radiocarbon dates came out within 150 years of the historians' figures.

Although it was apparently true that the proportion of radiocarbon to ordinary carbon is uniform in all living matter, regardless of its location, and although the tests on Egyptian wood seemed to show that this proportion has been the same during the last 4600 years, this was not the end. For it was not yet proved that the same held true for still more ancient times, more than 4600 years ago. To prove this it would be necessary to test still older material. But there is one difficulty about all such very old objects. Their dates are unknown; therefore, they provide nothing with which to compare the radiocarbon dates.

There was only one thing to do: get radiocarbon dates for a large number of samples, collected from many different geologic and archaeologic deposits, compare them, and see whether the dates were consistent with each other. If they were, it could be fairly concluded that the concentration of radiocarbon had been very nearly constant throughout the whole period represented by the samples.

The selection of all these samples, demanded the cooperation of a number of specialists. So the American Anthropological Association and the Geological Society of America were asked to name a committee of four specialists to collaborate with Libby and Arnold in the tests. As a result, more than 200 samples were collected and dated. Their ages ranged back more than 15,000 years.

The results were so generally consistent with each other as to leave little doubt that the radiocarbon calendar is reliable for dates as far back as it can reach. Unfortunately, however, its reach is limited, and for a simple reason. The rate at which radiocarbon disintegrates is comparatively rapid —so rapid that 5568 years after a plant or animal has died, the radiocarbon it contains is half gone. Although the rate gradually diminishes with time, still the proportion of radiocarbon remaining at the end of 20,000 years is so small as to make accurate laboratory counting very difficult. It is not likely that any refinement in technique will stretch the reach of the calendar much beyond 30,000 years.

Still, we must not ask for too much; during the last 20,000 to 30,000 years a great many things have happened, the dates of which science wants very much to know.

History of the Ancient Forest

At this point we can return to the ancient Wisconsin spruce forest from which we got the sample of wood we began with. The forest is buried beneath a thick layer of earth and stones plastered over it by the great glacial ice sheet that—not so long ago, as we have seen—flowed into the United States from Canada. Because of its burial, the forest would have been unsuspected if the ground had not been deeply cut into by the waves of Lake Michigan. In the course of time the Lake has created a bluff, or cliff, in which the ancient forest is exposed as a dark-colored layer of peat. Thanks to meticulous studies made by Professor L. R. Wilson of the University of Massachusetts, the peat has been described almost inch by inch. In it are spruce stumps still rooted in the ground beneath the peat. All the stumps are splintered. Many spruce logs lie in the peat or in the earth above it, all of them pointing southwest, the direction in which the invading glacier moved. Some of the logs are flattened, as though a great weight had pressed on them. Furthermore, their butts are splintered and twisted as only green, living wood splinters. From these facts it cannot be doubted that the trees were overwhelmed by the glacier, which snapped them off as a giant bulldozer might have done and then overrode the prostrate wreckage, smearing stony clay across its top.

When the logs were sawed through, they revealed confirmatory evidence. The oldest of the trees has 142 annual growth rings; so it is sure that the forest existed for at least 142 years before it was destroyed. The few outermost or youngest rings are closer together than the ones inside them, showing that during the last few years of their lives the trees were growing poorly. The reason for the poor growth is not far to seek. The upper part of the peat, unlike the peat layers beneath, consists only of mosses that live on wet ground or in water itself. The topmost moss plants are still upright, just as they grew, and their youngest branches are thin and scraggly. Water-laid clay overlies the plants and is packed around their stems, as could be the case only if it had been deposited while the mosses were still growing.

Clearly, what slowed the growth of the trees and what killed the mosses was a muddy lake, which must have flooded the spruce forest for several years before the final catastrophe. The lake could only have been created by the advancing glacier, which, by blocking the Strait of Mackinac, dammed up the water in the Lake Michigan basin and caused the lake gradually to rise. Meanwhile the glacier continued to flow southward, so that by the time our spruce forest had been standing in muddy water for several years, the moving wall of ice stood towering above it, ready to complete the task of destruction.

Following the track of this glacier, geologists have found the ridges of earth and stones that mark its outer limit only 25 miles southwest of the buried forest. They have traced this outer limit nearly as far south as Milwaukee, and, on the eastern side of Lake Michigan, to a point between Ludington and Muskegon. Between Milwaukee and Muskegon the glacier formed a great tongue that projected southward.

The radiocarbon calendar has given a date—11,400 years—for the destruction of the spruce forest. No buried wood has yet been found nearer than this to the outer limit of the glacier; so geologists have to resort, for the present at least, to their former method of estimating. From what is known about the rates of present-day glaciers, a reasonable estimate of the time required for the ice to flow from the forest site to its extreme and final limit is 400 years. Therefore we can say, provisionally, that the glacier stood at its final or "Milwaukee" line about 11,000 years ago, and from that line it began to melt away.

Matching up Ice-Age Events

Away down near St. Charles, Missouri, close to the point where the Missouri River empties into the Mississippi, a group of geologists found, in 1949, a log of wood imbedded in the base of a terrace, the top of which stands 50 feet above the river. They recognized the terrace as the remnant of a deposit made when the Mississippi and the Missouri were choked with sand and silt poured into them by melting ice from the glacier hundreds of miles to the north. Gradually, the rivers silted up their valleys with these "melt-water" sediments, until in time their beds stood 50 feet or more above their present positions. Sediment 50 feet thick, filling a valley more than a mile wide through a distance of hundreds of miles, forms a very bulky deposit.

When the glacier melted away, it ceased to pour extra sediment into the big valleys. The two great rivers, no longer burdened with superloads of sand and silt, could then cut down into the thick sediments they had been obliged to deposit on their floors during the glacial invasion. With their great volumes of water they succeeded in flushing much of the accumulation away, but in protected places along the valley sides flat-topped terraces of sand and silt remained as witnesses to the former great fills of glacial waste. St. Charles is one of those places.

All this was apparent to the group of geologists as they examined the log projecting from the terrace sand. They realized further that the log must have been floated in and deposited as driftwood during the time when the river was choked with glacial sand. They guessed that the ice invasion was probably the one in which the ice in the Lake Michigan basin stood along the "Milwaukee line," straddled the Mississippi at Minneapolis-St. Paul, and stood in the Missouri valley at Yankton, South Dakota.

All this seemed probable, though it wasn't proved. But thanks to radiocarbon, several months later it *was* proved. A sample of wood from the log was sent to the laboratory, reduced to carbon, put into the counter, and the automatic counting was begun. Result: the log proved to be just a few hundred years older than the Wisconsin spruce forest. Clearly the choking of the rivers had begun while the ice stood somewhere north of the "Milwaukee line," before, but not long before, it overwhelmed the forest. Radiocarbon had proved a close connection between two events that occurred more than 11,000 years ago at places hundreds of miles apart.

For the group of geologists who were trying to decipher the history of the Ice Age it was a godsend. To have established by ordinary geologic methods even an approximate relation in time between the driftwood log and the buried spruce forest would have required years of difficult and painstaking measurement of the terraces along the Mississippi River between Wisconsin and Missouri.

Prehistoric Man in America

One of the important ways in which the radiocarbon calendar can add to the history of America is by dating the races of prehistoric man that lived on this continent thousands of years before the coming of white Europeans, and even before the arrival of the present-day Indians.

Since 1926, archaeologists had been finding evidence in western North America that Stone Age people once lived there. The evidence consisted of a peculiar kind of arrowhead, or rather dart-head, made of quartz and unlike any ever seen before. In contrast with the later Indian arrowheads these dart-heads had no notch at the base, but they did have a long groove down each side, and they were beautifully and delicately made. As the first ones were found near the little town of Folsom, New Mexico, they came to be called *Folsom points,* and soon archaeologists were speaking of the Stone Age people who had made them as *Folsom Man.*

More Folsom points came to light. Implements of this sort were discovered in New Mexico, Texas, Nebraska, Colorado, Wyoming, Nevada, western Canada, and even Alaska. Furthermore, they were found in the same layers with the bones of an extinct kind of bison, larger than modern buffalo and quite different from them. Indeed, one point was found imbedded in the backbone of a bison—a bison that a prehistoric hunter probably killed to secure food and clothing. Clearly western North America had been the ancient hunting ground of a skillful and wide-ranging race of early Americans.

These discoveries led to all sorts of estimates of the time when the Folsom people lived, but the estimates remained estimates—until radiocarbon came into the picture. A group of geologists and archaeologists in Texas, sensing at once that the radiocarbon calendar could solve the problem, searched for Folsom material suitable for laboratory use. At Lubbock, Texas, they found it: pieces of burned bone, the cold leftovers of a Folsom meal. Off went the collected and ticketed bones to the University of Chicago. There they were reduced to carbon, placed in the Geiger counter, and in due course the time since their animal possessors died was calculated. The radiocarbon date came out at 9883 years, with a possible error of less than 5 per cent.

Ten thousand years ago, then, in round numbers, Folsom people were living and hunting bison in North America. Ten thousand years ago was only a thousand years later than the time when the last big flow of glacier ice reached its outer limit at Milwaukee and Minneapolis and Yankton.

There is a good deal of evidence that the period when the Folsom people left their dart-heads and dinner scraps lying about was a period when western North America had more lakes, bigger streams, and therefore more rain than most of it has today. This is exactly what we should expect if most of Canada and part of the United States were covered with a thick ice sheet some millions of square miles in area. So the radiocarbon calendar shows that the evidence collected by geologists and archaeologists is consistent and illuminates another chapter in the ancient history of the continent.

DAVID C. HOLMES

The Weather

•

I

THE OCEAN of atmosphere which surrounds us has provided a canopy over the earth for untold ages. It would appear logical that this sea of air would long ago have stabilized into a soft, still envelope so that now there would be no thunderstorms, no sudden gales, no lashing rain. The reasons why this is not so are simple, although they puzzled man for many centuries. It is only lately that we have become aware that our atmosphere is like soup steaming in a gigantic kettle, churned and boiled by the radiant heat from our sun.

When Einstein established the equivalence of mass and energy, he ushered in an age during which many startling facts have become evident. The sun, for example, daily radiates an enormous quantity of energy. Since energy and mass are mutually convertible, the solar radiation corresponds to a weight loss in the sun of nearly 5,000,000 tons per second. The earth absorbs 150 tons of this mass each day.

If the observers from other celestial systems know of our existence at all, it is only because of the radiated light from our beacon sun. The energy it releases permeates all the planets of the solar system and extends beyond them to the outer reaches of space to become a true link with infinity. The powerful effects of this radiant energy, unevenly absorbed by our planet, together with the rotation of the earth produce the great circulation patterns of the atmosphere and are the basic cause of all our weather.

The key to an understanding of weather phenomena lies in a knowledge of these circulation patterns. They are produced, not by the sun alone, but also by the rotating earth; however, in order to simplify the picture, let us first assume that the earth is stationary. Thus we will be able to see clearly just what effect the radiation of the sun has upon our atmosphere (see Figure 1).

The earth's rotation has suddenly stopped, and the sun appears stationary at a point above the equator. Most of the heat from its rays is absorbed by the atmosphere midway between the poles, and this causes the air near the equator to become much hotter than that of the poles. The hot air rises, producing low-pressure areas at the equator, and the cold dense air from the extremities of our sphere rushes in to fill the gap. Warmed by the equatorial heat, this air also expands and must rise. The result

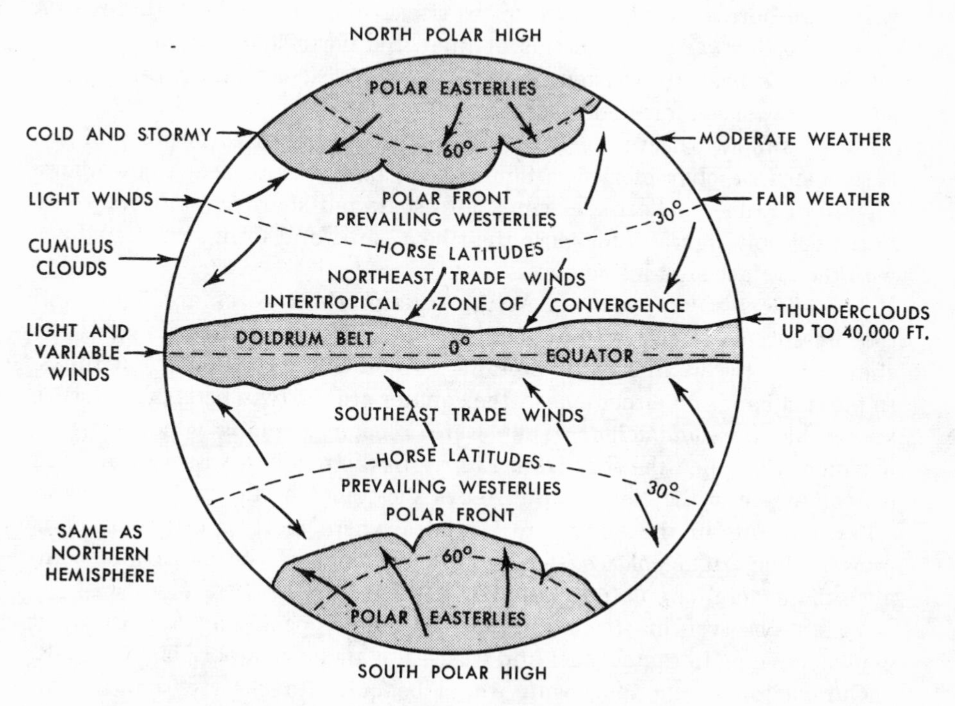

Fig. 1. The global circulation of air provides an infinite pattern of change and variety, bringing rain, sun, warmth and cold to refresh our world.

is a gigantic, invisible Ferris wheel of atmosphere which circulates toward the equator on the surface and away from it at high altitudes. As the upper air flows away from the sun, it is cooled and becomes heavy, causing a downward flow at the poles. Thus the cycle is complete, and the atmosphere is continually on the move, aerating our world, shifting constantly over the land, a flow without beginning and without end, enduring beyond time.

The rotating earth superimposes on these basic north-south air currents another and more complex flow. As the earth rotates, so does its atmosphere. And since outer space is wholly frictionless, the mass of atmosphere keeps pace with earth. At the equator, the earth's circumference is large, almost 25,000 miles. Thus the air at the equator turns through 25,000 miles of space each day. Since the earth likewise moves at this speed, there is little or no surface wind at the equator.

As the tropical air, heated by the sun, rises to take its long northward path to the poles, it maintains this rotary speed. But as it moves north, the earth's circumference rapidly grows less, and the earth's surface speed diminishes proportionately. Hence, as the equatorial air travels northward, it moves faster than the surface below, and the middle latitude winds are born. These winds seem to rise in the east and for this reason they are called *easterlies*. In this manner, the north-south winds produced by the sun's heat are twisted unevenly to the west, and the winds of the earth blow askew across her surface.

These simple patterns become enormously complicated because of the earth's surface characteristics. Some areas absorb more heat than others. Mountain ranges deflect the winds into scattered shards. Winds en route to the poles at high altitudes lose their heat, sink, and form swirling eddies with the surface streams below.

The effects are interesting. At the equator the breezes are light, and the pressure is relatively low. This belt is called the doldrums and was shunned by the weather-wise captains of sailing vessels in ancient times. In parallel bands on either side of the equator are the twin belts of easterlies known as the *trade winds*. They were a boon to many generations of mariners who plied the seas from Europe in their sailing vessels searching for commerce in the New World. Above the trades lie the *prevailing westerlies,* currents of the temperate latitudes where most of our weather is brewed. The frigid *polar easterlies* chill the ends of the earth and complete the general circulation pattern.

To an observer in space, the earth would appear ringed with wind bands, flowing alternately east and west—a planetary barber-pole.

Our visitor in the space suit would be quick to observe that in two localities, one just under the polar easterlies and the other at the equator, opposing flow systems meet. These boundary regions are called *fronts*—the *polar front* and the *inter-tropical zone of convergence* or *equatorial front.* Both are low-pressure areas created by the ascension of surface-heated air to higher altitudes, and both are characterized by storminess and bad weather. In the United States we are primarily concerned with the polar front. The waves and nodes which form in it during the winter months invade the temperate zone to produce most of our cold weather.

The radiant power of the sun, beating down on a polished earth, would produce the circulation effects we have noted. But the earth is not a mirror which reflects the sun's radiation uniformly. The great oceans absorb and store the atomically generated heat and act as a giant thermostat to dampen changes of temperature in our atmosphere. Sea currents carry tropical heat to wash lands which would otherwise be snow-encrusted. Half our world owes its mild climate to the heat-charged Gulf Stream. The land, the mountains, the valleys and the low, level plains—each mirrors the sun's energy differently and so shapes the local climate.

The earth's varying geography thus alters the balance of temperature throughout the world. Most of the land and the highest mountain ranges lie in the Northern Hemisphere. This unbalanced distribution, coupled with the fact that three-fourths of earth's surface is covered by water, produces major temperature changes. If the earth were an ideal sphere, the North and South Poles would be the coldest regions. In the Southern Hemisphere this is almost correct. However, the northern cold pole is located in Siberia, almost one-third the distance down toward the equator.

The heat equator, or band of hottest temperature, varies with the path of the sun. Generally it is attracted northward by the continents. During the summer months it extends into the United States through the California valleys and sometimes almost up to the Oregon border—forty per cent of the distance to the pole.

The net result of coalescing sun, atmosphere, rotation, and geography is inconsistency and fluctuation. Although the steady, ageless sun is the prime mover of our weather, it is forever different and unpredictable. Yet miraculously these weather variations are but the merest ripples in the galactic scale and always remain within the incredibly thin limits necessary to foster the fragile, jelly-like growth of life upon our planet.

II

The great wind patterns of the earth have been known and used by men for many centuries. Since these air currents carry our weather with them, it is surprising that the tricks and variations of the global weather patterns baffled the human mind for so many generations.

The problem persisted primarily because of false beliefs rather than false weather observations. The early weather prophets believed, reasonably enough, that the canopy of air covered the earth with a uniform, protective blanket having the same characteristics at the equator and the poles. Weather was assumed to be a local phenomenon and many false trails were generated in the search for the weather-producing elements. Primitive peoples explained weather even more simply. It was a form of punishment or reward visited upon them by the deities. Good weather came from benevolent, happy gods; storm and drought meant that these overlords were angry. The beating of drums and the offering of sacrifices were the best way of appeasing their wrath and changing the weather.

The early scientists and primitives were both wrong. The world-wide weather observations of today have established the fact that the air over us is divided into huge air masses—continents of air drifting with the winds of earth. These air masses have vastly differing characteristics and, as they drift over our part of the world, they vary our weather and cover us with alternate umbrellas of rain, snow, or sunshine.

Although the theory that the atmosphere is divided into distinct and separate islands of air is less than sixty years old, it has already solved many of the riddles of weather phenomena. It explains those cold days in May and why the southwestern air of the United States is so clear. It unravels the puzzle of the thunderstorm and the reason for the California fog, and it allows our weathermen to predict the weather with an average correctness of over eighty-five per cent. Like the clue in the detective story, it was there all the time.

Basically there are only two sources of weather air. The most important one is the polar region where the great winter air masses are born. These are, obviously enough, formed of cold air; since cold air does not hold much moisture, the polar air masses are relatively dry. The second source lies near the equator. This is the spawning ground of the *tropical air masses* which are warm, wet, and heavily cloud-laden.

The polar air masses are built in the higher altitudes above the fortieth parallel over the arctic seas or the ice-capped lands near the poles. Continental polar air masses originate primarily near the desolate tundras and wastelands of Canada and Siberia. The polar air mass called *maritime polar* forms over water. Most of the snow which falls on the United States comes from the clouds generated in one of these air masses.

The *tropical air masses,* as the name implies, are generally created in the equatorial regions between the Tropic of Cancer and the Tropic of Capricorn. Continental tropical air masses are not generally associated with our continent, but are formed over the arid lands of the Middle East, Africa, and Australia. Occasionally one of these air masses will form over Mexico and extend into our southwestern states. This type of air is bone-dry—so dry that it holds no further interest for us.

Maritime tropical air masses breed over the warm equatorial waters. As a consequence, these air masses absorb enormous quantities of water vapor and are extremely cloudy.

In addition to these basic types, meteorologists have given names to smaller air masses because of their extreme local importance. Those which bear mentioning are the *Arctic, Equatorial, Monsoon* and *Superior* air masses. In general these are only extreme cases of polar or tropical types. Arctic air is found on the Antarctic continent, Siberia, and occasionally Canada. It is the coldest air on the face of this earth. Equatorial and Monsoon air is found over southeast Asia and is extremely moist and warm.

Superior air is something of a mystery. It is very dry and usually exists at high altitudes. Superior air is believed to result from large-scale settling motions in the upper atmosphere. Occasionally such an air mass will overrun maritime air and come sliding down the damp slope to descend to ground level and cover the southwestern plains of the United States and Mexico. The visibility in this clear, dry air is remarkable; some Texans claim they can see over one hundred miles—which is not a very great distance in Texas.

Just as the nations of the world are separated by their boundaries, air masses are kept apart by barriers of their own making called *fronts*. Unlike the continents, however, air masses are usually on the move. In rare cases where geography constrains them, they may be walled-in like lakes, captured by the natural barriers of the land. An example of this is the North Pacific air mass which nestles against the Olympic Range, held there by the westerly flow of air aloft. A similar air mass, known as the Azores High, exists along the coast of Spain. But such cases are rare. Air masses may remain in the same region for long intervals, but eventually they begin to move, and in traveling they create the changing cycle of weather. The air mass is a chameleon with an ability to assume the temperature and humidity of the surface over which it flows. If this surface is different from its origin, if it is hotter or colder or more moist or dry, the air mass will accommodate itself to the new condition, sometimes losing, sometimes gaining moisture, and changing the weather below as a consequence.

A specific illustration is perhaps the best way of explaining the weather changes resulting from the voyage of an air mass. Continental polar air will often surge through the United States from its Canadian breeding ground. At the beginning of its journey, temperatures range downward to $-40°$, and the citizens of Minnesota huddle against their stoves. As the air travels southeast, it absorbs heat from the land. By the time it reaches the southern states, temperatures are rarely below freezing. At this point it is shunted seaward over the warm Gulf Stream by the tropical maritime masses swirling north from the equator. Once over the Gulf Stream, our wandering air mass picks up both heat and moisture. In the normal course of events, driven by the westerlies, it will reach the European continent. As it encounters that cool surface, it will lose heat. Since cold air cannot hold as much moisture as hot, clouds will form and the water vapor will condense into rain. Finally, our air mass will return to what it started out to be: cold, dry air.

In their surges across the fields and seas, the air masses do more than vary the weather. In a wholly automatic way they are the thermostats of earth, regulating its temperature by distributing its heat. The story of this heat distribution starts with the oceans. Water heats more slowly than land. It reflects about sixty per cent of the solar radiation, leaving only forty per cent to be absorbed. Land, on the other hand, absorbs ninety per cent and reflects only ten. Unlike the land, the oceans do not store their heat on the surface to be siphoned away by the winds; their mobile currents, however, distribute warmth to the vast ocean deeps. Moreover, it takes four times as much solar heat to raise water one degree than it does to heat an equivalent amount of earth.

For these reasons the oceans have become the great moderators, acting to narrow the extremes of temperature in the air masses which journey over them. These air masses, in turn, pass on the beneficial effects of the water to the land. An excellent example of this heat-transfer process can be found along the California coast during the summer. Cool maritime air from the Pacific flows over the sun-baked California farmlands, keeping them so cool that the range in temperature between summer and winter in San Diego is less than ten degrees.

Along the equator, tropical air masses absorb water and are heated by the seas below. This air expands, becomes lighter, and rises through the troposphere. As it reaches the higher levels, some of its heat is conducted away to the surrounding air. This leaves the rising air cooler and it can no longer retain the huge quantities of moisture which it carried aloft from the evaporating seas below. As a result, clouds form, blotting out the sun, and cooling rain falls. Thus the air masses serve to regulate the heat absorption so that the lush tropical regions are not burned by the direct rays of the sun. Another dissipation of solar energy occurs in the hurricanes created by the tropical air masses. Huge quantities of heat are transformed into the motion of the wind and waves during these storms.

In this way, the air masses air condition the earth in the same manner as the forced air systems in our houses heat and cool our rooms. Heat is brought to cool areas and moisture to dry regions. The total effect creates the miraculously narrow range of temperatures which exists on our globe and allows us to live in relative comfort on most of its surface.

Weathermen follow and report the passage of air masses over our country because it allows them to estimate the nature of weather to come. This information has become so much a part of our daily lives that we scarcely notice it, unless, of course, the prediction does not come true. To the aviator, however, weather knowledge is of more than casual interest, since he gambles his life on its accuracy.

We mentioned earlier that air masses are separated by fronts, the region where two air masses are competing for the same space. The aviator will make every effort to avoid such fronts, since they are the areas of great atmospheric turbulence and the source of most bad weather. However, there is a very considerable amount of bad weather which occurs within an individual air mass. Fogs, thunderstorms, easterly waves, and tropical cyclones are weather problems within an air mass which plague airmen and surface-dwellers alike.

Most formations of clouds occur when air masses of different temperatures collide. Sometimes the results of these collisions are quite spectacular. Large-scale mixing of air masses often occurs very rapidly. When this happens, the atmosphere somersaults and thunderstorms are born. These storms begin to form when huge quantities of water vapor are lifted into clouds which blot out the sun. Then, like the Monday-morning wash hung on the line to dry, the moisture is wrung out of them to drench the land in torrents. Static electricity is produced by the friction of the falling raindrops, the charges build up, ultimately to discharge into the ground below. This lightning causes a temporary vacuum in the atmosphere and the air, swirling in to fill the space, creates the thunderclap.

In addition to fog and thunderstorms as air-mass weather phenomena, we have two disturbances which originate in maritime tropical or equatorial air and travel in a direction opposite to that of storms in the temperate latitudes. These are called *easterly waves* and *tropical cyclones*.

An easterly wave is merely a bend in the inter-tropical front. It is oriented in a northeast-southwest line which moves irregularly westward. Along this line, thunderclouds build and showers have their maximum intensity; the wind may shift from northeast to southeast as the storm passes by.

The two types of easterly waves are mainly identified by their areas of rain shower activity. The stable wave is characterized by clear skies preceding the weather and then about twenty-four hours of light-to-moderate shower activity. This type has very few thunderstorms and is quite common around the tropical islands bordering the Caribbean.

The unstable easterly wave is usually an active disturbance with rainfall and thunderstorms occurring over a period of one to two days. There are usually showers on both sides of the line squall; in many cases an unstable wave is the parent of a full-fledged hurricane or typhoon. To cite an example, hurricane Carol of the 1952 season was spawned east of Miami within an unstable easterly wave which dumped eight inches of rainfall on Puerto Rico in a twenty-four-hour period.

Such storms may be transformed into raging tropical hurricanes by microscopic changes in the upper air-flow pattern. It is often true in nature that the difference between order and confusion is an extremely narrow containment of many complex variables. In the case of the hurricane, the difference between a sunny day and a violent storm is the consequence of only a minor shift in the atmospheric balance.

III

Although weathermen have many names for the fronts which drift across our sky, they are usually classified according to the temperature of the air behind them. Those fronts which are followed by colder air are logically called *cold fronts*. On the other hand, *warm fronts* are followed by warm air. Since these two air-borne battle lines cause more than three-fourths of our weather, we will examine them more carefully.

A cold front develops when the leading edge of a cold-air mass meets the trailing edge of a warm-air mass. Cold air, being heavier, will underrun warm air in the same manner that a shoe can be slid under a rug. Because of this undercutting movement of air, the front of the cold-air mass forms a gigantic, invisible wedge of atmosphere extending backward from the point of surface contact. The upward curve of this wedge constitutes the cold front and provides the battleground for the most violent skirmishes of the meteorological war. The cold front consists of turbid mixtures of hot and cold air, gusty winds, and vertically developed clouds which present a dark and foreboding appearance. Long lines of these cumulonimbus clouds extend throughout the length of the cold front, rising skyward like giant mountain peaks. They are characterized by an anvil-like tip which is shaped when their tops are flattened by the strong westerly winds of the upper atmosphere. Violent rain, squeezed out of the warm air as it is pushed upward, drenches the earth below, while thunder and lightning form a fitting backdrop for the battle.

The cold front is usually heralded by high cirrus clouds which ride like lacy messengers well ahead of the storm. These are blown from the tops of the billowing thunderclouds which are next to follow them. Finally, the surface front arrives with its vertical clouds, heavy rainstorms, and gusty winds. These winds generally blow from the south or southwest. As the surface front passes, the winds shift radically to the northwest and the temperatures drop, often with startling suddenness. The cold-front skies clear rapidly once the surface front passes, and the damp, cool atmosphere soon has a gemlike clearness.

A warm front develops when a warm-air mass follows a cold-air mass. The light, warm air rides up above the cold, producing a soft creamlike blanket over the heavier milk of the cold-air mass. Thus the warm front also forms a wedge of cold air over the earth. But unlike the cold-front wedge, the warm-front wedge extends ahead of the surface front. It is also much flatter than that of the cold front and its weather extends over a greater area. While the average cold front has a cross-sectional width of fifty miles, the warm front is usually about two hundred miles wide.

Warm-front boundaries are a zone of atmospheric battle resembling exactly those of the cold front. The moisture of the rising warm air condenses to form clouds and bring rain. However, there is a sharp difference in the form of these clouds. They lie over the land in a flat blanket which slopes downward toward the surface boundary. The rain falls in a monotonous drizzle which may last for days. There is very little thunder and lightning, and the winds are light instead of gusty. High cirrus clouds drift in softly ahead of the warm front and gradually thicken into a solid layer of gray clouds. As the surface front approaches, one has the feeling that one can reach up and touch the slow moving clouds. The visibility is poor; often thin fog forms in the steady rain ahead of the front. As with all frontal passages, the wind shifts, usually from easterly to southerly as the front passes. The air becomes noticeably warmer behind the front. Compared to the slashing, beachhead type of battle of the cold front, warm frontal passages have the personality of heavy slowness characteristic of a long winter campaign.

When accompanied by temperature changes, the sudden shifting of rain-laden winds is the surest indication of a frontal passage and of better weather to come. Weathermen, observing these facts on their recording instruments, earn an easy reputation for reliable forecasting while the local weather is still bad. We can forgive this easy victory, however, when we realize that these opportunities are rare and that for the most part they must study endlessly the areas hundreds of miles away in order to predict the local weather.

The passage of the giant storm fronts becomes a continuous cycle as the air masses travel their great atmospheric routes. They weave the local sky into a pattern of ceaseless change, providing variety and diversity to make the air alternately gray and blue, sunny and rain-swept, dappled with tufted cirrus, or darkly coated by low stratus. The result is distribution. Water is brought to the arid land and heat is delivered to the cold places of the sphere. Just as the ships carry their cargo from one nation to another over the world's waterways, so do the weather fronts transport the riches of the atmosphere over the earth. Areas neglected by the treasure-laden fronts are for the most part shunned by living things. Some lie as tranquil and barren deserts under the pitiless sun, while others are coated by eternal frost.

In the Northern Hemisphere, fronts move from west to east at an average speed of fifteen miles per hour, but they do not always travel at the same rate. The reason for this difference in velocity lies in the variation in the air pressure between the air masses. Some contain relatively dense air which rises to great heights over the earth; such air masses exert high pressure on the surface below them and result in what weathermen call *high-pressure zones*. Dips and valleys also occur in the atmosphere, and the air masses which contain them are said to possess *low-pressure zones*. High-pressure air tends to flow toward a low-pressure area much as water can be sucked from a "high-pressure" glass into a "low-pressure" mouth through a straw. For this reason high-pressure air following a low-pressure air mass will be sucked toward the low pressure and will gradually overtake the low-pressure air mass. The intervening body of air, caught between these two opposing pressure systems, will slowly be squeezed skyward until its forward and rear fronts meet. When this happens, the point of contact is called an *occlusion* and the new line of battle is termed an *occluded front*.

The blue waters of the Pacific are often the spawning grounds of these occluded fronts. As the warm and cold fronts sweep inland like ocean waves to cross the shoreline of our western states, they are delayed by the high coastal mountain ridges. Occasionally this delay permits later fronts to catch up and engage the straggler, producing occlusions and occluded fronts at the points of contact. Other occluded fronts are formed over the prairie plains of our mid-western states when the polar continental air mass sends icy fingers stabbing southward. Such protuberances in the polar front are called *waves*. As these waves deepen, their sides draw closer together. Eventually contact occurs between the leading and following edges —each a true front in its own right—and an occluded front is born. The early stages of this process, before the fronts meet, produces the weather phenomenon known as a *wave cyclone*.

Three types of air are involved in the occluded front. The central section just below the occlusion is covered by a blanket of warm air. The region just before the front is cool; the air following the front is definitely cold. The conversion of a wave cyclone into an occluded front takes place when the pressure drops in the warm middle sector, causing the two fronts on either side to approach and make contact like the claws of a lobster. When this takes place, the warm air is gripped between the cool- and cold-air masses and squeezed aloft. In a practical sense this means that occluded fronts cause cold surface weather since the warm air is lifted away from the earth's surface.

The exact characteristics of occluded fronts depend upon whether the coldest air lies ahead of, or behind, the occlusion. When the coldest air lies ahead of the surface front, the system is known as a warm-front type occlusion. Another type of occluded front occurs when the coldest air is behind the surface front. As before, the warm air is pushed aloft, but the front remaining on the surface is now the cold front which has nudged its way under the cool air.

From the standpoint of a ground-dweller, the passage of an occluded front does not appear significantly worse than the passage of a strong cold front. Both types of weather are equally bad. However, the appearance of an occluded front on his weather map is a matter of the greatest concern to an aviator. One of the most important features in his flight planning is the avoidance of that area near the occlusion where the warm and cold fronts meet, for here the meteorological combat is fought in its most violent form. Giant thunderstorms, extreme turbulence, slashing rain, and assaulting lightning greet the unhappy aviator who has the misfortune to blunder into these destructive areas. Of the two types of occluded fronts, the cold-front occlusion is the more violent, since the energy of the warm-front type is spread over a larger area. If either one lies in his path, however, the wise pilot plots a detour to avoid both types of ambush.

We have outlined the great air-borne global conflicts which produce our stormy weather. The knowledge of air masses and their meteorological battle lines which we call fronts is far from complete. Indeed, fifty years ago the term *front* did not exist. However, for the first time in the history of our earth we are no longer subject to the surprises and whims of the ever-changing, eternal cycle of weather. Although our present ability to forecast is limited to a few days ahead at best, our techniques will advance as more information becomes available. After reviewing the marvelous strides of the past half-century, who can tell where we will be in the next? This much is certain: each passing year will increase our independence from the elements of our planetary environment.

D. *Physical Law*

•

If astronomy is the oldest of the sciences, physics, which studies the basic properties and interrelationships of matter and energy, is the most fundamental. All other sciences as they become more exact attempt to explain their separate disciplines in terms of physical law. It was not until Galileo that physics itself began to loosen the shackles imposed by Aristotelian dogma. Galileo it was who first questioned the theories of dynamics which had been considered inviolate since they had been formulated by the Greeks. In so doing he paved the way for a genius of equal or even greater stature. The year of Galileo's death, 1642, saw the birth of a premature baby who was not expected to live but who grew to manhood, to discover the law of gravitation, the laws of motion and the composite nature of white light; and, simultaneously with Leibniz, to invent the calculus. Newton's invention of the reflecting telescope, which would have established the reputation of a lesser scientist, pales almost into insignificance compared with the towering nature of his other achievements. Newton, of course, owed a great debt to Galileo and to two other astronomers who lived in this same extraordinary period: Tycho Brahe, who first recorded accurately the motions of the planets, and Johannes Kepler, whose laws of planetary motion showed how these planets moved with relation to their central sun. On the foundations laid by these three, Newton built a conception of the world and the forces that guide it that was destined to hold undisputed place until the beginning of the twentieth century, and even at that distant date to undergo but minor modification. Newtoniana *tells us something of the man; while* Discoveries *gives us all too brief glimpes of the work that made him what he was.*

The history of physics is replete with oddities, and its practitioners have been men and women of the most diverse personalities. One of the most interesting, not in the first rank among physicists but certainly a great man, was the protean Ben Franklin, some of whose contributions are contained in Experiments and Ideas. *Franklin the scientist is best known for his electrical experiments. Less well known are his invention of bifocals, his discovery of the origin of northeast storms, his extraordinary prophecy of aerial invasion.*

An equally fascinating personality is described in The Discovery of Radium *by Eve Curie, daughter of Marie Curie, whose basic discoveries helped lay the foundations of modern physics. Nearly half a century before the explosion of the atomic bomb, Henri Becquerel observed a fogging of a carefully wrapped photographic plate which had accidentally been placed near a fragment of the same substance, uranium, which was to play a dominant part in later events. The next development, which came as a sequence to Becquerel's observation, was the work of Madame Curie, who kept house, brought up a family and discovered radium. It is a story which gains new meaning when it is related to the course of modern physics. From Becquerel's studies, to Marie Curie's discovery, through numberless experiments by men and women of many nationalities, to the final release of nuclear energy in the bomb, we can trace the development of one of the most intricately constructed edifices in science. Some of the details of this progression are contained in* From X-Rays to Nuclear Fission *by Henry D. Smyth, formerly a member of the Atomic Energy Commission and author of* Atomic Energy for Military Purposes, *the official account of the Manhattan Project.*

We offer three episodes in the history of the atomic bomb. In 1939 a group of scientists who had emigrated to the United States from the Axis countries, profoundly disturbed by the possibility that Hitler might obtain a bomb before the Allies, prevailed on Albert Einstein to write a letter to President Roosevelt, which is here reproduced. It would be satisfying to report that the letter had an immediate effect. Actually it was buried in bureaucratic files, and it was more than two years after its delivery that the decision was made to build the bomb. The incident is an interesting commentary on the importance of unifying the "two cultures" which, as C. P. Snow has pointed out, exist in contemporary civilization.

The second episode is described in the eyewitness account, The First Atomic Pile *by Allardice and Trapnell. In one of the most crucial experiments in the history of science, man initiated and controlled a self-sustaining nuclear chain reaction. The experiment, which was of course shrouded in the utmost secrecy, took place on December 2, 1942, "beneath the West Stands of Stagg Field in Chicago."*

The third episode, The Death of Louis Slotin *by Ralph E. Lapp, is a terrifying example of the silent and unseen danger of radiation. The implications for all mankind of the radiations threat are examined in further detail in an article by Warren Weaver which appears in a later section.*

In recent years developments in physics have helped us probe ever deeper into the problem of the ultimate constitution of matter. Examination of this problem is on an entirely different level from that on which a previous generation approached it. Yet the solution remains as evasive as in the past. Indeed, contemporary physicists are perhaps more bewildered than their predecessors: our knowledge of fundamental particles is in a state bordering confusion. O. R. Frisch, who with Lisa Meitner made a basic contribution to the achievement of atomic fission, describes our present knowledge in Fundamental Particles. *Although it is written for a scientific and technical audience, a reading of Glenn T. Seaborg's* Peaceful Uses of Atomic Energy *throws light on his remark that nuclear power technology "can provide in the future enough energy for all the peoples of the world—the energy that is central to the banishment of hunger, poverty and fear of the future."*

It was from his work on the theory of relativity that Einstein drew his famous equation on the equivalence of matter and energy, $E=mc^2$. *At first, the relationship seemed incomprehensible to all but a handful of specialists, yet it has since revolutionized man's most fundamental concepts of the universe. Relativity theory remains deep water indeed. Yet in* Space, Time and Einstein *Dr. Heyl, whose techniques for weighing the earth are famous, says interesting things about it which are not too difficult for the informed layman to understand.*

Newtoniana

•

"I DO NOT KNOW what I may appear to the world, but to myself I seem to have been only like a boy playing on the sea-shore, and diverting myself in now and then finding a smoother pebble and a prettier shell than ordinary, whilst the great ocean of truth lay all undiscovered before me."

—Sir Isaac Newton

"If I have seen farther than Descartes, it is by standing on the shoulders of giants."—Sir Isaac Newton

"Newton was the greatest genius that ever existed and the most fortunate, for we cannot find more than once a system of the world to establish."—Lagrange

"There may have been minds as happily constituted as his for the cultivation of pure mathematical science; there may have been minds as happily constituted for the cultivation of science purely experimental; but in no other mind have the demonstrative faculty and the inductive faculty co-existed in such supreme excellence and perfect harmony."

—Lord Macaulay

"Taking mathematics from the beginning of the world to the time when Newton lived, what he had done was much the better half."

—Leibniz

"Let Men Rejoice that so great a glory of the human race has appeared."—Inscription on Westminster Tablet

"The law of gravitation is indisputably and incomparably the greatest scientific discovery ever made, whether we look at the advance which it involved, the extent of truth disclosed, or the fundamental and satisfactory nature of this truth."—William Whewell

"Newton's greatest direct contribution to optic appears to be the discovery and explanation of the nature of color. He certainly laid the broad foundation upon which spectrum analysis rests, and out of this has come the new science of spectroscopy which is the most delicate and powerful method for the investigation of the structure of matter."
—Dayton C. Miller

"On the day of Cromwell's death, when Newton was sixteen, a great storm raged all over England. He used to say, in his old age, that on that day he made his first purely scientific experiment. To ascertain the force of the wind, he first jumped with the wind and then against it, and by comparing these distances with the extent of his own jump on a calm day, he was enabled to compute the force of the storm. When the wind blew thereafter, he used to say it was so many feet strong."—James Parton

"His carriage was very meek, sedate and humble, never seemingly angry, of profound thought, his countenance mild, pleasant and comely. I cannot say I ever saw him laugh but once, which put me in mind of the Ephesian philosopher, who laughed only once in his lifetime, to see an ass eating thistles when plenty of grass was by. He always kept close to his studies, very rarely went visiting and had few visitors. I never knew him to take any recreation or pastime either in riding out to take the air, walking, bowling, or any other exercise whatever, thinking all hours lost that were not spent in his studies, to which he kept so close that he seldom left his chamber except at term time, when he read in the schools as Lucasianus Professor, where so few went to hear him, and fewer that understood him, that ofttimes he did in a manner, for want of hearers read to the walls. Foreigners he received with a great deal of freedom, candour, and respect. When invited to a treat, which was very seldom, he used to return it very handsomely, and with much satisfaction to himself. So intent, so serious upon his studies, that he ate very sparingly, nay, ofttimes he has forgot to eat at all, so that, going into his chamber, I have found his mess untouched, of which, when I have reminded him, he would reply—'Have I?' and then making to the table would eat a bite or two standing, for I cannot say I ever saw him sit at table by himself. He very rarely went to bed till two or three of the clock, sometimes not until five or six, lying about four or five hours, especially at spring and fall of the leaf, at which times he used to employ about six weeks in his laboratory, the fires scarcely going out either night or day; he sitting up one night and I another till he had finished his chemical experiments, in the performance of which he was the most accurate, strict, exact. What his aim might be I was not able to penetrate into, but his pains, his diligence at these set times made me think he aimed at something beyond the reach of human art and industry. I cannot say I ever saw him drink either wine, ale or beer, excepting at meals and then but very sparingly. He very rarely went to dine in the hall, except on some public days, and then if he has not been minded, would go very carelessly, with shoes down at heels, stockings untied, surplice on, and his head scarcely combed.

"His elaboratory was well furnished with chemical materials, as bodies, receivers, heads, crucibles, etc. which was made very little use of, the crucibles excepted, in which he fused his metals; he would sometimes, tho' very seldom, look into an old mouldy book which lay in his elaboratory, I think it was titled Agricola de Metallis, the transmuting of metals being his chief design, for which purpose antimony was a great ingredient. He has sometimes taken a turn or two, has made a sudden stand, turn'd himself about, run up the stairs like another Archimedes, with an Eureka fall to write on his desk standing without giving himself the leisure to draw a chair to sit down on. He would with great acuteness answer a question, but would very seldom start one. Dr. Boerhave, in some of his writings, speaking of Sir Isaac: 'That man,' says he, 'comprehends as much as all mankind besides.' "—Humphrey Newton

"When we review his life, his idiosyncrasies, his periods of contrast, and his doubts and ambitions and desire for place, may we not take some pleasure in thinking of him as a man—a man like most other men save in one particular—he had genius—a greater touch of divinity than comes to the rest of us?"—David Eugene Smith

SIR ISAAC NEWTON

Discoveries

•

Concerning the Law of Gravitation

HITHERTO we have explained the phaenomena of the heavens and of our sea by the power of gravity, but have not yet assigned the cause of this power. This is certain, that it must proceed from a cause that penetrates to the very centres of the sun and planets, without suffering the least diminution of its force; that operates not according to the quantity of the surfaces of the particles upon which it acts (as mechanical causes used to do), but according to the quantity of the solid matter which they contain, and propagates its virtue on all sides to immense distances, decreasing always in the duplicate proportions of the distances. Gravitation towards the sun is made up out of the gravitations towards the several particles of which the body of the sun is composed; and in receding from the sun decreases accurately in the duplicate proportion of the distances as far as the orb of Saturn, as evidently appears from the quiescence of the aphelions of the planets; nay, and even to the remotest aphelions of the comets, if these aphelions are also quiescent. But hitherto I have not been able to discover the cause of those properties of gravity from phaenomena, and I frame no hypotheses; for whatever is not deduced from the phaenomena is to be called an hypothesis; and hypotheses, whether metaphysical or physical, whether of occult qualities or mechanical, have no place in experimental philosophy. In this philosophy particular propositions are inferred from the phaenomena, and afterwards rendered general by induction. Thus it was that the impenetrability, the mobility, and the impulsive force of bodies, and the laws of motion and gravitation were discovered. And to us it is enough that gravity does really exist, and act according to the laws which we have explained, and abundantly serves to account for all the motions of the celestial bodies, and of our sea.

Laws of Motion

Law I. *Every body perseveres in its state of rest, or of uniform motion in a right line, unless it is compelled to change that state by force impressed thereon.*

Projectiles persevere in their motions, so far as they are not retarded by the resistance of the air, or impelled downwards by the force of gravity. A top, whose parts by their cohesion are perpetually drawn aside from rectilinear motions, does not cease its rotation, otherwise than as it is retarded by the air. The greater bodies of the planets and comets, meeting with less resistance in more free spaces, preserve their motions both progressive and circular for a much longer time.

Law II. *The alteration of motion is ever proportional to the motive force impressed: and is made in the direction of the right line in which that force is impressed.*

If any force generates a motion, a double force will generate double the motion, a triple force triple the motion, whether that force be impressed altogether and at once, or gradually and successively. And this motion (being always directed the same way with the generating force), if the body moved before, is added to or subducted from the former motion, according as they directly conspire with or are directly contrary to each other; or obliquely joined, when they are oblique, so as to produce a new motion compounded from the determination of both.

Law III. *To every action there is always opposed an equal reaction; or the mutual actions of two bodies upon each other are always equal, and directed to contrary parts.*

Whatever draws or presses another is as much drawn or pressed by that other. If you press a stone with your finger, the finger is also pressed by the stone. If a horse draws a stone tied to a rope, the horse (if I may so say) will be equally drawn back towards the stone; for the distended rope, by the same endeavor to relax or unbend itself, will draw the horse as much towards the stone, as it does the stone towards the horse, and will obstruct the progress of the one as much as it advances that of the other. If a body impinge upon another, and by its force change the motion of the other, that body also (because of the equality of the mutual pressure) will undergo an equal change, in its own motion, towards the contrary part. The changes made by these actions are equal, not in the velocities but in the motions of bodies; that is to say, if the bodies are not hindered by any other impediments. For, because the motions are equally changed, the changes of the velocities made towards contrary parts are reciprocally proportional to the bodies.

The Dispersion of Light

In the year 1666 (at which time I applied myself to the grinding of optick glasses of other figures than spherical) I procured me a triangular glass prism, to try therewith the celebrated phaenomena of colours. And in order thereto, having darkened my chamber, and made a small hole in my window-shuts, to let in a convenient quantity of the sun's light, I placed my prism at its entrance, that it might be thereby refracted to the opposite wall. It was at first a very pleasing divertissement, to view the vivid and intense colours produced thereby; but after a while applying myself to consider them more circumspectly, I became surprised, to see them in an oblong form; which, according to the received laws of refraction, I expected should have been circular. They were terminated at the sides with straight lines, but at the ends, the decay of light was so gradual that it was difficult to determine justly, what was their figure; yet they seemed semicircular.

Comparing the length of this colour'd Spectrum with its breadth, I found it about five times greater, a disproportion so extravagant, that it excited me to a more than ordinary curiosity to examining from whence it might proceed. I could scarce think, that the various thicknesses of the glass, or the termination with shadow or darkness, could have any influence on light to produce such an effect; yet I thought it not amiss, first to examine those circumstances, and so try'd what would happen by transmitting light through parts of the glass of divers thicknesses, or through holes in the window of divers bignesses, or by setting the prism without, so that the light might pass through it, and be refracted, before it was terminated by the hole. But I found none of these circumstances material. The fashion of the colours was in all these cases the same.

The gradual removal of these suspicions led me to the Experimentum Crucis, which was this: I took two boards, and placed one of them close behind the prism at the window, so that the light might pass through a small hole, made in it for the purpose, and fall on the other board, which I placed at about 12 feet distance, having first made a small hole in it also, for some of the incident light to pass through. Then I placed another prism behind this second board, so that the light trajected through both the boards might pass through that also, and be again refracted before it arrived at the wall. This done, I took the first prism in my hand, and turned it to and fro slowly about its axis, so much as to make the several parts of the image cast, on the second board, successively pass through the hole in it, that I might observe to what places on the wall the second prism would refract them. And I saw by the variation of those places, that the light, tending to that end of the image, towards which the refraction of the first prism was made, did in the second prism suffer a refraction considerably greater than the light tending to the other end.

And so the true cause of the length of that image was detected to be no other, than that light is not similar or homogenial, but consists of *Difform Rays, some of which are more Refrangible than others;* so that without any difference in their incidence on the same medium, some shall be more Refracted than others; and therefore that, according to their *particular Degrees of Refrangibility,* they were transmitted through the prism to divers parts of the opposite wall.

On the Origin of Colours

The colours of all natural bodies have no other origin than this, that they are variously qualified, to reflect one sort of light in greater plenty than another. And this I have experimented in a dark room, by illuminating those bodies with uncompounded light of divers colours. For by that means any body may be made to appear of any colour. They have there no appropriate colour, but ever appear of the colour of the light cast upon them, but yet with this difference, that they are most brisk and vivid in the light of their own daylight colour. Minimum appeareth there of any colour indifferently, with which it is illustrated, but yet most luminous in red, and so bise appeareth indifferently of any colour, but yet most luminous in blue. And therefore minimum reflecteth rays of any colour, but most copiously those endowed with red, that is, with all sorts of rays promiscuously blended, those qualified with red shall abound most in that reflected light, and by their prevalence cause it to appear of that colour. And for the same reason bise, reflecting blue most copiously, shall appear blue by the excess of those rays in its reflected light; and the like of other bodies. And that this is the entire and adequate cause of their colours, is manifest, because they have no power to change or alter the colours of any sort of rays incident apart, but put on all colours indifferently, with which they are enlightened.

These things being so, it can be no longer disputed, whether there be colours in the dark, or whether they be the qualities of the objects we see, no nor perhaps, whether light be a body. For, since colours are the quality of light, having its rays for their entire and immediate subject, how can we think those rays qualities also, unless one quality may be the subject of, and sustain another; which in effect is to call it substance. We should not know bodies for substances; were it not for their sensible qualities, and the principle of those being now found due to something else, we have as good reason to believe that to be a substance also.

Besides, who ever thought any quality to be a heterogeneous aggregate, such as light is discovered to be? But to determine more absolutely what light is, after what manner refracted, and by what modes or actions it produceth in our minds the phantasms of colours, is not so easie; and I shall not mingle conjectures with certainties.

BENJAMIN FRANKLIN

Experiments and Ideas

•

As FREQUENT MENTION is made in public papers from Europe of the success of the Philadelphia experiment for drawing the electric fire from clouds by means of pointed rods of iron erected on high buildings, &, it may be agreeable to the curious to be informed, that the same experiment has succeeded in Philadelphia, though made in a different and more easy manner, which is as follows:

Make a small cross of two light strips of cedar, the arms so long as to reach to the four corners of a large thin silk handkerchief when extended; tie the corners of the handkerchief to the extremities of the cross, so you have the body of a kite; which being properly accommodated with a tail, loop, and string, will rise in the air, like those made of paper; but this being of silk, is fitter to bear the wet and wind of a thunder-gust without tearing. To the top of the upright stick of the cross is to be fixed a very sharp pointed wire, rising a foot or more above the wood. To the end of the twine, next the hand, is to be tied a silk ribbon, and where the silk and twine join, a key may be fastened. This kite is to be raised when a thunder-gust appears to be coming on, and the person who holds the string must stand within a door or window or under some cover, so that the silk ribbon may not be wet; and care must be taken that the twine does not touch the frame of the door or window. As soon as any of the thunder-clouds come over the kite, the pointed wire will draw the electric fire from them, and the kite, with all the twine, will be electrified, and the loose filaments of the twine will stand out every way, and be attracted by an approaching finger. And when the rain has wet the kite and twine, so that it can conduct the electric fire freely, you will find it stream out plentifully from the key on the approach of your knuckle. At this key the phial may be charged; and from electric fire thus obtained, spirits may be kindled, and all the other electric experiments be performed, which are usually done by the help of a rubbed glass globe or tube, and thereby the sameness of the electric matter with that of lightning completely demonstrated. *Letter to Peter Collinson, 1752*

Electrical Experiments and Electrocution

Your question, how I came first to think of proposing the experiment of drawing down the lightning, in order to ascertain its sameness with the electric fluid, I cannot answer better than by giving you an extract from the minutes I used to keep of the experiments I made, with memorandums of such as I purposed to make, the reasons for making them, and the observations that arose upon them, from which minutes my letters were afterwards drawn. By this extract you will see, that the thought was not so much "an out-of-the-way one," but that it might have occurred to any electrician.

"*November* 7, 1749. Electrical fluid agrees with lightning in these particulars. 1. Giving light. 2. Colour of the light. 3. Crooked direction. 4. Swift motion. 5. Being conducted by metals. 6. Crack or noise in exploding. 7. Subsisting in water or ice. 8. Rending bodies it passes through. 9. Destroying animals. 10. Melting metals. 11. Firing inflammable substances. 12. Sulphureous smell. The electric fluid is attracted by points. We do not know whether this property is in lightning. But since they agree in all particulars wherein we can already compare them, is it not probable they agree likewise in this? Let the experiment be made."

The knocking down of the six men was performed with two of my large jarrs not fully charged. I laid one end of my discharging rod upon the head of the first; he laid his hand on the head of the second; the second his hand on the head of the third, and so to the last, who held, in his hand, the chain that was connected with the outside of the jarrs. When they were thus placed, I applied the other end of my rod to the prime-conductor, and they all dropt together. When they got up, they all declared they had not felt any stroke, and wondered how they came to fall; nor did any of them either hear the crack, or see the light of it. You suppose it a dangerous experiment; but I had once suffered the same myself, receiving, by accident, an equal stroke through my head, that struck me down, without hurting me. And I had seen a young woman, that was about to be electrified through the feet, (for some indisposition) receive a greater charge through the head, by inadvertently stooping forward to look at the placing of her feet, till her forhead (as she was very tall) came too near my prime-conductor: she dropt, but instantly got up again, complaining of nothing. A person so struck, sinks down doubled, or folded together as it were, the joints losing their strength and stiffness at once, so that he drops on the spot where he stood, instantly, and there is no previous staggering, nor does he ever fall lengthwise. Too great a charge might, indeed, kill a man, but I have not yet seen any hurt done by it. It would certainly, as you observe, be the easiest of all deaths.

Letter to John Lining, 1755

Origin of Northeast Storms

Agreeable to your request, I send you my reasons for thinking that our northeast storms in North America begin first, in point of time, in the southwest parts: That is to say, the air in Georgia, the farthest of our colonies to the Southwest, begins to move southwesterly before the air of Carolina, which is the next colony northeastward; the air of Carolina has the same motion before the air of Virginia, which lies still more northeastward; and so on northeasterly through Pennsylvania, New-York, New-England, &c., quite to Newfoundland.

These northeast storms are generally very violent, continue sometimes two or three days, and often do considerable damage in the harbours along the coast. They are attended with thick clouds and rain.

What first gave me this idea, was the following circumstance. About twenty years ago, a few more or less, I cannot from my memory be certain, we were to have an eclipse of the moon at Philadelphia, on a Friday evening, about nine o'clock. I intended to observe it, but was prevented by a northeast storm, which came on about seven, with thick clouds as usual, that quite obscured the whole hemisphere. Yet when the post brought us the Boston newspaper, giving an account of the effects of the same storm in those parts, I found the beginning of the eclipse had been well observed there, though Boston lies N. E. of Philadelphia about 400 miles. This puzzled me because the storm began with us so soon as to prevent any observation, and being a N. E. storm, I imagined it must have begun rather sooner in places farther to the northeastward than it did at Philadelphia. I therefore mentioned it in a letter to my brother, who lived at Boston; and he informed me the storm did not begin with them till near eleven o'clock, so that they had a good observation of the eclipse: And upon comparing all the other accounts I received from the several colonies, of the time of beginning of the same storm, and, since that of other storms of the same kind, I found the beginning to be always later the farther northeastward. I have not my notes with me here in England, and cannot, from memory, say the proportion of time to distance, but I think it is about an hour to every hundred miles.

From thence I formed an idea of the cause of these storms, which I would explain by a familiar instance or two. Suppose a long canal of water stopp'd at the end by a gate. The water is quite at rest till the gate is open, then it begins to move out through the gate; the water next the gate is first in motion, and moves towards the gate; the water next to that first water moves next, and so on successively, till the water at the head of the canal is in motion, which is last of all. In this case all the water moves indeed towards the gate, but the successive times of beginning motion are the contrary way, *viz.* from the gate backwards to the head of the canal. Again, suppose the air in a chamber at rest, no current through the room till you make a fire in the chimney. Immediately the air in the chimney, being rarefied by the fire, rises; the air next the chimney flows in to supply its place, moving towards the chimney; and, in consequence, the rest of the air successively, quite back to the door. Thus to produce our northeast storms. I suppose some great heat and rarefaction of the air in or about the Gulph of Mexico; the air thence rising has its place supplied by the next more northern, cooler, and therefore denser and heavier, air; that, being in motion, is followed by the next more northern air, &c. &c., in a successive current, to which current our coast and inland ridge of mountains give the direction of northeast, as they lie N.E. and S.W. *Letter to Alexander Small, 1760*

A Prophecy of Aerial Invasion

I have this day received your favor of the 2d inst. Every information in my power, respecting the balloons, I sent you just before Christmas, contained in copies of my letters to Sir Joseph Banks. There is no secret in the affair, and I make no doubt that a person coming from you would easily obtain a sight of the different balloons of Montgolfier and Charles, with all the instructions wanted; and, if you undertake to make one, I think it extremely proper and necessary to send an ingenious man here for that purpose; otherwise, for want of attention to some particular circumstance, or of not being acquainted with it, the experiment might miscarry, which, in an affair of so much public expectation, would have bad consequences, draw upon you a great deal of censure, and affect your reputation. It is a serious thing to draw out from their affairs all the inhabitants of a great city and its environs, and a disappointment makes them angry. At Bourdeaux lately a person who pretended to send up a balloon, and had received money from many people, not being able to make it rise, the populace were so exasperated that they pulled down his house and had like to have killed him.

It appears, as you observe, to be a discovery of great importance, and what may possibly give a new turn to human affairs. Convincing sovereigns of the folly of wars may perhaps be one effect of it; since it will be impracticable for the most potent of them to guard his dominions. Five thousand balloons, capable of raising two men each, could not cost more than five ships of the line; and where is the prince who can afford so to cover his country with troops for its defence, as that ten thousand men descending from the clouds might not in many places do an infinite deal of mischief, before a force could be brought together to repel them?

Letter to Jan Ingenhousz, 1784

Daylight Saving

You often entertain us with accounts of new discoveries. Permit me to communicate to the public, through your paper, one that has lately been made by myself, and which I conceive may be of great utility.

I was the other evening in a grand company, where the new lamp of Messrs. Quinquet and Lange was introduced, and much admired for its splendour; but a general inquiry was made, whether the oil it consumed was not in proportion to the light it afforded, in which case there would be no saving in the use of it. No one present could satisfy us in that point, which all agreed ought to be known, it being a very desirable thing to lessen, if possible, the expense of lighting our apartments, when every other article of family expense was so much augmented.

I was pleased to see this general concern for economy, for I love economy exceedingly.

I went home, and to bed, three or four hours after midnight, with my head full of the subject. An accidental sudden noise waked me about six in the morning, when I was surprised to find my room filled with light; and I imagined at first, that a number of those lamps had been brought into it; but, rubbing my eyes, I perceived the light came in at the windows. I got up and looked out to see what might be the occasion of it, when I saw the sun just rising above the horizon, from where he poured his rays plentifully into my chamber, my domestic having negligently omitted, the preceding evening, to close the shutters.

I looked at my watch, which goes very well, and found that it was but six o'clock; and still thinking it something extraordinary that the sun should rise so early, I looked into the almanac, where I found it to be the hour given for his rising on that day. I looked forward, too, and found he was to rise still earlier every day till towards the end of June; and that at no time in the year he retarded his rising so long as till eight o'clock. Your readers, who with me have never seen any signs of sunshine before noon, and seldom regard the astronomical part of the almanac, will be as much astonished as I was, when they hear of his rising so early; and especially when I assure them, *that he gives light as soon as he rises*. I am convinced of this. I am certain of my fact. One cannot be more certain of any fact. I saw it with my own eyes. And, having repeated this observation the three following mornings, I found always precisely the same result.

This event has given rise in my mind to several serious and important reflections. I considered that, if I had not been awakened so early in the morning, I should have slept six hours longer by the light of the sun, and in exchange have lived six hours the following night by candle-light; and, the latter being a much more expensive light than the former, my love of economy induced me to muster up what little arithmetic I was master of, and to make some calculations, which I shall give you, after observing that utility is, in my opinion the test of value in matters of invention, and that a discovery which can be applied to no use, or is not good for something, is good for nothing.

I took for the basis of my calculation the supposition that there are one hundred thousand families in Paris, and that these families consume in the night half a pound of bougies, or candles, per hour. I think this is a moderate allowance, taking one family with another; for though, I believe some consume less, I know that many consume a great deal more. Then estimating seven hours per day as the medium quantity between the time of the sun's rising and ours, he rising during the six following months from six to eight hours before noon, and there being seven hours of course per night in which we burn candles, the account will stand thus;—

In the six months between the 20th of March and the 20th of September, there are

Nights	183
Hours of each night in which we burn candles	7
Multiplication gives for the total number of hours	1,281
These 1,281 hours multiplied by 100,000, the number of inhabitants, give	128,100,000
One hundred twenty-eight millions and one hundred thousand hours, spent at Paris by candle-light, which, at half a pound of wax and tallow per hour, gives the weight of	64,050,000
Sixty-four millions and fifty thousand of pounds, which, estimating the whole at the medium price of thirty sols the pound, makes the sum of ninety-six millions and seventy-five thousand livres tournois	96,075,000

An immense sum! that the city of Paris might save every year, by the economy of using sunshine instead of candles.

Letter to the Authors of "The Journal of Paris," 1784

Bifocals

By Mr. Dollond's saying, that my double spectacles can only serve particular eyes, I doubt he has not been rightly informed of their construction. I imagine it will be found pretty generally true, that the same convexity of glass, through which a man sees clearest and best at the distance proper for reading, is not the best for greater distances. I therefore had formerly two pair of spectacles, which I shifted occasionally, as in travelling I sometimes read, and often wanted to regard the prospects.

Finding this change troublesome, and not always sufficiently ready, I had the glasses cut, and half of each kind associated in the same circle.

By this means, as I wear my spectacles constantly, I have only to move my eyes up or down, as I want to see distinctly far or near, the proper glasses being always ready. This I find more particularly convenient since my being in France, the glasses that serve me best at table to see what I eat, not being the best to see the faces of those on the other side of the table who speak to me; and when one's ears are not well accustomed to the sounds of a language, a sight of the movements in the features of him that speaks helps to explain; so that I understand French better by the help of my spectacles.

Letter to George Whatley, 1785

EVE CURIE

The Discovery of Radium

•

AFTER ROENTGEN'S DISCOVERY of X-Rays, Henri Poincaré conceived the idea of determining whether rays like the X-ray were emitted by "fluorescent" bodies under the action of light. Attracted by the same problem, Henri Becquerel examined the salts of a "rare metal," uranium. Instead of finding the phenomenon he had expected, he observed another, altogether different and incomprehensible: he found that uranium salts *spontaneously* emitted, without exposure to light, some rays of unknown nature. A compound of uranium, placed on a photographic plate surrounded by black paper, made an impression on the plate through the paper. And, like the X-ray, these astonishing "uranic" salts discharged an electroscope by rendering the surrounding air a conductor.

Henri Becquerel made sure that these surprising properties were not caused by a preliminary exposure to the sun and that they persisted when the uranium compound had been maintained in darkness for several months. For the first time, a physicist had observed the phenomenon to which Marie Curie was later to give the name of *radioactivity*. But the nature of the radiation and its origin remained an enigma.

Becquerel's discovery fascinated the Curies. They asked themselves whence came the energy—tiny, to be sure—which uranium compounds constantly disengaged in the form of radiation. And what was the nature of this radiation? Here was an engrossing subject of research, a doctor's thesis! The subject tempted Marie most because it was a virgin field: Becquerel's work was very recent and so far as he knew nobody in the laboratories of Europe had yet attempted to make a fundamental study of uranium rays. As a point of departure, and as the only bibliography, there existed some communications presented by Henri Becquerel at the Academy of Science during the year 1896. It was a leap into great adventure, into an unknown realm.

There remained the question of where she was to make her experiments—and here the difficulties began. Pierre made several approaches to the director of the School of Physics with practically no results: Marie was given the free use of a little glassed-in studio on the ground floor of the school. It was a kind of storeroom, sweating with damp, where unused machines and lumber were put away. Its technical equipment was rudimentary and its comfort nil.

Deprived of an adequate electrical installation and of everything that forms material for the beginning of scientific research, she kept her patience, sought and found a means of making her apparatus work in this hole.

It was not easy. Instruments of precision have sneaking enemies: humidity, changes of temperature. Incidentally the climate of this little workroom, fatal to the sensitive electrometer, was not much better for Marie's health. But this had no importance. When she was cold, the young woman took her revenge by noting the degrees of temperature in centigrade in her notebook. On February 6, 1898, we find, among the formulas and figures: "Temperature here 6°25.[1] Six degrees . . . !" Marie, to show her disapproval, added ten little exclamation points

The candidate for the doctor's degree set her first task to be the measurement of the "power of ionization" of uranium rays—that is to say, their power to render the air a conductor of electricity and so to discharge an electroscope. The excellent method she used, which was to be the key to the success of her experiments, had been invented for the study of other phenomena by two physicists well known to her: Pierre and Jacques Curie. Her technical installation consisted of an "ionization chamber," a Curie electrometer and a piezoelectric quartz.

At the end of several weeks the first result appeared: Marie acquired the certainty that the intensity of this surprising radiation was proportional to the quantity of uranium contained in the samples under examination, and that this radiation, which could be measured with precision, was not affected either by the chemical state of combination of the uranium or by external factors such as lighting or temperature.

These observations were perhaps not very sensational to the uninitiated, but they were of passionate interest to the scientist. It often happens in physics that an inexplicable phenomenon can be subjected, after some investigation, to laws already known, and by this very fact loses its interest for the research worker. Thus, in a badly constructed detective story, if we are told in the third chapter that the woman of sinister appearance who might have committed the crime is in reality only an honest little housewife who leads a life without secrets, we feel discouraged and cease to read.

Nothing of the kind happened here. The more Marie penetrated into intimacy with uranium rays, the more they seemed without precedent,

[1] About 44° Fahrenheit.—Eds.

essentially unknown. They were like nothing else. Nothing affected them. In spite of their very feeble power, they had an extraordinary individuality.

Turning this mystery over and over in her head, and pointing toward the truth, Marie felt and could soon affirm that the incomprehensible radiation was an *atomic* property. She questioned: Even though the phenomenon had only been observed with uranium, nothing proved that uranium was the only chemical element capable of emitting such radiation. Why should not other bodies possess the same power? Perhaps it was only by chance that this radiation had been observed in uranium first, and had remained attached to uranium in the minds of physicists. Now it must be sought for elsewhere.

No sooner said than done. Abandoning the study of uranium, Marie undertook to examine *all known chemical bodies*, either in the pure state or in compounds. And the result was not long in appearing: compounds of another element, thorium, also emitted spontaneous rays like those of uranium and of similar intensity. The physicist had been right: the surprising phenomenon was by no means the property of uranium alone, and it became necessary to give it a distinct name. Mme Curie suggested the name of radioactivity. Chemical substances like uranium and thorium, endowed with this particular "radiance," were called *radio elements*.

Radioactivity so fascinated the young scientist that she never tired of examining the most diverse forms of matter, always by the same method. Curiosity, a marvelous feminine curiosity, the first virtue of a scientist, was developed in Marie to the highest degree. Instead of limiting her observation to simple compounds, salts and oxides, she had the desire to assemble samples of minerals from the collection at the School of Physics, and of making them undergo almost at hazard, for her own amusement, a kind of customs inspection which is an electrometer test. Pierre approved, and chose with her the veined fragments, hard or crumbly, oddly shaped, which she wanted to examine.

Marie's idea was simple—simple as the stroke of genius. At the crossroads where Marie now stood, hundreds of research workers might have remained, nonplussed, for months or even years. After examining all known chemical substances, and discovering—as Marie had done—the radiation of thorium, they would have continued to ask themselves in vain whence came this mysterious radioactivity. Marie, too, questioned and wondered. But her surprise was translated into fruitful acts. She had used up all evident possibilities. Now she turned toward the unplumbed and the unknown.

She knew in advance what she would learn from an examination of the minerals, or rather she thought she knew. The specimens which contained neither uranium nor thorium would be revealed as totally "inactive." The others, containing uranium or thorium, would be radioactive.

Experiment confirmed this prevision. Rejecting the inactive minerals, Marie applied herself to the others and measured their radioactivity. Then came a dramatic revelation: the radioactivity was a *great deal stronger* than could have been normally foreseen by the quantity of uranium or thorium contained in the products examined!

"It must be an error in experiment," the young woman thought; for doubt is the scientist's first response to an unexpected phenomenon.

She started her measurements over again, unmoved, using the same products. She started over again ten times, twenty times. And she was forced to yield to the evidence: the quantities of uranium found in these minerals were by no means sufficient to justify the exceptional intensity of the radiation she observed.

Where did this excessive and abnormal radiation come from? Only one explanation was possible: the minerals must contain, in small quantity, a *much more powerfully radioactive substance* than uranium and thorium.

But what substance? In her preceding experiments, Marie had already examined *all known chemical elements.*

The scientist replied to the question with the sure logic and the magnificent audaciousness of a great mind: The mineral certainly contained a radioactive substance, which was at the same time a chemical element unknown until this day: *a new element.*

A new element! It was a fascinating and alluring hypothesis—but still a hypothesis. For the moment this powerfully radioactive substance existed only in the imagination of Marie and of Pierre. But it did exist there. It existed strongly enough to make the young woman go to see Bronya one day and tell her in a restrained, ardent voice:

"'You know, Bronya, the radiation that I couldn't explain comes from a new chemical element. The element is there and I've got to find it. We are sure! The physicists we have spoken to believe we have made an error in experiment and advise us to be careful. But I am convinced that I am not mistaken."

These were unique moments in her unique life. The layman forms a theatrical—and wholly false—idea of the research worker and of his discoveries. "The moment of discovery" does not always exist: the scientist's work is too tenuous, too divided, for the certainty of success to crackle out suddenly in the midst of his laborious toil like a stroke of lightning, dazzling him by its fire. Marie, standing in front of her apparatus, perhaps never experienced the sudden intoxication of triumph. This intoxication was spread over several days of decisive labor, made feverish by a magnificent hope. But it must have been an exultant moment when, convinced by the rigorous reasoning of her brain that she was on the trail of new matter, she confided the secret to her elder sister, her ally always. Without exchanging one affectionate word, the two sisters must have lived again, in a dizzying breath of memory, their years of waiting, their mutual sacrifices, their bleak lives as students, full of hope and faith.

It was barely four years before that Marie had written:

> Life is not easy for any of us. But what of that? We must have perseverance and above all confidence in ourselves. We must believe that we are gifted for something, and that this thing, at whatever cost, must be attained.

That "something" was to throw science upon a path hitherto unsuspected.

In a first communication to the Academy, presented by Prof. Lippmann and published in the *Proceedings* on April 12, 1898, "Marie Sklodovska Curie" announced the probable presence in pitchblende ores of a new element endowed with powerful radioactivity. This was the first stage of the discovery of radium.

By the force of her own intuition the physicist had shown to herself that the wonderful substance must exist. She decreed its existence. But its incognito still had to be broken. Now she would have to verify hypothesis by experiment, isolate this material and see it. She must be able to announce with certainty: "It is there."

Pierre Curie had followed the rapid progress of his wife's experiments with passionate interest. Without directly taking part in Marie's work, he had frequently helped her by his remarks and advice. In view of the stupefying character of her results, he did not hesitate to abandon his study of crystals for the time being in order to join his efforts to hers in the search for the new substance.

Thus, when the immensity of a pressing task suggested and exacted collaboration, a great physicist was at Marie's side—a physicist who was the companion of her life. Three years earlier, love had joined this exceptional man and woman together—love, and perhaps some mysterious foreknowledge, some sublime instinct for the work in common.

The valuable force was now doubled. Two brains, four hands, now sought the unknown element in the damp little workroom in the Rue Lhomond. From this moment onward it is impossible to distinguish each one's part in the work of the Curies. We know that Marie, having chosen to study the radiation of uranium as the subject of her thesis, discovered that other substances were also radioactive. We know that after the examination of minerals she was able to announce the existence of a new chemical element, powerfully radioactive, and that it was the capital importance of this result which decided Pierre Curie to interrupt his very different research in order to try to isolate this element with his wife. At that time—May or June, 1898—a collaboration began which was to last for eight years, until it was destroyed by a fatal accident.

We cannot and must not attempt to find out what should be credited to Marie and what to Pierre during these eight years. It would be exactly what the husband and wife did not want. The personal genius of Pierre Curie is known to us by the original work he had accomplished before this collaboration. His wife's genius appears to us in the first intuition of discovery, the brilliant start; and it was to reappear to us again, solitary, when Marie Curie the widow unflinchingly carried the weight of a new science and conducted it, through research, step by step, to its harmonious expansion. We therefore have formal proof that in the fusion of their two efforts, in this superior alliance of man and woman, the exchange was equal.

Let this certainly suffice for our curiosity and admiration. Let us not attempt to separate these creatures full of love, whose handwriting alternates and combines in the working notebooks covered with formulae, these creatures who were to sign nearly all their scientific publications together. They were to write "We found" and "We observed"; and when they were constrained by fact to distinguish between their parts, they were to employ this moving locution:

> Certain minerals containing uranium and thorium (pitchblende, chalcolite, uranite) are very active from the point of view of the emission of Becquerel rays. In a preceding communication, *one of us* showed that their activity was even greater than that of uranium and thorium, and stated the opinion that this effect was due to some other very active substance contained in small quantity in these minerals.
>
> (Pierre and Marie Curie: *Proceedings of the Academy of Science,* July 18, 1898.)

Marie and Pierre looked for this "very active" substance in an ore of uranium called pitchblende, which in the crude state had shown itself to be four times more radioactive than the pure oxide of uranium that could be extracted from it. But the composition of this ore had been known for a long time with considerable precision. The new element must therefore be present in very small quantity or it would not have escaped the notice of scientists and their chemical analysis.

They began their prospecting patiently, using a method of chemical research invented by themselves, based on radioactivity; they separated all the elements in pitchblende by ordinary chemical analysis and then measured the radioactivity of each of the bodies thus obtained. By successive eliminations they saw the "abnormal" radioactivity take refuge in certain parts of the ore. As they went on, the field of investigation was narrowed. It was exactly the technique used by the police when they search the houses of a neighborhood, one by one, to isolate and arrest a malefactor.

But there was more than one malefactor here: the radioactivity was concentrated principally in two different chemical fractions of the pitchblende. For M and Mme Curie it indicated the existence of two new elements instead of one. By July 1898 they were able to announce the discovery of one of these substances with certainty.

"You will have to name it," Pierre said to his young wife, in the same tone as if it were a question of choosing a name for little Irène.

The one-time Mlle Sklodovska reflected in silence for a moment. Then, her heart turning toward her own country . . . answered timidly: "Could we call it 'polonium'?"

In the *Proceedings of the Academy* for July 1898 we read:

> We believe the substance we have extracted from pitchblende contains a metal not yet observed, related to bismuth by its analytical properties. If the existence of this new metal is confirmed we propose to call it *polonium*, from the name of the original country of one of us.

We find another note worthy of remark.

It was drawn up by Marie and Pierre Curie and a collaborator called G. Bémont. Intended for the Academy of Science, and published in the *Proceedings* of the session of December 26, 1898, it announced the existence of a second new chemical element in pitchblende.

Some lines of this communication read as follows:

> *The various reasons we have just enumerated lead us to believe that the new radioactive substance contains a new element to which we propose to give the name of* RADIUM.

HENRY D. SMYTH

From X-Rays to Nuclear Fission

•

SINCE THE END of the Second World War, science and scientists have been more in the public eye than ever before. It has become recognized that some of the most important national and international questions arise from scientific and technological developments. In fact, sometimes it is difficult to discern the dividing line between science and politics; at least we find physicists talking about world government, statesmen discussing nuclear fission, and everyone sure that we are on the threshold of an "atomic age." At such a time scientists in general and physicists in particular should study their methods and should examine the power and limitations of science. This article on 50 years of atomic physics is offered as a contribution along these lines. In it I shall not catalogue the great discoveries of the past half century, but I shall discuss some of those that seem to me most significant or most typical. I shall start with a description of the state of physics in 1895 and I must end, inevitably, with the chain reaction in 1942.

In the 300 years since Galileo, a large body of knowledge had been built up and a method firmly established. Much was known of mechanics, electricity and magnetism, heat, sound, and light; they persist in most modern textbooks.

The successes of the electromagnetic theory of light, of the kinetic theory of gases, and of the application of the fundamental laws of mechanics had been very great. The atomicity of matter had been firmly established by the chemists; it was possible to estimate the sizes and velocities of atoms and molecules, however roughly. The laws of electrolysis had suggested rather clearly that electrical charge was made up of atoms or particles of identical size. Yet most of physics was not concerned with these ideas. Except for the kinetic theory of gases, physics was all in terms of continuous media and continuous exchange of energy. No discontinuities appeared in the theoretical picture of electricity and magnetism nor in heat flow nor in light nor sound.

One might properly have expected that further knowledge about atoms and molecules and any discoveries about their structure—if they were in fact not indivisible—would come from chemistry rather than physics. Physics was supposed to be concerned with energy and its transformations, chemistry with matter. But it so happened that the study of electricity was a province of physics. For some 30 or 40 years physicists had studied the electrical discharges in gases at low pressure and they had also worked in the field of electrolysis, which the chemists had not yet taken over. It was out of the study of electrolysis that the suggestion of the atomicity of electricity had come, and it was out of the study of discharge in gases that many of the great discoveries of the modern period were to come, directly or indirectly.

Three Great Discoveries

The first of these great discoveries and, in many respects, the most sudden and startling was the discovery of X-rays by Roentgen in 1895. The second was the discovery of radioactivity by Becquerel in 1896; and the third, which was not nearly so clear-cut, was the discovery of the electron. I should like to examine these three discoveries as illustrations of the scientific method and the ways in which it is used to increase our knowledge.

In the fall of 1895 Roentgen was studying the passage of cathode rays through thin aluminum windows. This was one more experiment in the innumerable series on the discharge of electricity through gases which has been going on now for nearly a hundred years, has given us the fluorescent lamp and other devices, but has still not told us what happens in an electrical discharge in gases. On November 8, 1895, in the laboratory at Würzburg, it happened that a fluorescent screen was lying near the discharge tube, and Roentgen noticed that it became luminous when the discharge tube was running, even though the tube was covered with black paper. Roentgen had the imagination and the curiosity to look into the matter further, and within a week he had discovered many of the principal properties of X-rays. Nevertheless he kept his discovery to himself for nearly two months, during which he devised and carried out a whole series of experiments. He first announced his discovery in a paper submitted to the Physical Medical Society of Würzburg on December 28, 1895. This paper has deservedly become a scientific classic, both because of the importance of the discovery it announces and the clarity and modesty of its presentation. It describes the stepwise establishment of the properties of X-rays by a sequence of experiments logically following each other. It is a beautiful example of logical analysis of new data continually used to guide new experiments. Furthermore, the observed facts and suggested interpretations are clearly distinguished. At the end, Roentgen suggests that X-rays may be longitudinal vibrations in the ether, but he does not press the point and has provided enough information about the experiments to enable the reader to judge the validity of the interpretation for himself. I wish I could escape the feeling that if Roentgen were publishing today his paper would be entitled "Demonstration of Longitudinal Waves in the Ether," would begin by a mathematical treatment of the expected properties of these waves based on unstated assumptions, and would conclude by a report of experimental results which did not condescend to say what was actually done but gave graphs showing the comparison between theory and experiment.

The discovery of radioactivity by Becquerel was somewhat similar to that of X-rays. In the tubes used by Roentgen the X-rays came largely from the glass walls of the tube when cathode rays impinged on them. This phenomenon was accompanied by the fluorescence of the glass. Becquerel, who was a specialist in the phenomena of fluorescence and phosphorescence, had the idea that there might be a definite relation between the flourescence of the glass and the emission of X-rays and that he might find that phosphorescence was always accompanied by X-rays. Certain salts of uranium are phosphorescent, that is, they continue to glow in the dark for some time after they have been exposed to sunlight. Becquerel proved that a layer of uranium salt which had been exposed to sunlight could produce an effect on a photographic plate which had been wrapped in black paper so that no ordinary light got to it. This was fine evidence of the correctness of his hypothesis. Fortunately, he was a careful worker, and a lucky one. He prepared a combination of photographic plate, black paper, and uranium salt for confirming his experiments, but the sun went under a cloud after a short exposure so that he laid the combination away in a drawer, expecting to use it later. Fortunately, he decided to develop it without further exposure and found to his surprise that the blackening of the plate was very striking. In other words, the exposure of the uranium to sunlight in his previous experiments had nothing to do with the blackening of the photographic plate.

Notice that both the discovery of X-rays and of radioactivity were "accidental." In Roentgen's work there was no initial hypothesis, merely extremely shrewd observation followed by analysis and a logical series of successive experiments. There was a tentative final interpretation as there should have been. That this ultimately proved to be wrong was not so important as that it was sensible in terms of the evidence and that it suggested further experiments. In Becquerel's work there was a sensible working hypothesis, which appeared at first to be right but later turned out to be wrong.

The discovery of the electron was quite a different matter. In fact, one could probably say the discovery of the electron took nearly 50 years, from the discovery of Faraday's laws of electrolysis to the great measurement of the electronic charge by Millikan. Nevertheless, it is true that the experiments by J. J. Thomson and his students in the last five years of the nineteenth century did establish definitely the existence of small particles carrying identical negative electrical charges and having identical masses, about 1/2,000 of the mass of a hydrogen atom, and further established that these particles or electrons could be obtained from all kinds of matter. This then was the first definite evidence that there is at least one constituent common to every known chemical element, and suggested strongly that atoms have structure.

In terms of method, these experiments on electrons correspond more nearly than either of the other two to the conventionally accepted scientific procedure. From many sources the idea had grown up that electricity might be atomic. Cathode rays had been discovered many years earlier and studied by many investigators. Two different hypotheses had been advanced as to their nature: (1) that they were corpuscular; (2) that they were waves. The deflection of cathode rays by magnetic fields had been observed and in some cases measured. Thus a whole series of investigations had determined many of the properties of cathode rays more or less quantitatively, even before J. J. Thomson and his pupils did definitive experiments in measuring the ratio of mass to charge and preliminary experiments in measuring the charge alone. Evidently this was a much more cooperative and complicated process than the discovery of X-rays happened to be. It is not nearly so well suited to Sunday-supplement science, but is the only kind of discovery in which most scientists today, however able, can expect to participate. Men like Roentgen are not only very able but also very lucky.

The Nuclear Atom

As a result of the three major discoveries which we have been discussing, and many auxiliary experiments, physics at the beginning of the century was engaged on a totally new set of problems concerned with atomic structure. Not that the old problems were settled; the ether was still with us, and there still remained the need of accurate measurements. In the course of the next 15 years atomic physics developed along two separate lines; in modern terms we would call one nuclear physics and the other the electronic structure of the atom. Actually the study of nuclear structure was very largely a straight experimental study of the properties of radioactive substances and their radiations. To one who takes pleasure in simple and ingenious experiments which produce really significant results, a study of this period is most rewarding. Mathematical theory was almost nonexistent, but good logical analysis of experimental results and imaginative planning of new experiments were constantly in evidence. To cite only one example, consider the experiment of Rutherford and Royds on the production of helium by alpha particles from radium emanation.

This work was done in 1908 and the results published in 1909. By that time enough evidence of one sort and another had been accumulated to indicate that the alpha particles emitted by radioactive substances are in fact atoms of helium carrying two positive charges, or as we would now say, the alpha particles are helium nuclei. Nevertheless, this conclusion depended on an involved chain of deduction from the results of various experiments. In order to clinch the matter, Rutherford and Royds carried out an experiment which is described in Rutherford's Nobel lecture as follows:

> A cylinder of glass was made with walls so thin that alpha particles could shoot right through them, and yet the walls remained quite gas tight. Radium emanation was introduced into this cylinder and the alpha rays were collected in an outside tube, and compressed in another fine tube through which a spark was passed. After a time the spectrum of helium appeared, proving that the alpha particles were, or became, helium!

Notice that this experiment differs quite fundamentally from the other three we have described. It is a very direct and very elegant experimental proof of a conclusion that had already been drawn and generally accepted. Its value lay not in the novelty of the results, which were not at all surprising, but in the economy of logic required to draw a significant conclusion from experimental results. I believe that the most satisfactory experiments are the ones where the complexity comes in designing or performing the experiment, not in the interpretation of the result. The kind of experiment where a year is spent in setting up the apparatus, in getting every irrelevant variable eliminated, but only a day is spent in obtaining results of immediate significance, seems to me far nearer the scientific ideal than an experiment which takes reams of data over a period of many months, data which then require elaborate statistical analysis before any conclusions can be drawn.

To return to the general development in the field of nuclear physics, the significant results that had been obtained by 1912 from the studies of radioactivity were the establishment of the nature of the three radiations, the nature of the law of disintegration, and the so-called displacement laws. According to these laws, a radioactive atom has a certain probability of spontaneously emitting either an alpha particle or an electron. Either of these changes produces a new atom which is chemically different from the parent. The emission of an alpha particle corresponds to a displacement of two places to the left in the periodic table, that is, the daughter atom is chemically analogous to an element two places to the left of the parent element in the periodic table. In the emission of an electron, the daughter atom is displaced one column to the right in the periodic table. Thus it was established not only that atoms were divisible but that one chemical element could be made from another. Furthermore, the high velocities of some alpha and beta particles showed that very great energies might be released in such transformations.

Curiously enough the indirect contribution of radioactive materials to the knowledge of atomic structure was at least comparable in importance to the direct contribution. The alpha particles were probes for the study of atomic structure vastly superior to anything that could then be produced in the laboratory, and it was the study of alpha particle scattering that led Rutherford in 1911 to suggest the modern picture of a nuclear atom, just as the study of the scattering of cathode rays had led Lenard to suggest a similar structure in 1903. More specific knowledge of the nature of the charge on the nucleus and its relation to chemical properties was established by the work of Moseley on the X-ray spectra of various elements. This experiment or series of experiments is so well known that there is no need to describe it in detail.

Nor is it necessary to review the many experiments on the properties of electrons, on positive rays and gaseous discharge in general, which took place in the first 15 years of this century. In relation to the general development of atomic physics, they were hardly comparable in importance with the radioactivity experiments. In 1900 we were almost sure that all atoms had electrons in them and that there must be some positive charge to compensate them. By 1914 we knew that all atoms had electrons in them and that an equal positive charge was concentrated on a very small but very massive nucleus, that the amount of the positive charge and consequently the number of external electrons increased step by step from hydrogen up through the heavier elements; but we knew practically nothing about the arrangement of the electrons or the laws governing their behavior.

Other Developments Before 1913

I might mention, however, one interesting series of experiments which illustrates still a different kind of occurrence in the development of science, namely, the Zeeman effect. It was first found experimentally by Zeeman in 1896. In the so-called normal Zeeman effect an ordinary spectral line is split up into three when observed transversely to a magnetic field, the middle line being polarized parallel to the field and the outside lines at right angles to the field, but when a spectral line is observed longitudinally, that is, with the magnetic field parallel to the light path, it splits into two components, each of which is circularly polarized. This effect was beautifully explained by Lorentz in terms of an electron oscillating in a magnetic field, that is to say, in purely "classical" terms. This was before the quantum theory had ever been heard of. Unfortunately for the classical theory, further study of the Zeeman effect showed that in fact there was often a so-called anomalous effect in which a spectral line split into many components and this effect was never explained until the quantum theory had been well worked out. Here then we have an example of a satisfactory explanation of an incomplete observation by a theory that later proved inadequate, yet there is no doubt that both Zeeman's discovery and Lorentz' early explanation were of immense importance in stimulating both experimental and theoretical research in the field of atomic physics.

I have deliberately avoided discussing the developments in theoretical physics that occurred in the first decade of the century, not because they were unimportant, but because they did not have any very immediate influence on the contemporary experimental work here described. I would merely recall two major items: first, the development of the quantum theory, and second, the suggestion of the equivalence of mass and energy by Einstein in his famous paper on relativity.

Classical theory was able to make predictions about the radiation given off by hot bodies. These predictions were based solely on the existence of some sort of electromagnetic oscillators in solids and did not specifically say anything about atomic structure or even about the atomicity of matter. The predictions turned out to be at variance with experiments, in fact, at variance with common sense since they led to an infinite rate of radiation of energy. In an attempt to avoid this difficulty, Planck in 1900 introduced the hypothesis that the emission and absorption of radiant energy by matter does not take place continuously but in finite quanta of energy, each quantum containing $h\nu$ ergs of energy where h is a universal constant and ν is the frequency of the radiation. This hypothesis of Planck not only led to a reasonable theory of the emission of continuous radiation by matter but, as modified by Einstein, succeeded in explaining photoelectric phenomena which were completely contradictory to classical ideas. According to Einstein, electromagnetic radiation itself had a kind of atomicity, that is to say, when light was emitted it was emitted in packets of energy, each of which flew through space like a corpuscle traveling with the speed of light. These quanta of light, or photons as we now call them, when striking a metallic surface, released electrons photoelectrically and these electrons had kinetic energies determined by the frequency of the radiation rather than its average intensity.

Still another set of data must be cited which were woven into the fabric of atomic theory in the next period to be discussed. For many years it had been known that different substances gave out different spectra. The use of the diffraction grating had made the measurements of the frequencies characteristic of the different elements among the most accurate measurements ever made. For certain substances such as hydrogen and the alkali metals, the number of observed lines in the spectra was small and the lines appeared to be related in definite patterns. For other elements such as iron, thousands of spectral lines were found and the complexity of their relations defied analysis. Though the precision of measurement and apparent simplicity of some spectra had made it possible to establish very accurate empirical relations between the frequencies of some spectral lines, for example, to arrange them in series like the Balmer series for hydrogen, these relations proved to be of a totally novel kind, utterly different from the harmonics familiar in music and expected in terms of classical theory. Fortunately, earnest spectroscopists had continued to make measurements and to arrange their data as well as they could in definite empirical patterns. This is one of the many cases in physics where the empirical relations were still so complicated as to be very difficult to grasp. Only later, when the interpretation had been made by Bohr and his successors, did the whole scheme fall into an intelligible pattern, a pattern in fact that can be extended to include great amounts of material that would never have been correlated by purely empirical methods. For this reason no attempt will be made here really to describe a spectral series from an empirical point of view.

Energy Levels, Spectra, and Electron Impact

I now come to a period in physics extending roughly from 1913 to 1930, during which theory and experiment went along hand in hand in a remarkable way. In 1913 Niels Bohr published his work on the structure of the hydrogen atom, thereby laying the foundation for our present-day picture of the arrangement of the electrons around the nucleus. It is hardly necessary to recall the fundamental principles of this theory, which is now taught in every elementary course in physics and chemistry, but I should like to go back to the position of a student beginning graduate work in physics at the end of the First World War. Remember that there were still educational institutions in this country and probably abroad where the very existence of atoms was viewed with skepticism. Although these were rare, certainly the idea of a nuclear atom composed of positive and negative charges was still vague except in research centers. Yet here was Bohr proposing that electrons revolved in certain specified orbits around a positively charged nucleus but that they gave off no radiation except when they changed from one permitted orbit to another. This meant that a given kind of atom, say hydrogen, could absorb energy only in definite quanta, and would then re-emit energy in definite quanta of radiation, thereby producing a line spectrum. Thus according to Bohr's picture normal neutral hydrogen has one electron moving in the innermost permitted orbit. In terms of energy content of the atom as a whole this condition represents the lowest energy state of the hydrogen atom. If this one electron is completely ejected from the atom we have a hydrogen ion, and in returning the electron may pause in any one of the permitted orbits, that is, the atom may exist momentarily in any one of the permitted energy states. A less drastic disturbance than complete ejection of the electron will leave the atom in some energy state higher than normal, i.e., in a so-called excited state but still neutral. In changing from a state of energy E_1, to a one of lower energy E_2, a quantum of energy is given off as electromagnetic radiation of definite frequency determined by the equation $h\nu = E_1 - E_2$, where ν is the frequency and h is Planck's constant.

Since this theory was designed to explain the spectrum of hydrogen, it is not surprising that it did so. Attempts to extend it to more complex atoms ran into difficulties so far as details of electron orbits were concerned, but the concepts of permitted orbits and corresponding energy states did explain in a general way the whole set of empirical relations that had been observed in spectra. What had been called spectral terms now could be identified with energy states, and the fact that the wave numbers of spectral lines corresponded to differences between spectral terms immediately made sense.

Furthermore, a simple prediction could be made that many kinds of energy transfer to and from atoms could occur only in quanta. Specifically, electrons could not transfer energy to an atom except in quanta. This idea led to a whole series of experiments on the collisions of electrons in gases and vapors. Experiments of this kind were initiated by Franck and Hertz in 1914 following the general pattern of earlier experiments by Lenard. According to our theory, if an electron has less than a certain amount of energy, corresponding to a transition from normal to the lowest excited energy state of the target atom, it collides elastically with the atoms without any transfer of energy (since the electrons are so light compared to the atoms, an elastic collision transfers a negligible amount of energy). If, however, an electron has slightly more than the minimum exciting energy, it can transfer most of its energy to the target atom, leaving the atom in an excited state and the electron with little or no kinetic energy. Such effects can be detected in several ways. It was shown that the electrons do in fact suffer no loss of energy at low speeds but do lose this whole quantum of energy once they are above the threshold. It was shown that at a still higher energy, corresponding to complete removal of the orbital electron, positive ions are produced. But perhaps the most interesting experiment of all that was done in this period was one which demonstrated that atoms bombarded with electrons of just over the minimum exciting energy radiated one and only one spectral line instead of a whole series. Thus, from a combination of theoretical suggestions of Bohr, experimental work of Franck and Hertz and many others on electron collisions and on spectra, the idea of discrete energy levels in atoms was firmly established.

Attempts to extend Bohr's ideas of the motions of electrons around nuclei soon ran into difficulty. More and more quantum numbers were introduced and modifications of Bohr's original principles made. Finally, in 1925 and 1926, the work of Schrödinger and Heisenberg gave a new basis of quantum mechanics for the theory of electronic structure and of spectra. I wish I could discuss the philosophical implications of Heisenberg's and Schrödinger's ideas. Perhaps I can say just this much: that their work, and even more the later work of Dirac, represents an almost complete departure from the idea of a mechanical model of an atom. In that sense Bohr's theory is intermediate between the old vortex ring ether model mentioned earlier, and the point of view of Schrödinger and Heisenberg. Bohr introduced radical new laws but was still working with a model. Heisenberg, particularly, attempted to work only from observations, emphasizing the danger of extending to submicroscopic atomic phenomena the ideas of length, motion, and position which are familiar to us from ordinary experience.

There remained many problems to be solved about the details of spectra and the behavior of electrons in chemical reactions; nevertheless it is fair to say that by 1926 the principles governing the electronic structure of atoms and their chemical behavior had been laid down, and by 1930 the detailed application of these principles to the known elements was well advanced. Throughout this period, there had been a remarkable interplay between theory and experimental results, one example of which has been discussed in some detail.

Unfashionable Physics, 1913–1930

Returning to the subject of radioactivity and nuclear physics, I feel that the period from 1912 to 1930 was relatively unexciting. There were, however, a number of discoveries which foreshadowed the brilliant developments that have occurred in the past 15 years.

By 1913 a study of the radioactive elements had made it clear (Russell, Soddy, and Fajans) that many places in the periodic tables were multiply occupied. Apparently it was possible to have atoms of identical chemical properties but different radioactive properties. Such atoms were called isotopes. The existence of such isotopes in ordinary elements was first demonstrated by J. J. Thomson in 1913 and was investigated thoroughly by Aston in 1919 and subsequent years. The result of these studies showed that all atomic species have masses that are very nearly integral multiples of the mass of hydrogen. In the many cases where the atomic weights chemically determined are not integral, it was found that the elements are mixtures of two or more isotopes of different masses so that the atomic weight is merely an average depending on the proportions of different isotopes and their masses.

Without describing the mass spectrograph that Aston constructed nor elaborating on his results, it may be interesting to note that he is a rare example of a physicist whose work was of sufficient importance to earn him a Nobel Prize and yet was so narrow in scope that after his student days he worked with only three instruments, each a modification and improvement of the former one. Aston showed that the masses of most nuclei were nearly whole numbers but that they certainly were not exactly whole numbers. He not only proved this but was able to measure the amount by which the masses of nuclei differed from whole numbers.

The other work that was of first importance in the nuclear field in this period was Rutherford's discovery that he could break up the nuclei of atoms artificially by striking them with alpha particles. In 1919 Rutherford announced that he had succeeded in changing a few atoms of nitrogen into atoms of oxygen by bombarding them with alpha particles and that in the course of this change a hydrogen nucleus or proton was ejected.

Perhaps this is an appropriate point to discuss the very great importance of unfashionable physics. I do not know whether in the period from 1900 to 1914 the physicists of the world in general and this country in particular were as given to fads as they now seem to be, but the work in spectroscopy in that period was certainly not flashy or spectacular and I suspect it was somewhat difficult to interest graduate students in doing theses in that field. The techniques had already been well developed and the theory was almost nonexistent so that the work must have consisted largely in meticulously accurate measurements of the lines emitted by various sources. There was some doubt as to what lines came from what elements and there was some interest attached to the study of the Zeeman effect, the Stark effect, the Doppler effect, and so forth, but the major interest in this accumulated material did not come until the theory began to develop. However, the development of the theory would have been very much retarded had there not been this large body of dependable empirical data available. It was, to be sure, incomplete and had to be vastly extended in the period from 1915 to 1930 when spectroscopy was fashionable.

On the other hand, nuclear physics, or—speaking more strictly—radioactivity, fashionable in the earlier period, went rather out of style from 1915 to 1930. Similarly the study of isotopes was carried on almost exclusively by one man, Aston, although Dempster and one or two others did make significant contributions. Yet, as we have seen, it was Rutherford's work on the disintegration of nitrogen by alpha particles that experimentally foreshadowed the whole field of modern nuclear physics, while Aston's work on isotopes provided data of great value to the modern idea of transmutation of elements with the release of energy.

The State of Physics in 1930

Perhaps we should look at the state of physics in 1930 and compare it with that in 1895. More than two-thirds of the 50-year period I have chosen to consider had passed. There had of course been vast developments in the fields of technological and applied physics. Many of these, such as the thermionic effect, had gone almost directly from the laboratory to industrial use, and their industrial development had then been reflected in vastly strengthened laboratory techniques. For example, in so fundamental a problem as the measurement of small currents, the electrometer and the high sensitivity galvanometer had largely been replaced by amplifier systems of greater sensitivity and reliability.

Similarly high vacuum techniques developed in the laboratory had been carried over into the lamp and radio tube industry for further development. Thermionic and gas-filled rectifiers, added to the industrial development of transformers and high voltage technique, had also provided new tools for fundamental science.

Perhaps most important of all, the major problems in external atomic structure had been solved, so that young and imaginative physicists were looking for new fields to conquer. It had been a period of radical developments in our knowledge of physics and in our point of view. Both theory and experiment had established the nuclear atom on a firm basis and had specified the numbers of electrons in various atoms, their arrangement, and the laws that govern their behavior. The introduction of the theory of relativity and of the quantum theory had brought to an end the idea that the universe could be explained in terms of mechanical models. Little was known of nuclear structure, but there was at least good reason to suppose that nuclei did have structure and that all nuclei were made up from only a few kinds of particles.

Rutherford's experiment on the disintegration of nitrogen by alpha particles was well known, and a number of other similar experiments had been performed. Einstein's equation $E = mc^2$ was familiar and had been used to explain Aston's results and the great energies involved in radioactive processes. The idea of converting mass into energy had been discussed, but as a practical possibility it seemed remote. Few would have guessed that the discoveries of the next decade would have been as brilliant as those of Roentgen, Becquerel, Thomson, and the Curies. Yet in the one year, 1932, the neutron, the positron, and the deuteron were all discovered, and the first nuclear disintegration by artificially accelerated particles was achieved; even the discovery of artificial radioactivity in 1934 really belongs to this same group. In 1939 uranium fission was first observed, and in 1942 the first nuclear chain reaction using uranium fission went into operation. A very fruitful decade!

The Neutron

It should be interesting to compare the work of this last decade in our 50-year period with that of the first. We discussed the manner in which X-rays, radioactivity, and the electron were discovered as typical of the ways in which science progresses. For the sake of comparison we shall describe three major discoveries in this most recent period, namely, the neutron, uranium fission, and the chain reaction.

It is difficult to say when the idea of a neutral particle of about the mass of a proton was first suggested. It is the kind of thing that was probably talked about in groups of physicists almost as soon as the proton itself had been discovered; but the history of the actual experimental discovery of the neutron really began in 1930 when Bothe and Becker in Germany found that if the very energetic natural alpha particles from polonium fell on certain of the light elements, specifically beryllium, boron, or lithium, an unusually penetrating radiation was produced. At first this radiation was thought to be gamma radiation although it was more penetrating than any gamma rays known, and the details of experimental results were very difficult to interpret on this basis. The next important contribution was reported in 1932 by Irène Curie and F. Joliot in Paris. They showed that if this unknown radiation fell on paraffin or any other hydrogen-containing compound it ejected protons of very high energy. This was not in itself inconsistent with the assumed gamma-ray nature of the new radiation, but detailed quantitative analysis of the data became increasingly difficult to reconcile with such a hypothesis. Finally (later in 1932), J. Chadwick in England showed that the gamma-ray hypothesis was untenable. He suggested that in fact the new radiation consisted of uncharged particles of approximately the mass of the proton, and he performed a series of experiments verifying his suggestion. Thus the existence of neutrons was established.

Even in this very brief account of the discovery of the neutron, we see several points that are characteristic of the scientific method: first, the observation of an unusually penetrating radiation, i.e., of a new effect; second, the attempt to explain the effect in terms of what was already known, i.e., as gamma rays; third, further experiments correlating this unknown radiation with another observation, the ejection of protons of high energy; fourth, the gradual accumulation of quantitative data which disprove the previous explanation; fifth, a new hypothesis and experiments to test this hypothesis which in effect confirm it. I should say that this is a perfectly typical course of events in scientific discovery, exactly similar to those which led to discoveries in the seventeenth century or in the last decade of the nineteenth century. It was also typical in being based on a free interchange of information and on the assumption that that information was reliable. I should also like to point out that the contributions came from individuals or very small groups working independently because these problems interested them. There was no organization or direction.

Uranium Fission

One of the most striking characteristics of atomic and nuclear physics is the use of bombarding particles as instruments of investigation. We produce X-rays by bombarding targets with electrons, we study atomic structure by bombarding atoms with alpha particles and observing the scattered alpha particles, and so on. To a good physicist a new particle is potentially a new projectile, so it was natural that neutrons were seized upon for this purpose. Furthermore, they had the great advantage of being unchanged so that they would not be repelled by the nuclear charges of atoms as are the alpha particle, the proton, and the electron. This advantage appeared particularly favorable for the study of heavy atoms where the nuclear charge is greatest. Consequently, by 1934 Fermi, in Rome, had begun to study the results of bombarding uranium with neutrons. It was anticipated that effects would be observed similar to those of natural or artificial radioactivity, that is, that the neutron might be absorbed with the emission of an alpha particle, a proton, or an electron, and the consequent formation of a new element. But these elements would differ from uranium in mass and nuclear charge by only a small number of units. Such appeared to be the explanation of the initial results of Fermi and his group. As more data accumulated, this interpretation proved more and more difficult. The initial hypothesis was inadequate. It was not clearly wrong, as in Becquerel's discovery of radioactivity; it was more like Lorentz' theory of the Zeeman effect, or Roentgen's suggestion that X-rays were longitudinal waves in the ether.

By 1939 it was clear that we did not understand what happened when neutrons struck uranium nuclei. Note the relatively long preliminary period of confusion in contrast to the rapid clarification after the discovery of the neutron or of X-rays. The techniques are difficult, involving precise chemical separations in short times on a microscopic scale. Only the perfection and careful repetition of these methods by Hahn and Strassman gave the clue. In Berlin in 1939 they convinced themselves that at least one of the new elements formed from uranium by neutron bombardment was not a close neighbor in the periodic table, not a heavy element of nuclear charge one or two units greater or less than uranium; on the contrary, it was barium, whose nucleus has not much more than half the mass or charge of uranium.

On hearing this news, Meitner and Frisch in Copenhagen guessed that the bombarding neutrons were splitting the uranium nucleus into two parts each about half the size of the original nucleus. They also realized that an immense amount of energy would be released in the process. News of the discovery and its interpretation spread rapidly through the scientific world and within a few months the energy released by the "fission" process had been observed in many laboratories. Much as with X-rays, there were many laboratories equipped to observe the process once they knew what to look for. Curiously enough, although the two discoveries were approached very differently, their immediate world-wide verification, the recognition of their importance, and their rapid utilization were very similar.

In many respects the discovery of uranium fission marks the end of an era in scientific research. It was truly international, it was made by small groups working on a small scale, for the most part in university laboratories, and it was made in the atmosphere of freedom and frankness that had meant so much to science. It remains to be seen how fully we can return to such conditions.

The Chain Reaction

The chain reaction, the next and last discovery to be mentioned, was made under circumstances superficially very different—in secret, in a war laboratory, heavily financed by the United States Government. Even now, a complete quantitative description of the chain reaction has not been published. The beauty of Roentgen's paper or of those on fission was that anyone who read them could repeat the experiment. This is perhaps the most fundamental principle of science; its methods of operation are so objective and its means of communication so precise that its results are reproducible. Each man can build on the other's work so that knowledge is advanced by a vast cooperative movement.

But we cannot work in this way when atomic bombs may be the reproducible result in a world that is politically unable to control them. Hence, we have had to violate our fundamental principles and keep silent about numbers and dimensions. Yet I think enough has been published about the first chain reaction to indicate the way in which it was approached. Under other conditions, all the necessary scientific constants might have been measured first and the theory so well worked out that the experiment would have been analogous to that of Rutherford and Royds, a confirmation of an anticipated result, though on an engineering scale. In fact, the first nuclear reactor was approached very much in the scientific tradition, conceived, prepared, and carried out by a professor with the help of his students and colleagues working on a university campus. Furthermore, up to the time that Fermi and his group first made the nuclear chain reaction go, the uranium project was essentially scientific in aims and methods. December 2, 1942, represents the end of the epoch of atomic physics that began in 1895. Whether the effort of scientists in this period will ultimately benefit mankind or destroy it remains to be seen. The power of the scientific method has been amply demonstrated.

Conclusion

The thoughtful scientist reviewing his present situation is deeply concerned. He can take pride in the achievements of the past 50 years. They show what human beings can do when they work with disciplined minds and objective methods in a spirit of cooperation and freedom. Yet mankind has failed to understand the lesson of method. Apparently the method is inadequate for social and political problems, or its use conflicts too violently with long-cherished prejudices and shortsighted ideas of self-interest. The world ignores the free interchange of thought, the precision of method and communication, the disciplined imagination which science has proved so potent, and instead seizes on the power released by science, and with an eager impulse toward suicide turns it to purposes of destruction. Some scientists themselves, suddenly awakened to the world in which they live, make political pronouncements in a spirit of emotionalism and panic that would never be tolerated in their own field.

What can the scientist do? Each must act according to his judgment, as he would in his own field, listening to all sides, attempting to evaluate and criticize such evidence as is presented, opposing authoritarian dogma as he would in science. He must be humble, recognizing the limitations of his own knowledge and that of others. He must be restrained and infinitely patient. He must think imaginatively but not emotionally. He must be willing to alter his ideas to meet changing conditions. All these processes are part of his training and have proved themselves as powerful tools. They should be easier for the scientist than for many of his fellows. Perhaps he can help those whose concern with immediate human problems has prevented them from developing an objective attitude. To render such help he must learn from those familiar with the fields of politics and human affairs, of which he knows so little. He must remember that he has no monopoly on brains but is merely the lucky custodian of a method well designed to meet the kind of problems that have heretofore confronted him. It remains to be seen whether this method can be applied successfully to the larger field of human relations.

ALBERT EINSTEIN

Letter to President Roosevelt

August 2nd, 1939

F. D. Roosevelt
President of the United States
White House
Washington, D. C.

SIR:

Some recent work by E. Fermi and L. Szilard, which has been communicated to me in manuscript, leads me to expect that the element uranium may be turned into a new and important source of energy in the immediate future. Certain aspects of the situation which has arisen seem to call for watchfulness and, if necessary, quick action on the part of the Administration. I believe therefore that it is my duty to bring to your attention the following facts and recommendations.

In the course of the last four months it has been made probable through the work of Joliot in France as well as Fermi and Szilard in America—that it may become possible to set up a nuclear chain reaction in a large mass of uranium, by which vast amounts of power and large quantities of new radium-like elements would be generated. Now it appears almost certain that this could be achieved in the immediate future.

This new phenomenon would also lead to the construction of bombs, and it is conceivable—though much less certain—that extremely powerful bombs of a new type may thus be constructed. A single bomb of this type, carried by boat and exploded in a port, might very well destroy the whole port together with some of the surrounding territory. However, such bombs might very well prove to be too heavy for transportation by air.

The United States has only very poor ores of uranium in moderate quantities. There is some good ore in Canada and the former Czechoslovakia, while the most important source of uranium is the Belgian Congo.

In view of this situation you may think it desirable to have some permanent contact maintained between the Administration and the group of physicists working on chain reactions in America. One possible way of achieving this might be for you to entrust with this task a person who has your confidence and who could perhaps serve in an unofficial capacity. His task might comprise the following:

a) to approach Government Departments, keep them informed of the further development, and put forward recommendations for Government action, giving particular attention to the problem of securing a supply of uranium ore for the United States;

b) to speed up the experimental work, which is at present being carried on within the limits of the budgets of University laboratories, by providing funds, if such funds be required, through his contacts with private persons who are willing to make contributions for this cause, and perhaps also by obtaining the cooperation of industrial laboratories which have the necessary equipment.

I understand that Germany has actually stopped the sale of uranium from the Czechoslovakian mines which she has taken over. That she should have taken such early action might perhaps be understood on the ground that the son of the German Under-Secretary of State, von Weizsäcker, is attached to the Kaiser-Wilhelm Institute in Berlin where some of the American work on uranium is now being repeated.

Yours very truly,
(signed) *A. Einstein*
(ALBERT EINSTEIN)

CORBIN ALLARDICE AND
EDWARD R. TRAPNELL

The First Atomic Pile

•

ON DECEMBER 2, 1942, man first initiated a self-sustaining nuclear chain reaction, and controlled it.

Beneath the West Stands of Stagg Field, Chicago, late in the afternoon of that day, a small group of scientists witnessed the advent of a new era in science. History was made in what had been a squash-rackets court.

Precisely at 3:25 P.M., Chicago time, scientist George Weil withdrew the cadmium-plated control rod and by his action man unleashed and controlled the energy of the atom.

As those who witnessed the experiment became aware of what had happened, smiles spread over their faces and a quiet ripple of applause could be heard. It was a tribute to Enrico Fermi, Nobel Prize winner, to whom, more than to any other person, the success of the experiment was due.

Fermi, born in Rome, Italy, on September 29, 1901, had been working with uranium for many years. Awarded the Nobel Prize in 1938, he and his family went to Sweden to receive the prize. The Italian Fascist press severely criticized him for not wearing a Fascist uniform and failing to give the Fascist salute when he received the award. The Fermis never returned to Italy.

From Sweden, having taken most of his personal possessions with him, Fermi proceeded to London and thence to America where he remained.[1]

The modern Italian explorer of the unknown was in Chicago that cold December day in 1942. An outsider looking into the squash court where Fermi was working would have been greeted by a strange sight. In the center of the 30-by-60-foot room, shrouded on all but one side by a gray balloon-cloth envelope, was a pile of black bricks and wooden timbers, square at the bottom and a flattened sphere on top. Up to half of its height, its sides were straight. The top half was domed, like a beehive. During the

[1] Until his death in 1954.—Eds.

construction of this crude-appearing but complex pile (the name which has since been applied to all such devices) the standing joke among the scientists working on it was: "If people could see what we're doing with a million and a half of their dollars, they'd think we are crazy. If they knew why we were doing it, they'd be sure we are."

In relation to the atomic bomb program, of which the Chicago pile experiment was a key part, the successful result, reported on December 2, formed one more piece for the jigsaw puzzle which was atomic energy. Confirmation of the chain reactor studies was an inspiration to the leaders of the bomb project, and reassuring at the same time because the Army's Manhattan Engineer District had moved ahead on many fronts. Contract negotiations were under way to build production-scale nuclear chain reactors, land had been acquired at Oak Ridge, Tennessee, and millions of dollars had been obligated.

Three years before the December 2 experiment it had been discovered that when an atom of uranium was bombarded by neutrons, the uranium atom sometimes was split, or fissioned. Later it had been found that when an atom of uranium fissioned, additional neutrons were emitted and became available for further reaction with other uranium atoms. These facts implied the possibility of a chain reaction, similar in certain respects to the reaction which is the source of the sun's energy. The facts further indicated that if a sufficient quantity of uranium could be brought together under the proper conditions, a self-sustaining chain reaction would result. This quantity of uranium necessary for a chain reaction under given conditions is known as the critical mass, or more commonly, the "critical size" of the particular pile.

Further impetus to the work on a uranium reactor was given by the discovery of plutonium at the Radiation Laboratory, Berkeley, California, in March, 1940. This element, unknown in nature, was formed by uranium 238 capturing a neutron, and thence undergoing two successive changes in atomic structure with the emission of beta particles. Plutonium, it was believed, would undergo fission as did the rare isotope of uranium, U^{235}.

Meanwhile at Columbia Fermi and Walter Zinn and their associates were working to determine operationally possible designs of a uranium chain reactor. Among other things, they had to find a suitable moderating material to slow down the neutrons traveling at relatively fast velocities. In July, 1941, experiments with uranium were started to obtain measurements of the reproduction factor (called "k"), which was the key to the problem of a chain reaction. If this factor could be made sufficiently greater than 1, a chain reaction could be made to take place in a mass of material of practical dimensions. If it were less than 1, no chain reaction could occur.

Since impurities in the uranium and in the moderator would capture neutrons and make them unavailable for further reactions, and since neutrons would escape from the pile without encountering uranium 235 atoms, it was not known whether a value for "k" greater than unity could ever be obtained.

Fortunate it was that the obtaining of a reproduction factor greater than 1 was a complex and difficult problem. If Hitler's scientists had discovered the secret of controlling the neutrons and had obtained a working value of "k," they would have been well on the way toward producing an atomic bomb for the Nazis.

One of the first things that had to be determined was how best to place the uranium in the reactor. Fermi and Leo Szilard suggested placing the uranium in a matrix of the moderating material, thus forming a cubical lattice of uranium. This placement appeared to offer the best opportunity for a neutron to encounter a uranium atom. Of all the materials which possessed the proper moderating qualities, graphite was the only one which could be obtained in sufficient quantity of the desired degree of purity.

The study of graphite-uranium lattice reactors was started at Columbia in July, 1941, but after reorganization of the uranium project in December, 1941, Arthur H. Compton was placed in charge of this phase of the work, under the Office of Scientific Research and Development, and it was decided that the chain reactor program should be concentrated at the University of Chicago. Consequently early in 1942 the Columbia and Princeton groups were transferred to Chicago, where the Metallurgical Laboratory was established.

At Chicago, the work on sub-critical size piles was continued. By July, 1942, the measurements obtained from these experimental piles had gone far enough to permit a choice of design for a test pile of critical size. At that time, the dies for the pressing of the uranium oxides were designed by Zinn and ordered made. It was a fateful step, since the entire construction of the pile depended upon the shape and size of the uranium pieces.

It was necessary to use uranium oxides because metallic uranium of the desired degree of purity did not exist. Although several manufacturers were attempting to produce the uranium metal, it was not until November that any appreciable amount was available.

Although the dies for the pressing of the uranium oxides were designed in July, additional measurements were necessary to obtain information about controlling the reaction, to revise estimates as to the final critical size of the pile, and to develop other data. Thirty experimental sub-critical piles were constructed before the final pile was completed.

Meantime, in Washington, Vannevar Bush, Director of the Office of Scientific Research and Development, had recommended to President Roosevelt that a special Army Engineer organization be established to take full responsibility for the development of the atomic bomb. During the summer, the Manhattan Engineer District was created, and in September, 1942, Major General L. R. Groves assumed command.

Construction of the main pile at Chicago started in November. The project gained momentum, with machining of the graphite blocks, pressing of the uranium oxide pellets, and the design of instruments. Fermi's two "construction" crews, one under Zinn and the other under Herbert L. Anderson, worked almost around the clock. V. C. Wilson headed the instrument work.

Original estimates as to the critical size of the pile were pessimistic. As a further precaution, it was decided to enclose the pile in a balloon-cloth bag which could be evacuated to remove the neutron-capturing air.

The bag was hung with one side left open; in the center of the floor a circular layer of graphite bricks was placed. This and each succeeding layer of the pile was braced by a wooden frame. Alternate layers contained the uranium. By this layer-on-layer construction a roughly spherical pile of uranium and graphite was formed.

Facilities for the machining of graphite bricks were installed in the West Stands. Week after week this shop turned out graphite bricks. This work was done under the direction of Zinn's group, by skilled mechanics led by millwright August Knuth. In October, Anderson and his associates joined Zinn's men.

Describing this phase of the work, Albert Wattenberg, one of Zinn's group, said: "We found out how coal miners feel. After eight hours of machining graphite, we looked as if we were made up for a minstrel. One shower would remove only the surface graphite dust. About a half hour after the first shower the dust in the pores of your skin would start oozing. Walking around the room where we cut the graphite was like walking on a dance floor. Graphite is a dry lubricant, you know, and the cement floor covered with graphite dust was slippery."

Before the structure was half complete, measurements indicated that the critical size at which the pile would become self-sustaining was somewhat less than had been anticipated in the design.

Day after day the pile grew toward its final shape. And as the size of the pile increased, so did the nervous tension of the men working on it. Logically and scientifically they knew this pile would become self-sustaining. It had to. All the measurements indicated that it would. But still the demonstration had to be made. As the eagerly awaited moment drew nearer, the scientists gave greater and greater attention to details, the accuracy of measurements, and exactness of their construction work.

At Chicago during the early afternoon of December 1, tests indicated that critical size was rapidly being approached. At 4 P.M. Zinn's group was relieved by the men working under Anderson. Shortly afterward the last layer of graphite and uranium bricks was placed on the pile. Zinn, who remained, and Anderson made several measurements of the activity within the pile. They were certain that when the control rods were withdrawn, the pile would become self-sustaining. Both had agreed, however, that should measurements indicate the reaction would become self-sustaining when the rods were withdrawn, they would not start the pile operating until Fermi and the rest of the group could be present. Consequently, the control rods were locked and further work was postponed until the following day.

That night the word was passed to the men who had worked on the pile that the trial run was due the next morning.

About 8:30 on the morning of Wednesday, December 2, the group began to assemble in the squash court.

At the north end of the squash court was a balcony about ten feet above the floor of the court. Fermi, Zinn, Anderson, and Compton were grouped around instruments at the east end of the balcony. The remainder of the observers crowded the little balcony. R. G. Noble, one of the young scientists who worked on the pile, put it this way: "The control cabinet was surrounded by the 'big wheels'; the 'little wheels' had to stand back."

On the floor of the squash court, just beneath the balcony, stood George Weil, whose duty it was to handle the final control rod. In the pile were three sets of control rods. One set was automatic and could be controlled from the balcony. Another was an emergency safety rod. Attached to one end of this rod was a rope running through the pile and weighted heavily on the opposite end. The rod was withdrawn from the pile and tied by another rope to the balcony. Hilberry was ready to cut this rope with an ax should something unexpected happen, or in case the automatic safety rods failed. The third rod, operated by Weil, was the one which actually held the reaction in check until withdrawn the proper distance.

Since this demonstration was new and different from anything ever done before, complete reliance was not placed on mechanically operated control rods. Therefore a "liquid-control squad," composed of Harold Lichtenberger, W. Nyter, and A. C. Graves, stood on a platform above the pile. They were prepared to flood the pile with cadmium-salt solution in case of mechanical failure of the control rods.

Each group rehearsed its part of the experiment.

At 9:45 Fermi ordered the electrically operated control rods withdrawn. The man at the controls threw the switch to withdraw them. A small motor whined. All eyes watched the lights which indicated the rods' position.

But quickly the balcony group turned to watch the counters, whose clicking stepped up after the rods were out. The indicators of these counters resembled the face of a clock, with "hands" to indicate neutron count. Nearby was a recorder, whose quivering pen traced the neutron activity within the pile.

Shortly after ten o'clock, Fermi ordered the emergency rod, called "Zip," pulled out and tied.

"Zip out," said Fermi. Zinn withdrew "Zip" by hand and tied it to the balcony rail. Weil stood ready by the "vernier" control rod which was marked to show the number of feet and inches which remained within the pile.

At 10:37 Fermi, without taking his eyes off the instruments, said quietly:

"Pull it to 13 feet, George." The counters clicked faster. The graph pen moved up. All the instruments were studied, and computations were made.

"This is not it," said Fermi. "The trace will go to this point and level off." He indicated a spot on the graph. In a few minutes the pen came to the indicated point and did not go above that point. Seven minutes later Fermi ordered the rod out another foot.

Again the counters stepped up their clicking, the graph pen edged upwards. But the clicking was irregular. Soon it leveled off, as did the thin line of the pen. The pile was not self-sustaining—yet.

At 11 o'clock, the rod came out another six inches; the result was the same: an increase in rate, followed by the leveling off.

Fifteen minutes later, the rod was further withdrawn and at 11:25 was moved again. Each time the counters speeded up, the pen climbed a few points. Fermi predicted correctly every movement of the indicators. He knew the time was near. He wanted to check everything again. The automatic control rod was reinserted without waiting for its automatic feature to operate. The graph line took a drop, the counters slowed abruptly.

At 11:35, the automatic safety rod was withdrawn and set. The control rod was adjusted and "Zip" was withdrawn. Up went the counters, clicking, clicking, faster and faster. It was the clickety-click of a fast train over the rails. The graph pen started to climb. Tensely, the little group watched and waited, entranced by the climbing needle.

Whrrrump! As if by a thunderclap, the spell was broken. Every man froze—then breathed a sigh of relief when he realized the automatic rod had slammed home. The safety point at which the rod operated automatically had been set too low.

"I'm hungry," said Fermi. "Let's go to lunch."

Perhaps, like a great coach, Fermi knew when his men needed a "break."

It was a strange "between halves" respite. They got no pep talk. They talked about everything else but the "game." The redoubtable Fermi, who never says much, had even less to say. But he appeared supremely confident. His "team" was back on the squash court at 2:00 P.M. Twenty minutes later, the automatic rod was reset and Weil stood ready at the control rod.

"All right, George," called Fermi, and Weil moved the rod to a predetermined point. The spectators resumed their watching and waiting, watching the counters spin, watching the graph, waiting for the settling down, and computing the rate of rise of reaction from the indicators.

At 2:50 the control rod came out another foot. The counters nearly jammed, the pen headed off the graph paper. But this was not it. Counting ratios and the graph scale had to be changed.

"Move it six inches," said Fermi at 3:20. Again the change—but again the leveling off. Five minutes later, Fermi called: "Pull it out another foot."

Weil withdrew the rod.

"This is going to do it," Fermi said to Compton, standing at his side. "Now it will become self-sustaining. The trace will climb and continue to climb. It will not level off."

Fermi computed the rate of rise of the neutron counts over a minute period. He silently, grim-faced, ran through some calculations on his slide rule.

In about a minute he again computed the rate of rise. If the rate was constant and remained so, he would know the reaction was self-sustaining. His fingers operated the slide rule with lightning speed. Characteristically, he turned the rule over and jotted down some figures on its ivory back.

Three minutes later he again computed the rate of rise in neutron count. The group on the balcony had by now crowded in to get an eye on the instruments, those behind craning their necks to be sure they would know the very instant history was made. In the background could be heard William Overbeck calling out the neutron count over an annunciator system. Leona Marshall (the only girl present), Anderson, and William Sturm were recording the readings from the instruments. By this time the click of the counters was too fast for the human ear. The clickety-click was now a steady brrrrr. Fermi, unmoved, unruffled, continued his computations.

"I couldn't see the instruments," said Weil. "I had to watch Fermi every second, waiting for orders. His face was motionless. His eyes darted from one dial to another. His expression was so calm it was hard. But suddenly, his whole face broke into a broad smile."

Fermi closed his slide rule——

"The reaction is self-sustaining," he announced quietly, happily. "The curve is exponential."

The group tensely watched for twenty-eight minutes while the world's first nuclear chain reactor operated.

The upward movement of the pen was leaving a straight line. There was no change to indicate a leveling off. This was it.

"O. K., 'Zip' in," called Fermi to Zinn, who controlled that rod. The time was 3:53 P.M. Abruptly, the counters slowed down, the pen slid down across the paper. It was all over.

Man had initiated a self-sustaining nuclear reaction—and then stopped it. He had released the energy of the atom's nucleus and controlled that energy.

Right after Fermi ordered the reaction stopped, the Hungarian-born theoretical physicist Eugene Wigner presented him with a bottle of Chianti wine. All through the experiment Wigner had kept this wine hidden behind his back.

Fermi uncorked the wine bottle and sent out for paper cups so all could drink. He poured a little wine in all the cups, and silently, solemnly, without toasts, the scientists raised the cups to their lips—the Canadian Zinn, the Hungarians Szilard and Wigner, the Italian Fermi, the Americans Compton, Anderson, Hilberry, and a score of others. They drank to success—and to the hope they were the first to succeed.

A small crew was left to straighten up, lock controls, and check all apparatus. As the group filed from the West Stands, one of the guards asked Zinn:

"What's going on, Doctor, something happen in there?"

The guard did not hear the message which Arthur Compton was giving James B. Conant at Harvard, by long distance telephone. Their code was not prearranged.

"The Italian navigator has landed in the New World," said Compton.

"How were the natives?" asked Conant.

"Very friendly."

RALPH E. LAPP

The Death of Louis Slotin

•

ORDINARY or natural uranium is quite harmless for it will not, by itself, sustain chain reaction. Only when it is embodied in an enormous matrix of some light element like graphite or heavy water does it sustain a slow chain reaction. Were this to be allowed to run out of control, it would not in general produce anything like an explosion. Heat would be produced and some inner parts of the reactor might melt, but it would not qualify as a bomb.

Enriched uranium or plutonium is quite different from ordinary uranium. Assemble too much of it in one place, and the chain reaction will automatically run away. Thus, it was rather important for the people at Oak Ridge and at Hanford to know how much was "enough" so that safety precautions could be taken. At Los Alamos the experts refined their calculations as to the size of the critical mass, but it was essential to have experimental measurements.

The man who headed up the "critical assembly" group was a good friend of mine. I knew Louis Slotin while an undergraduate at the University of Chicago and liked him very much for his pleasant manner and friendly advice. He was never too busy to help out a Ph.D. aspirant, and I remember that he gave me valuable pointers on making Geiger counters. On my visits to Los Alamos I used to stop by to see Slotin and give him the news of Chicago. He was a short, wiry youth with dark hair and soft sad eyes. Somehow or other, he always ended up doing jobs nobody else wanted. He never complained, and I respected the cheerful way that Slotin did dirty work.

Slotin had nerves of iron and he needed them for his critical experiments with the "nukes." Here is essentially what he did in making a critical assembly, or in "tickling the dragon's tail," as we called it. He would set up a table with a neutron counter and a rack. On the rack he would place two pieces of bomb stuff, each one being somewhat less than a critical amount.

Then he would push the two pieces, often in the form of hemispheres the size of a split baseball, toward each other. As the gap narrowed between the pieces, he would measure the buildup of the chain reaction inside the assembly. He used a small source of neutrons to amplify the effect, rather than waiting for stray neutrons to come from cosmic rays or from the material itself. He determined the tempo of the buildup by listening to the clicks in an amplifier connected to the neutron counter and in watching a recorder trace out a jagged red line on a moving roll of graph paper.

As the hemispheres came closer and closer, more and more of the neutrons would tend to be caught within the bomb stuff and fewer would be lost through the narrowing air gap. The chain reaction would build up, and, just before it was ready to rip, Slotin would calmly stop the experiment, measure the separation and deduce just how big the critical mass was. He grew quite adept at the experiment for he repeated it fifty times or more. His nonchalance amazed Fermi who once warned him, "Keep doing that experiment that way and you'll be dead within a year." Some of Slotin's colleagues tried to get him to build in automatic safety devices, like powerful springs, which could be triggered to hurl the two hemispheres apart when the neutrons built up too fast. He turned aside this suggestion with this retort: "If I have to depend upon safety devices I am sure to have an accident."

Slotin was asked to repeat the experiment "just one more time" to demonstrate the technique to others in the laboratory. So he gathered the group of six people behind him in the sunlit room where he did his work. One man, Dr. Alvin Graves, had his hand almost on his shoulder as Slotin proceeded to demonstrate his technique. He used two hemispheres that he had worked with before and holding a screwdriver he moved the two pieces of bomb material together to form a "nuke" or nuclear core. Slowly, at first, then more quickly the counters clicked away and the red line moved upward on the white paper chart.

Suddenly the counters screamed and the red ink indicators swung off scale. There had been an accident! The chain reaction was running away. Almost as if by reflex action Slotin hurled himself forward and tore the reacting mass apart with his bare hands. The others gasped and, turning around, Slotin, his face whitely reflecting his terror, motioned them to leave the room.

Slotin telephoned the hospital and said that there had been an accident. Then he telephoned his close friend, Phil Morrison. He was nauseated but, always the true scientist, paused in the hallway and drew a pencil sketch of the room and marked everyone's position, putting a big X for himself. Then he scribbled the time, 3:20 P.M., and hustled the group off to the hospital, all of them jamming into two jeeps.

The big question in the mind of everyone was: how much dose did Slotin get? The neutrons and X-rays which flashed through his body before he tore the assembly apart caused biological damage to his body. This we measure in certain units—called roentgens or r-units. A total of about 400 r over the entire body is considered the lethal amount for most people. This deadly amount does not produce immediate effect but takes time . . . weeks . . . or days . . . depending on the dose.

Phil Morrison, gifted theoretical physicist, worked feverishly to reconstruct the accident and to learn how serious was his friend's plight. Slotin's very blood had been made radioactive by the burst of neutrons which riddled his body, and a small sample of his blood gave a clue to the dose. Of course, Slotin was hospitalized and became ill rather soon, but during the first few days he was cheerful and would ask when visited by Morrison, "Well, what's the dose?" Nobody really knew and it took a long time to find out. Before they did, the tide had changed in Slotin's reaction to the radiation. His differential blood cell count told the story—a picture so hopeless that the attending Army nurse, hardened to hospital routine, broke down and sobbed when she saw the results.

Slotin had been most severely irradiated around the hands and arms. These parts of his pain-ridden body swelled grotesquely and the skin sloughed off. The nation's best doctors were flown to the Army hospital at Los Alamos but they could do little for the weakening patient. Nor could we do much more today.

Technicians strung a telephone connection into the bare hospital room and Slotin talked with his mother in Winnipeg, Canada. The next day, his parents were flown to New Mexico by special Army plane, and they stayed at their son's bedside until he breathed his last. The end came early on the morning of the ninth day after the accident.

The man who stood behind Slotin, Dr. Graves, was severely injured by the accident but he recovered and went on to become associate director of Los Alamos in the postwar period. He had this to say of Slotin: "I can perhaps tell you as much about his personality and character as I could in very many words if I merely quote to you his first statement when we were alone together in the hospital room. He said, 'I'm sorry I got you into this. I am afraid that I have less than a fifty-fifty chance of living. I hope you do better than that.' "

Slotin was not destined to be a great or a famous man. He was one of the many scientists who worked devotedly and unselfishly throughout the war. The young scientist gave his life, just as did many of his comrades in arms.

Slotin's experiment was outlawed at Los Alamos. With the development of television and remote-control gadgetry, it became possible to do the critical assembly operations with no one within a quarter of a mile. White-coated technicians, principally women, control the assembly and make all their observations without the slightest danger to themselves.

O. R. FRISCH

Fundamental Particles

•

FOR GOOD REASONS this has become known as the atomic age. Power from atomic nuclei is about to transform our world—and threatens to destroy it. Nearly every recent advance in engineering is based on what we know about the structure of atoms—high-strength alloys and plastics no less than fluorescent lamps, transistors and ferrites. Even in the study of the phenomena of life the stage has been reached when we examine effects of single atoms on living organisms. For these reasons alone it is clear that the study of atoms and their components is of great importance, but physicists have another strong incentive: curiosity. They just want to know what the world is made of, what are the smallest particles and how they behave. Therefore they are always ahead of practical applications; there always exists some knowledge about which people can ask "what is it good for?" It is true that at the moment nobody can see any use for the "newer" particles—mesons and hyperons—yet in the past any new discovery has invariably, within a few decades, found some practical use, or at least has become such an indispensable part of our knowledge that many practical advances would have been impossible without it. I do not foresee meson guns or hyperon boilers, but if applications for these particles are ever found, it is unlikely that even the most imaginative of present science-fiction writers will have envisaged them correctly.

Let me first recapitulate what is known about the structure of atoms. Since 1911 we have known that each atom consists of a heavy core or nucleus, surrounded by a number of much lighter particles called electrons. Different atoms have different kinds of nuclei, but electrons are all alike. Electrons all weigh the same and have the same negative electric charge, but an atom is not electrically charged, the negative charge of its electron being offset by an equal positive charge of the nucleus. To put it the other way round: the nucleus has a positive charge which is Z times the charge of an electron; hence Z electrons become arranged around the nucleus to form an atom, which is electrically neutral (i.e., uncharged).

The number (Z) of positive electronic charges on the nucleus, or of electrons around it, is called the atomic number. This determines the chemical properties of the atom, and many of its physical properties as well. Thus atoms with the same Z will stay together and not become separated when passed through chemical reactions; they belong to the same chemical element. Each element is characterized by its Z. Thus hydrogen has $Z = 1$; carbon, 6; copper, 29; and uranium, 92.

Though all the nuclei in any one element carry the same positive charge, they have not necessarily the same mass. Nuclei with the same charge but different mass are said to belong to different isotopes of the same element. The mass of a nucleus is, however, always very nearly an exact multiple of a certain unit mass which, in turn, is very nearly equal to the mass of the lightest nucleus of all, that of the lightest (and most common) hydrogen isotope, the proton. For instance, a gold nucleus weighs about 197 of these units; we say its mass number A is 197. But it cannot consist simply of 197 protons, for then its atomic number Z would also be 197 whereas it is really only 79. We now know that it consists of 197 nuclear particles of which, however, only 79 are protons; the other 118 have practically the same mass but no electric charge and are called neutrons. Protons and neutrons are known collectively as nucleons.

Free neutrons were first recognized in 1932 when it was found that they could be knocked out of certain light nuclei. For the knocking, nature had very kindly supplied fast-moving helium nuclei which are emitted from the nuclei of certain heavy elements such as uranium, radium, and others; when they were first observed their nature was not known and they were labelled "alpha particles," a name which has stuck. An alpha particle—like any other fast-moving charged particle—damages the atoms in the air through which it passes, leaving behind a trail of atoms or molecules which are electrically charged by having either lost an electron or gained an extra one, and which are called "ions." But a neutron slips through the air without making ions and is for that reason hard to detect; that is why it was discovered so late. A neutron can only be detected if it happens to strike a nucleus, which is then sent flying or is broken up; in that case, fast charged particles are formed which makes ions and can thereby be detected.

C. T. R. Wilson's cloud chamber provides a means of making trails of ions visible as fine tracks of water droplets, and so of detecting neutrons at second hand. It was a very convincing proof of the existence of neutrons to see the track of a nucleus, struck by a neutron somewhere in the gas of a cloud chamber, and sent flying by the collision. By placing the chamber between the poles of a strong electromagnet one can deflect the nuclei so that their tracks become curved; the degree of curvature indicates the speed of the nucleus, and hence the speed of the neutron that sent it flying. This kind of technique has been used again and again in the discovery and study of new particles.

When the neutron was first found, its study amounted to research of the purest kind. Nobody could have foreseen any practical use for it. This changed dramatically with the discovery (1939) that neutrons could cause the fission of uranium nuclei, with the liberation of more neutrons; in this way a chain reaction became possible, and this is now our chief source of neutrons and an increasingly important source of power. Today neutrons are an industrial commodity; several tons have been produced (and immediately consumed) in the brief history of atomic energy. Only ten years elapsed between the discovery of the neutron and the operation of the first atomic pile!

A nucleus always weighs a little less than the sum of the neutrons and protons—or briefly the nucleons—of which it consists. The reason is that the nucleons are bound together by strong forces; it would require a considerable energy, the so-called binding energy, to take a nucleus completely to pieces. Now it is one of the consequences of Einstein's theory of relativity that an energy E possesses a mass m, the exchange rate being given by the famous formula $E = mc^2$ where c is the speed of light. The energy contained in 700 domestic units of electricity corresponds to only one millionth of an ounce; so in ordinary life the mass equivalent of energy can be neglected. But if you could assemble a nucleus out of its nucleons the energy liberated would amount to about one per cent of the total mass. Even minor rearrangements inside nuclei cause changes in mass which can be accurately measured with the so-called mass spectrometer. The energies, too, can be measured, and in this way Einstein's equation has been checked and confirmed many times. Energies here are measured in MeV (the energy gained by an electron on being accelerated through one million volts). 1 MeV corresponds to one 940th of the mass of a proton, or to twice the mass of an electron.

Almost at the same time as the neutron, the positive electron or positron was discovered, and this opened entirely new vistas. Physicists had often asked themselves why protons were always positively and electrons negatively charged when the fundamental laws of electricity were quite symmetrical in respect of charge. Indeed the quantum theory of the electron, as developed in England by P. A. M. Dirac (1928), showed that positively charged electrons must be possible, and that one could produce a positron and a (negative) electron "out of nothing," provided the necessary energy was supplied. Indeed when gamma rays pass through matter, positrons are produced, and when they pass through a cloud chamber the production of pairs of electrons, one deflected to the right, the other to the left, by the electromagnet, can be clearly seen. The opposite process also occurs: when a positron meets an electron both disappear in a flash of gamma radiation—they are said to annihilate each other. This is the reason why positrons are so rare: they disappear within less than a millionth of a second in contact with matter although they can exist indefinitely in a perfect vacuum. The positron is said to be the "antiparticle" of the electron, and vice versa.

As soon as the positron was discovered, it was realized that there ought to be an antiproton as well, a proton of negative charge, capable of annihilating itself with an ordinary proton. But since the proton is 1,836 times heavier than the electron, 1,836 times as much energy is needed to produce a proton-antiproton pair than to produce a positron-electron pair. For the latter process one needs about 1 MeV energy, an amount possessed by many ordinary gamma rays, but gamma rays of several thousand MeV are rare even in the cosmic radiation, the fine rain of very fast particles that comes to us from outer space. An accelerator was therefore built (in Berkeley, California) which could accelerate protons to about 6,000 MeV, and a determined and successful search was made for antiprotons. Particles were found which weighed as much as protons, but had the opposite (*i.e.*, negative) charge, and which suffered annihilation in the expected manner on passing through matter.

In the same experimental setup, antineutrons were also found. It may sound surprising that the neutron, which has no charge, should possess an "electric mirror image" different from itself; but a neutron is a little magnet, which spins about its magnetic axis rather like the earth. In the antineutron, not only the electric but also the magnetic properties are reversed; so a neutron and an antineutron, spinning in the same direction, have their magnetic poles pointing in opposite ways, and this makes them different particles. Nobody has yet thought of a way of measuring the magnetic properties of antineutrons (though with neutrons it can be done); but there can be little doubt that the neutral particles, created under the same conditions as antiprotons and suffering annihilation in the same way, are indeed antineutrons.

Let us take stock. We have mentioned six particles so far: the electron, the proton, and the neutron, each with its respective antiparticle—protons and neutrons are jointly called nucleons. The positive electron (the "antielectron") is usually called the positron. The word electron is sometimes used for both kinds and sometimes just for the negative kind, the meaning being usually clear from the context.

There is also the photon, the quantum of electromagnetic radiation, recognized by Einstein as early as 1905. Its existence, he realized, was a necessary consequence of the quantum theory of Planck (Germany) who had concluded—from a rather subtle argument about heat radiation —that radiation must be emitted and absorbed, not continuously as people had previously thought, but in packets whose energy-content was proportional to the frequency of oscillation in the radiation in question.

For instance, the main part of the gamma radiation of radium has an oscillation frequency of $5 \cdot 3 \times 10^{20}$ per second, which corresponds with an energy content of $2 \cdot 2$ MeV for photons of this radiation. The frequency of visible light is a million times lower and hence its photons have a million times less energy; on the other hand, photons a million times more energetic are found in the cosmic radiation. But all these photons are basically the same thing, only endowed with different energies.

Next on my list is the neutrino, which has an extremely interesting story. Almost as soon as radioactivity was discovered at the turn of the century the various types of radiation were roughly classified and labelled with Greek letters. I have already mentioned alpha particles, the least penetrating of all, which were later identified as fast helium nuclei, and gamma rays, which are energetic photons rather like X-rays. Those with intermediate penetrating power were called beta rays and were soon recognized as fast electrons. Some of them were just atomic electrons which had been set in motion by energy from the nucleus, but others came out of the nucleus itself, created in the transformation of a neutron into a proton. It was possible to calculate with what energy they ought to come out (from the mass of the nucleus before and after), and the disturbing fact emerged that they never came out with this full energy, but with a distribution of lower energies. In each of these "beta transformations" a random amount of energy was missing and could not be traced.

Where something is missing there must be a thief, so a young Austrian theoretician, W. Pauli (in Switzerland), suggested that the beta transformation consisted in the emission, not of one particle (the electron that is observed), but of two, which share the available energy, one of them escaping unobserved. Lead blocks had been set up by the experimental physicists in order to trap the missing energy, and Pauli had to assume that his unobserved particle was penetrating enough to go through all of them, and so had to be electrically neutral. It was therefore soon nicknamed *il neutrino* (the little neutral one) by the Italian physicist E. Fermi, who was the first to take it seriously. It also had to be very light in order to give good agreement with the observed distribution of the electrons. More and more indirect evidence accumulated that Pauli had been right, but the neutrino itself escaped all the traps set for it and proved to be millions of times more elusive than Pauli had assumed at first. Even so it could not be completely elusive: if atomic nuclei could send out neutrinos they must also be capable of stopping them. It was possible to calculate the minimum stopping-power and it turned out to be extremely small: the chance that out of a million neutrinos traversing the entire earth a single one could be stopped was about one in a million. Yet, in 1956, it was announced that neutrinos had been caught and their existence definitely confirmed. This was made possible by using an atomic pile as a source (some 10^{18} neutrinos every second) and by means of special counters capable of detecting any neutrino that got stopped in a cubic yard or so of scintillating fluid. Even so only a few neutrinos per hour were recorded, but this was enough to identify them, and a proud day it was for Pauli, whose bold guess of some twenty-five years ago had at last been fully vindicated.

There are also antineutrinos, and the difference between the two is that they spin in opposite directions about their direction of flight, like bullets fired respectively from barrels with right-handed and left-handed rifling. You may object that this is not a valid distinction, that to an observer overtaking a neutrino its direction of flight would appear the opposite, and hence it would look like an antineutrina to him. But a neutrino always travels with the speed of light and thus cannot be overtaken since no object can move faster. It looks like a neutrino to any observer, whichever way and however fast he moves.

You may feel that I have spent too much time on an almost unobservable particle, but in the first place I think the physicists can be proud of having traced and pinned down something so elusive, and secondly, the effects of neutrinos in the large are not negligible. The energy of the sun and stars comes from nuclear transformations in their interiors. Some of the transformations are of the beta kind, and it has been computed that no less than about one-sixth of the energy is taken away by neutrinos and lost in the depths of space.

How did the photon, the quantum of electromagnetic radiation, come to be discovered? Our eyes being sensitive to a particular kind of radiation which we call light, once the laws of the quantum theory were understood, the existence of light quanta or photons followed of necessity. But would a race of blind physicists have discovered the photon? I think they would. They could have discovered electric and magnetic phenomena and found out their laws, and from these laws they would have deduced the existence of electromagnetic waves (just as Maxwell did in 1864, though the fact that he knew light might have helped him!) and hence of photons. Once the electromagnetic forces were known the rest could be done by mathematics alone.

Do other forces also give rise to waves and hence to associated quanta? What about gravity? One can show that gravitational waves (if they exist; the mathematicians are still arguing) would have exceedingly small effects, and so the "graviton"—the quantum of gravitational waves—is hardly even a matter of speculation. It is certainly not identical with the neutrino though some people suspect the neutrino may in some way be connected with gravity.

A force that is much more important on the subatomic scale is the so-called nuclear force which holds the protons and neutrons together in nuclei. Some twenty-five years ago a young Japanese mathematician,

H. Yukawa, began to wonder whether the nuclear forces might be capable of wave motion, and what the associated quanta would be like. His task was difficult, for very little was known about the nuclear force, except that it fell off much more rapidly with distance than the electric force, *i.e.*, much more rapidly than the inverse square of the distance. But just how it fell off was not known. Yukawa did what mathematicians do in such cases: he made the simplest assumption that was compatible with the scanty experimental data, and went ahead. His result was startling: the quanta would be heavy, with a mass about 300 times that of the electron, and they might carry a positive or negative electric charge, equal to that of a proton. They would be very different from the photon which has no charge and no intrinsic mass.

Yukawa's heavy quanta were eventually (1947) discovered in cosmic rays and they are now usually called pions (brief for pi-mesons; the term meson denoting the fact that their mass is intermediate between that of the electron and the proton). They are rare in nature, at least at sea level, but high up in the stratosphere they are quite common because there they are constantly produced by the impact of the fast cosmic ray particles entering the atmosphere from outside. As Yukawa had guessed, they are unstable and live on an average for only one forty-millionth of a second. Yukawa expected that they would break up into an electron and a neutrino, and indeed occasionally they do so, but mostly they break up into a neutrino and a particle that was quite unexpected: the muon (brief for mu-meson). Muons are much tougher than the original pions. They live about 80 times longer and can go clean through atomic nuclei: they can travel right down through the atmosphere, and every square inch of ground is struck by several muons a minute. Muons are lighter than the pions—only 207 times as heavy as electrons, while pions are 273 times as heavy. Being much more common near sea level, they were the first to be discovered (1937) and at first were thought to be Yukawa's "heavy quanta." The war interfered with their study, and it was not until 1947—when pions were discovered—that the confusion was straightened out.

In some ways the muon is the most mysterious of the new particles—precisely because it is so commonplace. Apart from its instability (it breaks up into an electron, a neutrino, and an antineutrino) it is just an overweight electron. It resembles the electron completely in every respect, that is, spin, magnetic properties, and indifference to nuclear forces. Why the electron should exist in two sizes we do not know. None of the other particles do.

A few years ago there was a brief flutter of excitement when, for a short time, it looked as if the muon might be tremendously important as the key that would unlock the energy of fusion reactions. By a fusion reaction is meant a collision between two light nuclei that results in the formation of a heavier nucleus from them. This process, on a grand scale, keeps the sun hot. Explosively it is the energy source in the hydrogen bomb. In each case, a temperature of millions of degrees is needed to make the process start, but the muon, it was found, could cause "cold fusion." In an ordinary hydrogen molecule, two hydrogen nuclei are held together by the electrons that circle around them; the distance—about one 400-millionth of an inch—is too large for fusion to happen. But when a muon is present it will tend to take the place of one of the electrons and will describe an orbit 200 times smaller; the two nuclei will be pulled 200 times closer together, and fusion will quickly occur. (One nucleus must be "heavy hydrogen," that is, the hydrogen isotope of mass number 2; two protons will not fuse.) The muon merely acts as a catalyst; when fusion occurs it finds itself loose again and immediately starts to round up another pair of nuclei. Unfortunately each fusion requires about a millionth of a second, and so the muon in its short life cannot do the trick more than once or twice, and the energy it liberates is much less than that needed to produce the muon in the first place. The excitement passed, but the episode showed again how the most academic type of research can perhaps lead to important industrial applications. If the muon had happened to have a lifetime a few thousand times longer, the process would have worked.

Both the pion and the muon exist with either a positive or a negative charge, one the antiparticle of the other. In addition, there is a neutral pion; this has an extremely short life, millions of times shorter even than that of the charged pion, and breaks up into two photons. A photon must be considered to be its own antiparticle, for if all the electromagnetic fields that make up a photon are reversed, the photon is still the same as it was before. The neutral pion must also be considered its own antiparticle since it breaks up into two photons almost at once.

We have now dealt with 14 particles: the electron, proton, neutron, muon, and neutrino, each with its antiparticle; and the photon and the three pions—positive, neutral, and negative.

There are 16 others which are called the "strange particles." Some are heavier than protons and are called hyperons; there are six of these—each with its antiparticle, not all of which have as yet been observed. The others are about half the weight of a proton and are called kaons (brief for *K*-mesons). They were called strange particles because when they were first discovered (from 1948 onward) their behavior was very puzzling. Even now, though we have a scheme that accounts for most of their properties, they are still pretty mysterious.

Consider, for example, the particle that comes next in mass after the proton. It is called the Lambda particle (written with a capital L because its symbol is Λ, Greek capital lambda). It is uncharged and after a short life of about one 10,000-millionth of a second it breaks up into a proton and a negative pion. It can also be made by what looks like the inverse process, namely by hitting a proton with a negative pion; but the pion needs much more energy than one would compute—by Einstein's formula—from the masses of the particles concerned. It turns out that together with a Lambda, a kaon is always produced; neither a kaon nor a Lambda can be produced alone. The present explanation is that the kaon and the Lambda are saddled with opposite amounts of something which has been given the slightly facetious name "strangeness" and which cannot be created in the brief space of a collision. In this it resembles an electric charge, but it has no other physical effect, and it is not permanent. It takes some time to disappear and therefore delays the break-up of the Lambda which otherwise ought to occur much more rapidly. The same scheme applies also to the three Sigma particles (positive, neutral, and negative), which come next in mass. But beyond these are two particles—the Xi particles (negative and neutral) which have twice as much "strangeness." They are therefore produced with two kaons (not one) and they break up in two steps with the Lambda as an intermediate stage. The theory accounts for a great many complexities of behavior, but what "strangeness" is we do not know.

There are two kaons (positive and neutral) and six hyperons. If we add the antiparticles, we have 16 strange particles. Together with the 14 mentioned earlier that makes 30. Are they all fundamental? We are fairly certain that none of them is "simply" a compound of two or more. But perhaps some are more fundamental than others. There is still a great deal we do not know.

GLENN T. SEABORG

Peaceful Uses of Atomic Energy[1]

•

WE HAVE MET in approximately the twenty-fifth anniversary year of the great discoveries which are responsible for our being here today.

Fittingly, we have seen during this year the first ripening of our labors. We have achieved economic nuclear power in limited but important areas. I believe this conference marks the beginning of the age of nuclear power. We can now foresee the end of the specter of an energy shortage which has haunted the world since the beginning of the Industrial Revolution. As nuclear power technology progresses, I believe we can provide in the future enough energy for all the peoples of the world—the energy that is central to the banishment of hunger, poverty and fear of the future.

The magnitude of the accomplishment in this quarter of a century can be appreciated by retrospection. Some of you will recall, as I do, the impact on nuclear science laboratories around the world of the startling reports in late 1938 and early 1939 of the discovery in Germany of nuclear fission by Otto Hahn and Fritz Strassman, with elucidations by Lise Meitner and Otto Frisch. In only four years the late Enrico Fermi and his colleagues had operated a reactor. Even so, in this period many of us would have been content with the thought that nuclear power might be an economic reality by the turn of the century. I find it astonishing that so much has been accomplished in only twenty-five years.

The contribution of the three International Conferences to this stage of progress has been immeasurable. The first conference, in 1955, dropped the shrouds of secrecy from many aspects of nuclear energy, and began a renewal of the channels of communication between nuclear scientists and engineers of the world. In the second conference, communications and international cooperation were further expanded, and fusion research was

[1] Excerpts from Dr. Seaborg's summary of the Third United Nations International Conference on the Peaceful Uses of Atomic Energy, at Geneva.

removed from the pale of secrecy. The third conference, bringing us to the borders of the age of nuclear power, might be called the Conference of Fulfillment.

And, now, let me turn to my attempt to distill the essence of this notable conference.

Let me begin with some general remarks, directed, first, to world power needs and the place of nuclear power in meeting those needs, and, second, to what some of us consider to be the three phases of nuclear power.

The conference has dramatized the fact that the world will require huge increases in available energy during the remainder of the century. The accelerating pace of the Scientific Revolution in the developed countries will require huge new increments of power. At the same time, we sorely need to bring the developing nations into the orbit of today's technologies.

What does the conference seem to have concluded regarding the advantages of nuclear power in meeting these energy needs? These advantages go beyond economics alone. They permit nations to manage wisely their fossil fuel resources as irreplaceable raw materials rather than as sources of heat. As President Emelyanov observed in his opening remarks, "The raw materials for the chemical industry, for the manufacture of plastics, fabrics, artificial leather and similar products are natural gas, oil and coal. Organic fuels provide the chemical industry with its primary materials. If we go on using oil at the present rate, all our oil will soon be burnt up, and the chemical industry will be deprived of a most important source of raw material."

Let us see how the conference gives us reason to believe in the future of nuclear power. The installed nuclear capacity of the world has grown from 1955 with only 5 megawatts, to 1958 with 185 megawatts, [to] 1964 [with] almost 5,000 megawatts. The future prognosis is also excellent. By 1970 the total world nuclear power capacity will be about 25,000 megawatts, and by 1980 this will have increased to 150,000 to 200,000 megawatts.

The testimony of the technical papers to the fact that nuclear power has come of age is supported by corridor discussions reflecting the entry of commercial competition. This in itself is a strong sign of economic arrival, and I should like to think that it has not intruded too heavily on the conference. However, competition—sometimes aggressive—inevitably accompanies economic development. It plays an important role in driving costs down. Perhaps it is pertinent to note the observation of one delegate which implied that aggressive techniques are sometimes needed to convince conservative financiers and utility engineers of the value of new developments.

It should be noted that our projections of nuclear power development do not assume a breakthrough in controlled thermonuclear fusion or in any now unknown energy source. Such a development is not on the horizon, but could occur. If fusion does become practical and if it is economical, I am sure there will be time to modify our plans.

The subject of my second group of general remarks deals with reactors as we find them today and as we expect them to be in the future. Many of the delegates to this conference view nuclear power in three phases.

The first is that phase reached in the past year or so—the coming of age of the three types of presently economic reactors: the graphite-moderated, gas-cooled reactor; the heavy-water-moderated, heavy-water-cooled reactor; and the light-water-moderated, light-water-cooled reactor.

The second phase of nuclear power is the improved or advanced converter reactors, including near breeders. This phase of nuclear power development promises to bring greater fuel utilization, preparation of fuel for breeders at a faster rate and potentially even lower-cost power than today's reactor.

The third and somewhat concurrent phase of nuclear progress is the development of breeder reactors. These breeder reactors promise to extend by an order of a magnitude and more the fuel utilization of our uranium and thorium resources since they will produce more fissionable material than they consume. In essence they are our key to unlocking the energy stored in the nonfissionable but extremely abundant isotopes, uranium-238 and thorium-232.

Clearly, the conference shows that the aim of all the nations is to achieve abundant economic nuclear power. We are approaching this goal in different ways, and it appears fortunate that alternative explorations are being made. It seems likely that the nuclear power base for some time to come will consist of a number of different systems paralleling each other in time. We are not likely to find a sharp cut-off point at which one type of reactor will cease to be useful. Moreover, it is unlikely that any advanced system—converter or breeder—will be widely adopted if it does not become economical.

The proceedings here show that the focus of competitive nuclear power today is on large-size plants, 500 megawatts of electrical output each. Plans for even larger plants of 1,000 electrical megawatts each are just quietly assumed. When we look toward the reactors for desalination of sea water—one of the most remarkable future human benefits spilling from the cornucopia of nuclear energy—we consider plant sizes as large as 2,000 electrical megawatts. But the other end of the spectrum also requires attention. Not all countries can use power in these large blocks. Hopefully, one of the outcomes of the large-scale development of nuclear power will be the advent of economic power reactors in the smaller sizes more suited to many of the developing countries. Nuclear reactors below the 500 megawatt size are already showing economic benefits in certain of these countries.

Technical and economic factors have restricted the number of reactor types that are well enough developed to be candidates at present for immediate large-scale power programs. The reported prospective construction costs for the three predominant reactor types in the world today are only about one-half the cost of the first large reactors.

Those systems which tend to have high capital costs tend to have low running costs. Those countries which have concentrated on the development of those reactors with low fuel costs have low fixed-charge rates which tend to offset their higher capital costs. In fact, the lower carrying charges were one of the main reasons which made these lines of reactor development attractive. It speaks well for these reactor technologies that they have been able to become more competitive with other fuels in their own countries.

All of these reactor types still have considerable potential for economic improvement through increased unit size, through multiple-unit stations, through large-scale production by replication of designs, and through steady engineering improvements of the type that have substantially reduced the cost of conventional stations over the years.

Experience has also demonstrated that many power reactors can operate safely at power levels considerably higher than their initial design ratings, thus substantially reducing unit costs.

The reports have further indicated that there remains the possibility of substantial reductions in fuel cycle costs. The reductions to be achieved between the costs we have experienced up to now and future costs may be much greater than the 50 percent by which we have already reduced capital costs, although the amount of reduction will vary with reactor type. Fuel fabrication costs will be greatly reduced as we go on to develop improved fabrication techniques and increase the scale of the fabrication industry.

As far as the fuel itself is concerned, possibilities for further savings have also been reported. Natural uranium prices are currently well below the cost of the uranium used in the early cores of existing reactors.

Perhaps the most encouraging economic evidence of all lies in the projections of nuclear power growth given to this conference. It appears quite certain that nuclear energy will play an increasing role in meeting the electrical power needs of many countries. The role of nuclear power will vary from country to country depending upon the extent and costs of their conventional resources. In this respect the relative economics of nuclear power will be most important. The managers of the electric utility systems in all countries seem to insist, quite naturally, on economic competitiveness before they will engage in large-scale nuclear power programs.

The present economic types of power reactors, as developed to date, represent only a first step. Many reports on forthcoming converter reactors promise substantial improvements in economics and fuel utilization. While opinions may differ on the paths to pursue for the future, there is substantial agreement that we must develop improved converters to obtain these greater efficiencies. In looking toward ultimate future needs, there is almost general agreement as to the need to develop breeder reactors, although some nations, like Canada, support the opinion that breeder reactors may not be required for many years in the future, if at all.

A variety of forthcoming converter reactor types is presently under study and development in many parts of the world. Some of the potential advantages of these reactors include such properties as high conversion ratios, high specific power with resultant lower fuel inventory, high temperature and high thermal efficiency, larger single unit capacity, more efficient use of natural uranium, thorium and plutonium, and a potential contribution toward ultimate breeding systems. In some cases these converter concepts are based on design variations and extensions and combinations of the technology of the proven reactor types. In other cases they represent innovations in technology.

As an example, heavy-water-moderated reactors, long of interest to many countries, represent a class of reactors which has good long-range as well as short-range potential.

Advanced reactor versions of graphite-moderated, gas-cooled reactors are also being developed leading to high temperature, improved fuel cycle and higher conversion ratio performance.

We find substantial development effort is also being directed toward sodium-cooled reactors. The development of the sodium-cooled, graphite-moderated reactor in the U.S. has as its prime objective an economic, high-temperature, large-power reactor system; in addition it is also contributing a significant amount of sodium technology which is applicable to the sodium-cooled, fast reactor systems.

The reports reflect that emphasis on fast reactors has increased spectacularly in many countries during recent years.

Although the main effort in breeder reactor development is in the direction of sodium-cooled, fast reactors, the reports here reflect that other fast reactor systems are being studied which would use other coolants such as gas and steam.

Work in several countries which may lead to thermal breeder reactors also was presented, utilizing the thorium and uranium-233 fuel cycle.

Let me turn to the manufacture of nuclear fuel, one of the most important technologies of nuclear power, where we find steady improvement through a diversity of approaches. These improvements have accounted for a slow, steady gain in fuel performance over the years, with operating temperatures becoming higher and an increase in the heat output of each unit of fuel throughout.

While the number of separate reactor concepts being pursued has narrowed somewhat in recent years, there are still many options as to the fuel cycle. Uranium metals are being used in low-exposure, natural-uranium systems. For water-cooled reactors the presently favored and most thoroughly proven fuel material is uranium oxide. The results reported to this conference indicate that this fuel will continue to be favored in these reactors because it can sustain high irradiation exposures. A number of papers suggest that carbide fuels appear quite promising for nonwater reactors. Other compounds such as nitride, sulfide, silicide, etc., are also being studied, and have been reported upon.

Stainless-steel and zirconium alloys continue to be prominent cladding materials, although fuel designers continue to be concerned about high-temperature embrittlement under irradiation. Magnesuim alloys have been serving well in low-temperature, gas-cooled systems. For future applications ceramic fuel particles coated with carbon offer a good fuel for high-temperature, gas-cooled reactors.

A number of fluid-fuel concepts such as molten salt, molten plutonium and aqueous slurries are also being studied and developed. These concepts are very intriguing and offer the promise of significant reductions in fuel cycle economics, but they pose formidable problems that are still to be solved.

An important factor in the development and application of the peaceful uses of atomic energy, as important as its economic and scientific impact, has been the continuing concern for and emphasis on nuclear safety and reasoned management of nuclear wastes. The public has manifested concern in many countries over the safety considerations associated with the location, design, construction and operation of nuclear power plants. This is only natural in view of the increasing number of nuclear plants actually being operated, built or planned, and the desire to place these plants nearer to the centers of the electrical power needs.

Fortunately, safety has been a foremost consideration, from the start, in developing the peaceful applications of nuclear energy. The remarkable safety record is powerful and eloquent testimony to the high degree of importance attached to this aspect of nuclear energy activities, and to the efforts and accomplishments by workers in this field throughout the world. For example, the U.S. has been operating reactors of various types for about twenty years—with an accumulation of over 1,200 reactor years of operating experience. In that time there has been no known instance of public injury or even public inconvenience outside an immediate plant site that can be attributed to a reactor operation or accident.

From the reports given at this conference it would appear that throughout the world a general safety philosophy is developing for these nuclear reactor systems. The approach appears to be based largely on two related but separated and conservative paths. First, to prevent accidents, the reactor is generally designed conservatively, taking into account the kinetic or neutronic behavior of the system, the characteristics of the materials used in its construction, and the incorporation of redundant instrumentation and control systems made as fail-safe as possible. In addition, reactor operators are carefully trained and detailed plant operating procedures are carefully followed. Second, most power reactors are equipped with a variety of engineered safeguards and emergency systems to minimize the consequences of an accident should it somehow occur. For example, in some countries it is common practice to enclose the entire reactor system in a containment structure built to withstand considerable pressure and with a high degree of leaktightness.

The concern for nuclear safety does not cease with the continued safe operation of nuclear reactors. Their radioactive wastes must also be disposed of safely. Significant advances during the past years have been made and reported on here by Czechoslovakia, France, India and the U.S. in the handling of radioactive waste products from nuclear energy operations, including power reactor installations. There has been a strong impetus throughout the world for vigorous waste management research and development programs directed at further reduction in the quantities of radioactive materials being discharged into the environment.

The disposal of certain types of solid and liquid low-level waste effluents to the ground has proven to be safe and acceptable in many countries including Canada, the U.S. and the U.S.S.R. The growth of land burial or storage sites, with the resulting economies, have essentially eliminated ocean disposal as an important waste management operation in many nations of the world with available land area.

More than fifteen years' experience in the U.K., the U.S. and the U.S.S.R. with the improving methods of handling highly radioactive liquid waste from fuel reprocessing by storage in special underground tanks has shown such storage to be a safe and practical interim measure. The long-term usefulness of this method is limited, however, by the long effective life of the waste (hundreds of years) and the comparatively short life of storage tanks, estimated at several decades. Accordingly, a number of countries are developing means to convert high-level liquid waste to stable solids.

After high-level liquid wastes are converted to solids, there still exists a requirement for permanent storage of these solid wastes. Man-made structures may not be adequate to last for the hundreds of years that must pass before the wastes become relatively harmless; underground salt formations appear to offer an attractive alternate site for solids and concentrated liquid wastes because of their unique geologic characteristics. Salt formations are dry, impermeable, have good structural strength and thermal conductivity, and are not associated with usable ground water sources, and they exist in many parts of the world. Future developmental work in several countries along this path will be watched with interest.

With continued attention to this area of nuclear safety and waste management, I firmly believe that we can achieve the potential benefits of nuclear power and at the same time protect, or even improve, our general standards of public health and safety. The increasing use of nuclear power may indeed help to lessen atmospheric pollution, largely a result of the widespread use of fossil fuels.

To this point my remarks have been largely limited to nuclear reactors providing the heat for the conventional generation of electricity. The progress reported to this conference in the area of energy conversion techniques, nuclear thermoelectric and thermionic conversion and magnetohydrodynamics is opening new vistas for power generation.

The conversion of the heat of nuclear fission directly into electrical energy by means of the thermionic emission of electrons has been demonstrated as a practical concept in the short time since the last conference.

The generation of electric power using magnetohydrodynamics techniques is being actively pursued in a number of laboratories. Here again the high temperatures of the plasmas required pose serious materials problems, although the use of an inert working fluid would reduce the severity of the materials problem.

Generation of electric energy by the direct thermoelectric conversion of the decay heat of radioisotopes has become an established technology since the last conference. The technology is now being demonstrated not only in space but also in a number of terrestrial applications including weather stations, navigation buoys and lighthouses.

The direct thermoelectric conversion of the fission heat of a nuclear reactor has been demonstrated.

As I indicated earlier, one of the more exciting new applications for nuclear reactors is the desalting of sea water. We have heard reports of the U.S. studies for combination power and desalting application, and the studies by Israel and Tunisia. The U.S.S.R. and France have presented interesting data on reactors for process applications such as would be the case in desalting situations. We are encouraged by these reports and expect that one or more combination nuclear power and desalting installations producing millions of gallons per day of fresh water will be constructed and in operation within the next four to eight years.

The potential of nuclear energy in the combined production of power and water is not only a very fascinating peaceful use of the atom, but one which can provide tremendous benefits for all mankind. The availability of economic power and water could open new frontiers throughout the world for population growth and industrialization.

The studies that have been undertaken to date indicate that combination nuclear installations in the next few decades will be able to produce fresh water and electric power at costs which may be attractive for many municipal and industrial needs throughout the world. The water from these combination plants may even find economic potential for selected agricultural use when compared with other alternatives in specific situations.

As the nations of the world develop and populations increase, the economic natural water sources are likely to become depleted, especially in some geographical regions. Other areas, already deficient in water, will need water for development to support larger populations. Thus desalting of sea water by nuclear energy will become more and more important. What today is a matter of interest could well become tomorrow's necessity.

Another immediate application of nuclear reactors is to supply power and heat in remote locations. We have heard a Soviet Union report on the ARBUS organic cooled and moderated package plant. This plant consists of nineteen packages, each weighing not more than twenty tons. Soviet scientists have also described a pressurized water plant arranged on four large, tracked vehicles. U.S. scientists described their portable pressurized water reactors, using compact cores of UO_2-stainless steel cermet fuel, together with details of operating experience at several sites in the U.S., the Arctic and the Antarctic. The success of these plants, which generate up to 2,000 Kw(e) in addition to a substantial quantity of space heat, provides the technology which is applicable to any small nuclear power plant for remote installations, be it for mining, a scientific mission or for other needs.

The hopeful outlook for the maritime application of nuclear power expressed in the 1958 conference can now be supported by successful operating experience with two nuclear-powered vessels. The icebreaker *Lenin* has demonstrated the advantages of nuclear power for this important service. The *N.S. Savannah* is also meeting its expectations.

Thus, with successful operating experience amounting to well over 100,000 miles for the two existing ships and with new, prospectively economic projects already under way or planned, we can have confidence in the ultimate success of expanding nuclear propulsion programs for the merchant marine.

The U.S. Plowshare program for developing peaceful uses of nuclear explosives has received considerable attention. Despite the fact that this program is in an early stage of development and many data are needed before useful projects can be undertaken, the potential for use of nuclear explosives in excavation, mining, recovery of gas and oil and as a research tool appears promising.

Although nuclear power has been the major focus of interest at this conference, there has been considerable discussion of research and high-flux reactors and their associated programs, and the applications of radioisotopes to the physical and life sciences.

Conference papers suggest that the uses of newer research reactors fall plainly into three main kinds of activity: First, there is the continuing examination of radiation effects on materials for the construction, moderaation and fueling of reactors. Second is the more fundamental and better controlled kind of physical research made possible by reactors designed to meet more specific research needs. A good example of this is the work reported on pulse reactors. The third area is the production of radioisotopes for medical therapy, for tracer uses and now for the production of relatively large quantities of transplutonium elements, in both the U.S. and U.S.S.R.

My personal bias is evident when I say that the prospect of performing basic and exploratory research on gram quantities of californium isotopes, hundreds of milligrams of berkelium, milligrams of einsteinium and up to a milligram of fermium produced in these reactors is one of the most exciting to which we have been exposed in decades.

Another area exemplifying the research applications of nuclear energy is that of radioisotopes. It is clear that the technology associated with isotopes now permeates every scientific and engineering field. It is indeed one outgrowth of the atomic age that can be employed by all countries regardless of size or state of technological advancement.

In my judgment, an outstanding technical accomplishment of the past few years—as represented by papers presented here—has been the effort to produce, separate and purify radioisotopes in quantities sufficient to permit consideration of their use as sources of thermal and radiation power. Several countries have reported major progress in this area. Another area of outstanding achievement has been the use of radioisotopes in medicine, to alleviate man's suffering; this is the domain of what I have called "The Humane Atom."

From the scientific viewpoint, the widespread establishment of neutron activation analysis as a standard technique for measurement of trace quantities of almost every element in the periodic table represents a contribution of immeasurable value to medicine, agriculture and the physical sciences. In fact, its application has been extended even to law enforcement.

Ionizing radiation—whether the source be radioisotopes, machines or reactors—is finding a place in the processing of organic chemicals, plastics and other materials, in sterilization of medical supplies and in the preservation of foods.

The disclosure of previously classified research on controlled thermonuclear reactions was one of the main features of the 1958 conference. That year and its aftermath was an age of innocence for this intriguing field of research—a field of research which could lead to the extraction of an inexhaustible supply of energy from the oceans. The papers presented here show that we have learned a great deal in the intervening years. Plasma physicists now know well the hard scientific and engineering realities of suspending, squeezing and holding in space gases with temperatures of the order of those found in the stars. They have learned that the prospects for an easy engineering short-cut to controlled fusion are not bright. They have demonstrated, to the satisfaction of themselves and the nuclear community, that controlled fusion is one of the most difficult scientific and engineering problems ever encountered.

The early optimism followed by the sobering experience of the last six years should not, however, blind us to the truly significant progress that has been made. An important and exciting new area of fundamental science in plasma physics has grown up. Whereas in 1958 plasma scientists were only on the verge of producing fusion reactions with thermal neutrons truly attributable to the reaction, a number of laboratories today regularly produce plasma with ion-energies exceeding the so-called minimum ignition temperature.

A usable controlled fusion reaction would require nuclear reactions of adequate duration, temperature and density of plasma. Today one machine may best approach the production of the required temperature, another the required duration of nuclear reactions and another the desirable density of plasma, but no one machine is capable of meeting all three requirements. The aim now—and it is a long-range one—is to achieve reactions combining all of these factors satisfactorily in a single machine.

We cannot be absolutely sure that controlled thermonuclear power can be developed, although the general feeling at the conference is that this will be accomplished at some time—perhaps before the end of the century. Certainly the benefit—essentially unlimited power for the earth's population for all time—is one we cannot overlook. Indeed, I agree with expressions of some of the delegates that the approximately one hundred million dollars spent world-wide each year in the nuclear fusion field is too low an investment for research with such vast potential benefit.

In my closing remarks I should like to depart again from the form of the technical progress summary. I wish to review some of the human implications, especially in an international context of what has been said here. The degree of international cooperation in the development of this coming major energy resource in the last decade is surely unusual, and perhaps unique, in world history.

The conference has demonstrated many reasons why international cooperation must be continued and strengthened. The free flow of information, not only in science but also at the more restrictive technological level, is the key to the most rapid technical progress for all people.

This international collaboration practiced so successfully in nuclear energy gives further strength to the thesis that science can serve as a common ground among all nations of the world. A uranium or plutonium atom knows no nationality. Through international conferences such as this, and other broader and more intensive progress of exchange and collaboration, science may be a leading factor in resolving the differences which still remain between countries.

If we are to implement the major conclusion of this conference—that nuclear power will become an increasingly powerful force in the world's work—it will be necessary to evolve as rapidly as possible an appropriate world body of nuclear law. Considerable progress has been made in the development and application of appropriate safeguards under the aegis of the International Atomic Energy Agency.

By the turn of the century, our conference suggests, more than half of the world's electricity will be generated by nuclear energy. Nuclear energy is, therefore, the hope of the peoples of the world for a good life. If the major future energy base of the world continues to evolve in an environment of international development and law, nuclear energy can be an important unifying force in a world of peace, security and human well-being.

PAUL R. HEYL

Space, Time and Einstein

•

WHETHER WE UNDERSTAND IT or not, we have all heard of the Einstein theory, and failure to understand it does not seem incompatible with the holding of opinions on the subject, sometimes of a militant and antagonistic character.

In 1905, Einstein published his first paper on relativity, dealing principally with certain relations between mechanics and optics. Since that time a new generation has grown up to whom pre-Einstein science is a matter of history, not of experience. Eleven years after his first paper Einstein published a second, in which he broadened and extended the theory laid down in the first so as to include gravitation. Thirteen years later, in a third paper, Einstein broadened his theory still farther so as to include the phenomena of electricity and magnetism.

The general interest taken in this subject is frequently a matter of wonder to those of us who must give it attention professionally, for there are in modern physical science other doctrines which run closely second to that of Einstein in strangeness and novelty, yet none of these seems to have taken any particular hold on popular imagination.

Perhaps the reason for this is that these theories deal with ideas which are remote from ordinary life, while Einstein lays iconoclastic hands on two concepts about which every intelligent person believes that he really knows something—space and time.

Space and time have been regarded "always, everywhere and by all," as independent concepts, sharply distinguishable from one another, with no correlation between them. Space is fixed, though we may move about in it at will, forward or backward, up or down; and wherever we go our experience is that the properties of space are everywhere the same, and are unaltered whether we are moving or stationary. Time, on the other hand, is essentially a moving proposition, and we must perforce move with it. Except in memory, we cannot go back in time; we must go forward, and at the rate at which time chooses to travel. We are on a moving platform, the mechanism of which is beyond our control.

There is a difference also in our measures of space and time. Space may be measured in feet, square feet or cubic feet, as the case may be, but time is essentially one-dimensional. Square hours or cubic seconds are meaningless terms. Moreover, no connection has ever been recognized between space and time measures. How many feet make one hour? A meaningless question, you say, yet something that sounds very much like it has (since Minkowski) received the serious attention of many otherwise reputable scientific men. And now comes Einstein, rudely disturbing these old-established concepts and asking us to recast our ideas of space and time in a way that seems to us fantastic and bizarre.

What has Einstein done to these fundamental concepts?

He has introduced a correlation or connecting link between what have always been supposed to be separate and distinct ideas. In the first place, he asserts that as we move about, the geometrical properties of space, as evidenced by figures drawn in it, will alter by an amount depending on the speed of the observer's motion, thus (through the concept of velocity) linking space with time. He also asserts in the second place that the flow of time, always regarded as invariable, will likewise alter with the motion of the observer, again linking time with space.

For example, suppose that we, with our instruments for measuring space and time, are located on a platform which we believe to be stationary. We cannot be altogether certain of this, for there is no other visible object in the universe save another similar platform carrying an observer likewise equipped: but when we observe relative motion between our platform and the other it pleases our intuition to suppose our platform at rest and to ascribe all the motion to the other.

Einstein asserts that if this relative velocity were great enough we might notice some strange happenings on the other platform. True, a rather high velocity would be necessary, something comparable with the speed of light, say 100,000 miles a second; and it is tacitly assumed that we would be able to get a glimpse of the moving system as it flashed by. Granting this, what would we see?

Einstein asserts that if there were a circle painted on the moving platform it would appear to us as an ellipse with its short diameter in the direction of its motion. The amount of this shortening would depend upon the speed with which the system is moving, being quite imperceptible at ordinary speeds. In the limit, as the speed approached that of light, the circle would flatten completely into a straight line—its diameter perpendicular to the direction of motion.

Of this shortening, says Einstein, the moving observer will be unconscious, for not only is the circle flattened in the direction of motion, but the platform itself and all it carries (including the observer) share in this shortening. Even the observer's measuring rod is not exempt. Laid along the diameter of the circle which is perpendicular to the line of motion it would indicate, say, ten centimeters; placed along the shortened diameter, the rod, being itself now shortened in the same ratio, would apparently indicate the same length as before, and the moving observer would have no suspicion of what we might be seeing. In fact, he might with equal right suppose himself stationary and lay all the motion to the account of our platform. And if we had a circle painted on our floor it would appear flattened to him, though not to us.

Again, the clock on the other observer's platform would exhibit to us, though not to him, an equally eccentric behavior. Suppose that other platform stopped opposite us long enough for a comparison of clocks, and then, backing off to get a start, flashed by us at a high speed. As it passed we would see that the other clock was apparently slow as compared with ours, but of this the moving observer would be unconscious.

But could he not observe our clock?

Certainly, just as easily as we could see his.

And would he not see that our clock was now faster than his? "No," says Einstein. "On the contrary, he would take it to be slower."

Here is a paradox indeed! *A*'s clock appears slow to *B* while at the same time *B*'s clock appears slow to *A!* Which is right?

To this question Einstein answers indifferently:

"Either. It all depends on the point of view."

In asserting that the rate of a moving clock is altered by its motion Einstein has not in mind anything so materialistic as the motion interfering with the proper functioning of the pendulum or balanced wheel. It is something deeper and more abstruse than that. He means that the flow of time itself is changed by the motion of the system, and that the clock is but fulfilling its natural function in keeping pace with the altered rate of time.

A rather imperfect illustration may help at this point. If I were traveling by train from the Atlantic to the Pacific Coast it would be necessary for me to set my watch back an hour occasionally. A less practical but mathematically more elegant plan would be to alter the rate of my watch before starting so that it would indicate the correct local time during the whole journey. Of course, on a slow train less alteration would be required. The point is this: that a timepiece keeping local time on the train will of necessity run at a rate depending on the speed of the train.

Einstein applies a somewhat similar concept to all moving systems, and asserts that the local time on such systems runs the more slowly the more rapidly the system moves.

It is no wonder that assertions so revolutionary should encounter general incredibility. Skepticism is nature's armor against foolishness. But there are two reactions possible to assertions such as these. One may say: "The man is crazy" or one may ask: "What is the evidence?"

The latter, of course, is the correct scientific attitude. To such a question Einstein might answer laconically: "Desperate diseases require desperate remedies."

"But," we reply, "we are not conscious of any disease so desperate as to require such drastic treatment."

"If you are not," says Einstein, "you should be. Does your memory run back thirty years? Or have you not read, at least, of the serious contradiction in which theoretical physics found itself involved at the opening of the present century?"

Einstein's reference is to the difficulty which arose as a consequence of the negative results of the famous Michelson-Morley experiment and other experiments of a similar nature. The situation that then arose is perhaps best explained by an analogy.

If we were in a boat, stationary in still water, with trains of water-waves passing us, it would be possible to determine the speed of the waves by timing their passage over, say, the length of the boat. If the boat were then set in motion in the same direction in which the waves were traveling, the apparent speed of the waves with respect to the boat would be decreased, reaching zero when the boat attained the speed of the waves; and if the boat were set in motion in the opposite direction the apparent speed of the waves would be increased.

If the boat were moving with uniform speed in a circular path, the apparent speed of the waves would fluctuate periodically, and from the magnitude of this fluctuation it would be possible to determine the speed of the boat.

Now the earth is moving around the sun in a nearly circular orbit with a speed of about eighteen miles per second, and at all points in this orbit light waves from the stars are constantly streaming by. The analogy of the boat and the water-waves suggested to several physicists, toward the close of the nineteenth century, the possibility of verifying the earth's motion by experiments on the speed of light.

True, the speed of the earth in its orbit is only one ten-thousandth of the speed of light, but methods were available of more than sufficient precision to pick up an effect of this order of magnitude. It was, therefore, with the greatest surprise, not to say consternation, that the results of all such experiments were found to be negative; that analogy, for some unexplained reason, appeared to have broken down somewhere between mechanics and optics; that while the speed of water-waves varied as it should with the speed of the observer, the velocity of light seemed completely unaffected by such motion.

Nor could any fault be found with method or technique. At least three independent lines of experiment, two optical and one electrical, led to the same negative conclusion.

This breakdown of analogy between mechanics and optics introduced a sharp line of division into physical science. Now since the days of Newton the general trend of scientific thought has been in the direction of removing or effacing such sharp lines indicating differences in kind and replacing them by differences in degree. In other words, scientific thought is monistic, seeking one ultimate explanation for all phenomena.

Kepler, by his study of the planets, had discovered the three well-known laws which their motion obeys. To him these laws were purely empirical, separate and distinct results of observation. It remained for Newton to show that these three laws were mathematical consequences of a single broader law—that of gravitation. In this, Newton was a monistic philosopher.

The whole of the scientific development of the nineteenth century was monistic. Faraday and Oersted showed that electricity and magnetism were closely allied. Joule, Mayer and others pointed out the equivalence of heat and work. Maxwell correlated light with electricity and magnetism. By the close of the century physical phenomena of all kinds were regarded as forming one vast, interrelated web, governed by some broad and far-reaching law as yet unknown, but whose discovery was confidently expected, perhaps in the near future. Gravitation alone obstinately resisted all attempts to coordinate it with other phenomena.

The consequent reintroduction of a sharp line between mechanics and optics was therefore most disturbing. It was to remove this difficulty that Einstein found it necessary to alter our fundamental ideas regarding space and time. It is obvious that a varying velocity can be made to appear constant if our space and time units vary also in a proper manner, but in introducing such changes we must be careful not to cover up the changes in velocity readily observable in water waves or sound waves.

The determination of such changes in length and time units is a purely mathematical problem. The solution found by Einstein is what is known as the Lorentz transformation, so named because it was first found (in a simpler form) by Lorentz. Einstein arrived at a more general formula and, in addition, was not aware of Lorentz's work at the time of writing his own paper.

The evidence submitted so far for Einstein's theory is purely retrospective; the theory explains known facts and removes difficulties. But it must be remembered that this is just what the theory was built to do. It is a different matter when we apply it to facts unknown at the time the theory was constructed, and the supreme test is the ability of a theory to predict such new phenomena.

This crucial test had been successfully met by the theory of relativity. In 1916 Einstein broadened his theory to include gravitation, which since the days of Newton had successfully resisted all attempts to bring it into line with other phenomena. From this extended theory Einstein predicted two previously unsuspected phenomena, a bending of light rays passing close by the sun and a shift of the Fraunhofer lines in the solar spectrum. Both these predictions have now been experimentally verified.

Mathematically, Einstein's solution of our theoretical difficulties is perfect. Even the paradox of the two clocks, each appearing slower than the other, becomes a logical consequence of the Lorentz transformation. Einstein's explanation is sufficient, and up to the present time no one has been able to show that it is not necessary.

Einstein himself is under no delusion on this point. He is reported to have said, "No amount of experimentation can ever prove me right; a single experiment may at any time prove me wrong."

Early in the present year Einstein again broadened his theory to include the phenomena of electricity and magnetism. This does not mean that he has given an electromagnetic explanation of gravitation; many attempts of this kind have been made, and all have failed in the same respect —to recognize that there is no screen for gravitation. What Einstein has done is something deeper and broader than that. He has succeeded in finding a formula which may assume two special forms according as a constant which it contains is or is not zero. In the latter case the formula gives us Maxwell's equations for an electromagnetic field; in the former, Einstein's equations for a gravitative field.

E. *Chemistry*

•

As physics and chemistry continue to advance, it becomes harder to decide where one ends and the other begins. In the eighteenth century, when men first began to comprehend the distinction between the atoms themselves and their combinations into molecules, there was little connection. In Some Early Pioneers in Chemistry *the British scientists Stephen and L. M. Miall discuss early discoverers and discoveries and relate them to the thousands of experiments which have produced our chemical civilization.*

In The Rise of Organic Chemistry *John Read defines the two main divisions of the science, inorganic and organic, and describes milestones in the growth of the latter. Practically all modern industrial processes depend on chemical reactions. The organic chemist, by his manipulation of atoms and molecules, has created substances unknown in nature which can be tailored to fit specific environmental conditions, including extremes of heat, cold, pressure and the like.*

The early history of chemistry dealt in large part with the alchemists, the legendary medieval experimenters whose goal was the transmutation of the elements. After a lapse of centuries the contemporary chemist finds himself concerned with the same problem. Now, however, his theories are based not on vague speculation and on attempted manipulation of earth, air, fire and water but on a solid background of experimental data. He has even advanced beyond the concept of transmutation to that of creation, and has manufactured elements which have not been discovered in nature. The leading figure in this search, who is also Chairman of the Atomic Energy Commission, Glenn T. Seaborg, describes it in The Man-made Chemical Elements Beyond Uranium.

The importance of chemistry in contemporary civilization is the subject of Where Is Chemistry Today? *by Lawrence P. Lessing. Its possibilities for growth are almost boundless. Revolutions in the production of food, clothing and shelter, in the conquest of disease, in the understanding of basic biological processes have resulted and will continue to result from chemical research.*

STEPHEN MIALL
AND L. M. MIALL

Some Early Pioneers in Chemistry

•

MANY of the familiar materials with which we are so well acquainted—wood, paper, bone, meat, and coal—are very complicated substances: so complicated that a hundred or fifty years ago it seemed that it was almost hopeless to understand how they were built up. The early chemists spent much of their time in studying not the familiar complicated substances, but the simpler, if rarer, ones; they investigated the elements, gold, silver, iron, lead, tin, carbon, sulfur, phosphorus, and others that had been known for a very long time. They found that the air is a mixture of oxygen and nitrogen, that water is a compound of oxygen and hydrogen; they discovered the yellowish gas, chlorine, an element that is contained in common salt. With these elements, and twenty or thirty others less familiar to most of us, they performed all kinds of experiments that have very greatly enlarged our knowledge; but most of the substances made by them are unknown outside chemical laboratories or chemical books. Their very names are unfamiliar and repellent; many of these names are long, and are meaningless except to those who possess the clues and know the chemists' language.

It is convenient, because many of the simpler compounds were investigated first, to take a few instances from the early days, when chemistry began to be a science, so as to lead us on to a consideration of the more complicated structures and the singular processes that cannot easily be explained until we have grasped some of the most fundamental articles in the chemist's creed.

It was recognized that chemical elements existed, and that by the combination of these a multitude of compounds could be formed. The elements copper, silver, gold, sulfur, carbon, mercury, tin, lead, iron, arsenic, antimony, and phosphorus had long been known. Platinum, zinc, cobalt, and nickel were fairly recent discoveries; during the last half of the eighteenth century the gases hydrogen, oxygen, chlorine, and nitrogen were prepared and recognized as elements, and a very few years later attempts were made to prepare other elements which had not been isolated, although compounds of them were known. Among these elements we may mention silicon, boron, aluminum, calcium, sodium, potassium, and magnesium.

These were the elementary substances available for the use of chemists. Chemical experiments were then carried out in a room that more resembled an old-fashioned kitchen than a modern laboratory; most of the chemical appliances were very crude; coal-gas was not available, and the electric current did not come into use until after Volta's discovery of a primitive electric cell in 1791. There was no systematic teaching of chemistry, and the early workers had to make a good deal of their primitive apparatus; petrol and paraffin were unknown, and there were very few chemicals that you could buy at a shop, although drugs had, of course, been readily purchased for many years.

The first stage in the science of chemistry was to recognize the distinction between elements and compounds. The next stage was to recognize that one element could only combine with another in certain definite proportions. Thus hydrogen combines with oxygen, in the proportion of one part of hydrogen by weight and eight parts of oxygen, to form water, and in the proportion of one part of hydrogen by weight and sixteen parts of oxygen to form hydrogen peroxide. It will be seen that the difference in the relative weights of hydrogen and oxygen can cause great differences in the properties of the compound. It is the extra atom of oxygen that gives hydrogen peroxide its bleaching and disinfecting properties. When charcoal or coke burns in an open grate, twelve parts by weight of carbon combine with thirty-two parts by weight of oxygen to form carbon dioxide. The exhaust gases from a motorcar frequently contain a certain quantity of the poisonous carbon monoxide, in which twelve parts by weight of carbon are combined with sixteen parts of oxygen. These two are the only compounds of carbon with oxygen that have ever been prepared in a large quantity, though unusual compounds in which carbon and oxygen are combined in different proportions have been prepared in very minute quantities; in general, one element will only combine with another in one, two, or three proportions comparable with those we have already mentioned in the case of hydrogen and oxygen, and carbon and oxygen.

One of the early workers who helped to establish the science was Cavendish. Henry Cavendish was born in 1731, the elder son of Lord Charles Cavendish, and grandson of the Duke of Devonshire; he was educated at Cambridge, and until he was about forty he lived on a very moderate allowance from his father and contracted very economical habits. He lived at Clapham Common and had a library in Soho, and his methodical nature was such that he never took a volume from his own shelves without entering it in a book. He became a Fellow of the Royal Society in 1760, and dined regularly with the club that was formed from the fellows; otherwise he shunned society and was painfully shy when strangers were introduced to him; he ordered his dinner by a note placed on his table, and his women servants were instructed to keep out of his sight; his dress was old-fashioned and often shabby, for his whole interest was in science; he hated to be bothered about business matters, and on one occasion his bankers found that he had something like £80,000 in their hands, and they sent a representative to see Mr. Cavendish; the following conversation ensued:

"We have a very large balance in hand of yours, and wish for your orders respecting it."

"If it is any trouble to you, I will take it out of your hands. Do not come here to plague me."

"Not the least trouble to us, sir, not the least; but we thought you might like some of it to be invested."

"Do so, and don't come here and trouble me, or I will remove it."

Occasionally he invited a few friends to dine at his house, and invariably gave them a leg of mutton; once he asked four scientific men to dine with him, and when the housekeeper asked what should be got for dinner Cavendish replied:

"A leg of mutton."

"Sir, that will not be enough for five."

"Well, then, get two."

This anecdote illustrates not merely the oddity of the man, but also the small size of the leg of mutton in those days. Since the enclosure of pastures and the use of fertilizers the size of sheep and cattle has increased very much. One leg of mutton would now be ample for a large party of scientific men.

Cavendish was one of the very great men of science; he was remarkably exact and painstaking, and his experiments were of great importance. He prepared hydrogen and studied its density and its power of combining with oxygen so as to form water, and was the first man to have a clear idea of the composition of water. He determined with great exactness the composition of the air, proving that it was a mixture of about one part of oxygen with four parts of nitrogen, and that, in addition to these gases, the air contained about 1 per cent of a different gas; he collected a little of this and recorded the experiment carefully; he did not know what it was, but a hundred years later Rayleigh and Ramsay discovered the gas argon, that is now often used to fill electric light bulbs, and they found that the gas that had been collected by Cavendish was argon. Cavendish also studied the production of oxides of nitrogen—compounds of oxygen and nitrogen—produced by means of electric sparks in a mixture of oxygen and nitrogen, and he worked at the composition of nitric acid. Nitric acid is a compound of nitrogen, oxygen, and hydrogen; it is a very useful liquid in chemical research and of great commercial importance, playing a part in the manufacture of explosives and fertilizers. In these ways Cavendish laid an exact foundation for the study of some of the most important elementary gases. He studied heat and electricity, and determined the density of the earth, and busied himself with geology; but these researches are rather outside our subject. He died in 1810, leaving a fortune of a million pounds and having enriched chemistry by two researches of great value, the composition of air and the composition of water.

His contemporary Joseph Priestley was as unlike Cavendish as it would be possible to be; he was born in 1733 of parents in humble circumstances, and at the age of seven was taken charge of by an aunt and educated to become a nonconformist minister. In his young days he studied electricity, and wrote a history of it which secured his election to the Royal Society in 1766. He became a Unitarian minister in Leeds, and was attracted to the study of chemistry, partly because he lived next door to a brewery where quantities of carbon dioxide were produced in the ordinary course of the fermentation of beer. He became librarian to Lord Shelburne in 1772, and remained with him for seven years investigating the different kinds of gases and making important discoveries. He was the first to prepare and publish an account of sulfur dioxide, ammonia gas, nitrous oxide (laughing gas), and oxygen. The last gas he discovered in 1774, not knowing that two years or so earlier the Swedish chemist Scheele had also discovered it, but had not published his discovery. Priestley proved that green plants are able in sunlight to absorb carbon dioxide from the air and make use of it, and to restore to the air a certain quantity of oxygen. He played a part in the discovery of the composition of the air, and invented soda water, which has nothing to do with soda, but is merely water in which a considerable quantity of carbonic acid (carbon dioxide) has been dissolved. Priestley at one time learned many foreign languages and taught them; he was a vehement theologian and a keen politician, with radical or revolutionary views; he was acrimonious in his writings, and made himself very unpopular in Birmingham, where he went to live in 1780. During the riots of 1791 the mob wrecked his house and burned his books, and he was in some danger of losing his life. Finally he emigrated to America, and died there in the year 1804.

Carl Wilhelm Scheele was another early pioneer; he was born in 1742 at Stralsund in Pomerania; when he was fourteen he was apprenticed to an apothecary at Göteborg in Sweden, and then began to experiment in chemistry. He became an apothecary at Malmö, then at Stockholm, Upsala, and Köping in Sweden, and spent all his spare time in his chemical studies. He independently discovered oxygen, ammonia, and hydrochloric acid gas (a compound of hydrogen and chlorine); he was the first to isolate the gas chlorine and to discover hydrofluoric acid, a compound of hydrogen and fluorine used for etching glass; he was also the first to prepare lactic acid, oxalic acid, citric acid, and tartaric acid; lactic acid occurs in sour milk, oxalic acid in the sorrel herb, citric acid in lemon juice, tartaric acid is made from grape juice. He was the first to isolate glycerine, and the sugar contained in milk, now called lactose, an important component of some infant foods; he prepared prussic acid (hydrocyanic acid, a compound of hydrogen, carbon, and nitrogen), and he found out the nature of prussian blue, and made many other discoveries of great interest. Cavendish, Priestley, and Scheele made so many discoveries of new elements and new compounds, and did their work so thoroughly, that chemistry was enriched by a great mass of facts that enabled later workers to deduce from them the laws of chemical combination. Had Scheele lived to be an old man the volume of his discoveries would doubtless have been much greater, but he died in 1786, in his forty-third year.

None of these three men knew much about the theory of chemistry, or the laws of chemical combination, or the nature of chemical compounds, and the little they guessed about these matters was confused and inaccurate. It was a young French chemist, Lavoisier, with that logical kind of mind that is commoner in France than elsewhere, who reduced the chaos to order, distinguished exactly between chemical elements and chemical compounds, gave names to most of the new elements, and devised a logical system of names for the numerous chemical compounds.

Antoine Laurent Lavoisier was born in Paris in 1743, and studied chemistry at the Jardin du Roi; he was the first to show that the "setting" of plaster of Paris is due to the taking up of water by the dry plaster and the combination of the two; he was soon elected a member of the French Academy. In 1768 he became a member of the *Ferme-générale,* a company to which the Government, in return for a fixed sum of money, granted the right of collecting many of the taxes. He was later made a Powder Commissioner, and effected improvements in the manufacture of saltpeter and gunpowder; he drew up reports on the cultivation of flax and potatoes and prepared a scheme for establishing experimental farms; he introduced the cultivation of the beet, and helped to improve the breed of sheep; he was secretary of the Commission of Weights and Measures that devised the metric system, and he rendered great services to his country and to chemistry. But his connection with the *Ferme-générale* and the Government was his ruin. In 1793 the Convention ordered the arrest of the twenty-eight members of the *Ferme-générale*; they were arrested and sentenced to death; a plea was made on behalf of Lavoisier, but the officer in charge declared: *"La république n'a pas besoin des savants."* Lavoisier was executed and his property confiscated.

Some of those who experimented in the new pastime, or science, improved the apparatus that was employed in it. Each such improvement enabled new discoveries to be made. That has remained true to some extent ever since; at any rate, it is due to the extraordinary improvements in our scientific instruments that many of our greatest modern discoveries have been made. Modern science depends upon accurate observation far more than ingenious speculation, and our knowledge of matter and its ways would be very imperfect but for our marvelous microscopes, spectroscopes, vacuum pumps, thermostats, and the other triumphs of physics and engineering. Cavendish and his contemporaries had none of these; they would have done far more if they had.

Like so much else in this world, our knowledge of the structure of matter and the nature of chemical change is very largely a case of evolution. Our alphabet is the result of three or four thousand years of gradual experiment and development, the steam engine the result of effort and improvement spread over a few hundred years. Almost every scientific discovery depends upon observations made by earlier workers. When a sufficient mass of facts has been accumulated someone comes along and finds out a law connecting them. It was necessary to have the careful work of the four great men we have mentioned, and many men of less fame, before an adequate explanation of the facts could be given; as is usual, the explanation was not due to one man alone: several of his predecessors had made guesses with some foundation of truth in them; but until all the necessary data had been collected, the theory could neither be proclaimed nor proved. The man who gave us the first great chemical theory, one that is still universally accepted, was John Dalton, not such a scientific genius as Cavendish or Lavoisier, not such a prolific discoverer as Priestley or Scheele, but a simple-minded, clear-thinking man who tried to interpret the facts simply and clearly, and succeeded in doing so. To Dalton we owe the modern theory of atoms; it was he who first believed that every element was composed of infinitely small atoms each exactly like the other, and that the atoms of any one element were different from the atoms of any other element. He pictured an atom of element *A* combining with one, or perhaps two, or perhaps three, atoms of another element *B*, and that this was the basis of all chemical combination. This is now as obvious to us all as the existence of petrol, but it was not obvious a hundred and thirty years ago. Now when we read Darwin's *Origin of Species* it seems that he devoted several hundred pages to prove what needed only to be stated in one page; his conclusions are obvious to everybody. It is hard for us to put our minds in the state that was characteristic of educated people some seventy or eighty years ago.

John Dalton, the son of a Quaker weaver, was born in Cumberland in 1776, and became a teacher in Manchester and afterwards the Secretary of the Manchester Philosophical Society. He studied meteorology, and experimented on gases, in particular some of the gaseous compounds of carbon and hydrogen, one of which is known as methane or marsh gas; he was also the first person to call attention to color blindness; he himself had this peculiarity to a marked extent. After a life free from excitement, he died in 1844.

Dalton's views did not encounter the opposition that met Darwin's views; the ancient Jewish writers had not speculated on the constitution of matter, and no question of religion or orthodoxy was involved. Dalton's Atomic Theory was published in 1807, and has never since been disputed, though modern research on the structure of the atom has led to some interesting developments. Dalton was able to calculate very roughly the relative weights of many of the various atoms. Water, he knew, consisted of one part by weight of hydrogen and eight parts by weight of oxygen. The chemists of that period did not know whether they ought to regard water as composed of two atoms of hydrogen with a relative weight of 1 each and one atom of oxygen with a relative weight of 16, or whether they ought to regard water as composed of one atom of hydrogen with a relative weight of 1 and one atom of oxygen with a relative weight of 8. Either supposition fitted in with the facts.

The chemists during the next twenty or thirty years did not spend much time in discussing such a supposition; perhaps they thought it was too metaphysical, or at any rate a waste of time, and they busied themselves in discovering new elements and new compounds, in analyzing all kinds of minerals and substances of animal or vegetable origin, in studying the shapes of crystals, and in scores of other ways. They improved very greatly their methods of analysis, and Dalton's crude list of the relative combining weights of a few elements was superseded by a far more accurate list of atomic weights—that is, relative atomic weights—of some fifty or sixty elements.

Various scientific men investigated the density of different gases and the effect of compression on them and the way in which a gas expands when it is heated, and after the publication of Dalton's Atomic Theory an Italian physicist named Avogadro published his views on the nature of the ultimate small particles of elementary or compound gases. Translating his notions into modern terms, we may say that he regarded hydrogen as a gas containing a practically infinite number of molecules, each molecule consisting of two atoms of hydrogen; chlorine gas consisted of molecules each of which was made up of two atoms of chlorine. When hydrogen combines with chlorine to form hydrogen chloride (or hydrochloric acid gas), we consider that a molecule of each gas splits up into its two atoms, and that two new molecules are formed, each containing one atom of hydrogen and one atom of chlorine. Avogadro stated in 1811 that whatever gas you considered, a cubic foot of it would contain the same number of molecules as a cubic foot of any other gas, assuming that the temperatures and the pressures were the same. This was really a considerable development, but it attracted very little attention among chemists; it was not in advance of the facts, but in advance of the ideas of scientific men.

We can mention another and more recent instance of a molecular theory being in advance of its time. We now believe that every gas consists of a number of molecules dashing about at a great speed, and when these hit the side of a vessel containing the gas, they exert a pressure on it. It is possible to calculate the number of molecules in a given volume of any gas, at a fixed temperature and pressure, to calculate the average speed of the molecules, and other such matters. This theory was developed by Waterston in 1845, by Clausius in 1857, and by Clerk Maxwell in 1859. Waterston, who anticipated the views of Clausius and Clerk Maxwell, was several years in advance of scientific opinion, and when his paper was submitted to the Royal Society, the referee said: "The paper is nothing but nonsense, unfit even for reading before the Society." But although Waterston's paper was not published at the time, it was preserved, and in 1892 Lord Rayleigh called attention to its importance, and at last arranged for its publication with a suitable explanation.

A very versatile man of science was William Hyde Wollaston, the son of a clergyman, who had fifteen children who grew up to manhood. After studying at Cambridge, Wollaston became a country doctor, but he soon came to London to devote himself to science. He engaged in the manufacture of platinum, and made a considerable sum of money in this way; he invented an instrument, the reflecting goniometer, to enable the angles of crystals to be measured accurately, and he made slide-rules as ready appliances for making calculations.

In the year 1812 Wollaston discovered that certain compounds of similar metals crystallized in similar crystals having angles very nearly but not quite the same; thus calcium carbonate (calcite), a compound of calcium, carbon, and oxygen, crystallizes in rhombohedra with an angle of 105°5′, calcium magnesium carbonate (dolomite), a compound of calcium, magnesium, carbon, and oxygen, crystallizes in rhombohedra with an angle of 106°15′, and iron carbonate (siderite) crystallizes in rhombohedra with an angle of 107°0′. This phenomenon is called isomorphism. Wollaston discovered a new silvery-looking element called palladium, but he was an odd man and he made his discovery known in an odd and provoking way. Having prepared a small quantity of this metal, he thought he would play a practical joke on another chemist named Chenevix. He arranged in April, 1803, that a printed notice should come into the hands of Chenevix stating that a new metal called palladium was to be sold at Mrs. Forster's of Gerrard Street, Soho. Chenevix, believing that it was a fraud, bought up the whole stock, and after investigating it came to the conclusion that the substance was not a new metal, but that it was an alloy of platinum and mercury with peculiar properties. In May, 1803, Chenevix sent a paper to this effect to the Royal Society, which was read to the meeting by Wollaston, who was then one of the secretaries. Shortly afterwards an anonymous advertisement appeared in which a handsome reward was offered to anyone who should prepare any of the new substance. No one succeeded in preparing any, and in the following year Wollaston announced that he was the discoverer of palladium, and described its preparation. He intimated at the same time his discovery of another new element, rhodium, now used for the electroplating of silver. Wollaston was the first to see the dark lines in the solar spectrum that have subsequently given us a knowledge of the nature of that wonderful body the sun, and helped us to understand the structure of atoms. He was also the first to explain why it is that when we look at a portrait hanging on a wall in a room, the eyes seem to look at us, in whatever part of the room we happen to be. The explanation is really a mathematical deduction from the rules of perspective.

Chevreul, a French chemist, made a very careful examination of fats between the years 1811 and 1823; he found that the fats are mixtures of compounds of glycerine, the principal of which are the stearates, the oleates, and palmitates of glycerine. Chevreul was born in the year 1786, and died in 1889, in his hundred and third year. A writer who visited him in 1874 stated: "He is constantly at work, allowing only ten minutes for each of his meals, of which he has but two a day . . . except a small loaf at noon, which he eats standing and by the side of his alembics." He said on this occasion: "I am very old, and I have a great deal to do, so I do not wish to lose my time in eating."

The advancement of scientific knowledge is the work of many men. Dalton could not have discovered the broad highway of atomic theory along which we advanced so rapidly had not Cavendish, Priestley, Scheele, and Lavoisier preceded him. Lavoisier could not have reduced the infant science to a state of order had not a score of diligent workers collected a number of facts from which his genius deduced a system. After the time of Dalton we understood chemical combination far better than ever before. We think of sulfur as composed of an almost infinite number of sulfur atoms, of equal size and weight, each one weighing 32 times as much as a hydrogen atom. The atoms of sulfur are nearly indivisible, and until comparatively recent years it was thought that they could not be divided. We now know that by elaborate instruments it is possible to break up such an atom and to produce particles far too small to be seen with the help of a microscope, but these particles are not sulfur. When sulfur combines with hydrogen we picture one sulfur atom as attracting to itself two hydrogen atoms so firmly that they cannot escape. In this way is formed a molecule of hydrogen sulfide or sulfurated hydrogen, an unpleasant-smelling substance that is given off by rotten eggs, and can be made in other ways. A fragment of iron is an assembly of a number of atoms of iron, each atom weighing 55.8 times as much as an atom of hydrogen. So too every other element is composed of atoms of that element. The various chemical compounds are formed of almost infinitely small molecules consisting of two, three, or four atoms of different elements in the case of the simple compounds and twenty, fifty, or a hundred, or many more atoms belonging to perhaps eight or ten elements in the case of the more complex compounds. Many of the substances found in living creatures seem to be even more complex than this. Thousands of experiments carried out day by day for more than a hundred years by a multitude of chemists in many countries have proved beyond doubt the essential truth of Dalton's theory; that in a few details it requires a little modification need cause us no surprise.

JOHN READ

The Rise of Organic Chemistry

•

THROUGHOUT THE AGES, so far back as the Islamic period of alchemy, it had been recognised that some kinds of matter were of a mineral nature while others were associated with living organisms. Much later, the iatrochemists, among them le Febure, Glaser, and Lemery, arranged their pharmaceutical preparations according to their derivation from animal, vegetable, or mineral sources. During the eighteenth century this division became more distinct, and eventually the great and growing number of substances known to the chemist were arranged in two main classes. One class comprised substances which were either found in, or could be prepared from, inanimate mineral matter: these were called inorganic substances. In the other class were placed all substances obtained from plant and animal organisms, and which therefore owed their origin to living or "organised" matter: these were called organic substances.

For a long time it was supposed that none of these plant and animal substances could be made artificially, and that their production was dependent upon life-processes and the operation of an imagined "vital force." This belief, which was held in the phlogistic era and accepted later by Berzelius and others, was undermined in 1828 by an experiment of Wöhler, who prepared in the laboratory, without the intervention of any living matter, the substance urea. This white crystalline solid, discovered in urine by Rouelle in 1773, is one of the most typical products of animal metabolism. "I can prepare urea," wrote Wöhler to his venerated master, Berzelius, "without requiring a kidney or an animal, either man or dog." Soon afterwards the English chemist, Hennell, prepared alcohol artificially. Laboratory syntheses of many other natural organic substances followed, and the idea of a vital force was abandoned.

However, these substances are so exceedingly numerous that it was found advantageous to keep the term "organic" in use, especially as it began to be realised that the element carbon could give rise to further multitudes of compounds unknown in Nature. The term Organic Chemistry now comprehends all compounds containing carbon, whether of natural occurrence or solely man-made, such as, for example, chloroform, picric acid, synthetic dyes, sulpha drugs, and artificial fibres and plastics. "Organic Chemistry" is thus the chemistry of the carbon compounds.

The Beginnings of Organic Chemistry

Organic chemistry, as a science, may be said to date from about the time of Wöhler's artificial production of urea, in 1828. Of course, many natural organic materials were used from the earliest times, since they include all the fundamental types of foodstuffs and clothing materials (from fig-leaves onwards), as well as oils, gums, waxes, resins, perfumes, spices, dyes, and so forth, in lavish variety. Such typical processes of organic chemical technology as soapmaking and dyeing were known to the early civilisations, as were also baking and brewing. Much later, by virtue of their special outlook, the iatro-chemists were much concerned with the chemistry of plant and animal products. Nevertheless, until the Atomic Theory had been put forward and established, early in the nineteenth century, the progress of organic chemistry had consisted of little more than a stumbling by chance upon new naturally occurring organic compounds; and here Scheele made a notable contribution by his discovery of many new organic acids and of glycerine. Without an inspiring theory, no further development of the chemistry of these natural organic substances could well be expected.

Even after the establishment of the Atomic Theory, it was held for some time to be dubious whether some of the leading consequences of the theory could be applied to organic compounds; whether, for example, organic compounds conformed, like inorganic ones, to definite formulae. There seemed, indeed, to be a wide gulf separating the two divisions of compounds. In the first place, organic compounds were found to be invariably destroyed by heat, whereas many inorganic compounds sustained even the temperature of the blowpipe. Secondly, there came the question of number and diversity. The considerable number of inorganic compounds can be ascribed to the fact that they embrace (at the present time) all the derivatives of all the ninety-odd elements, except those of carbon. On the other hand, as Lavoisier and Berthollet showed, organic compounds are for the most part built up from four elements, namely, carbon, hydrogen, oxygen, and nitrogen. It follows that their enormous number and variety must be ascribed to the occurrence in organic chemistry of great molecular complexity.

A third great difference between organic and inorganic compounds was found in a phenomenon known as "isomerism." Wöhler's preparation of artificial urea in 1828 had a twofold significance; for, besides what has been said above, it confirmed a somewhat earlier observation that two distinct substances can have the same percentage composition, and, as was shown later, the same molecular formula. Urea, in fact, has the same percentage composition as ammonium cyanate: these two substances, which have widely different physical and chemical properties, have the common formula, CH_4ON_2. Berzelius, in 1830, coined the word "isomerism" to denote this kind of relationship. Thus, isomers are compounds having the same molecular formula. Their number is often very large, and may theoretically in some cases run into millions. For example, there are more than 120 distinct substances now known which have the common molecular formula, $C_{10}H_{16}O$, and many others are theoretically possible; one of them is camphor, a crystalline solid, another is citral, a fragrant liquid found in oil of lemon. Isomerism, which is exceedingly rare in inorganic chemistry, is one of the most typical features of organic chemistry.

In spite of the above differences between inorganic and organic compounds, it began to be evident, through the analytical work of Berzelius from about the year 1815, that organic compounds, like inorganic ones, could be represented by definite formulae. Some years later, in 1831, Liebig introduced adequate experimental methods of determining the exact percentage composition of a large number of organic substances, and thenceforward organic chemistry attained the status of an exact science capable of mathematical treatment, as inorganic chemistry had become earlier through the work of Lavoisier.

The frequent occurrence of isomerism among organic compounds showed the overwhelming importance of molecular structure in this branch of chemistry; for it became evident thereby that the mode of arrangement of the atoms in the organic molecular edifice was no less important than their number and nature. Indeed, it is quite conceivable that the same assortment of atoms assembled in one way might furnish a beneficent drug, and when rearranged in another way might lead to the formation af a toxic compound.

The riddle of organic molecular structure took a long time and much work and thought to solve, particularly as the significance of Avogadro's Hypothesis was not realised until 1858. However, long before this, the "dawn of a new day," as Berzelius called it, came with a publication of Wöhler and Liebig in 1832, in which they showed that a common association of atoms, which they termed a radical, could survive unaltered a series of changes brought about by chemical means in the other parts of the molecules concerned.

Liebig (1803-73) and Wöhler (1800-82), who became great friends, exerted a profound influence upon the progress of organic chemistry during this period. It has been said that, about 1840, ambitious young chemists from all parts of the world "used to flock either to the laboratory of Liebig at Giessen or to that of Wöhler at Göttingen." Both of these pioneering schools offered the novelty of public instruction in the practice of experimental chemistry and original chemical research, thereby inaugurating a new epoch in chemistry. "We worked from break of day till nightfall," wrote

Liebig. "Dissipation and amusement were not to be had at Giessen. The only complaint which was continually repeated was that of the attendant, who could not get the workers out of the laboratory in the evening when he wanted to clean it."

Work on the elucidation of organic molecular structure went through various developments until about 1850, by which time these delicate studies of molecular anatomy were bringing a definite molecular pattern into focus. Then came, in 1852, the "Theory of Valency," put forward by Sir Edward Frankland. This theory regarded each kind of atom as possessing a certain valency, or capacity of combining with other atoms, which was expressible as a definite number; but the theory could not meet with a proper application until the significance of Avogadro's Hypothesis was recognised, in 1858. Before that time, if the atomic weight of carbon were taken as 6, the molecular formula of methane appeared to be C_2H_4, or perhaps CH_2, instead of CH_4, as it was seen to be when carbon was assigned its correct atomic weight of 12. It was in that year, 1858, charged with significance for the whole future of organic chemistry, that Kekulé advanced the fundamental theory of organic molecular structure.

Organic Molecular Structure

Among the numerous pupils of Liebig who became famous in the world of chemistry, Friedrich August Kekulé (1829-96) takes high rank. The notes of Liebig's lectures which he took as a student are still extant; they cover 346 pages of a neat and closely written script, with many small illustrative sketches. They are entitled "Experimentalchemie vorgetragen von Prof. Dr. v. Liebig, 1848," with the flyleaf inscription, "A. Kekulé stud. chem." Kekulé had entered the University of Giessen as a student of architecture, but Liebig's brilliant lectures turned him from the gross to the molecular aspect of this art. Later, Kekulé migrated as a junior teacher to Bunsen's laboratory at Heidelberg, where, like Berzelius, he rigged up a laboratory in a room and kitchen of his dwelling.

Here, Kekulé carried out experimental work and concentrated his thoughts upon the fascinating and elusive problem of molecular structure. It has been written of him, that, as Mark Twain might have said, although young, Kekulé was industrious: he lectured on organic chemistry; he prepared the experiments and also most of the lecture specimens himself; and often he closed his arduous day by sweeping out the classroom in readiness for the morrow's lectures. Kekulé's "long, long thoughts" culminated in a flash of inspiration during a visit to London. As he related in a speech many years afterwards:

One fine summer evening I was returning by the last omnibus, "outside," as usual, through the deserted streets of the metropolis, which are at other times so full of life. I fell into a reverie (*Träumerei*), and lo, the atoms were gambolling before my eyes! Whenever, hitherto, these diminutive beings had appeared to me, they had always been in motion; but up to that time I had never been able to discern the nature of their motion. Now, however, I saw how, frequently, two smaller atoms united to form a pair; how a larger one even embraced two smaller ones; how still larger ones kept hold of three or even four of the smaller; whilst the whole kept whirling in a giddy dance. I saw how the larger ones formed a chain, dragging the smaller ones after them. . . . The cry of the conductor: "Clapham Road," awakened me from my dreaming; but I spent a part of the night in putting on paper at least sketches of these dream forms. This was the origin of the *Structurtheorie*.

These "dream forms" led Kekulé to the *Theory of Molecular Structure*, which he published in 1858. It made two fundamental postulates: the quadrivalency of the carbon atom; and a capacity of carbon atoms for linking together to form chains. The organic molecule was thus pictured as containing a backbone of linked carbon atoms, to which other atoms, or groups of atoms (sometimes called radicals) could be attached, provided that each carbon atom maintained its valency of 4. The simplest examples are provided by hydrocarbons, which are composed of carbon and hydrogen only. Thousands of these substances are known, the two simplest being methane, CH_4, and ethane, C_2H_6.

By a curious coincidence, to which there are a number of parallels in the history of science, in the year of Kekulé's publication (1858) a paper appeared in a French scientific journal expressing essentially the same view of organic molecular structure. This paper remained practically unnoticed, and the identity of the author was not discovered for half a century. The author proved to be a young, obscure Scottish assistant in Lyon Playfair's laboratory at Edinburgh, who suffered a sudden breakdown in health and was forced to retire from scientific work. "In the history of organic chemistry," wrote the German chemist Anschütz, in 1909, "the sorely tried Archibald Scott Couper deserves a place of honour beside his more fortunate fellow-worker, Friedrich August Kekulé." The theory of organic molecular structure is now often called the theory of Kekulé and Couper.

The theory accounted satisfactorily for the molecular structure of the numerous open-chain substances; but it failed to embrace the whole field of organic chemistry. There remained an important and growing body of substances, referable at that time mainly to the coal-tar hydrocarbon, benzene, C_6H_6, which could not be brought into line with the original theory. This large division became known as "aromatic" compounds, because many of its members were found to occur in various fragrant oils and aromatic spices, obtained from plants. The fundamental problem upon which Kekulé brooded for a further seven years was to extend his theoretical conception so as to devise structural formulae accounting for certain peculiar chemical characteristics of these aromatic compounds. By this time (1865) Kekulé had been appointed to the chair of chemistry in the University of Ghent, and he was deeply immersed in writing a textbook of organic chemistry when a second revealing vision came to him. This is how he described it:

> I was sitting, writing at my text-book; but the work did not progress; my thoughts were elsewhere. I turned my chair to the fire and dozed. Again the atoms were gambolling before my eyes. This time the smaller groups kept modestly in the background. My mental eye, rendered more acute by repeated visions of the kind, could now distinguish larger structures of manifold conformation: long rows, sometimes more closely fitted together; all twining and twisting in snake-like motion. But look! What was that? One of the snakes had seized hold of its own tail, and the form whirled mockingly before my eyes. As if by a flash of lightning I awoke; and this time also I spent the rest of the night in working out the consequences of the hypothesis.

And so this vision of the Ouroboros Serpent, the "tail-eater" of Greece and ancient Egypt, a symbol "half as old as time," brought across the wide ocean of time to Kekulé the solution of one of the most baffling and most important problems of organic chemistry. "Let us learn to dream, gentlemen," Kekulé used to say to his students, "and then perhaps we shall learn the truth"; but he was careful to add the note of warning conveyed in the words, "let us beware of publishing our dreams before they have been put to the proof by the waking understanding." Adolf Baeyer, one of the greatest of all organic chemists, who had been a pupil of Kekulé at Heidelberg, and later succeeded Liebig at Munich, had caught from Kekulé one of the secrets of success in chemistry; for he used to say to his students: "*So viele Chemiker haben nicht genügend Phantasie* (So many chemists suffer from a lack of imagination)."

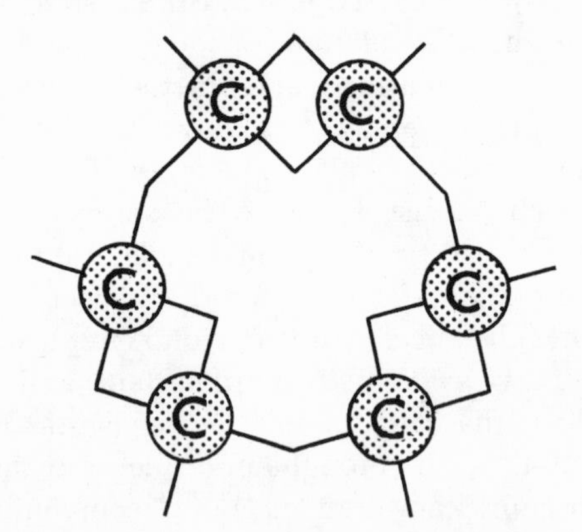

Fig. 1. Kekulé's Original Representation of the Benzene Ring.

Kekulé's imagination had led him to the conception that assemblages of linked carbon atoms could form rings as well as open chains. The benzene ring consists of six carbon atoms, each carrying one hydrogen atom, and linked to its two adjacent carbon atoms, according to Kekulé's original idea, by alternate single and double bonds. Figure 1 shows Kekulé's original representation of the benzene ring, as given in his *Lehrbuch der organischen*

Chemie (1866): the outstanding bonds provide for the attachment of one hydrogen atom to each carbon atom, and the alternate single and double connections between adjacent carbon atoms represent the single and double bonds of his formula. Nowadays, the benzene molecule, C_6H_6, is usually represented in one of the ways shown in Figure 2, the plain elongated hexagon being merely an abbreviated version of the fuller formula.

The conception of the benzene ring has been called the crowning achievement of the linking of carbon atoms. Kekulé's fundamental ideas of organic

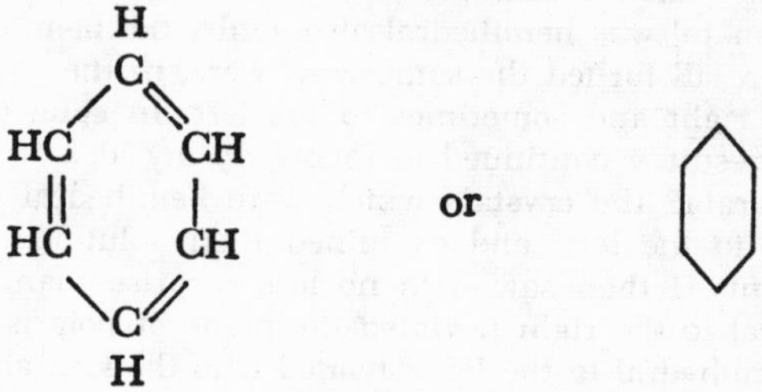

H
C
HC CH
or
HC CH
C
H

Fig. 2.

molecular structure, that is to say, his conceptions of the carbon-chain and the carbon-ring, have led to developments in pure and applied chemistry that stand unsurpassed in the whole history of science.

Later researches have shown that all organic Nature is based upon the carbon-chain and the carbon-ring, and that life itself depends upon the capacity of carbon atoms to link together so as to form the molecular chains and rings of acyclic and cyclic compounds, respectively.

Chemistry in Space

In Paris, in the year 1848, there was working at the École Normale a young assistant named Louis Pasteur, who had been born of humble parents at Dôle, near Dijon, in 1822. He had undergone a training in science, and in order to perfect himself in crystallography he undertook the repetition of some work on the crystalline forms of certain salts of tartaric acid. At that time, tartaric acid was known in two isomeric forms, both obtainable from grape juice, and related in a close and very puzzling way. One of them, ordinary tartaric acid (as also its salts), had the power of turning the plane of a beam of polarised light in a right-handed direction. The isomeric acid, known as racemic acid, did not possess this property of "optical activity," since it allowed polarised light to pass through its solutions without deviation. It was already known that well-developed crystals of tartrates (salts of tartaric acid) were characterised by the occurrence upon their surface of small subsidiary facets, known as hemihedral facets. Moreover, an eminent German chemist, Mitscherlich, had stated in 1844 that sodium ammonium tartrate and sodium ammonium racemate had the same crystalline form; but that the tartrate deviated the plane of polarisation, while the racemate remained indifferent, or optically inactive. This statement aroused the attention of Pasteur, who thought that the crystals of the racemate should differ from those of the tartrate by lacking the hemihedral facets; because throughout his work on the tartrates he had come to correlate their property of optical activity with the presence of these facets. The sequel is best related in Pasteur's own words:

> I hastened therefore to re-investigate the crystalline form of Mitscherlich's two salts. I found, as a matter of fact, that the tartrate was hemihedral, like all the other tartrates which I had previously studied, but, strange to say, the paratartrate [racemate] was hemihedral also. Only, the hemihedral faces which in the tartrate were all turned the same way, were, in the paratartrate inclined sometimes to the right and sometimes to the left. In spite of the unexpected character of this result, I continued to follow up my idea.
>
> I carefully separated the crystals which were hemihedral to the right from those hemihedral to the left, and examined their solutions separately in the polarising apparatus. I then saw with no less surprise than pleasure that the crystals hemihedral to the right deviated the plane of polarisation to the right, and that those hemihedral to the left deviated it to the left; and when I took an equal weight of each of the two kinds of crystals, the mixed solution was indifferent towards the light in consequence of the neutralisation of the two equal and opposite individual deviations.

This observation, one of the most classical in the whole history of science, was at first received by the scientific world with astonishment, mingled with incredulity. The Académie des Sciences delegated the veteran physicist, Biot, the greatest authority of that time upon polarised light and its effects, to examine the plausible but startling story of this unknown young man. Again, Pasteur has left an account of what happened:

> He [M. Biot] sent for me to repeat before his eyes the several experiments. He gave me racemic acid which he himself had previously examined and found to be quite inactive to polarised light. I prepared from it in his presence the sodium ammònium double salt, for which he also desired himself to provide the soda and ammonia. The liquid was set aside for slow evaporation in one of the rooms of his own laboratory, and when 30 or 40 grams of crystals had separated he again summoned me to the Collège de France, so that I might collect the dextro- and laevo-rotatory [right- and left-handed] crystals before his eyes, and separate them according to their crystallographic character, asking me to repeat the statement that the crystals which I should place on his right hand would cause deviation to the right, and the others to the left. This done, he said that he himself would do the rest.
>
> He prepared the carefully weighed solutions, and, at the moment when he was about to examine them in the polarimeter, he again called me into his laboratory. He first put the more interesting solution, which was to cause rotation to the left, into the apparatus. Without making a reading, but already at the first sight of the colour-tints presented by the two halves of the field in the Soleil saccharimeter, he recognised that there was a strong laevo-rotation. Then the illustrious old man, who was visibly moved, seized me by the hand, and exclaimed: "My dear child, I have so loved the sciences throughout my life that this makes my heart leap with joy! (*Mon cher enfant, j'ai tant aimé les sciences dans ma vie que cela me fait battre le coeur!*)."

It is strange that this dramatic scene, so charged with human interest, and rarely paralleled in the history of science, has never attracted the attention of any great artist.

Pasteur ascribed the existence of this so-called "optical isomerism" to the formation of enantiomorphous molecules, that is to say, of molecules capable of existing in right- and left-handed forms, and related to each other as an object to its non-coincident image. His researches in this field of chemistry, carried out mostly between 1848 and 1858, led to the conclusion that organic molecules, like tangible objects, fall into two categories, designated by the terms symmetric and asymmetric (or non-symmetric).

A symmetric object, such as a teapot, gives a coincident image; also it possesses a plane of symmetry, that is to say, it can be divided into two similar halves, in this particular example by a plane directed at right angles to its base and bisecting the teapot midway through the spout and the handle. An asymmetric object, such as a hand, foot, glove, or shoe, gives a non-coincident image, that of a right hand being a left hand; also it has no plane of symmetry.

Pasteur's work showed that organic molecules also fall into these two categories. It follows that molecules must be three-dimensional, and not two-dimensional, or flat, as Kekulé's structural formulae might suggest. A flat object could not exist in right- and left-handed forms. If such a being as a Flatlander could exist, his two hands would be identical, and he would be unable to tell one from the other, or to decide whether a clock-hand were moving around in a "clockwise" or "counter-clockwise" direction. Since some molecules can exist in right- and left-handed forms, it follows that they must be extended in space of three dimensions; and therefore, by implication, that all molecules, whether asymmetric or symmetric, are three-dimensional and not flat.

Pasteur (1822-95) became the founder of stereochemistry, or "chemistry in space," which deals with the spatial arrangement of atoms in combination. After a century of research many of the ramifications and implications of this wide field of science, of equal importance in organic chemistry and biochemistry, still remain to be explored. But the foundation of stereochemistry was only the beginning of Pasteur's scientific work. Outstanding in his versatility, this great paladin of science passed on from stereochemistry to fermentation, pathology, and in particular to investigations showing that diseases could be induced by micro-organisms; and this work led in turn to the introduction of antiseptic surgery and immunisation. Huxley said that, in terms of money alone, Pasteur's work would have met the whole cost of the French war indemnity of 1870 (amounting to 5000 million francs). Louis Pasteur was one of the greatest of Frenchmen and one of the most human and humane of scientists. He gained the affection and reverence of all. Osler said to him that he was the most perfect man that ever entered into the kingdom of science.

Pasteur was unable to interpret his ideas in precise terms of molecular structure, because Kekulé's theory of molecular structure, leading to structural formulae, had not been put forward at the time of Pasteur's experiments. With the advent of Kekulé's theory it became possible, during the next few years, to examine the structural formulae of the tartaric and lactic acids, and of various other substances which were known to exhibit optical activity. As a result, it was found that the molecules of all these substances assumed an asymmetric form—or "configuration," as it is usually called—when their flat Kekuléan representations were converted into three-dimensional ones, according to a very simple principle. This principle was embodied in the "Theory of Molecular Configuration," which was advanced independently and almost simultaneously in 1874 by the French chemist Le Bel and the Dutch chemist van 't Hoff.

It has been pointed out that the expansion of the Atomic Theory into the theories of molecular structure and molecular configuration provides a classical example of the gradual evolution of a scientific theory. Each successive stage of the theory defined a limited problem, the solution of which opened the way for a further advance; until the attainment of a position from which (to adapt some words of Sir John Herschel, referring to the still unseen planet, Neptune) we see the organic molecule "as Columbus saw America from the shores of Spain. Its movements have been felt trembling along the far-reaching line of our analysis with a certainty hardly inferior to ocular demonstration."

The Onward March

In no branch of science has theory determined practical progress to a greater extent than in chemistry. In organic chemistry the structural and spatial molecular theories came at psychological moments in the history of the science, when the way lay open for rapid advances. It was in 1856, only two years before the publication of Kekulé's theory, that an eighteen-year-old English student, William Henry Perkin (1838-1907), working during the Easter vacation in his "rough laboratory at home," discovered mauveine, the first synthetic organic dyestuff to be applied in dyeing. His synthesis of this dye from impure aniline, as a result of experiments aimed at the synthesis of quinine, became the herald of an enormous stream of work on synthetic coal-tar derivatives which ran in full flood throughout the next half-century. Without the guidance of Kekulé's benzene theory this great expansion would have been impossible.

From a few "primaries" actually present among the numerous constituents of coal-tar—notably benzene, toluene, phenol, naphthalene, and anthracene—were prepared, as time went on, several hundred "intermediates," such as aniline from benzene, benzaldehyde from toluene, and phthalic acid from naphthalene. These led in turn, through the application of a variety of chemical processes, to the synthesis of thousands of purely artificial dyes, drugs, explosives, disinfectants, photographic developers, and many other useful kinds of fine chemicals. During the half-century after Perkin's day, certain coal-tar constituents were brought into use in the manufacture of nylons, synthetic rubbers and plastics, and other substitutes for age-old natural materials such as wood, glass, and natural fibres. Such work still continues. Moreover in the present century the great deposits of natural petroleum are undergoing an increasing development not only as sources of energy for heating, lighting, and locomotion, but also as a raw material for the manufacture of useful chemicals.

Concurrently with the synthesis of purely artificial substances from coal-tar, much work was being devoted to unravelling the molecular constituents of natural organic substances. Here, the German chemist, Emil Fischer (1852-1919), stood out as a master. His elucidation of the molecular constitution of sugars and their subsequent synthesis, which he began in the 1880's, would have been impossible without the guidance of the Space Theory of Le Bel and van 't Hoff. From this classical series of investigations he passed on to explore the molecular mysteries of the uric acid derivatives and the proteins. Other organic chemists took up similar work upon such important groups of compounds as the natural dyes, the pigments of flowers and leaves, the alkaloids, hormones, vitamins, constituents of the fragrant oils of plants, and many series besides.

With the deciphering of so many natural molecular types, it became possible to discern certain general structural relationships, including the recognition of standard patterns forming variations upon fundamental molecular themes. Hand in hand with the molecular diagnosis of natural organic compounds went very often their artificial synthesis. These advances opened the way in turn to the practical realisation of artificial variations upon natural molecular themes, leading to improved dyes, drugs, rubbers, and many other useful substances with molecular structures based upon natural models. The artificial rubber, neoprene, for example, is a synthetic variant of natural rubber in which the methyl group ($—CH_3$) attached to every fourth carbon atom in a long chain of thousands of such linked atoms in the natural molecule is replaced by a chlorine atom ($—Cl$) in the artificial molecule. As another example, the macromolecules of the purely artificial fibrous nylons are variants of structures found in natural protein molecules.

Well may the organic chemist, in his advance from the simple molecule of methane to the gigantic macromolecules of natural celluloses and proteins and of artificial fibres and plastics, exclaim with the poet:

> All experience is an arch wherethro'
> Gleams that untravell'd world, whose margin fades
> For ever and for ever when I move.

GLENN T. SEABORG

The Man-made Chemical Elements Beyond Uranium

•

AN EXCITING BRANCH of science, which started as recently as World War II and has a clearly discernible future of great promise, is that of the "transuranium elements." These are the man-made chemical elements with atomic numbers greater than that of the heaviest natural element, uranium, which has the atomic number 92.

The transuranium elements are, for all practical purposes, synthetic in origin and must be produced by transmutation, starting in the first instance with uranium. The key to the discovery of these essentially "synthetic" elements was their position in the Periodic Table.

Prior to the discovery of these transuranium elements, the relationship of the heaviest, naturally occuring elements—actinium, thorium, protactinium, and uranium—in the Periodic Table was not clearly understood. Recognition of the fact that the transuranium elements represented a whole new family of "actinide" elements analogous to the rare-earth series of elements, "lanthanides," permitted the discoverers to predict the chemical properties of the unknown transuranium elements and thereby enabled them to discover and separate them from all the other elements in the Periodic Table.

At the present time, eleven transuranium elements have been created and identified, with a total of nearly one hundred isotopes. All these new transuranium elements are unstable and therefore radioactive. The half-life of the various isotopes decreases in general with increasing atomic number—meaning that, as we create heavier and heavier elements, they exist for shorter and shorter periods, making their production, separation, and identification progressively more difficult.

It may still be possible to synthesize, separate, and identify a half-dozen or so more of the transuranium elements; but barring unknown experimental breakthroughs or unknown regions of stability in these heavier elements, the end should come somewhere in the region of element 110. The elements up to and including einsteinium, element 99, have isotopes sufficiently long-lived to be isolated in macroscopic (that is, weighable) quantities, but this does not seem to be true beyond einsteinium. Unfortunately for the prospect of studying ever-higher elements, the longest-

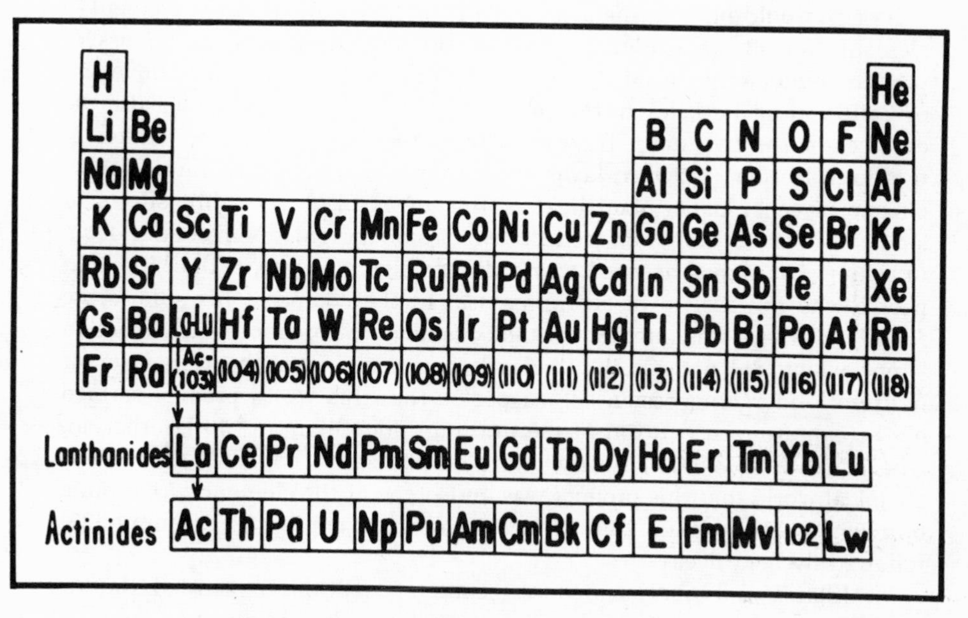

Fig. 1. The Periodic Table of the Elements.

lived isotopes that can be made beyond element 104 or 105 will probably not exist long enough for conventional chemical identification.

However, the prediction of the chemical properties of the yet undiscovered chemical elements is quite straightforward. Lawrencium, the element with the atomic number 103, completes the actinide series, and it is expected that elements 104, 105, 106, etc., will be fitted into the Periodic Table under hafnium, tantalum, tungsten, etc., and have analogous chemical properties.

Let me briefly review some of the history of these discoveries. The efforts of men to change one element into another, or to create new elements, date back to the time of the alchemists. But the first serious scientific attempts to go beyond the heaviest of the known elements—uranium—and explore the transuranium region were those of Enrico Fermi, Emilio Segrè, and their co-workers in Rome, in 1934, shortly after the existence of the neutron had been discovered. This group bombarded uranium with slow neutrons in the hope that the uranium nucleus would capture a neutron. A number of radioactive products were found. In the immediately following years, many more such radioactive species were observed. However, chemical investigations led to the discovery that these were not transuranium elements, but, rather, products of the fission process. The discovery of fission by Otto Hahn and F. S. Strassman, in December, 1938, which led to the "atomic age," was in a sense a dividend or by-product of man's quest for the transuranium elements.

As fate would have it, the discovery of the first transuranium element—element No. 93, *neptunium*—was a by-product of studies of the fission process conducted by E. M. McMillan. McMillan, working at the University of California at Berkeley in the spring of 1940, was trying to measure the energies of the two main fragments from the neutron-induced fission of uranium. He placed a thin layer of uranium compound (uranium oxide) on one piece of paper; and next to this he stacked very thin sheets of similar paper to stop and collect the fission fragments recoiling from uranium. The paper he used was ordinary cigarette paper, the kind that people who roll their own cigarettes use. In the course of his studies, he found that there was another unstable, radioactive product of the reaction—one which did not recoil sufficiently to escape from the thin layer of uranium undergoing fission. He suspected that this was a product formed by the capture of a neutron in the uranium. McMillan and P. H. Abelson, who joined him in this research, were able to show, on the basis of their chemical work, that this product was an isotope of the element with atomic number 93—neptunium-239, formed by neutron capture in uranium-238, followed by beta decay.

McMillan's and Abelson's investigation of neptunium showed that it resembles uranium—not rhenium, as predicted—in its chemical properties. Therefore, analogous to uranium, which was named after the planet Uranus, element 93 was named neptunium, after the next planet, Neptune. This was the first definite evidence that an inner electron shell (the $5f$ electron shell) is filled in the transuranium region.

The element with the atomic number 94, *plutonium*, was next to be discovered. By bombarding uranium with deuterons raised to a high energy in the 60-inch cyclotron at the University of California at Berkeley, E. M. McMillan, J. W. Kennedy, A. C. Wahl, and I, in late 1940, succeeded in preparing a new isotope of neptunium, neptunium-238, which decayed to plutonium-238.

The first bombardment of uranium oxide with the 16-MeV deuterons was performed on December 14, 1940. Alpha radioactivity was found to grow into the chemically separated element 93 fraction during the following weeks, and this alpha activity was chemically separated from the neighboring elements (especially elements 90 to 93, inclusive) in experiments performed during the next two months. These experiments constituted the positive identification of plutonium.

The chemical properties of elements 93 and 94 were studied by the so-called "tracer method" at the University of California for the next year and a half. This meant that invisible amounts of these elements were followed in chemical studies by their telltale radioactivity. These first two transuranium elements were referred to by the group simply as "element 93" and "element 94," or by code names, until the spring of 1942, at which time the first detailed reports on them were written. The early work, even in those days, was carried on under a self-imposed cover of secrecy. Throughout 1941, element 94 was referred to by the code name of "copper," which was all right until it was necessary to introduce the element copper into some of the experiments. This posed the problem of distinguishing between the two. For a while, plutonium was referred to as "copper" and the real copper was "honest-to-God copper." This seemed clumsier and clumsier as time went on, and element 94 was finally christened "plutonium," after the planet Pluto and analogous to uranium and neptunium.

The plutonium isotope of major importance is the one with mass number 239. The search for this isotope, as a decay product of neptunium-239, was being conducted by the same group, with the collaboration of E. Segrè, simultaneously with the experiments leading to the discovery of plutonium. The isotope plutonium-239 was identified, and its possibilities as a nuclear energy source established, during the spring of 1941.

Using neutrons produced by the 37-inch cyclotron at the University of California, the group first demonstrated, on March 28, 1941, with a sample containing 0.5 microgram of plutonium-239, that this isotope undergoes slow-neutron-induced fission with a cross section even larger than that of uranium-235. A fission cross section for plutonium-239 some 50 percent greater than that for uranium-235 was found, in remarkable agreement with the accurate values which were determined later.

The realization that plutonium-239 could serve as the explosive ingredient of a nuclear weapon, and that it might be created in quantity in a nuclear reactor, followed by chemical separation from uranium and the highly radioactive fission products, made it imperative to carry out chemical investigations of plutonium with weighable quantities, even though only microgram quantities could be produced using the cyclotron sources of neutrons available at that time. In August, 1942, B. B. Cunningham and L. B. Werner, at the wartime Metallurgical Laboratory of the University of Chicago, succeeded in isolating about a microgram of plutonium-239—less than one ten-millionth of an ounce—which had been prepared by cyclotron irradiations. Thus, plutonium was the first man-made element to be obtained in visible quantity. The first weighing of this man-made element took place on September 10, 1942, and was performed by investigators Cunningham and Werner.

These so-called "ultramicrochemical" studies conducted by the research workers on plutonium were remarkable. It was possible to perform many significant studies with almost invisible amounts of material—work that was carried out under a microscope. If extremely small volumes are used, even microgram quantities of material can give relatively high concentrations in solution; and with the development of balances of the required sensitivity, micrograms were also sufficient for gravimetric analysis. Liquid volumes in the range of 10^{-1} to 10^{-5} cc were measured with an error of less than one percent by means of finely calibrated capillary tubing. Chemical glassware, such as test tubes and beakers, was constructed from capillary tubing and was handled with micromanipulators.

Plutonium is now produced in much larger quantities than any other synthetic element. The large, wartime chemical-separation plant at Hanford, Washington, was constructed on the basis of the investigations performed on the tracer and ultramicrochemical scale; the scale-up between ultramicrochemical experiments to the final Hanford plant corresponds to a factor of about one billion—surely, a scale-up of unique proportions.

After the completion, at the wartime Metallurgical Laboratory, of the most essential part of the investigations concerned with the chemical processes involved in the production and separation of plutonium, attention turned to the problem of synthesizing and identifying the next-heavier transuranium elements. As my collaborators in this endeavor, there were A. Ghiorso, R. A. James, and L. O. Morgan.

There followed a period during which the attempts to synthesize and identify elements 95 and 96 bore no fruit. The unsuccessful experiments were based on the premise that these elements should be much like plutonium, in that it should be possible to oxidize them to a higher oxidation state and utilize this oxidation in the chemical-isolation procedures. It was not until the middle of the summer of 1944, upon the first recognition that these elements were part of an "actinide" transition series (i.e., were chemically very similar to the element actinium and to the long-known rare-earth elements) that any advance was made; and then progress came quickly. Incidentally, this element-by-element analogy in chemical properties between the actinide and lanthanide (rare-earth) elements has been the key to the chemical identification—and, hence, discovery—of the subsequent transuranium elements.

When it was recognized that these elements could be oxidized only with extreme difficulty, if at all, the identification of an isotope then thought to be due to element 95 or 96 followed immediately. Thus, the isotope of element 96—now known to be *curium-242*—was produced in the summer of 1944 as a result of the bombardment of plutonium-239 with 32-MeV helium ions in the cyclotron at Berkeley.

The identification of element 95, *americium,* in the form of the isotope americium-241 followed, during late 1944 and early 1945, as a result of the bombardment of plutonium-239 with neutrons in a nuclear reactor.

Some comments should be made concerning the rare-earth-like properties of these two elements. Our hypothesis that they should greatly resemble the rare-earth elements in their chemical properties proved to be so very true that, for a time, it appeared to be unfortunate. The better part of a year was spent in trying, without success, to separate chemically the two elements from each other and from the rare-earth elements; and although we felt entirely confident, on the basis of their radioactive properties and the methods of production, that isotopes of elements 95 and 96 had been produced, the chemical proof was still undemonstrated. The elements remained unnamed during this period of futile attempts at separation (although one of our group referred to them as "pandemonium" and "delirium," in recognition of our difficulties). The key to their final separation, and the technique which made feasible the separation and identification of these and subsequent transuranium elements, was the so-called ion-exchange technique. The elements were named americium, after the Americas, and curium, in honor of Pierre and Marie Curie, by analogy to the naming of their rare-earth counterparts, europium (after Europe) and gadolinium (after Finnish chemist Gadolin).

The most important prerequisite to the process for making the transcurium elements was that sufficiently large amounts of americium and curium be manufactured to serve as starting materials. Because of the intense radioactivity of these starting substances, even in milligram or submilligram amounts, it was necessary to develop extremely efficient chemical separation methods in order to obtain the enormous separation factors needed for the isolation of the new elements from the starting material, so that it would be possible to detect the very small amounts of radioactivity due to the new transcurium elements. The dangerous radioactivity of the source material made it necessary to develop complicated methods for remote control operation to keep the health hazards at a minimum.

These production, separation, and protection problems were solved, and successful experiments were performed at the end of 1949 and the beginning of 1950. Americium for target material was prepared in milligram amounts by intense neutron bombardment of plutonium over a long period of time, and curium target materials were prepared in microgram amounts as the result of the intense neutron bombardment of some of this americium. Both of these neutron bombardments took place in a high-power reactor having a high neutron flux.

Element 97, *berkelium,* was discovered by S. G. Thompson, A. Ghiorso, and myself, in December, 1949, as a result of the bombardment of the milligram amounts of americium-241 with 35-MeV helium ions accelerated in the 60-inch cyclotron of the University of California at Berkeley. The first isotope produced has the mass number of 243 and decays with a half-life of 4.5 hours.

Element 98, *californium*, was first produced and similarly identified by Thompson, Ghiorso, K. Street, and myself in February, 1950, at the University of California at Berkeley. The first isotope produced is now assigned the mass number 245 and decays with a half-life of 44 minutes. This isotope was produced by the bombardment of microgram amounts of curium-242 with 35-MeV helium ions accelerated in the 60-inch cyclotron. It is interesting to note that this identification of element 98 was accomplished with a total of only some 5000 atoms; someone remarked at the time that this number was substantially smaller than the number of students attending the university.

The naming of elements 97 and 98 presents an interesting story. Element 97 was called berkelium after the city of Berkeley, California, where it was discovered. Element 98 was named californium after the university and state where the work was done.

Upon learning about the naming of these elements, the "Talk of the Town" section of *The New Yorker* magazine had the following to say:

> New atoms are turning up with spectacular, if not downright alarming, frequency nowadays, and the University of California at Berkeley, whose scientists have discovered elements 97 and 98, has christened them berkelium and californium, respectively. While unarguably suited to their place of birth, these names strike us as indicating a surprising lack of public relations foresight on the part of the university, located, as it is, in a state where publicity has flourished to a degree matched perhaps only by evangelism. California's busy scientists will undoubtedly come up with another atom or two one of these days, and the university might well have anticipated that. Now it has lost forever the chance of immortalizing itself in the atomic tables with some such sequence as universitium (97), ofium (98), californium (99), berkelium (100).

The discoverers sent the following reply:

> "Talk of the Town" has missed the point in their comments on naming of the elements 97 and 98. We may have shown lack of confidence but no lack of foresight in naming these elements "berkelium" and "californium". By using these names first, we have forestalled the appalling possibility that after naming 97 and 98 "universitium" and "ofium", some New Yorker might follow with the discovery of 99 and 100 and apply the names "newium" and "yorkium".

The answer from *The New Yorker* staff was brief:

> We are already at work in our office laboratories on "newium" and "yorkium". So far, we just have the names.

In 1958, S. G. Thompson and B. B. Cunningham at Berkeley succeeded in isolating, for the first time, macroscopic amounts of berkelium (as berkelium-249) and californium (as a mixture of californium-249 and -252), which were synthesized by the long-term irradiation of plutonium-239 and its transmutation products with neutrons. The first pure compound of californium was isolated in 1960, by B. B. Cunningham and

J. C. Wallmann, using californium-249 obtained via the decay of previously separated berkelium-249. Three tenths of a microgram of californium oxychloride (CfOCl) was isolated. Later, the trichloride and oxide were also prepared. This experimental work was the first carried out on a "submicrogram" scale as a result of new techniques and represented an order-of-magnitude scaledown from the ultramicrochemical work done during the last war.

The discovery of elements 99 and 100, *einsteinium* and *fermium*, represents an outstanding example of the unexpected in science. The seventh and eighth transuranium elements were discovered in debris from "Mike," the first, large thermonuclear or "hydrogen-bomb" explosion which took place in the Pacific in November, 1952. Debris from the explosion was collected, first on filter papers attached to airplanes which flew through the clouds and later, in more substantial quantity, gathered up as fallout materials from the surface of a neighboring atoll. This debris was brought to the United States for chemical investigation in a number of laboratories.

Initial investigation of this debris at the Argonne National Laboratory in Chicago, and at the Los Alamos Scientific Laboratory of the University of California in New Mexico, led to the unexpected observation of heavy isotopes such as plutonium-244 and plutonium-246. At that time, the heaviest known isotope of plutonium was plutonium-243. Since this pointed to the capture of many successive neutrons by the uranium-238 in the device, and thus the presence of neutron-rich isotopes in greater abundance than expected, a group at the University of California Radiation Laboratory undertook to look for isotopes of transcalifornium elements in this material. Ion-exchange experiments of the type previously mentioned immediately demonstrated the existence of a new element. Later, in order to secure a larger amount of source material, it was necessary to work up many hundreds of pounds of coral from one of the atolls adjoining the explosion area. Eventually, such coral was worked up by the ton in a pilot-plant operation which went under the name of "Paydirt."

Without going into the details, it may be pointed out that the experiments by the research groups at the three laboratories led to the positive identification of isotopes of elements 99 and 100. A radioactive isotope with the mass number 253, which decays with a half-life of twenty days, was identified as an isotope of einsteinium—element 99—named in honor of Albert Einstein. A radioisotope with the mass number 255 and a half-life of about 20 hours was identified as an isotope of fermium, element 100, named in honor of Enrico Fermi.

The successive instantaneous capture of many neutrons by uranium-238 had led to heavy uranium isotopes. The heavy uranium isotopes then decayed into the isotopes appearing beyond them in the Periodic Table, hence increasing the atomic number in successive steps. The first identification of element 100 was made with only about 200 atoms. The most striking previous accomplishment in this category was the positive identification of element 98 with a total of about 5000 atoms.

Before declassification and the subsequent announcement of the original discovery experiments could be accomplished, isotopes of elements 99 and 100 were produced by a number of other methods. Chief among these was that of successive neutron capture as the result of intense neutron irradiation in a nuclear reactor of high neutron density. The difference between this method of production and that of the "Mike" thermonuclear explosion is one of time as well as of starting material. In a reactor, it was necessary to bombard gram quantities of plutonium for several years; thus, the short-lived, intermediate isotopes of the various elements had an opportunity to decay. Unfortunately, this process cannot be used to prepare the elements beyond fermium (element No. 100), because some of the intermediate isotopes which must capture neutrons have half-lives so short as to preclude their presence in the appreciable concentrations that are necessary. In the thermonuclear device, larger amounts of uranium were subjected to an extremely high neutron irradiation for a period of a few millionths of a second; the subsequent decay of the ultraheavy isotopes of uranium formed the products of high atomic number found in the debris. In principle, this method could be used to prepare elements beyond fermium.

It was not until 1961 that sufficient einsteinium had been produced through intense neutron bombardment in reactors to permit separation of a macroscopic and weighable amount. B. B. Cunningham, J. C. Wallmann, L. Phillips, and R. C. Gatti, at Berkeley, were able to separate—working on the new, submicrogram scale—an extraordinarily small amount of einsteinium-253. A total amount of only a few hundredths of a microgram, a billionth of an ounce, was weighed on a special, magnetic-type balance.

The discovery of *mendelevium,* element 101, by A. Ghiorso, B. G. Harvey, G. R. Choppin, S. G. Thompson, and myself, was in many ways the most dramatic of them all. It was decided to make an attempt in a situation which would have been regarded by most sensible people as very premature. All of the previous discoveries of transuranium elements had been based on starting with weighable amounts of target materials; however, it was thought that techniques had advanced to a point where it might be possible to identify the element with the atomic number 101 in a target of unweighable amount.

The plan of attack involved the bombardment of the maximum available quantity of einsteinium, element 99, in the form of the isotope einsteinium-253, with helium ions in the Berkeley 60-inch cyclotron. On gathering together all of the available einsteinium-253 which had been produced in nuclear reactors, we found that it amounted to about 10^9 atoms. In order to assay the possibilities, the number of atoms of element 101 which could reasonably be expected from the bombardment could be deduced from simple considerations. The calculation showed that we could expect to produce approximately one atom in each experiment!

Adding immeasurably to the complexity of the experiment was the absolute necessity for the chemical separation of the one atom of element 101 from the 10^9 atoms of einsteinium in the target and its ultimate, complete chemical identification by separation with the now-familiar ion-exchange method. This separation and identification would presumably have to take place in a period of hours, or even one hour or less, because the expected half-life was of this order of magnitude.

The requirements indicated the desperate need for new techniques, together with some luck; and fortunately both were forthcoming. The new technique involved the separation of the element 101 by the recoil method from the einsteinium in the target. The einsteinium was placed on a gold foil in an invisibly thin layer. The helium-ion beam produced in the cyclotron was sent through the back of the foil so that the atoms of element 101, recoiling due to the reaction with the impinging helium ions, could be caught on a second thin gold foil. This second gold foil which contained the recoiled atoms, and which was relatively free of the einsteinium target material, was dissolved, and the chemical operations performed began with this solution.

The earliest experiments were confined to looking for short-lived, alpha-particle-emitting isotopes of element 101. No alpha activity was observed that could be attributed to element 101, even when the time between the end of bombardment and the beginning of the alpha particle analyses had been reduced to five minutes.

However, the experiments were continued and, in one of the subsequent bombardments, a single large pulse caused by spontaneous fission was observed in the detection apparatus. With probably unjustified self-confidence, we thought that this might be a significant result. Although such an attitude might ordinarily have been considered foolish, it must be recalled that rapid decay by spontaneous fission was up until that time confined to only a few isotopes, none of which should have been introduced spuriously into the experiment or produced in the experiment. In addition, background counts due to this mode of decay should be essentially zero in proper equipment.

The major question, of course, was whether the experiment could be repeated. In a number of subsequent bombardments, spontaneous fission events were observed in some, while none was observed in others. This, of course, was to be expected, because of the statistical fluctuation inherent in the production of the order of one atom per experiment. Furthermore, more advanced chemical experiments seemed to indicate that the spontaneous fission counts, when they did appear, came in about the element 100 or 101 chemical fractions. At about this time, a huge fire bell was hung in the hall of the chemistry building, connected to the counting circuit in such a manner that a loud "clang" rang out on each occasion when one of these rare spontaneous fission events registered. However, this sport was put to a justifiable end when it came to the attention of the fire department.

The definitive experiments were performed in a memorable, all-night session. Three successive, three-hour bombardments were made, and, in turn, their transmutation products were completely and quickly separated by the ion-exchange method. A total of five spontaneous fission counts was observed in the element 101 chemical fraction, while a total of eight spontaneous fission counts was also observed in the fermium (element 100) chemical fraction; no such counts were observed in any other chemical fraction.

The spontaneous fission activity in both the element 101 and fermium (element 100) fractions decayed with a half-life of about three hours. This and other evidence led to the interpretation that the mendelevium, element 101, isotope has the mass number 256 and decays by electron capture—with a half-life of the order of one hour—to the isotope fermium-256, which is responsible for the spontaneous fission decay. Element 101 was named mendelevium in honor of the Russian scientist, Dmitri Mendeleev, who was the first to use the Periodic Table of the Elements to predict the chemical properties of undiscovered elements, a principle which has been the key to the discovery of so many of the transuranium elements.

In 1957, a team of scientists from Argonne National Laboratory in the United States, the Atomic Energy Research Establishment in England, and the Nobel Institute for Physics in Sweden announced the discovery of an isotope of *element 102*, as a result of research performed at the Nobel Institute. The name "nobelium" for element 102 was suggested by this group. Unfortunately, neither experiments using the facilities at the University of California in Berkeley nor related experiments performed in the USSR have confirmed this Stockholm research.

In 1958, a group at the University of California—A. Ghiorso, B. G. Harvey, G. R. Choppin, S. G. Thompson, and I—reported the positive identification of the isotope of element 102 with the mass number 254 as a product of the bombardment of curium-246 with carbon-12 ions accelerated in the new, heavy-ion linear accelerator (HILAC) at Berkeley, specially constructed for experiments on the transuranium elements. The new element was detected by the chemical identification of its known daughter, fermium-250. Although the name "nobelium" for element 102 will undoubtedly have to be changed, these investigators have not yet made their suggestion for a new name.

The removal of the element 102 isotope from the target material and the separation of the daughter element from the parent element 102 were performed by the use of a new method in which two physical recoil separations were utilized. This experiment bears resemblance to that of the discovery of mendelevium, except that the half-life involved was even shorter, necessitating more sophisticated techniques. Here, again, the amount involved required identification of the new element, atom by atom. The target consisted of curium deposited on a thin nickel foil and was enclosed in a container filled with helium gas. The curium was bombarded with carbon-12 ions, and the transmuted atoms were knocked into the helium gas to absorb their recoil energy. With a sufficient electric field strength, it was found that practically all of these positively charged atoms could be attracted to a moving, negatively charged metallic belt placed directly beneath the target. The belt was then passed under a foil (the catcher foil), which was charged negatively relative to the belt. Approximately half of the element 102 atoms undergoing radioactive decay by alpha-particle emission would cause their daughter atoms to recoil from the surface of the belt to a catcher foil.

The catcher foil was cut transversely to the direction of the belt motion into five equal-length sections—after a time of bombardment suited to the half-life of the daughter atom to be examined—and each section was analyzed simultaneously in counters. All of the desired measurements could be made for identifying the daughter atoms caught on the catcher foils, and thus the half-life of the parent of the recoiling atoms could be determined. It was found that fermium-250 could be collected on the catcher foil in accordance with a parent half-life (i.e., half-life for element 102 with a mass number 254) of three seconds. Changing the belt speed was found to change the distribution of the fermium-250 on the catcher foil in a manner conforming to a three-second parent.

In later experiments, the recoil atoms of element 102 were caught on a belt which was quickly pulled inside an alpha-particle counter in order to measure directly the energy of the alpha particles and the half-life of the isotope in question.

In the spring of 1961, after almost three years of work which began shortly after the discovery of element 102, a group at the University of California, Berkeley (A. Ghiorso, T. Sikkeland, A. Larsh, and R. Latimer) found an isotope of element 103, *lawrencium,* named in honor of Ernest O. Lawrence.

The method used to produce and identify lawrencium was similar to that used in the later, direct-counting experiments performed in connection with the discovery of element 102. About three micrograms of a mixture of californium isotopes were bombarded with boron ions accelerated in the heavy-ion linear accelerator (HILAC). The atoms of lawrencium recoiled from the target into an atmoshpere of helium, where they were electrostatically collected on a copper conveyor tape. This tape was then periodically pulled into place before radiation detectors to measure the emission rate and the energy of the alpha particles being emitted. By this means, it was possible to identify the lawrencium isotope, lawrencium-

257, with a half-life of eight seconds. At present, because of the short half-life and the lack of a suitable daughter isotope, available in the case of element 102, it has not been possible to perform a chemical identification, and the discovery rests solely on nuclear evidence.

The search is continuing for even heavier transuranium elements with atomic numbers beyond 103. These may be discovered by means of bombardments with heavy ions, as in the case of elements 102 and 103. This requires the use of isotopes of the heaviest available elements, especially californium and einsteinium, as target materials. Present supplies of californium are limited to microgram quantities, and in the case of einsteinium to submicrogram quantities.

Up to this point, I have addressed myself, for the most part, to the exciting search for and discovery of the transuranium elements. The work was clearly basic research—a quest for further knowledge and understanding of the atomic and nuclear properties of the elements. However, as history would have it, this basic research work has developed into an outstanding example of the great and important effect that such discoveries can have on people everywhere. The most immediate effect that the discovery of a transuranium element had on the world was, of course, the use of plutonium in nuclear weapons.

Historically, after the fission reaction was discovered in uranium, effort was devoted to separating the fissionable uranium isotope, uranium-235, from the bulk of the uranium, uranium-238; that is, to the enrichment of uranium-235. During the crucial and urgent war years, an alternative method was vital for concentrating a fissionable isotope that might be used in atomic weapons. After the discovery of plutonium-239 and its fissionable properties, an intensive program of research and development engineering was sponsored to produce significant quantities of this new element for military uses.

Whereas the separation of uranium isotopes required exceedingly refined techniques, since chemically the two isotopes were nearly indistinguishable, the separation of plutonium from the uranium in which it was formed, and the huge quantities of radioactive fission products present with it, presented a more straightforward chemical problem. One could rely on the chemical properties of the new element to separate it from uranium and all the other elements present—if one knew the chemical properties of this new element. Fortunately, it was possible to determine the chemical properties of plutonium by means of the tracer and ultramicrochemical experiments. The engineering of large-scale separation plants based on this knowledge was of course successful.

The first large facility (constructed during World War II) for the production of plutonium was the Hanford Engineer Works at Hanford, Washington. Some years after World War II, a second plutonium production site, the Savannah River Plant, was constructed south of Aiken, South Carolina, on the Savannah River.

However, the uses of these transuranium elements are not limited to military applications. They also have very important peaceful applications. Plutonium-239 and -241, which are both fissionable by thermal neutrons, can also be used in power reactors. As more power reactors come into operation, more plutonium will be produced; and it will become more feasible to use this increasing supply of plutonium for fueling reactors. This will increasingly supplement ordinary fuels in meeting the large, growing need for additional electrical power in the future.

The really important role of plutonium in peaceful applications lies in the fact that a fissionable isotope can be produced from uranium-238, a nonfissionable isotope. In any analysis of the world's energy resources, it is evident that gas, oil, and coal are not inexhaustible, but eventually limited. Studies to date would indicate that the energy locked in available supplies of uranium is factors greater than that in these available fossil fuels. However, this relative excess of energy is calculated on the basis of our being able to utilize not only the several tenths of a percent of uranium-235 present, but the inherent energy in the abundant uranium-238 that can be converted into fissionable plutonium.

It is quite evident that even further applications will be found for these transuranium elements. Of even greater importance is the contribution which the study of their chemical and nuclear properties has made to the store of human knowledge. The time and effort devoted to the basic research involved in their discovery and investigation have been more than compensated for by the results.

LAWRENCE P. LESSING

Where Is Chemistry Today?

•

As WE MOVE OUT into the last half of the twentieth century we find chemistry underlying nearly all aspects of man's endeavors. Its enormous range runs from the revolutionizing of many industries to the broadening, understanding, and extension of life itself. Indeed, chemistry, reflecting this universality of application, is rapidly absorbing or being absorbed into nearly all industry, that basic index of human life and activity. There is nothing strange in this. As man has become more adept at changing and molding matter to his use, which is industry, he inevitably has moved deeper and deeper into chemistry, which is the controlled transformation of matter through science or knowledge.

The growth of the chemical industry itself in only the first half of this century, based on this revolution in knowledge, has been something phenomenal. In the United States it has been growing at a yearly rate about double the average of all manufacturing industry. From being only a minor part of all industry at the turn of the century, the chemical industry today has suddenly become the most dynamic, if not the premier, industry of the century. It is so big and expanding so fast in so many directions that its boundaries are difficult to place at any given moment. If the chemical industry is confined simply to the basic producers of chemicals and allied products—which is the core of the industry that orderly statisticians find most manageable—it ranks about fourth in size, preceded only by petroleum, primary metals, and transportation equipment, in that order. This is one of the sharpest ascents of any new industry in history. If petroleum refining is included, a process merging more and more indistinguishably into chemistry year by year, then this total chemical industry is unquestionably the new colossus.

But even this chemical-petroleum combination does not wholly cover all the permeations and penetrations of chemical industry into the broad reaches of the industrial scene. For instance, the steel, copper, zinc, and other ancient metal industries are on their way—through a new infusion of chemical research into the basic chemical elements, processes, and alloys with which they deal—to becoming strongly based metallurgical chemical industries. This transformation is likely to be seen completed well before the century is out. And the electrical-electronics industry is already deep in chemistry, developing and manufacturing many of the special materials, plastics, metal combinations, and ceramics upon which a steady stream of new, improved, or, in some instances, revolutionary products is based. Altogether, if all the products created by chemistry could be traced, they might be found to constitute well over 30 per cent, if not close to half, of the total output of goods in the country. Moreover, in ever-mounting volume, the basic chemical industry feeds materials and semifinished products into all sixty-eight industrial categories into which the U.S. Department of Commerce and economists generally divide the total industry of the nation.

To be a chemist, therefore, is to be at the very center of a driving new force expanding the economy of life and of the country. Moreover, the chemist has a large hand in some of the vital developments for the future. It is characteristic of this science and this industry that it engages in or supports more research, stretching from all manner of industrial laboratories into the universities, than almost any other category of science or human endeavor. In sheer numbers, chemists today constitute the largest scientific professional group in the country, exceeded only by medical practitioners. Yet the role of chemistry is expanding so rapidly that these numbers are judged to be not nearly enough. It is estimated that unless many more chemists and chemical engineers are created by 1965, there will be a deficit of 93,000 in meeting the needs of the chemical industry alone, not to mention the other growing areas in which chemists are needed.

The reasons for this large, urgent, and even fantastic-sounding need for more chemists and technologists, particularly of the creative variety, lie deep in the many revolutions that the chemical industry has under way in the most vital areas of human interest, and the many more revolutions that are on the verge of opening out into the future.

Consider first those basic areas of food, clothing, and shelter that underlie the life of man on earth. In the production of food, chemistry has had a big and still growing hand in an agricultural revolution that may well be a leading factor in the continuation of life on earth. A good part of the remarkable rise in yield per acre, true of nearly all United States crops, must be credited to farm mechanization and to the plant geneticists' creation of new, constantly improved plant varieties. But a large part of this jump in productivity also is to be credited to the endlessly mounting new fertilizers, insecticides, fungicides, weed killers, and growth agents which, cutting crop losses and labor, have brought a chemical revolution to the farm. As biochemistry unravels more of the inner secrets of plant and animal life, more and more subtle chemical products are in the offing that will build resistance to disease and attack right into crop structures and, beyond this, control growth itself to almost any desired limits.

Chemistry's success, along with other factors, in thus helping to raise United States farm productivity is already so great that it has brought on continuing large and embarrassing surpluses, despite all efforts at economic or political control. But this is only a local and passing situation in a world still largely underfed and expected to double in population before the end of the century. By then, if it is not already long overdue now, the greater dissemination and adoption of the knowledge, materials, and techniques accumulated and proved in this country will be imperative to human survival on the planet.

In clothing and textiles, of course, chemistry is in the midst of one of the most glamorous and far-reaching of its revolutions. Synthetic and semi-synthetic fibers, taking in all types of man-made fabrics, today constitute one-quarter of all United States textile production. In little more than another decade, at their present rate of growth, they will account for close to half of all production. Actually, for the first time in history, the man-made materials are beginning to make inroads on that king of natural fibers, cotton, which in turn is being modified by more and more chemical surface treatments, adding such desirable new characteristics as crease-resistance and the ability to be "drip-dried," in order to retain its markets. Meanwhile, still newer synthetic fibers are steadily being developed to add to the spectrum of textile properties, a spectrum which in its breadth, versatility, and blending qualities is something new on earth. This, too, has its relation to the world situation, for, as population pressures increase, the pressure for more food-growing acreage will inevitably squeeze the production of natural fibers, so that to clothe itself the world must turn more and more to the products of the test tube and chemical ingenuity.

In shelter, the chemical revolution has only begun, though about 15 per cent of United States chemical production already is going into the building industry in such important materials as paints and surface coatings, adhesives, plywood, insulation, floor tiling, interior finishing, furnishing, trim, and plastic paneling. Plastics alone in building now account for over $500 million a year, in such items as lighting fixtures, hardware, plastic pipe, roofing and glazing materials, but this is still only a drop in the immense construction bucket. As plastics gain in experience, development, and economy, they will be ready to move into major structural uses, still largely untouched. The all-plastic bathroom, for instance, molded in a single unit with all-plastic fixtures attached, appears to be just around the corner. And other integrally designed structural units, such as whole wall and roof sections, sandwich-panel members, and ingenious combinations of metals and molded plastics, promise new advances in prefabricated building construction.

The great moldability of plastics, and the almost endless range of different properties that can be built into them through molecular engineering—a range constantly being strengthened and added to by research into materials able to stand up to such extreme conditions as are encountered in high-speed aircraft and rocket flight—portend a structural revolution in human shelter that will take many forms.

Still, all these things do not by any means exhaust the revolutions under way through chemistry. Perhaps the most portentous of developments for the future are those revolving around the tapping or transformation of new energy sources and the betterment of human health, the already visible extension of life on earth.

In the field of energy, where again the populous world is moving toward inevitable stringencies unless new sources can be unlocked economically and usefully, chemistry is pursuing many exciting lines of development. On chemistry's back shelf are such processes as hydrogenation or partial oxidation of carbon compounds, which, by various routes, can turn coal, peat, lignite, or a large range of lighter organic waste materials, such as certain farm wastes, into gasoline and other liquid fuels to keep the wheels moving as soon as petroleum resources begin to falter.

There are also a whole new group of strange metal combinations and intermetallic compounds in development that have the property of transforming heat or light from whatever sources directly into electricity, at generally low efficiencies as yet, but steadily rising. Some of these may hold a coming revolution in self-powered appliances, such as radios, television sets, and other small household equipment. Others will eventually be used to convert the enormous heat of atomic reactions directly and more efficiently into power, rather than go through the old roundabout route of steam turbines. Researchers at Los Alamos National Laboratory in New Mexico invented a so-called plasma thermocouple device which, through an ingenious combination of a metal (tantalum) and a gas (cesium vapor), converts heat to electricity at a theoretical efficiency of 30 per cent or better. This would be equal to some of the present steam turbines.

Perhaps the most far-reaching of the energy revolutions ahead lies in the field of solar power, in which intensive work is going on in many directions, most of them employing chemistry or chemical materials as the unlocking key. The solar-heated house, already experimentally in existence with chemical heat collectors and heat-pump devices, is probably less than a decade away as a practical matter. Further off, but growing more feasible by the year, are methods for chaining the biochemical process of plant photosynthesis to the production of heat, light, or energy through such routes as the electrolysis of water to cheap hydrogen. The hydrogen would then be fed into chemical fuel cells, already in forward development, where, by oxidation or catalytic processes, the hydrogen would be converted to heat or directly to electrical energy. Thus, long before conventional fossil fuels run out, chemistry and nuclear physics will be ready to supply energy from a great variety of sources more limitless than any heretofore.

The deepest of all chemical revolutions is taking place, however, in the biochemical laboratories of the universities and the pharmaceutical industry, where the fight to control diseases and extend the human life span has marked up so many recent and well-known victories. Pharmaceuticals, indeed, constitute one of the most notable examples of an ancient industry almost completely revolutionized by chemistry within the last quarter of a century. From being an industry based largely on the extraction of drugs and nostrums from botanical and other natural sources, pharmaceuticals now compose an industry based strongly on creative chemistry, with research programs exceeding those of almost any other industry of equal size. The success of this transformation of medical science through chemistry is written in the bluntest of vital statistics. In only the last fifty years, for the first time in recorded history, the average of human life expectancy has been raised, in countries cultivating the new science, from 42 years to about 70 years, a figure composed of an average life expectancy of 67 years for men and 73 for women.

Nearly all this heart-warming record has been achieved by a panoply of new chemical drugs, antibiotics, and vaccines for beating back the infectious diseases, the diseases caused by invading organisms, bacteria or viruses. Except for the more elusive viral diseases, on which a great deal of work still remains to be done, nearly all the once common diseases that took off so many men in their youth are now largely conquered or held closely in check. The remaining area to be vanquished, and a large one in which the broadest and most exciting research attacks are going forward, is the area of the so-called organic diseases, products of still largely obscure changes or maladjustmnets in the complex biological mechanisms of the human body itself. These are mainly the diseases of middle and old age, of which the major ones are heart diseases, cancer, mental diseases, and senility itself.

If these organic diseases can be brought under any large measure of control—and there is no indication that they cannot—the human life span may be usefully lengthened, according to all present theoretical calculations, to an average life expectancy of 110 years. And this may be accomplished by the year 2000, if the scientific research attack is deep and broad and well supported. Or at least that is now the exciting goal.

This quickening of prospects has come about through the finer and finer explorations into the body's cell structures and chemical mechanisms, and into the origins and generation of life itself. Every advance into life's hidden recesses, in nearly every direction in which investigators have probed, illustrates ever more clearly that the secret of health or illness lies in the extremely delicate balance of complex, interlocking chemical systems that make up the life processes, processes which, with more knowledge, chemistry may come to control.

Chemical factors more complex than a single protein may be found to be behind schizophrenia and other true brain diseases, but it is now generally accepted that research is on the right track. "No twisted thought without a twisted molecule," is the way a noted brain biologist, Ralph Gerard, has succinctly put it. And what is chemically caused may someday soon be chemically changed. It may take a long and arduous effort to trace the cause of these crooked molecules back to their source, for the chemistry of the blood, brain, and nervous system is exceedingly complicated, but that a solution will eventually be found to the age-old mystery of most mental illnesses and the human brain is now as nearly certain as that their causes are physical, and therefore graspable by the tools of science.

Some evidence of chemistry's ability to deal with such matters is already at hand in the pharmaceutical industry's recent development of the so-called tranquilizing drugs, based on the rediscovery by Western medicine of an old East Indian remedy made from the root of a native plant, *Rauwolfia serpentina.* From this starting point, the industry has gone on to synthesize, by chemistry's now well-known molecular juggling, a whole group of related, steadily improved, and more potent tranquilizers, which have had some marked success in quieting violently psychotic patients, alleviating mental disturbances, and treating many other ills in which emotional tensions are a factor. Conversely, the industry also has developed a group of so-called energizing drugs—derived oddly but not inappropriately from the rocket fuel, hydrazine—which have had some remarkable results in bringing deeply depressed, withdrawn schizophrenics to life and making them more amenable to treatment.

These psychiatric drugs open a whole new area of research, extending up into the very functioning of the human brain, and demonstrate forcefully that chemistry can be employed to good effect in dealing even with mental processes. Psychiatrists now foresee the day when such chemical agents, greatly improved and extended, will be used not only to cure the mentally deranged, but also to raise healthy brains to powers still dormant within them. And this, too, may be a necessary measure not unrelated to the continued survival of the human race.

In much the same way as seen in mental illness, another abnormal molecule in the blood stream has been found to be a major concomitant of the most prevalent of all so-called heart diseases, called atherosclerosis. This is the constriction of arteries by irregular fatty deposits on the interior walls of blood vessels, accompanied by high blood pressure or hypertension, the danger of blood-clot formations or thromboses, and other organic ills.

Meanwhile, though atherosclerosis cannot yet be attacked directly, pharmaceutical research has raised a whole army of new drugs to control or turn aside some of its deadly effects: a group of highly useful anticoagulant drugs to prevent formation of possibly dangerous blood clots, a steadily growing range of drugs to lower blood pressure, and new drugs that show promise of reducing the level of cholesterol in the blood. Without a doubt. the day is drawing ever closer when the root causes of so-called heart attacks, today's No. 1 killer of men in their prime, will be exposed and brought under control.

Cancer is even more complex than the heart diseases, but no less hopeful of eventual solution in the current strides of research. What makes it particularly difficult is that there are so many different complex forms of cancerous growth, many hard to diagnose, some not even readily identifiable as yet. The leading indication now is that many cancers are triggered by still unknown viruses, acting in ways different from known viruses, and probably in concert with certain malfunctions in the body cells' genetic and growth-controlling material, setting off that wild, uncontrolled reproduction of misshapen cells that is cancer. A few rare types of cancer recently have proved to be eradicable by specific antibiotics, indicating that the virus theory is not unfounded. But a great deal more needs to be learned about the intricate chemistry of the body's cells, their control and reproductive mechanisms, and the interplay of viruses with these materials, before enough is known to design the specific agents needed to control the range of cancers effectively.

As this research moves deeper into what essentially is the mystery of life itself—the living cell—the probability is that these studies will also have a profound effect on the understanding of old age and senility, and the aging process, which is the reverse of cancer—for old age without other complications is simply the running down of the cell's reproductive, growth, and repair facilities.

Nothing is simple in this area of life's chemistry, and achievements will not come easily. But in this direction lies the greatest of all challenges to the adventurous and inquiring mind, the bold explorer of the future. In this chemical research area arises the proposition that not only the gross material surroundings of man can be improved but that human life itself can be greatly improved in its inner being, both mental and physical.

All this opens new philosophic vistas of understanding. "The chemical descent of man," says Robert R. Williams, isolator and synthesizer of vitamin B_1, "extends his perception of kinship and his sense of the trends of evolution through far greater ranges than the anatomical evidence with which Darwin had largely to be content. It tells us where we have come from and, if we read it wisely and well, I believe it may tell us much about where we are going."

Part III

•

THE WORLD OF LIFE

A. *The Study of Life*

•

We have scanned the skies, wandered over the earth, penetrated the atom. We now examine "The World of Life." Our first selection deals with the environment to which such life must adjust itself. The earth, which is neither too hot nor too cold, which is surrounded by a suitable atmosphere and contains the necessary constituents, is extraordinarily well fitted for life as we know it. Or to rephrase the statement more precisely in accordance with evolution theory, life has managed to adjust with great success to the earthly environment. Paul B. Sears, the noted ecologist, discusses the subject in The Fitness of Earth.

An important aspect of the study of biology is the organization of our knowledge about it. We must divide plants and animals into groups which have characteristics in common or we will be so overwhelmed by unorganized information that it will be virtually useless. In The Naming of Organisms and the Catalogue of Nature *Marston Bates elaborates on the need for such classification, introduces us to the Swedish biologist Carl Ingemarsson, who later adopted the name Linnaeus and whose basic system has survived for two hundred years, and describes the main divisions of the plant and animal kingdoms.*

Linnaeus did not come to grips with the distinction between life and nonlife, a question that has concerned natural philosophers and theologians for millennia. Historically, it has centered on the problem of the spontaneous origin of life. The initial scientific contribution to this problem was made during the seventeenth century by Anton Leeuwenhoek, the first to see those tiny "animalcules" which we call microbes. The more he looked, the more he found—in the tissues of a whale, the scales of his own skin, the head of a fly, the sting of a flea. He watched them attack mussels and so realized that life lives on life. It is doubtful whether he knew that life, under conditions as they now exist normally on earth, must always come from life or that microbes play a dominant role in disease.

Those were discoveries that were to take centuries and test the abilities of men like Spallanzani, Redi, Pasteur, Tyndall, Koch. We no longer think that eels develop spontaneously in stagnant pools, that kittens without parents spring from piles of dirty clothes. The idea is so foreign to us that we hardly believe men could have thought it possible. Yet the classic experiments which disproved once and for all the doctrine of spontaneous generation were performed no earlier than the last century by Pasteur and Tyndall. Pasteur showed that water boiled in flasks to which the dust-filled air was not admitted would never generate life. He showed that flasks opened in Paris contained numerous microbes, while those opened in the Jura Mountains contained few or none. "There is no condition known today," he wrote, "in which you can affirm that microscopic beings came into the world without germs, without parents like themselves."

Yet somewhere along the road, if we may believe the latest researches, life and nonlife seem to merge. One attack on the problem has been made by experimenters like Wendell Stanley. Working with filterable viruses, composed of molecules made up of millions of atoms far smaller than anything observed by Leeuwenhoek, they seem to have reached a point on the scale bridging the gap between the living and the dead. Are these viruses alive? It depends on our definition of life. By some standards they are, by others they are not.

There is still another recent approach to the problem, an approach which gives aid and comfort to the spontaneous generationists, but in a way never dreamed of by them. Workers like Oparin in Russia (the originator of the theory) and Harold Urey, S. L. Miller and George Wald in the United States have suggested, with experiments to back their theories, that under conditions as they probably existed when our planet was young, and with the assistance of electrical discharges such as lightning, spontaneous generation did occur during the long stretches of early geologic time. It is this attempt to "bring the origin of life within the realm of natural phenomena" which Professor Wald of Harvard discusses in The Origin of Life.

Although it is only in recent years that we have begun to have an inkling of the method through which life began, it has long been possible to observe life processes as they exist in nature. Biology as a science has concerned itself with structure, chemical change, action and reaction, adaptation to environment, reproduction, development and death. Karl von Frisch, the famous Viennese biologist, introduces us to this study in Of Cells, Life and Death.

PAUL B. SEARS

The Fitness of Earth

•

THE EARTH on which we live is one of nine major planets in our solar system. That system, in turn, is but one of an uncounted number in the astronomical universe. For this reason, and on the basis of probability alone, it is believed that the universe contains many planets enough like Earth to be favorable to the existence of living beings. For life as we know it is an expression of energy and the organization of matter which appears to require special conditions for its origin and survival. These conditions are definitely absent from most of the planets in our own system, uncertainly present on one other, Mars.

However interesting the possibilities thus opened up, our primary concern is with the planet on which we are living. Through the knowledge we now have and the uses we make of it, we have a large measure of responsibility for what happens on Earth, and consequently what happens to us. Man, through the power of his numbers, intelligence, machines and the command of energy that drives them, at length has become a geological force. It is essential that he not only know his own strength, but guide it wisely. He has already made some of his surroundings less livable than they once were, supporting fewer people and affording less opportunity for satisfactory existence. Thus in parts of Asia Minor, where agriculture began between ten and fifteen thousand years ago, great civilizations have flourished. But there the slow deterioration of the irrigation works, and their final destruction by invaders in the twelfth century, have ended one of the great adventures of mankind.

Treasure is best protected by those who understand its value. What are the peculiar advantages of Earth for us? In essence they are matters of its physics and chemistry, and of subsequent developments which these properties have made possible. We are indebted to the mass of the earth, its distance from the sun, the states of matter here, and their chemical composition. Conceivably some or all of these conditions might vary considerably and still sustain life, but not our own.

If the mass of earth were as great as that of Jupiter, its gravitational force would surround it with an atmosphere too dense to permit the sun's energy to reach the surface of the planet directly. It is this radiant energy which, on our own planet, is stored by green plants in the form of carbon compounds that sustain all life. Astronomers tell us that, in fact, the atmosphere of Jupiter is a kind of mush, far more opaque than the densest storm clouds that sometimes darken our own surface.

On the other hand, if the mass of earth were much less than it is, it would be unable to hold, by gravitation, such an atmosphere as we have. Not only does our own gaseous envelope contain chemical elements—oxygen, carbon and nitrogen—that are essential to the kind of life we know, but their physical condition as gases immensely increases the ease with which they can be moved in and out of living bodies and take part in necessary chemical changes. One of the most critical of the many problems in space travel is that of devising a safe and sustained gaseous environment for men who are sent outside of our atmosphere. Any explorer of the moon will have to take his atmosphere along with him.

Again, the distance between earth and sun is a most favorable circumstance. The temperatures thus insured are generally those which enable water to exist in the liquid state—neither too hot nor too cold. By comparison the surface temperature of Jupiter (which helps account for its mushy atmosphere) is about a frigid −129° C.

The revival of a wilting plant when given water shows how important liquid water is in maintaining form in living cells, while our own experience of recurring thirst is further evidence that water is constantly needed to sustain life. Not only is water an essential part of living matter, but its power to dissolve other substances and allow them to move about as invisible particles in solution makes possible an infinite number of the chemical reactions so necessary to the continuing existence of living organisms.

This is not all. Water is chemically neutral. It can also pass into the form of gas or vapor at temperatures below its boiling point, the rate depending upon atmospheric conditions. As vapor it can be lifted from sea and land, be transported in the clouds above us and be redistributed in the form of rain or snow. Flowing back to the sea from higher ground or soaking into the land it becomes available to sustain vegetation and animal life, while its presence as vapor in the air helps prevent too rapid evaporation from living bodies. On its way to the oceans water becomes the active tool of gravity, molding earth's surface by dissolving, carving, moving and depositing rock particles.

It is true that during what we call glacial ages a great amount of water is converted into masses of continental ice, lowering sea level some hundreds of feet. But even then much of the earth remains free from ice, and at temperatures that enable plant and animal life to carry on. The moving glaciers themselves shift and mix rock materials of varying composition, often leaving a deposit of highly fertile soils when they retreat as they have in our Midwest. They also leave numerous lake basins that still furnish us with fishing and other forms of recreation. From Minnesota to Maine and northward through Canada these lakes contribute to the economy, an advantage broadcast on the slogans invitingly placed on more than one state's car license.

Since water heats and cools much more slowly than solid earth, water is also an important stabilizer of temperature. On lands near to oceans or large lakes extremes of hot and cold are tempered. Such climates do not show the abrupt and violent contrasts between day and night, winter and summer temperatures that are found in the drier continental interiors. These and other benefits result from the presence of water on this planet and of temperatures that enable so much of it to remain in its liquid state. Yet familiar as it is, we still have much to learn about this marvelous substance, so important a part of our environment. Despite its abundance, one of our gravest concerns has been and still is to insure an adequate supply where and when we need it.

One of the distinguishing characteristics of living organisms is their ability to draw upon environment for chemical substances, break them down and remold them into new combinations that serve the business of survival. There are, generally speaking, two kinds of chemical change—one that requires energy and stores it in the resulting compounds, the other that releases such stored energy. The manufacture of sugar in cane or beet leaf is an example of the first, the use of a candy bar by the mountain climber an example of the second. Such reactions depend upon a ready supply of the necessary chemical elements present in the materials around us as well as upon the energy that flows continuously from the sun. For example, two elements, the gases oxygen and hydrogen, are combined to form the water just discussed, and both are used in and by living substance.

Another invaluable element is carbon. Present in our atmosphere as less than half (by weight) of the .3 of 1 per cent (by volume) of the carbon dioxide found in the air, it nevertheless furnishes more than half of the dry weight of average plant material. It was this paradoxical fact that so much solid material could be derived from an invisible gas that delayed our understanding of plant nutrition until after the discovery of oxygen and the consequent beginning of modern chemistry about the time of the American Revolution.

Since we know carbon as charcoal, graphite and diamond, we suspect that it is an element with some interesting, perhaps unique, properties. This is true. Its atom or chemical unit has four bonds, compared with the two of oxygen and one of hydrogen. Thus this four-armed carbon atom can join with four of hydrogen, as it does in methane, or two of oxygen, as it does in carbon dioxide. Also it can link up with other atoms of its own kind in chains of varying length or form a ring like youngsters on the playground. But because the carbon atom has four "arms" instead of two it can form side chains and patterns of the most varied shape. When this happens many links are left free to pick up other elements, or groups of them.

Not surprisingly, then, enormous numbers of carbon compounds, or, as they are often called, organic compounds, can and do exist, and their variety is intimately associated with the variety of plant and animal life and what goes on inside of living things. More than this, energy is required to put carbon compounds together and the energy so used is stored within them. Under proper conditions, usually when oxygen is present, as it is in the burning of a log or the breathing of an athlete, this stored energy can be released as the carbon compounds break down. It is in this way, rather than directly from sunlight, that plants and animals get energy for the work they do in the business of living, the solar energy having first been stored by green plants in the form of carbon compounds.

Fortunately again, as we consider the fitness of our planet, oxygen is present abundantly in the air and is constantly being released by green plants as they build carbon compounds out of carbon dioxide, water and minerals from the earth. These simple materials are in turn being given back to the reservoir from which they came, released by plants and animals when they, in the course of living, break down the compounds that give them energy.

Although more than a hundred elements are known, some twenty at most seem to be necessary to sustain life. Had any of these been lacking, perhaps living organisms might have "learned" to use others instead. In theory silicon might have served the purpose now served by carbon, and sulfur filled the role of oxygen. But from what we know, the arrangement might have been considerably more awkward than the one we have. To say the least, the chemical substances present in earth, air and water have proved highly usable in the processes that keep life going.

While we rely upon physiology and biochemistry to explain the roles played by various chemical elements, the fact that they are essential can and should be understood by all. The wise farmer knows that he must supply his fields with materials to make up for those he sells as grain or meat. He limes his land with calcium in the form of crushed rock. The bags of fertilizer he buys are labeled with numbers such as 5-10-5 or 10-5-10, to show the proportions of nitrogen, potassium and phosphorus, respectively. Often they bear the legend "Trace Elements Added" to show the presence of necessary elements needed in very small amounts. A few ounces of borax to the acre may determine the health of the clover crop. But our point now is that what plants and animals need for living they must get from earth, air and water.

Needed materials are not, however, equally abundant everywhere. Nor are they always directly available, even though present. All life must have nitrogen for without it there would not be protein substance, and living stuff itself is largely protein. Compounds containing phosphorus carry on much of the intricate chemical work in living stuff. Yet many soils are formed of rocks which contain neither.

Now there is plenty of nitrogen in the air—some 80 per cent in fact. But nitrogen is a stubborn element, not easily formed into compounds that can be taken in by green plants and so built into proteins for plant and animal to use. Some measure of the energy required to form nitrogen compounds—and therefore stored in them—can be appreciated by recalling that some of these compounds are high explosives.

However, simple nitrogen compounds can be formed by lightning discharge and so reach the earth in rain. More important, there are some very small and ancient forms of plant life that "fix" nitrogen, as we say. Presumably this process was in full swing long before the development of higher plants and animals, making possible their activities. So we must add, in speaking of the fitness of environment on earth, the early appearance of simple nitrogen-fixing microorganisms. Incidentally, the seas were the first cradle of life and it must have been here that this process was first started. Sea water today contains all of the elements necessary to living organisms, curiously enough in about the same concentration as the salts in our blood.

Without discussing all of the score of elements needed by plants and animals, a word about phosphorus is in order. Although present in the oceans, this element is absent in many of the rocks from which soil has been formed, yet found, though not always abundantly, in most soils. The likeliest explanation is that in the millions of years before fences, railroads and other barriers existed, phosphorus was distributed by the movement to and fro of animals. These animals, in turn, had obtained it from feeding where it was abundant or from mineral springs or salt licks to which they resorted in great numbers. Great quantities of phosphorus were also brought inland by ocean fish migrating upstream to spawn and die.

Thus life has, throughout the ages, enhanced the capacity of the environment to sustain life by making available needed mineral nutrients. But the service of life to life goes further. Very early in the game there developed types of microscopic organisms that lived by breaking down the remains and waste products of other forms of life, so that the substances that had been built up out of the raw materials of earth, air and water could be returned to the source whence they came and so be recycled, that is, used over and over again by succeeding generations.

Within and without the living body chemical processes are essential to life. Chemical change is an inseparable part of the business of living. Indeed it was that circumstance which was involved in Priestley's discovery of oxygen and the birth of those unidentical but inseparable twins, modern chemistry and physiology. Presently it became evident that measurable energy is involved in chemical process and, since energy is one of the most familiar manifestations of life, the bond between the physical and biological sciences has become a permanent one. Modern techniques now enable us to probe deeply into minute and delicate physical and chemical processes within living bodies, and even to conjecture reasonably as to how such processes may have led, through vast reaches of time, to the origin of life itself.

So fruitful and challenging are the problems thus opened up that they overshadow, for the time being, the larger-scale and no less important aspects of biological science, especially those which give it continuity in time and space, and which deal with the organism entire in its form, behavior and relation to the outer world. It is these larger aspects which have raised the problems for and still give perspective to the brilliant detailed research that has become possible.

No consideration of the fitness of Earth is complete without thinking of time and continuity of process. One of the marks of Darwin's greatness was his discouragement at Lord Kelvin's calculation of the age of our planet. On the basis of heat change as he understood it, the great physicist decided that the earth could not be much more than about twenty million years old. Darwin knew that this was not sufficient to have permitted the development of life into its present highly organized expression, nor was he ever reassured. Had he lived until today, he would have been. The techniques of measuring radioactivity in rocks give the earth an antiquity of three—perhaps even five—billion years and assign an age of at least half that time to the oldest known fossils. Time there has been, enough and to spare.

And because the earth is so old, the time available for necessary changes is no less a part of its fitness than the favorable conditions already discussed. Thanks to time, kinds and patterns of life ever more favorable to life itself have developed. Life has come to participate increasingly in geological process, giving rise to concentrations of certain mineral ores and such sources of energy as coal, oil and gas. Also in the course of time great cycles of activity that renew, replenish and stabilize the environment have become established. Without the circulation of water, purification of the atmosphere and formation of soil, for example, Earth would be far less capable of sustaining life than it now is. And so far as man is concerned, he runs perhaps a greater risk in disrupting such great natural cycles than in depleting the materials he needs.

Measured in Earth-time, man is a parvenu, dangerously combining power with responsibility for its use—dangerously yet hopefully. He emerged onto a scene populated with warm-blooded vertebrates, staged amidst a vegetation of seed-bearing plants that sustained them. Not much knowledge of previous evolution, nor much imagination is required to picture his chances in an earlier world, even had such a biological absurdity been possible, for he too is a warm-blooded vertebrate.

The animals he found furnished him with convenient food, service and companionship, while he, like other animals, found sustenance and materials from vegetation. Meanwhile, the abundant survival of earlier and simpler organisms such as bacteria, fungi and invertebrates added to the fitness of his surroundings. Some of them, the yeasts, mushrooms, various insects and shellfish, for example, served directly to supply nourishment. Others, as we have already noted, served as middlemen to insure the breakdown and reuse of materials. Wherever this process stops, as it does on unfertilized or sterile land, production slows down or stops. Still other forms of life added, it is true, to his difficulties. Some became his rivals for food or space; some, by poison or appetite, his enemies. Others produced disease and shortened his life span. But the end result—his survival and increase—is eloquent proof that the general picture was in his favor.

There were other benefits, too. We have only a few dim and slit-like windows into the long past of the landscape, but so far as we can judge from what we know, the plants and animals which occupied it at the time of our advent (and since, for they have not changed greatly) form a more varied, flexible and closer-knit pattern than those which preceded them. Be that as it may, the *system* of life had become beautifully integrated with environment and, in general, tended to produce maximum effectiveness in utilizing the energy and materials available to it. The short and ready proof of this is in the high fertility of virgin soils and in the oft-mentioned "balance of nature."

Darwin noted the fact that in nature the populations of organisms remained fairly even in their proportions through the years. And Dr. and Mrs. Fairfield Osborn, newly returned from a visit to the great African wild-life preserves, were profoundly moved by the relative peace and harmony —almost Eden-like—among hunter and hunted. Lions and their prey lived side by side, the latter no more in panic than the steers of our pastures who face a certain end in the packing house, or the poultry in an old-fashioned chicken yard who have often seen their companions dispatched on the chopping block.

Among the many phases of biology, the science of life, is one called ecology. Much, perhaps most, of the work of the ecologist centers about the study of living communities of plants and animals—what is in them, where they are, how they develop and change, and how they have come to be what and where they are. One cannot draw neat lines around such problems as these as he can in mathematical and experimental problems. In one direction they lead to the individual organism, its requirements, its equipment and its adventures in surviving. Carried still further this path leads down into the most minute and intricate details of living things.

In another direction—or shall we say outward—the ecologist moves into the study of the nonliving world of earth, air and water. He concerns himself with the differences between communities and tries to explain their differences in space and time. He becomes geographer and historian as well as climatologist and geologist. Obviously he cannot master all of these fields, but he must know enough about them so that he knows where to look for help when he needs it. He is usually a pretty good specialist himself in one of the relevant fields, but in other respects he is like the general practitioner in medicine, obliged at times to consult specialists in various fields.

So numerous and varied are the problems in ecology that many workers—foresters, fisheries experts, students of population, even some economists—are really ecologists under another name. Even scientists working on the most minute and exacting laboratory experiments can be reckoned as ecologists, providing they are aware of the importance of what they are doing in relation to the whole scheme of nature. On the other hand, there probably are some students of living nature who have little or no perspective with regard to what they do. They may accumulate useful information, but it is hard to consider them as true ecologists, no matter what they may call themselves.

MARSTON BATES

The Naming of Organisms and the Catalogue of Nature

•

I

CHARLES ELTON has remarked that there is little use in making observations on an animal unless you know its name. The first step in a survey of natural history, then, should be the acquisition of some familiarity with the system of names and the system of classification, with the word equipment used by naturalists.

Many animals and plants have vernacular names that everyone learns in childhood, or that form parts of special vocabularies such as those of farmers, woodsmen or hunters. It is surprising, though, how quickly we exhaust this supply of names. It works well enough for large mammals such as bobcats, deer, foxes, raccoons. But if we start to make observations on field mice, we soon find that there are no common names for all of the kinds that we find; in the ordinary course of events, these different kinds of field mice simply don't come to our attention.

Vernacular names have a limited usefulness for biological purposes. In the first place, they are available for only a few kinds of organisms. Second, they are apt to be very local. Some names have gained wide currency in English because of literary practices, so that we do not realize how regional a vocabulary can be. Even in English an animal like the puma may have quite a list of aliases—catamount, cougar, lion, panther. In some countries all of the names for common animals and plants may change from valley to valley. Third, at very best the names are limited to one language, and science must necessarily strive after internationalism in its vocabulary as well as its ideas. There are millions of different kinds of organisms, and the invention of parallel millions of names in each language is unthinkable. Fourth (I think this list is about long enough), it is hard to give a vernacular name a precise meaning. Words like oak, or pine, or rabbit, may cover a variety of things, and we can't always straighten this out with adjectives like white oak or cottontailed rabbit. We often (perhaps better, usually) discover inconspicuous but significant differences that would make it necessary to name the Florida cottontailed rabbit, or Wilberforce's eastern white oak, and in the end we would find ourselves completely and hopelessly tangled up in our vernacular vocabulary, which started out looking so simple.

Biological nomenclature forms a beautiful system. I doubt whether it is fully appreciated even by the average biologist, who is apt to be irritated by the trivial inconsistencies that turn up through daily contact.

The Linnaean System

Biological nomenclature is the invention of Carolus Linnaeus. Linnaeus, a Swede, was born in 1707 and died in 1778. He was the eldest son of a peasant, Nils Ingemarsson, who became a pastor and who, with this rise in the world, adopted Linnaeus as a family name from a huge linden tree that grew near his home. Carl apparently even in childhood showed a great interest in plants, which in those days meant mostly medicinal herbs, and he resolved to study medicine. He was very poor, and probably would have been unable to survive at all at the university of Upsala except for the help that he early received from Celsius, the dean. Linnaeus must have been a very engaging fellow, since all through his life he received the admiration, friendship and sympathy of his fellow scientists. Perhaps he had that quality of infectious enthusiasm. Certainly he must have been an enthusiast, since otherwise he could never have carried through the tremendous volume of work for which he is responsible.

He received various grants and honors at Upsala, but in order to obtain the degree of doctor of medicine, he had to leave Sweden. So, with funds advanced by his future father-in-law, he went to Holland where he stayed for several years. With the assistance of patrons in Amsterdam and Leyden, he published the first edition of his great book, the *Systema naturae*. This was in 1735, when he would have been twenty-eight years old. He published other books in fairly rapid succession. He returned to Sweden in 1738 and continued to live there the rest of his life, refusing honors and appointments abroad. He was, however, held in high esteem in his own country, where he had the professorship first of medicine and then of botany at Upsala, and where he was given noble rank. He had large numbers of students whom he sent on collecting expeditions to various parts of the world.

Linnaeus divided the Empire of Nature into the Kingdoms of Animals, Plants and Minerals. The Kingdom of Animals he divided into six classes: the mammals, the birds, the amphibia, the fish, the insects and the worms.

The classes were in turn divided into orders, the orders into genera, and the genera into species. Linnaeus went through the material of all of the museums available to him, studied all of the books on animals and plants that he could get hold of, and gave a name in his system to every kind of organism thus known to exist, with a brief description of its characteristics, and with page references to the places where it was mentioned in the various natural history books. Even aside from the invention of the system of naming, this represents a remarkable achievement of industry, of critical evaluation, by a brilliantly systematic mind. His classifications have, of course, undergone many changes over the years: but after all, his great contribution to science was the invention of a system that permitted orderly change with the constant increase in knowledge.

The Concept of Species

The basis of biological names is the *species*. A species is a kind of animal or plant. Nowadays we like to think of a species as a population which includes all of the individuals that, in the natural course of events, could mate or are likely to mate with one another. All of the bizarre varieties of domestic dog still constitute one species, because they all belong to a general inter-mating population, as everyone knows who has tried for a few days to protect a bitch in heat. Red foxes and timber wolves do not look more different than many varieties of domestic dog; but red foxes do not normally mate with timber wolves, hence they are different species.

I shouldn't have got started with dogs, because they are a rather special case. No one is sure what wild dog, or which wild dogs, formed the ancestry of our household pets, and domestic dogs can be crossed with various kinds of wolves in different parts of the world. The important point is that they do not habitually cross, so that in actual fact we have a number of reproductively isolated populations, which we can safely call species.

As long as we stick to one place, we can be fairly sure about species. If, around our town, we have three kinds of mice and two kinds of squirrels; if we observe that there aren't any or very many intermediates among these mice and squirrels; and that they have somewhat different habits as well as a different appearance, we can be pretty sure that we have three species of mice and two of squirrels. If the males of one kind of squirrel took to chasing after the females of the other kind with any frequency or with any success, the place would soon be overrun with a lot of intermediates so that our two original kinds would no longer be easy to distinguish—in fact, they wouldn't be two species.

The naturalist is most apt to get in difficulties when he has a squirrel from Utah and another one from Massachusetts. Maybe one has more gray in his tail than the other, a few more whiskers, and slightly bigger ears. Now, are these two species, or are they varieties of the same species? You can't always get a live male from Utah and a live female from Massachusetts to try it out; and anyway, behavior in captivity is rather uncertain. Sometimes animals that wouldn't touch each other with a ten foot pole in the wild will carry on scandalously if they are put in a cage together. Other animals that have no inhibitions out in the woods take monastic vows as soon as they are put in a cage. So it isn't easy to use the experimental method.

What the naturalist actually does is to try to get a lot of specimens of squirrels from country in between Utah and Massachusetts. If the change seems to be gradual, the tail getting grayer, the whiskers more luxuriant, the ears bigger as Massachusetts is approached, he calls his two squirrels subspecies (geographical varieties, or races if you will). If the change is abrupt somewhere about Missouri, he assumes that they are two species: that somewhere along the line they have had a chance to mix and have turned up their respective noses. If the naturalist can't get enough specimens to decide about this, he guesses, and some of his colleagues agree with him, and others don't.

The Grouping of Species

Names for species can hardly be handled unless they are classified into groups. The more natural the groups are, the easier will be the problem of handling the names (and the corresponding organisms). Just as individual plants and animals fall into discontinuous populations, species, so the populations fall into groups the members of which are more or less similar to each other in various ways. The problem is to arrange a hierarchy of groupings that will be both convenient and natural.

Linnaeus, as I said before, grouped his species into genera, those into orders, and the orders into classes. We have added several categories to this hierarchy. Among animals, we have now the species, the genus, the family, the order, the class and the phylum, sometimes with the tribe placed between the genus and family. Each of these may be further split with "sub" or "super" categories, so that we can have subfamily, family, superfamily, suborder, order and so forth.

The Genus

A group of similar species forms a genus. There is no rule about how a genus is formed and the genus doesn't correspond to any definite thing in nature—it and all of the higher categories are primarily conveniences, necessary filing systems for our accumulating information. The word genus keeps its Latin plural in English, so we write about various genera. Generic names are always capitalized, and the generic and specific names go together to make up the name formula for the organism. Take cats, for instance. The cats in general form the genus Felis. The house cat is Felis domestica. Like all of our domestic animals, its origin is a little uncertain, but it is supposed to be descended from the Egyptian wild cat (Felis ocreata) and the European wild cat (Felis catus), various breeds perhaps having different proportions of each ancestry.

Keeping Felis domestica, F. catus and F. ocreata separate is pretty much merely a convenience if the first is a mixture of the other two. But biologists are not always consistent, and the separation of these three kinds of cat really is a convenience. The domestic animals, in any case, present rather special nomenclatorial problems. The lion, Felis leo, and the tiger, Felis tigris, are better examples of species. The American jaguar (Felis onca) is another. The North American bobcat is generally placed in another genus, and called Lynx rufus; the cheetah (Cynaelurus jubatus) is another example of a cat that is usually considered not to belong to the genus Felis. All of these cats, though, are put together in the family Felidae, which takes us another step up in the hierarchy.

The genus and the species form the basic name formula of an animal or plant and these two names are all that is needed to look up what has been written about the organism (if the library is big enough). This is accomplished by two rules: that a particular name can be used for a species only once in a particular genus; and that a particular name can only be used once in the animal or plant kingdom for a genus. A word used for a genus of plants can be used again as a generic name in animals, and there are a few cases; but since it is usually fairly obvious whether a plant or an animal is being discussed this causes a minimum of confusion. It is customary to capitalize the generic name but not the specific name, though some botanists capitalize specific names if they are proper nouns.

A genus may have any number of species. Sometimes a genus has only one species, if that particular kind of animal or plant has no close relatives. Thus man is put in a genus by himself among living animals, though various fossils have been described as other species of Homo. Some genera have hundreds of species: groups of animals or plants where there are many slightly different kinds, some perhaps quite different from others, but with such gradual connecting links that it is most sensible to include them all together under one generic name.

Families and Higher Groups

Families, groups of genera, are also sometimes very big and sometimes very small. Among animals, family names are always made by adding the letters "idae" to some genus name. Thus the dogs are Canidae; the cats, Felidae. In plants family names are usually formed by adding "ceae" to a genus: Rosaceae for the rose family and Liliaceae for the lily family. Scientists are apt to differ considerably in defining the extent of families, and again it seems to me that the final decision should rest on convenience as well as on the expression of natural relationships. Families are rarely covered by words in the common language. It is necessary to refer to the Canidae, for instance, as the "dogs and their relatives," the Procyonidae as the "raccoons and their relatives," the Liliaceae as the "lilies and their relatives" (including the onions).

Groups of families make orders, and orders are more apt to be covered by common words than are families. Thus the beetles and flies represent orders of insects, the primates, carnivores and marsupials orders of mammals. Plant orders, curiously, are rarely recognized as discrete groups by the nonspecialist. Orders again are grouped into classes, and the distinction of the classes is often obvious enough so that the groups are recognized in the common vocabulary. Thus the classes of chordates (vertebrates) include the mammals, birds, reptiles, amphibia and fish. Insects form a class of arthropods. The major subdivisions of the animal kingdom, such as the chordates (vertebrates), the arthropods, the molluscs and the protozoa, are called phyla. The term "phylum" is less often used by the botanists.

The classification of organisms is built up from individuals. The specimens studied by the naturalist are grouped into species—which is already an abstract "concept," though one that corresponds to a fairly concrete phenomenon insofar as species are interbreeding populations. Genera are defined on the basis of a study of the characteristics of many different species, families on genera, orders on families, and so forth.

In using this classification, however, the naturalist works the other way. Given a particular organism to identify, the first question is, what phylum does it belong to? Is it a seed plant, a moss, a vertebrate, a protozoan, or what? The next question is, what class does it belong to? Which, among the vertebrates, means deciding whether it is a fish, amphibian, reptile, bird or mammal. Within the class, the order, then the family, then the genus, and finally the species must be determined. When these last two have been found, the name formula of the organism is at hand.

II

It is usually estimated that about a million species of organisms have been described and given names. No one can make more than a rough guess as to how much progress this represents toward the goal of getting all of the kinds of organisms named. In a few groups, like the birds, almost all of the very distinct kinds have surely been found and catalogued. In other groups, like some of the smaller and less conspicuous insects, only a small percentage of existing kinds have been given names. Charles Brues has estimated that there may be ten million species of insects alive in the world today, and that only a half a million or so of these have as yet been catalogued. Whatever the exact figures, it is clear that the job of naming is enormous. Biology has got swamped by the problem, and the end is nowhere in sight.

The catalogue of nature interests us, as naturalists, chiefly because it is a necessary tool. Without some such filing system, the accumulation and indexing of observations and experiments would be impossible. Imagine trying to keep notes that apply to one kind of thing out of a million or so, without any generally accepted filing system.

Plants Versus Animals

In forming a classification, we have to list our groups in a lineal order. This, inevitably, is arbitrary. We may compare the evolutionary history of organic types to a branching tree, the various sorts living today being the end result of this historical process of divergence, the final twigs on branches that have spread in different directions during the slow course of growth in geological time. To make a list of groups, we have to cut these branches and lay them out in a straight order, A, B, C, D and so forth, which may give a very misleading impression.

The basic fork in our tree, for instance, involves the separation of plants and animals, a divergence that started somewhere very far back in geological time. It is easiest to deal, in the list, first with all of the plant groups, then with all of the animal groups. But this doesn't mean that plants are "lower" than animals, or ancestral to animals. They are two forks of a tree that have got to be dealt with somehow in a list on a sheet of paper. The order on the list is a pure convenience, except that we try to group the branches in a way that reflects our idea of how the main trunks probably diverged on this hypothetical tree back in geological time.

The difference between plants and animals is taken as a matter of course with complex organisms—anyone can tell a horse from an oak tree. But with relatively simple organisms, the differences are not so obvious, and in some cases it is really impossible to decide whether a given organism should be called a plant or an animal.

We think of plants as fixed, growing in one position, and of animals as capable of various kinds of movement, of locomotion. There are many exceptions, however. Quite a variety of marine animals have fixed positions —corals, sponges, sea anemones, for instance. Moving plants are less common, but the slime molds creep, and some microscopic algae are very active.

The basic difference between plants and animals is one of food economy and habits. Plants are the key industry organisms on which the development of other forms of life depends. They can start with carbon dioxide from the air and through the magic of green chlorophyl build up starch and from this the complex molecules of carbohydrates and proteins, which animals must get by eating plants or by eating other animals that in their turn have eaten plants. As for habits, animals eat their food, while plants absorb their food in water solutions through cell surfaces. Even microscopic animals live by engulfing pieces of food, while microscopic plants lie in the sun and absorb the surrounding liquids. The exceptions to this are mostly parasitic animals that have taken to absorbing predigested food in the intestines or blood stream of other animals. It is largely this difference in food habit that leads biologists to include the great group of fungi, including bacteria, among plants, since these organisms do not possess chlorophyl, though they are capable of some astonishing chemical tricks.

Viruses

But if we start the filing system with structurally simple organisms, the first group is one that I have never seen even the most enthusiastic zoologist list among the animals, or the most grasping botanist among the plants. This group includes the viruses, things so small and so simple that no one understands them. So small that they cannot be seen through a microscope with visible light because they would fall between the light waves; so simple that it is possible to have long arguments about whether they are really alive at all or not.

W. M. Stanley was given a Nobel Prize because he was able to turn one virus—the cause of a plant disease called tobacco mosaic—into crystals, and to prove that his crystals were really the virus, the cause of the disease. Now many complex chemicals may be crystallized, but it is hard to imagine turning a living organism into a crystal, or a lot of crystals, and then turning it back again. Yet this tobacco mosaic undoubtedly grows, multiplies, reproduces itself, acts in most ways like a living thing.

All known viruses are parasites. Let us assume now that everyone knows what a parasite is—an organism that gets inside of another organism, or hangs on the outside, and makes it sick, or at least makes it want to scratch. The trouble is, we would have no way of catching a virus if it were not a parasite, because we can recognize the virus only by the symptoms that it causes. Virus particles can be photographed with an electron microscope, but it would have been impossible to get the stuff to make the picture if the virus hadn't made some animal or plant sick to start with. There may be all kinds of harmless viruses lying around in the mud, but we have absolutely no way of finding them, of distinguishing them from all of the dead contents of the mud.

Bacteria

Which brings us to the bacteria. The botanists always list the bacteria very boldly among the plants, calling them Class Schizomycetes of the fungi. The botanists, however, pay little attention to bacteria beyond listing them in the books; and the bacteriologists, who actually study bacteria, usually shrug off the problem of whether to call them plants or animals with a few paragraphs pointing out ways in which they might be classed as either, or neither. They are very special organisms that must be studied by very special methods.

Bacteria are defined as organisms built of a single cell, without apparent nucleus. Bacteria are of all sizes and shapes—though a microscope is still needed to see the biggest one. They may include various kinds of things that are not really related, and one group (the corkscrew spirochaetes of syphilis) gets pushed back and forth between the animal and plant kingdoms, each author having his own ideas on the subject. The bacteria as a whole have got a bad name because of a few kinds that cause diseases, but the vast majority are busy doing their microscopic but tremendously important part to maintain the economy of nature. Bacteria are the basic organisms in the process of rotting the corpses of dead animals and plants into simple materials that can be used again. Many kinds of bacteria have taken up partnership with big organisms, aiding in the digestion of things like cellulose in return for a safe and protected refuge in the intestine or in other parts of the big partner.

Perhaps the most important bacteria of all are the ones that keep the supply of nitrogen constantly going for the rest of the organic world. Some of these live free lives in the soil, others have taken up partnership with bean plants (legumes); each kind has specialized for a particular operation in the nitrogen chain, and the whole thing makes a fascinating story that has been fairly well publicized, though perhaps not well enough. There are other bacteria that can do tricks with sulphur and phosphorus; bacteria that get along very nicely without air; bacteria that are not killed even if they are boiled. There are bacteria everywhere; they are by far the most ubiquitous and numerous of organisms.

Fungi

Next come the fungi—mushrooms and a lot of poor relations. The poor relations have been coming up in the world, though, since one of the molds, Penicillium, got into the newspapers and the medicine cabinets. Bacteria are usually considered as a division of the fungi, which would then include all plants that do not have chlorophyl; other divisions of the fungi are the yeasts, molds, and various types that do not have definite group names in the common language. I have been looking up several classifications while writing this chapter, and I can't find any two classifications of the "lower plants" that agree in details of arrangement.

The phenomenon of sexual reproduction appears with the fungi. Knowledge of sex among the bacteria is scanty and indirect, but with fungi sex is well studied and has all kinds of complications, since the same plant may look quite different when it is in a sexual phase from the way it looks when going about its ordinary asexual business of rotting shoes. Yeasts are the simplest fungi, but even they present sexual problems—in fact the book open before me lists four different ways in which yeasts can multiply.

Algae

The algae are the simplest of the green plants, the plants with chlorophyl that can make starch out of air and water and sunlight. Almost all algae live in water, and every housewife knows the kinds that form a green film over the sides of a glass vase that has been left too long with water. Aquarium enthusiasts have rubber gadgets for scraping the algae off the sides of their aquaria so that they can see the fish. Even the clearest of natural waters have a considerable population of algae, all microscopically busy making starch, waiting to be eaten by microscopic animals that will in turn be eaten by somewhat larger animals, and so on up the scale of size. They make up, then, the pastures of our ponds and seas.

The algae, like the fungi, are of all sizes and shapes. They are normally divided into a variety of classes depending on their predominant color—the green, the yellow-green, the brown, the red and the blue-green algae. A great many are single-celled organisms, and some of them, because of their active movement and close similarity to the animal protozoa, are claimed by both the botanists and the zoologists. Some microscopic algae (diatoms and desmids) have beautiful symmetrical forms, like the patterns of snow crystals, and are favorite objects of the microscopists. In others the cells group together in various ways, first forming simple chains of cells, then forming more complex communities with different cells taking on different functions, until finally many algae form definite, obvious plants, the seaweeds, including the largest plant of them all, the giant kelp of the Pacific Ocean. These algae never, however, achieve the special structures of the seed plants, roots, stems, leaves, flowers, though certain parts of the plant may look like a stem, a root, or a leaf.

Lichens

I suppose lichens should be the next group of plants on our list. We all know them, making splotches of color on rocks, or forming tiny Japanese gardens on old logs. These lichens are always listed as a special class of plants, though in reality they are not a single class of plants at all, but a queer natural partnership in which the thing that we see is a mixture of two completely different kinds of organisms, of a fungus and an alga. Each kind of lichen is made up of a particular kind of fungus and a particular kind of alga. The fungus provides support, salts and water, while the alga carries on its business of manufacturing starch. This partnership results in a curious naming situation: each kind of lichen has a name as a lichen, yet it is composed of a particular kind of fungus with a name in the fungus system, and of alga with a name in the algal system. In some cases the algae and fungi are of kinds that can also live separately, but mostly the fungus, at least, cannot live without its alga. The partnership involves quite different types of algae and fungi in different types of lichen, and must thus be a very ancient agreement. Perhaps the different types of fungi discovered quite independently the advantages of having some kind of alga always close at hand.

Mosses and Liverworts

The mosses, with their cousins the liverworts, form the next great group of plants, called Bryophyta by the botanists. The algae are essentially aquatic organisms, and the mosses are the simplest green plants that have solved the problem of how to grow on land. Even with mosses, the problem has been only imperfectly solved, since they are pretty much limited to damp places. In this group the tissues of the plant have become differentiated into root and stem, and the liverworts have leaves, but these parts do not really correspond to the highly specialized root, stem and leaves of the higher plants. I haven't said anything about reproduction in fungi and algae, beyond mentioning that these plants get involved with sex, because the reproductive mechanisms are very diverse and complex. With mosses, too, reproduction is a complicated business of spore formation and alternation of sexual and asexual generations.

Ferns

With the next major group, the ferns, we meet plants in the conventional sense of the word. The ferns have solved the main problems of growing on land, but they still have a reproductive cycle reminiscent of the algae and fungi. They do not have flowers or seed, but produce millions of spores, of special cells, usually in little brown packets on the under side of the leaves. These spores are produced in tremendous quantities, so that a single plant may release fifty million or so in a season. The spores grow into tiny, scale-like plants, called prothalli, which produce sexual cells, the sperm cells swimming through water from dew or rain to reach the female cells, a lingering trace of the aquatic life of the simpler plants. The fern that we know grows from this fertilized female cell, so that there is an alternation of generations of sexual and asexual plant forms.

Ferns were dominant in the landscapes of the geological past, and there are many fossils in different kinds of rocks and in coal beds. Eight or nine thousand species have been found growing in the world today, some of them, in the tropics, making fair-sized trees.

Seed Plants

There is probably no use in writing very much about the seed plants here. We all know the more obvious things about them, remembered from high school botany or picked up at meetings of the Garden Club. My chief object has been to put the seed plants in perspective, by listing them with the other major plant groups.

The seed plants, which may be called phanerogams or spermatophyta, are divided into three classes, all with jawbreaker names, the gymnosperms, the monocotyledons and the dicotyledons. The gymnosperms do not have flowers in the usual sense, the seeds being borne on scale-like leaves—in a pine cone, for instance. These plants were abundant in the geological past, and the coniferous forests still cover a respectable part of the earth's surface.

From a grain of corn, a single leaf-like shoot comes out of the ground; from a bean, a nicely balanced pair of seed leaves first spread out. This difference is the key to the division between the monocotyledons and the dicotyledons. Both are divided into a series of orders and numerous families, the monocots leading up through grasses, palms, lilies and bananas to reach their most complex development in the orchids. The dicots include all of our ordinary herbs, shrubs and trees, and are considered to reach their highest or most complex development in the composites, plants of the daisy family.

Protozoa

I started out this catalogue with the tree analogy, whereby diverging evolutionary history can be compared with a branching process. In that figure, we have got clear out at the tip of the plant fork with the orchids and the daisies, and we now have the problem of climbing back down to the main trunk, to the world of single-celled organisms, to start out again on the animal fork with the protozoa.

The protozoa are one-celled animals. Among plants the bacteria and many types of algae are single-celled, and of course all organisms pass through a stage—the fertilized egg—in which the individual consists of only one cell.

In complex animals, the cells are specialized for particular functions, but in protozoa all of the functions of the animal—digestion, excretion, locomotion, respiration—must be carried out within the cell, and various special functions may be performed by special parts of the cell which act like organs of complex animals, and so are called "organelles" or "organoids." The fact that protozoa are single-celled, then, does not necessarily mean that they are simple; and it has often been suggested that they would better be called noncellular animals rather than unicellular animals.

The protozoa, from any point of view, are fascinating creatures, and their study (protozoology) has become an extensive and specialized science. Most protozoa are quite active, some creeping in a manner that has become familiar to us all as "amoeboid movement," some moving rapidly with the beating action of numerous tiny hairs (cilia), or by the lashing of a single long tail (flagellum). Many have become parasitic and several important human diseases (malaria, African sleeping sickness, amoebic dysentery, for instance) are caused by animals of this group.

Some protozoa live together as colonies, each cell still a complete and potentially independent organism, but bound together with its brothers to live a communal life. It is an easy step from such colonies of protozoa to the members of the next great animal phylum, the sponges (porifera).

Sponges

The sponges are generally considered to lie outside of the main lines of animal evolution, to represent a group of collared protozoa that learned to live together as colonies sufficiently well so that they were able to build up into great congregations of cells held together by a mass of fibers, spicules and secreted jelly. The fibrous skeleton of one type of sponge is familiar to all of us. The sponge cells cooperate to form a supercellular organism in which water moves along regular channels through the pores so that food materials can be extracted and waste materials thrown out, and the cells may be specialized in various ways. Parts of a sponge will reproduce the whole animal, like cuttings from plants. Reproduction is usually by a ball-like group of undifferentiated cells called a "gemmule" which swims away from the parent sponge, but there is also a complicated form of sexual reproduction.

Coelenterates

In almost all arrangements of the animal kingdom the third phylum, after the protozoa and the sponges, is that of the coelenterates—the jellyfish, corals and sea anemones. About five thousand species are known, all marine except for a few inconspicuous fresh-water forms. The coelenterates are definite, organized, multicellular animals. The individual cells are organized into tissues with distinctive functions, and one may recognize digestive, muscular, nervous, sensory and even skeletal systems, though respiratory, excretory and circulatory systems are lacking. Distinctive sex cells, with reproductive functions, are produced, but they do not form part of a reproductive system as in more complex animals.

Coelenterates show radial symmetry, with all of the different parts disposed within the circular form centering on the mouth opening, instead of the bilateral symmetry of most animals. There are two body types, the medusa form (jellyfish), which is free swimming; and the polyp form (sea anemone), which has a fixed position. Within the same species, there may be an alternation of generations, the fixed polyp producing free-swimming medusae which, through sexual reproduction, give rise to a generation of polyps again. The same animal (or the same kind of animal) may thus look very different at different stages in its life history.

The coelenterates also tend to form colonies. Coral reefs are examples of gigantic colonies of polyps formed by the skeletons of the innumerable individual animals. The Portuguese man-of-war—familiar to people who have done much swimming in tropical waters—is an example of another kind of colony, made up of several individual animals with different forms and functions. A whole collection of animals hangs under the bright, bladderlike cell—some individuals specialized as feelers and stingers, others for eating, others for protection, and still others for reproduction.

Nematodes

The next major phylum, skipping several obscure but very interesting groups, is that of the nematodes. These might also be considered to be obscure, in that they have attracted relatively little attention from most people, but they are incredibly numerous and apt to be important in any consideration of the general economy of nature.

The nematodes are worms. This word "worm" is used for a variety of different kinds of animals—anything that is long, round, wriggly, and too small to be called a snake. Thus, all sorts of insect larvae are worms, though they would perhaps more properly be called grubs, caterpillars or maggots. The true worms, that is, animals that keep the worm form all of their life, include two very different phyla, the nematodes and the annelids. The nematodes include the worms with a smooth, unsegmented body, for which biologists have attempted to establish the vernacular terms "roundworm" or "threadworm," while the annelids include the segmented worms, such as the common earthworm.

Something like 80,000 different kinds of nematodes have been described and probably only a small proportion of the existing kinds have been named. They are much more highly organized animals than the coelenterates: the digestive system is a complete tract, with both mouth and anus, and the sex cells are associated with a reproductive system, so that individual animals may have male or female characteristics. Most nematodes are very small, microscopic, and while many kinds live in salt and fresh water, others live in soil or other moist situations. A great many also are parasitic, living in plants, insects and vertebrates. Some of the human intestinal worms are nematodes. Filaria, forming a large group of vertebrate parasites, are also nematodes; they include the organism that causes "elephantiasis" in man.

Annelids

It would perhaps be best to discuss the segmented worms, or annelids, next, even though in classifications they are usually placed quite separately from the nematodes. The annelids include the earthworms, the leeches and a variety of marine worms. They are considerably more complex organisms than the nematodes, with all of the main organ systems of the higher animals. Their chief characteristic is the *segmentation* of the body, which affects not only the outward form, but the internal arrangement of the organs.

Only about a tenth as many annelids (8,000) as nematodes have been described, but the earthworms, at least, are numerous enough as individuals to be very important in the economy of nature. Charles Darwin became interested in the abundance and activity of earthworms, and wrote a book about them in which he showed that they were primarily responsible for the maintenance of favorable soil conditions, at least in temperate latitudes.

Echinoderms

The echinoderms include the starfish, sea urchins, sea lilies and their relatives. The adult organisms show a radial symmetry, like the coelenterates, but it is generally considered that this symmetry is a secondary development, since larval forms are bilaterally symmetrical. Indeed, because of the structure of these larval forms, it is thought that the echinoderms may lie close to the main stem of the ancestry of the most complex animals, the vertebrates.

About 6,000 living species of echinoderms have been described. They form a distinctive group because of the radial symmetry (almost always with five basic segments), the horny skin (formed by intracellular spicules of lime) and the curious hydraulic apparatus that furnishes power for movement.

Molluscs

The shellfish and their relatives—the snails, clams, slugs, squids and octopuses—make up the next phylum, that of the molluscs. These are mostly marine, but many forms have invaded fresh water and many snails have become completely adapted to terrestrial existence. The molluscs form a very large group, numerous both in species and individuals, with something like 80,000 described living kinds. They show a complex development of organ systems like that of higher animals, but they lack the jointed appendages of arthropods and chordates. The shell is of course the most obvious characteristic of the molluscs, though this is shared with another phylum of marine organisms, the brachiopods, which has not been described here because there are so few living species. And many molluscs, such as the slugs of the garden and the squids and octopuses of the sea, have no obvious shell. One mollusc, the giant squid, is the largest of all invertebrates, attaining a length of more than fifty feet.

Arthropods

Two phyla remain to complete this review of the animals—the arthropods and the chordates. These include the dominant land animals of our geological time—the arthropods because of the insects, which are the most numerous of visible organisms both in number of kinds and in number of individuals; and the chordates because of the mammals, which gain a peculiar interest because we are one of them.

The arthropods include five main classes, the crustacea (crabs, shrimps, barnacles), the arachnids (spiders, scorpions, ticks), the onychophora (the few queer species of Peripatus that have survived into our time from the dim geological past), the myriapods (millipedes and centipedes) and the insects.

To us as mammals, the insects seem to belong to some topsy-turvy world almost outside the reach of our understanding. They have the skeleton on the outside of the body, the main nervous system below the digestive tract; they carry air to the cells of the body through a complicated system of tubes and use blood (body fluid) only for the transport of food materials. They may become involved in amazingly complex patterns of behavior, but these patterns seem to depend on some inherited fixed instinct, not in any way involving a learning process. And the insects are endlessly abundant, endlessly prolific, endlessly obtruding themselves on human consciousness either in the garden or the kitchen. We, in our snobbish pride, consider this to be the Age of Mammals; but even we sometimes stop to wonder whether these ubiquitous insects are not destined, after all, to inherit the earth.

Chordates

Biologists now speak of the chordates, instead of the vertebrates, because they have recognized the cousinly relationships of some marine organisms which, although they have no backbone, must definitely be included in the family circle. These nonvertebrate chordates, however, include only a few inconspicuous species of little concern in a broad survey of natural history. For our purposes we can still think of the chordates as composed of five main classes, the fish (sharks and rays should really be a separate class), the amphibians, the reptiles, the birds and the mammals.

The chordates, with 70,000 or so species, make a respectable showing in comparison with any phylum except the arthropods; and they include organisms that dominate, in size if not in numbers of individuals, in all of the major types of environments.

Mammals are the most complex, the most specialized, the most recently evolved of the classes of chordates. With pride, we place man at the tip of the mammalian line, adducing his brain size, his manual skill, his geologically recent intrusion on the landscape, as arguments for this position. With the mammals and with man we thus come to the end of the survey of the animal branch of organic nature, having reached an extreme of specialization comparable to that of the orchids and the daisies on the plant branch.

GEORGE WALD

The Origin of Life

•

ABOUT A CENTURY AGO the question, how did life begin? which has interested men throughout their history, reached an impasse. Up to that time two answers had been offered: one that life had been created supernaturally, the other that it arises continually from the nonliving. The first explanation lay outside science; the second was now shown to be untenable. For a time scientists felt some discomfort in having no answer at all. Then they stopped asking the question.

Recently ways have been found again to consider the origin of life as a scientific problem—as an event within the order of nature. In part this is the result of new information. But a theory never rises of itself, however rich and secure the facts. It is an act of creation. Our present ideas in this realm were first brought together in a clear and defensible argument by the Russian biochemist A. I. Oparin in a book called *The Origin of Life*, published in 1936. Much can be added now to Oparin's discussion, yet it provides the foundation upon which all of us who are interested in this subject have built.

The attempt to understand how life originated raises a wide variety of scientific questions, which lead in many and diverse directions and should end by casting light into many obscure corners. At the center of the enterprise lies the hope not only of explaining a great past event—important as that should be—but of showing that the explanation is workable. If we can indeed come to understand how a living organism arises from the nonliving, we should be able to construct one—only of the simplest description, to be sure, but still recognizably alive. This is so remote a possibility now that one scarcely dares to acknowledge it; but it is there nevertheless.

One answer to the problem of how life originated is that it was created. This is an understandable confusion of nature with technology. Men are used to making things; it is a ready thought that those things not made by men were made by a superhuman being. Most of the cultures we know contain mythical accounts of a supernatural creation of life. Our own tradition provides such an account in the opening chapters of Genesis. There we are told that beginning on the third day of the Creation, God brought forth living creatures—first plants, then fishes and birds, then land animals and finally man.

The more rational elements of society, however, tended to take a more naturalistic view of the matter. One had only to accept the evidence of one's senses to know that life arises regularly from the nonliving: worms from mud, maggots from decaying meat, mice from refuse of various kinds. This is the view that came to be called spontaneous generation. Few scientists doubted it. Aristotle, Newton, William Harvey, Descartes, van Helmont, all accepted spontaneous generation without serious question. Indeed, even the theologians—witness the English Jesuit John Turberville Needham—could subscribe to the view, for Genesis tells us, not that God created plants and most animals directly, but that He bade the earth and waters to bring them forth; since this directive was never rescinded, there is nothing heretical in believing that the process has continued.

But step by step, in a great controversy that spread over two centuries, this belief was whittled away until nothing remained of it. First the Italian Francesco Redi showed in the seventeenth century that meat placed under a screen, so that flies cannot lay their eggs on it, never develops maggots. Then in the following century the Italian abbé Lazzaro Spallanzani showed that a nutritive broth, sealed off from the air while boiling, never develops microorganisms, and hence never rots. Needham objected that by too much boiling Spallanzani had rendered the broth, and still more the air above it, incompatible with life. Spallanzani could defend his broth; when he broke the seal of his flasks, allowing new air to rush in, the broth promptly began to rot. He could find no way, however, to show that the air in the sealed flask had not been vitiated. This problem finally was solved by Louis Pasteur in 1860, with a simple modification of Spallanzani's experiment. Pasteur too used a flask containing boiling broth, but instead of sealing off the neck he drew it out in a long, S-shaped curve with its end open to the air. While molecules of air could pass back and forth freely, the heavier particles of dust, bacteria and molds in the atmosphere were trapped on the walls of the curved neck and only rarely reached the broth. In such a flask the broth seldom was contaminated; usually it remained clear and sterile indefinitely.

This was only one of Pasteur's experiments. It is no easy matter to deal with so deeply ingrained and common-sense a belief as that in spontaneous generation. One can ask for nothing better in such a pass than a noisy and stubborn opponent, and this Pasteur had in the naturalist Félix Pouchet, whose arguments before the French Academy of Sciences drove Pasteur to more and more rigorous experiments. When he had finished, nothing remained of the belief in spontaneous generation.

We tell this story to beginning students of biology as though it represents a triumph of reason over mysticism. In fact it is very nearly the opposite. The reasonable view was to believe in spontaneous generation; the only alternative, to believe in a single, primary act of supernatural creation. There is no third position. For this reason many scientists a century ago chose to regard the belief in spontaneous generation as a "philosophical necessity." It is a symptom of the philosophical poverty of our time that this necessity is no longer appreciated. Most modern biologists, having reviewed with satisfaction the downfall of the spontaneous generation hypothesis, yet unwilling to accept the alternative belief in special creation, are left with nothing.

I think a scientist has no choice but to approach the origin of life through a hypothesis of spontaneous generation. What the controversy reviewed above showed to be untenable is only the belief that living organisms arise spontaneously under present conditions. We have now to face a somewhat different problem: how organisms may have arisen spontaneously under different conditions in some former period, granted that they do so no longer.

To make an organism demands the right substances in the right proportions and in the right arrangement. We do not think that anything more is needed—but that is problem enough.

The substances are water, certain salts—as it happens, those found in the ocean—and carbon compounds. The latter are called *organic* compounds because they scarcely occur except as products of living organisms.

Organic compounds consist for the most part of four types of atoms: carbon, oxygen, nitrogen and hydrogen. These four atoms together constitute about 99 per cent of living material, for hydrogen and oxygen also form water. The organic compounds found in organisms fall mainly into four great classes: carbohydrates, fats, proteins and nucleic acids. The fats are simplest, each consisting of three fatty acids joined to glycerol. The starches and glycogens are made of sugar units strung together to form long straight and branched chains. In general only one type of sugar appears in a single starch or glycogen; these molecules are large, but still relatively simple. The principal function of carbohydrates and fats in the organism is to serve as fuel—as a source of energy.

The nucleic acids introduce a further level of complexity. They are very large structures, composed of aggregates of at least four types of unit—the nucleotides—brought together in a great variety of proportions and sequences. An almost endless variety of different nucleic acids is possible, and specific differences among them are believed to be of the highest importance. Indeed, these structures are thought by many to be the main constituents of the genes, the bearers of hereditary constitution.

Variety and specificity, however, are most characteristic of the proteins, which include the largest and most complex molecules known. The units of which their structure is built are about 25 different amino acids. These are strung together in chains hundreds to thousands of units long, in different proportions, in all types of sequence, and with the greatest variety of branching and folding. A virtually infinite number of different proteins is possible. Organisms seem to exploit this potentiality, for no two species of living organism, animal plant, possess the same proteins.

Organic molecules therefore form a large and formidable array, endless in variety and of the most bewildering complexity. One cannot think of having organisms without them. This is precisely the trouble, for to understand how organisms originated we must first of all explain how such complicated molecules could come into being. And that is only the beginning. To make an organism requires not only a tremendous variety of these substances, in adequate amounts and proper proportions, but also just the right arrangement of them. Structure here is as important as composition—and what a complication of structure! The most complex machine man has devised—say an electronic brain—is child's play compared with the simplest of living organisms. The especially trying thing is that complexity here involves such small dimensions. It is on the molecular level; it consists of a detailed fitting of molecule to molecule such as no chemist can attempt.

One has only to contemplate the magnitude of this task to concede that the spontaneous generation of a living organism is impossible. Yet here we are—as a result, I believe, of spontaneous generation. It will help to digress for a moment to ask what one means by "impossible."

With every event one can associate a probability—the chance that it will occur. This is always a fraction, the proportion of times the event occurs in a large number of trials. Sometimes the probability is apparent even without trial. A coin has two faces; the probability of tossing a head is therefore ½. A die has six faces; the probability of throwing a deuce is ⅙. When one has no means of estimating the probability beforehand, it must be determined by counting the fraction of successes in a large number of trials.

In such a problem as the spontaneous origin of life we have no way of assessing probabilities beforehand, or even of deciding what we mean by a trial. The origin of a living organism is undoubtedly a stepwise phenomenon, each step with its own probability and its own conditions of trial. Of one thing we can be sure, however: whatever constitutes a trial, more such trials occur the longer the interval of time.

The important point is that since the origin of life belongs in the category of at-least-once phenomena, time is on its side. However improbable we regard this event, or any of the steps which it involves, given enough time it will almost certainly happen at least once. And for life as we know it, with its capacity for growth and reproduction, once may be enough.

Time is in fact the hero of the plot. The time with which we have to deal is of the order of two billion years. What we regard as impossible on the basis of human experience is meaningless here. Given so much time, the "impossible" becomes possible, the possible probable, and the probable virtually certain. One has only to wait: time itself performs the miracles.

This brings the argument back to its first stage: the origin of organic compounds. Until a century and a quarter ago the only known source of these substances was the stuff of living organisms. Students of chemistry are usually told that when, in 1828, Friedrich Wöhler synthesized the first organic compound, urea, he proved that organic compounds do not require living organisms to make them. Of course it showed nothing of the kind. Organic chemists are alive; Wöhler merely showed that they can make organic compounds externally as well as internally. It is still true that with almost negligible exceptions all the organic matter we know is the product of living organisms.

The almost negligible exceptions, however, are very important for our argument. It is now recognized that a constant, slow production of organic molecules occurs without the agency of living things. Certain geological phenomena yield simple organic compounds. So, for example, volcanic eruptions bring metal carbides to the surface of the earth, where they react with water vapor to yield simple compounds of carbon and hydrogen. The familiar type of such a reaction is the process used in old-style bicycle lamps in which acetylene is made by mixing iron carbide with water.

Harold Urey, Nobel laureate in chemistry, has become interested in the degree to which electrical discharges in the upper atmosphere may promote the formation of organic compounds. One of his students, S. L. Miller, performed the simple experiment of circulating a mixture of water vapor, methane (CH_4), ammonia (NH_3) and hydrogen—all gases believed to have been present in the early atmosphere of the earth—continuously for a week over an electric spark. The circulation was maintained by boiling the water in one limb of the apparatus and condensing it in the other. At the end of the week the water was analyzed by the delicate method of paper chromatography. It was found to have acquired a mixture of amino acids! Glycine and alanine, the simplest amino acids and the most prevalent in proteins, were definitely identified in the solution, and there were indications it contained aspartic acid and two others. The yield was surprisingly high. This amazing result changes at a stroke our ideas of the probability of the spontaneous formation of amino acids.

A final consideration, however, seems to me more important than all the special processes to which one might appeal for organic syntheses in inanimate nature.

It has already been said that to have organic molecules one ordinarily needs organisms. The synthesis of organic substances, like almost everything else that happens in organisms, is governed by the special class of proteins called enzymes—the organic catalysts which greatly accelerate chemical reactions in the body. Since an enzyme is not used up but is returned at the end of the process, a small amount of enzyme can promote an enormous transformation of material.

Enzymes play such a dominant role in the chemistry of life that it is exceedingly difficult to imagine the synthesis of living material without their help. This poses a dilemma, for enzymes themselves are proteins, and hence among the most complex organic components of the cell. One is asking, in effect, for an apparatus which is the unique property of cells in order to form the first cell.

This is not, however, an insuperable difficulty. An enzyme, after all, is only a catalyst; it can do no more than change the *rate* of a chemical reaction. It cannot make anything happen that would not have happened, though more slowly, in its absence. Every process that is catalyzed by an enzyme, and every product of such a process, would occur without the enzyme. The only difference is one of rate.

Once again the essence of the argument is time. What takes only a few moments in the presence of an enzyme or other catalyst may take days, months or years in its absence; but given time, the end result is the same.

Indeed, this great difficulty in conceiving of the spontaneous generation of organic compounds has its positive side. In a sense, organisms demonstrate to us what organic reactions and products are *possible*. We can be certain that, given time, all these things must occur. Every substance that has ever been found in an organism displays thereby the finite probability of its occurrence. Hence, given time, it should arise spontaneously. One has only to wait.

It will be objected at once that this is just what one cannot do. Everyone knows that these substances are highly perishable. Granted that, within long spaces of time, now a sugar molecule, now a fat, now even a protein might form spontaneously, each of these molecules should have only a transitory existence. How are they ever to accumulate; and, unless they do so, how form an organism?

We must turn the question around. What, in our experience, is known to destroy organic compounds? Primarily two agencies: decay and the attack of oxygen. But decay is the work of living organisms, and we are talking of a time before life existed. As for oxygen, this introduces a further and fundamental section of our argument.

It is generally conceded at present that the early atmosphere of our planet contained virtually no free oxygen. Almost all the earth's oxygen was bound in the form of water and metal oxides. If this were not so, it would be very difficult to imagine how organic matter could accumulate over the long stretches of time that alone might make possible the spontaneous origin of life. This is a crucial point, therefore, and the statement that the early atmosphere of the planet was virtually oxygen-free comes forward so opportunely as to raise a suspicion of special pleading. I have for this reason taken care to consult a number of geologists and astronomers on this point, and am relieved to find that it is well defended. I gather that there is a widespread though not universal consensus that this condition did exist. Apparently something similar was true also for another common component of our atmosphere—carbon dioxide. It is believed that most of the carbon on the earth during its early geological history existed as the element or in metal carbides and hydrocarbons; very little was combined with oxygen.

This situation is not without its irony. We tend usually to think that the environment plays the tune to which the organism must dance. The environment is given; the organism's problem is to adapt to it or die. It has become apparent lately, however, that some of the most important features of the physical environment are themselves the work of living organisms. Two such features have just been named. The atmosphere of our planet seems to have contained no oxygen until organisms placed it there by the process of plant photosynthesis. It is estimated that at present all the oxygen of our atmosphere is renewed by photosynthesis once in every 2,000 years, and that all the carbon dioxide passes through the process of photosynthesis once in every 300 years. In the scale of geological time, these intervals are very small indeed. We are left with the realization that all the oxygen and carbon dioxide of our planet are the products of living organisms, and have passed through living organisms over and over again.

In the early history of our planet, when there were no organisms or any free oxygen, organic compounds should have been stable over very long periods. This is the crucial difference between the period before life existed and our own. If one were to specify a single reason why the spontaneous generation of living organisms was possible once and is so no longer, this is the reason.

We must still reckon, however, with another destructive force which is disposed of less easily. This can be called spontaneous dissolution—the counterpart of spontaneous generation. We have noted that any process catalyzed by an enzyme can occur in time without the enzyme. The trouble is that the processes which synthesize an organic substance are reversible: any chemical reaction which an enzyme may catalyze will go backward as well as forward. We have spoken as though one has only to wait to achieve syntheses of all kinds; it is truer to say that what one achieves by waiting is *equilibria* of all kinds—equilibria in which the synthesis and dissolution of substances come into balance.

In the vast majority of the processes in which we are interested the point of equilibrium lies far over toward the side of dissolution. That is to say, spontaneous dissolution is much more probable, and hence proceeds much more rapidly, than spontaneous synthesis. For example, the spontaneous union, step by step, of amino acid units to form a protein has a certain small probability, and hence might occur over a long stretch of time. But the dissolution of the protein or of an intermediate product into its component amino acids is much more probable, and hence will go ever so much more rapidly. The situation we must face is that of patient Penelope waiting for Odysseus, yet much worse: each night she undid the weaving of the preceding day, but here a night could readily undo the work of a year or a century.

How do present-day organisms manage to synthesize organic compounds against the forces of dissolution? They do so by a continuous expenditure of energy. Indeed, living organisms commonly do better than oppose the forces of dissolution; they grow in spite of them. They do so, however, only at enormous expense to their surroundings. They need a constant supply of material and energy merely to maintain themselves, and much more of both to grow and reproduce. A living organism is an intricate machine for performing exactly this function. When, for want of fuel or through some internal failure in its mechanism, an organism stops actively synthesizing itself in opposition to the processes which continuously decompose it, it dies and rapidly disintegrates.

What we ask here is to synthesize organic molecules without such a machine. I believe this to be the most stubborn problem that confronts us—the weakest link at present in our argument. I do not think it by any means disastrous, but it calls for phenomena and forces some of which are as yet only partly understood and some probably still to be discovered.

At present we can make only a beginning with this problem. We know that it is possible on occasion to protect molecules from dissolution by precipitation or by attachment to other molecules. A wide variety of such precipitation and "trapping" reactions is used in modern chemistry and biochemistry to promote syntheses. Some molecules appear to acquire a degree of resistance to disintegration simply through their size. So, for example, the larger molecules composed of amino acids—polypeptides and proteins—seem to display much less tendency to disintegrate into their units than do smaller compounds of two or three amino acids.

Such molecular aggregates, of various degrees of material and architectural complexity, are indispensable intermediates between molecules and organisms. We have no need to try to imagine the spontaneous formation of an organism by one grand collision of its component molecules. The whole process must be gradual. The molecules form aggregates, small and large. The aggregates add further molecules, thus growing in size and complexity. Aggregates of various kinds interact with one another to form still larger and more complex structures. In this way we imagine the ascent, not by jumps or master strokes, but gradually, piecemeal, to the first living organisms.

Where may this have happened? It is easiest to suppose that life first arose in the sea. Here were the necessary salts and the water. The latter is not only the principal component of organisms, but prior to their formation provided a medium which could dissolve molecules of the widest variety and ceaselessly mix and circulate them. It is this constant mixture and collision of organic molecules of every sort that constituted in large part the "trials" of our earlier discussion of probabilities.

The sea in fact gradually turned into a dilute broth, sterile and oxygen-free. In this broth molecules came together in increasing number and variety, sometimes merely to collide and separate, sometimes to react with one another to produce new combinations, sometimes to aggregate into multimolecular formations of increasing size and complexity.

What brought order into such complexes? For order is as essential here as composition. To form an organism, molecules must enter into intricate designs and connections; they must eventually form a self-repairing, self-constructing dynamic machine. For a time this problem of molecular arrangement seemed to present an almost insuperable obstacle in the way of imagining a spontaneous origin of life, or indeed the laboratory synthesis of a living organism. It is still a large and mysterious problem, but it no longer seems insuperable. The change in view has come about because we now realize that it is not altogether necessary to *bring* order into this situation; a great deal of order is implicit in the molecules themselves.

The epitome of molecular order is a crystal. In a perfect crystal the molecules display complete regularity of position and orientation in all planes of space. At the other extreme are fluids—liquids or gases—in which the molecules are in ceaseless motion and in wholly random orientations and positions.

Lately it has become clear that very little of a living cell is truly fluid. Most of it consists of molecules which have taken up various degrees of orientation with regard to one another. That is, most of the cell represents various degrees of approach to crystallinity—often, however, with very important differences from the crystals most familiar to us. Much of the cell's crystallinity involves molecules which are still in solution—so-called liquid crystals—and much of the dynamic, plastic quality of cellular structure, the capacity for constant change of shape and interchange of material, derives from this condition. Our familiar crystals, furthermore, involve only one or a very few types of molecule, while in the cell a great variety of different molecules come together in some degree of regular spacing and orientation—i.e., some degree of crystallinity. We are dealing in the cell with highly mixed crystals and near-crystals, solid and liquid. The laboratory study of this type of formation has scarcely begun. Its further exploration is of the highest importance for our problem.

In a fluid such as water the molecules are in very rapid motion. Any molecules dissolved in such a medium are under a constant barrage of collisions with water molecules. This keeps small and moderately sized molecules in a constant turmoil; they are knocked about at random, colliding again and again, never holding any position or orientation for more than an instant. The larger a molecule is relative to water, the less it is disturbed by such collisions. Many protein and nucleic acid molecules are so large that even in solution their motions are very sluggish, and since they carry large numbers of electric charges distributed about their surfaces, they tend even in solution to align with respect to one another. It is so that they tend to form liquid crystals.

Recently several particularly striking examples have been reported of the spontaneous production of familiar types of biological structure by protein molecules. Cartilage and muscle offer some of the most intricate and regular patterns of structure to be found in organisms. A fiber from either type of tissue presents under the electron microscope a beautiful pattern of cross striations of various widths and densities, very regularly spaced. The proteins that form these structures can be coaxed into free solution and stirred into completely random orientation. Yet on precipitating, under proper conditions, the molecules realign with regard to one another to regenerate with extraordinary fidelity the original patterns of the tissues.

We have therefore a genuine basis for the view that the molecules of our oceanic broth will not only come together spontaneously to form aggregates but in doing so will spontaneously achieve various types and degrees of order. This greatly simplifies our problem. What it means is that, given the right molecules, one does not have to do everything for them; they do a great deal for themselves.

Oparin has made the ingenious suggestion that natural selection, which Darwin proposed to be the driving force of organic evolution, begins to operate at this level. He suggests that as the molecules come together to form colloidal aggregates, the latter begin to compete with one another for material. Some aggregates, by virtue of especially favorable composition or internal arrangement, acquire new molecules more rapidly than others. They eventually emerge as the dominant types. Oparin suggests further that considerations of optimal size enter at this level. A growing colloidal particle may reach a point at which it becomes unstable and breaks down into smaller particles, each of which grows and redivides. All these phenomena lie within the bounds of known processes in nonliving systems.

We suppose that all these forces and factors, and others perhaps yet to be revealed, together give us eventually the first living organism. That achieved, how does the organism continue to live?

We have already noted that a living organism is a dynamic structure. It is the site of a continuous influx and outflow of matter and energy. This is the very sign of life, its cessation the best evidence of death. What is the primal organism to use as food, and how derive the energy it needs to maintain itself and grow?

For the primal organism, generated under the conditions we have described, only one answer is possible. Having arisen in an oceanic broth of organic molecules, its only recourse is to live upon them. There is only one way of doing that in the absence of oxygen. It is called fermentation: the process by which organisms derive energy by breaking organic molecules and rearranging their parts. The most familiar example of such a process is the fermentation of sugar by yeast, which yields alcohol as one of the products. Animal cells also ferment sugar, not to alcohol but to lactic acid. These are two examples from a host of known fermentations.

The yeast fermentation has the following over-all equation: $C_6H_{12}O_6 \rightarrow 2\,CO_2 + 2\,C_2H_5OH +$ energy. The result of fragmenting 180 grams of sugar into 88 grams of carbon dioxide and 92 grams of alcohol is to make available about 20,000 calories of energy for the use of the cell. The energy is all that the cell derives by this transaction; the carbon dioxide and alcohol are waste products which must be got rid of somehow if the cell is to survive.

The cell, having arisen in a broth of organic compounds accumulated over the ages, must consume these molecules by fermentation in order to acquire the energy it needs to live, grow and reproduce. In doing so, it and its descendants are living on borrowed time. They are consuming their heritage, just as we in our time have nearly consumed our heritage of coal and oil. Eventually such a process must come to an end, and with that life also should have ended. It would have been necessary to start the entire development again.

Fortunately, however, the waste product carbon dioxide saved this situation. This gas entered the ocean and the atmosphere in ever-increasing quantity. Some time before the cell exhausted the supply of organic molecules, it succeeded in inventing the process of photosynthesis. This enabled it, with the energy of sunlight, to make its own organic molecules: first sugar from carbon dioxide and water, then, with ammonia and nitrates as sources of nitrogen, the entire array of organic compounds which it requires. The sugar synthesis equation is: $6\,CO_2 + 6\,H_2O +$ sunlight $\rightarrow C_6H_{12}O_6 + 6\,O_2$. Here 264 grams of carbon dioxide plus 108 grams of water plus about 700,000 calories of sunlight yield 180 grams of sugar and 192 grams of oxygen.

This is an enormous step forward. Living organisms no longer needed to depend upon the accumulation of organic matter from past ages; they could make their own. With the energy of sunlight they could accomplish the fundamental organic syntheses that provide their substance, and by fermentation they could produce what energy they needed.

Fermentation, however, is an extraordinarily inefficient source of energy. It leaves most of the energy potential of organic compounds unexploited; consequently huge amounts of organic material must be fermented to provide a modicum of energy. It produces also various poisonous waste products—alcohol, lactic acid, acetic acid, formic acid and so on. In the sea such products are readily washed away, but if organisms were ever to penetrate to the air and land, these products must prove a serious embarrassment.

One of the by-products of photosynthesis, however, is oxygen. Once this was available, organisms could invent a new way to acquire energy, many times as efficient as fermentation. This is the process of cold combustion called respiration: $C_6H_{12}O_6 + 6\,O_2 \rightarrow 6\,CO_2 + 6\,H_2O$ + energy. The burning of 180 grams of sugar in cellular respiration yields about 700,000 calories, as compared with the approximately 20,000 calories produced by fermentation of the same quantity of sugar. This process of combustion extracts all the energy that can possibly be derived from the molecules which it consumes. With this process at its disposal, the cell can meet its energy requirements with a minimum expenditure of substance. It is a further advantage that the products of respiration—water and carbon dioxide—are innocuous and easily disposed of in any environment.

It is difficult to overestimate the degree to which the invention of cellular respiration released the forces of living organisms. No organism that relies wholly upon fermentation has ever amounted to much. Even after the advent of photosynthesis, organisms could have led only a marginal existence. They could indeed produce their own organic materials, but only in quantities sufficient to survive. Fermentation is so profligate a way of life that photosynthesis could do little more than keep up with it. Respiration used the material of organisms with such enormously greater efficiency as for the first time to leave something over. Coupled with fermentation, photosynthesis made organisms self-sustaining; coupled with respiration, it provided a surplus.

The entry of oxygen into the atmosphere also liberated organisms in another sense. The sun's radiation contains ultraviolet components which no living cell can tolerate. We are sometimes told that if this radiation were to reach the earth's surface, life must cease. That is not quite true. Water absorbs ultraviolet radiation very effectively, and one must conclude that as long as these rays penetrated in quantity to the surface of the earth, life had to remain under water. With the appearance of oxygen, however, a layer of ozone formed high in the atmosphere and absorbed this radiation. Now organisms could for the first time emerge from the water and begin to populate the earth and air. Oxygen provided not only the means of obtaining adequate energy for evolution but the protective blanket of ozone which alone made possible terrestrial life.

This is really the end of our story. Yet not quite the end. Our entire concern in this argument has been to bring the origin of life within the compass of natural phenomena. It is of the essence of such phenomena to be repetitive, and hence, given time, to be inevitable.

This is by far our most significant conclusion—that life, as an orderly natural event on such a planet as ours, was inevitable. The same can be said of the whole of organic evolution. All of it lies within the order of nature, and apart from details all of it was inevitable.

Astronomers have reason to believe that a planet such as ours—of about the earth's size and temperature, and about as well lighted—is a rare event in the universe. Indeed, filled as our story is with improbable phenomena, one of the least probable is to have had such a body as the earth to begin with. Yet though this probability is small, the universe is so large that it is conservatively estimated at least 100,000 planets like the earth exist in our galaxy alone. Some 100 million galaxies lie within the range of our most powerful telescopes, so that throughout observable space we can count apparently on the existence of at least 10 million million planets like our own.

What it means to bring the origin of life within the realm of natural phenomena is to imply that in all these places life probably exists—life as we know it. Indeed, I am convinced that there can be no way of composing and constructing living organisms which is fundamentally different from the one we know—though this is another argument, and must await another occasion. Wherever life is possible, given time, it should arise. It should then ramify into a wide array of forms, differing in detail from those we now observe (as did earlier organisms on the earth) yet including many which should look familiar to us—perhaps even men.

We are not alone in the universe, and do not bear alone the whole burden of life and what comes of it. Life is a cosmic event—so far as we know the most complex state of organization that matter has achieved in our cosmos. It has come many times, in many places—places closed off from us by impenetrable distances, probably never to be crossed even with a signal. As men we can attempt to understand it, and even somewhat to control and guide its local manifestations. On this planet that is our home, we have every reason to wish it well. Yet should we fail, all is not lost. Our kind will try again elsewhere.

KARL VON FRISCH

Of Cells, Life and Death

•

THERE ARE very different ways of looking at the world. A small boy who goes into the country with his dog on a holiday and sits under an oak tree may simply enjoy being out of doors in the sunshine without noticing what is stirring around him. Perhaps he watches a robin and feels sorry for the fat earthworm that the bird has taken out of the ground for its breakfast. Or he may watch a butterfly flitting from flower to flower in search of nectar, or a slug that is inconspicuously looking for food in its own way. Maybe he will be a biologist some day, and is having early thoughts about the riddle of life, which has puzzled men's minds since time immemorial.

There are fundamental characteristics of life: growth, individual organization, capacity for propagation, and inevitable death; and man shares them with the bird and the worm, with the oak and with all the plants in the fields. Yet something within us rebels against the idea that we have a great deal in common with these creatures and that we are intimately related to them. This is partly true if we think of a cat or a bird, and more so if we think of a tree or a delicate and perishable plant. How different their ways and the tenor of their lives are from ours!

One of the greatest discoveries in the history of science was therefore the recognition that all living organisms, whether high or low, whether animal or plant, actually have one common feature in their inherent structure. Their bodies are all composed of minute building blocks, called cells, and these are by no means as different from one another in a plant or an animal, or even in a worm or a man, as might be guessed from the external differences between these creatures. On the contrary! They correspond so extensively in structure, down to the finest details, that they may be considered as common and essential features of all life.

Cells as the Visible Building Blocks of Which Organisms Are Made

If, as in the heading of this section, cells are called "visible building blocks," one reservation must be added. The cells, because of their small size, are not visible directly. Their discovery had to await the invention of the microscope.

The microscope, like other great achievements, cannot be attributed to one definite inventor. The development of this kind of modern instrument requires the work of generations. Simple magnifying glasses (hand lenses) were already available in the 13th century. These were used with a small stand which served as a holder for the object that was to be magnified, and with other accessory devices. Such an instrument is called a simple microscope. It permits magnification from 2 to 300 times. The compound microscope was devised around 1590; in this instrument, the enlarged image of the object cast by one lens, the objective, is again enlarged by a second lens, the ocular. Later, the polishing of lenses was improved, and special kinds of glass were used for their manufacture. Good illumination was provided, and other optical and mechanical improvements were made. In this way the modern microscope was achieved, with its capacities for magnifying objects up to 2000 times. For several centuries people believed that this was as far as progress could go. For when objects to be viewed are as small as light waves themselves—the physicists can demonstrate it convincingly—light rays cannot form a microscopic image of them. A brilliant discovery in 1934 permitted this barrier to be surmounted.

In the electron microscope the image of the object is formed by beams of electrons instead of by rays of light. With this apparatus, objects 1000 times smaller than before can be perceived. The smallest building blocks of many objects—their molecules—are visible and can be photographed. Although the electron microscope is very expensive, and also fussy and difficult to operate, it is now indispensable in many biological investigations. It will never of course replace the usual microscope, but it will become an effective supplement to it.

Around 1600, Antony van Leeuwenhoek, a Dutch draper and an amateur scientist who was untrained but nonetheless enthusiastic, skillful, and persevering, became one of the first people to use the microscope for the study of natural science. A contemporary of his, the English naturalist Robert Hooke, observed with his compound microscope the cellular structure of plant tissue in a small piece of cork. But more than a century passed before the significance of this discovery became clear. Then, in 1838, the German naturalist M. J. Schleiden announced that the whole body of the plant is composed of microscopic elements, the cells. A year later Theodor Schwann, who was also German, asserted the same thesis for the animal body.

It was not accidental that cells were first found in plants. Most plant cells are very sharply delimited from one another by a cellulose wall, and they can therefore be seen even with crude microscopes. Animal cells have no cellulose membrane; their surface covering is very delicate and often invisible.

It cannot be held against the discoverers of the cells that they considered the cell walls the main thing, because these were all they saw. Much later, Max Schultze (1861) was the first to recognize that the most important part of each plant cell is what is included within the cellulose wall, the protoplasm, the bearer of life. Like the pit in the fruit of the cherry, the nucleus lies in the interior of the protoplasm. And Schultze found that animal cells also consist of protoplasm and nucleus. He was farsighted enough to recognize cells as the common feature in the structure of the whole world of life, but he could not of course foresee how convincingly later investigations would confirm his bold generalization.

The Size and Shape of Cells

Many cells are spherical, with the cell nucleus in the center—here the analogy we have already made is appropriate, with the cherry and its pit. Generally, however, they look quite different. Many pages of this book could be filled with illustrations showing the varied modifications of shapes of animal cells, from their simplest basic forms to their strangest complicated shapes; and the plant kingdom is not far behind the animal kingdom in this respect.

The size of cells bears no relationship to body size. The mighty elephant is not composed of larger, but only of more cells than the little mouse; and the cells of the California giant Sequoia, which rears its head more than 300 feet above the ground, are no larger than those of a snowdrop.

The diameter of cells is usually, if they are roundish in shape, between $1/10$ and $1/100$ mm. Yet there are deviations from these limits, downward as well as upward. Under favorable circumstances, cells $1/10$ mm. in diameter can just be seen with the naked eye, the black pigment cells, for instance, in the skin of fishes, where they stand out clearly against their lighter background in some places. Many nerve cells are true giants; their threadlike extensions become more than a yard long, but they are so slender that they still can be seen only with the microscope. The elongated tubular cells that contain the milk-sap in spurges are as much as several yards long in the treelike representatives of this genus. Many cells also become wide, rather than long, so big and plump that they can easily be seen with the naked eye. When a roundworm is dissected, storage cells as big as grains of millet can be found. Furthermore, the yolk of a hen's or an ostrich's egg is a single cell, although this, to be sure, owes its unusual size only to the storage materials it contains.

The Contents of a Cell

Protoplasm, under the microscope, generally looks colorless and transparent. Often it has small granules or droplets embedded in it. Sometimes its structure appears foamy like lather, only here what looks like foam is not made of gas bubbles in small compartments of fluid, but rather of two fluids which do not mix with each other, whose contents and partitions form the vesicles. . . . Fluids? What is the physical state of protoplasm?

If a cell is pricked with a fine needle, its contents flow out. Even in the absence of injury the streaming of plasma can be observed in many objects, best perhaps in an ameba, or in the stamen hairs of the spiderwort, often a modest and unassuming ornament in our collection of house plants. In these cases there is a convincing impression of a streaming liquid. Yet it cannot be maintained that plasma is generally liquid.

According to physical laws, a drop of liquid suspended in another liquid always becomes spherical. Yet many one-celled organisms, for instance the beautiful heliozoan, do not fulfill this expectation, but are radiate, or some other shape.

This contradiction results from a special feature of protoplasm. Let us consider the following example. Sugar, dissolved in water, breaks down into its molecules. These are so small that they are far below the limits of visibility of our microscopes. They are also too small to be retained by the best of our usual filters, and during filtration they slip through the finest of pores along with the water. Solutions of this kind are called true or crystalloid solutions. If we mix fine sand with water, what we obtain is not a solution but a suspension. The solid particles can be seen with a magnifying glass, and with a suitable filter they can be separated out of the liquid again. Colloidal solutions are intermediate between true solutions and suspensions. It is not the kind of particles, but their size, that is characteristic of colloidal solutions. Their upper and lower limits are arbitrary; solutions are called true solutions when the individual molecules consist of 1000 atoms or less, and suspensions when they are composed of a billion or more.

One peculiarity of colloidal solutions is that under certain conditions they can change from a fluid to an increasingly viscid, jellylike, and finally solid state. It is a characteristic of colloidal solutions that under certain circumstances even slight changes of salt concentration, of acidity, of temperature, or of other factors can cause the transition from the fluid to the solid condition, from sol to gel, and the reverse. Similar changes also take place in the plasma of cells.

The chemical constituents of protoplasm are carbohydrates, fats, lipoids (fatlike substances) and—as the special bearers of life—proteins with their highly complex molecules. The structural formula of clupein may give an idea of how complicated proteins may be. In it, 179 atoms of carbon, 405 atoms of hydrogen, 69 of oxygen, and 99 of nitrogen take part in the construction of the molecule. Moreover, the molecules of clupein are less complete than the protein molecules that compose protoplasm. These are at least 4 times, and occasionally even several hundred times, as large. As a rule they contain, in addition to C, H, O, and N, several atoms of S (sulfur) and P (phosphorus) also.

In addition to the organic substances we have named, protoplasm always contains salts and a great deal of water. How much water is present becomes evident only if it is all removed; almost nothing remains of many living creatures if they are completely dried out. A jellyfish contains up to 98 percent water; and two-thirds of the weight even of the human body is only water, in spite of its hard weighty bones.

It is primarily protoplasm that is responsible for the organization of cells and for their manifestations of life, insofar as we can know them. The nucleus seems to be an unconcerned bystander. When the plasma streams, the nucleus is carried along with it; but it seems not to worry about what is going on around it. It is delimited from its environment by a firm membrane, the nuclear membrane. Internally it contains a framework of granules that is very hard to see in the living nucleus, even under the highest magnification; these granules are only perceptible at all because their ability to refract light is greater than that of the nuclear sap which fills the spaces between them. They look, therefore, like tiny ice crystals in a drop of water. It was important for their discovery that they very readily take up certain dyes, for instance red or blue ones, from their surroundings. Because of this characteristic the framework of granules is called chromatin. It consists principally of complex proteins called nucleoproteins, which differ from some of the proteins of the protoplasm in their association with nucleic acids.

From time to time the nucleus temporarily gives up its apparently indifferent behavior. It seems suddenly to become enterprising. Its activity will be taken up in the next section.

Cell Division and Nuclear Division

The ability to grow and to multiply is one of the basic characteristics of the living substance. The phenomenon is expressed in a primitive way in the growth and multiplication of cells. What happens in general is that a cell that has reached about double its size divides into two cells and thus again restores the earlier size. The nucleus divides first in this process, and then the protoplasm. The protoplasmic division takes place without any conspicuous ceremony. The nucleus, however, goes about its business very elaborately, which indicates that its contents must be divided carefully, and in a very specific manner, between the two new nuclei.

The preparation for division begins with the division into two of an insignificant-looking granule, the central body (centrosome), which has been lying next to the nucleus in the protoplasm. In the inside of the nucleus the fragments of chromatin, which seem to be irregularly arranged, line up and finally form little sausage-shaped bodies called chromosomes. The two central bodies back away from each other and they become surrounded, like tiny suns, by halos of very fine plasma fibers, the asters. The nuclear membrane dissolves. The boundary between nucleus and protoplasm therefore vanishes, and as the central bodies move away from each other they become the poles of a spindle about whose equator the chromosomes group themselves. A longitudinal division appears in each chromosome, and the two split halves shift away from each other to opposite poles of the spindle. It looks as though the spindle fibers shorten and in this way draw away from each other the chromosomes to which they seem to adhere. When they have arrived at the two poles, they become surrounded by a new nuclear membrane; the asters disappear, and next the cell body is constricted into two parts. The division of the plasma has followed the nuclear division. The chromatin gradually loosens to form the irregular network that distinguishes the nucleus in its resting state; during this process the particles that make up any particular chromosome preserve their connection with each other, and in the next division they unite again to form the same chromosome. In many cells the whole division process occupies only a few minutes, while in others it requires hours.

If these processes are compared in different animal and plant cells, a real variety becomes apparent in some respects. The number of chromosomes is the same in all the nuclei of a particular plant or animal species, but it often differs in different species. It can be as low as 2, or as high as over 100. Generally, however, between 4 and 60 chromosomes can be identified in the nucleus. Their shape may be spherical, rodlike or threadlike. Centrosomes are found in most animal cells, while they are absent from the cells of higher plants. Also, the spindle is not always demonstrable.

What is always evident in cell division, however, whether in plant or animal cells, whether in a worm or in the cells of the human body, is the careful ordering of the chromosomes, their longitudinal division, and the distribution of the split halves to the daughter nuclei. There is a reason for this regularity: each chromosome thread contains tiny particles, packed into long rows, whose behavior and arrangement determine the organization and the functions of the cells to be produced by later divisions. Before the nucleus divides, these hereditary factors have duplicated themselves, and when the chromosomes split longitudinally the doubled factors separate as the chromosome halves shift away from each other, with the result that they are passed on, unaltered in composition, to the two daughter nuclei.

The kind of nuclear division just described is called mitotic division, or in brief, mitosis. It is also called indirect nuclear division, because of the intricate preparation for the process. Amitotic (direct) division also occurs in a number of cell types; in this, the nucleus constricts into two halves without first forming chromosomes. Amitotic division is relatively rare, and takes place only in nuclei and cells that have, so to speak, no future before them and that are entering into an early decline.

Although the nucleus behaves in essentially the same way during division in both animals and plants, the division of the protoplasm is different, because in the plant kingdom most of the cells are surrounded by a rigid membrane and therefore cannot be easily constricted. Here, a "cell plate" and a new cell membrane are formed in the equatorial plane of the spindle, in the middle of the cell, and the protoplasm is divided in this way, each half with its nucleus forming its own chamber.

Cells as Living Building Blocks

The microscope has thus made us aware that the bodies of plants and animals are composed of cells that are delimited from one another. They seem to be building blocks of organisms in the same way that bricks are building blocks of houses. A house is certainly as far from alive as are the bricks of which it is built. Are the cells alive? Are the countless cells of which the body of a man or a frog is constructed their living building blocks, or is life present only when these are combined into the higher unity of the body?

When a snake is run over and cut in two in the middle, it dies. It seems as though only its body as a whole is capable of life. Yet if we dig up the earth and cut through an earthworm with a spade, we do not kill it at all. On the contrary, we make two living worms out of one. Thus the question is not so simple as it seems. Also, separated parts of lower animals can remain alive. We can go further than this: fragments removed even from a frog can be kept alive if they are handled correctly and provided with everything they need. If the heart is removed from an anesthetized frog, and placed in a fluid which is as close as possible to blood in its composition, it will continue to beat for days. But the heart is still an assemblage of many hundred thousand cells. We can proceed even further. If a fragment the size of a poppy seed is cut out of fresh living tissue and placed in a drop of fluid that contains all the substances necessary for its growth, then it not only remains alive but its cells multiply so vigorously that they soon grow out from the fragment on all sides. If they were left to themselves, they would soon overrun each other and suffocate. But if they are taken care of attentively, and a few cells repeatedly removed and placed in a fresh drop where they have more room and fresh nourishment, then they live for years, continuing to grow and divide.

There can be no doubt that here cells detached from a higher organism carry on their life independently. The cells are living building blocks of the animal body.

We might add that the cell, consisting of protoplasm and nucleus, is the smallest unit independently capable of life. It is no overwhelmingly difficult task, with modern technical equipment, to dissect the nucleus out of a microscopically small cell. The protoplasm deprived of the nucleus perishes after a short time, and in the same way the nucleus without protoplasm is condemned to death. Both are dependent on one another.

On Ameba, and on the Characteristics of Life

By now we have said so much about life that it is time to give an account of its fundamental distinguishing characteristics. When man's faithful companion the dog boisterously and noisily greets his master on his return home, no one doubts that the dog is alive. But have not some of us heard of pitiful occasions in bygone days when people thought to be dead were buried alive? Where is the essence of life concealed, if we can be so horribly deceived?

Movement seems to us to be one of the most certain signs of life. Even a sleeping bird shows breathing movements, and its heart, although less easily seen, beats uninterruptedly. We consider an animal dead when it is completely motionless, internally and externally—yet this is not quite correct. Feigned death is not a rare occurrence in lower animals, and it is a trustworthy method of weathering unfavorable times. The ameba, *Ameba proteus,* bears the name *proteus* for a Greek god, because of its mobility and its ability to change its shape. But when the pond in which it lives dries out in hot weather, it can round up into a sphere, secrete a stiff protective covering, and wait motionless for the return of the water. It is not dead, because the first abundant rainfall ends its sleep. The movement in plants—for instance the streaming of protoplasm inside the cells, or the turning of leaves and blossoms toward the light—are usually so narrowly limited that it is only rarely that they strike the eye as signs of life. Thus even a motionless cell can be alive.

On the other hand, lifeless objects sometimes move and creep like an ameba. A kind of foam can be made from oil lather, and if a small drop of this is placed in fluid this so-called artificial ameba immediately begins to stretch out pseudopods and to withdraw them, and sometimes it creeps around for days. No one will wish to maintain that when we have made this oil lather that crudely imitates the foamy structure often seen in living protoplasm, we have actually created life. Rather, the movements originate as a result of local changes in surface tension that cause the pseudopods to flow out in regions where the tension is decreased.

The creeping of a real ameba is not so simple a process. For it can be affected in many ways by outside influences, by stimuli of different kinds. If a pseudopod that is streaming forward, for instance, is touched with the point of a needle, the movement stops; it can even change to the opposite direction. Not only will the ameba withdraw the pseudopod that has been touched, but it will stretch out another one on the other side where nothing has happened. The impulse to do this comes from the touched region. The excitation has been passed on through the protoplasm; conduction has taken place. Such irritability is not very easily imitated in model experiments, although it has been done successfully to some degree.

Perhaps the engulfing of food can be imitated by a lifeless model, but its digestion and utilization cannot be, and here we have arrived at an infallible characteristic of life. The ameba must feed in order to grow. It has no mouth; any region of its body surface can rapidly form pseudopods which flow around something edible, for instance a one-celled alga, and thus take it into the interior of the protoplasm. In this way a food vacuole arises; inside this, the material which has been taken in is digested, that is, it is broken down into simpler, easily soluble compounds which pass into the protoplasm. The undigested remains are expelled, and then the food vacuole disappears again.

The intake of food and its transformation to body substances leads to the growth of the ameba, but only within limits. When the ameba has reached a certain size, it undergoes a relatively simple process of propagation by nuclear and cell division. From one ameba, there are formed two of half size. This goes on continually, and there would be more and more amebas if hungry enemies or adverse conditions were not constantly doing away with untold numbers of them. Thus the driving force of life is everywhere challenged by counterforces. This is the destiny, and often the excitement too, of life.

Division of Labor and Differentiation Within Cells

In the fresh-water forms [of ameba], the protoplasm contains also a contractile vacuole, so-called because at regular intervals of several minutes it contracts and empties its fluid contents to the outside. It works like a pump in a leaky ship. A local differentiation has occurred, the formation of a specific structure with a specific assignment.

So far zoologists have identified several thousand different species of one-celled animals (Protozoa) in the United States alone that live in water or moist soil. These delicate creatures are not made to live in dry places. Many are similar to the ameba, but many look remarkably different. There are many kinds of differentiation within their cells by which their accomplishments are improved in one way or another. A few examples will show this.

The ameba creeps by means of its pseudopods. Nothing of this kind occurs in the flagellates. Instead, a delicate protoplasmic filament, the flagellum, arises from the front end of the rather longish body, and this drives the water backward, by rapidly beating whiplike movements. In this way it draws the body along behind it, as a propeller does an airplane. In ciliates (Infusoria) there are also delicate and movable plasma outgrowths, the cilia, but these are shorter and are present in greater numbers. In *Paramecium,* for instance, they cover the whole body surface. *Paramecium* swims through the water as a result of the orderly beating of these cilia, and, driven by a thousand tiny oars, it attains a far greater speed than an ameba. To be sure, by human standards neither of them moves very fast. If an ameba is in a hurry, it takes it about two to three hours to cover the distance of a centimeter. *Paramecium* speeds over the same course in about 10 seconds. We are indeed in the world of microscopic dimensions, and need only look through the microscope to convince ourselves of the "racing" speed of the small oarsman.

Cilia are also used to acquire food. *Paramecium* cannot feed, like the ameba, with whatever part of its surface it wishes, for its outer plasma layer is a rather firm membrane. It owes its fixed shape to this. On one side it has a deep "gullet"; at the bottom of this the plasma is soft and can take in solid food particles (bacteria, algae, and similar things). The beat of more powerful cilia causes a stronger flow of water in the gullet, and this carries in the food.

Paramecium even has weapons. Small, short, transparent little rods, the trichocysts, lie closely lined up near the surface of the plasma. As a result of certain stimuli they can suddenly explode and shoot out long hairs; it looks as though a forest of hairs has grown out in a fraction of a second. They serve for defense, and perhaps also for offense in predatory infusorians.

These are remarkable achievements for cells microscopic in dimension. But even they are surpassed in two other ciliates with whose consideration we shall conclude this hasty expedition into a vast field.

Stentor has the habit of fixing itself to water plants and such objects by its slender hind end, and its widened trumpetlike front end stretches out hungrily for prey. A spiral of strong cilia proceeds from the rim of the trumpet opening to the sunken cell mouth, creating a vigorous stream of water which sweeps along all sorts of things to eat. If it wishes to change its position, the *Stentor* can swim away and establish itself in a new place.

Vorticella cannot do this, for its long thin stalk has grown fast to the substrate. The rest of the cell sits on the stalk like a small head; it too is provided with a spiral of cilia for the acquisition of food. In the interior of the stalk, the protoplasm forms a cord; this, wound in a loose spiral, reaches from the lower end of the stalk to the upper end, where it radiates into the plasma of the little head. The cord can suddenly shorten, like the fibers in our muscles, and when it does so it causes the little head to whiz back to the protection of the substrate. The wall of the stalk, which does not shorten, is twisted together into the shape of a corkscrew; it gradually, as a result of its elasticity, again regains its former length.

What a contrast to that dab of flowing protoplasm, the ameba! Extensive differentiation has brought about a division of labor between different regions of the cell plasma. We see within one and the same cell one part of the outer plasma layer forming a kind of skeleton, cilia procuring food, a muscle cord effecting movement as fast as lightning, the contractile vacuole providing for the correct salt concentration, defense weapons serving for protection; and we find very widely distributed food vacuoles that are the equivalent of small stomachs for these miniature creatures. In this way, parts of cells have been specialized for particular assignments, and they can perform them all the more successfully.

Another Form of Specialization: The Many-Celled Condition and the Division of Labor Among Cells

We asume that creatures built as simply as an ameba have remained similar to the most primitive and the oldest animals ever to have been formed on our earth, and that in the course of the earth's history, during inconceivably long periods of billions of years, more highly organized forms have gradually developed from progenitors of this kind. We have just followed one such path of development, which leads from the ameba to the highly differentiated cell of a *Stentor* or a *Vorticella.* It seems to us miraculous that so much has been conjured into the plasma droplet of a single cell. And yet these infusorians occupy only a small modest spot in the realm of the whole great world of life. Their small delicate bodies are ill-adapted to leaving the dampness or to resisting the brutality of their neighbors.

Another path of development has proved more successful. This leads to the many-celled condition, and to the division of labor between the various cells which constitute the body. Certain one-celled organisms considered to be transitional between one-celled and many-celled organisms show how this could have come about.

The flagellates, of which there are many different kinds, usually propagate in the same way as does the ameba. The daughter cells originating from their division separate and go their own ways. The genus *Eudorina* is remarkable in that the daughter cells, which are formed by several divisions that succeed rapidly upon one another, remain together in a jointly secreted envelope, from which only the flagella project. We call this a cell colony. It is a very superficial union of cells that do not enter into any close relationships with one another. Indications of a beginning division of labor are, however, found in another significantly larger colony, that of *Volvox.* In *Volvox,* many hundred or even several thousand individuals that have been produced by division are united in a single jelly envelope. These organisms like light, and they are found, when conditions are suitable for them, in brightly illuminated parts of ponds. Like many other flagellates, they can move toward the light because they have a light-sensitive "eye-spot" at the front end of their body. When the *Volvox* colony swims, one particular pole of the sphere is always directed forward. The flagellates at this pole are responsible for steering the colony, and in order better to accomplish this mission they are provided with eye-spots more impressive than those of their sister cells at the back of the colony, which cannot see where the colony is heading and are therefore unimportant for purposes of steering.

From modest beginnings of this kind may have developed the more extensive division of labor among different types of cells that we see today in the bodies of all many-celled animals. It is characteristic of even the lowest forms, of which we may single out the fresh-water polyp as an example to consider more closely.

This insignificant-looking inhabitant of our ponds is called *Hydra*. It is a giant compared to the protozoans; its body is several millimeters long, but a microscope is still needed to understand its structure. The whole animal resembles a double-walled sack closed at one end. Movable tentacles surrounding the mouth opening serve to seize prey, for this harmless-looking sack is a dangerous predator that catches and poisons small crustaceans (water fleas) if they come too near it; it transports them through its small but very extensible mouth into its digestive cavity and digests them. The inner layer of the double-walled sack is the lining of the digestive system, its outer layer is the skin. The tentacles are only hollow extensions of the sack, like the fingers of a glove, and they are similar in structure to the rest. The animal is thus really no more than an alimentary canal covered with skin. No other body parts are present.

Microscopic examination of the finer structure of both layers reveals, however, unsuspected variety within them. The skin and the alimentary sack are each formed mainly by a close union of cylindrical cells (epithelial cells). The cells of the gut epithelium have flagella to stir the contents of the gut. The movement of the body is brought about by contractile fibers. These, in contrast to the muscle fibers of other animals, are not independent cells, but are developed as appendages of epithelial cells. In the outer skin layer all the muscle fibers run lengthwise in the body, so they shorten it when they contract, or bend it if they contract on one side of the body. In the gut layer, they are perpendicular to these (circular) so that when they contract they make the animal long and thin. Large-bellied cells, the gland cells, whose secretion decomposes the food, are scattered between the epithelial cells of the gut. Slender sensory cells are particularly sensitive to chemical and mechanical stimuli. They come in contact with long fiberlike extensions of nerve cells; these serve to conduct impulses and are connected by additional extensions to other nerve cells, or to muscle cells which they can cause to contract. The most remarkable are the stinging cells, a specialty of *Hydra* and its relatives. The large jellyfish of the sea are among these; many bathers have been stung by them. *Hydra* is too small and delicate to be dangerous to man. But it owes its ability to overpower the water flea to the stinging cells crowded onto its tentacles. Inside of each stinging cell is a capsule in which a long, very fine thread is coiled up. The capsule is hollow, like the finger of a glove turned outside in, and it contains a strong poison. When the tenacle is touched, the capsules explode, the threads are unwound and are slung out with lightning speed; they injure the prey, and at the same time inject the deadly poison into it.

Thus the body of this animal consists of cells that have been differentiated along different directions and specialized for different tasks. None of these cells has really learned anything completely new. Motility, irritability, the ability to conduct an impulse and to secrete digestive juices are primitive characteristics of living protoplasm. But in sensory cells irritability is developed to a higher degree of perfection than in other cells at the cost of other capacities; similarly, the ability to transmit excitation is especially developed in nerve cells, as is motility in muscle cells, and the ability to secrete in gland cells. The shape of the cell changes and takes the form most suited to its function. A cell devoted to one single thing accomplishes more than the cell of *Vorticella* with all its different assignments. But this kind of differentiation is successful only if there is unified and meaningful teamwork on the part of the various differentiated cells. *Hydra* is only a beginner at this, compared with other animals. We search vainly in *Hydra* for a brain, the center for an organized cooperation of parts.

On Death

Every day we read in the newspapers that accidents or illnesses have taken the lives of a number of people. Even the most careful and the healthiest people do not escape their fate. When they have lived for 80 to 100 years they eventually become senile and die. We see the same thing in animals and plants, even though the appointed times for their existences have different lengths. Every lover of dogs knows that after about 12 years he must take leave of his pet forever. Whoever finds this too sad might do better to keep a parrot, that can become more than a hundred years old. Anyone who wishes to spare his children and his grandchildren also, the pain of farewell should choose a giant Galápagos tortoise for a house pet. It can attain an age of 200 to 300 years. Many worms live only a few days; chickens live 15 to 20 years. Many plants are granted only a few months of life; the mighty giant Sequoia trees in California have lived thousands of years—but sooner or later old age and the end of life come for them all.

Must this be? Is death the necessary consequence of life? The ameba teaches us otherwise. When it divides, both halves live on, with the same rights and privileges, and with the same outlook for the future. They will divide again, and so it continues from generation to generation. Obviously a misfortune can strike them. Many are eaten, or dried up; or they may fall victim to the germs of disease, by no means a privilege of mankind alone. But they do not die from inner necessity, they undergo no senile decrepitude, they have no natural time for death. They bear within themselves the capacity for life unlimited. One-celled organisms are potentially immortal.

In order to understand correctly this contrast between one-celled and many-celled organisms, let us return once more to the colonies of flagellates that we mentioned as showing the first beginnings of division of labor among the cells. When we spoke of them before, we did not take up the question of their mode and manner of propagation. In *Eudorina* a new colony arises from each cell as a result of several successive divisions. These small young daughter colonies become independent; they leave the old jelly envelope, and when they have grown up, the same method of multiplication is repeated. The members of the group differ as little in their capacity for differentiation as in their appearance. *Volvox* is different. Here only a few, perhaps a dozen, cells out of the many hundred individuals of the colony are able to produce new colonies by division. These daughter colonies move inward and constrict themselves off; they then lie inside the mother colony until this finally ruptures and releases the young. The life of the mother is then at an end, and its dead body shows us that here a new division of labor of critical importance has come into being. In this group of cells, only relatively few, the germ cells, remain immortal like the ameba and other one-celled creatures. The great majority of body cells have a limited length of life and even if they do not meet with accidental destruction they die naturally at a given time.

This separation into body cells (also somatic cells, or, in the aggregate, soma for short) and germ cells is characteristic of all many-celled animals and plants. The germ cells have kept the potential immortality of the protozoans, and they can pass life on to unlimited successions of generations. On the other hand, the life of the body cells has a definite time limit. But why, really?

The fact that the daughter colonies of *Volvox* can become free only when the mother bursts is a sufficient reason for the mother colony's death. But when a cat or a water flea can give birth to offspring without suffering harm, why must the mother's life eventually come to an end? This has been explained by saying that the narrowly specialized body cells are gradually worn out and exhausted by the performance of their duties. The traces of such damage are often visible under the microscope. The nerve cells of a feeble old dog are so full of drossy waste substances that it is not surprising that they finally refuse to do any more work. It is, however, thoroughly conceivable that such damage might not occur if cells could be relieved of their labors and transferred to conditions highly favorable for their life. This has been tried with tissue culture; in one experiment, cells from the heart of a chick went on multiplying happily, without cessation, from 1911 to 1940, when the culture was abandoned for extraneous reasons. Thus they lived for 29 years, although if they had been left alone in the heart of the chicken they would have survived at most for only 15 to 20 years.

Perhaps some day highly skilled surgeons will learn to wash the wastes away from body cells while these are still in their normal places, and thus lengthen their lives unlimitedly. But we do not need to rack our brains as to whether or not that would be a blessing.

B. *The Evolution of Life*

•

Like the origin of life, the origin of the different species has been one of the fundamental problems of biology. Until the end of the eighteenth century, the Biblical account of creation was generally accepted. At that time vague stirrings of discontent began to appear. It seemed obvious that if the theories of the early geologists had any merit, the Book of Genesis could not be accepted literally. Bound up with the geological problem was that of biology. Were all species of plants and animals created at the same time and were they in fact fixed and immutable? Those who dared express their doubts were considered crackpots not only by laymen and the clergy but also by many reputable scientists.

This was the intellectual atmosphere into which a book which ranks with that of Copernicus, "The Origin of Species" by Charles Darwin, was launched in 1859. Its publication created an unprecedented scientific and religious controversy, even greater than that which surrounded Galileo. The foundations of the Christian religion were considered under attack. Man himself, according to the theory, was descended from some lower form. The clergy reacted with fury. Darwin himself avoided personal involvement, but T. H. Huxley rose to his defense. Only gradually did the tide of battle turn in favor of the evolutionists. Indeed, as late as 1925 a young schoolteacher named Scopes was convicted of teaching this "false doctrine" in the state of Tennessee.

The impact of the theory of evolution has been enormous. It has changed our way of thinking not only in biology but in almost every other aspect of man's intellectual activity. In Darwinisms *we are offered brief insights into the character of its chief proponent, and in* Darwin and "The Origin of Species" *by Sir Arthur Keith we learn why Darwin's masterpiece is essentially as fresh and significant as when it was first written. Darwin recognized that variation in nature is the means by which natural selection can operate. How variation occurred he did not know. It remained for others to analyze the problem further: de Vries and Bateson discovered that plants and animals are subject not only to small variations but also to large and sudden "mutations." And as a result of these inherited mutations, new varieties are bred.*

It was then that biologists rediscovered the work of a forgotten Austrian monk. Hugo Iltis, one of his compatriots, describes Gregor Mendel and His Work *with clarity and charm. Mendel's work has served as the foundation of the modern science of genetics. The mathematical relationships which he uncovered were explained with the discovery of the genes and chromosomes. More recently, with the discovery of DNA, biologists have begun to understand reproduction and inheritance in exact biochemical terms. John Pfeiffer's article on* DNA *is a careful analysis of the progress that has been made.*

The story is brought up to date in Darwin's Theory in the Light of Modern Genetics *by Karl von Frisch, a clear and illuminating discussion of one of the most important subjects in modern biology.*

Darwinisms

•

Darwin's Father:

"YOU CARE for nothing but shooting, dogs, and rat-catching, and will be a disgrace to yourself and all your family."

T. H. Huxley on The Origin of Species:

"It is doubtful if any single book, except the 'Principia,' ever worked so great and so rapid a revolution in science, or made so deep an impression on the general mind."

Darwin:

"I think that I am superior to the common run of men in noticing things which easily escape attention, and in observing them carefully. My industry has been nearly as great as it could have been in the observation and collection of facts."

"Accuracy is the soul of Natural History. It is hard to become accurate; he who modifies a hair's breadth will never be accurate. Absolute accuracy is the hardest merit to attain, and the highest merit."

"Facts compel me to conclude that my brain was never formed for much thinking."

"I have steadily endeavored to keep my mind free so as to give up any hypothesis, however much beloved (and I cannot resist forming one on every subject), as soon as the facts are shown to be opposed to it."

"If I am wrong, the sooner I am knocked on the head and annihilated so much the better."

"I had, also, during many years followed a golden rule, namely, that whenever a published fact, a new observation or thought came across me, which was opposed to my general results, to make a memorandum of it without fail and at once; for I had found by experience that such facts and thoughts were far more apt to escape from the memory than favorable ones."

"I am very poorly to-day, and very stupid, and hate everybody and everything. One lives only to make blunders."

"I have been speculating last night what makes a man a discoverer of undiscovered things; and a most perplexing problem it is. Many men who are very clever—much cleverer than the discoverers—never originate anything. As far as I can conjecture, the art consists in habitually searching for the causes and meaning of everything which occurs."

"I think I can say with truth that in after years, though I cared in the highest degree for the approbation of such men as Lyell and Hooker, who were my friends, I did not care much about the general public. I do not mean to say that a favorable review or a large sale of my books did not please me greatly, but the pleasure was a fleeting one, and I am sure that I have never turned one inch out of my course to gain fame."

"I look at it as absolutely certain that very much in the *Origin* will be proved rubbish; but I expect and hope that the framework will stand."

"It is a horrid bore to feel as I constantly do, that I am a withered leaf for every subject except Science. It sometimes makes me hate Science, though God knows I ought to be thankful for such a perennial interest, which makes me forget for some hours every day my accursed stomach."

"I do not believe any man in England naturally writes so vile a style as I do."

"Now for many years I cannot endure to read a line of poetry: I have tried lately to read Shakespeare, and found it so intolerably dull that it nauseated me."

"What a book a devil's chaplain might write on the clumsy, wasteful, blundering, low, and horribly cruel works of nature!"

SIR ARTHUR KEITH

Darwin and "The Origin of Species"

•

WHEN H.M.S. BEAGLE, "of 235 tons, rigged as a barque, and carrying six guns," slipped from her moorings in Devonport harbour on 27 December, 1831, the events which were to end in the writing of *The Origin of Species* were being set in train. She had on board Charles Darwin, a young Cambridge graduate, son of a wealthy physician of Shrewsbury, in the rôle of naturalist. On the last day of February 1832 the *Beagle* reached South America and Darwin, just entered on his twenty-fourth year, stepped ashore on a continent which was destined to raise serious but secret doubts in his mind concerning the origin of living things. He was not a naturalist who was content merely to collect specimens, to note habits, to chart distributions, or to write accurate descriptions of what he found; he never could restrain his mind from searching into the reason of things. Questions were ever rising in his mind. Why should those giant fossil animals he dug from recent geological strata be so near akin to the little armour-plated armadillos which he found still alive in the same place? Why was it, as he passed from district to district, he found that one species was replaced by another near akin to it? Did every species of animal and plant remain just as it was created, as was believed by every respectable man known to him? Or, did each and all of them change, as some greatly daring sceptics had alleged?

In due course, after surveying many uncharted coasts, the *Beagle* reached the Galápagos Islands, five hundred miles to the west of South America. Here his doubts became strengthened and his belief in orthodoxy shaken. Why was it that in those islands living things should be not exactly the same as in South America but yet so closely alike? And why should each of the islands have its own peculiar creations? Special creation could not explain such things. South America thus proved to be a second University to Charles Darwin; after three and a half years spent in its laboratories he graduated as the greatest naturalist of the nineteenth century. It had taken him even longer to obtain an ordinary pass degree from the University of Cambridge.

The first stage in the preparation of *The Origin of Species* thus lies in South America. The second belongs to London. The *Beagle* having circumnavigated the world returned to England in October 1836, and by his [twenty-eighth] birthday, 12 February, 1837, Darwin was ensconced in London with his papers round him working hard at his *Journal and Reports,* but at the same determined to resolve those illicit doubts which had been raised by his observations in South America and which still haunted him, concerning the manner in which species and animals had come into the world. He knew he was treading on dangerous ground; for an Englishman to doubt the truth of the Biblical record in the year 1837 was to risk becoming a social outcast; but, for Darwin, to run away from truth was to be condemned by a tender conscience as a moral coward. He was a sensitive man, reflective, quiet, warm-hearted, ever heeding the susceptibilities of his friends. Added to this he was also intensely modest and as intensely honest, fearing above all things even the semblance of a lie in thought or in act. The facts he had observed in South America merely raised his suspicions. They suggested to him that animals and plants might become, in the course of time, so changed as to form new species. At first they were but suspicions, but as he proceeded to collect evidence in London, the suspicions deepened. More particularly was this the case when he inquired into the methods employed by breeders to produce new varieties of pigeons, fowls, dogs, cattle and horses. He soon realised that for the creation of new domestic breeds two factors were necessary—first there must be a breeder or selector, and secondly the animals experimented on must have in them a tendency to vary in a desired direction. Given those two factors, a new breed, having all the external appearances of a new species, could be produced at will.

Having satisfied himself on this point, he turned again to animals and plants living in a state of nature and found that they too tended to vary. "But where," he had to ask himself, "is Nature's selector or breeder?" At this juncture he happened to read an *Essay* written by the Rev. T. R. Malthus, first published in 1798, *On the Principle of Population,* and as he read, realised that the breeder he was in search of did exist in Nature. It took the form, he perceived, of a self-acting mechanism—a mechanism of selection. Among the individuals of every species, there goes on, as Malthus had realised, a competition or struggle for the means of life, and Nature selects the individuals which vary in the most successful direction. The idea that living things had been evolved had been held by many men before Darwin came on the scene; it was already well known that animals tended to vary in form and in habit, but the realisation that Nature had set up in the world of living things an automatic breeder, which utilised variations as a means of progress, was entirely Darwin's discovery.

And thus it came about that during his second year in London (1838) and before he had completed the [twenty-ninth] of his life, Darwin had wrested from Nature one of her deepest secrets—a secret which gave him a clue to one of her many unsolved mysteries. Great ideas, if they are to come at all, usually come before a man is thirty and it was so in Darwin's case. In South America he had merely had doubts about the orthodox belief; the revelation which came in London convinced him that the real story of creation was quite different from the one usually told and accepted. With the discovery of the law of Natural Selection in 1838 *The Origin of Species* entered its second stage of preparation, and it is convenient to regard this stage as ending in January 1839, when Darwin married his cousin Emma Wedgwood.

The third stage opened in September 1842, when he resolved to find peace for study and for health by removing his family from London to Down in the chalky uplands of Kent, where he lived until his death on 14 April, 1882. He had inherited money and resolved to devote his life to the solution of the old problem of creation, instead, as is so often the case with men of his class, to leisure and to sport. On his arrival at Down he believed he was in possesion of a secret of momentous import—and so unholy that he determined to say nothing of it until he had attained complete certainty. He had at that time many researches in hand and, as he worked at them, he was ever on the outlook for evidence to prove the truth or untruth of his theory. We know that, just before he left London, he had permitted himself the luxury of seeing what his theory looked like when reduced to paper; that sketch, written in June 1842, is really the first outline of *The Origin of Species*, but it then filled only thirty-five pages of manuscript. It was not until 1844, when he had been two years at Down, and had amassed much additional evidence, that he committed to writing a complete exposition of his theory; this time he succeeded in filling 230 pages of manuscript. This third stage—the stage of accumulating evidence—continued with many intermissions until 1854, when the preparation of *The Origin of Species* entered its fourth stage.

In 1854 he completed his research on Barnacles—a seven years' task, and was thus free to set in systematic order the immense amount of evidence he had accumulated—all of it bearing upon the problem of transmutation or evolution of every form of life. This he now proceeded to do, but there were many interruptions. From time to time, while busy with many inquiries and experiments and sadly hindered by indifferent health, a chapter of his projected work was written and as his self-imposed task proceeded it became apparent to him it was to be a big book—three volumes at least. And so he went along until the summer of 1858 was reached, when on a day early in June the rural postman pushed into his letter-box a missive which gave him the shock of his life and brought his projected book to a sudden end. The postmark showed that the missive had been dispatched from an address in the Celebes Islands. In this sudden manner we pass from the fourth to the fifth and final stage in the preparation of *The Origin of Species*.

In the history of Science there is no episode so dramatic as that which compelled Charles Darwin to pass so abruptly to the fifth and final stage in the preparation of *The Origin of Species*. He was no longer a young man; he was in his fiftieth year. Let us look for a moment at the staging of this drama and the actors who took part in it. In February 1858, when Darwin, in his study at Down, was suffering from his "accursed stomach" and struggling painfully with his proofs of transmutation, another Englishman, Alfred Russel Wallace, was lying in the small island of Ternate, in the Malay Archipelago, suffering from bouts of malarial fever, and puzzling over the same problem as engaged Darwin's attention at Down. The writer has experienced these bouts of ague and knows how vivid is the imagery that then races through the brain and how nimbly the mind hunts along a train of ideas. Such a bout brought Wallace his revelation. He was fourteen years Darwin's junior. He was also a poor man, being dependent for a livelihood on the collections he made as a travelling naturalist. He, too, had visited South America just as Darwin had, and it was while collecting on the Amazon that he became impressed by the tendency of animals and plants to vary. Soon after his arrival in Borneo he had read, just as Darwin had done eighteen years before him, Malthus's *Essay on Population*. He had, before then, begun to suspect that species were not immutable, and as his brain raced along during his attack of fever in Ternate it stumbled across the idea which came to Darwin in London—the idea that the struggle would favour those individuals which tended to vary in an advantageous direction and that such individuals might continue to change until a new species was brought into existence. As soon as the attack of fever was over and his temperature had returned to normal he began to write, and at one sitting finished an account of his discovery—an idea which would explain the origin of new species without calling in the aid of any supernatural agency whatsoever. Having written his sketch, he thereupon addressed it to a man who was almost a stranger to him—*Charles Darwin Esq., F.R.S., Down House, Down, Kent*, where it duly arrived in the third week of June 1858.

On opening this missive Darwin found that the fears of his best friends, Sir Charles Lyell and Dr. Joseph Hooker of Kew Gardens, had come only too true; he had been forestalled. By a curious stroke of fate, the favourite child of his brain, which he had nursed and tended in secret for over twenty years, was suddenly deprived of that which is so dear to the heart of a father—the birthright of priority. Wallace's sketch, he found, was almost a replica of the one he himself had penned after his arrival at Down; and how much had he discovered and added to the original sketch in the intervening years! Darwin knew that if he acted rationally, and he was as nearly rational as men are made, he ought to welcome Wallace's communication. It was a confirmation of his own conclusions. He was ashamed to find himself troubled at heart over this paltry matter of priority. It is a long way from Kent to the land of Moriah and from Darwin's day to that of Abraham, but distant as are the places and the times, they are linked together by the same human nature. Abraham with his knife and bundle of faggots was resolved to make the supreme sacrifice and so was Darwin, and he would have done it had not his friends Lyell and Hooker intervened. They exercised a judgment worthy of Solomon; justice was to be done to both authors by a conjoint communication to a learned society. They asked Darwin to supply them with a brief abstract of his theory and this, with Wallace's sketch, they sent to the Linnean Society of London. The two papers were read at a meeting held on 1 July, 1858, and caused no great commotion.

This communication having been made, Lyell and Hooker insisted that Darwin must now prepare for publication, and he then began to work on *The Origin of Species* as we now know it. He set himself to abstract and to condense what he had already written. The opening chapters were finished in September 1858 but it took him fully twelve months of toil and tribulation before he could write *finis*. On 24 November, 1859, the book was published and thus ended the fifth stage in the preparation of *The Origin of Species*.

The publishers apparently did not expect a big demand for *The Origin;* at least they printed only 1250 copies. A second edition was called for in 1860—one of 3000 copies. A third appeared in 1861, a fourth in 1866, a fifth in 1869 and a sixth and final edition in 1872. Darwin lived for ten years after the issue of the sixth edition, but so thoroughly had he winnowed his data, so fully had he met the expert criticism of his time, that he did not feel called upon to make any further alteration in its text.

Such is a brief account of how *The Origin of Species* came to be written. Its preparation occupied, from first to last, a period of forty years, for its foundation was being laid in 1832 when Darwin began his researches in South America, and its building was not finished until the last edition appeared in 1872. The book came into being during a period when Europe was in a state of intense intellectual activity, and the effect it produced was immediate and profound. The generation which felt its first shock is dying or dead. The generation which has grown up, like every new generation, is passing the household gods inherited from its predecessor through the fiery furnace of criticism. How is *The Origin of Species* to emerge from this ordeal? Having served its day and generation is it now dead? Or does it possess, within itself, the seeds of eternal youth and is it thus destined to become one of the world's perpetual possessions? The latter, I am convinced, is its destiny. On the foundations laid by Darwin in this book his successors have erected a huge superstructure which will be infinitely extended and modified as time goes on. Yet I feel certain that as long as men and women desire to know something of the world into which they have been born, they will return, generation after generation, to drink the waters of evolutionary truth at the fountain-head.

The Origin of Species is still freely abused and often misrepresented, just as it was when Darwin was alive. In his final edition he entered a mild protest—a luxury he rarely indulged in—against a misrepresentation to which his theory was persistently subjected. "But as my conclusions have lately been much misrepresented," he wrote, "and it has been stated that I attribute the modification of species exclusively to *natural selection*, I may be permitted to remark, that in the first edition of this work, and subsequently, I placed in a most conspicuous position—namely, at the close of the Introduction—the following words: *I am convinced that natural selection has been the main, but not the exclusive means of modification.* This has been of no avail. Great is the power of steady misrepresentation, but the history of science shows that fortunately this power does not long endure."

The power of error to persist is more enduring than Darwin thought; the misrepresentation of which he complained is being made now more blatantly than ever before. It is being proclaimed from the housetops that *The Origin of Species* contained only one new idea, and that this idea, the conception of natural selection, is false. Natural selection, some of his modern critics declare, is powerless to produce new forms of either plant or animal. Darwin never said it could. In his book the reader will find him giving warning after warning that by itself selection can do nothing. To effect an evolutionary change two sets of factors, he declared, must be at work together—those which bring about variations or modifications in animal or in plant and those which favour and select the individuals which vary or become modified in a certain direction. Why should so many critics continue to misunderstand the essentials of Darwin's theory of evolution?

Men do not wilfully persist in misrepresentation; there must be some explanation of their error. The truth is that Darwin himself was at fault; the full title he gave to his book was *The Origin of Species by Means of Natural Selection.* Plainly such a title was a misnomer, his book was and is much more than such a title implies; it was much more than a mere demonstration of the action of natural selection, it was the first complete demonstration that the law of evolution holds true for every form of living thing. It was this book which first convinced the world of thoughtful men and women that the law of evolution is true. Long before Darwin's time men had proclaimed the doctrine of evolution, but they failed to convince their fellows of its truth, both because their evidence was insufficient and because they had to leave so much that was unexplained. Darwin, on the other hand, brought forward such an immense array of facts in this book and set them in such a logical sequence that his argument proved irresistible. He never resorted to any kind of special pleading, but permitted facts to speak for themselves. However longingly his readers clung to age-long beliefs, Darwin compelled them to face facts and draw conclusions, often at enmity with their predilections. We all desire to be intellectually honest, and sooner or later truth wins. It was this book which won a victory for evolution, so far as that victory has now been won. When it appeared in the nineteenth century the Why and the How of evolution were immaterial issues. What had to be done then was to convince men that evolution represented a mode of thinking worthy of acceptation and in that *The Origin of Species* succeeded beyond all expectation. Nor has it finished its appointed mission. No book has yet appeared that can replace it; *The Origin of Species* is still the book which contains the most complete demonstration that the law of evolution is true.

This, then, is Darwin's essential service to the world—not that he discovered the law of Natural Selection—but that he succeeded in effecting a complete revolution in the outlook of mankind on all living things. He wrought this revolution through *The Origin of Species*. Darwin himself formed a true estimate of what the nature of this revolution was. In the last paragraph of his Introduction, he writes, "Although much remains obscure and will long remain obscure, I can entertain no doubt, after the most deliberate study of dispassionate judgment of which I am capable, *that the view which most naturalists until recently entertained, and which I formerly entertained—namely, that each species has been independently created—is erroneous. I am firmly convinced that species are not immutable.*" From this statement we see that Darwin's aim was to replace a belief in special creation by a belief in evolution and in this he did succeed, as every modern biologist will readily admit. No one was in a better position to measure what Darwin succeeded in doing than his magnanimous contemporary and ally Alfred Russel Wallace. Writing to Professor Newton of Cambridge in 1887, five years after Darwin's death, he penned the following passage: "I had the idea of working it out [the theory of natural selection], so far as I was able, when I returned home, not at all expecting that Darwin had so long anticipated me. I can truly say now, as I said many years ago, that I am glad it was so, for I have not the love of work, experiment and detail that was so preeminent in Darwin and without which anything I could have written *would never have convinced the world.*" Darwin succeeded in convincing the world not only by his superabundance of proof but by the transparently honest way in which he presented his case. No one can read *The Origin of Species* without feeling that Darwin had the interests of only one party at heart—his client, Truth.

HUGO ILTIS

Gregor Mendel and His Work

•

IN A SMALL VILLAGE on the northern border of what was called Austria at that time, a boy was born in a farmer's house who was destined to influence human thoughts and science. Germans, Czechs and Poles had been settled side by side in this part of the country, quarreling sometimes, but mixing their blood continually. During the Middle Ages the Mongolic Tatars invaded Europe just there. Thus, the place had been a melting pot of nations and races and, like America, had brought up finally a splendid alloy. The father's name was Anton Mendel; the boy was christened Johann. He grew up like other farmers' boys; he liked to help his father with his fruit trees and bees and retained from these early experiences his fondness for gardening and beekeeping until his last years. Since his parents, although not poor compared with the neighbors, had no liquid resources, the young and gifted boy had to fight his way through high school and junior college (Gymnasium). Finally he came to the conclusion, as he wrote in his autobiography, "That it had become impossible for him to continue such strenuous exertions. It was incumbent on him to enter a profession in which he would be spared perpetual anxiety about a means of livelihood. His private circumstances determined his choice of profession." So he entered as a novice the rich and beautiful monastery of the Augustinians of Bruenn in 1843 and assumed the monastic name of Gregor. Here he found the necessary means, leisure and good company. Here during the period from 1843 to 1865 he grew to become the great investigator whose name is known to every schoolboy to-day.

On a clear cold evening in February, 1865, several men were walking through the streets of Bruenn towards the modern school, a big building still new. One of those men, stocky and rather corpulent, friendly of countenance, with a high forehead and piercing blue eyes, wearing a tall hat, a long black coat and trousers tucked in top boots, was carrying a manuscript under his arm. This was Pater Gregor Mendel, a professor at the modern school, and with his friends he was going to a meeting of the Society of Natural Science where he was to read a paper on "Experiments in Plant Hybridization." In the schoolroom, where the meeting was to be held, about forty persons had gathered, many of them able or even outstanding scientists. For about one hour Mendel read from his manuscript an account of the results of his experiments in hybridization of the edible pea, which had occupied him during the preceding eight years.

Mendel's predecessors failed in their experiments on heredity because they directed their attention to the behavior of the type of the species or races as a whole, instead of contenting themselves with one or two clear-cut characters. The new thing about Mendel's method was that he had confined himself to studying the effects of hybridization upon single particular characters, and that he didn't take, as his predecessors had done, only a summary view upon a whole generation of hybrids, but examined each individual plant separately.

The experiments, the laws derived from these experiments, and the splendid explanation given to them by Mendel are to-day not only the base of the modern science of genetics, but belong to the fundamentals of biology taught to millions of students in all parts of the world.

Mendel had been since 1843 one of the brethren of the beautiful and wealthy monastery of the Augustinians of Bruenn, at that time in Austria, later in Czechoslovakia. His profession left him sufficient time, and the large garden of the monastery provided space enough, for his plant hybridizations. During the eight years from 1856 to 1864, he observed with a rare patience and perseverance more than 10,000 specimens.

In hybridization the pollen from the male plant is dusted on the pistils of the female plant through which it fertilizes the ovules. Both the pollen and the ovules in the pistils carry hereditary characters which may be alike in the two parents or partly or entirely different. The peas used by Mendel for hybridization differed in the simplest case only by one character or, better still, by a pair of characters; for instance, by the color of the flowers, which was red on one parental plant and white on the other; or by the shape of the seeds, which were smooth in one case and wrinkled in the other; or by the color of the cotyledons, which were yellow in one pea and green in the other, etc. Mendel's experiments show in all cases the result that all individuals of the first generation of hybrids, the F_1 generation as it is called to-day, are uniform in appearance, and that moreover only one of the two parental characters, the stronger or the dominant one, is shown. That means, for instance, that the red color of the flowers, the smooth shape of the seeds or the yellow color of the cotyledons is in evidence while the other, or recessive, character seems to have disappeared. From the behavior of the hybrids of the F_1 generation, Mendel derived the first of the experimental laws, the so-called "Law of Uniformity," which is that all individuals of the first hybrid generation are equal or uniform. The special kind of inheritance shown by the prevalence of the dominant characters in the first hybrid generation is called alternative inheritance or the pea type of inheritance. In other instances, however, the hybrids show a mixture of the parental characteristics. Thus, crossing between a red-flowered and a white-flowered four o'clock (*Mirabilis*) gives a pink-flowered F_1 generation. This type of inheritance is called the intermediate, or Mirabilis, type of inheritance.

Now, Mendel self-fertilized the hybrids of the first generation, dusting the pistils of the flowers with their own pollen and obtained thus the second, or F_2, generation of hybrids. In this generation the recessive characters, which had seemingly disappeared, but, which were really only covered in the F_1 generation, reappeared again and in a characteristic and constant proportion. Among the F_2 hybrids he found three red-flowered plants and one white-flowered plant, or three smooth-seeded and one wrinkled-seeded plant, or three plants with yellow cotyledons and one with green ones. In general, the hybrids of the F_2 generation showed a ratio of three dominant to one recessive plants. Mendel derived from the behavior of the F_2 generation his second experimental law, the so-called "Law of Segregation." Of course, the characteristic ratio of three dominant to one recessive may be expected only if the numbers of individuals are large, the Mendelian laws being so-called statistical laws or laws valid for large numbers only.

The third important experimental law Mendel discovered by crossing two plants which distinguished themselves not only by one but by two or more pairs of hereditary characters. He crossed, for instance, a pea plant with smooth and yellow seeds with another having green and wrinkled seeds. The first, or F_1, generation of hybrids was of course uniform, showing both smooth and yellow seeds, the dominant characters. F_1 hybrids were then self-fertilized and the second hybrid, or F_2, generation was yielded in large numbers, showing all possible combinations of the parental characters in characteristic ratios and that there were nine smooth yellow to three smooth green to three wrinkled yellow to one wrinkled green. From these so-called polyhybrid crossings, Mendel derived the third and last of his experimental laws, the "Law of Independent Assortment."

These experiments and observations Mendel reviewed in his lecture. Mendel's hearers, who were personally attached to the lecturer as well as appreciating him for his original observations in various fields of natural science, listened with respect but also with astonishment to his account of the invariable numerical ratios among the hybrids, unheard of in those days. Mendel concluded his first lecture and announced a second one at the next month's meeting and promised he would give them the theory he had elaborated in order to explain the behavior of the hybrids.

There was a goodly audience, once more, at the next month's meeting. It must be admitted, however, that the attention of most of the hearers was inclined to wander when the lecturer became engaged in a rather difficult algebraical deduction. And probably not a soul among the audience really understood what Mendel was driving at. His main idea was that the living individual might be regarded as composed of distinct hereditary characters, which are transmitted by distinct invisible hereditary factors—to-day we call them genes. In the hybrid the different parental genes are combined. But when the sex cells of the hybrids are formed the two parental genes separate again, remaining quite unchanged and pure, each sex cell containing only one of the two genes of one pair. We call this fundametal theoretical law the "Law of the Purity of the Gametes." Through combination of the different kinds of sex cells, which are produced by the hybrid, the Law of Segregation and the Law of Independent Assortment can be easily explained.

Just as the chemist thinks of the most complicated compound as being built from a relatively small number of invariable atoms, so Mendel regarded the species as a mosaic of genes, the atoms of living organisms. It was no more nor less than an atomistic theory of the organic world which was developed before the astonished audience. The minutes of the meeting inform us that there were neither questions nor discussions. The audience dispersed and ceased to think about the matter—Mendel was disappointed but not discouraged. In all his modesty he knew that by his discoveries a new way into the unknown realm of science had been opened. "My time will come," he said to his friend Niessl.

Mendel's paper was published in the proceedings of the society for 1866. Mendel sent the separate prints to Carl Naegeli in Munich, one of the outstanding biologists of those days, who occupied himself with experiments on plant hybridization. A correspondence developed and letters and views were exchanged between the two men. But even Naegeli didn't appreciate the importance of Mendel's discovery. In not one of his books or papers dealing with heredity did he even mention Mendel's name. So, the man and the work were forgotten.

When Mendel died in 1884, hundreds of mourners, his pupils, who remembered their beloved teacher, and the poor, to whom he had been always kind, attended the funeral. But although hundreds realized that they had lost a good friend, and other hundreds attended the funeral of a high dignitary, not a single one of those present recognized that a great scientist and investigator had passed away.

The story of the rediscovery and the sudden resurrection of Mendel's work is a thrilling one. By a peculiar, but by no means an accidental, coincidence three investigators, in three different places in Europe, de Vries in Amsterdam, Correns in Germany, Tschermak in Vienna, came almost at the same time across Mendel paper and recognized at once its great importance.

Now the time had arrived for understanding, now "his time had come" and to an extent far beyond anything of which Mendel had dreamed. The little essay, published in the great volume of the Bruenn Society, has given stimulus to all branches of biology. The progress of research since the beginning of the century has built for Mendel a monument more durable and more imposing than any monument of marble, because not only has "Mendelism" become the name of a whole vast province of investigation, but all living creatures which follow "Mendelian" laws in the hereditary transmission of their characters are said to "Mendelize."

As illustrations, I will explain the practical consequences of Mendelian research by two examples only. The Swede, Nilsson-Ehle, was one of the first investigators who tried to use Mendelistic methods to improve agricultural plants. In the cold climate of Sweden some wheat varieties, like the English square-hood wheat, were yielding well but were frozen easily. Other varieties, like the Swedish country wheat, were winter-hard but brought only a poor harvest. Nilsson-Ehle knew that in accordance with the Mendelian Law of Independent Assortment, the breeder is able to combine the desired characters of two different parents, like the chemist who combines the atoms to form various molecules or compounds. He crossed the late-ripening, well-yielding, square-hood wheat with the early-ripening, winter-hard, but poor-yielding Swedish country wheat. The resulting F_1 generation, however, was very discouraging. It was uniform, in accordance with Mendel's first law, all individuals being late-ripe and poor-yielding, thus combining the two undesirable dominant characters. In pre-Mendelian times the breeder would have been discouraged and probably would have discontinued his efforts. Not so Nilsson-Ehle, who knew that the F_1 generation is hybrid, showing only the dominant traits, and that the independent assortment of all characters will appear only in the F_2 generation. Self-fertilizing the F_1 plants he obtained an F_2 generation showing the ratio of nine late-ripe poor-yielding to three late-ripe well-yielding, to three early-ripe poor-yielding, to one early-ripe, well-yielding wheat plants. The desired combination of the two recessive characters, early-ripe, well-yielding, appeared only in the smallest ratio, one in sixteen —but because recessives are always true-breeding, or as it is called "homozygote," Nilsson-Ehle had only to isolate these plants and to destroy all others in order to obtain a new true breeding early-ripe and well-yielding variety which after a few years gave a crop large enough to be sold. Thus, by the work of the Mendelist, Nilsson-Ehle, culture of wheat was made possible even in the northern parts of Sweden and large amounts heretofore spent for imported wheat could be saved.

Another instance shows the importance of Mendelism for the understanding of human inheritance. Very soon after the rediscovery of Mendel's paper it became evident that the laws found by Mendel with his peas are valid also for animals and for human beings. Of course, the study of the laws of human heredity is limited and rendered more difficult by several obstacles. We can't make experiments with human beings. The laws of Mendel are statistical laws based upon large numbers of offspring, while the number of children in human families is generally small. But in spite of these difficulties it was found very soon that human characters are inherited in the same manner as the characters of the pea. We know, for instance, that the dark color of the iris of the eye is dominant, the light blue color recessive. I remember a tragi-comic accident connected with this fact. At one of my lecture tours in a small town in Czechoslovakia, I spoke about the heredity of eye color in men and concluded that, while two dark-eyed parents may be hybrids in regard to eye color and thus may have children both with dark and blue eyes, the character blue-eyed, being recessive, is always pure. Hence two blue-eyed parents will have only blue-eyed children. A few months later I learned that a divorce had taken place in that small town. I was surprised and resolved to be very careful even with scientifically proved statements in the future.

Even more important is the Mendelian analysis of hereditary diseases. If we learn that the predisposition to a certain disease is inherited through a dominant gene, as diabetes, for instance, then we know that all persons carrying the gene will be sick. In this case all carriers can be easily recognized. In the case of recessive diseases, feeblemindedness, for instance, we know that the recessive gene may be covered by the dominant gene for health and that the person, seemingly healthy, may carry the disease and transmit it to his children.

With every year the influence of Mendel's modest work became more widespread. The theoretical explanation given by Mendel was based upon the hypothesis of a mechanism for the distribution and combination of the genes. To-day we know that exactly such a mechanism, as was seen by the prophetic eye of Mendel, exists in the chromosome apparatus of the nucleus of the cells. The development of research on chromosomes, from the observations of the chromosomes and their distribution by mitosis to the discovery of the reduction of the number of chromosomes in building the sex cells and finally to the audacious attempt to locate the single genes within the chromosomes, is all a story, exciting as a novel and at the same time one of the most grandiose chapters in the history of science. A tiny animal, the fruit-fly, Drosophila, was found to be the best object for genetical research. The parallelism between the behavior of the chromosomes and the mechanism of Mendelian inheritance was studied by hundreds of scientists, who were trying to determine even the location of the different genes within the different chromosomes and who started to devise so-called chromosome maps.

From 1905 to 1910, I tried by lectures and by articles to renew the memory of Mendel in my home country and to explain the importance of Mendelism to the people. This was not always an easy task. Once I happened to be standing beside two old citizens of Bruenn, who were chatting before a picture of Mendel in a book-seller's window. "Who is that chap, Mendel, they are always talking about now?" asked one of them. "Don't you know?" replied the second. "It's the fellow who left the town of Bruenn an inheritance!" In the brain of the worthy man the term "heredity" had no meaning, but he understood well enough the sense of an inheritance or bequest.

JOHN PFEIFFER

DNA

•

COILED like tiny springs in the nucleus of a fertilized human ovum are molecules of a remarkable compound known to biochemists as deoxyribonucleic acid or, more conveniently, as DNA. Half of this DNA has come from the sperm and half from the ovum itself. The sum of the two halves is exceedingly small, about two ten-trillionths of an ounce. But this minuscule total is enough to start a process in which a single cell, barely visible to the naked eye, will develop into a mature human, a highly organized community that contains some ten thousand billion cells.

DNA represents the master substance of life and growth. It is hereditary material, the stuff out of which genes are made. Studying its structure and how it works has become a major objective of present-day research in the life sciences. In fact, biology is currently in the midst of a breakthrough as significant and exciting as the recent advances in physics that led to the release of nuclear energy—and one important reason for this is science's new insight into the nature of these complex molecules.

The more we learn about the DNA molecule, the better we shall understand the subtlest reactions of living matter. It is the basic unit, the "atom," of biology. For one thing, this nucleic acid has the notable property of being able to reproduce. The cells in our bodies contain—in their genes—faithful copies of the DNA molecules in the ovum that was each of us at the time of conception. In other words, the DNA we inherited from our parents has duplicated itself many billions of times and it will continue to do so in our children and theirs for endless generations.

Reproduction is an enormously complex process, even at the molecular level, and DNA is an enormously complex substance. The discovery of its basic structure, one of the key advances of modern biology, came during the winter of 1953 at Cambridge University, England. The setting was hardly glamorous—not a shining laboratory but an unimpressive office in a sort of barracks, a space just large enough to hold two desks and a bookcase. Francis Crick and James Watson, a physical chemist and a biologist, found themselves engaged in a curious scientific game.

The game was a kind of jigsaw puzzle in three dimensions. The pieces of the puzzle were metal sheets, about the size of ordinary playing cards, but cut into different shapes representing six different substances. First was the sugar deoxyribose (the D in DNA); second was a phosphate group; and finally came four nitrogen-containing compounds known as "bases." Previous studies had shown that the DNA molecule is made up entirely of these six "building blocks," although it includes many of each type. Other evidence had hinted that the over-all shape of the molecule might be some sort of intricate coil.

Crick and Watson started fastening their metal sheets together in various ways, using rotatable joints to serve as analogies of chemical bonds. They spent more than two weeks getting nowhere before they finally hit upon a new way of putting the metal parts together to form a structure that checked with the known evidence. Then, in three days and nights of work, they assembled a model that resembled a piece of abstract modern sculpture—a double helix, two interconnected molecular strands twisted into a spiral-staircase structure.

Each strand in the model has as its foundation a long chain of alternate sugar and phosphate units. Attached to this chain at regular spacings, and sticking out as side groups, are sequences of four nitrogenous bases: adenine (A), guanine (G), cytosine (C), and thymine (T). Each strand has thousands of base side groups.

Now, each such DNA strand has a similar structure, and each is attached to another strand by its side groups. In other words, the nitrogenous bases not only pair off with one another, but pair according to a definite rule that Crick and Watson deduced while building their model. Adenine can combine only with thymine, and guanine only with cytosine. So two attached strands look something like a ladder or a zipper.

Investigators have been quick to note the biological implications of this model. Among other things, it offers possible mechanisms for the self-duplicating capacity of DNA. Imagine a solution containing the six essential DNA building blocks, including ample quantities of the four bases, plus a "parent" molecule of DNA. The two strands of the parent molecule have separated by some sort of unraveling process. Now this solution, like most solutions, is a site of frenzied activity. Particles of the four bases are moving at high speeds in all directions. They collide with one another and with the unraveled DNA strands and, as a rule, they bounce apart after colliding. A free adenine particle, for example, bounces off an adenine or guanine or cytosine group attached to a DNA strand; but if it collides with an attached thymine group, there may be a neat fit and it may stick in place.

A sufficient number of appropriate collisions will thus produce two new DNA strands, formed on and fastened to each strand of the original molecule. In short, each of the original strands is complementary to the other and effectively serves as a template or mold that determines the other's structure. The result is reproduction at the molecular level: two DNA molecules exist where only one existed before. Furthermore, because there are billions and billions of collisions in the solution every second, reproduction may occur swiftly. Arthur Kornberg of Stanford University has prepared special solutions in which DNA "breeding" actually takes place (in 1959 he won a Nobel Prize for his work in this field).

The Crick-Watson model is of outstanding significance for another reason. Since DNA is the main hereditary material—the physical basis of genes—differences in its makeup determine the differences among species. These differences are extremely subtle. In fact, the similarities are far more obvious than the dissimilarities.

DNA exists throughout the vast range of living things. It comes in "packets," molecules of about 15,000 "base-pairs" each. At the lowest end of the scale of life, DNA is found in many viruses—for instance, in certain tadpole-shaped viruses so tiny that more than five billion of them would fit comfortably into a sphere no bigger across than the period at the end of this sentence. DNA is also found in bacteria, and in all higher organisms from amoeba to man. It is generally constructed according to the same pattern, a double helix of paired strands, made of the same half-dozen building blocks. Moreover, the foundation chains of alternate sugar and phosphate groups are always identically the same in basic structure.

The subtle essential differences that characterize different species involve the base-pairs connecting the chains, the "rungs" of the DNA ladder. Broadly speaking, the higher we move up the evolutionary scale, the greater the number of base-pairs—that is, the longer the ladder. There may be packets containing some half-million base-pairs in the molecules of virus DNA, ten million in bacterial DNA, and perhaps ten billion or more in mammalian DNA. This is what we would expect, assuming that a more advanced species needs more DNA to transmit its hereditary traits to future generations, and this rule appears to hold in general.

But a word of warning is in order. Part of the excitement and challenge of biological research is its exquisite complexity; living things frequently fail to obey our most "logical" rules. To cite a stubborn fact, it apparently takes more DNA to form a lungfish than to form a man, and this striking exception to an orderly increase of complexity has yet to be explained.

Suppose that we had complete information (which we emphatically do *not*); that we knew the exact and complete sequence of base-pairs in the DNA molecules of every species on earth. Suppose, further, that the molecule packets of each species were attached and then untwisted to form a single continuous strip. In that case, we would have about two million such strips, because the earth houses an estimated two million species.

Now, according to current concepts, each one of these strips would be unique, in that its base-pairs would be arranged in a different *order* from those of every other species. In theory, we could study the strips one by one and "read off" their base-pairs as though we were reading information from a vast collection of ticker tapes. No two strips would read in exactly the same way and each would represent a distinct species. Thus a sequence starting "A-T, G-C, T-A, C-G, A-T . . ." might represent a jellyfish, "G-C, A-T, C-G, G-C, T-A . . ." might represent an elephant, and "T-A, A-T, G-C, C-G, G-C . . ." a man.

Getting down to finer detail, a complete reading of base-pair sequences—theoretically—would make it possible not only to distinguish strains within each species but also individuals within strains. This is another way of saying that, excepting identical twins, no two individuals have exactly the same order of base-pairs and every individual represents a unique combination of inherited traits. The differences between all living things, in this conception, may be traced to the four bases in their DNA molecules: the language of heredity may be written with an alphabet of only four letters.

The amazingly precise organization of hereditary material reflects the job it has to do. DNA contains the plan or blueprint for the construction of living cells and systems of living cells. Its specifications are expressed in a code of base-pairs that includes all the information needed to synthesize that all-important class of biological compounds, the proteins. The contraction of each muscle, the action of growth-promoting hormones, the flash of electrical signals along nerve fibers, digestion and the regulation of body temperature, even thought itself—in these and all other functions, the proteins play a fundamental role.

The cycles of life would be impossible if it were not for enzymes—substances that speed the rates of biochemical reactions without themselves undergoing changes in the process. And all of the six hundred to seven hundred known enzymes are proteins. So are the thousands of different antibodies, each designed to combat the effects of a different infective agent. Yet, for all their variety, proteins—like DNA—are composed of a relatively small number of building blocks, some twenty organic units known as amino acids. Here, again, DNA has a vital and complex part to play.

DNA controls the intricate "assembly line" that links amino acids together into the giant molecules we know as proteins. There is still a great deal to learn about how this assembly line works, but considerable progress in understanding has been made during the past decade or so. The following account is based on experiments conducted at laboratories throughout the world and particularly on the work of Jean Brachet of the University of Brussels, of Torbjörn Caspersson of Stockholm, and of Mahlon Hoagland and his associates at the Massachusetts General Hospital in Boston. It involves numerous recent studies of the anatomy and physiology of cells.

The metabolic chores of any cell are guided, by remote control, by the DNA in its nucleus. DNA molecules may be compared to expensive master dies required to manufacture the precision parts for machinery. Such dies are too valuable to be used directly in the making of parts. The actual work, the shaping of metal, calls for cheaper, expendable working dies formed from the master dies. Similarly, DNA is too valuable to take part in the everyday business of protein synthesis and must be preserved for special duties. Locked like a treasure in the cell's nucleus, it makes "working dies" of itself—molecules of another substance, known to biochemists as ribosenucleic acid, or RNA for short.

The RNA molecules then migrate out of the nucleus. We do not know for certain how or in what form they do this, but we do know where most of them end up. With the electron microscope, we see that the region surrounding the cell's nucleus contains a tangled network of threadlike tubes, hollow vessels that are believed to serve as the cell's circulatory system. Attached to these tubes, like berries on vines, are thousands of small dense granules with diameters of from 100 to more than 150 angstroms (or some 2 to 2.5 millionths of an inch). These granules play a central role in our story. Known as "ribosomes," they contain 80 per cent of all RNA.

Let us take an imaginary look at the protein assembly line. Think of a single ribosome, magnified so many million times that it appears as a sphere about the size of a grapefruit. We may picture an RNA molecule, fastened securely to the surface of this sphere and containing—by way of mechanical analogy—a succession of "notches" or indentations. The working die, so to speak, is now in place.

Now, the solution in which the ribosome is constantly bathed includes large numbers of special "carrier" molecules. There are twenty different types of these molecules, and each type has attached itself to a different one of the twenty amino-acid units that, in combination, make up all proteins. Each carrier type is also designed to fit into a particular notch of the RNA working die.

In the solution, the carrier molecules are moving at high speeds. They collide with one another and with the RNA working die: usually, they bounce off. But when the right molecules collide in just the right way with the right notches, they fit into place as snugly as a key in a lock and stay put.

What has been accomplished? As the "keys" fit their "locks," a series of amino acids, attached to their individual carrier molecules, have been put into a particular alignment. Lined up in this order, the amino acids then link to one another and separate from their carrier molecules as a completely assembled protein.

It should be emphasized that this is both a highly simplified and a highly mechanistic analogy for what is believed to be happening in protein synthesis. Proteins contain anywhere from fifty to more than ten thousand amino acids, and actual RNA working dies are correspondingly long. Furthermore, research is continuing at such a pace that our ideas about what happens will certainly be modified in the light of increased knowledge.

The picture is a dynamic one. Every one of the thousands of proteins in the body has its specific working die, a length of RNA that occupies a key position in the amino-acid assembly line. Every second, protein molecules are being manufactured by the billion in the legions of ribosomes of tissue cells. These new molecules appear at just the right rate to replace old, broken-down molecules that have served their purposes. Life is a delicate balance between synthesis and demolition. Survival depends ultimately on the structure and function of the self-duplicating master substance in the cell nucleus—DNA molecules passed on from parents to offspring, generation after generation.

Research on the structure and the function of DNA and RNA has created a new, rapidly expanding area of biochemical genetics. It has brought a new understanding of mutation and the possibility of new treatments for hereditary diseases. But it has done a great deal more than that. The implications of current studies are so far-reaching that they concern not only the mechanisms of heredity but every major field of biology.

To cite one example, consider what happens when a virus enters the system. Certain defense cells go into action and produce antibodies that are designed to neutralize the effects of the virus. The significant point is that a unique antibody is produced for each invader. According to one estimate, the average person may be exposed to some ten thousand different kinds of infective agents during a lifetime—and in consequence manufactures some ten thousand different kinds of antibodies.

As we have already mentioned, antibodies are proteins. This means that the vast store of RNA working dies in the tissues includes a supply of special antibody dies ready for emergency use. It also means that our genetic reserves are enormous, because each working die is shaped according to the precise specifications of its DNA "master die" in the nuclei of antibody-producing cells. So an intimate connection exists between one's resistance to infection and one's own hereditary material.

The connection goes far deeper. Viruses are hereditary material on the loose; mobile genes, themselves DNA or RNA molecules enclosed in protein shells. A struggle is going on in the body of a patient suffering from polio, smallpox, influenza, or any other virus disease. The invading virus may take over the genetic apparatus of an infected cell, so that the cell no longer manufactures its own proteins. Instead it manufactures hundreds or thousands of new viruses, and infection may spread swiftly. This is a war to the death, a war of competing hereditary materials. The importance of understanding such processes becomes even clearer in the light of recent work suggesting that viruses may be involved in some forms of human cancer.

In an entirely different area, nucleic acids may have something to do with the highest workings of the mind. We hold enough information in our brains to fill hundreds of thousands of volumes, and investigators are speculating about the possibility that our records of times past may be stored in the form of appropriately patterned protein molecules. If so, DNA-RNA research and brain biochemistry may bring a dramatic increase in our knowledge of memory and learning.

Finally, biologists now believe that the development of animate from inanimate matter may have involved DNA-like molecules. Perhaps such self-duplicating molecules, taking shape in primeval waters some three billion years ago, were the biochemical ancestors of all modern organisms. Perhaps these naked genes mutated, developed protective shells to become bodies like viruses, and gradually gave rise to cells—so that the great march of evolution was under way. Ideas like these, and the accumulating body of solid evidence behind them, are signs that we are on the verge of breakthrough along the entire front of biology. Basic research on this hereditary material is providing the biologists of today with fresh insights into the fundamental nature of life itself.

KARL VON FRISCH

Darwin's Theory in the Light of Modern Genetics

•

DARWIN HAD assumed that the hereditary characteristics of animals and plants are variable and that the variations are heritable. Only half a century later geneticists checked up thoroughly on whether these premises of Darwin's theory of natural selection were valid. For a time it looked as if this meant the end of Darwinism. But Darwin prevailed.

Let Us Take a Closer Look at the Variability of Hereditary Characteristics

If one wants to study variability carefully it is best to choose a trait that can be measured accurately. The size of bean seeds, for example, can be easily measured. It is in fact a hereditary characteristic, because there are varieties of bean plants that produce large seeds and others that produce small seeds. The development of hereditary characteristics depends not only on the make-up of the gene but on environmental conditions as well. This can be observed very clearly in our example. Beans belong to those plants that can be continuously reproduced by self-fertilization. Thus one can prevent hybridization with other varieties and obtain material of completely uniform characteristics.

The progeny produced by selfing individuals with uniform hereditary characteristics are called true breeders. Yet despite this uniformity each of the plants produces seeds of unequal size. If the length of all the beans produced is measured, it will be found that one size is the most common, say seeds five eighths of an inch long. Besides, one gets larger and smaller beans in decreasing numbers.

[In] the kind of variability just described, a size of five eighths of an inch is the heritable characteristic as it appears under normal circumstances. All seeds of the plant, as it were, try to reach this size. The environmental conditions, however, have an important say in the matter. Some seeds develop on a sunny, others on a strongly shaded, part of the plant, some on a strong, others on a weak branch, some are favorably, others unfavorably joined to the nourishing flow of sap. Thus the hereditary factor for size can be strengthened or inhibited for all kinds of reasons. As a rule the stimulating and the inhibiting conditions will balance equally; occasionally one will have more influence than the other. Large aberrations from the normal median size are therefore relatively rare. In other varieties of beans, seed length will have a different average value (for example, three eighths of an inch).

Artificial Selection That Ends in Failure

If we use the largest and the smallest seeds of a true-breeding bean plant, we get, under identical conditions, plants that produce seeds similar to those of the mother plant, because the factors for size are the same in both. This result cannot be altered, even if for generations one uses always the largest seeds for breeding. Hence, selection from true breeders proved ineffective. This was discovered several decades ago by the Danish botanist and geneticist Wilhelm Ludwig Johannsen, much to the surprise of his fellow scientists. It had not been found before, because uniform animal or plant material had not been used in breeding but always a mixture of varieties.

If seeds are collected from a whole field of beans grown from a number of varieties with differing hereditary characteristics—for simplicity's sake let us assume three—the harvest will again consist of middle-sized, large, and small beans. However, the range of sizes will be wider, because the effect of environmental conditions is added to that of heredity. The curves representing the variability within each of the three races overlap. The actual number of beans of different sizes is obtained by summation of the three curves. Selection from these seeds will be successful. If especially large seeds are chosen, there is a chance that it is their hereditary factor that made them larger, and the average size of bean seeds can be raised accordingly from half an inch to nearly five eighths of an inch.

Of course there are limits to successful outbreeding. When one has picked out the seeds with the factor for the largest size, further selection fails to produce an increase in size. Johannsen therefore concluded that artificial selection can do nothing but disentangle a mixture of variations and that a continual elaboration of a character as it was postulated by Darwin is impossible. Consequently Johannsen, and many other geneticists with him, believed that this was the end of the theory of natural selection.

Mutations and Their Significance

Such reasoning, however, is correct only if hereditary characteristics are once and for all immutable. For quite some time odd cases had become known in which hereditary factors suddenly changed. They are now called mutations, from the Latin *mutare*—to change. Thus within a herd of sheep, which so far had produced nothing but normal straight-legged members, normal parents might one day produce a bowlegged lamb, the progeny of which would show inherited bowleggedness. This variety was for a time greatly favored by farmers, because such sheep could not jump over low fences. The German sausage dog, or dachshund, very likely came about in a similar way. In another instance a hornless variety of cattle originated just as suddenly and unexpectedly. Sometimes one finds among flies with well-developed wings some specimens that have a pair of short stumps only. If one breeds from them, this special characteristic proves to be heritable. All such sudden changes are brought about by mutations of genes and are therefore heritable.

The importance of mutations as a source for new species may seem to be negligible. But certain adaptations found through them a simple and immediately convincing explanation, for example the frequent occurrence of stunted-winged or wingless insects on remote islands. One of the best examples is Kerguelen, an isolated windswept island. All indigenous insects such as the nine species of flies living there, even the butterflies and beetles, are incapable of flight. On the continent this would be of great disadvantage; on this island underdeveloped wings are an asset. Those insects that rise into the air are in these stormy regions all too easily carried out over the open sea, where they perish. Among artificially bred flies the sudden appearance of forms with heritably stunted wings has been repeatedly observed. It is reasonable to assume that the flies of Kerguelen have their origin in corresponding mutations, which on account of the peculiar living conditions on the island superseded the original winged forms in the struggle for existence.

Although mutations might explain such isolated instances, they seemed of too rare occurrence to explain the innumerable and complicated adaptations necessary for the evolution of different species. According to Darwin an abundance of hereditary variations is necessary from which in the struggle for existence the most suitable ones are selected and promoted. It was a second turning point in the history of Darwinism when geneticists found that in fact mutations occur very frequently.

Striking aberrations like bowlegs or stunted wings are of course rare. The new data came to light after careful and laborious studies. During several decades the American geneticist Morgan and his research workers bred millions of little fruit flies and studied them in the greatest detail. They found that the obvious mutations, like white-eyed flies among normal red-eyed ones or the already mentioned animals with vestigial (stunted) wings, were relatively rare, but that on the other hand small but nevertheless heritable mutations were very frequent, differing only in degree. They might involve any organ or function of the body, be it the color of the eyes or the wing pattern, the number and distribution of bristles on the body, the segmentation of the body, its color and patterning and many other characteristics. Within a short time Morgan's breeding experiments presented us with hundreds of new mutations, and of course hundreds of thousands may have occurred unnoticed in Nature within the same brief span of time. Nature works with an abundance of material compared to laboratory experiments.

When the German botanist Baur started to cultivate snapdragons his experiences were similar, but with a significant difference: he found that mutations occurred far more frequently than in the fruit fly. Even within most carefully selected true-breeding strains of this plant, his discerning eye discovered on the average among ten plants at least one that showed a heritable mutation. This might relate to the color of the flower, the way of branching, the earlier or later time of flowering, the color of the leaves, and many other characteristics. Many of these mutations do not affect the life of the plant and many are even disadvantageous. On the other hand, many are of a kind that might under certain conditions be of advantage in the struggle for existence. There are certainly many animals and plants the hereditary characteristics of which are less liable to mutate than those of the fruit fly or the snapdragon. But it is just as certain that countless mutations will be discovered in all kinds of organisms if looked for with the assiduity of a Morgan, a Baur and a Muller.

Yet another important discovery, which strongly supports Darwin's theory, was made in modern genetics. It was found that the rate of mutations can be increased artificially by external influence. If, for example, fruit flies are exposed to X rays, as they were by the American Nobel Prize winner Hermann J. Muller or bred at an exceptionally high temperature, the germ cells so maltreated show greater numbers of heritable mutations. Now, it is quite possible that during the earlier periods of the earth's history, high temperatures or other extraordinary external conditions favored the production of mutations, and that this high variability of organisms led to the formation of new species and natural selection.

The appearance of a great number of mutations is welcome so long as the less favored ones are eradicated by ruthless natural selection and only the more viable ones preserved.

Nowadays we have entered an age of developments that is uniquely threatening. Human health is endangered by exposure to large amounts of Man-made radiation, without the balancing safeguard of natural selection.[1]

The first warning signs appeared some time ago. The early pioneers working with X rays were shocked to find that they as well as their patients began to suffer from the effects of this radiation. It happened quite unexpectedly and often a long time after exposure.

Nowadays radiation is produced not only for use in laboratories, for X-raying broken limbs, for healing growths, or for experiments on the induction of great mutability in organisms. We now split atoms, a process that releases not only an undreamed-of amount of energy but at the same time radioactivity, the effect of which is considerably stronger than that of X rays. It is horrifying to realize that people who were within the active radius of the first atom bomb dropped over Hiroshima are dying a slow death.

These ruinous if delayed effects on our health are the dreadful price we pay for the achievements of modern technology. But still more sinister because more far-reaching is the harm done by radiation to the germ cells, the carriers of heredity. This concerns the future of mankind.

Whenever an atom bomb is exploded, the atmosphere is likely to be polluted with a great quantity of radioactive substances, which remain active for long periods of time and reach plants and ultimately animals through fallout and in rain and snow. The exploitation of atomic energy for peaceful purposes also results in radioactive waste, the removal and safe storage of which are still great problems, quite apart from the possibility of their liberation by accident. It has been established that all over the earth Man is exposed to a considerably higher degree of radiation than was the case before the time of atomic explosions, and it is obvious that this will get worse and worse if human reason does not prevail. It is known that germ cells can be influenced by the smallest doses, in proportion to the applied radiation, and that the effects of exposure are cumulative. This means that the germ cells, which cannot "forgive and forget," hand on the sum of our sins against them from one generation to the next. They live on, while the individual dies. As time goes on a great increase in mutations may take place, and we know that most of them will be unfavorable. This ever-present danger now threatens the life of future generations, although years or even several decades may pass between the first exposure to radiation and the appearance of the final dire consequences in the human race.

Man is aware of the danger, but it is questionable whether he has the sense to keep it in check. Incredibly careless about the welfare of future generations, he plays with fire before he has learned how to control it.

Let us summarize: Modern genetics has clarified the meaning of the term "variability." There are on the one hand changes in the external ap-

[1] See *Radiations and the Genetic Threat* by Warren Weaver, page 710.—Eds.

pearance of an organism, brought about by the influence of the environment, that are not heritable and that are called modifications. Quite different on the other hand are those changes in the external appearance that are due to heritable alterations of the genes within the germ cells, known as mutations. The frequent occurrence of natural mutations had at first escaped the notice of geneticists. Now they can increase the number artificially by special manipulation. Thus the existence of mutations is one of the most powerful arguments in favor of Darwin's theory, however much it seemed initially to contradict it. In studying Darwin's work, we find that he was quite aware of the existence of heritable and nonheritable variations and that he believed in the possibility of the inheritance of so-called acquired characteristics, provided the external causes get the chance to exert their influence for a long enough time to work the change.

C. *The Spectacle of Animal Life*

•

As we have seen in previous sections, all living things have basic characteristics in common. All, for example, are composed of cells, ingest food, reproduce and die. But there are obvious differences. We are immediately struck by the variation in size among the single-celled animal, the grasshopper and the elephant. Size profoundly influences an animal's activities. J. B. S. Haldane explains its importance in On Being the Right Size. *Haldane, who died in 1964, was one of the world's outstanding geneticists. He was also an able writer, as is evidenced by the present article.*

T. H. Huxley was a polemicist of great skill, a zoologist of standing and one of the greatest popularizers of science in history. His article A Lobster; or, The Study of Zoology *helps us "to see how the application of common sense and common logic to the obvious facts it presents, inevitably leads us into all the branches of zoological science." Huxley shows us the unity of plan and diversity of execution which characterize all animals, whether they swim, crawl, fly, swing from trees or walk on the ground.*

The following article, A Living Fossil *by Professor J. L. B. Smith of Rhodes University College in South Africa, recounts one of the most extraordinary happenings in the study of natural history—the discovery of a fish which had previously been thought to be extinct for fifty million years. The find was "a confirmatory link in the chain of evidence upon which the theory of evolution is based."*

A striking example of the interrelationships of all the sciences, and of how seemingly diverse investigations can be correlated, is contained in Space Tracks *by Dwain W. Warner. In it is shown how a space satellite, traveling at 18,000 miles per hour at an altitude of 200 miles, can furnish new information about the migration of birds.*

Finally, Frank A. Beach, the well-known student of animal psychology at Yale University, discusses the question Can Animals Reason? *by describing a variety of illuminating experiments on both animals and men.*

J. B. S. HALDANE

On Being the Right Size

•

THE MOST OBVIOUS DIFFERENCES between different animals are differences of size, but for some reason the zoologists have paid singularly little attention to them. In a large textbook of zoology before me I find no indication that the eagle is larger than the sparrow, or the hippopotamus bigger than the hare, though some grudging admissions are made in the case of the mouse and the whale. But yet it is easy to show that a hare could not be as large as a hippopotamus, or a whale as small as a herring. For every type of animal there is a most convenient size, and a large change in size inevitably carries with it a change of form.

Let us take the most obvious of possible cases, and consider a giant man sixty feet high—about the height of Giant Pope and Giant Pagan in the illustrated *Pilgrim's Progress* of my childhood. These monsters were not only ten times as high as Christian, but ten times as wide and ten times as thick, so that their total weight was a thousand times his, or about eighty to ninety tons. Unfortunately the cross sections of their bones were only a hundred times those of Christian, so that every square inch of giant bone had to support ten times the weight borne by a square inch of human bone. As the human thigh-bone breaks under about ten times the human weight, Pope and Pagan would have broken their thighs every time they took a step. This was doubtless why they were sitting down in the picture I remember. But it lessens one's respect for Christian and Jack the Giant Killer.

To turn to zoology, suppose that a gazelle, a graceful little creature with long thin legs, is to become large, it will break its bones unless it does one of two things. It may make its legs short and thick, like the rhinoceros, so that every pound of weight has still about the same area of bone to support it. Or it can compress its body and stretch out its legs obliquely to gain stability, like the giraffe. I mention these two beasts because they happen to belong to the same order as the gazelle, and both are quite successful mechanically, being remarkably fast runners.

Gravity, a mere nuisance to Christian, was a terror to Pope, Pagan, and Despair. To the mouse and any smaller animal it presents practically no dangers. You can drop a mouse down a thousand-yard mine shaft; and, on arriving at the bottom, it gets a slight shock and walks away, provided that the ground is fairly soft. A rat is killed, a man is broken, a horse splashes. For the resistance presented to movement by the air is proportional to the surface of the moving object. Divide an animal's length, breadth, and height each by ten; its weight is reduced to a thousandth, but its surface only to a hundredth. So the resistance to falling in the case of the small animal is relatively ten times greater than the driving force.

An insect, therefore, is not afraid of gravity; it can fall without danger, and can cling to the ceiling with remarkably little trouble. It can go in for elegant and fantastic forms of support like that of the daddy-longlegs. But there is a force which is as formidable to an insect as gravitation to a mammal. This is surface tension. A man coming out of a bath carries with him a film of water of about one-fiftieth of an inch in thickness. This weighs roughly a pound. A wet mouse has to carry about its own weight of water. A wet fly has to lift many times its own weight and, as everyone knows, a fly once wetted by water or any other liquid is in a very serious position indeed. An insect going for a drink is in as great danger as a man leaning out over a precipice in search of food. If it once falls into the grip of the surface tension of the water—that is to say, gets wet—it is likely to remain so until it drowns. A few insects, such as water-beetles, contrive to be unwettable; the majority keep well away from their drink by means of a long proboscis.

Of course tall land animals have other difficulties. They have to pump their blood to greater heights than a man, and therefore, require a larger blood pressure and tougher blood-vessels. A great many men die from burst arteries, especially in the brain, and this danger is presumably still greater for an elephant or a giraffe. But animals of all kinds find difficulties in size for the following reason. A typical small animal, say a microscopic worm or rotifer, has a smooth skin through which all the oxygen it requires can soak in, a straight gut with sufficient surface to absorb its food, and a single kidney. Increase its dimensions tenfold in every direction, and its weight is increased a thousand times, so that if it is to use its muscles as efficiently as its miniature counterpart, it will need a thousand times as much food and oxygen per day and will excrete a thousand times as much of waste products.

Now if its shape is unaltered its surface will be increased only a hundredfold, and ten times as much oxygen must enter per minute through each square millimetre of skin, ten times as much food through each square millimetre of intestine. When a limit is reached to their absorptive powers their surface has to be increased by some special device. For example, a part of the skin may be drawn out into tufts to make gills or pushed in to make lungs, thus increasing the oxygen-absorbing surface in proportion to the animal's bulk. A man, for example, has a hundred square yards of lung. Similarly, the gut, instead of being smooth and straight, becomes coiled and develops a velvety surface, and other organs increase in complication. The higher animals are not larger than the lower because they are more complicated. They are more complicated because they are larger. Just the same is true of plants. The simplest plants, such as the green algae growing in stagnant water or on the bark of trees, are mere round cells. The higher plants increase their surface by putting out leaves and roots. Comparative anatomy is largely the story of the struggle to increase surface in proportion to volume.

Some of the methods of increasing the surface are useful up to a point, but not capable of a very wide adaptation. For example, while vertebrates carry the oxygen from the gills or lungs all over the body in the blood, insects take air directly to every part of their body by tiny blind tubes called tracheae which open to the surface at many different points. Now, although by their breathing movements they can renew the air in the outer part of the tracheal system, the oxygen has to penetrate the finer branches by means of diffusion. Gases can diffuse easily through very small distances, not many times larger than the average length travelled by a gas molecule between collisions with other molecules. But when such vast journeys—from the point of view of a molecule—as a quarter of an inch have to be made, the process becomes slow. So the portions of an insect's body more than a quarter of an inch from the air would always be short of oxygen. In consequence hardly any insects are much more than half an inch thick. Land crabs are built on the same general plan as insects, but are much clumsier. Yet like ourselves they carry oxygen around in their blood, and are therefore able to grow far larger than any insects. If the insects had hit on a plan for driving air through their tissues instead of letting it soak in, they might well have become as large as lobsters, though other considerations would have prevented them from becoming as large as man.

Exactly the same difficulties attach to flying. It is an elementary principle of aeronautics that the minimum speed needed to keep an aeroplane of a given shape in the air varies as the square root of its length. If its linear dimensions are increased four times, it must fly twice as fast. Now the power needed for the minimum speed increases more rapidly than the weight of the machine. So the larger aeroplane, which weighs sixty-four times as much as the smaller, needs one hunderd and twenty-eight times its horsepower to keep up. Applying the same principle to the birds, we find that the limit to their size is soon reached. An angel whose muscles developed no more power weight for weight than those of an eagle or a pigeon would require a breast projecting for about four feet to house the muscles engaged in working its wings, while to economize in weight, its legs would have to be reduced to mere stilts. Actually a large bird such as an eagle or kite does not keep in the air mainly by moving its wings. It is generally to be seen soaring, that is to say balanced on a rising column of air. And even soaring becomes more and more difficult with increasing size. Were this not the case eagles might be as large as tigers and as formidable to man as hostile aeroplanes.

But it is time that we pass to some of the advantages of size. One of the most obvious is that it enables one to keep warm. All warm-blooded animals at rest lose the same amount of heat from a unit area of skin, for which purpose they need a food-supply proportional to their surface and not to their weight. Five thousand mice weigh as much as a man. Their combined surface and food or oxygen consumption are about seventeen times a man's. In fact a mouse eats about one quarter its own weight of food every day, which is mainly used in keeping it warm. For the same reason small animals cannot live in cold countries. In the arctic regions there are no reptiles or amphibians, and no small mammals. The smallest mammal in Spitzbergen is the fox. The small birds fly away in winter, while the insects die, though their eggs can survive six months or more of frost. The most successful mammals are bears, seals, and walruses.

Similarly, the eye is a rather inefficient organ until it reaches a large size. The back of the human eye on which an image of the outside world is thrown, and which corresponds to the film of a camera, is composed of a mosaic of "rods and cones" whose diameter is little more than a length of an average light wave. Each eye has about a half a million, and for two objects to be distinguishable their images must fall on separate rods or cones. It is obvious that with fewer but larger rods and cones we should see less distinctly. If they were twice as broad two points would have to be twice as far apart before we could distinguish them at a given distance. But if their size were diminished and their number increased we should see no better. For it is impossible to form a definite image smaller than a wave-length of light. Hence a mouse's eye is not a small-scale model of a human eye. Its rods and cones are not much smaller than ours, and therefore there are far fewer of them. A mouse could not distinguish one human face from another six feet away. In order that they should be of any use at all the eyes of small animals have to be much larger in proportion to their bodies than our own. Large animals on the other hand only require relatively small eyes, and those of the whale and elephant are little larger than our own.

For rather more recondite reasons the same general principle holds true of the brain. If we compare the brain-weights of a set of very similar animals such as the cat, cheetah, leopard, and tiger, we find that as we quadruple the body-weight the brain-weight is only doubled. The larger animal with proportionately larger bones can economize on brain, eyes, and certain other organs.

Such are a very few of the considerations which show that for every type of animal there is an optimum size. Yet although Galileo demonstrated the contrary more than three hundred years ago, people still believe that if a flea were as large as a man it could jump a thousand feet into the air. As a matter of fact the height to which an animal can jump is more nearly independent of its size than proportional to it. A flea can jump about two feet, a man about five. To jump a given height, if we neglect the resistance of the air, requires an expenditure of energy proportional to the jumper's weight. But if the jumping muscles form a constant fraction of the animal's body, the energy developed per ounce of muscle is independent of the size, provided it can be developed quickly enough in the small animal. As a matter of fact an insect's muscles, although they can contract more quickly than our own, appear to be less efficient; as otherwise a flea or grasshopper could rise six feet into the air.

T. H. HUXLEY

A Lobster; or, The Study of Zoology

•

CERTAIN BROAD LAWS have a general application throughout both the animal and the vegetable worlds, but the ground common to these kingdoms of nature is not of very wide extent, and the multiplicity of details is so great, that the student of living beings finds himself obliged to devote his attention exclusively either to the one or the other. If he elects to study plants, under any aspect, his science is botany. But if the investigation of animal life be his choice, the name generally applied to him will vary according to the kind of animals he studies, or the particular phenomena of animal life to which he confines his attention. If the study of man is his object, he is called an anatomist, or a physiologist, or an ethnologist; but if he dissects animals, or examines into the mode in which their functions are performed, he is a comparative anatomist or comparative physiologist. If he turns his attention to fossil animals, he is a palæontologist. If his mind is more particularly directed to the specific description, discrimination, classification, and distribution of animals, he is termed a zoologist.

For the purpose of the present discourse, however, I shall recognise none of these titles save the last, which I shall employ as the equivalent of botanist, and I shall use the term zoology as denoting the whole doctrine of animal life, in contradistinction to botany, which signifies the whole doctrine of vegetable life.

Employed in this sense, zoology, like botany, is divisible into three great but subordinate sciences, morphology, physiology, and distribution, each of which may, to a very great extent, be studied independently of the other.

Zoological morphology is the doctrine of animal form or structure. Anatomy is one of its branches; development is another; while classification is the expression of the relations which different animals bear to one another, in respect of their anatomy and their development.

Zoological distribution is the study of animals in relation to the terrestrial conditions which obtain now, or have obtained at any previous epoch of the earth's history.

Zoological physiology, lastly, is the doctrine of the functions or actions of animals. It regards animal bodies as machines impelled by certain forces, and performing an amount of work which can be expressed in terms of the ordinary forces of nature. The final object of physiology is to deduce the facts of morphology, on the one hand, and those of distribution on the other, from the laws of the molecular forces of matter.

Such is the scope of zoology. But if I were to content myself with the enunciation of these dry definitions, I shall ill exemplify that method of teaching this branch of physical science, which it is my chief business to-night to recommend. Let us turn away then from abstract definitions. Let us take some concrete living thing, some animal, the commoner the better, and let us see how the application of common sense and common logic to the obvious facts it presents, inevitably leads us into all these branches of zoological science.

I have before me a lobster. When I examine it, what appears to be the most striking character it presents? Why, I observe that this part which we call the tail of the lobster, is made up of six distinct hard rings and a seventh terminal piece. If I separate one of the middle rings, say the third, I find it carries upon its under surface a pair of limbs or appendages, each of which consists of a stalk and two terminal pieces.

If I now take the fourth ring, I find it has the same structure, and so have the fifth and the second; so that, in each of these divisions of the tail, I find parts which correspond with one another, a ring and two appendages; and in each appendage a stalk and two end pieces. These corresponding parts are called, in the technical language of anatomy, "homologous parts." The ring of the third division is the "homologue" of the ring of the fifth, the appendage of the former is the homologue of the appendage of the latter. And, as each division exhibits corresponding parts in corresponding places, we say that all the divisions are constructed upon the same plan. But now let us consider the sixth division. It is similar to, and yet different from, the others. The ring is essentially the same as in the other divisions; but the appendages look at first as if they were very different; and yet when we regard them closely, what do we find? A stalk and two terminal divisions, exactly as in the others, but the stalk is very short and very thick, the terminal divisions are very broad and flat, and one of them is divided into two pieces.

I may say, therefore, that the sixth segment is like the others in plan, but that it is modified in its detail.

The first segment is like the others, so far as its ring is concerned, and though its appendages differ from any of those yet examined in the simplicity of their structure, parts corresponding with the stem and one of the divisions of the appendages of the other segments can be readily discerned in them.

Thus it appears that the lobster's tail is composed of a series of segments which are fundamentally similar, though each presents peculiar modifications of the plan common to all. But when I turn to the forepart of the body I see, at first, nothing but a great shield-like shell, called technically the "carapace," ending in front in a sharp spine on either side of which are the curious compound eyes, set upon the ends of stout movable stalks. Behind these, on the under side of the body, are two pairs of long feelers, or antennae, followed by six pairs of jaws folded against one another over the mouth, and five pairs of legs, the foremost of these being the great pinchers, or claws, of the lobster.

It looks, at first, a little hopeless to attempt to find in this complex mass a series of rings, each with its pair of appendages, such as I have shown you in the abdomen, and yet it is not difficult to demonstrate their existence. Strip off the legs, and you will find that each pair is attached to a very definite segment of the under wall of the body; but these segments, instead of being the lower parts of free rings, as in the tail, are such parts of rings which are all solidly united and bound together; and the like is true of the jaws, the feelers, and the eye-stalks, every pair of which is borne upon its own special segment. Thus the conclusion is gradually forced upon us, that the body of the lobster is composed of as many rings as there are pairs of appendages, namely, twenty in all, but that the six hindmost rings remain free and movable, while the fourteen front rings become firmly soldered together, their backs forming one continuous shield—the carapace.

Unity of plan, diversity in execution, is the lesson taught by the study of the rings of the body, and the same instruction is given still more emphatically by the appendages. If I examine the outermost jaw I find it consists of three distinct portions, an inner, a middle, and an outer, mounted upon a common stem; and if I compare this jaw with the legs behind it, or the jaws in front of it, I find it quite easy to see, that, in the legs, it is the part of the appendage which corresponds with the inner division, which becomes modified into what we know familiarly as the "leg," while the middle division disappears, and the outer division is hidden under the carapace. Nor is it more difficult to discern that, in the appendages of the tail, the middle division appears again and the outer vanishes; while, on the other hand, in the foremost jaw, the so-called mandible, the inner division only is left; and, in the same way, the parts of the feelers and of the eye-stalks can be identified with those of the legs and jaws.

But whither does all this tend? To the very remarkable conclusion that a unity of plan, of the same kind as that discoverable in the tail or abdomen of the lobster, pervades the whole organisation of its skeleton, so that I can return to the diagram representing any one of the rings of the tail and by adding a third division to each appendage, I can use it as a sort of scheme or plan of any ring of the body. I can give names to all the parts of that figure, and then if I take any segment of the body of the lobster, I can point out to you exactly, what modification the general plan has undergone in that particular segment; what part has remained movable, and what has become fixed to another; what has been excessively developed and metamorphosed and what has been suppressed.

But I imagine I hear the question, How is all this to be tested? No doubt it is a pretty and ingenious way of looking at the structure of any animal; but is it anything more? Does Nature acknowledge, in any deeper way, this unity of plan we seem to trace?

Happily, however, there is a criterion of morphological truth, and a sure test of all homologies. Our lobster has not always been what we see it; it was once an egg, a semifluid mass of yolk, not so big as a pin's head, contained in a transparent membrane, and exhibiting not the least trace of any one of those organs, the multiplicity and complexity of which, in the adult, are so surprising. After a time, a delicate patch of cellular membrane appeared upon one face of this yolk, and that patch was the foundation of the whole creature, the clay out of which it would be moulded. Gradually investing the yolk, it became subdivided by transverse constrictions into segments, the forerunners of the rings of the body. Upon the ventral surface of each of the rings thus sketched out, a pair of bud-like prominences made their appearance—the rudiments of the appendages of the ring. At first, all the appendages were alike, but, as they grew, most of them became distinguished into a stem and two terminal divisions, to which, in the middle part of the body, was added a third outer division; and it was only at a later period, that by the modification, or absorption, of certain of these primitive constituents, the limbs acquired their perfect form.

Thus the study of development proves that the doctrine of unity of plan is not merely a fancy, that it is not merely one way of looking at the matter, but that it is the expression of deep-seated natural facts. The legs and jaws of the lobster may not merely be regarded as modifications of a common type—in fact and in nature they are so—the leg and the jaw of the young animal being, at first, indistinguishable.

These are wonderful truths, the more so because the zoologist finds them to be of universal application. The investigation of a polype, of a snail, of a fish, of a horse, or of a man, would have led us, though by a less easy path, perhaps, to exactly the same point. Unity of plan everywhere lies hidden under the mask of diversity of structure—the complex is everywhere evolved out of the simple. Every animal has at first the form of an egg, and every animal and every organic part, in reaching its adult state, passes through conditions common to other animals and other adult parts; and this leads me to another point. I have hitherto spoken as if the lobster were alone in the world, but, as I need hardly remind you, there are myriads of other animal organisms. Of these, some, such as men, horses, birds, fishes, snails, slugs, oysters, corals, and sponges, are not in the least like the lobster. But other animals, though they may differ a good deal from the lobster, are yet either very like it, or are like something that is like it. The cray fish, the rock lobster, and the prawn, and the shrimp, for example, however different, are yet so like lobsters, that a child would group them as of the lobster kind, in contradistinction to snails and slugs; and these last again would form a kind by themselves, in contradistinction to cows, horses, and sheep, the cattle kind.

But this spontaneous grouping into "kinds" is the first essay of the human mind at classification, or the calling by a common name of those things that are alike, and the arranging them in such a manner as best to suggest the sum of their likenesses and unlikenesses to other things.

Those kinds which include no other subdivisions than the sexes, or various breeds, are called, in technical language, species. The English lobster is a species, our cray fish is another, our prawn is another. In other countries, however, there are lobsters, cray fish, and prawns, very like ours, and yet presenting sufficient differences to deserve distinction. Naturalists, therefore, express this resemblance and this diversity by grouping them as distinct species of the same "genus." But the lobster and the cray fish, though belonging to distinct genera, have many features in common, and hence are grouped together in an assemblage which is called a family. More distant resemblances connect the lobster with the prawn and the crab, which are expressed by putting all these into the same order. Again, more remote, but still very definite, resemblances unite the lobster with the woodlouse, the king crab, the water flea, and the barnacle, and separate them from all other animals; whence they collectively constitute the larger group, or class, *Crustacea*. But the *Crustacea* exhibit many peculiar features in common with insects, spiders, and centipedes, so that these are grouped into the still larger assemblage or "province" *Articulata;* and, finally, the relations which these have to worms and other lower animals, are expressed by combining the whole vast aggregate into the sub-kingdom of *Annulosa.*

If I had worked my way from a sponge instead of a lobster, I should have found it associated, by like ties, with a great number of other animals into the sub-kingdom *Protozoa;* if I had selected a fresh-water polype or a coral, the members of what naturalists term the sub-kingdom *Cœlenterata,* would have grouped themselves around my type; had a snail been chosen, the inhabitants of all univalve and bivalve, land and water, shells, the lamp shells, the squids, and the sea-mat would have gradually linked themselves on to it as members of the same sub-kingdom of *Mollusca;* and finally, starting from man, I should have been compelled to admit first, the ape, the rat, the horse, the dog, into the same class; and then the bird, the crocodile, the turtle, the frog, and the fish, into the same sub-kingdom of *Vertebrata.*

And if I had followed out all these various lines of classification fully, I should discover in the end that there was no animal, either recent or fossil, which did not at once fall into one or other of these sub-kingdoms. In other words, every animal is organised upon one or other of the five, or more, plans, the existence of which renders our classification possible. And so definitely and precisely marked is the structure of each animal, that, in the present state of our knowledge, there is not the least evidence to prove that a form, in the slightest degree transitional between any of the two groups *Vertebrata, Annulosa, Mollusca,* and *Cœlenterata,* either exists, or has existed, during that period of the earth's history which is recorded by the geologist. Nevertheless, you must not for a moment suppose, because no such transitional forms are known, that the members of the sub-kingdoms are disconnected from, or independent of, one another. On the contrary, in their earliest condition they are all similar, and the primordial germs of a man, a dog, a bird, a fish, a beetle, a snail, and a polype are, in no essential structural respects, distinguishable.

Turning from these purely morphological considerations, let us now examine into the manner in which the attentive study of the lobster impels us into other lines of research.

Lobsters are found in all the European seas; but on the opposite shores of the Atlantic and in the seas of the southern hemisphere they do not exist. They are, however, represented in these regions by very closely allied, but distinct forms—the *Homarus Americanus* and the *Homarus Capensis:* so that we may say that the European has one species of *Homarus;* the American, another; the African, another; and thus the remarkable facts of geographical distribution begin to dawn upon us.

Again, if we examine the contents of the earth's crust, we shall find in the latter of those deposits, which have served as the great burying grounds of past ages, numberless lobster-like animals, but none so similar to our living lobster as to make zoologists sure that they belonged even to the same genus. If we go still further back in time, we discover, in the oldest rocks of all, the remains of animals, constructed on the same general plan as the lobster, and belonging to the same great group of *Crustacea;* but for the most part totally different from the lobster, and indeed from any other living form of crustacean; and thus we gain a notion of that successive change of the animal population of the globe, in past ages, which is the most striking fact revealed by geology.

Consider, now, where our inquiries have led us. We studied our type morphologically, when we determined its atonomy and its development, and when comparing it, in these respects, with other animals, we made out its place in a system of classification. If we were to examine every animal in a similar manner, we should establish a complete body of zoological morphology.

But you will observe one remarkable circumstance, that, up to this point, the question of the life of these organisms has not come under consideration. Morphology and distribution might be studied almost as well, if animals and plants were a peculiar kind of crystals, and possessed none of those functions which distinguish living beings so remarkably. But the facts of morphology and distribution have to be accounted for, and the science, the aim of which it is to account for them, is Physiology.

Let us return to our lobster once more. If we watched the creature in its native element, we should see it climbing actively the submerged rocks, among which it delights to live, by means of its strong legs; or swimming by powerful strokes of its great tail, the appendages of the sixth joint of which are spread out into a broad fan-like propeller: seize it, and it will show you that its great claws are no mean weapons of offence; suspend a piece of carrion among its haunts, and it will greedily devour it, tearing and crushing the flesh by means of its multitudinous jaws.

Suppose that we had known nothing of the lobster but as an inert mass, an organic crystal, if I may use the phrase, and that we could suddenly see it exerting all these powers, what wonderful new ideas and new questions would arise in our minds! The great new question would be, "How does all this take place?" the chief new idea would be, the idea of adaptation to purpose—the notion, that the constituents of animal bodies are not mere unconnected parts, but organs working together to an end. Let us consider the tail of the lobster again from this point of view. Morphology has taught us that it is a series of segments composed of homologous parts, which undergo various modifications—beneath and through which a common plan of formation is discernible. But if I look at the same part physiologically, I see that it is a most beautifully constructed organ of locomotion, by means of which the animal can swiftly propel itself either backwards or forwards.

But how is his remarkable propulsive machine made to perform its functions? If I were suddenly to kill one of these animals and to take out all the soft parts, I should find the shell to be perfectly inert, to have no more power of moving itself than is possessed by the machinery of a mill when disconnected from its steam-engine or water-wheel. But if I were to open it, and take out the viscera only, leaving the white flesh, I should perceive that the lobster could bend and extend its tail as well as before. If I were to cut off the tail, I should cease to find any spontaneous motion in it; but on pinching any portion of the flesh, I should observe that it underwent a very curious change—each fibre becoming shorter and thicker. By this act of contraction, as it is termed, the parts to which the ends of the fibre are attached are, of course, approximated; and according to the relations of their points of attachment to the centres of motions of the different rings, the bending or the extension of the tail results. Close observation of the newly-opened lobster would soon show that all its movements are due to the same cause—the shortening and thickening of these fleshy fibres, which are technically called muscles.

Here, then, is a capital fact. The movements of the lobster are due to muscular contractility. But why does a muscle contract at one time and not at another? Why does one whole group of muscles contract when the lobster wishes to extend his tail, and another group when he desires to bend it? What is it originates, directs, and controls the motive power?

Experiment, the great instrument for the ascertainment of truth in physical science, answers this question for us. In the head of the lobster there lies a small mass of that peculiar tissue which is known as nervous substance. Cords of similar matter connect this brain of the lobster, directly or indirectly, with the muscles. Now, if these communicating cords are cut, the brain remaining entire, the power of exerting what we call voluntary motion in the parts below the section is destroyed; and, on the other hand, if, the cords remaining entire, the brain mass be destroyed, the same voluntary mobility is equally lost. Whence the inevitable conclusion is, that the power of originating these motions resides in the brain and is propagated along the nervous cords.

In the higher animals the phenomena which attend this transmission have been investigated, and the exertion of the peculiar energy which resides in the nerves has been found to be accompanied by a disturbance of the electrical state of their molecules.

If we could exactly estimate the signification of this disturbance; if we could obtain the value of a given exertion of nerve force by determining the quantity of electricity; or of heat, of which it is the equivalent; if we could ascertain upon what arrangement, or other condition of the molecules of matter, the manifestation of the nervous and muscular energies depends (and doubtless science will some day or other ascertain these points), physiologists would have attained their ultimate goal in this direction; they would have determined the relation of the motive force of animals to the other forms of force found in nature; and if the same process had been successfully performed for all the operations which are carried on in, and by, the animal frame, physiology would be perfect, and the facts of morphology and distribution would be deducible from the laws which physiologists had established, combined with those determining the condition of the surrounding universe.

There is not a fragment of the organism of this humble animal whose study would not lead us into regions of thought as large as those which I have briefly opened up to you; but what I have been saying, I trust, has not only enabled you to form a conception of the scope and purport of zoology, but has given you an imperfect example of the manner in which, in my opinion, that science, or indeed any physical science, may be best taught.

And if it were my business to fit you for the certificate in zoological science granted by this department, I should pursue a course precisely similar in principle to that which I have taken to-night. I should select a fresh-water sponge, a fresh-water polype or a *Cyanœa*, a fresh-water mussel, a lobster, a fowl, as types of the five primary divisions of the animal kingdom. I should explain their structure very fully, and show how each illustrated the great principles of zoology.

J. L. B. SMITH

A Living Fossil

•

SCIENTIFIC DISCOVERY rarely follows a smooth and orderly course. Like most natural processes it proceeds spasmodically, and important results frequently come only after long drawn-out, exhausting, and apparently fruitless endeavor, sometimes even almost by what appears to be a lucky chance. Scientific discoveries may roughly be divided into two main classes: Those which affect the material welfare of mankind (*e.g.*, the existence and action of bacteria) and those which represent merely an addition to knowledge. There has recently been discovered near East London, South Africa, a very remarkable fish which represents an event of the latter sort. The interest it has aroused is on account of the great scientific importance which attaches to it. It is a living link with a past so remote as to be almost beyond the grasp of the ordinary mind.

Beyond the disputes of scientists about minor points is the fact that fishes of sorts were the ancestral forms from which all other vertebrate creatures have originated. We have no record of any accepted "link" between these fishes and their most likely invertebrate ancestors. The fishes are suddenly there, some three hundred and seventy million years ago, in numbers, and in a diversity of weird forms. From the fishes originated the amphibia, and from them the reptiles. These were all cold-blooded creatures, very much at the mercy of sudden climatic changes. The first amphibia appeared about three hundred and twenty million years ago, and reptiles evolved from them by ninety to one hundred million years later. Those sluggish creatures were produced by Nature in great diversity of size and shape, but most of the larger forms in specialized groups have become extinct.

The call for greater activity and mobility produced from the reptiles the warm-blooded birds and mammals, the latter class, as typified by man, being now dominant on the earth.

To return to those early fishes, some of which were our ancestors: Many of them, the only ones we know, left traces in the rocks. Usually the skeleton and any teeth, spines or hard skins are preserved, in rare cases perfectly. The "soft parts" are largely unknown, and the reconstructed outlines of extinct forms are to some extent guesswork, but no more guesswork than the diagnosis of appendicitis by a physician. In each case visualization of the hidden condition is based upon experience and knowledge.

The primitive ancestral form of fishes was almost certainly something rather sharklike, without any true bone in its make-up. From those creatures in a relatively brief period of time evolved a multiplicity of types which are generally divided by scientists into four main groups. These are known as the Placoderms (clumsy "armor-plated fishes"), the Marsipobranchs (jawless, sucking-mouthed fishes), the Selachians (fishes with cartilaginous skeletons), and Pisces (fishes with bony skeleton). Many were experimental forms which found competition too severe and so vanished. All of the first group are extinct. They were too clumsy. There are a few miserable remnants of the Marsipobranchs still alive today (hagfish and lampreys). The Selachii are the sharks and rays which have remained vigorous and numerous, and which are one of the great forces in the waters of the earth. In the vast periods of time since their ancestors first spread terror in prehistoric waters, the sharks have changed perhaps less than any other creatures. Many became extinct, but the line was carried on by forms of vigor and activity. Under Pisces are grouped the vast majority of living fishes, and a number who have vanished, some of great significance in the ancestral line.

The immediate importance of this recent discovery lies in the information it affords us about the developmental processes which have led to the typical forms of fishes, and which have been the subject of much research and speculation. This Coelacanthid specimen (*Latimeria chalumnae*) sheds a great deal of light upon many of those questions, since for some reason parts of this fish are in a condition which may be termed arrested metamorphosis. That is, it bears certain structures which are in process of changing from one thing to another, but the change has not gone to completion. Many of the outer bones of the head in fishes are supposed to have been derived from scales. Also the teeth in the jaws of fishes are believed to be merely scales that have migrated inward and have been changed into tooth-bearing structures. Fins are regarded as having originated from continuous folds of skin developed as stabilizers along the long axis of the body. On these and on many other points the present specimen affords a great deal of important evidence.

The main outline of the evolution of various types belonging to the two chief groups of fishes is shown by the accompanying diagram, which is not to scale. Branches which reach the line 1939 represent groups and forms which have survived to the present day. The others represent extinct forms. The cross-hatched line shows the addition to this scheme necessitated by the recent discovery.

One of the great main branches of the evolutionary tree was the group of the Crossopterygii (or fringe-finned). They were mostly large, active, predaceous fishes that probably dominated the extensive areas in which they occurred. Like most fishes they originated in fresh water and later migrated to the sea. A large number of forms developed, and were characterized by this peculiar "fringe-finned" state, and by the heavy bony

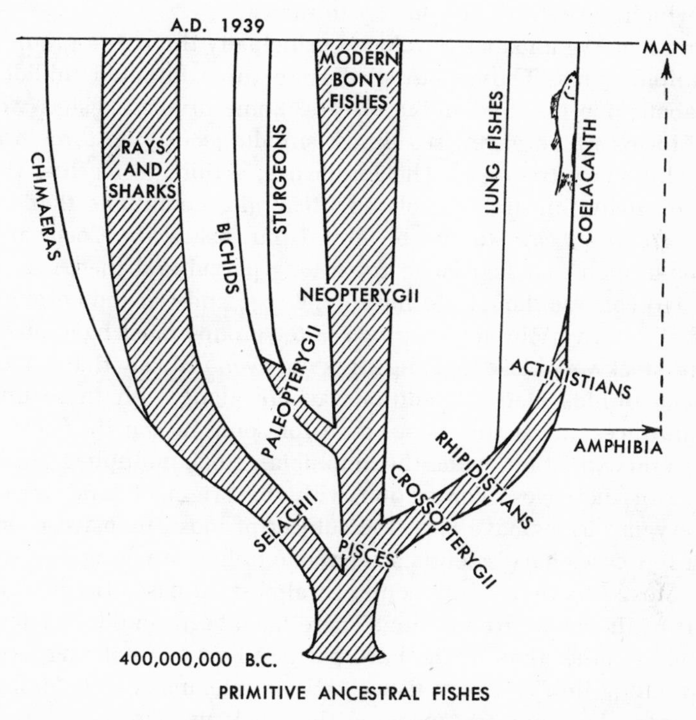

Fig. 1. Diagram illustrating the main lines of the evolution of fishes belonging to the groups Selachians and Pisces.

armature of the head. Many were not very different from the primitive sharklike ancestor, since their inner skeleton consisted at least partly of cartilage or gristle. They were covered by heavy scales, the outer surface of which was ornamented by an enamel-like substance known as ganoin. Many of them possessed most peculiar tails, known as gephyrocercal, which are really two tails in one, the extreme tip being a remnant of the original true tail which degenerated. Their pectoral and pelvic fins had developed so as to be very like limbs. It had been supposed that the sole living representatives of this ancient line were the few rather scarce species of "lungfishes" living in fresh water in America, Australia and Africa. They are degenerate forms which are but feeble shadows of their active and predaceous ancestors.

After having lived and flourished for some two hundred and fifty million years all those other numerous and vigorous Crossopterygian fishes had been supposed to have become extinct by fifty million years ago. The record in the rocks showed how, after having occurred in great numbers over a wide area, they diminished, until finally all traces ceased before the end of the Mesozoic Era—fifty million years ago. Those important fishes had all vanished. Important because they were a link between the early, intriguing creatures about whose structure we know so little, and the later vertebrates which have given rise finally to man.

The Crossopterygian stock developed principally through a group known as the Rhipidistians. Those flourished from three hundred million years ago for about one hundred million years. Some of those fishes were the ancestral forms which gave rise, almost simultaneously, to three branches of the evolutionary tree: (a) The lungfishes, a thin, feeble line that has survived by living an almost isolated life under conditions that scarcely any other creatures can stand; (b) Actinistian fishes, the Coelacanthidae, a vigorous branch that flourished for a long period and then just petered out long ago (or was thought to have done so); and (c) the amphibia, the origin of all land vertebrates. The two latter groups must have budded off the parent stock very close together. It is not even unlikely that some of the early Coelacanthids made expeditions ashore along with those unknown amphibian ancestors. If that is so, then for some reason they returned to the water and extinction, while the amphibian stock multiplied and throve.

Because of their close connection with the origin of land vertebrates, Crossopterygian fishes have been the subject of most intensive researches. In only a few cases have scientists been able to find anything like complete remains. Mostly there are fragments. In almost all cases the bones of the front part of the snout are missing. There have been found very few clues as to structure other than of the hard parts. All very tantalizing, especially as that "missing link" between fishes and amphibians is still missing.

Now, suddenly, there has appeared this great five-foot fish, bearing the full panoply of his early Mesozoic forebears, but larger than any of them. He is neither puny nor degenerate like the lungfishes, but a great, robust animal prepared and fitted to face all the risks in the sea (except a trawl net!). It is as if a fish of one hundred and fifty millon years ago had suddenly come to life. In that incomprehensibly long stretch of time this species has remained virtually unchanged, evidently completely satisfied with itself. In every way this is a true Coelacanthid from that remote past. For at least one hundred and fify million years this representative of that ancient but vigorous line has lived in such obscurity as never to have left any known traces of its existence.

The discovery of this Coelacanth is a confirmatory link in the chain of evidence upon which the theory of evolution is based. It stands as a high tribute to the reconstructional ability of scientists who have had to work chiefly with distorted and fragmentary remains. Although the scientists engaged in such work have been reasonably confident that their reconstructions were fairly close to the truth, there naturally remained a certain element of doubt. It appeared that there could never be any possibility of comparing their efforts with actual specimens, and many people regarded those reconstructions as mere fantasies. This Coelacanth shows that the scientists, in this case at least, have been remarkably accurate in their reconstructions.

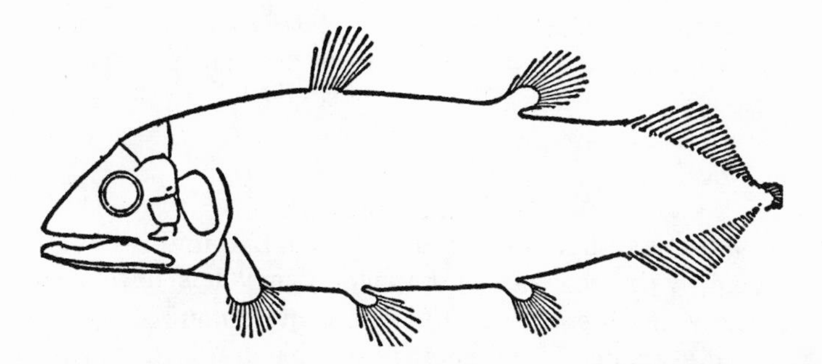

Fig. 2. The outline of a Coelacanthid fish, which lived about 150 million years ago, as reconstructed from fossil remains.

Naturally enough, this fish will fill in many of the gaps in our knowledge of those earlier forms. What is as important is that the discovery makes it at least possible that there may be other primitive creatures, believed long since extinct, lurking unsuspected in the depths of the ocean. It is more than likely that there is a real "sea serpent." So many reliable persons have testified independently to having seen that creature (or those creatures) that it cannot all be fabrication. We know almost nothing about what may be present in the depths of the ocean.

I have been asked where this fish is likely to have lived. My opinion—it can be only a guess—is that the species lived among rocky ledges where trawlers cannot operate, and at depths greater than that at which line fishing is practicable. But a number of factors incline me to believe that it does not live at very great depths. Probably one hundred to two hundred fathoms, along the outer ledges where rocky slopes fade down into the abyss, these Coelacanths lead a "coney-like" existence. Our specimen is probably a stray. We hope that the advent of other specimens will not long be delayed. We may even expect other species.

Editors' Note: After this article was written, as the result of an intensive campaign of education among native fishermen, a number of other specimens, including one which was captured alive but died shortly thereafter, became available for study.

DWAIN W. WARNER

Space Tracks

•

IN THE FEW YEARS that have passed since the mid-twentieth century, we have come to the realization that technology has developed both the power and the tools to enable us to explore more completely earth, space, our moon, and even nearby planets. These explorations, which at present fall almost entirely in the realm of the physical sciences and engineering, are proceeding at a rate few can grasp.

Meanwhile, biologists, and especially those concerned with the studies of the natural environment and its faunal and floral elements, have noted the marvelous accomplishments with mixed feelings. Particularly they have envied the new tools for making numerous measurements, which, in turn, have led to the accumulation of vast amounts of information. It has, after all, been only very recently that instrumentation, more sensitive and comprehensive than the human senses, has been developed for the acquisition of these data.

The field biologist, unfortunately, still depends largely on sight and hearing for his information and, as a result, often obtains data too discontinuous and incomplete for reliable statistical analyses and interpretation. Yet the same biologist also recognizes that, except almost peripherally, the information and measurements that have been collected by the physical scientists, who have always been considered to be in a discipline apart, have been but little concerned with life of the earth on which we live. Interdisciplinary research between biologists, physical scientists, and engineers in studies of our natural environment has been slow and pursued with reluctance.

Years ago the conflict was expressed by the pioneer ecologist Dr. Royal Chapman, who, in 1929, wrote: "Ecology is bound to become quantitative. Many of us are observing this inevitable tendency with regret. There is a feeling that the wonders of observational natural history are to be brushed aside by the cold, dry calculations of a mechanistic mathematics." But he added: "The urgent needs are, first, more accurate measurements of environmental factors and the populations which make up the natural associations, and, in the second place, better methods of evaluating the measurements of the factors."

A brief review of one aspect of man's efforts in these directions may help to put into better perspective the need for expanding interdisciplinary research and the real potential of modern instrumentation as applied to the biological sciences.

Watching and listening to the passage of animals in their seasonal migrations has been a necessary and traditional behavior pattern of man since he began to evolve as a flesh eater. Increasing curiosity about bird migrations led to the first markings of individual birds several hundred years ago, and the usefulness of this technique was well recognized by the early 1900's. As a result, several million birds and many other animals have been tagged with numbered metal bands or with various color markers. But this method has told us only a little of an animal's movements, for band recoveries represent a relatively small percentage of the marked animals. Usually, all we have available as an end result are two points on a map—the banding and recovery sites—with a known time interval, but an unknown distance of travel, between. This information has served many purposes, but has contributed little or nothing to understanding the animal's orientation, navigation, or relationship and responses to its environment.

Now, observations of an animal's motility are of utmost significance. Since the environment apparently furnishes or modifies the stimuli that trigger basic responses, the ability to follow an animal's movements continuously and to attempt to correlate these movements with environmental factors seems essential to advances in ecology. Nearly all animals are at some time motile, and since the causal factors of this motivation are not adequately understood, major emphasis must be given to a continuous recording of animal movements correlated with other environmental events.

For instance, various researchers have gathered data indicating a relationship between the distribution and abundance of animal populations and the biotic and physical forces of their natural environments. However, only a few quantitative studies on an individual animal's motile responses to environmental factors have been attempted *under natural conditions*. In addition, relatively few studies have been concerned with accurate measurements of the effects animals have on their immediate microclimates and the relation of these alterations to the animal itself and to its habitat.

In order to describe a total energy balance of an environment, the following physical properties must be measured: absorbed and reflected solar and infrared radiations, air and soil temperatures, relative and absolute humidities, precipitation, and barometric pressures. Current advances in instrumentation give hope for truly accurate measurements of these and many other variables. Movement studies have previously been based almost entirely on visual observation, recapture of previously trapped and marked individuals, and on the appearance in study areas of species intermittently captured by various sampling methods. From these studies have come current interpretations of an animal's response to environmental change, the area of its "home range," its migration and dispersal, population structure, and even conclusions on systematics, evolution, and biogeography. Under conditions of adversity, an animal adjusts physiologically (for example, by changing heart, respiratory, and metabolic rates). It may shift its position in its normal home area; it may migrate; it may die. If we use the problem of animal movements and environmental stimuli as a basis for interdisciplinary research among engineers, physical scientists, and biologists, what can we present as a feasible technique to give us results of an importance equal to those being obtained by researchers in the physical sciences?

Of course, the development of radar during World War II enabled us to "see" movements of birds both day and night on the radarscope and on film, and also to see weather and the responses of the birds to it. It now allows us to measure, within the radar's range, the magnitude and speed of the migration, altitudes of flight, and direction. But we cannot very accurately identify the birds as species nor can we distinguish between individuals and flocks. The use of radar by biologists evolved slowly—partly as a result of wartime security measures and partly because of the reluctance of many scientists to recognize that nearly all "angels" seen on radar were really birds.

Within the last three or four years, the availability of transistors and other miniature and now microminiature electronic components has given us another major breakthrough in the study of animal movements in their natural environment. Tiny radio transmitters, which, together with batteries, weigh less than an ounce and which can transmit a signal for weeks, have been placed on or in animals. To date, ruffed grouse, cottontail rabbits, woodchucks, porcupines, skunks, and other animals have been located intermittently by portable receivers carried in the field. Some of the results obtained have been both surprising and dramatic. The transmitted signal was sometimes altered by the behavior of the animal and by one or more body functions. Changes in the signal have distinguished between a resting and a flying bird, and analyses of signals recorded on paper have given accurate measurements of rates of respiration and wingbeats of a mallard duck.

Information on relatively sedentary animals and their environments may become known by telemetering data to ground stations over short distances. But radio signals do not curve, as does the earth's surface, so signals traveling along the ground are largely lost over any considerable distance. Thus, tracking long migratory movements would require either numerous receiving antennas on ground towers or planes equipped with receivers—a tremendous expenditure in equipment and manpower.

Is there any way we could obtain line-of-sight radio signals from numerous animals that may be moving over vast distances in all three of the earth's mediums? How can we find out where the penguins of Antarctica go after mating season; the routes of the wandering albatross or the Caribbean turtle; the forces governing caribou movements; the track of the Canada goose? One tool in common current use by physical scientists and engineers could answer all these and many more. That is the artificial satellite.

Satellites carrying various instruments have already relayed back to earth vast amounts of data in the fields of geophysics, radiation, and other aspects of space studies. Most instruments for making these measurements and the radio transmitters for sending the data back to earth are very small in size and light in weight. What equipment would be needed to locate animals on various parts of the globe and to transmit the information back to earth? To obtain the necessary calculations, I went to my colleague, William W. Cochran, Director of the Bioelectronics Laboratory in the Minnesota Museum of Natural History. His experience in designing and building both satellite- and animal-borne radio transmitters and receivers places the following instrumentation plan beyond the realm of mere speculation.

A satellite could be instrumented to receive and relay to ground stations the signals obtained from many transmitter-carrying animals in various parts of the world. Initially, we might seek only the migration track of each animal. Cloud patterns and perhaps other meteorological and geophysical data that might affect animal movements might be available from measurements made by other projects (such as the Tiros satellites) for areas of the earth over which birds would fly or swim. These data could provide information on such phenomena as disorientation of birds under conditions of overcast. Successes in tracking animals on a world-wide basis by the use of a satellite would, it is hoped, provide guides for further instrumentation to telemeter one or two physiological or environmental factors concurrently with migratory movements.

Species that could be tracked by this method, and for which prototype transmitters have already been designed and tested, are larger birds such as geese, swans, cranes, penguins, and albatrosses. Among mammals the caribou could be tracked in its migrations. Species in these groups are large enough to carry currently available transmitters and power supply. Their modes and regions of travel and their physical environments are so different that numerous tests of the instrumentation could be provided effectively. For example, Canada, blue, and snow geese, which could be tagged with radio transmitters, pass over much of continental North America, moving mostly in long, direct flights. The longest nonstop flight is probably from James Bay, the southern extension of Hudson Bay, to the Gulf Coast of Louisiana and east Texas.

Instrumentation to track these and other animals on a global basis could be relatively simple. Each animal could carry a transmitter and power supply weighing approximately 20 to 60 grams. The satellite, traveling 18,000 miles an hour (approximately 5 miles per second), at an altitude of 200 miles above the earth in a polar orbit, would cross over both North and South Poles once about every 103 minutes. In the satellite would be a receiver and transmitter weighing a few pounds—or possibly even less—and occupying perhaps half of one cubic foot of space. Power supply, of course, would require additional weight and space, but might be supplied by solar cells, which would extend the period of tracking and would weigh less than batteries.

In order to receive the signals from the satellite at all times in its flight, twenty-four optimum-located receiving stations would cover the earth. A tape recorder at each of the ground stations would record the signal that had been received in the satellite from an animal-borne transmitter and then retransmitted to earth. This tape would then be run through an audiospectrometer. The spectrograph would be displayed and used—together with the known positions of the satellite in its passes within the line-of-sight range of each animal—to locate the animals at various times of day and night.

Some of the details of this tracking method might be of interest. The tiny radio transmitters, each with its battery pack, would be no larger than the first joint of a thumb. It would be attached to a simple harness with the antenna forming one of the harness loops around the body of the animal. It has already been found that animals adjust relatively easily to harnesses that do not inconvenience their habits. The entire unit would weigh some two ounces or less, depending on the number of batteries. The transmitter signals would be intermittent pulses, each at a slightly different frequency and, therefore, distinguishable from others in the same part of the earth.

To help determine just how far a signal from these small instruments could be received, we attached half-ounce transmitters to high-altitude balloons used for research by the University of Minnesota Department of Physics. With radiated power of much less than one microwatt—far less than that anticipated for satellite tracking—we received the signal from 270 miles away when the balloon was 25 miles above the earth. And this was in a region of considerable radio interference. The test adequately demonstrated the feasibility of long-range line-of-sight transmission.

The proposed satellite would contain a receiver and an antenna tuned to that band of frequencies occupied by the transmitters. Satellite transmitter power should be at 500 microwatts to assure a satisfactory signal (distinguishable from all types of interference) at the ground receiving sites.

A lower power could be used at the expense of adding or utilizing more ground receiving sites.

Signals from the satellite could be received by a ground station at a distance of about 1,200 miles. The satellite could receive signals from the animals up to about 800 miles. Thus, the effective coverage area for one receiving station for tracking animals is about 4,000 miles across. Therefore, one station near the center of North America could conceivably cover most of the United States and Canada. A few stations in the Pacific islands could cover the whole Pacific area.

Each orbit of the satellite "scans" a swath about 1,600 miles across. The overlap in successive orbits depends upon the latitude. No overlap would occur at the Equator and a 100 per cent overlap would occur at the Poles. Thus, in equatorial zones, two locations, or fixes, a day are possible. In temperate latitudes, four fixes daily are possible, and in arctic latitudes about 14 (the number of orbits per day). Of course, to get a fix the satellite must be in range of both the receiver and the tag on the animal, so the maximum number of fixes is obtained only if ideal receiver distribution is used—that is, at least one receiver must be so located as to cover any point in the area concerned. This means that receivers can be spaced somewhat less than 1,200 miles apart, depending on the area covered, whereas for computing coverage, a radius of about 2,000 miles from the ground station can be used. For example, on Pass No. 1 of the satellite both ground station receiver and tag transmitter are in range of the satellite; but on Pass No. 2 only the tag is in its range. Thus, if we consider the number of fixes as the number of times the satellite is in tag range, we must provide receiver coverage to insure satellite contact. This means an overlap of receiver circles, each 1,200 miles across. But if we are only interested in complete coverage and not in the maximum number of fixes possible each day, we can use overlapping 2,000-mile circles. This is of importance at low latitudes, where tagged animals could be missed.

We have said that if a satellite is 200 miles high it will cross over both the North and South Poles once every 103 minutes, or every one hour and 43 minutes. (Its orbit can be thought of as fixed in space, for although precession occurs, it is very gradual.) If the satellite passes northward over point A on the Equator at noon, it will pass southward over the Equator 51½ minutes later, approximately on the opposite side of the earth. About 103 minutes later it will return, but this time it will pass over the Equator about 1,600 miles *east* of point A, which is the distance earth (and point A) has rotated during the satellite's orbit. The next orbit will be about 3,200 miles east of point A, and seven orbits after crossing A the first time, the satellite will pass over the Equator about 11,500 miles east of A, going north. Since 11,500 miles is about one half the circumference of the earth, point A will see the satellite coming south on its seventh orbit.

In each ground receiving system would be one FM receiver and associated antenna tuned to the frequency of the satellite transmitter, one two-channel tape recorder, and one highly accurate clock to emit sound signals that could be recorded as time intervals. Equipment of this type is available and relatively simple to operate.

Each time the satellite passed within the 1,200-mile maximum range, a recording would be made of the modulation of the satellite transmitter and, on the tape's other recorder channel, of time tones accurate to one second (the satellite travels at five miles per second). These recordings are preserved for analysis. The signals from the transmitters may or may not be heard through the interference, depending on the distance between the satellite and the tagged animals. Much of the time they will be inaudible. The tapes will be run through an audiospectrometer with a resolution of 10 cycles per second, or less; the narrow band width of the spectrometer will reduce the interference to a point at which the signals are discernible.

When these signals are plotted on paper as radio frequency versus time, they form an S-shaped curve. This curve is a result of the Doppler effect between the satellite receiver and the signals coming to it from the transmitter on the animal. This curve, plus accurate time, plus the satellite's path would enable us to locate the animal in a circle 50 miles or less across.

Paradoxically, extracting these data from a single tape may take some 1,500 times as long (about four continuous days) as the four minutes or so required to record it initially, although the time could be reduced by using audiospectrometers with higher capacity.

The complexity of analyzing the tapes suggests the creation and use of analysis centers. And here still another field is introduced into this cooperative scientific venture—the adequate utilization of computers whose programming could be jointly undertaken by physical and biological scientists. The information that could conceivably be transmitted by, say, a goose equipped with a transmitter, is almost limitless, at least theoretically. The speed, altitude, and direction of flight are only a few examples. Wing-beat and respiration rates could be sent to the satellite, as could temperatures of the individual animals. These factors could be examined by one of the new biological computers, which have been specially designed to extract desired information from tapes that might be cluttered with various kinds of interference. Latitude and longitude calculations to locate the animals could also be determined to accuracies of about 20 miles, depending on the particular electronic system utilized.

With knowledge of long-distance movements of only a few kinds of animals in various parts of the world, we could begin to obtain data on environmental and physiological factors, and with these to work with, answers could be found to such questions as: does the wandering albatross really wander, or does it follow routes prescribed by environmental factors?

To plan ways in which to use natural resources properly and to expand basic knowledge of the earth's ecologies, more extensive information on the biological and physical factors of the natural environment is obviously essential. It is also essential that such information be acquired at a rapid enough pace to bridge the gap that now exists in data acquisition between the physical and the biological sciences. If biologists are ever to participate in the application, to their own disciplines, of measurement programs such as those demonstrated during the I.G.Y., they must be ready to understand and to use advanced instrumentation whenever it is expedient.

On the last day of October, 1962, seven Canada geese, each wearing one of our transmitters, were monitored at their mid-continent stopover before the last leg of their fall migration flight from the Arctic to Texas. For twelve days previously they had done little flying—they ate, preened, walked, or swam. At that time we could determine the location of each goose and what each was doing by the signals from their transmitters. During those twelve days we learned much about goose behavior at a stopover area. When they left on that last long flight, and disappeared over the horizon, their signals were lost to us on the ground. But those signals were still being sent out into space where, if a satellite had been "listening," it might have been able to tell us infinitely more than we know now about the mysteries of migration.

FRANK A. BEACH

Can Animals Reason?

TWENTY-THREE HUNDRED years ago, Aristotle wrote that many animals can learn and remember but that only man can reason. This conclusion has been challenged many times during the ensuing centuries, and even today men still ask, "Can animals reason?" The query is nonsensical and therefore unanswerable. What "animals" are meant? Peacocks? Protozoa? Amoebae? Ants? Monkeys? Men? (After all, men *are* animals.) And what is meant by "reason"? Before engaging in wordy debates about reasoning in other creatures, we would do well to think seriously about reasoning as it occurs in human beings.

From the time of Heraclitus (500 B.C.), man has gratuitously described himself as *the* rational animal. What does the title imply? Philosopher John Dewey regarded reasoning as a special method of problem solving. (This is helpful. We should, then, begin our search for reasoning in other animals by observing them when they are trying to solve problems.) Dewey added that reasoning usually occurs in four steps.

(1) First comes the analysis of the problem. (2) Next, the reasoner formulates one or more hypotheses that represent possible solutions. (3) Then these hypotheses are tested, one after another. Man usually tests his hypotheses by talking to himself about them. He may not speak aloud, but his thought takes the form of unuttered language. In the course of his subvocal soliloquy, man exposes the weakness in each erroneous hypothesis and then discards it. (4) The final step comes when he hits upon one that stands the test. He puts it into action, and his problem is solved.

If this emphasis upon the importance of words in reasoning is justified, it is difficult to see how lower animals, lacking a language, could employ reasoning. But we need not surrender without a fight. Let's look at the facts. Is there anything in the behavior of other animals that resembles, however slightly, Professor Dewey's description of human reasoning?

Yes, there is. He mentions the formulation of hypotheses as a first step, and one psychologist has written a whole series of erudite scientific articles on the general subject of "Hypotheses in Rats." His experiments show that when a rat finds itself confronted with a novel problem, it does not attack the situation blindly. Instead, the rodent "progresses through a series of systematic behavior patterns, adopting one form of response, persisting in it for some time, dropping that habit [if it is unsuccessful] and adopting another, and so on."

In other words, the languageless rat cannot test his "hypotheses" by discussing them with himself, but he can put them into practice to see if they will work. The significant point established by these studies is that when the lowly rat is trying to solve a problem, he displays adaptability. If one approach proves fruitless, he has enough sense to drop it and try another. This ability by itself certainly is not reasoning, but there could be no reasoning without it. Perhaps it is not too much to say that variability of approach is one evolutionary forerunner of human reasoning.

As a matter of fact, the ability to give up an unsuccessful line of attack and shift to a new one is fundamental to practical reasoning. The failure to do this frequently keeps people from developing their reasoning powers to the fullest extent. Some men who pride themselves on their stick-to-itiveness are in reality confusing perseverance with pigheadedness.

Another psychologist has scientifically analyzed human reasoning in the experimental laboratory. He devised a series of rather simple mechanical problems for college students and then watched while various solutions were tried. After everyone had either solved the problems or given up in disgust, the group was given some practical suggestions on how to reason.

The advice was surprisingly simple. (1) When you are confronted with a problem, try to locate at least one of the difficulties, and then try to overcome it. If you fail, get that difficulty completely out of mind for the time being and look for an entirely different one. (2) Do not be a creature of habit and stay in a rut. Keep your mind open for new meanings. (3) The solution pattern appears suddenly. You cannot force it. Keep your mind open for new combinations and do not waste time on unsuccessful attempts.

At the conclusion of the lecture, a second series of problems was presented. This time the average score for the group was twice as high as it had been before. These results agreed with John Dewey's belief that reasoning does not depend entirely upon native ability. The efficiency with which we reason can often be improved through deliberate training.

The psychologist's advice about staying out of a rut and keeping an eye open for new solutions recalls the white rats and their "hypotheses." Their behavior involves a great deal more than one is apt to realize. The ability to give up one line of attack and shift to a new one is quite limited in animals that belong below the mammals in the evolutionary scale.

If a hungry octopus is separated from his prey by a sheet of clear glass, he behaves in a curious fashion. The tentacles writhe over the surface of the barrier, and the animal's body strains against the smooth surface. There on the opposite side, just out of reach, is a delectable prawn. The octopus expends every effort to force his way to the food *by the most direct route*. It would seem a simple matter for him to move sideways just far enough to pass around the edge of the glass. But this would mean moving away from the visible prey, and such a course of action is too much for the cephalopod "mind."

When hunting under normal conditions, the octopus will follow a prawn that has disappeared behind a rock. Yet when his dinner is in plain sight but mysteriously unobtainable, only one "solution" occurs to the octopus, and he cannot discard it even though it always results in failure. Ophelia, the "Clever Octopus," first *accidentally* knocked the lid off the food jar. It learned by trial and error. Here we are talking about reasoning.

Some birds aren't a great deal different from the octopus. A hen separated from a pile of corn by a short length of chicken wire fence is a pathetic sight. She stands in the center of the fence directly opposite the grain and clucks nervously. The unobstructed pathway around the end of the fence is plainly visible, but to turn away from the food is very difficult. The bird runs a few steps toward the end of the fence, hesitates, and then dashes back to the spot just opposite the corn. Another halfhearted excursion toward the end, and another loss of courage and return. Feathered indecision! Gradually the trips away from the food grow longer, and finally the hen rounds the turn and comes flying down the homestretch to gobble up the corn. Here is an example of extremely limited variability of behavior. One would not expect reasoning in such a creature.

Mammals are much more adaptable. Anyone who knows dogs will tell you that they can master this type of problem with little or no difficulty. In one experiment, a Canary Island bitch was put on the inside of a U-shaped wire fence. Then a nice, meaty bone was dropped just outside the fence at the base of the U. For a few minutes the dog tried to push her paw through the wire. Then she dug at the dirt in an unsuccessful effort to tunnel under the fence. Briefly she gnawed at the wire; but within a few minutes she gave up these fruitless forms of attack, sat down, and looked about her. In a flash she was on her feet, running, and in less time than it takes to tell, she had the bone in her teeth.

For human children such a situation hardly deserves to be called a problem. A little four-year-old girl was led into the same U-shaped enclosure, and her favorite doll was laid on the grass just outside the fence. The child looked at the wire, followed it with her eyes, and immediately saw the opening behind her. A happy smile lit her face; she began to laugh and ran swiftly around the fence to where the doll lay.

Such experiments are too simple to reveal the existence of reasoning, but they do tell us where to look for it. Any animal species that shows limited variability in simple types of problem solving is quite unlikely to demonstrate reasoning ability in any situation.

The inability to respond adaptively when faced with a problem may be due to lack of the necessary brain development. This would seem to be the explanation in the case of the octopus and, to a slightly lesser degree, in the case of birds. Unfortunately, however, possession of a well-developed cerebrum is no guarantee that reason will rule behavior. In man himself there are many foes to reason, and one of the most powerful is habit. In some life situations, man's habits make him behave like an octopus when it comes to meeting new problems.

Many times we encounter a new problem that bears a superficial resemblance to some we have solved before. Then we are likely to persist in attacking the novel difficulty with unsuitable methods just because they have been successful on other occasions. More than once an amateur has solved some problem that had been puzzling experts for years. Sometimes the amateur is smarter than the experts, but more often his only advantage is a fresh point of view. His approach to the solution is not cluttered up with habitual modes of thought about the problem.

From early childhood we build up habits that guide our day-to-day lives. It is fortunate that we can do so. Imagine how troublesome life would be if we had to solve every problem anew each time it arose. In 99 out of every 100 of the little problems confronting us each day, habit provides a ready-made answer, and original thinking is unnecessary. It is only when old habits will not serve that reasoning makes its appearance.

Habit limits our ways of thinking about a problem, and this is a very real block to successful reasoning. The trouble may originate at a very simple level. The way in which we "see" a problematical situation has a powerful control over our ability to solve it.

In one experiment young men and women were led into a bare room and given three candles, three thumbtacks, and three small boxes. They were instructed to fasten the candles to the door of the room in any way they could devise. There was only one solution, but it was simple. The boxes were fastened to the door with thumbtacks, and one candle was placed upright in each box. Every individual discovered this method in short order.

Then the experiment was repeated, using a new group of "subjects" and introducing one minor change in the setup. This time the little boxes contained matches. The rest of the equipment was the same, and so were the experimenter's instructions. Less than half of the men and women solved the problem.

What was the difference? In the second version of the experiment, most of the people "saw" the boxes as containers for matches, and this blocked the possibility of "seeing" them as potential candleholders. Furthermore, the presence of matches was distracting, for they suggested (due to habit) lighting the candles. A burning candle suggested (due to habit) the use of hot wax as an adhesive. The problem had been designed so that no such solution would work. Nevertheless, the force of habit was so strong that many individuals stubbornly persisted in trying minor variations on the same futile theme. The little boxes continued to be match containers, and the thumbtacks were simply ignored.

By simple trial and error man and many other animals are capable of learning how to meet certain types of problems. This is something different from reasoning, which always consists of devising the solution to a problem which has not previously been mastered through learning. Therefore, it is almost impossible to decide whether or not a particular item of behavior viewed in isolation is or is not the product of reasoning.

The difficulty is beautifully illustrated in the story of Congo, the fastidious gorilla. Congo had the distinction of being the first of her species to serve as the subject of psychological experiment. In one test she learned to reach through the bars of her cage and operate various mechanical devices. One sunny morning after a rainy night, the scientist approached her cage and called Congo to work. The ape emerged from her living quarters and lumbered over to the bars, apparently anxious to begin the day's tests.

The floor on which she usually sat was still wet. Congo looked at the puddles for a moment, scratched her head, and then deliberately turned her back on the experimenter and returned to her bedroom. Shortly she emerged, her arms loaded with dry straw from her sleeping couch. Carefully depositing the straw on the damp cement, the fastidious gorilla sat down on her improvised cushion and gestured to indicate that she was now comfortable and ready to work.

Now does this performance reflect reasoning? It is impossible to decide. There is always the possibility that puddles on the floor was an old problem for Congo and that in going after the straw she was acting by habit. You may say that she must have solved the problem some time, or the habit wouldn't be there. That is true, but we don't know *how* that first solution was achieved. It might have been the chance product of a very long process of trial-and-error behavior and, as such, not due to reasoning.

This attitude may seem overcautious, but it really underlies our everyday judgments of human beings. I once won the reputation of possessing great reasoning powers because I solved a conundrum that baffled every other guest at a party. When I finally confessed that I had heard it before, my stock as a reasoner dropped a long way. The original, flattering reaction had been based upon the assumption that the problem was a new one and that therefore my solution was original. Another puzzle was presented, and more than half of the guests finally found the answer, but this time everyone made many mistakes, advanced many incorrect answers, and finally reached the solution by a laborious process of elimination. No one was crowned King of Reason. Why? Because unconsciously we do not accept trial-and-error behavior as true reasoning.

If asked, "Can animals reason?" one is apt to say "Yes" without a second thought. By refusing to take the answer for granted, we have tried to straighten out a few of the difficulties that might have prevented us from a correct evaluation. We have seen several ways in which reasoning can be defined and ways in which human reasoning can be observed in action. Yes, there is some evidence to suggest that lower animals can reason, but lest we expect too much from the experimental results, let us hold firmly in mind the fact that men and women employ reason only at infrequent intervals, and even then reasoning does not come easily to The Reasoning Animal.

It is sad but true that grown men and women often become quite emotional when they find themselves in life situations demanding analysis and the formulation and testing of hypotheses. When anger or fear or anxiety (and some old sourpuss is sure to add, love) comes in the door, reason flies out the window.

The same shows up very clearly in other animals. Many psychologists have observed that the chimpanzee resembles man in emotional make-up more closely than in general intelligence; and these anthropoid apes sometimes behave in distressingly human fashion when called upon to solve problems.

Koko, a three-year-old male, was one of a number of animals used in a series of experiments on problem solving. He was obviously gifted, but like some human geniuses, Koko was a slave to temperament. If his first few attempts to solve a new problem failed, the little ape was apt to fly into a rage, hopping first on one foot and then on the other, pommeling the floor with his fists, and screaming at the top of his lungs to express his frustration. Since the reasoning tests occurred with regularity, Koko was usually in a state of chronic indignation. Of course, his performance suffered, because these emotional outbursts totally prevented any objective analysis of the problematical situation.

Every chimpanzee had a distinct personality. Nueva, a young female, had a trusting disposition and winning ways that almost made one forget that she was partly bald and, by human standards, decidedly ugly. She showed a childlike attachment to the experimenting psychologist. Then, too, there were Sultan and Konsul, both males and as different as night and day, but each in his own way was an efficient problem solver.

All of the problem-solving tests were quite simple. They were based upon two principles. First, the animal must be offered some truly attractive reward, but the simple, direct approach to the prize must be blocked. Second, the ape must be in a position to survey the entire situation, to study the various possible solutions without actually having to go through the motions of trying them out.

In one test a banana was suspended from a high limb, and the little chimpanzees were prevented from climbing the tree. In order to earn their reward, they had to work out an original solution. Discarded packing boxes of various sizes were left scattered about the enclosure in which the animals worked. Would any ape hit upon the idea of moving a box under the dangling fruit and using it as a stepladder?

Some chimpanzees did exactly this without any human guidance. Several of the animals became so adept that they could reach bananas that hung several yards above the ground. The simian engineers sometimes employed as many as four boxes stacked one upon another. The completed structures wouldn't have conformed to any sort of safety regulations. They were so rickety that they swayed perilously to and fro as their intrepid builder clambered toward the top. But the apes had implicit confidence in the products of their own labors, and if the movement became too violent, the climber merely took off with a flying leap, snatched the banana at the peak of his jump, and then fell to the earth. Landing on your feet is easy if your ancestors have spent most of their lives in trees.

Does this sort of behavior show reasoning ability? Well, the problem was novel. These apes, reared in captivity, had never before found their meals hanging in mid-air. Much less had they been taught to build their own stepladders! On some occasions, at least, the solution was hit upon suddenly, with little or no fumbling, no trial-and-error behavior. But before passing judgment, let's examine more of the evidence.

There is a story, possibly apocryphal, about a psychologist who shut a chimpanzee in a soundproof room filled with dozens of mechanical toys. Eager to see which playthings the ape would choose when he was all alone in this treasure house, the scientist bent down on his knees and put his eye to the keyhole. What he saw was one bright eye peering through from the other side of the aperture. If this anecdote isn't true, it certainly ought to be, for it illustrates the impossibility of anticipating exactly what an animal will do in a test situation.

After Koko, Konsul, and the others had been stacking boxes for a few days, they began to generalize from their experience in surprising fashion. One day, when a bit of fruit was hanging on high and no boxes were available, the psychologist strolled into the enclosure. One enterprising young ape rushed over to the experimenter, seized him by the seat of his pants, and dragged him to a spot directly beneath the banana. Then the chimpanzee pushed and pulled in an effort to get the scientist to bend over so that his back would provide a platform to be used in lieu of a box.

Such surprises occurred more than once. The animals sometimes solved their problems in ways that had not occurred to the experimenter when he designed the test. On one occasion the bait was suspended in a basket from a high limb, and the apes were given long bamboo poles. The psychologist expected the poles to be used to tip the basket and spill its contents, but he had forgotten that chimpanzees are natural-born acrobats. Placing a pole upright but unsupported directly beneath the basket, they swarmed rapidly up the stick and snatched the reward before the pole could fall over.

Sociologists sometimes describe man as a "tool-using animal," as though this were a unique characteristic. As a matter of fact, many animals from wasps to walruses are said to use rocks or twigs as a man uses a sledge or lever, and some apes have shown the ability to manufacture tools to suit their needs.

Favorite types of food were thrown on the ground several feet outside an ape's cage, and a short length of bamboo was left lying within reach. Some chimpanzees grasped the solution with little delay. Picking up one end of the bamboo, they extended an arm to full reach through the cage bars and raked in their reward.

Sultan went a step further. He was given two pieces of bamboo, neither of which was long enough to reach the food. After trying various possible solutions to his personal problem, Sultan began to gnaw on the end of one of the sticks. When he had reduced its size sufficiently, the ape forced the tapered end of the smaller stick into the hollow end of the larger one, thus constructing a new tool long enough to meet his needs.

These experiments with chimpanzees represent a very small part of the work that comparative psychologists have conducted in their attempts to plumb the animal mind. Other tests have been used, and many species from birds to baboons have been studied. Some of the methods have resembled those described here, and others have been quite different; but in nearly every case the underlying assumptions concerning the nature and manifestations of reasoning have been very much the same. Reason has been regarded throughout as a special way of solving new problems.

As a matter of fact, it is more nearly correct to say that science studies problem solving rather than reasoning. It is doubtful if any professional experimentalist would stick his neck out so far as to answer a simple "Yes" or "No" to the question, "Can animals reason?" Instead he asks himself what kinds of problems animals can solve, and he places particular emphasis upon the ways in which animals work out their solutions. Birds have shown practically no indication of being able to solve problems in any way except by simple learning. Even the fabled crow turns out to be incapable of passing a reasoning test. All mammals demonstrate the ability to vary their attack, but the only problems that cats, horses, and even dogs can solve without previous practice are so elementary that the behavior scarcely qualifies as reasoning. Monkeys and apes turn in a better score, and some of their reactions are so reminiscent of human reasoning that the same term seems applicable to both.

Some readers are probably saying to themselves, "Humph! My dog does things that are much more intelligent than simply piling boxes. Those skeptical psychologists in their artificial laboratories know very little about the natural behavior of animals. Why, if they could only see Towser . . .!"

Perhaps so. No experimentalist is so presumptious as to think he knows all there is to know about animal psychology. However, before you jump to conclusions, remember the limits of our definition of reasoning. Be sure that the situation represents a genuine problem that has never been encountered before. If Towser looks the matter over and then proceeds at once to the correct solution, score him high. If he keeps making the same mistakes over again and only gradually improves, credit him with learning but not with reasoning. Between these extremes the analysis is not so easy. The fewer the trials and the more sudden the success, the better the excuse for invoking reasoning. It is not a hard and sharp line, but from the scientific point of view, the most parsimonious interpretation is the correct interpretation. Problem solving that can be explained as a product of chance or of gradual learning is not to be accepted as evidence for any higher type of mental activity such as reasoning.

And there is one more precaution, so important that it can hardly be repeated too often. *Keep your heart out of this.* Always try to form your opinions on the basis of what you actually see, not what you want to believe. Those of us who love pets frequently let our emotions rule our thinking in judging animal intelligence. A dramatic illustration of this human failing came to light a few years ago when one of the mass-circulation magazines published an article about the intelligence of horses.

The article was short and obviously intended as a humorous piece. The author simply described one funny incident after another which, so he claimed, proved that under some circumstances a horse can behave as though it hasn't a brain in its head. Just why this offended a lot of readers is difficult to understand. Surely the same thing could be said of human beings. Nevertheless, offend them it did, and indignant letters poured into the editorial offices.

The editor became alarmed and published a nicely worded apology to horses and horse lovers everywhere. "The article," he explained sorrowfully, "had been printed as entertainment, not as an official pronouncement on the limitations of the equine mind." He was very, very sorry that so many people (and horses?) had had their feelings hurt. In closing, he printed a few of the letters he had received. The point of this story is that not a single one of the published letters contained any evidence whatsoever to refute the original claim that horses are unreasoning beasts.

The letter writers pointed out the obvious fact that horses have served man for thousands of years. They called to mind the horse's willingness to labor faithfully for the meanest kind of reward. Inevitably there was a letter about the horse who saved his master's life by finding the way home through the traditional blizzard. A young girl complained fretfully that the original article had brought ruin to her plans for owning a pony. By months of feminine strategy, she had convinced her father of the feasibility of her scheme, but when he read of the low intelligence of horses, he withdrew his consent.

Now what in the world do these letters tell us about the horse's *intelligence*? Not a thing! Suppose you were a schoolteacher and found it impossible to promote some child simply because he lacked the native intelligence to do the daily assignments. Would you revise your opinion just because someone pointed out that he was a good boy, never told lies, always said his prayers, and minded his mother? In forming daily judgments about the reasoning ability of our human friends and associates, we do not ordinarily rely on character references, but that is exactly what these letter writers offered to back up their fond belief that horses are highly intelligent. Just character references.

We know that men and women can be faithful, friendly, trustworthy, and lovable without being highly intelligent. In fact, we consciously differentiate between these qualities and sheer intellectual ability. Surely it is neither unfair nor unfeeling to insist that the same discrimination be observed when it comes to judging our animal friends. And when this essential rule is followed, we are likely to arrive at the conclusion that man can reason and does so occasionally, that his primate cousins show faint glimmerings of similar ability when they are carefully tested, but that for all other species the record is not encouraging. At this stage of knowledge, it would be premature to state flatly that dogs, horses, cats, and other four-footed creatures are totally incapable of reasoning; but thus far every objective search for scientifically acceptable proof has been completely unsuccessful.

D. *The Spectacle of Plant Life*

•

In previous articles the distinction between plants and animals has been made. In the present section we turn to a more extended discussion of the plant kingdom. A. J. Sharp is professor of botany at the University of Tennessee. In The Compleat Botanist *he explains why trained botanists with a broad background and a liberal education are urgently needed by society.*

In his introduction to The Web of Life, *the book by John H. Storer from which the following selection is taken, Fairfield Osborn points out that "there are some truths, even fundamental ones, that are apt to elude us. The most basic truth regarding our Earth-home is that all living things, in some manner, are related to each other." Storer is highly qualified to discuss the subject through his decades of activity as an observer and photographer of wildlife. His article helps us understand the intimate connection between ecology and conservation.*

In Feeding in Plants *by John Tyler Bonner and* The Pollination of Flowers *by Verne Grant some of the specific functionings of plants as they cope with their environment are analyzed. And in* Strangler Trees *Theodosius Dobzhansky and João Murça-Pires describe "a most remarkable adaptation in the vegetable world which illustrates . . . a step-by-step evolution" of the strangling habit.*

A. J. SHARP

The Compleat Botanist

•

My INDEBTEDNESS, in this speculation about the "compleat" botanist, is clear. Izaak Walton was not a professional biologist, but he did realize that to be a thorough fisherman one had to understand every angle of angling. It is my thesis that to be a "compleat" botanist one must have a breadth of perspective that permits him to see beyond his own specialty and understand the importance of relationships which exist today not only between the fragments (and I use the word advisedly) of botany but also between plants and the everyday life of man. What I have to say about botany is equally true of zoology and biology in general. And I will be so bold as to substitute often in my discussion the word *biology* for *botany*.

It is wise and healthful in any discipline for the practitioners periodically to re-examine it and themselves. It is clear that the time has come for botanists to take a fresh look at themselves and their science. The current, bitter, internecine discussion concerning the relative validity of modern or molecular biology and classical or organismal biology indicates that we are confused. Worse yet, it is confusing our clients and benefactors, many of whom can see no reason why classical and modern approaches should not supplement each other.

As I understand it, botany is the study of plants in the broadest terms, and ranges from the chemistry of the DNA molecule in the nucleus of a plant cell to the spatial relationship of individuals in a desert community. It includes studies of the chemical and physical natures of the materials and processes in the cells, of the organization of the cells into tissues, of tissues into organs, of the movement of materials into, through, and out of plants. In addition, the botanist is concerned with the plants of the past, with phylogeny, with modern floras, with the relations of plants to all phases of their environment, not excluding man.

It should be clear to the most naïve of our fellowmen that his welfare is dependent upon plants, but man seems far from understanding that fact.

Although he is a "newcomer" relative to plants, he constantly treats them without comprehension of their significance. Sears reminds us that the earth, around 5 billion years old, produced life only after a period of about 3 billion years and that, from that life, man came after another 2 billion years. This upstart Man (not over 1 million years old and, in civilized form, much younger than that) mistreats the vegetation of the earth and irreparably damages the macrocosm in which his children and grandchildren must live.

I do not have to remind professional biologists that plants are directly or indirectly the source of all food, most clothing, much fuel, and all lumber; that they exert a tremendous control over water supplies, soil, and the production of fish and wildlife; and that, as McKinley emphasizes, esthetically and spiritually they are of great value to man. Plants have an enormous impact on man's culture and well-being.

Man, at the same time, exerts an influence on vegetation. He is rapidly destroying large, and often virgin, plant communities, even in remote areas, by saw, bulldozer, fire, and distribution of destructive pests. This devastation must be seen to be understood. The destruction of vegetation and its natural environment is usually accomplished without any thought of what will happen to water supplies, soils, or the general economy of the region affected. Unfortunately, man usually thinks of plants and vegetation only in terms of food, clothing, lumber, and raw materials for manufacture.

Men in primitive groups seem to understand the relationships between themselves and plants better than men in more advanced cultures. There seems to be an almost direct correlation between the degree of civilization of a society and its failure to appreciate this dependence. This lack of awareness is further enhanced when a civilized society changes from an agricultural to an industrial economy. Industrialization seems effectively to isolate the individual from the primary plant resources on which his culture is based. This fact may help explain why botany and botanists are still much more revered in the less industrialized Latin American countries than in more industrialized countries.

I have been trying to emphasize the significance of plants to man and, thus, the great importance of botanists to society. Regardless of the recognition afforded us, we can be proud to be botanists. An awareness of our role in our culture imposes a heavy obligation on us to try to do something about the deteriorating balance between human populations and the plant resources upon which they depend. This could be the focal point of much of our teaching and of our research, which should include basic studies as well as applications.

Long ago research at the general level raised more questions than it answered, and, as the sciences of geology, chemistry, physics, and mathematics developed or matured, botanists turned to them for assistance. No one should infer from this that we have exhausted the problems of a general nature; for most of them remain, demanding constantly more refined and complex answers, not only to the question "How?" but, as recently emphasized by Simpson, to the legitimate but more fundamental question "What for?" Nor should the great amount of information already accumulated in certain areas mislead one into thinking that research in these areas is no longer needed. As Laetsch has indicated, most of the fads (and there seem to be as many fads in biology and botany as there are in women's clothes) subside without solutions being found to most of the basic problems.

As an example of the continuing need for research in the *classical* phases of botany, one could cite the considered opinion of most taxonomists that something less than half of the plant species in the world have been detected and described. It is probable that in this reservoir of "unknowns" are entities of high significance to the understanding of phylogeny and other basic phases of botany. There certainly will be found species of economic importance which may give us new antibiotics or other valuable chemicals, or breeding stock for better or more disease-resistant crops. To discover these species before it is too late is a critical matter, because in the next decade most of the natural areas and many of their species will have been eliminated. I agree with Corner that more understanding of the phylogeny and evolution of the flowering plants is to be gained from a study of existing tropical floras than has been given us by paleobotany. Shall we permit this reservoir of evidence to slip through our fingers because we deprecate exploratory and field botany?

Certainly few botanists know the American tropics as does Bassett Maguire, of the New York Botanical Garden, who warns us that the pristine tropical forests of South America are being put to the torch with devastating effect. The destruction, which, he says, at least matches that accomplished by the bulldozer in the United States, is the result of man's migrating from population centers and clearing land for farming. This land is useful for only 2 to 3 years at best. At the end of that time man starts burning new areas. Even the Andes, Maguire says, has felt the impact of this destruction; so far, only the inaccessible Lost World section has escaped. This is the region where four countries meet: British Guiana, Surinam, Brazil, and Venezuela.

The paramount need in the world of biology today, according to Maguire, is an inventory of tropical vegetation. The largest and most important zones not yet inventoried, he notes, are in the tropical areas of the New World. Whereas the flora in the temperate areas have been studied exhaustively, the surface has been only scratched in the hemisphere's tropics, despite repeated expeditions.

Research is still needed in plant geography, where recent findings in regard to geomagnetism and discussions concerning the origins and structure of continents and mountains cause us to question our previous postulates. These findings and discussions, plus our increased knowledge of the mechanics of evolution, make it clear that a re-examination of phytogeography is in order.

I am sure that in every phase of biology there are unsolved problems, whether the field be classical or modern or something in between, such as physiology or genetics. Any investigation of breadth seems to call upon a galaxy of related sciences. How, for example, can one understand physiology without knowledge of anatomy, chemistry, physics, and, in many cases, even taxonomy and ecology? There is a great need for the "generalist," as explained by Storer. At one time or another each part of biology makes use of nearly all the other phases. And there is no need to use "descriptive" as a derogatory epithet. Each science, even chemistry and physics, describes phenomena, processes, and products, and these descriptions are no less useful for being neither analytical nor mathematical in form.

Should any botanist think he has the final technique or the final answer, may I remind him that science has taught us nothing more clearly in this century than that there are no absolutes and that everything is relative and can be predicted only within certain statistical limits. The nature of living material, because of its tremendous complexity and constant change in a variable environment, is extremely difficult to study and understand. We may remove bits of material from cells, or cells from organisms, and place them in a test tube or a culture dish for analysis or observation, but findings have diminished meaning when the material is dissociated from the organism as a whole, and maximum significance only when the organism is in its natural environment. As Simpson has indicated, to our knowledge of physical, chemical, and mechanical principles must be added understanding of the adaptive usefulness of structures and processes to the whole organism and to the species to which the organism belongs, and, in addition, understanding of the ecological function of the species in the communities to which it belongs.

Our knowledge of the recycling of elements or of the capture and routing of energy in a biotic community is relatively in its infancy, and there is demand for new techniques to facilitate such studies. There are many other problems waiting to be solved in all phases of botany and biology. In fact the problems are far more numerous than the available workers.

In conclusion, let me summarize my feeling about botanists and the position of botany today.

The "compleat" botanist, because his materials are the prime resources of man, must be not only a specialist but at the same time a generalist of the most expansive kind. He must lead the members of a reluctant society to a realization of the importance of plants in most aspects of their daily living. He must teach his students in such a way that they, too, can see the woods *and* the trees, and carry on where he left off. If he must be a specialist, he should see the implications of his specialty for society, and its relation to other bodies of knowledge. He should understand teaching and demand of his charges a healthy skepticism, a tolerance of other points of view, and resistance to the temptation to conform when independent decision is called for. He must realize that diverseness does not imply perverseness and that in biology both classical and modern approaches are necessary. Botany needs taxonomists who understand and utilize chemistry, and biochemists who appreciate taxonomy. Intramural strife is enfeebling to both botanists and botany and must not be allowed to flourish. These are some of the challenges facing botanists today, and they cannot be ignored without penalty.

Finally, I realize, of course, that there is no "compleat" botanist, nor can there be, because no single individual can master all of botanical fact and philosophy. The easy way out is to specialize and "let the rest of the world go by." But we must strive toward the ideal of a broad perspective, for botanists above all others should understand the relationship between society and plant resources and should be able to impart to their students and to society at large the importance of this relationship to the present and future welfare of man.

JOHN H. STORER

The Web of Life

•

I

THERE IS A SPOT in the woodlands of southeastern Tennessee that can never be forgotten by one who has seen it. To reach it, one may travel for a hundred miles through forest-covered hills, rich with laurel, azalea, and rhododendron, and along springs and brooks and ravines which sometimes open up into green meadows where cattle graze.

Suddenly this green world disappears. The forest gives way to a hundred square miles of desert as dead as the Sahara. The rolling hills are cut into rows of low, steep-sided ridges, sterile and bare of any life. The soil is dry, the springs and brooks are gone. In this area the annual rainfall is less than in the surrounding country. The winds are stronger. It is hotter in summer and colder in winter. Here and there on this desert there stand in rows the dead skeletons of small trees, planted by people who hoped to start a new forest.

The soil in the nearby woodland is dark, rich, and spongelike. That on the desert is coarse, hard, and yellow. This desert was once covered by a forest and by rich forest soil. But today that soil lies five miles down the valley at the bottom of a reservoir and the shoals of coarse desert soil grow deeper, year by year, as every rain washes its fresh quota down to the reservoir.

This all happened because, many years ago, a copper smelter was built here, and the fumes from the smelter killed the surrounding trees, thereby setting in motion a train of events that finally produced the desert. The owners of the smelter have long since learned to control these fumes, which no longer poison the air so seriously; but the harm has been done.

After the fumes were controlled, many attempts were made to restore the forest. Desert grasses were planted, in the hope of furnishing some green cover to hold the soil in place; for rich soil cannot exist without the help of plants to build and protect it. But with the killing of the forest, the living soil that gave it life had also been killed. There were no roots to hold the soil in place, no litter to absorb the rain, and the grass seeds washed away, the seedling trees withered, and the dead soil continues to this day to wash from the desert and drift down to fill the reservoir.

At prohibitive cost, this desert land could be restored to life. Modern machinery might partly substitute for some of nature's processes. It could fill the eroding gullies and build level ridges along the hillsides to stabilize the soil, hold the rain, and so give moisture to the earth. Then seeds of grass and trees could take root and find a chance for life. Given such a start, nature could take over and slowly rebuild the ruined soil, organizing once again the community of living things that makes life possible for the forest. This might even be accomplished in a human lifetime.

Without man's help, nature might take ten times as long, for her processes are slower. Life cannot make a new start here until the earth offers it a firm foothold, and this must wait until nature has washed away the dead soil down to a solid foundation. This is the process that we call erosion. In this particular area it has been going on for many years and the land still looks as dead as the excavations for any city building. Eventually nature will find a solid footing, whether it be a ledge or a rock heavy enough to withstand the movement of the soil. Here seeds may find shelter for a foothold.

Or life may even start on the face of the rock itself. The rock offers little in the way of food and moisture, but on its secure base the pioneer plants may gain a foothold. Such plants must have the ability to dry up and lie dormant through times of dry weather, then waken to absorb the moisture of every rain or heavy dew. Many species of lichen have this power. Their rootlike fibres secrete an acid which dissolves minerals from the rock. Eating their way into it, they prepare an entrance for moisture which may later freeze and crack off rock particles—the beginnings of soil. The lichen thus offers to other more delicate plants a seedbed with moisture, a foothold on the rock, and mineral solutions for food.

In this seedbed, mosses, annual "weeds," and hardy ferns may grow and die, adding their substance to it, catching wind-blown dust, and building a deeper bed where seedling trees and other plants may find food and moisture. The growing community of plants will slowly spread over the rock till at last the roots reach into the soil around it. Here the tiny hairs will fill the earth so compactly that they may touch every soil particle, tying it firmly into place, making a secure foundation for a further spread of new plants around the parent rock.

A root system is a really incredible thing. Many studies have been made of its extent. In one study, a plant of winter rye grass was grown for four months in a box with less than two cubic feet of earth. In that time the plant grew twenty inches high, with about 51 square feet of surface above the ground. But underground the root system had developed 378 miles of roots and an additional 6,000 miles of root hairs! This meant an average growth of three miles of roots and 50 miles of root hairs for each day of the four-month growing season. The growth rate varies with different plants, of course, but this gives us some idea of the activity that goes on under the surface of a quiet-looking meadow, while the grass prepares food that will later become milk and meat and butter for us.

But these growing roots are doing far more than just binding together the rock particles that form the soil. They are taking the first step toward creating an entirely new kind of soil.

A Flemish physician who lived in the 17th century gave an interesting picture of this when he tried growing a willow sprout in a tub of earth. For five years nothing was added except rainwater, and the willow grew into a small tree. At the end of the five years the tree was weighed. It had gained more than 164 pounds in weight, while the soil in the tub had lost only two ounces. Actually the soil weight must by now have included millions of microscopic root hairs from the tree, but the figures are accurate enough to show that those 164 pounds of tree must have come from somewhere outside the soil.

If we divide up the plant into the various elements that form its substance, we find that only 5 percent of its weight comes originally from the soil. The elements in a mature corn plant include carbon, 44.58 percent; oxygen, 43.79 percent; hydrogen, 6.26 percent; nitrogen, 1.43 percent. These all come originally from the air and water, and together they form more than 95 percent of the plant. A good proportion of them comes to the plant through the roots, by way of the soil, after earlier plants have fitted the soil to receive them. The rest of the plant, that which comes from the soil itself, includes potassium, 1.62 percent; calcium, .59 percent; silicon, .54 percent; magnesium, .44 percent; phosphorus, .25 percent; chlorine, .20 percent; sodium, .15 percent; iron, .10 percent, and sulphur, .05 percent. Some plants contain very small amounts of other elements, such as copper, boron, and cobalt. These we call trace elements.

All these elements are built together into a living plant through the agency of the chlorophyll, the green coloring matter that is carried in the leaves. This building process is the essential first step that prepares the way for all the life that exists on earth. Chlorophyll has not yet yielded to man all the secrets, either of its composition or of the magic by which it transforms inert building blocks into living material, but we do know certain basic facts. The essential first step consists of building sugar out of sunlight, carbon dioxide, and water. This is called photosynthesis (putting together). To make one molecule of sugar the chlorophyll produces the union of six molecules of water and six of carbon dioxide. With them it binds the energy from the sunlight, and in the process six molecules of free oxygen taken from the water and the carbon dioxide are released into the air.

As the roots spread through the soil, they fill it with this new living substance built from air, sunlight, and water. But this has not yet become a part of the soil. It is not until the plant itself dies that the dramatic change takes place in the soil. For now the dead plant's roots and leaves offer food to the small organisms that are among the most important factors in the whole cycle of life, the bacteria, the molds, and the rest, most of them too small for the eye to see. Their most important function lies in decomposing the remains of the higher plants and animals, changing them into new chemical combinations that can be used again by succeeding plant generations for food.

The decomposing plant attracts a host of small creatures that help to break it up. Earthworms eat it, mixing it with the soil particles that pass through their bodies, digesting the whole, and casting it up on the surface, a revitalized and richer soil. The number of earthworms in the soil depends largely on its chemistry and on the amount of plant material they find in the earth.

Myriads of small creatures spend parts of their lives in the soil: ants, beetles, wasps, spiders, and many others. Some of these come to eat the plants, and many meat eaters come to eat the plant eaters. Among these the shrews and moles play a very important part. In favorable locations there may be as many as 100 shrews to the acre, and each shrew may eat the equivalent of its own weight in other living things each day. All this activity combines to carry on the work of plowing, mixing, and fertilizing as the creatures add their remains to the land.

This hive of living things in the soil, the eaters and the eaten, adds up to incredible numbers. The bacteria alone may range from comparatively few up to three or four billion in a single gram of dry soil. At the Rothamsted Experiment Station in England it has been estimated that in good soil the bacterial matter, living and dead, may weigh as much as 5,600 pounds per acre. Does this seem like a small amount in a whole acre? At the rate of even one billion to a gram of soil the total body surface of the bacteria in an acre, if spread out flat, would equal 460 acres.

The fungi may add up to a million in a gram of dry soil, weighing over 1,000 pounds to the acre.

Each of these small living things adds its tiny bit to the building of the living earth until, in the average acre of good topsoil, with 4 percent organic matter, there are stored about 80,000 pounds of such organic matter from plants and animals, containing energy from the sunlight equal to that in 20 to 25 tons of anthracite coal.

But while the soil lives, this stored-up energy is constantly being used for food by the teeming life it supports and, as we have seen, it must be constantly renewed by the plants in order to maintain this life. For good soil is actually a living thing, and its health is a matter of life and death to the plants and animals that live on its surface. We ourselves are as dependent on its health as the smallest of its creatures.

II

As the plant roots and fungi grow into the soil, tying its rock particles together into a firm mass, opening the way for other living things, and filling the earth with organic matter containing packaged energy from the sunlight, a subtle change occurs.

The root tips release carbon dioxide, the source of carbonic acid. This reinforces the action of the rain, helping to dissolve minerals from the soil particles, making them available to the plants for food. As the bacteria decompose the dead plant matter, they too release carbon dioxide and contribute their share toward the enrichment of the soil. The decomposition of animal and vegetable residues by microorganisms produces many other acids, besides carbonic acid, including citric, tartaric, oxalic, and malic. Such acids are probably of even greater importance in making minerals available.

In raw, unprepared soil these dissolved minerals might be carried away by every heavy rain, and the supply left for the plants would be rather precarious. But here the bacteria provide another service, for in the decomposition process there are some parts of the dead plants and animals that are more resistant than others. These stay in the soil for a long time, forming a dark, spongy, very absorbent material called humus. Humus stores rainwater, with its dissolved minerals, holding both as a reservoir for plants to draw on. It also stores minerals drawn up from below by the deeper roots.

As the bacteria use up their food supply, billions of them die of hunger or become inactive, and the life process in the soil slows down until further stores of food are added by plants or animals. As in most of nature's activities, this whole life cycle in the soil becomes a self-regulating system —an organized community, adjusting its numbers to the food supply so long as it is undisturbed by outside forces.

On the surface this community may appear to be merely a blanket of dead leaves and litter from last season, but under these lie the decaying remains of their predecessors from earlier seasons. As we go deeper these become mixed with soil particles and with the roots of plants living and dead. Through this material run myriads of passageways left by insects, mammals, and decayed roots. The whole makes a perfect protection for the earth and a sponge to check the runoff of rain, which it absorbs and introduces slowly into the soil reservoir below.

This power of the organized topsoil to store water and minerals is the key to the next step for the developing plant community.

For, while minerals form a very small part of the whole plant—only one twentieth—that small fraction, together with sunlight, is the key that makes the whole function.

This has been proved in some very interesting tests in Kansas. In the western part of that state, where there is comparatively low rainfall, the good earth is able to store most of the rain and the dissolved minerals near the surface. But in the eastern part of the state the heavier rainfall supplies more water than the soil can hold. Here a good share of the mineral solutions is leached away. In a comparative analysis of wheat raised in the eastern and western parts of the state, it was found that grain raised in the dry western part contained nearly 50 percent more protein than the same kind of wheat raised in the eastern part. An important cause of this difference appears to lie in the leaching of the eastern soil which, having lost a share of its dissolved minerals, was unable to give the wheat a proper supply of them.

Obviously the quality of the soil has a great effect on the quality of the plant. And since animals, including man, require roughly the same elements as plants, the health of man and the lower animals must depend on the ability of the soil to supply those needed elements in the right proportions.

Some interesting experiments have been performed to show this interrelationship. At the University of Missouri, Dr. William A. Albrecht grew hay on three plots of ground which were somewhat deficient in minerals. One of these plots was given no fertilizer. It produced about 1,700 pounds of hay per acre. The second plot was fertilized with nitrogen. It produced nearly double the crop of hay—more than 3,200 pounds. But when fed to rabbits, a pound of this nitrogen-fed hay produced less rabbit meat per pound than the much less luxuriant unfertilized hay. The nitrogen had forced its growth, but the soil had been unable to supply enough of the other minerals to keep pace with it.

The third plot was given a balanced fertilizer with the essential minerals that the grass needed. This produced less hay than the nitrogen-fed plot—about 2,400 pounds—but when fed to rabbits it produced nearly twice as much meat per pound of hay. So it appears that the minerals can be of greatest value only when the plant can use them and when they are in the right proportions. The actual figures shown in these tests were:

Unfertilized hay	23 lbs. of hay produced 32 oz. of rabbit growth.
Nitrogen-fertilized hay	23 lbs. of hay produced 26 oz. of rabbit growth.
With balanced fertilizer	23 lbs. of hay produced 49 oz. of rabbit growth.

From all this it might seem that plants grow best in a land where there is little rain. Actually, most food plants need huge amounts of water. For example, a single corn plant uses about 50 gallons in 100 days of growth.

In one test made in Kansas, an acre of corn containing 6,000 plants used 325,000 gallons of water in 100 growing days. This equalled the amount of water it would have taken to cover the acre 11 inches deep—or 11 acre inches.

To demonstrate the value of water, in an experiment in Utah some corn was raised on dry land without irrigation. It produced 26 bushels per acre. On another plot the corn was irrigated with 15 acre inches of water. It produced 53 bushels to the acre, more than double the other crop.

Hence, to raise a good crop of plants, the land must carry an adequate supply of moisture. It must have the ability to store it without leaching. To do so it must be well supplied with humus built into it by the plants and small living creatures of the soil. It is commonly estimated that nature may normally take as much as 500 years to build an inch of topsoil. This topsoil is one of the keys to man's existence on earth.

III

As the tree draws in its raw materials from the air, water, soil, and sunlight, these are carried to the leaves. Here, through the miracle of chlorophyll, they are woven together and transformed into sugar. And, from this sugar, into the vast number of chemical combinations that form the living substance of the tree—into roots and leaves and branches, into flowers with their male and female parts that together produce seeds. In the seeds are stored food and microscopic cells for the production of new life. The cells will later develop, each one producing its own assigned part of a new plant, with vitamins, enzymes, and other chemical combinations that interact to make it function as a living thing.

Some soil fungi form a direct partnership with trees and other plants. Beeches and pines apparently cannot make healthy growth unless there exists an active association between their roots and certain kinds of fungi. This partnership—or symbiotic association—is known as mycorrhiza. Its exact function is something of a mystery, but it evidently plays an important part in the transfer of food from the soil to the root system.

One group of plants, the legumes—clovers, beans, locust trees, and other pod bearers—joins forces with bacteria to form a sort of chemical laboratory in the earth. When nature has built up the soil's chemistry to a condition that will support them, these legumes take their place in the plant community. They offer a home in the soil to nitrogen-fixing bacteria which enter their roots and cause them to swell into lumps called nodules, where the bacteria live in colonies of many millions. Taking their energy from the sugar in the plant roots, the bacteria gather nitrogen from the air to form nitrogen compounds, which they store in the nodules.

When the roots die, the nitrogen is left in the soil, and with this enrichment the plant community bursts into full life.

Near Athens, Ohio, a plantation of cedar trees was set out on an area of very poor soil by the U. S. Forest Service, who wanted to test different trees in an effort to start a successful forest. In one part of the cedar plantation they set out among the cedars a number of locust trees (legumes), which carried the nitrogen-fixing bacteria on their roots. Eleven years later the cedar trees, planted alone, stood on the average about thirty inches high, while those among the locusts averaged perhaps seven feet. Between the small cedars the ground carried a thin, sickly cover of low grass, an occasional white poplar, and very few other plants. The soil was dry and offered little food or shelter for living creatures. Here nature might need a century or more to establish a healthy forest.

Under the locust-and-cedar combination, however, one can step across from a semi-desert into a rich young embryo natural forest. The ground is covered with a lush growth of grasses, weeds, and vines. On its surface a litter of dead leaves and stems has begun to collect from plants of past seasons. Beneath the litter the soil is cool and damp with moisture stored from the rains. Shaded from the strong sunlight, it is a good natural bed for the seeds of forest trees to make their start in. And here are growing seedlings of tulip trees, red oak, red maple, white ash, and others.

Both groups of trees were given an equal start on the same kind of soil, and the entire area has been bombarded every year by millions of seeds of many kinds, brought in by wind and bird and mammal.

Why are the white poplars the only new trees to start among the cedars? Why are there so few poplars under the taller cedar-and-locust combination, while many other trees have started here whose seeds failed to survive out in the open? The answer is that each plant is a specialist, adapted by its own habit of growth and its own special requirement for light and moisture to grow best in its own preferred environment.

Among the cedars only the seedlings of the grasses and the poplars have been able to withstand the drying heat of the strong sunlight. Under the taller, richer, cedar-and-locust growth the white poplars have had to meet the competition of red oak, red maple, white ash, and tulip, whose seedlings thrive in partial shade.

These trees are growing here, not because the ground received any more of their seed, but because the seeds that fell found the added moisture, soil quality, and just the amount of protection from sunlight that they must have to get a secure start in life.

This young forest will now follow the normal steps of forest development; for the seedlings will eventually grow up to overtop the earlier trees that nursed them and finally crowd them out by robbing them of sunlight and the free space they need for growth. And with the early trees will go many of the small plants of the ground surface which also need sunlight.

For many years this slow development will continue, until the crowding newcomers start to battle among themselves. At length their towering crowns will cut off the sunlight that their own seedlings must have in order to live. Then they will have destroyed their own power to reproduce. But among the many kinds of seedlings now struggling in the forest shadows, there are some that do thrive under these new conditions. Chief among them are the hemlock, beech, and sugar maple. These will now outgrow the others in their race to reach upward for the sunlight, finally touching their tops against the forest canopy. Here they will await their turn until a windstorm strikes some giant oak or tulip to the ground, making room for one of the newcomers. And at last these will take over to form the enduring climax forest. They are now the dominant trees. Nothing else can compete with them under the conditions they have established.

In some forests, where the soil is shallow, as the great trees fall, one by one, tearing their spreading root systems out of the shallow soil, the roots lift huge balls of earth with them. In this process the ground is plowed up into series of small ridges and depressions that fill with water after rains or melting snow. The water is held until it sinks slowly into the ground, and thus is reinforced the absorption system already established by humus and insects.

The trees have even made a new climate for all the lives that exist under them. Near Cleveland, Ohio, measurements were taken to compare the climate in a climax forest of this type with that in an open field nearby. It was found that on a bright day there was 750 times more light in the open field than in the forest. The trees slowed down the speed of the wind until, 1,000 feet inside the forest border, its velocity in summer was only one tenth as much as in the open field. In winter, when the leaves that slowed it had fallen, wind speed increased to one quarter that in the field.

These differences influence the moisture in the forest; for the wind and the sun's heat have a great effect on evaporation. It was found in the Cleveland tests that average moisture evaporation was 55 percent less in the forest than in the field during the summer, and 38 percent less in the winter.

With its blanket of humus and its protection from chilling winds, the forest soil is guarded against freezing in the winter, while the branches shade the snow against rapid melting. The soil is always ready to receive the water from the slowly melting snow, while the surface of the hard-frozen field nearby sheds the snow water like a roof, as it floods off under the warmth of the spring sun.

So, in addition to building timber and providing environment for many kinds of plants and animals, the forest also builds a reservoir to catch and store the huge amounts of water it must have for growth. From this reservoir, springs break out and brooks slowly cut their channels and bring life-giving water to land outside the forest.

IV

Our growing community of plants now provides three essentials needed by other forms of life—shelter, water, and a dependable supply of food. Many forms of plant eaters now come to use it—insects, mammals, birds, and other creatures.

Here, for example, is a caterpillar busily transforming the organic substance of a leaf into the juices and organs that make up the parts of a caterpillar. Eventually this substance will go through a series of further changes until at last it turns into a flying insect. Then it will repay with interest the damage it has done to the plant, for it becomes a partner in the plant's life process, carrying pollen to fertilize the blossoms. Insects make possible the continued existence of many plants.

Nearly all the fruits and vegetables used by man are directly dependent on this partnership with insects. But this partnership requires the most exacting regulation to meet two fundamental laws of nature on which all life is based.

First, insects, like all living things, must have the power to multiply faster than their normal death rate to insure against the catastrophes of disease and weather. Without this insurance no species could survive.

But this power carries with it tremendous danger; for insects, if allowed to multiply unchecked, would soon destroy all the leaves and kill all the plants that support them. We can see that danger most clearly in the life of soil bacteria, the smallest and simplest of all living things. One of these invisible cells may seem like a very insignificant part of the living community. But watch its numbers grow.

Each individual multiplies by dividing into two complete new ones. Under favorable conditions this may happen about twice each hour. Even if it happened but once each hour, and if each one lived, the offspring from a single individual would number 17,000,000 in 24 hours. By the end of six days the 17,000,000 would have increased to a bulk larger than the earth, and every living thing on earth would have become engaged in a suffocating struggle for food and air and life.

The useful bacteria that, kept in their place, make possible the higher forms of life, would have turned into irresistible destroyers.

The same principle, of course, applies to all forms of life. The insects that pollinate the blossoms are no exception.

To control them, nature uses a highly organized police force of flesh eaters—bacteria, insects, mammals, and birds, each one a specialist designed for a particular role to which it is best adapted.

For example, in this small community we are watching, the larvae of many insects spend parts of their lives in the upper soil. Shrews hunt them under the leaf mold. Other insects and some of the molds and bacteria also feed on them. On the surface many ground-nesting birds, such as the towhee, turn up the leaves to find them. The brown thrasher hunts here too, and continues the search among the bushes where it makes its nest. It is joined by the warblers and vireos, which extend their search up to the tree tops.

As the trees grow larger and their lower branches die, the fungi may decompose and soften the wood in the knotholes, offering favorable nesting sites for woodpeckers and nuthatches, whose feet are adapted for hunting on the bark of tree trunks. The woodpeckers go one step further, drilling holes through the bark to catch the insects hidden within.

Some of the insect pupae that survive this search emerge to fly over the tree tops, but here they are met by the swifts and swallows by day, by the nighthawks at dusk, and later by the bats that are equipped with nature's radar systems to hunt in the darkness. Each one of these controlling predators must in turn hold its own place against others that are larger or stronger or more active—hawks, owls, foxes, weasels.

It may seem a haphazard collection of predators and prey. Actually it has evolved into a very highly organized, beautifully controlled form of community government, with its own automatic system of farming, pedigree breeding, sanitation, policing, and insurance.

There may be many niches in the forest that give shelter for a nest or a den, but their inhabitants can survive only if the surrounding territory can offer enough food to support life. They can rear their families and multiply only if they can find the added food to feed their young, and many young birds and animals require the equivalent of their own weight in meat or insects each day.

As the growing families move out to forage for themselves, they find the best hiding places occupied, the best feeding areas already in use. They are then more exposed to their enemies. The first to tire or weaken in the search for food become the easiest prey. In this way nature removes the unlucky and the weaklings, saving in the long average the best, hardiest, or quickest to learn, to live and to carry on their kind. Thus their population is adjusted to the number that the land can safely support.

Sometimes a predator species may be wiped out for a time by weather or disease. But nature carries insurance against such catastrophes, for the area patrolled by each species will slightly overlap that of its neighbors. For example, if the ground-dwelling towhee disappears, the thrasher, the shrew, and the white-footed mouse all are ready to step into reap the more abundant food supply and, with more plentiful food, to rear larger families.

This local police organization is backed up by a more mobile patrol force; for the larger predators, requiring more food, must manage to cover more ground than their smaller prey, taking off the season's increase wherever it is easiest to catch. They concentrate on areas of heaviest production.

In this hierarchy of life some creatures may fall prey to smaller predators, the disease germs that attack them from within, or the parasites from without: the worms that live in their bodies, for example, or the fleas and lice living on their skins. As a rule, these take only a small share of the daily product of energy from their host, for the body of a healthy animal can usually control such parasites. But when the host is weakened by hunger or age, the parasites may prove fatal. Then the sanitary agents of the community step in to do their share. Vultures eat the dead flesh, bacteria and the larvae of flies feed in it and decompose it. Burying beetles carry its remains underground to enrich the soil, laying the foundation for a new cycle of life on a cleansed surface.

So the living community in the forest grows, the plants drawing in elements from air and water, and energy from the sun, building them into life—the plant eaters staying near their stationary food supply, turning its energy into meat, passing this on to the larger, more mobile flesh eaters. These in turn pass on their energy to the hierarchy of larger forms, each group in its turn becoming fewer in number as the larger creatures require more small bodies to feed them, with each successive individual forced to cover the larger area of ground needed to raise these many smaller lives. But in the end even the largest succumb to the bacteria and beetles that complete the circle in the earth.

All these creatures live and flourish completely unaware that they owe their existence to a few hundred million invisible bacteria living in the roots of some locust trees which pioneered the way for them many years ago.

JOHN TYLER BONNER

Feeding in Plants

•

I

SOME SPECIES of bacteria have an extraordinary ability to obtain energy from an extraneous chemical reaction which they encourage, and with this energy they perform all their life functions. These so-called chemosynthetic bacteria vary in the kind of chemical reaction they can stimulate, but in each case the method is basically similar. The bacteria are little salesmen in that they promote the reaction of buying and from this they grab their commission, the energy upon which they live.

A good example is the bacteria that live in the soil and burn ammonia. That is, they encourage the combination of ammonia and oxygen to give nitrous acid and water, and this reaction liberates energy which they capture and use to convert carbon dioxide and water into sugars. This is an uphill reaction, from water and carbon dioxide to sugar, but with a good measure of energy the push up the hill is sufficient, and now in the form of sugar the bacterium has real high-energy fuel. By burning this sugar in respiration it can in turn get energy to build its proteins and all the other substances it needs; its minute machine can run smoothly provided it has sugar to burn, and it makes its own sugar by promoting and parasitizing the oxidation of ammonia. These bacteria are incidentally very important in the soil for they fix the gaseous nitrogen of ammonia into nitrogen compounds, so necessary in the fertility of the land.

Ammonia is not the only substance attacked. For example another species will turn nitrous acid into nitric acid, and again energy is captured. In many swamps, especially where the mud is black, hydrogen sulphide is burned to sulphur by chemosynthetic bacteria, and again sulphur may be oxidized to sulphuric acid, and in each reaction the bacterium gains energy. The iron bacteria provide another example, one which is of considerable nuisance value, for they encourage the rusting of iron pipes.

Ferrous carbonate combines with water and oxygen to form ferric hydroxide or rust, and here the gain of the bacterium is the loss of the water company. Sometimes simple organic compounds are attacked, and another swamp species of bacterium oxidizes methane gas to get its energy. But perhaps the most remarkable of all is the species which burns hydrogen gas, giving water, and from the energy derived from this simple combination it can build up all the complicated chemical constituents of protoplasm.

All that is required, in the case of these chemosynthetic organisms, is the presence of some source of energy, carbon dioxide, and water, and from then on, except for the materials necessary to make various types of internal compounds, they are entirely self-sufficient. But the fact that they must always be near some chemical substance that they can break down to get energy means that their position in the environment is strictly limited, and it is in this respect that photosynthetic plants have a great advantage. An iron bacterium must always be near iron, but a green plant need only be in the light, and light is far more abundant and widespread on the surface of the earth than deposits of iron or sulphur.

Basically the process of photosynthesis is the same as that of chemosynthesis. With energy, carbon dioxide and water combine to produce carbohydrates. The real difference lies in the source of the energy. The green pigment of plants absorbs light; it is hit by light particles and after the collision the chlorophyll molecule becomes violently agitated; it is bumped from a sluggish low-energy state to an active high-energy state. In this condition the chlorophyll can release its energy and force carbon dioxide and water to become carbohydrate and oxygen. The details of the chemical process are extremely complex and involve a number of steps still not wholly understood, but the principle of the process and the end result are crystal clear.

It will be noted that this photosynthesis reaction is the reverse of respiration and the two may be written together in one equation:

$$\text{Carbon Dioxide and Water} \underset{\text{respiration}}{\overset{\text{light-chlorophyll}}{\rightleftharpoons}} \text{Sugar and Oxygen}$$

This means that if the plant is to use sugar as a fuel, which it does constantly so that it can make new substances and keep its machinery going, it will undo the advantage gained by photosynthesis and destroy all the valuable sugars. Obviously, if the plant is to grow, or for that matter exist at all, the total amount of photosynthesis must exceed the respiration so that the plant can accumulate energy. During the night respiration alone takes place and there is considerable degradation; during the day both take place, but photosynthesis outstrips respiration by a great amount and the bank account of fuel is always kept in excellent condition. In fact photosynthesis is so effective that not only does it provide for the plant, but the whole animal world is ultimately dependent on this source of available or "free" energy. The struggle for existence, the physicist Ludwig Boltzmann said many years ago, is the struggle for free energy, and the free energy of the whole living world comes almost entirely from photosynthesis, the insignificant group of chemosynthetic bacteria being the exception.

Since animals are dependent on plants for their fuel it is obvious that plants must have arisen first in the early history of the earth. But it is not at all certain whether chemosynthesis or the more successful photosynthesis was the first method of energy capture.

Whatever the beginnings, the photosynthetic organisms were soon overwhelmingly successful and evolved into many different forms and shapes, finding their way into every conceivable kind of environment: salt and fresh water, swamp, desert, mountain, plain. The only places they could not enter, at least without modifying their mode of nutrition, were places devoid of light. And with this expansion of habitat came a broadening of their structure; the simple bacterial cells were succeeded by filaments of the algae which in different ways produced larger and larger cell masses, achieving great size in the marine kelp, and then with the reconquering of land of the larger forms came the evolution of all our higher land plants, all the trees small and large, the bushes and the grasses. In every step of this evolution, in every modification of form that arose in the course of natural selection, one factor was never neglected and that is that the plant must capture light as efficiently as possible. This was no problem at all with the minute single-celled algae and even the small colonies such as Volvox, but when the size of the organism and its thickness increased, the problem of getting enough light for photosynthesis became serious and limiting. The problem is easy to visualize in terms of two hypothetical plants shaped like round balls, one a fraction of a millimeter thick and the other the size of a basketball. The sun striking on these green balls will penetrate to all the cells in the small one, but only a small proportion of cells in the larger one will get enough light, just the cells on the outside layer. So if plants are to compete successfully with their neighbors and still retain a large size, they must devise more and more efficient ways of capturing sunlight.

The obvious course is to produce structures that are not spherical balls, but that are flat and thin so that the sun can efficiently reach all parts. In early forms such as the kelp, which is a large alga, the main body is a flat blade supported by a long stem. The stem is attached at one end to the rocky bottom by a root-like holdfast and the other leads to the broad blade which floats horizontally under the surface of the water. The blade then has not only the advantage of its position in relation to the sun, but its thin broad structure is effective in the absorption of the light.

The leaves of higher plants—the oval leaf of the elm, the webbed leaf of the maple, the checkered leaf of ivy, the thin blade of grass—all these are sun-catchers. They are beautifully designed for their purpose, not only in their external shape but in their internal construction and their position in relation to the whole plant. On a tree or a bush the leaves are arranged so that they do not cast too much shadow upon one another. This is especially striking in ivy where the leaves dovetail one into another at different heights, allowing many leaves to grow close together, yet each may feel the sun. In many cases the leaves even move with the sun during the course of the day by growing more on one side than the other, so that the flat of the blade remains exposed to the warm rays.

The internal anatomy of a leaf is marvelously constructed for its function. On the top surface of the leaf there is a layer of thin transparent cells and below there is a layer of pencil-shaped cells all standing on end and pointing directly toward the sun. Each cell is lined with many chloroplasts, the small structures that contain the chlorophyll, and this palisade layer, as it is called, captures the light. This construction and orientation of the palisade cells is efficient in capturing light, just as an ear-horn is good at capturing sound. Below the palisade layer there is a spongy layer of cells where the movement of carbon dioxide, oxygen, and water vapor is made easy because of the large spaces between the cells. On the undersurface itself there is again a layer of covering cells, pierced with holes or stomata, and these holes can be opened or closed by two guard cells to control the amount of gas exchange and water vapor loss to the outside environment.

In a small single-celled plant all the processes take place together in one location, but in a large oak tree photosynthesis takes place only at the tips of the branches, yet the roots deep underground must grow and spread in proportion to the branches. Obviously there must be some method for conducting food to the various parts. This not only involves the transport of sugars but also proteins and vitamins and other protoplasmic substances which are synthesized in the leaves. For this water and salts are required, substances which are taken in by the roots and must be taken from the roots to the other parts. The transportation of dissolved substances in plants takes place in the vascular bundles, the bundles which are seen in the leaves as veins and in the stem as the hard parts. In a thick woody stem such as the trunk of an oak tree the whole interior wood is really a vascular conducting system, as well as a thin and more delicate layer of living cells (the bast) lying just under the bark. In all cases the conducting elements are cells, often, as in the case of wood, dead cells, which have become hollow and pipe-like to carry water with its dissolved food or salts. Many experiments have been performed to find if any particular pathways are taken by particular substances and it is now known that the water and salts absorbed by the roots go up through the internal wood, while the sugar manufactured in the leaves by photosynthesis comes down the outside bast. This latter phenomenon is especially easy to demonstrate, for if a tree is ringed so that the bast is severed in a circle around the trunk, a bulge of growth will appear above the ring because the carbohydrate is caught in its downward movement and the presence of this high-energy food in that region causes a local excess in growth. Very often the roots are involved in the storage of the food produced by the leaves, like an underground hoard, and this is the reason why carrots and beets and radishes are so plump and nutritious.

Photosynthesis and chemosynthesis are by no means the only ways in which plants feed. The great group of colorless fungi, for instance, absorbs food directly from the environment. A mold will send its delicate thread-like filaments into decaying wood or rotten meat or a host of other potential food sources and sop up the food directly. They may even parasitize a living plant or animal; and in all cases they do not synthesize their energy, they simply grow in or about the energy source. There are even cases of higher plants such as the dodder or mistletoe that rely in great part on the substance they can drain from the vascular system of the tree they parasitize, even though they also indulge simultaneously in some photosynthesis.

Most remarkable of all are the predacious forms which actually capture and devour live animals. In the fungi there are numerous examples where the fine filaments can capture small round worms or rotifer worms or amoebae. Some species of these ferocious fungi have a loop that is really a snare and if a worm should squeeze into it the loop will tighten and the fungal filament will then penetrate and grow into the dying worms until finally all its flesh has been absorbed. Other species have small adhesive pegs that appear especially desirable to worms, but once some unfortunate worm has surrounded this lethal knob with its mouth, it becomes stuck and the knob grows out into the body of the worm, drinking in its substance. Still other species have spores that are eaten by the worm or the amoeba, and once inside it suddenly germinates and the sprouts soon devour the innards of the unsuspecting prey.

There are predacious plants among the higher forms too, most of them thriving on a diet of insects. The tropical pitcher plant has a leaf in the form of a saxophone and, attracted by the foul-smelling odors, insects wander in the open end only to find that they are trapped in the watery soup of digestive enzymes at the base, and any escape is prevented by a vicious array of downward-pointed spike-like hairs. The fly's substance after it has been digested is absorbed directly by the plant and is distributed by its vascular system. Sundews, which we find in our temperate zones, are small plants that look like a pin cushion, the head of each pin having a drop of viscous honey on it. This apparently harmless, attractive plant entices insects which become trapped by the sticky substance, and then the pin-like hairs move and hold the fly as digestive juices perform their destruction. Perhaps the most dramatic of all is Venus's-flytrap which, like the iron maiden, has a hinge and large spikes. As a fly lands on the open leaf it snaps shut like a clam with extraordinary rapidity and the fly can be heard buzzing in a frenzy within its cage. There are further examples, but in each case the plant has forsaken any total dependence on photosynthesis and taken on an essentially animal mode of nutrition, for even the digestive enzymes involved are similar to those of animals. These predacious plants are curious anomalies and are important in that, since they feed like animals, they show the essential sameness of all living forms. Energy is needed for all life, and it makes no difference whether the energy comes in by chemosynthesis, by photosynthesis, or by feeding on dissolved substances or on live animals. However each of these types of energy-intake imposes certain restrictions on the organism and often involves specific specialized structures; a chemosynthetic organism, for instance, is unlikely to achieve the size of a tree, or a photosynthetic organism of large size cannot have the shape of a sphere.

VERNE GRANT

The Pollination of Flowers

•

WHAT IS A FLOWER? How is it constructed, how does it function, how did flowers originate and evolve into the more than 150,000 species found on earth? Although man has always lived in a world of flowers, as recently as two hundred years ago their biological meaning was still a mystery. Today, thanks to two centuries of patient work by many botanists all over the world, the mystery is to some extent explained.

Flowers are the reproductive structures of plants. The structures consist of pollen-bearing stamens (the male organs) and carpels (the female organs) containing pollen-catching stigmas and ovules, the plant's "eggs." A union of the pollen with the ovules produces seeds. Most of the flowers with which we are familiar have both organs, the stamens and the carpels, in the same flower. It would be most convenient for the plant if each flower's pollen fertilized its own ovules, but many flowers cannot pollinate themselves. They are fertilized by pollen from other individuals of the same species. From an evolutionary standpoint this has advantages, for it produces a combination of different heredities and yields more variable and more flexible progeny.

In the animal kingdom this kind of union poses no special difficulties. Impelled by an urge to mate, the male and female swim, crawl, walk or fly until they find each other. But the union of two flowering plants, anchored by their roots to separate spots of ground, presents a problem that can be solved only by the intervention of some third party. The pollen of one plant must be carried to the ovules of the other by an external agent—the wind, water currents, an insect or some other animal. Obviously if this is to occur it must be advantageous or inevitable for that agent to carry the pollen. In the long course of evolution the flowers of plants have become adapted through natural selection to the characteristics of their pollinators. Thus the various species of flower owe their structure, shape, color, odor and other attributes to the particular agents that cross-pollinate them. The flowers of the earth group themselves into several broad classes depending on how they are pollinated. There are the bee flowers, the moth flowers, the fly flowers, the beetle flowers, the bird flowers, the bat flowers, the wind flowers and so on.

The bee flowers include some orchids, verbena, violets, blue columbine, larkspur, monkshood, bleeding heart, many members of the mint, snapdragon and pea families, and a host of others. All of them offer nectar as a reward to the bees, and all advertise their presence by showy, brightly colored petals and a sweet fragrance. The bee flowers and bees are beautifully adapted to each other in biological construction and habits. Most bee flowers are blue or yellow or some mixture of these two colors, and experiments show that the vision of bees is mainly in this part of the spectrum; they are color-blind to red. Bees respond to sweet, aromatic or minty odors and apparently are not stimulated at all by foul ones. Bees fly only by day, and bee flowers are always open in the daytime but often closed at night.

In visiting a flower a bee habitually alights first on a petal. Many bee flowers provide a protruding lip as a landing platform. The bee then pushes its way into the region of nectar and pollen. Bee flowers secrete nectar from special glands which often lie at the base of a tube of petals. Bees, with their long, slender tongues, can reach the nectar, but most other insects cannot. As the bee takes the nectar, its body hairs inevitably pick up pollen from the flower's stamens. In some bee flowers the stamens have special lever, trigger or pistol devices for dusting pollen on some particular spot of the bee's body. When the bee has finished working on one flower, it flies rapidly on to another. Bees have an instinct to confine their attention to flowers of one species at a time; they recognize a species by its characteristic odor, form and color. This is very convenient for the flower, since it assures that the bee will deliver its load of pollen where it will do the most good, namely, to another flower of the same species which the pollen can fertilize. Since the stamens and carpels are grouped together in the flower, the bee simultaneously delivers its load of pollen to the carpels and picks up a new load from the stamens at each visit. It delivers enough pollen grains to fertilize a large number of ovules, and most bee-pollinated plants do in fact ripen numerous seeds in each flower.

Bee flowers thrive best in arid, semiarid and sunny parts of temperate regions, where bees find the climatic conditions to their liking. Many bee-pollinated plants are unable to reproduce themselves in areas that lack certain kinds of bees. For example, monkshood, a bumblebee flower, does not occur naturally outside the range of bumblebees in the Northern Hemisphere. Alfalfa, a cultivated species of the pea family, is often infertile in California, where the proper kinds of bees are scarce under the highly artificial conditions of cultivation.

The moth and butterfly flowers also are very numerous. Among the moth flowers are the morning-glory, tobacco, yellow columbine, datura, white catchfly, yucca, phlox, some evening primroses and many orchids; the butterfly flowers include the carnation, red catchfly and many lilies. Nearly all species of moths and butterflies have a long tongue for sucking nectar; in tropical hawk moths the tongue is sometimes as long as 25 centimeters. Unlike bees, moths do not settle down on the flower during feeding; they hover above it with the tongue inserted in the nectar. They are guided to flowers by a combination of sight and smell, but most of them fly during dusk and at night, so that moth flowers run mainly to white shades and a very heavy fragrance. Many of these flowers open only in the late afternoon or evening and are closed through the hours of bright sunshine. Butterfly flowers, on the other hand, often have red or orange colors, since butterflies feed during the daytime and some of them, unlike bees, can see the red part of the spectrum.

The nectar of moth and butterfly flowers is secreted at the base of a long slender floral tube, where it is accessible only to the long-tongued moths and butterflies. In many species of moths the length of the tongue closely matches the length of the spur in the particular flower that the species visits. Like bees, moths and butterflies tend to feed on one kind of flower at a time.

Moth flowers apparently are most plentiful in tropical and warm temperate latitudes. They are common at high elevation on temperate mountain ranges, but are absent from the Arctic and Antarctic.

The flies that feed on flowers fall into two classes: long-tongued and short-tongued. The long-tongued flies, such as the syrphids and bombylids, in the main visit the same types of flowers as bees, since they are well adapted in bodily structure, habits and sensory perception to live on the nectar of these flowers. The truly distinctive fly flowers, therefore, are those that feed the short-tongued flies.

These flies, consisting of some thirty or more families, are a diverse and miscellaneous lot, and they have no particular specializations for feeding on flowers. Most of them probably derive their nourishment mainly from other sources, notably carrion, dung, humus, sap and blood. The flowers that attract them are those that carry similar odors. Unlike bees, moths and long-tongued flies, the short-tongued flies are attracted to flowers not primarily by vision but by their sense of smell. The fly flowers, consequently, are generally dull of color and rank in odor. Rafflesia, a large-blossomed fly flower of Malaysia, smells like putrefying flesh; another fly flower, black arum, has the odor of human dung; another, the lily *Scoliopus bigelovii,* smells like fish oil; there is a species of Dutchman's-pipe that smells like decaying tobacco and another that smells like humus.

The performance of short-tongued flies on flowers may be described as unindustrious, unskilled and stupid. Some fly flowers obtain the pollen-carrying services of flies without even expending any nectar on them. In this case the flies are attracted to special glistening streaks, spots or bodies, as in the saxifrage *Parnassia,* the lily *Paris* and the orchid *Ophrys.* The flowers of Dutchman's-pipe and arum not only fail to feed the flies but actually imprison them for a day or two in a floral trap while they are being doused with pollen. After the flies escape from one flower, they may go on to a second trap, where they have the opportunity during another day or two to pollinate the flower thoroughly.

The American floral ecologist John Lovell once remarked upon the vast difference in the reception that flowers accord to bees and to flies. The efficient and constant bees are offered nectar, pollen, shelter, a landing platform, bright colors and sweet odors, and their competitors for the limited supply of food are excluded from the floral chamber. For the stupid flies, on the other hand, there are pitfalls, prisons, pinch-traps and deceptive nectaries!

Fly flowers are especially common in plants of Arctic and high mountainous regions where other types of animal-pollinated flowers are infrequent or absent. They are also found in shady woods of the temperate and tropical zones.

The beetle flowers also attract their insect pollen-carriers chiefly by odor rather than by sight. The flower-visiting beetles, with rare exceptions, are not especially adapted for feeding on flowers and may derive most of their nourishment from other sources, such as sap, fruit, leaves, dung and carrion. They may be attracted to flowers by fruity, spicy or sweet odors. There are two general types of beetle flowers. One group has large, solitary blossoms; among them are magnolias, pond lilies, California poppy, sweet shrub (Calycanthus) and wild rose. The other has clusters of small flowers; examples in this group are dogwood, elder, spiraea, buckthorn and some members of the arum, parsley and sunflower families.

The beetles not only lap up the nectar and other juices of a flower but also feed on the tissues of petals and stamens. To protect the ovules from the chewing jaws of their pollinators, most beetle flowers have the ovules well buried beneath the floral chamber. Several of them hold the beetles in a trap while the stigmas receive pollen and the stamens sprinkle a fresh supply onto the bodies of the prisoners. They then open up an exit by which the beetle escapes. On the other hand, many beetle flowers have a shallow, open basin freely accessible to all comers. This makes them a common camping ground for many other kinds of small insects, including flies, wasps, bugs and bees.

Beetle flowers are most abundant in tropical latitudes and diminish toward the colder parts of the earth.

It is a common notion that insects are by far the most important animal pollinators of flowers, but actually in some parts of the world, particularly in the tropics and Southern Hemisphere, birds may be even more important. Hummingbirds in North and South America, sunbirds in Africa and Asia, honey eaters and lorikeets in Australia, honey creepers in the Hawaiian Islands, and several other groups of birds regularly visit flowers to feed on nectar, flower-inhabiting insects and pollen.

Birds have powerful vision and only a feebly developed sense of smell, so the bird flowers rely mainly on color to attract their pollinators. Most of them are large and colorful, and many are odorless. The sensitivity of the bird's eye, like that of man, is great in the red end of the spectrum but relatively weak in the region of blue and violet. Hence the most frequent colors in bird flowers are red and yellow. The bird flowers include red columbine, fuchsia, passionflower, eucalyptus, hibiscus and members of the pea, cactus, pineapple, banana and orchid families. Bird flowers are most common in tropical and south temperate latitudes.

Hummingbirds commonly suck the nectar of flowers on the wing, and the flowers that they visit most often are of the hanging type. On the other hand, the flowers favored by sunbirds, which settle on the plant, usually stand erect and provide a landing platform. A bird probes into the chamber of a flower with its sharp-pointed bill. This frequently causes considerable mechanical damage to the inner floral parts. The bird flowers therefore put their ovules out of harm's way in an ovary under the floral chamber, behind a sheath or at the end of a special stalk.

The petals of bird flowers are fused into a tube which holds copious quantities of thin nectar. The proportions of the tube often correspond to the length and curvature of the bird's bill. The stamens are usually brightly colored, numerous and turned out so that they touch the bird on the breast or head while it feeds. The pollen adheres in sticky masses or threads. A single pollinating visit thus suffices for the fertilization of scores or hundreds of ovules. The importance of birds as pollinators is indicated by the fact that the Mexican century plant, whose pollen is carried by hummingbirds, is barren when transplanted to Europe, where hummingbirds are absent. The flower is abundantly visited by bees, but without hummingbirds it remains sterile.

The bat flowers are pollinated by certain species of tropical bats with long, slender muzzles, extensile tongues and shortened or missing front teeth—all adaptations that enable the bats to feed on flowers. They feed at night and are probably guided to the flowers chiefly by their well-developed sense of smell. They clamber on the flower, hold on with their claws and extract nectar or small insects from the floral chamber with the tongue or chew the pollen or succulent petals. The tree-borne bat flowers of the tropics are large, frequently dirty white in color and open only at night. They attract the bats by a fermenting or fruitlike odor, which is given off at night. Examples of bat flowers are calabash, sausage tree, candle tree and some other members of the trumpet-vine family, various members of the sapodilla family and areca palm.

The flowers pollinated by the wind need no bright colors, special odors, nectar or other attractions, and they have none. Most of them even lack petals. Instead, their stamens and stigmas are exposed as freely as possible to the air currents, and they provide huge masses of light, smooth-skinned pollen that can be scattered far and wide. Pollen grains borne by the wind have been collected in air traps over the middle of the Atlantic, hundreds of miles from their source.

Because there is no special need for the stamens and carpels to be grouped together in the same flower, as in animal-pollinated plants, the sexes in the wind flowers are often separated into staminate and carpellate flowers, which are borne either in different parts of the plant or on different individual plants. The stigma is feathery, brushy or fleshy, so that the wind-carried pollen will stick to it. The pollen grains of a wind flower, being borne in a highly dispersed condition in the air, land singly on a stigma and not in masses as with animal-borne pollen. Each act of pollination thus leads to the fertilization of one ovule in each flower. Most wind-pollinated plants have become adapted to this condition by producing single-seeded fruits: the oak flower produces one acorn, the grass flower one grain.

Wind flowers are most common in the Arctic and Antarctic, where most insects cannot live, and also play a very important role in the cold temperate zones. Among the wind-pollinated plants are the grasses, sedges, rushes, cattail, dock, goosefoot, hemp, nettle, plantain, alder, hazel, birch, oak and poplar.

Many flowers, of course, fall into more than one category. Some European heaths are pollinated by bees in the spring but cease producing nectar and commit their pollen to the wind at the end of the season. Phlox flowers, normally pollinated by moths, may occasionally be fertilized by thrips. Some European species of gentian and violet are pollinated by bees in the lowlands and by butterflies in the Alps. Similar variations occur among the different species in a genus or family of plants. In short, the classes are not static. Changes in the method of pollination have occurred with considerable frequency during the history of the earth. These changes, and the nice adaptation of the flowers to their pollinators, show clearly that the pollinating agents have played a major active role in the evolution of flowers.

The fossil record indicates that flowers originated sometime during the middle of the Mesozoic Era, about 150 million years ago. Flowers, like mammals and birds, probably made their first humble appearance in the age of conifers, cycads, dinosaurs and beetles. Most of the seed-bearing plants of that age were probably pollinated by wind. They possessed the same kinds of reproductive structures, including separate sexes and in some instances winged pollen, that are associated with wind pollination in their modern survivors. The ovules were borne in cones or on leaves and exuded drops of sap. In the course of time beetles, feeding on the sap and resin of stems and on leaves, must have discovered that the liquid droplets from the ovules and the pollen in the male cones were nutritious foods. Some of these beetles, returning regularly to the newly found source of food, would have accidentally carried pollen to the ovules. For some Mesozoic plants this new method of pollination may have represented a more efficient method of cross-pollination than the releasing of enormous quantities of pollen into the air. Through natural selection they would develop adaptations to the potentialities of beetle pollination. The ovules, first of all, must be placed behind some protective wall to prevent their being chewed up by the beetles. One means of accomplishing this would be to fold the ovule-bearing leaf or branch into a hollow, closed carpel. The pollen-collecting function would then have to be transferred from the individual ovules to a central stigma serving all the ovules in the carpel.

The beetles could be drawn to the stigma by a special secretion of nectar which would replace the droplets previously given off by the individual ovules. A beetle visiting the stigma would be apt to leave behind sufficient pollen for the fertilization of numerous ovules. The number of seeds formed in a single pollination would no longer be one, as in the wind-pollinated ancestor, but ten or twenty. So the transition from wind to beetle pollination would increase the fertility of the plant.

The chances that the beetle would bring pollen to the stigma would be increased if the stamens were in close proximity to the carpels. The stamens and carpels might even be advantageously grouped within the same cone. The stamens would have to be present in large numbers so that they would not all be devoured by the beetles. In the course of time the outer stamens might become sterilized and pigmented and transformed into a set of showy petals. When these conditions had been fulfilled, there would have come into existence a structure possessing all the essentials of a modern flower.

This, in all likelihood, is how the evolution of flowers began. The most primitive type of flower of which we have any knowledge is a beetle flower, and it seems altogether probable that the selective factor that called flowers into being was beetle pollination. From these early flowers probably are descended most of those modern families that have separate petals and open nectar. When the bees, moths, butterflies and long-tongued flies arrived on the earth at the beginning of the Tertiary Period, some seventy million years ago, the evolution of flowers was greatly broadened.

In flowers pollinated by the long-tongued insects the petals became fused into a tubular corolla with a supply of nectar concealed at its base. The carpels were similarly fused into a compound ovary with a more localized and centralized stigma. The tubular structure of the corolla tended to screen out the beetles and small flies and restrict visitors to those insects —the bees, moths and long-tongued flies—that fly regularly from flower to flower of the same species. This was a great step forward in floral design: it marked a transition from promiscuous pollination by miscellaneous unspecialized insects to restricted pollination by specialized and flower-constant animals.

THEODOSIUS DOBZHANSKY
AND JOÃO MURÇA-PIRES

Strangler Trees

•

PERHAPS the most troublesome problem in the theory of evolution today is the question of how the haphazard process of chance mutation and natural selection could have produced some of the wonderfully complicated adaptations in nature. Consider, for instance, the structure of the human eye—a most intricate system composed of a great number of exquisitely adjusted and coordinated parts. Could such a system have arisen merely by the gradual accumulation of hundreds or thousands of lucky, independent mutations?

Some people believe that this is too much to ask natural selection to accomplish, and they have offered other explanations. One school of thought suggests that evolution is directed not by natural selection but by some inner urge of organisms—an inscrutable something called "psychoid." Another theorist proposes that the marvelous gifts of evolution to the living world came to birth through sudden and drastic "systemic mutations," which created "hopeful monsters" that were later polished down to the final product by evolutionary selection. But these theories amount only to giving more or less fancy names to imaginary phenomena: no one has even observed the occurrence of a "systemic mutation," for instance.

Actually we do not need to meander so far from the Darwinian theory. There is no necessity for assuming that the human eye, for example, had a sudden birth or that cruder forerunners of it could not have been useful to their possessors before the eye acquired its final perfection. The ancestors of the human species had eyes to see with, though they may have been less elaborate than ours. In short, the eye could have developed gradually from a very simple organ which in its earliest form gave some kind of "sight" or other useful ability to the animal that possessed it.

We shall consider in this article a most remarkable adaptation in the vegetable world which illustrates such a step-by-step evolution. In some exuberant rain forests of the tropics there grows a strange variety of plant known as strangler trees. Such a plant starts by seeding itself and growing like a vine on the trunk or branches of an ordinary forest tree. Climbing over its host, the strangler enfolds it in a thick mass of roots, strangles it to death and finally stands on its own as an independent tree!

The reason for the origin of the strangler trees (of which there are a number of species) is plain. In the dense tropical forest the competition for sunlight is keen. A young plant sprouting on the dark forest floor has a poor chance of survival unless it can somehow break through the canopy overhead. The stranglers have solved the problem by climbing on other trees. And the whole life history of these outlandish trees seems beautifully contrived to accomplish their objective: to seize a place in the sun in the midst of a dense tropical forest. How could this singular adaptation have arisen? Here is an extraordinary example of just the kind of complex adjustment that seems to justify some esoteric explanation such as "psychoids" or "systemic mutations." Let us see, however, whether a simpler explanation may suffice.

We need to consider first the life history of a strangler tree. Among the most common stranglers are certain fig trees (genus *Ficus*) of Brazil. The seeds of the strangler fig usually sprout high on the branches of a tall tree; just how they get there is not known, but there are reasons to believe that they may be carried by birds and fruit-eating bats. The young seedling produces roots of two kinds. One kind grows around the branch or the trunk of the supporting tree; the other descends toward the forest floor, either along the trunk or hanging in the air. The stem of the strangler sprouts leaves and grows upward to catch the sunlight. The young plant gets its water and its mineral food from accumulations of dirt and organic matter in crevices of the tree's bark. At this stage the future strangler is not a parasite, for it derives no nourishment from the living tissues of its host. It is an epiphyte, i.e., a plant which grows on another plant.

As soon as the descending roots take hold in the soil of the jungle floor, the growth of the strangler quickens. Its roots rapidly thicken and harden, and they put out many new branches and leaves. It is often difficult to tell from the forest floor which foliage belongs to the strangler and which to the host tree. New roots are formed, and they begin to branch on the surface of the supporting trunk. Eventually they form a mesh which envelops the host tree with an ever-hardening strangle hold. The appearance of a gigantic forest tree caught in the deadly embrace of a strangler is weird in the extreme. It makes one think of some of the grotesque creations of surrealist art, but it has the nobility and the purposefulness of life.

Now the strangler proceeds to kill the supporting tree. It does so not merely by preventing the trunk from expanding but actually by squeezing the tree. This is indicated by the fact that the strangler fig often kills palm trees, whose trunks grow steadily in length but little or not at all in thickness.

While the fig is throttling its host, its roots go on growing and hardening, until they completely or almost completely cover the trunk. They also form buttresses which enable the fig to stand on its own feet. By the time the supporting tree dies, the strangler has become an independent tree, with its own crown of branches and leaves. Many specimens of these figs reach colossal dimensions, rivaling in height and girth some of the giants of the tropical forest.

At the final stage of its development the strangler may or may not show outward signs of its murderous past. Its "trunk," which in reality is a mass of fused roots, often has a bizarre shape, owing to the many cable-like or planklike buttresses. But it may also attain an almost regular cylindrical shape. In either case its true nature is readily exposed if one cuts through the mass of roots: inside there is a cavity which contains the more or less decomposed remains of the victim. Near Belém at the mouth of the Amazon stands a gigantic fig tree which has grown on the tall chimney of a brick factory abandoned some 70 years ago. The chimney is now all but invisible.

The Brazilian figs, which belong to the mulberry family, are one of many kinds of stranglers. Strangling trees are common not only in Brazil but also in rain forests of Australia, New Zealand and other places. But now, from the point of view of how the strangler trees evolved, we note a highly significant fact. There are many stranglerlike plants which do not strangle their hosts. A notable example is the Brazilian tree called *Clusia*. Some species of this genus behave like the strangler figs in every respect except one: they seldom if ever kill their support. We have seen thousands of jungle trees attacked by Clusia, and all of them were alive. High up in the forest canopy the large, leathery, dark-green leaves and the showy, rose-colored flowers of Clusia mingle with the foliage of the host tree. Adolfo Ducke, the leading authority on Amazonian flora, has informed us that he cannot remember having seen a Clusia that caused the death of its host tree.

Clusia may, then, illustrate an important stage in the evolution of the strangling habit. It is well adapted to use other tree species for support, and it is able to cling quite firmly to its host. But it stops short of killing the host tree and taking its place. When the host tree dies, Clusia presumably perishes with it, although further observations on this point are needed.

Still earlier stages in the evolution of the strangling habit may be seen in Brazil in three genera of plants of the mulberry family, to which the figs also belong. These genera are *Coussapoa, Pouroma* and *Cecropia.* Unlike the strangler figs or even Clusia, they may start in the soil on the forest floor and grow for their entire lifetime without climbing on other trees; they often grow in this independent fashion in forest clearings. The three genera show varying degrees of epiphytism: Coussapoa acts as a strangler frequently, Pouroma less often and Cecropia only occasionally.

It is remarkable that the strangling adaptation has evolved independently in several quite unrelated families of plants. The forests of New Zealand have no strangler figs or other stranglers of the mulberry family, but there is a strangler there called "rata" which is a member of the myrtle family. A rata kills and replaces its supporting tree in just the same way as a strangling fig. Yet a species closely related to rata grows on trees like a vine without strangling them.

E. J. Godley and L. J. Dumbleton of New Zealand have called our attention to still other plants in New Zealand forests which furnish striking illustrations of the probable stages of the evolution of the strangling habit. These plants belong to several different families: *Weinmannia* of the family *Cunoniaceae, Schefflera* of the *Araliaceae, Melicitus* of the *Violaceae* and *Griselinia* of the *Cornaceae*. Yet these rather remotely related plants are all capable of growing either as stranglers or as independent trees from the soil. It is not difficult to see how their versatility has evolved. The trees that they victimize most often are tree ferns—whose beautiful feathery fronds are so characteristic of the New Zealand forests. The trunks of the tree ferns are covered with a spongy mass of fibers, which in the rainy climate of many parts of New Zealand provides an inviting medium for seeds. Various species of plants have seized the opportunity and evolved adaptations which permit them to grow on such trees. After a time the evolving climber may lose the ability to start its life without the support of another tree; it is no longer a facultative strangler but an obligatory strangler. On the other hand, some members of the same plant genus or family keep their ability to grow independently.

Evidences of such evolution can be seen not only in New Zealand but also in the forests of Brazil. Some of the fig species there grow into huge trees without ever resorting to the strangling techniques of their relatives.

To summarize, a comparative study of the strangler trees shows that these amazing representatives of the plant kingdom possess quite a variety of adaptations for life under the exacting conditions of the tropical forest. The origin of these adaptations can easily be visualized as being due to nature's selection of useful hereditary modifications. This view is in accord with the modern theory of evolution, which considers selective responses of the organism to opportunities in the environment to be the primary driving force of the evolutionary process.

Part IV

•

THE WORLD OF MAN

A. *From Ape to Civilization*

•

From Ape to Civilization is in some ways a long road, although the transition has taken place during a small segment of geologic time. And while man himself may live in skyscrapers, he has cousins today who spend their lives in trees. Man's relationship to the apes, like all Darwinian theory, remains a subject of controversy even at present, but Darwin suggests the evidence which resolves it in his classic The Evidence of the Descent of Man from Some Lower Form. *In Darwin's time the path of development which culminated in Homo sapiens was barely understood. Today the main outlines of the story are much more precise, although there are still formidable gaps in our knowledge. In* Evolution Revised *Ruth Moore explains what recent research has uncovered.*

One of the most distinguished historical anthropologists is L. S. B. Leakey, formerly the curator of the Cornydon Museum in Nairobi, Kenya. Leakey's instinct for the rewarding digging site seems at times to have bordered on the miraculous. He describes one such occasion in The Discovery of *Zinjanthropus*, *based on a previous article which Dr. Leakey has kindly brought up to date for the present edition.*

The interrelationship of the various scientific disciplines is illustrated in dramatic fashion in The Great Piltdown Hoax *by William L. Straus, Jr., professor of anatomy at the Johns Hopkins Medical School. Without such techniques as the fluorine test this hoax may have gone undetected. The history of science contains only a few instances of deliberate fraud. This is one of the most devious and interesting.*

Anthropology, the study of man, deals not only with the evolution of Homo sapiens but also with the cultures, from the most primitive to the most sophisticated, which he has developed. In The Growth of Culture *an outstanding modern anthropologist, Ruth Benedict, explains what can be learned from its study. A fascinating branch of this study is archaeology, which investigates the remains of ancient man. Sir Leonard Woolley has made some of the greatest discoveries in this field, including evidence of the existence of lost civilizations in Mesopotamia. In* Digging Up the Past *he discusses the aims and techniques of the field archaeologist.*

CHARLES DARWIN

The Evidence of the Descent of Man from Some Lower Form

•

The Bodily Structure of Man

IT IS NOTORIOUS that man is constructed on the same general type or model as other mammals. All the bones in his skeleton can be compared with corresponding bones in a monkey, bat, or seal. So it is with his muscles, nerves, blood-vessels and internal viscera. The brain, the most important of all the organs, follows the same law, as shewn by Huxley and other anatomists. Bischoff, who is a hostile witness, admits that every chief fissure and fold in the brain of man has its analogy in that of the orang; but he adds that at no period of development do their brains perfectly agree; nor could perfect agreement be expected, for otherwise their mental powers would have been the same. But it would be superfluous here to give further details on the correspondence between man and the higher mammals in the structure of the brain and all other parts of the body.

It may, however, be worth while to specify a few points, not directly or obviously connected with structure, by which this correspondence or relationship is well shewn.

Man is liable to receive from the lower animals, and to communicate to them, certain diseases, as hydrophobia, variola, the glanders, syphilis, cholera, herpes, etc., and this fact proves the close similarity of their tissues and blood, both in minute structure and composition, far more plainly than does their comparison under the best microscope, or by the aid of the best chemical analysis.

Man is infested with internal parasites, sometimes causing fatal effects; and is plagued by external parasites, all of which belong to the same genera or families as those infesting other mammals.

The whole process of that most important function, the reproduction of the species, is strikingly the same in all mammals, from the first act of courtship by the male, to the birth and nurturing of the young. Monkeys are born in almost as helpless a condition as our own infants: and in certain genera the young differ fully as much in appearance from the adults, as do our children from their full-grown parents. It has been urged by some writers, as an important distinction, that with man the young arrive at maturity at a much later age than with any other animal: but if we look to the races of mankind which inhabit tropical countries the difference is not great, for the orang is believed not to be adult till the age of from ten to fifteen years. Man differs from woman in size, bodily strength, hairiness, etc., as well as in mind, in the same manner as do the two sexes of many mammals. It is, in short, scarcely possible to exaggerate the close correspondence in general structure, in the minute structure of the tissues, in chemical composition and in constitution, between man and the higher animals, especially the anthropomorphous apes.

Embryonic Development

Man is developed from an ovule, about the 125th of an inch in diameter, which differs in no respect from the ovules of other animals. The embryo itself at a very early period can hardly be distinguished from that of other members of the vertebrate kingdom. At this period the arteries run in arch-like branches, as if to carry the blood to branchiae which are not present in the higher vertebrata, though the slits on the sides of the neck still remain, marking their former position. At a somewhat later period, when the extremities are developed, "the feet of lizards and mammals," as the illustrious Von Baer remarks, "the wings and feet of birds, no less than the hands and feet of man, all arise from the same fundamental form." "It is," says Prof. Huxley, "quite in the later stages of development that the young human being presents marked differences from the young ape, while the latter departs as much from the dog in its developments, as the man does. Startling as this last assertion may appear to be, it is demonstrably true."

After the foregoing statements made by such high authorities, it would be superfluous on my part to give a number of borrowed details, shewing that the embryo of man closely resembles that of other mammals. It may, however, be added, that the human embryo likewise resembles in various points of structure, certain low forms when adult. For instance, the heart at first exists as a simple pulsating vessel; the excreta are voided through a cloacal passage; and the os coccyx projects like a true tail, "extending considerably beyond the rudimentary legs." In the embryos of all air-breathing vertebrates, certain glands, called the corpora Wolffiana, correspond with, and act like the kidneys of mature fishes. Even at a later embryonic period, some striking resemblances between man and the lower animals may be observed. Bischoff says that the convolutions of the brain in a human foetus at the end of the seventh month reach about the same stage of development as in a baboon when adult. The great toe, as Prof. Owen remarks, "which forms the fulcrum when standing or walking, is perhaps the most characteristic peculiarity in the human structure," but in an embryo, about an inch in length, Prof. Wyman found "that the great toe was shorter than the others; and, instead of being parallel to them, projected at an angle from the side of the foot, thus corresponding with the permanent condition of this part in the quadrumana." I will conclude with a quotation from Huxley, who after asking, Does man originate in a different way from a dog, bird, frog, or fish? says, "The reply is not doubtful for a moment; without question, the mode of origin, and the early stages of development of man, are identical with those of the animals immediately below him in the scale: without a doubt in these respects, he is far nearer to apes than the apes are to the dog."

Rudiments

Not one of the higher animals can be named which does not bear some part in a rudimentary condition; and man forms no exception to the rule. Rudimentary organs are eminently variable; and this is partly intelligible, as they are useless, or nearly useless, and consequently are no longer subjected to natural selection. They often become wholly suppressed. When this occurs, they are nevertheless liable to occasional reappearance through reversion—a circumstance well worthy of attention.

Rudiments of various muscles have been observed in many parts of the human body; and not a few muscles, which are regularly present in some of the lower animals, can occasionally be detected in man in a greatly reduced condition. Every one must have noticed the power which many animals, especially horses, possess of moving or twitching their skin; and this is effected by the *panniculus carnosus*. Remnants of this muscle in an efficient state are found in various parts of our bodies; for instance, the muscle on the forehead, by which the eyebrows are raised.

Some few persons have the power of contracting the superficial muscles on their scalps; and these muscles are in a variable and partly rudimentary condition. M. A. de Candolle has communicated to me a curious instance of the long-continued persistence or inheritance of this power, as well as of its unusual development. He knows a family, in which one member, the present head of the family, could, when a youth, pitch several heavy books from his head by the movement of the scalp alone; and he won wagers by performing this feat. His father, uncle, grandfather, and his three children possess the same power to the same unusual degree. This family became divided eight generations ago into two branches; so that the head of the above-mentioned branch is cousin in the seventh degree to the head of the other branch. This distant cousin resides in another part of France; and on being asked whether he possessed the same faculty, immediately exhibited his power. This case offers a good illustration how persistently an absolutely useless faculty may be transmitted.

The sense of smell is of the highest importance to the greater number of mammals—to some, as the ruminants, in warning them of danger; to others, as the carnivora, in finding their prey; to others, again, as the wild boar, for both purposes combined. But the sense of smell is of extremely slight service if any, even to savages, in whom it is much more highly developed than in the civilized races. It does not warn them of danger, nor guide them to their food; nor does it prevent the Esquimaux from sleeping in the most fetid atmosphere, nor many savages from eating half-putrid meat. Those who believe in the principle of gradual evolution, will not readily admit that this sense in its present state was originally acquired by man, as he now exists. No doubt he inherits the power in an enfeebled and so far rudimentary condition, from some early progenitor, to whom it was highly serviceable, and by whom it was continually used. We can thus perhaps understand how it is, as Dr. Maudsley has truly remarked, that the sense of smell in man "is singularly effective in recalling vividly the ideas and images of forgotten scenes and places"; for we see in those animals, which have this sense highly developed, such as dogs and horses, that old recollections of persons and places are strongly associated with their odour.

Man differs conspicuously from all the other Primates in being almost naked. But a few short straggling hairs are found over the greater part of the body in the male sex, and fine down on that of the female sex. There can be little doubt that the hairs thus scattered over the body are the rudiments of the uniform hairy coat of the lower animals.

It appears as if the posterior molar or wisdom-teeth were tending to become rudimentary in the more civilised races of man. These teeth are rather smaller than the other molars, as is likewise the case with the corresponding teeth in the chimpanzee and orang; and they have only two separate fangs. They do not cut through the gums till about the seventeenth year, and I have been assured by dentists that they are much more liable to decay, and are earlier lost, than the other teeth. It is also remarkable that they are much more liable to vary both in structure and in the period of their development, than the other teeth. In the Melanian races, on the other hand, the wisdom-teeth are usually furnished with three separate fangs, and are generally sound; they also differ from the other molars in size less than in the Caucasian races. Prof. Schaffhausen accounts for this difference between the races by "the posterior dental portion of the jaw being always shortened" in those that are civilised, and this shortening may, I presume, be safely attributed to civilised men habitually feeding on soft, cooked food, and thus using their jaws less.

With respect to the alimentary canal, I have met with an account of only a single rudiment, namely the vermiform appendage of the caecum. The caecum is a branch or diverticulum of the intestine, ending in a cul-de-sac, and is extremely long in many of the lower vegetable-feeding mammals. In the marsupial koala it is actually more than thrice as long as the whole body. It is sometimes produced into a long gradually-tapering point and is sometimes constricted in parts. It appears as if, in consequence of changed diet or habits, the caecum had become much shortened in various animals, the vermiform appendage being left as a rudiment of the shortened part. That this appendage is a rudiment, we may infer from its small size, and from the evidence which Prof. Canestrini has collected of its variability in man. It is occasionally quite absent, or again is largely developed. The passage is sometimes completely closed for half or two-thirds of its length, with the terminal part consisting of a flattened solid expansion. In the orang this appendage is long and convoluted; in man it arises from the end of the short caecum, and is commonly from four to five inches in length, being only about the third of an inch in diameter. Not only is it useless, but it is sometimes the cause of death, of which fact I have lately heard two instances; this is due to small hard bodies, such as seeds, entering the passage, and causing inflammation.

The os coccyx in man, though functionless as a tail, plainly represents this part in other vertebrate animals. At an early embryonic period it is free, and projects beyond the lower extremities. In certain rare and anomalous cases, it has been known to form a small external rudiment of a tail.

The bearing of the three great classes of facts now given is unmistakable. But it would be superfluous here fully to recapitulate the line of argument given in detail in my *Origin of Species*. The homological construction of the whole frame in the members of the same class is intelligible, if we admit their descent from a common progenitor, together with their subsequent adaptation to diversified conditions. On any other view, the similarity of pattern between the hand of a man or monkey, the foot of a horse, the flipper of a seal, the wing of a bat, &c., is utterly inexplicable. It is no scientific explanation to assert that they have all been formed on the same ideal plan. With respect to development, we can clearly understand, on the principle of variation supervening at a rather late embryonic period, and being inherited at a corresponding period, how it is that the embryos of wonderfully different forms should still retain, more or less perfectly, the structure of their common progenitor. No other explanation has ever been given of the marvellous fact that the embryos of a man, dog, seal, bat, reptile, &c., can at first hardly be distinguished from each other. In order to understand the existence of rudimentary organs, we have only to suppose that a former progenitor possessed the parts in question in a perfect state, and that under changed habits of life they became greatly reduced, either from simple disuse, or through the natural selection of those individuals which were least encumbered with a superfluous part.

Thus we can understand how it has come to pass that man and all other vertebrate animals have been constructed on the same general model, why they pass through the same early stages of development, and why they retain certain rudiments in common. Consequently we ought frankly to admit their community of descent; to take any other view, is to admit that our own structure, and that of all the animals around us, is a mere snare laid to entrap our judgment. This conclusion is greatly strengthened, if we look to the members of the whole animal series and consider the evidence derived from their affinities or classification, their geographical distribution and geological succession. It is only our natural prejudice, and that arrogance which made our forefathers declare that they were descended from demi-gods, which leads us to demur to this conclusion. But the time will before long come, when it will be thought wonderful, that naturalists, who were well acquainted with the comparative structure and development of man, and other mammals, should have believed that each was the work of a separate act of creation.

RUTH MOORE

Evolution Revised

•

SINCE 1950 the scientific evidence has pointed inescapably to one conclusion: man did not evolve in either the time or the way that Darwin and the modern evolutionists thought most probable. The physicists and the geologists by 1950 had clearly shown that the world is older and man is younger than anyone had dared to estimate before.

By that year, the men who dug into the ancient caves of South Africa and China, looking for traces of early man and the part-human, part-anthropoid races that had preceded him had succeeded. All the major steps in the evolution of man were for the first time filled in; the hitherto missing link had been found. But the missing link was not what science had expected; no one had imagined a being with the head of an ape and the body of a man.

What did it mean? The physicists and geologists and fossil-hunters did not say. They merely presented their dates and materials.

The question was insistent: what did it mean? How did the new findings affect the theory of evolution? This was a problem for the anthropologist, because the whole problem of man's origins and evolution was affected.

The surprising and almost unbelievable fossils that came from the banks of the Solo River, from Dragon-bone Hill, and from the Sterkfontein caves indicated that man had developed according to a new pattern, and that the pattern given him in the past—and currently, in many cases—was wrong at some critical points. The fossils supplied disconcerting proof that the development of the body came first, and the typical development of the brain later—that we had human bodies long before we reached human intelligence.

At first, such evidence from the ground was disbelieved, as evidence is likely to be when it runs counter to what the world has always thought.

When Eugène Dubois found an undeniably human leg bone close by the skull of *Pithecanthropus erectus*, the world cried: "They can't belong together." The ridicule heaped upon his find, in large part because of this "discrepancy," drove the Dutch physician into his thirty-year retreat.

The skepticism was almost as strong when similar bones began to be found with the skulls of Peking man.

Broom ran headlong into the same feeling that there must be a mix-up when he discovered human-like pelvic bones in the same deposits with the unquestionably apelike skulls of the South African ape-men. For a number of years his work was not taken seriously because he ventured to claim that the apelike creatures could have had near-human bodies and walked like men.

The conviction that the first men began as replicas, however crude and primitive, of modern man was so deep that the evidence to the contrary long was discredited and discounted. It was a staple belief that many millions of years ago some of the anthropoids developed better brains, and that as they became smarter they came down out of the trees and gradually evolved into modern man.

Weidenreich was perhaps the first authority to point out that man's development might have followed a different course. In 1941 he wrote: "Little is known about the development of other parts of the skeleton, but it can be taken as definitely established that the erect posture and all that is connected with its adoption were attained long before [man reached his definitive form]. Thus the subsequent change of the skull, and above all that of the brain case, morphologically viewed, crowns the transformation in the true sense of the word, both in time and position."

But not until after the publication of the South African monographs and their appraisal by Clark of Oxford did science generally begin to grant that the body reached human form long before the brain.

Laboratory work at the University of Chicago and Harvard strongly confirmed the new pattern of evolution evinced by the fossils. Washburn, at Chicago, made a close study of the different body "complexes" of man and the apes.

In the arms, the ribs, and the shoulder girdle, he found, the two are very much alike. The important middle part of the body had changed little; there men still were essentially apes. The specialist can recognize the technical differences between the shoulders of modern man and the ape, but the differences are not great.

The most noticeable departure is in the hands, which Washburn, from the long-term anthropological viewpoint, considered of lesser importance. Once man had become a biped, selection inevitably would have favored a hand differing from that of the tree-living primates. Even so, the human hand still shows a remarkable amount of the primitive grasping adaptation, particularly in the long fingers and nails.

As Washburn and a number of other anthropologists now see it, the middle part of the body, the trunk and arms, began to take on its essential form millions of years ago, at the time when the earliest primates climbed up into the trees. The story as they trace it goes, in outline, something like this:

The development of the ability to grasp with the hands and feet set the

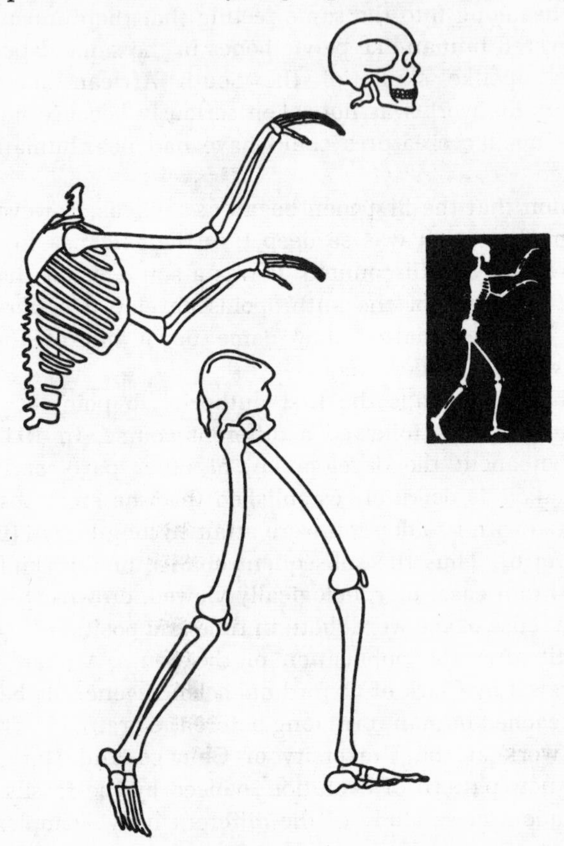

Fig. 1. The three great body "complexes" of man. Each section evolved differently. The arms, shoulders, and ribs changed little. In this middle part of the body men still are essentially apes. There was an all-important change in the pelvis-legs complex, a change that started man on a distinctive line of evolution. Man's ability to walk on two legs freed his hands for the use of tools. The development of the third great complex, the brain, then followed.

first primates apart from all the other primitive mammals and made it possible for them to take up life in the trees.

But life on the leafy green world above was a fairly restricted, though safe one. Without the great ranges over which the ground-living animals could move and mix, the tree-living primates came to differ widely from each other.

Many of them developed differences in the senses. In the trees, the strong sense of smell, the ears that cocked to the least sound, the hair that stood on end at the threat of danger, were not so important as on the ground. The monkeys with better eyes and color vision were the ones that survived and left descendants. And gradually the brain changed from a primitive "smell-brain" to a more advanced "sight-brain."

The monkeys thus equipped became abundantly successful in the Old World tropical forests and in most of the other areas to which they spread.[1]

Some of the numerous small bands of primates then began to develop a different mode of locomotion. Instead of hopping or running along, holding fast with hands and feet, they would swing along with a different motion of the arms. This mode of progress, called brachiation, involved anatomical changes in the wrist, elbow, shoulder, and thoracic region.

Fig. 2. Gibbon, gorilla, and spider monkey. The gibbon and gorilla swing along with the arms—brachiate. The monkey, lacking the arm and shoulder development of the apes and man, runs along holding fast with arms, legs, and, in the case of the New World monkey, with the tail too.

And thus the apes arose, and the shoulder, arms, and ribs took on the form that has been carried along almost unchanged to all of us today.

Washburn points out that no monkey anywhere in the world has such arms and shoulders. But every essential detail of this "complex" is shared by man and ape. He believes this crucially important change occurred about ten million years ago.

As the eons rolled by, the story continues, some of the apes that ate a more varied diet came down to the ground to live. Like their present-day

[1] The older, more primitive forms soon were replaced, except where they were especially protected, or retreated to a nocturnal life. The safety of the island of Madagascar, for example, saved the lemurs. The tarsiers, the lorises, and the galagos found refuge in the night.

descendants, they could take a few steps upright, perhaps even while holding a stick in one hand. But when they wanted to cover space, down went the knuckles and they proceeded on all fours.

About a million years ago there came what many anthropologists regard as the most important of all the changes in the evolution of man, a change that forever afterward was to set man apart from his anthropoid ancestors.

Some of the big ground-living apes were born with a different kind of pelvis. It meant that they could walk upright, on two legs! For the first time, in all of time, the hands were free. They no longer had to be used for locomotion. These ground-living apes could use tools, for any implement held in the hands no longer had to be dropped every time more than a few steps were taken. "The fact that we number more than a few thousand bipeds living in the Old World tropics is due to the development of tools," said Washburn.

From this point on, all was changed; a new future had been cast. Natural selection was on a new basis, for a premium had been placed on brains as well as brawn. The most intelligent of the biped man-apes, the ones that could most effectively use sticks and stones to beat off their enemies and kill their food, were the survivors and the parents of the next generation.

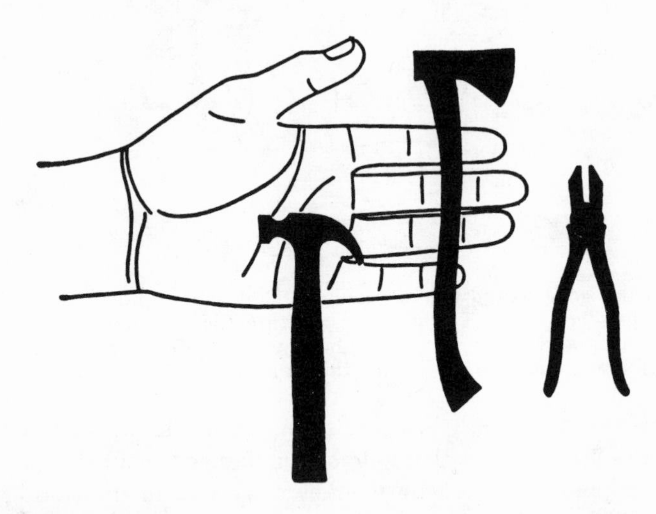

Fig. 3. Hand and tools.

The pelvis that precipitated this all-important turn in evolution has a number of functions. It not only connects the legs and trunk in such a way that it controls gait; it gives origin to many muscles and serves as a bony birth canal.

How any living creature stands and moves depends in large part upon the length of the pelvis and upon the angle at which it is inclined. In the apes it is long and slanting. In the ape-men and men it is shorter and more nearly upright.

Washburn believes that this whole vital evolutionary cycle began when some of the big ground-living apes were born with a shorter pelvis. When this basic genetic change occurred, the bone had to take on a more upright position to assure a safe birth for offspring. And once the pelvis is brought into such a position, the thigh muscles that attach to it are directly affected. These muscles make the human step.

Washburn argues that the real difference between the walk of man and ape is not in the extent of the motion, but in the ability to finish the step

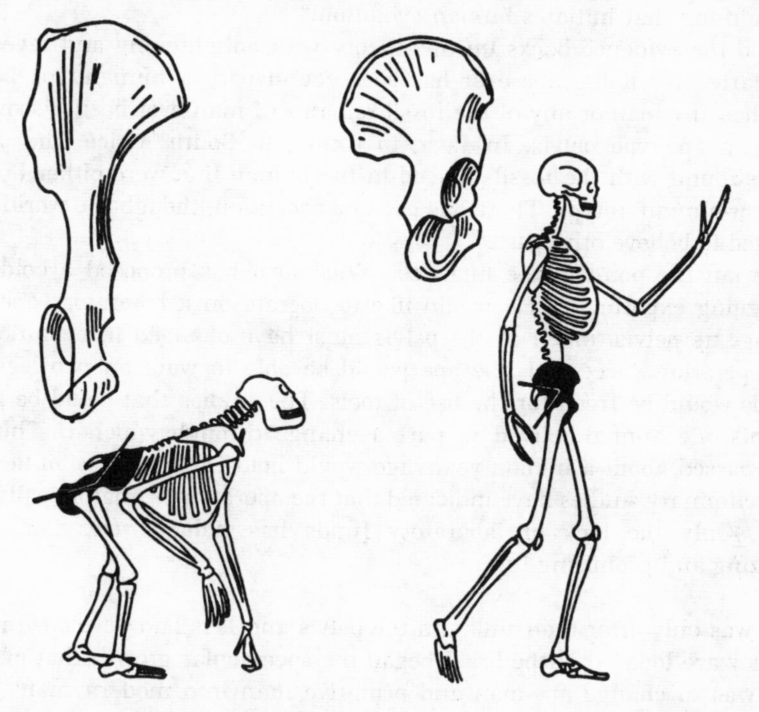

Fig. 4. One great change forever afterward set man apart from his anthropoid ancestors. It was in the pelvis. The apes, with their long slanting pelvis, can take only a few steps upright and must use the hands for locomotion. Man, because of his shorter pelvis, can stand and walk—he is a biped. His hands are then no longer needed for locomotion.

with a drive. The muscle that provides the drive and swings back the thigh is the gluteus maximus, the muscle that arises from the posterior part of the pelvis. When the pelvis is short, the gluteus maximus pulls hard. A vigorous step is possible.

Negative evidence of how important the muscle is appears when it becomes paralyzed. A man who suffers such an accident cannot walk normally, though he can get around easily with a flexed gait similar to that of the apes. "The paralysis of the single muscle makes the human type of bipedal locomotion impossible," wrote Washburn. "It shows that the form and function of this particular muscle is critical in the evolution of man's posture and gait."

Carrying this argument back to what may have happened when the first apes began to walk, Washburn maintains that since selection is for function, the animals able to walk and use the hands most freely were the ones to survive. Hence selection favored the new type of pelvis. "It is my belief," the Chicago anthropologist concludes, "that this single change is the thing that initiates human evolution."

And the evidence backs up this theory with enlightening and revealing regularity. No living ape ever has been found with a human-type pelvis. Nor has any man or any of the fossil remains of man ever been discovered with an ape-type pelvis. In Java, in China, in South Africa, the pelvic bones found with the fossils placed in the human line were either human or near-human in type. There has been no exception, though the world long wanted to believe otherwise.

To put the point to the final test, Washburn has proposed a bold and intriguing experiment. He would like to operate on a laboratory ape and change its pelvis, much as the pelvis must have changed in evolution. If the operation succeeded, the ape would be able to walk on two legs! Its hands would be freed for the use of tools. The studies that could be made as this one animal relived in part a change through which the human race passed about a million years ago would hold exciting possibilities.

Preliminary studies have indicated that the operation is anatomically possible. Only the lack of laboratory funds has halted work along this amazing and promising line.

It was only after the trunk and the pelvis and legs had developed much as they are today that the brain began the spectacular growth that eventually was to change ape-man and primitive man into modern man. Until the 1940's science assumed that the growth of the brain always came first. Supposedly it was a better brain that enabled the mammals as a group to triumph over the reptiles when the two contended for the control of that ancient world of sea and jungle.

And supposedly it was again the development of a better brain that led the apes out of the trees and onto the ground. Even Weidenreich, though he recognized that the evolution of the body came first, rated the development of the brain as the primary factor in evolution.

The faith in the priority of the brain was first seriously jarred in 1948. In that year Professor Tilly Edinger of Harvard showed that the growth of the brain tends to follow in evolution. The earliest mammals, their fossil remains revealed, had brains no more advanced than those of the reptiles. Only later, after they had become typical mammals, did the brains reach modern mammalian proportions.

The horse, which Edinger studied in particular, attained its characteristic form, its long legs and teeth, well before the brain reached its final size.

And so it was with man A comparison of the brain capacity of man and his forerunners sharply etches the pattern:

	CRANIAL CAPACITY
Chimpanzee and gorilla	325–650 (cubic centimeters)
South African ape-man	450–650
Java man	790–900
Peking man	900–1,200
Neanderthal man	1,100–1,500
Modern man	1,200–1,500

The South African ape-men—despite their near-human bodies and upright posture—were in the brain range of the apes.

"There is no doubt that all human fossils described so far have human pelves and limb bones, and the man-apes were remarkably human in these features," Washburn emphasized. "Therefore it appears that the differences in the brain between apes and man, just as those in dentition, were attained after full human status had been achieved in the limbs and trunk."

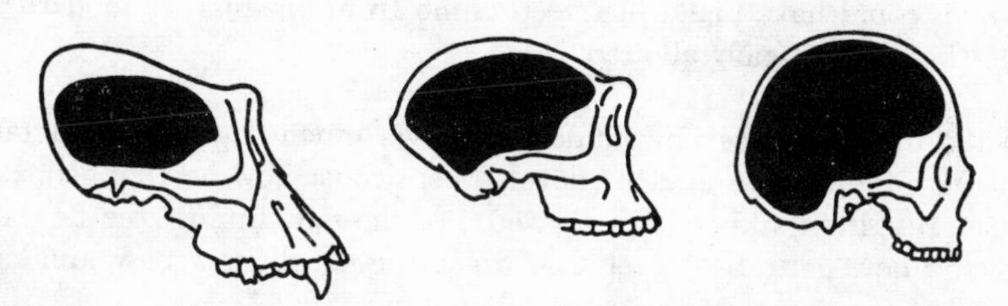

Fig. 5. The growth of the brain. The ape (left) had a brain averaging less than 650 cubic centimeters. Primitive man reached a brain of 900 to 1,200 cubic centimeters. Modern man often has a brain of 1,500 cubic centimeters. The brain of modern man is more than twice as large as that of the ape.

In its final evolution the change in the brain was large. Between the ape-men and the emergence of modern man, the brain more than doubled in size. It grew from the 650-cubic-centimeter maximum of the apes and the ape-men to the modern top average of 1,500 cubic centimeters.

Some of the Neanderthal men of Europe had a brain capacity very close to that of modern man, if not within the modern range. They were well along the way.

It should be remembered, too, that many of the other fossils that once appeared to date man with a fairly sizable brain back in the shadowy stretches of 500,000 to 1,000,000 years ago were also shown to belong to comparatively recent years.

How recent? Here Carbon-14 comes in. Although the Carbon-14 dates do not go beyond 25,000 years, they clearly indicate that the ice last extended down from the north both in Europe and in North America about 11,000 years ago. If this is correct, geologists hold that the final glaciation began less than 50,000 years ago.

This is an assumption, unsupported as yet by any absolute system of dating, but with the date of the final advance of the ice fixed, and with all the wealth of evidence left by the glaciers themselves, there is little speculation in estimating the duration of the last ice age.

The new timing indicates therefore that humans who had the requisite intelligence to be called men did not reach that high status until about 50,000 years ago. Modern man, then, is only about 50,000 years old.

The 50,000 years, of course, are approximate. But even if this estimate should later be enlarged to 75,000 or 100,000 years, modern man still would be the veriest of newcomers by all evolutionary standards. And if our 50,000-year tenure of the earth must be adjusted, the chances are that it will be shortened. Unpublished work and studies now going on in a number of universities are tending to pull the time of man's emergence as man even closer to today.

At the same time, the work of Washburn, Dice, and others demonstrated that man could have made the steep climb from ape-man to modern man in the shorter time now allotted.

In the light of these new understandings, much that has been taught about the time of man and his development must now be changed. Books must be rewritten and courses revised. For the new timing, the new fossil finds, the new pattern of evolution are bringing about a new and major revision in the theory of the origin of man.

The theory of evolution is not being weakened by this correction of past errors and misconceptions and by the opening of new understandings. On the contrary, the basic truths developed by Darwin and the brilliant succession of evolutionists who came after him are strengthened.

L. S. B. LEAKEY

The Discovery of *Zinjanthropus*

•

THE TEETH were projecting from the rock face, smooth and shining, and quite obviously human.

To my wife Mary and myself, who had long been looking for just such a clue, those bits of fossilized matter represented a priceless discovery, and the end of a 28-year search.

For there in the rock of that remote, sun-baked gorge in East Africa lay the remains of the most ancient near-man ever found.

In order to understand how the remains of *Zinjanthropus* can be buried in these ancient rocks, let us go back to an imaginary scene by the ancient lake shore about one and three-quarter million years ago. Rain had fallen for many days and the lake level was rising ominously. Camped by the shore, the little band of primitive humans realized by dawn that they would have to retreat.

Collecting together their few family possessions, the family moved away from the rising floods. Behind them they left the body of a youth who had tried to intrude into their camp during the night, and whose visit had resulted in his death.

Inexorably the lake crept over the camp-site, engulfing the body, together with many stone tools abandoned in the flight and the bones of small animals the hunters had eaten. Higher and higher the water rose, depositing a layer of silt over all.

Again and again during what we call pluvial periods—eras of increased rainfall probably coinciding with ice ages farther toward the poles—the lake rose and fell, adding layer on layer of silt and sand on top of the camp-site. Finally the water vanished, leaving the body entombed under several hundred feet of sediments that had hardened to rock.

There our story might have ended but for one of those quirks of nature that sometimes seem to do man's work for him.

Some 100,000 years ago—when the bones of our near-man had lain buried for one and three-quarter million years—violent earthquakes convulsed the area, fracturing and reshaping the land on a vast scale. When the tremors had ceased, an immense new chasm had appeared as part of the Great Rift Valley, which stretches from Jordan south through Kenya and Tanganyika.

For thousands of years Africa's sands drifted restlessly, filling an erosion-created side valley. Then what erosion had begun, torrential rains finished. The water, sweeping into the Great Rift, cut through the sands and deepened the channel through the cliffs. This is today's Olduvai Gorge, and it slices precisely through the beds of that long-vanished Stone Age lake.

Olduvai is a fossil-hunter's dream, for it shears 300 feet through stratum after stratum of earth's history as through a gigantic layer cake. Here, within reach, lie countless fossils and artifacts which but for the faulting and erosion would have remained sealed under thick layers of consolidated rock.

I have long believed that it would be at Olduvai that we would find evidence of human life earlier than that represented by Peking Man or Java Man. We now know that this was so, for both *Zinjanthropus* and *Homo habilis* (discovered subsequently) are older than Far Eastern man by more than a million years.

Why did I first go to Olduvai Gorge, and what made me so positive it would yield new secrets concerning the development of early man? Olduvai Gorge was discovered in 1911 by a German entomologist named Kattwinkel. The discovery nearly cost him his life: Chasing a butterfly, he almost followed his quarry over a cliff. When he climbed down over the edge, he came upon fossil bones, which he took back with him to Berlin.

German scientists sent an expedition to Olduvai in 1913 under the leadership of my old friend, the late Professor Hans Reck. World War I, however, interrupted the work, and in the postwar years Reck wrote to me in Kenya. He could not resume his initial operations in the gorge. Would I like to take over the exploration?

Early results had been promising. The shores of a vanished inland lake are always likely places to search for evidence of fossil man, since in his earlier stages he had no vessels to carry water and invariably stayed close to an ample source.

I wrote to Reck accepting the offer and made plans for an expedition. Lack of money caused endless delays, and it was not until 1931 that Reck and I finally set out from Nairobi for Olduvai.

The present route via the town of Arusha leads past the rim of the Ngorongoro Crater, then across the part of the Great Rift Valley known as the Balbal Depression to the gorge, at the edge of the Serengeti Plain. The 320-mile journey from Nairobi can nowadays be accomplished in any car in 8 hours or so. The first time I went down there it took us seven days by truck and car, using a different driving route. Only a few years ago, the journey needed a Land-Rover and took 13 or 14 hours.

There was no lack of wildlife along the way; we saw elephants, giraffes, rhinoceroses, zebras, wildebeests, and smaller game such as Thomson's gazelles and dik-diks, delightful little antelopes only about 14 inches tall. Most of the animals showed scant fear; we could often approach as close to them as 20 feet.

The first night, camped at the edge of the gorge, we met neighbors who have visited us regularly ever since. After darkness fell, I went out to see what was moving around the camp and switched on a large electric torch. In the blackness I was able to pick out the green eyes of eleven lions, some near, some far.

They had come from different parts of the gorge to investigate this invasion of their territory, for these animals are the most curious of all the cat family; they just seem to have to know what is happening. To this day they always greet us on our return to Olduvai, but they have never bothered us. Needless to say, we extend the same courtesy to them.

Not all the animals that visited us over the years were as unobtrusive as our lion neighbors. Giraffes, hyenas, and even rhinos have wandered into our camp. The explanation is simple: water.

From the beginning, lack of water has been the great hardship at Olduvai. Because of the predominance of clay in the soil, excavation during rainy months is difficult. When the rain stops, however, water simply vanishes, and we have to haul every precious drop by trailer from a spring 28 miles away on the rim of the Ngorongoro Crater—a laborious and expensive process. Up to now the cost has limited us to a working season of about seven weeks.

We have tried several ways to overcome this water shortage, sometimes with disastrous results.

One year Mary and I decided to visit the gorge toward the end of the wet season to avoid having to carry water. It worked beautifully for a week; with runoff from the mountains, we even had running water at the bottom of the gorge for a daily bath. But when the streams dried up, we found ourselves reduced to a single water hole.

Unfortunately for us, two rhinos discovered the hole about the same time, and not content to drink from it, they turned it into a wallow. To those who have never lived on rhino bath water, I can only offer my congratulations. The taste stayed with Mary and me for months afterward.

Another time, when we had a cloudburst, we made funnels out of the tent flaps and collected enough water for everybody to have plenty. We forgot that tent canvas is often treated with copper sulphate for protection against insects. Luckily it made us ill before we drank enough to kill us outright.

During Olduvai's dry spell, the animals often scent the camp's water supply, and this has led to some strange encounters.

One evening as our cook was preparing a meal in the kitchen—merely a fireplace of rocks surrounded by a low brushwood fence—he was interrupted by a faint growl. He glanced up, expecting to see the usual timid hyena sniffing for scraps, and found himself face to face with a leopard.

It is hard to say which bolted faster, our screaming cook or the startled leopard, but supper that evening was an understandably sketchy affair.

On another occasion a pair of rhinos ambled into camp and, unnoticed, approached one of my assistants. The nearsighted brutes practically stumbled over him before he gave out a great yell and put them to rout.

Strangely enough, the rhinos are terrified of our two Dalmatians, Sally and Toots. It is a ludicrous picture to see two tons of rhinoceros lumbering across the plain with a 50-pound dog yapping in hot pursuit.

We have discovered that Toots and Sally are our best protection against animals and snakes. They sense danger long before we do, and this saves us the bother of carrying guns, a definite handicap on an expedition.

That first season at Olduvai convinced me that the gorge was one vast storehouse of Stone Age relics, a fossil museum such as existed perhaps nowhere else in the world. The animal remains alone were staggering. Since we started digging at Olduvai, we have uncovered fossil remains of more than 180 different extinct beasts, some of them nightmarish in their size and make-up.

For example there was *Afrochoerus*, a prehistoric pig as big as a rhinoceros. This pig had tusks so gigantic that a German scientist once attributed one to an elephant!

There was *Pelorovis*, a gigantic sheep-like creature that towered six feet at the shoulder and had a horn span of three to four yards, and *Libytherium*, a burly, short-necked giraffe with broad antlers like those of the modern moose.

But perhaps the most fantastic and dreadful creature of all was the giant baboon, *Simopithecus jonathani*, named for my son Jonathan, who has spent several seasons with us and who discovered the beast's jaw in 1957. Jonathan's baboon dwarfs a gorilla and, indeed, any primate previously known.

A prize for astonishing discoveries goes to our youngest son, Philip. From time to time in our digging at Olduvai, we had come across bits of fossilized eggshell whose thickness suggested a bird of giant proportions. We theorized that the owner had been a member of the ostrich family, but whether it actually corresponded in size to its enormous eggs we could not tell.

In 1961 Philip staked out an area for excavation which he guarded with all the pride of a 12-year-old fossil-hunter. Walking across the plot one day, Mary stumbled over a bone barely sticking from the ground. Philip, as proprietor of the site, directed the digging, and together he and Mary began to unearth the specimen.

As more and more of the fossil came to light, they could scarcely believe their eyes. It was a femur—the upper bone of a leg—but what a femur! In size it matched the leg bone of a giraffe. Quickly the two hauled it to where I was working.

"Why, it's a bird," I exclaimed, "but what a giant!"

Now at last the riddle of the eggs was solved. Examination proved that the bird was indeed a giant, a member of the ostrich family. In size it may well have overshadowed the extinct moa of New Zealand, the largest of which stood about 12 feet high.

Indeed it became clear that Olduvai was a site such as none other in all the world. That is why, ever since 1931, we have gone back again and again, certain that sooner or later we would find, as we have done, evidence of the earliest men, and perhaps the remains of the men themselves.

To give a clear picture of our work, I should describe the Olduvai diggings briefly. They are scattered over the floor and slopes of the gorge, some sites separated from others by a mile or more.

The earliest fossils—those among which *Zinjanthropus* came to light—are in what I call Bed I, the bottommost layer. From 40 to 100 feet above the floor of this stratum, and representing a later era, lies Bed II. Still higher layers, of course, record more recent ages.

Our method of searching is simple and, to say the least, uncomfortable. It consists of crawling up and down the slopes of the gorge with eyes barely inches from the ground, stopping at the slightest fragment of a fossil bone or stone implement and delicately investigating the clue with a fine brush or a dental pick. All this in heat that sometimes reaches 110° F.

To nonscientists the procedure seems agonizingly endless and slow, and it is true that Mary and I often feel we have spent more of our lives on our hands and knees than on our feet. Fossil hunting is an exacting business, one that demands patience and endurance above all. Still, the rewards are great; the gorge has already contributed much to man's knowledge of his beginnings. And we have made only a start.

Olduvai has always promised to bear out Charles Darwin's prophecy, made nearly a century ago, that Africa would be revealed as the cradle of mankind. In 1931 and 1932 we were already uncovering crude stone implements from the dawn of the Paleolithic Age, predating the stone hand axes of the Chellean culture.

I named this well-defined new culture Oldowan, and over the years we came to learn more and more about its primitive artisans. But the men themselves, the fossil remains by which we could reconstruct those dim figures of a distant age, continued to evade us.

The long quest ended on July 17, 1959.

That morning I woke with a headache and a slight fever. Mary was adamant.

"I am sorry," she said, "but you just cannot go out this morning, even though you want to. You're not fit for it, and you'd only get worse. We cannot risk having to go back to Arusha so soon."

I recalled the harrowing drive we had once taken to the hospital there, when one of our staff had suddenly developed appendicitis. Reluctantly I agreed to spend the day in camp.

With one of us out of commission, it was even more vital for the other to continue the work, for our precious seven-week season was running out. So Mary departed for the diggings with Sally and Toots in the Land-Rover, and I settled back to a restless day off.

Some time later—perhaps I dozed off—I heard the Land-Rover coming up fast to camp. I had a momentary vision of Mary stung by one of our hundreds of resident scorpions or bitten by a snake that had slipped past the dogs.

The Land-Rover rattled to a stop, and I heard Mary's voice calling over and over: "I've got him! I've got him! I've got him!"

Still groggy from the headache, I couldn't make her out.

"Got what? Are you hurt?" I asked.

"Him, the man! *Our* man," Mary said. "The one we've been looking for. Come quick. I've found his teeth!"

Magically the headache departed. I somehow fumbled into my work clothes while Mary waited.

As we bounced down the trail in the car, she described the dramatic moment of discovery. She had been searching the slope where I had found the first Oldowan tools in 1931, when suddenly her eye caught a piece of bone lodged in a rock slide. Instantly she recognized it as part of a skull—almost certainly not that of an animal.

Her glance wandered higher, and there in the rock were two immense teeth, side by side. This time there was no question: They were undeniably human. Carefully, she marked the spot with a cairn of stones, rushed to the Land-Rover, and sped back to camp with the news.

The gorge trail ended half a mile from the site, and we left the car at a dead run. Mary led the way to the cairn, and we knelt to examine the treasure.

I saw at once that she was right. The teeth were premolars, and they had belonged to a human. I was sure they were larger than anything similar ever found, nearly twice the width of modern man's.

I turned to look at Mary, and we almost cried with sheer joy, each seized by that terrific emotion that comes rarely in life. After all our hoping and hardship and sacrifice, at last we had found a well-preserved fossil remains of a manlike creature, side by side with tools of the earliest known Stone Age culture.

Somehow we waited until the next day before doing anything further. Des Bartlett, a professional photographer sent by our friend and Nairobi neighbor, film producer Armand Denis, was on his way to Olduvai, and it was essential that we have proper photographs of the teeth just as they had been found.

Then, very gingerly, we began the work of uncovering the find with delicate camel's-hair brushes and dental picks. In the end it took us 19 days.

We soon discovered that the nearly complete skull—minus only the lower jaw—was imbedded in the soft rock, although expansion and contraction of the rock had cracked the fossil into more than 400 fragments. In order not to lose a single precious scrap, we had to remove and sift tons of scree below the find.

Once we got the skull back to camp, we faced the problem of reassembly, a feat somewhat akin to putting together a complex three-dimensional jigsaw puzzle. The task is now virtually complete, save for the missing lower jaw, and already the skull has begun to tell us a fascinating story.

First of all, how do we know we are dealing with a human being? What distinguishes *Zinjanthropus* from *Proconsul africanus*, the small apelike creature of at least 25 million years ago, whose skull was found by Mary and me in Kenya in 1948? Proconsul already foreshadowed man in a number of ways, but also retained some purely monkey characteristics.

What distinguishes *Zinjanthropus* from the ape men—or "near-men," as I prefer to call them—found in the Transvaal in South Africa by anthropologists Robert Broom, Raymond A. Dart, and J. T. Robinson?

To answer these questions, while *Zinjanthropus* has a number of characteristics which distinguish him as a species from the South Africa near-men, he must as well rank as slightly more like man as we know him today than they did. While, at first, we thought of him as a man by definition, although not in his physical appearance, this view no longer can be accepted. It was a view based upon a long-standing definition of man by anthropologists as a maker of tools of a set and regular pattern.

Soon after the discovery of *Zinjanthropus*, we found that another type of early man, also present, might have made the tools. Moreover, as a result of Jane Goodall's study of chimpanzees, we know that apes make simple tools regularly so that the definition of man has had to be revised.

But that is only part of the story. Let us turn to the skull itself and see what it tells us. Oddly enough, the most revealing feature is the one that led to the discovery—the teeth.

Zinjanthropus has the largest molars ever found in a human skull. But what is most important is that only the molars and premolars are extraordinarily large. The incisors and canines—the teeth used for cutting and tearing—are relatively small compared to the huge teeth behind them. Here is the key to our man's way of life, and even to his development as a tool-maker.

When we find such enormous flat-crowned molars, we can be reasonably certain that the owner fed mainly on coarse vegetation. Although broken bones associated with stone tools were found on the living floor where *Zinjanthropus* was discovered, we can no longer be certain that this site was his home, but suspect he was an intruder, and that Homo habilis lived here. Consequently, we can no longer discuss the nature of *Zinjanthropus*' animal food.

So much for *Zinjanthropus'* teeth. What about the structure of his skull? In some respects this new Stone Age skull more closely resembles that of present-day man than it does the skulls of the gorilla or of the South African near-men.

The curvature of the cheek region shows a facial architecture comparable to that of present-day man. It suggests a lower jaw with muscle attachments like those which in humans control movements of the tongue and are linked with speech. I shall indeed be surprised if the lower jaw, when we find it, does not exhibit the form characteristic of speaking man.

Another similarity to present-day man is to be found in the mastoid processes, part of the temporal bones behind the ear holes. In our Olduvai fossil near-man these have a shape and size often seen in present-day man but quite unlike those of the gorilla and near-man.

The base of the skull also shows us that this man held his head erect, possibly even more erect than man's carriage today. To some extent this fact may be linked with a very large and deep lower jaw, which he must have had in harmony with his long face.

He is a very odd-looking creature by our standards with his very long face and no brow. This supports a view that he was a near-man, but he is perhaps a little more like man as we know him than were some of his South African cousins.

Human that he is, *Zinjanthropus* clearly stands a long way from the state of development seen in Homo sapiens of the present day. For example, the portrait shows a very flat cranium, which probably housed a brain little more than half the size of ours. He had a sagittal crest, a bony ridge crowning the skull, that is seen in certain of the lower primates and some near-men. The same crest appears in carnivores like the lion and hyena. In *Zinjanthropus* it must have developed independently and served as an anchor for his powerful jaw muscles.

I have put the age of death of our particular specimen at 18 because his wisdom teeth, the third molars, show no signs of wear, while on the other hand the basioccipital suture—a seam between two bones of the skull—is closed, showing that the individual was more than 16 years old.

There is no riddle about the date of *Zinjanthropus*. The evidence of potassium argon dating, as well as that of glass fission track dating on all the fossil animals found at the same horizon, all point to a date of Lower Pleistocene, around about one and three-quarter million years ago.

In the present state of our knowledge, we can suggest that *Zinjanthropus* represents a near-man who is contemporary with true man Homo habilis at Olduvai. We are no longer certain whether *Zinjanthropus* made the tools found near him, or whether Homo habilis was responsible for them, or whether both hominids were tool-makers. Only future discoveries will finally settle this problem.

As to our new near-man's name, the full title is *Zinjanthropus boisei*,

Zinj being the ancient Arabic word for eastern Africa, where he was discovered, and *anthropus* of course from the Greek term for man. *Boisei* honors Charles Boise of London, who since 1948 has shown steadfast faith in our work by helping to finance the expeditions to Olduvai. (Our work there, ever since 1931, has been aided by a number of institutions and individual sponsors.)

Earlier I mentioned Bed II at Olduvai, and here in time we shall uncover a sequel to the story of *Zinjanthropus*. For lying somewhere, a mere 20 feet higher in the wall of the gorge than our earliest human, must be a more advanced man, whose slightly improved stone tools enabled him to hunt the giant animals with which his predecessor could not cope.

Already we know this later man used the bola, a weapon consisting of triple stones connected by a thong or vine to be hurled at a quarry with the hope of entangling and disabling it. Primitive Eskimos and Patagonians use modified bolas to this day. The size and weight of the stones at Olduvai indicate their owners were extraordinarily strong, even by present standards.

Since the finding of *Zinjanthropus*, a great deal more work has been done at Olduvai. Many important new discoveries have been made, and we hope that much more work will show others as our expeditions continue to search the gorge. We are working with generous grants from the National Geographic Society Research Committee and discoveries are made at very frequent intervals now. We are convinced that still more light on human evolution will be found.

WILLIAM L. STRAUS, JR.

The Great Piltdown Hoax

•

WHEN Drs. J. S. Weiner, K. P. Oakley, and W. E. Le Gros Clark recently announced that careful study had proved the famous Piltdown skull to be compounded of both recent and fossil bones, so that it is in part a deliberate fraud, one of the greatest of all anthropological controversies came to an end. Ever since its discovery, the skull of "Piltdown man"—termed by its enthusiastic supporters the "dawn man" and the "earliest Englishman"—has been a veritable bone of contention. To place this astounding and inexplicable hoax in its proper setting, some account of the facts surrounding the discovery of the skull and of the ensuing controversy seems in order.

Charles Dawson was a lawyer and an amateur antiquarian who lived in Lewes, Sussex. One day, in 1908, while walking along a farm road close to nearby Piltdown Common, he noticed that the road had been repaired with peculiar brown flints unusual to that region. These flints he subsequently learned had come from a gravel pit (that turned out to be of Pleistocene age) in a neighboring farm. Inquiring there for fossils, he enlisted the interest of the workmen, one of whom, some time later, handed Dawson a piece of an unusually thick human parietal bone. Continuing his search of the gravel pit, Dawson found, in the autumn of 1911, another and larger piece of the same skull, belonging to the frontal region. His discoveries aroused the interest of Sir Arthur Smith Woodward, the eminent paleontologist of the British Museum. Together, during the following spring (1912), the two men made a systematic search of the undisturbed gravel pit and the surrounding spoil heaps; their labors resulted in the discovery of additional pieces of bone, comprising—together with the fragments earlier recovered by Dawson—the larger part of a remarkably thick human cranium or brain case and the right half of an apelike mandible or lower jaw with two molar teeth in situ. Continued search of the gravel pit yielded, during the summer of 1913, two human nasal bones and fragments of a turbinate bone (found by Dawson), and an apelike canine tooth (found by the distinguished archeologist, Father Teilhard de Chardin). All these remains constitute the find that is known as Piltdown I.

Dawson died in 1916. Early in 1917, Smith Woodward announced the discovery of two pieces of a second human skull and a molar tooth. These form the so-called Piltdown II skull. The cranial fragments are a piece of thick frontal bone representing an area absent in the first specimen and a part of a somewhat thinner occipital bone that duplicates an area recovered in the first find. According to Smith Woodward's account, these fragments were discovered by Dawson early in 1915 in a field about two miles from the site of the original discovery.

The first description of the Piltdown remains, by Smith Woodward at a meeting of the Geological Society of London on December 18, 1912, evoked a controversy that is probably without equal in the history of paleontological science and which raged, without promise of a satisfactory solution, until the studies of Weiner, Oakley, and Clark abruptly ended it. With the announcement of the discovery, scientists rapidly divided themselves into two main camps representing two distinctly different points of view (with variations that need not be discussed here).

Smith Woodward regarded the cranium and jaw as belonging to one and the same individual, for which he created a new genus, *Eoanthropus*. In this monistic view toward the fragments he found ready and strong support. In addition to the close association within the same gravel pit of cranial fragments and jaw, there was advanced in support of this interpretation the evidence of the molar tooth in the jaw (which were flatly worn down in a manner said to be quite peculiar to man and quite unlike the type of wear ever found in apes) and, later, above all, the evidence of a second, similar individual in the second set of skull fragments and molar teeth (the latter similar to those imbedded in the jaw and worn away in the same un-apelike manner). A few individuals (Dixon, Kleinschmidt, Weinert), moreover, have even thought that proper reconstruction of the jaw would reveal it to be essentially human, rather than similar. Reconstructions of the skull by adherents to the monistic view produced a brain case of relatively small cranial capacity, and certain workers even fancied that they had found evidences of primitive features in the brain from examination of the reconstructed endocranial cast—a notoriously unreliable procedure; but subsequent alterations of reconstruction raised the capacity upward to about 1,400 cc.—close to the approximate average for living men.

A number of scientists, however, refused to accept the cranium and jaw as belonging to one and the same kind of individual. Instead, they regarded the brain case as that of a fossil but modern type of man and the jaw (and canine tooth) as that of a fossil anthropoid ape which had come by chance to be associated in the same deposit. The supporters of the monistic view, however, stressed the improbability of the presence of a hitherto unknown ape in England during the Pleistocene epoch, particularly since no remains of fossil apes had been found in Europe later than the Lower Pliocene. An anatomist, David Waterston, seems to have been the first to have recognized the extreme morphological incongruity between the cranium and the jaw. From the announcement of the discovery he voiced his disbelief in their anatomical association. The following year (1913) he demonstrated that superimposed tracings taken from radiograms of the Piltdown mandible and the mandible of a chimpanzee were "practically identical"; at the same time he noted that the Piltdown molar teeth not only "approach the ape form, but in several respects are identical with them." He concluded that since "the cranial fragments of the Piltdown skull, on the other hand, are in practically all their details essentially human . . . it seems to me to be as inconsequent to refer the mandible and the cranium to the same individual as it would be to articulate a chimpanzee foot with the bones of an essentially human thigh and leg.'

In 1915, Gerrit Miller, then curator of mammals at the United States National Museum, published the results of a more extensive and detailed study of casts of the Piltdown specimens in which he concluded that the jaw is actually that of a fossil chimpanzee. This view gradually gained strong support, e. g., from Boule and Ramström. Miller, furthermore, denied that the manner of wear of the molar teeth was necessarily a peculiarly human one; he stated that it could be duplicated among chimpanzees. That some other workers (Friederichs, Weidenreich) have ascribed the jaw to a fossil ape resembling the orangutan, rather than to a chimpanzee, is unimportant. What is important, in the light of recent events, is that the proponents of the dualistic theory agreed in pronouncing the jaw that of an anthropoid ape, and unrelated to the cranial fragments. Piltdown II remained a problem; but there was some ambiguity about this discovery, which was announced after the death of Dawson "unaccompanied by any direct word from him." Indeed, Hrdlička, who studied the original specimens, felt convinced that the isolated molar tooth of Piltdown II must have come from the original jaw and that there was probably some mistake in its published history.

A third and in a sense neutral point of view held that the whole business was so ambiguous that the Piltdown discovery had best be put on the shelf, so to speak, until further evidence, through new discoveries, might become available. I have not attempted anything resembling a thorough poll of the literature, but I have the distinct impression that this point of view has become increasingly common in recent years, as will be further discussed. Certainly, those best qualified to have an opinion, especially those possessing a sound knowledge of human and primate anatomy, have held largely—with a few notable exceptions—either to a dualistic or to a neutral interpretation of the remains, and hence have rejected the monistic interpretation that led to the reconstruction of a "dawn man." Most assuredly, and contrary to the impression that has been generally spread by the popular press when reporting the hoax, *Eoanthropus* has remained far short of being universally accepted into polite anthropological society.

An important part of the Piltdown controversy related to the geological age of the *Eoanthropus* fossils. As we shall see, it was this aspect of the controversy that eventually proved to be the undoing of the synthetic Sussex "dawn man." Associated with the primate remains were those of various other mammals, including mastodon, elephant, horse, rhinoceros, hippopotamus, deer, and beaver. The Piltdown gravel, being stream-deposited material, could well contain fossils of different ages. The general opinion, however, seems to have been that it was of the Lower Pleistocene (some earlier opinions even allocated it to the Upper Pliocene), based on those of its fossils that could be definitely assigned such a date. The age of the remains of "Piltdown man" thus was generally regarded as Lower Pleistocene, variously estimated to be from 200,000 to 1,000,000 years. To the proponents of the monistic, "dawn-man" theory, this early dating sufficed to explain the apparent morphological incongruity between cranium and lower jaw.

In 1892, Carnot, a French mineralogist, reported that the amount of fluorine in fossil bones increases with their geological age—a report that seems to have received scant attention from paleontologists. Recently, K. P. Oakley, happening to come across Carnot's paper, recognized the possibilities of the fluorine test for establishing the relative ages of bones found within a single deposit. He realized, furthermore, that herein might lie the solution of the vexed Piltdown problem. Consequently, together with C. R. Hoskins, he applied the fluorine test to the *Eoanthropus* and other mammalian remains found at Piltdown. The results led to the conclusion that "all the remains of *Eoanthropus* . . . are contemporaneous"; and that they are, "at the earliest, Middle Pleistocene." However, they were strongly indicated as being of late or Upper Pleistocene age, although "probably at least 50,000 years" old. Their fluorine content was the same as that of the beaver remains but significantly less than that of the geologically older, early Pleistocene mammals of the Piltdown fauna. This seemed to increase the probability that cranium and jaw belonged to one individual. But at the same time, it raised the enigma of the existence in the late Pleistocene of a human-skulled, large-brained individual possessed of ape-like jaws and teeth—which would leave *Eoanthropus* an anomaly among Upper Pleistocene men. To complete the dilemma, if cranium and jaw were attributed to two different animals—one a man, the other an ape—the presence of an anthropoid ape in England near the end of the Pleistocene appeared equally incredible. Thus the abolition of a Lower Pleistocene dating did not solve the Piltdown problem. It merely produced a new problem that was even more disturbing.

As the solution of this dilemma, Dr. J. S. Weiner advanced the proposition to Drs. Oakley and Clark that the lower jaw and canine tooth are actually those of a modern anthropoid ape, deliberately altered so as to resemble fossil specimens. He demonstrated experimentally, moreover, that the teeth of a chimpanzee could be so altered by a combination of artificial abrasion and appropriate staining as to appear astonishingly similar to the molars and canine tooth ascribed to "Piltdown man." This led to a new study of all the *Eoanthropus* material that "demonstrated quite clearly that the mandible and canine are indeed deliberate fakes." It was discovered that the "wear" of the teeth, both molar and canine, had been produced by an artificial planing down, resulting in occlusal surfaces unlike those developed by normal wear. Examination under a microscope revealed fine scratches such as would be caused by an abrasive. X-ray examination of the canine showed that there was no deposit of secondary dentine, as would be expected if the abrasion had been due to natural attrition before the death of the individual.

An improved method of fluorine analysis, of greater accuracy when applied to small samples, had been developed since Oakley and Hoskins made their report in 1950. This was applied to the Piltdown specimens. The results of these new estimations, based mainly on larger samples, are given in the first and second columns of the accompanying table. Little elaboration is necessary. The results clearly indicate that whereas the Piltdown I cranium is probably Upper Pleistocene in age, as claimed by Oakley and Hoskins, the attributed mandible and canine tooth are "quite modern." As for Piltdown II, the frontal fragment appears to be Upper Pleistocene (it probably belonged originally to Piltdown I cranium), but the occipital fragment and the isolated molar tooth are of recent or modern age. The foregoing conclusions are supported by evidence concerning the organic content of the specimens, as determined by analysis of their nitrogen content. This method is not as conclusive as fluorine analysis; but its results, given in the third column of the accompanying table, provide additional support for the conclusions arrived at by the fluorine-estimation method. In general, as would be expected, the nitrogen content decreases with age; the only specimen that falls out of line is the occipital of Piltdown II.

Weiner, Oakley, and Clark also discovered that the mandible and canine tooth of Piltdown I and the occipital bone and molar tooth of Piltdown II had been artificially stained to match the naturally colored Piltdown I cranium and Piltdown II frontal. Whereas these latter cranial bones are all deeply stained, the dark color of the faked pieces is quite superficial. The artificial color is due to chromate and iron. This aspect of the hoax is complicated by the fact that, as recorded by Smith Woodward, "the colour of the pieces which were first discovered was altered a little by Mr. Dawson when he dipped them in a solution of bichromate of potash in the mistaken idea that this would harden them." The details of the staining, which confirm the conclusions arrived at by microscopy, fluorine analysis, and nitrogen estimation, need not be entered into here.

In conclusion, therefore, the *disjecta membra* of the Piltdown "dawn man" may now be allocated as follows: (1) the Piltdown I cranial fragments (to which should probably be added Piltdown II frontal) represent

TABLE 1.—*Fluorine content, ratio of fluorine to phosphorus pentoxide, and nitrogen content of the bones and teeth of the so-called Piltdown I and Piltdown II skulls, compared with those of various Upper Pleistocene and Recent bones and teeth.* (From Weiner, Oakley, and Clark, rearranged.)

	% F	$\frac{\% F \times 100}{\% P_2O_5}$	% N
Upper Pleistocene:			
Bones (local) (minimum F content)	0.1	0.4	
Teeth, dentine (minimum F content)	0.1	0.4	
Bone (London)			0.7
Equine molar, dentine (Piltdown)			1.2
Human molar, dentine (Surrey)			0.3
Recent:			
Neolithic bone (Kent)			1.9
Fresh bone			4.1
Chimpanzee molar, dentine	<0.06	<0.3	3.2
Piltdown I:			
Cranium	0.1	0.8	1.4
Mandible, bone	<0.03	<0.2	3.9
Mandibular molar, dentine	<0.04	<0.2	4.3
Canine	<0.03	<0.2	5.1
Piltdown II:			
Frontal bone	0.1	0.8	1.1
Occipital bone	0.03	0.2	0.6
Isolated molar, dentine	<0.01	<0.1	4.2

a modern type of human brain case that is in no way remarkable save for its unusual thickness and which is, at most, late Pleistocene in age; (2) Piltdown I mandible and canine tooth and Piltdown II molar tooth are those of a modern anthropoid ape (either a chimpanzee or an orangutan) that have been artificially altered in structure and artificially colored so as to resemble the naturally colored cranial pieces—moreover, it is almost certain that the isolated molar of Piltdown II comes from the original mandible, thus confirming Hrdlička's earlier suspicion; and (3) Piltdown II occipital is of recent human origin, with similar counterfeit coloration.

Weiner, Oakley, and Clark conclude that "the distinguished palaeontologists and archaeologists who took part in the excavations at Piltdown were the victims of a most elaborate and carefully prepared hoax" that was "so extraordinarily skilful" and which "appears to have been so entirely unscrupulous and inexplicable, as to find no parallel in the history of palaeontological discovery."

It may be wondered why 40 years elapsed before the hoax was discovered. Two factors enter here: first, there was no reason at all to suspect the perpetration of a fraud, at least, not until fluorine analysis indicated the relative recency of all the specimens, thus making the association of a human cranium and an anthropoid-ape jaw, either anatomically or geologically, hardly credible; and, second, methods for *conclusively* determining whether the specimens were actual fossils or faked ones, short of their wholesale destruction, were developed only in recent years (it will be recalled that even the fluorine-estimation method used by Oakley and Hoskins a few years ago was inadequate for detecting a significant difference between brain case and jaw). It is of interest to note that Dawson, in his original report, stated:

> A small fragment of the skull has been weighed and tested by Mr. S. A. Woodhead, M.Sc., F.I.C., Public Analyst for East Sussex & Hove, and Agricultural Analyst for East Sussex. He reports that the specific gravity of the bone (powdered) is 2.115 (water at 5° C. as standard). No gelatine or organic matter is present. There is a large proportion of phosphates (originally present in the bone) and a considerable proportion of iron. Silica is absent.

This statement obviously refers to the brain case alone; for, in both the title and text of the original report the authors spoke of "skull *and* mandible" (italics mine). One cannot help but wonder what might have come to pass if samples of the jaw and teeth had also been submitted to chemical analysis, even though the present, more refined methods were not then available.

The ready initial acceptance of the Piltdown discovery at its face value, at least by a majority of interested scientists, can probably be attributed to the philosophical climate that invested the problem of human evolution at that time. In September 1912, before the announcement of the discovery of "Piltdown man," the distinguished anatomist Elliot Smith, in an address before the Anthropological Section of the British Association for the Advancement of Science at Dundee, expressed a prevailing point of view when he developed the theory that the brain led the way in the evolution of man and that modification of other parts of the body followed. Thus the stage was set for the ready acceptance of the Piltdown fragments as constituting a single individual, a "dawn man" possessing a human cranium housing a human brain, but with phylogenetically laggard, hence simian, jaws and teeth. To quote the paleontologist Sollas:

> The surprise which was first excited by what appeared to be a monstrous combination disappears on further reflection. Such a combination had, indeed, been long previously anticipated as an almost necessary stage in the course of human development. . . . In *Eoanthropus Dawsoni* we seem to have realised

precisely such a being . . . , one, that is, which had already attained to human intelligence but had not yet wholly lost its ancestral jaws and fighting teeth.

And, as Sir Arthur Keith, perhaps the most vocal champion of *Eoanthropus*, argued in supporting this view:

. . . before the anthropoid characters would disappear from the body of primal man, the brain, the master organ of the human body, must first have come into its human estate. Under its dominion the parts of the body such as the mouth and hands, the particular servants of the brain, became adapted for higher uses. Looking at the problem from this point of view, we cannot reject the Piltdown mandible because as regards the mylo-hyoid ridge it is simian and not human in character.

Recent finds of fossil men and other primates, however, indicate that it is the brain that was the evolutionary laggard in man's phylogeny; indeed, the studies of Tilly Edinger of the phylogeny of the horse brain suggest that this may well be a general rule in mammalian evolution. It was such concepts as this, leading to a change in philosophical climate, that evoked an increasing skepticism toward the validity of the monistic interpretation of the Piltdown fragments and led in turn to what appears to have been the prevailing recent opinion, namely, that the fragments should, as expressed in 1949 by Le Gros Clark, "be laid aside without further comment until more evidence becomes available." This view, enhanced by the redating of the remains by Oakley and Hoskins, provided the proper psychological setting for the coup de grâce delivered by Weiner, Oakley, and Clark.

As the three latter point out, the solution of the Piltdown enigma greatly clarifies the problem of human evolution. For *Eoanthropus*, both morphologically and geologically, just simply did not fit into the picture of human evolution that has gradually been unfolding as the result of paleontological discoveries throughout the world.

The Piltdown story is a significant one in the history of ideas, more particularly as it bears on the concept of the precise course of human evolution. For, if man's biological history be likened to a book, it is seen to be composed of both blank and written pages and, by those who note them carefully, many if not most of the written ones will be seen to be in the nature of palimpsests—pages that have been rewritten after their original writing has been rubbed out. Of this, the Piltdown affair is a striking demonstration. It is a demonstration, furthermore, that the palimpsest nature of the pages of man's history is not always due directly to new fossil discoveries but can also result from changes in the philosophical climate of the science. That this phenomenon is peculiar to anthropology, however, is seriously to be doubted.

RUTH BENEDICT

The Growth of Culture

•

All books with such titles as *Progress*, or *The History of Civilization*, or *The Growth of the United States*, or *Modern Finance*, or *Modern Warfare* are books about some aspect of the growth of culture. When we talk about such subjects, they are parts of this great story, whether we are speaking about ancient Greece or contemporary Iowa. Even when we read about how the Romans destroyed Greece or the Goths destroyed Rome, we are learning about cultural growth. For destruction and growth go on together. As culture grows, it also destroys, and as it destroys, new growths appear.

The history of the human race is a wonderful story of progress. Archaeologists tell us that for thousands of years men made the same flint tools by striking stones together, took shelter in caves, and tied a skin around themselves for warmth. Only a few families could live near one another, for each grown man had to stalk wild animals to get food for his women and children. He could only accept nature's supplies as he found them; he knew no way to increase them. It was thousands of years later that his descendants discovered that plants could be sown and tended and harvested or that animals could be domesticated.

The human race is unique among all animal species because of the progress it has made from that day to this. Man alone has constantly enriched his way of life by invention and by complex learning. The fabulous growth of culture is his great achievement, and no other mammal has made this kind of progress. When we examine the growth of culture in human history, we are examining the basis for man's pre-eminence. Members of the human race have a right to just one great boast: that they have an endless capacity to invent and learn. They can learn not merely as other mammals do, from imitation and from individual experience, but from experience passed down to a present generation from thousands of forebears now dead and gone.

The growth of culture in human history has created for the human race a man-made environment quite unlike the environment nature provided. Man took the wild grasses and developed them into wheat and barley and corn, which were productive enough to sustain his great cities. Even primitive people with no writing and no schools terraced mountains to make rice or corn fields and irrigated them by diverting water in regions where rainfall was not adequate.

Primitive tribes have invented tools and learned how to make pottery and baskets. They have fashioned elaborate traps and fish-weirs to make the food quest easier. They have made musical instruments to please their senses and lavished their craftsmanship on beautiful objects. Everywhere, also, among all races of men, they have expanded their world of known and seen human contacts to include also spirits and gods whom they call upon for help.

Human Beginnings

This long story of man's creativeness in the growth of culture begins far back with the first appearance of the human species. To us today Stone Age man seems culturally poverty-stricken because we have gone so far beyond him. But he had begun the distinctively human process of making inventions and transmitting them by teaching to his descendants. By the middle of the Stone Age, for instance, he had domesticated fire. No one will ever know what happened to make some men or women first put to domestic use that terrifying and destructive force. And what were the circumstances which led them to make water boil over a fire? They could hardly have seen in nature any boiling water. And how did they discover that they could kindle fire at will by rubbing two pieces of wood together? At any rate at least by the middle of the Stone Age in Europe man had not only learned these things; he had made them a part of this transmitted culture. He had learned ways of making and keeping fire and he could use it to warm himself when he was cold, and to cook and preserve his food. It was a complex invention which foreshadowed man's continuing career as a great inventor.

Perhaps even more significant was the gradual invention of language. It involved arranging things in the environment in different categories or classes; it involved creating verbs to show how these things could act and be acted on. Originally it involved something else too: the long, slow development of the muscles used in articulation. Early Stone Age man had less specialized tongue muscles and his speech was certainly hampered by this fact. We cannot know when human speech first became a complex set of symbols; it was certainly a long slow process. Today, however, there is no primitive tribe, no matter how poor in material culture, which has not a complicated language and a vocabulary of words with fine shades of meaning.

Stone Age man was remarkable for his skill in making implements out of flint. By striking brittle stone he could fashion tools and weapons. Stone Age man in different areas and periods made different tools. Their shape was so standardized that experts can name the period and area from which most worked flints in a modern museum come. Although their stability over centuries reminds us of the distinctive nest a robin or a crow inevitably builds this distinctiveness depends on learning, not instinct. The human race, even at that early date, had to transmit even this basic industry by teaching each generation what man could learn by experience. When he had learned, he could transmit the new technique.

Stone Age man had begun the process of creating man-made cultural environments. Because nothing is left of his works except those made of durable materials, we do not know what he had invented in social organization, in rules of marriage, or in religion and folklore. It may have been much; it may have been surprisingly little. We do know that with his handling of fire, language, and flint implements he had adopted unique human methods of invention and learning. From that day to this, man has followed this path.

Obstacles to the Adoption of New Traits

The growth of culture has not been as continuous and as purposeful, however, as we often imagine when we talk of progress. Our ideas of progress are themselves cultural inventions of restless modern man avid for improvements. In the modern world in one generation we adopt and learn to manipulate the automobile or the aeroplane or the telephone or the radio or the techniques of mass factory production. We do not pray: "Oh Lord, keep us as our fathers were." Even in finance or art, we invent freely, with our eyes on the future rather than on the past. We even create new religious cults by the dozen. It is easy, therefore, for us to picture human progress as if man had always reached out for a new idea or a new invention and had adopted it whenever he saw it.

History is full of examples of apparently simple discoveries that were not made even when they would be surpassingly useful in that culture. Necessity is not necessarily the mother of invention. Men in most of Europe and Asia had adopted the wheel during the Bronze Age. It was used for chariots, as a pulley wheel for raising weights, and as a potter's wheel for making clay vessels. But in the two Americas it was not known except as a toy in any pre-Columbian civilization. Even in Peru, where immense temples were built with blocks of stone that weighed up to ten tons, these huge weights were excavated, transported, and placed in buildings without any use of wheels.

The invention of the zero is another seemingly simple discovery which was not made even by classic Greek mathematicians or Roman engineers.

Only by the use of some symbol for nothingness can the symbol 1 be used so that it can have the value either of 1 or 10 or 100 or 1000. It makes it possible to use a small number of symbols to represent such different values as 129 and 921. Without such inventions figures cannot be added or subtracted by writing them one above another, and multiplication and division are even more difficult. The Romans had to try to divide CCCLVIII by XXIV and the difficulty was immense. It was not the Egyptians or the Greeks or the Romans who first invented the zero, but the Maya Indians of Yucatán. It is known that they had a zero sign and positional values of numbers by the time of the birth of Christ. Quite independently the Hindus made these inventions in India some five to seven centuries later. Only gradually was it adopted in medieval Europe, where it was known as Arabic notation because it was introduced there by the Arabs.

Necessity is not only not the inevitable mother of invention; it is not possible to assume that a people will adopt new inventions or accept discoveries others make. The technique of making bronze was established in Europe and Asia a couple of thousand years before iron ores were worked, and even after ways of forging and tempering iron were known, bronze remained for centuries the favorite metal. It was prettier though not nearly so good for tools. Yet iron ores are abundant and not difficult to extract, and tools of iron can be readily made at little outdoor primitive forges, as they are among African tribes today.

Primitive tribes in the modern world often continued to practice some old and back-breaking custom even when they were in contact with some other primitive peoples who had admirably solved that particular technological problem. The Chukchee, a reindeer herding tribe of eastern Siberia, carried on trade with Eskimo tribes who built themselves snow huts. These houses are dome-shaped; blocks of firm snow are cut out with knives and slanted inward until the final block at the top seals the dome. A single man can build one for himself in half-an-hour for a shelter, and large interconnecting ones are built for short-term winter dwellings. They can be heated with blubber lamps and can keep the inmates warm in Arctic winters. The Chukchee, however, stuck to their great skin tents, inside which they set up a smaller skin sleeping tent. Every morning the frost had to be beaten out of the skins, for the moisture from the breath and from perspiration froze, and, if left in, would make the skins crack. This daily beating of the tent covering was exhausting physical labor; in addition, the great bundle was heavy and cumbersome to transport, and erecting the tents on a new site was laborious. But the Chukchee never adopted the snow house of the Eskimo, no matter what difficulties they had with skin tents in the Arctic.

There was another side to this picture. The Chukchee were reindeer herders. They became rich through breeding and rearing these animals and they harnessed them to their sleds. The Eskimos, however, did not adopt reindeer herding. On the American continent, where the barren-ground caribou were available in large numbers and apparently might have been domesticated as the Siberian reindeer were, no domestication at all took place. The trait was not borrowed by American Eskimos or Indians, even though many other cultural inventions diffused across Bering Strait.

Even in modern western civilization, where we pride ourselves on our efficiency, each nation excludes some existing inventions. One might suppose that in great civilizations where counting and measurement are as important as they are in Europe and the United States, all nations would adopt systems with convenient units. The metric system, in which the integer is multiplied first by 10, then by 100 and 1000, can be applied to measurements of volume, of length, and of weight, and can be used to count money. In France the decimal system is used for every kind of counting and measuring. In the United States we use it to reckon money, but not volume, length, or weight. In England it is not even used in reckoning money and no measurements of length or weight or volume are reckoned by metric systems.

The growth of culture, therefore, has not had the kind of history an armchair student would imagine. It cannot be reconstructed logically and by deduction. Sometimes some obvious and simple thing was not discovered or accepted at all even when there was great logical need for it. Sometimes very complicated things were invented in the simplest primitive societies. This is true not merely in technology. It is true in social organization, in legal systems, in religion, and in folk philosophy. To understand the growth of culture in all these aspects, it is necessary to describe more fully how partial all men become to the special man-made environment they have created by their own cultural inventions and arrangements.

The habits of any culture fit the people who learn to use them like well-worn gloves. This fit goes very deep, for their ideas of right and wrong, their selection of human desires and passions, are part and parcel of their whole version of culture. They can react to another people's way of conducting life with a supreme lack of interest or at least of comprehension. Among civilized peoples this often appears in their depreciation of "foreign ways"; it is easy to develop a blind spot where another people's cherished customs are concerned. Among primitive peoples this lack of interest in "progress" has been proverbial. And for good reason. Every primitive tribe has its own elaborate cultural arrangements which ensure its survival, either technologically, or in their forms of social organization, or by ceremonies and offerings to the gods. Even though they may be eager for some things the white man brings—perhaps guns, perhaps beads, or whiskey, or empty tin cans out of which to make a knife— they do not generally look on the white man's culture as a solution of life's problems which is "better" than the one they have. They may be culturally uninterested even in laborsaving devices. Often the value they put on time is extremely low and "wisdom" is far more valued than efficiency. Our cultural system and theirs are oriented around different ideals.

Some primitive cultures have not been able to accommodate themselves to contact with the white man. Their whole way of living, when they were brought into contact with modern civilization, has fallen down like a house of cards. The Indians of the United States have most of them become simply men without a cultural country. They are unable to locate anything in the white man's way of life which is sufficiently congenial to their old culture. When the white man first came, the Plains tribes had a short-lived cultural upsurge when they enthusiastically incorporated the horse into their way of life, and the Northwest Coast Indians had a veritable renaissance of wood carving when they got metal. But closer contact laid bare the great gap between white and Indian values. The Indian cultures could not survive the white man's interference with their tribal war paths and the buffalo herds and salmon fisheries on which they depended. Acquaintance with the strange white customs of working for wages and paying for land and conducting private enterprise broke down their old social arrangements without putting anything intelligible in their place. The white man, for his part, was equally unable to see the cultural values which the Indian tribes cherished and which were being broken down and lost forever. Each side was blind to cultural ideals which to the other were the most real things in the universe.

In all such cases of contact between western civilization and other cultures, the white man is usually sure that he is of superior intelligence because he has the knives, the guns, the cigarettes, the metal skillets that the simpler people do not have. He judges that the others would have these things, along with reading, writing, and arithmetic, if they were not stupid. Actually careful observations and tests have shown that the matter is not so simple. Neither the intelligence nor the senses of primitive people need be inferior even when their manner of life is very simple. In western civilization we are heirs of inventions that have been made all over the face of the earth. All we have and know are items of our social inheritance. We were simply born into it by the accident of our birth. It is highly unlikely that any one of us has invented one single process. Just so, an American Indian was an heir of *his* culture. It was rather more likely than in the complex western civilization that he had individually had an opportunity to make some contribution to tribal ways or that he had had a chance to take leadership in some activities important to his people. His ways of life satisfied him because they solved human problems in ways he had been reared to understand. It had not crossed his mind to want the things the Europeans wanted. He had used his brains on a different set of activities that were more congenial.

No one has ever developed an objective scale of values according to which all different cultural goals may be graded as better or worse. Western civilization, for instance, is organized to extend its power widely over the earth. A valid case, however, can be made for the value of a cultural goal which has no place for conquest or financial domination. Every people value most the drives and emotions to which they are accustomed, and they usually condemn people who lack them. They are right in valuing their own way of life, but their depreciations of other cultures are often based on misunderstandings.

Internal Growth of Culture

Because all peoples defend their own way of life, it is easy to understand that one way in which cultures have grown richer and more complex has been by elaborating and multiplying their own most cherished customs. They carry further and further their favorite customs. Simple trading habits may be worked up into great tribal ceremonies. The potlatches of the Northwest Coast Indians were such ceremonies, in which chiefs tried to defeat other chiefs by giving them so many blankets and other goods that they would be unable to return the interest on them. Such tribes took the main theme of their cultural life from the situation of the creditor and the debtor, and they elaborated their ceremonies around this theme till their potlatches became systems of intricate cultural complexity.

In other tribes the most cherished observances are hospitable entertainments of the gods. In the Southwest pueblo of Zuñi, the spirits are thought to be happiest when they are given the opportunity to come to the world of the living and dance. Therefore men put on spirit masks and impersonate them. These Indians "dance" their corn, too, to make it happy, and put on elaborate welcoming rites for the carcass of a deer after the hunt. They greet and honor even little pine branches they cut for their ceremonies. Since the sun, and spirits of rain, and spirits of animals, and spirits of enemies, and spirits of curing all have to be honored, Zuñi has a staggering mass of ceremonial addressed to this end. Both on the Northwest Coast and in Southwest pueblos, the local process of cultural growth has been, just as in other parts of the world, a kind of industrious weaving of a more and more complex cocoon. But the threads of this cocoon are still old, chosen, and simple habits, even when they are fashioned into such complex observances. They are valued in their congeniality.

Cultures tend to develop in this way, and it is therefore possible to understand the different lines along which, for instance, eastern civilizations like China and India have developed as contrasted to western civilizations. Unless one is to the manner born, the elaborations of another culture often seem superfluous. But the whole history of the growth of culture is full of superfluities to which people of that tribe or nation have displayed deep attachment and loyalty.

Growth of Culture Through Diffusion

Besides this kind of internal elaboration of preferred traits cultures have grown mightily by borrowing techniques and ideas from one another. This borrowing is technically called the diffusion of cultural traits.

Western civilization itself is based on inventions which have been borrowed from every part of the world. Many of them were made by people of simple culture who did not share in western traditions. The alphabet was invented by Semitic peoples in the area north of the Red Sea and carried by Phoenicians to Greece and Rome. Over centuries it spread throughout Europe and into India. Paper—and gunpowder too—are old inventions made in China. The true arch, with its keystone, was a great architectural invention made in Babylonia thirty centuries before Christ; but ancient Greek architecture is not based on it. The great monuments and temples of Peru and Central America were built without any knowledge of it. Gradually, however, the Babylonian invention was adopted in ancient Etruria and in Rome, and became basic in Gothic cathedrals. Modified into a dome, it is used in modern public buildings.

Man has constantly enriched his food supply by introducing grains and fruits which were originally domesticated on the opposite side of the globe. Coffee was brought into cultivation in Abyssinia, but today we associate it particularly with Brazil and Java. Potatoes are roots first tended and harvested by South American Indians, and Bolivian Indians cultivated 240 varieties. But we call our white potatoes "Irish." Bananas come to us today from Central America, but wild varieties were first brought into cultivation in south Asia, and Polynesian peoples had carried them over immense areas of the Pacific before European navigators made their voyages of discovery. The banana in the New World is post-Columbian; it was borrowed from the Old World. Maize, an American Indian crop, is today a staple of many primitive tribes of Africa, and tobacco, also an American Indian crop, has been adopted in all parts of the world.

The diffusion of cultural traits from one people to another has constantly enriched human ways of life. Every little tribe is indebted to its neighbors for various inventions which the latter have borrowed farther afield and which they themselves modified after they copied it and perhaps improved.

Recasting of Borrowed Traits

Whatever traits tribes borrow from one another, they are likely to recast them to make them congenial to their own way of life. Sometimes this recasting has been drastic, sometimes not so drastic. But as a student follows any one cultural trait through tribe after tribe, he finds strange new meanings and uses given to it, or strange new combinations into which it has entered. The wheel, when it was invented in the Old World, spread rapidly in the period around 3000 B.C. into Assyria and Iran and India, and later into Egypt. These were regions of the world where pottery was at that time very important and when the wheel spread into Egypt, it was as a potter's wheel. Not until much later was it used as a chariot wheel. However, when after 200 B.C. the wheel was borrowed by peoples of northern Europe, they used it for wheeled vehicles for nearly a thousand years before they utilized it in pottery making. Tribes and nations could not put the wheel to use in a horse-drawn, wheeled chariot unless they had domesticated animals that could be trained to the harness, and they could not use it for pottery unless they had pottery industry and cared about making it more rapidly. So the wheel became a part of quite different arts of life as it diffused over the world.

New meanings are given to borrowed traits as they pass from area to area. This is just as conspicuous in traits of social organization, political arrangements, and religious practices as it is in traits of material culture. A religious ceremony, for instance, may be shared by all tribes over a great area. All may erect the same kind of house or enclosure on sacred ground, have the same kinds of torture or trance communication or order of march, and use the same insignia for officers and the same type of prayers. All these characteristics may have been spread from tribe to tribe in the area. Nevertheless, in spite of all these diffused traits, a widespread ceremony like the Sun Dance of the Plains Indians has been recast in tribe after tribe. In one tribe the whole ceremony is put on by someone who has had a vision of the Thundergod and desires to honor this spirit which has honored him; in another it is vowed by one man who proposes to avenge the death of a relative on the war path; in another it is a way of giving thanks for escape from danger or disease; in another it is a ceremony for the initiation of priests or shamans. These different meanings of the ceremony, of course, led to changes in the rites themselves and eventually the whole ceremony in one tribe comes to have its own special character, which it does not share with any other tribe.

This recasting of borrowed traits occurs in the same way in social organization and folklore and in any other field of life. A good example is the varied meaning of cannibalistic practices. Cannibalism did not occur in all parts of the world, but where it did it had the most contrasted meanings. It was used in some tribes as a way of ensuring the birth of children; only young children were eaten and only the immediate family participated; afterwards they believed that a child would be born again to the family. In other tribes, the hearts only of brave enemies were eaten; it was done in order to increase the bravery of the eaters. Sometimes cannibalism was a lusty enjoyment of good food; sometimes it was a proof that a man could face anything in the world if he could dare to swallow a portion of human flesh. Each tribe and area had taken this piece of behavior and used it in its own special way.

It is the same with adolescent ceremonies, with kinship systems, and with the institutions of kingship. People borrow, and, when they have adopted the trait, it has already become something else from the thing they borrowed. The process of diffusion has therefore not only allowed people all over the world to share in each others' creations and inventions; it has also increased the rich variety of human cultures.

Evolution

The history of man from the Stone Age to the present is a wonderful story of cultural growth. The social inheritance of man has been enriched by multitudinous inventions and arts. In spite of terrible periods of devastation and destruction, the human race has built for itself a cultural environment which is capable of almost infinite richness.

Although a large part of the history of any given culture is due to accident, an evolutionary process may be traced. The growth of culture has not been haphazard. That is, certain earlier inventions, whether of tools or of institutions or of ideas, have been necessary before other inventions could take place. In primitive tribes courts which administered tribal justice could not evolve until there was some organization of the tribal state. Kingdoms could not arise till certain political inventions had been made which brought many neighboring communities into mutual relations with one another. Standing armies in the service of chiefs required a preceding elaborate division of labor and the existence of centralized power.

Evolution can be well illustrated in two fields, the technological and the political. In technology, modern man has built upon the unplanned discoveries of the human race which began with man-made flint tools and the utilization of fire and later the invention of agriculture and herding. Modern man, however, has not left his inventions to chance. At long last, with the modern growth of scientific knowledge, man has arrived at the point where he consciously invents. That is, he sets up for himself a problem he wants to solve and tries all sorts of experiments and combinations until he solves it. He tests and retests till he is sure his solution works.

We are so used to this kind of problem-solving that it is hard for us to realize that most cultural advances have been chance discoveries rather than conscious invention. These discoveries, made without benefit of a previously imagined goal, thus were, strictly speaking, accidental. Even today primitive tribes are found sometimes in regions that have no agriculture but who have nevertheless dumped their garbage near their homes until seeds have sprouted in the enriched earth close to their houses. They had not planned to fertilize these patches and plant seed within easy reach of their camp fires. When they saw what had happened they did not think about their discovery and go straightway and plant new plots. But they picked the seeds and vegetables which sprouted on their dump heaps, and found them handy. They had accidentally stumbled upon an experience out of which the practice of agriculture could grow—probably the same experience which men stumbled upon in the New Stone Age when the human race first began systematically to exploit the possibilities of purposeful cultivation and planting of the soil.

The great upward curve of progress in technology of which man can rightly boast, therefore, is an evolution from unplanned discovery to planned invention. Man has learned purposefully to set his goal and then to check and recheck the experiments that he sets up to achieve that goal. Methods of curing diseases are a good example of this change. During most of the history of the human race, men accepted their traditional curing practices on faith. To treat eye troubles some peoples chose plants which had an "eye" on their fruit or blossom; "like," they said, "cured like." Some of these plants were actually beneficial, but others, we know by chemical analysis, could even cause blindness. Nevertheless the dangerous plant was used. Some tribes had cure-alls for the most unrelated human ailments. It might be "baking," which meant putting the sick person over a bed of buried hot stones and keeping him warm for days or weeks. This was good for certain aches and pains, but they used it also for broken bones without trying to reset them. Starving or bleeding might be their cure-all, but neither of these were good cures for tuberculosis. Nevertheless they did not experiment and they continued to use their cure-alls. In the practice of medicine we have come a long way.

In man's technological progress, therefore, it required tens of centuries to arrive at the idea of scientific planning and checking. Man made his latest great step forward when he said, "Just what is it I want to do?" and then tested his results to see if he had attained his object. In this way he discovered that planning could unlock the previously unknown.

A second great evolution in human culture is man's increasing ability to live together in large numbers. In early times and among the simpler societies only a few hundred people, or at most, on special occasions, a few thousands could be organized into a community. Man had to make inventions in social organization and in distribution of goods and in the political field before large organized states were possible. As man made more and more of these inventions, he was able to live in larger communities and to achieve law and order over larger and larger areas. Trade and ceremonies brought people together peacefully, and ideas circulated. Men were stimulated to think and build and create.

The growth of greater human communities is, therefore, in spite of all the devastation these large groups have often visited upon one another, one of the major themes of human progress. It has changed the human topography of the modern world. In earlier times small communities of a few hundred souls might be the only "in-group" these people knew; all the rest were "out-groups." An in-group is a group of people with loyalties and rights and obligations which they hold in common. Out-groups are all other communities. The primitive in-group might be an economically self-sufficient community within which each person was necessary to the livelihood and well-being of the tribe. Out-group people were annoyances or out-and-out enemies. Everywhere such tribes had one system of ethics to regulate their dealings with in-group members, and a different and often opposite one for out-group people. Stealing, for instance, was very frequently unknown within the in-group, but it was a virtue if a man stole from an out-group. Generosity was often a prime virtue within the tribe, but it did not extend to out-group people.

The advantages of extending the in-group to include millions of people who can profit by mutual security and mutual trade in material goods and mutual exchange of ideas is too obvious to need comment. Mankind has gone far in this kind of progress. We can project this upward curve into the future and recognize that some day mankind will organize the whole world so that he can reap the maximum benefits of security and commerce and exchange of ideas. We have not done it yet. We keep the old primitive contrast between in-group and out-group ethics in our distinction between killing a man of one's own country—which is murder and a major crime—and killing an enemy in war—which is a duty for which we honor the successful soldier. We keep the primitive contrast, too, in our hair-trigger suspiciousness of other sovereign nations—just as they keep them about us. We set up mechanisms of law and order within each nation, but, just as in primitive times, there are no such lawful mechanisms binding sovereign nations together. There is temporary alliance, but essentially there is still the old anarchy that has been traditional in the relations of out-groups to one another. In this world which has grown so small because of modern technological inventions in commerce and finance and armament and communication and trasportation, it is just as necessary today to organize the world community for the secure enrichment of human life as it was in earlier times to organize a dozen little in-groups scattered a few miles from each other along a river course.

SIR LEONARD WOOLLEY

Digging Up the Past

•

Before I begin to describe the methods of Field Archaeology it might be as well to say something about its aims. Nobody supposes that the digging up of antiquities is in itself a scientific end, and though there is always a thrill attending the discovery of buried treasure the ever-growing interest of the public in archaeological work is by no means limited to its dramatic accidents; behind the mere romance there is something of real and enduring value.

In these days natural science is unfolding before us a panorama which to our great-grandfathers seemed in its beginnings blasphemous: to them it undermined the foundations of belief, to us it establishes thought upon a base broader and more rational. Science reckons time in millions of years and stretches space to infinity; the wider outlook is there, part of our consciousness, and the more it is explored the better can we understand ourselves. Archaeology is doing the same thing in a smaller field: it deals with a period limited to a few thousand years and its subject is not the universe, not even the human race, but modern man. We dig, and say of these pots and pans, these beads and weapons, that they date back to 3000 or 4000 B.C., and the onlooker is tempted to exclaim at their age, and to admire them simply because they are old. Their real interest lies in the fact that they are new. If mere age be the standard, all that we unearth is insignificant compared to the dinosaur's fossil egg, and, for that matter, what is six thousand years in the life of the human race when we have to calculate that in terms of geological periods? The importance of our archaeological material is that it throws light on the history of men very like ourselves, on a civilization which is bound up with that of to-day.

The political thinker of a hundred years ago would cite his parallels and draw his arguments from the Roman or Greek world, finding that cognate with his own, but there he stopped short; Greek civilization presented itself to him as something born full-grown with no history behind it, giving little opportunity for observing development and cause. To-day we can see that modern man did not begin his career in 500 B.C., nor even perhaps in 5000 B.C.; from the flower of Attic culture we can work back and find the roots spreading far afield, and sending up perennial blossoms all differing with the nature of the soil and the tending they have received, but all of one stock, and in the light of such knowledge we can better judge and control the present and the future growth. And this enlightenment is not merely for the specialist, for the research student in history. The opening-up of the world affects us all, becomes part of the general intellectual inheritance, and the justification of archaeology is that it does in the end concern everyone. Its direct appeal is due to the fact that, compared with natural science, it comes with simpler introductions. Its subject is modern man, not a universe which resolves itself more and more into an intellectual abstraction, and its material is the work of man's hands. We see the elaborate drainage-system of Knossos and at once feel at home; the cosmetics found in an ancient grave strike us as pathetically up to date; the surprise which a visitor to a museum expresses at the age of a given object is in exact proportion to his recognition of the object's essential modernity—it is the surprise of one who sees his horizon suddenly opening out; and the advantage of archaeology is that it offers Darien peaks so many and so easy to climb.

I was led to write the above by being told that the first question which the reader would like to have answered might be, "Why does anyone dig?" and that came as a shock, for it had seemed to me so obvious that the purpose of archaeology is to illustrate and to discover the course of human civilization, which is certainly an end worth while. But if the historian uses as his material those relics of the past which the field archaeologist does, as a matter of fact, bring to light, could not the material be produced by casual digging? Is there any justification for a person who claims to be an expert in a specialized branch of science and then does in an elaborate way what the laborer could do much more cheaply?

If that is what the question means—and it could mean a good many things—it betrays complete ignorance of what Field Archaeology is. In its essence Field Archaeology is the application of scientific method to the excavation of ancient objects, and it is based on the theory that the historical value of an object depends not so much on the nature of the object itself as on its associations, which only scientific excavation can detect. The casual digger and the plunderer aim at getting something of artistic or commercial value, and there their interest stops. The archaeologist, being after all human, does enjoy finding rare and beautiful objects, but wants to know all about them, and in any case prefers the acquisition of knowledge to that of things; for him digging consists very largely in observation, recording and interpretation. There is all the difference in the world between the purpose and the methods of the scientific worker and those of the robber; it remains to be seen whether there is a corresponding difference in the value of the work done.

Supposing that a peasant somewhere or other unearths a marble statue or a gold ornament; he sells it, and it passes from hand to hand until from a dealer's shop it makes its way into a museum or a private collection. By this time nobody knows where it was found or how; it has been torn from its context and can be judged only as a thing in itself; its quality as a work of art does not suffer, but how about its historical value? Experts have to guess, from such knowledge as they already possess, to what country and age it belongs, and if they agree the statue or cup is assumed to illustrate further that particular known phase of art; very likely they will not agree, and it becomes merely a bone of contention for the learned, and a source of confusion for the layman. If the object found be, for instance, a clay pot having no claim to artistic merit, then, stripped of any significance it might have possessed as a historical document, it becomes absolutely valueless; if the finding of an important object be incorrectly reported it becomes a positive stumbling-block to science. Some Arabs, digging in the ruins of a Syrian church, discover by chance a silver goblet adorned with figures in relief, amongst them some which can credibly be identified as Christ and his apostles. Through various hands it passes to America. The dealers are ready with the story that it was discovered at Antioch, and "the disciples were called Christians first at Antioch"; and the world is assured that here is the Holy Grail, the actual chalice of the Last Supper, bearing contemporary portraits of the apostles of Christ; and though the goblet was, in fact, found more than a hundred miles away from Antioch, and though, judging from its style, it must have been made at least 300 years after Christ's death, it is hard to dispel an error which has already gained a hearing and has so dramatic a ring. In this particular case the harm done to science was less because the story told was demonstrably false, and the purpose of it was clearly interested; many people might be deceived, but the expert was not obliged to recast his knowledge, gained from innumerable dated objects, of the art of the first four Christian centuries; but where the background of definite knowledge is slight an object robbed of its context may be a snare even to the expert. I recall the case of a bronze figure of a lion purchased in China; presumably it was Chinese, but to a certain scholar it seemed to present analogies with the very few such monuments that we possess of Hittite art; he declared it to be Hittite, and then made it the criterion for judging other works of art whose Hittite origin could not be disputed. Here subjective criticism based on too partial knowledge was to blame, but had anything been recorded as to the conditions of the lion's finding we should have been spared so much confusion in the history of Near Eastern art.

On the other hand an object of no value in itself may become a historical document of the greatest importance just because its associations have been properly observed. The great stone ruins of Zimbabwe, in Rhodesia, had long been a puzzle, and the wildest theories were current about them —they had been built by the Phoenicians, they were the Ophir from which Solomon obtained his gold, they were an outpost of ancient Egypt; and observe that if any one of these theories had been proved correct we should have had to revise very thoroughly our views of ancient history. A worthless scrap of Chinese porcelain found in the foundations of the buildings, but found in the course of a scientific excavation properly controlled, proves that the so-called temple is Medieval in date, and must be native African in authorship. A speculator digging for profit would never have bothered about that little potsherd nor, if he had, would anyone have paid any attention to it, for the very good reason that his method would not have been such that his discoveries could have been accepted as scientific evidence: found as it was, it not only knocked falsehood on the head but opened up a new chapter in African history.

Treasure-hunting is almost as old as Man, scientific archaeology is a modern development, but in its short life of about seventy years it has done marvels. Thanks to excavation, thousands of years of human history are now familiar which a hundred years ago were a total blank, but this is not all, perhaps not even the most important part. The old histories, resting principally on written documents, were largely confined to those events which at every age writers thought most fit to record—wars, political happenings, the chronicles of kings—with such side-lights as could be gleaned from the literature of the time. The digger may produce more written records, but he also brings to light a mass of objects illustrating the arts and handicrafts of the past, the temples in which men worshipped, the houses in which they lived, the setting in which their lives were spent; he supplies the material for a social history of a sort that could never have been undertaken before. Until Schliemann dug at Mycenae, and Sir Arthur Evans in Crete, no one guessed that there had been a Minoan civilization. Not a single written word has been found to tell of it,[1] yet we can trace the rise and fall of the ancient Minoan power, can see again the splendours of the Palace of Minos, and imagine how life was lived alike there and in the

[1] In the destroyed palace at Minos Sir Arthur had discovered several thousand documents inscribed on clay—the writing paper of this civilization. These documents were not completely published until 1952, after Sir Arthur's death. Meanwhile, the American archaeologist, Carl Blegen, had found some hundreds of tablets using the same script in a Mycenaean building in the Peloponnesus, the southern peninsula of the Greek mainland. These tablets had been used until 1200 B.C., although the Knossos script ceased in use in 1400 B.C., when the palace had been sacked. In 1952 a young Englishman, Michael Ventris, using this material, succeeded in proving that Linear B was a form of Greek, thus linking the origins of Greek and Minoan civilization. This discovery was one of the great archaeological feats, rivaling that of Champollion in deciphering the Rosetta Stone of Egypt. These developments took place after Woolley's article was published.—Eds.

crowded houses of the humbler folk. The whole history of Egypt has been recovered by archaeological work, and that in astonishing detail; I suppose we know more about ordinary life in Egypt in the fourteenth century before Christ than we do about that of England in the fourteenth century A.D. To the spade we owe our knowledge of the Sumerians and the Hittites, great empires whose very existence had been forgotten, and in the case of other ancient peoples, the Babylonians and the Assyrians, the dry bones of previously known fact have had life breathed into them by the excavation of buried sites. It is a fine list of achievements, and it might be greatly expanded; all over Europe, in Central America, in China and in Turkestan excavation is supplementing our knowledge, and adding new vistas to our outlook over man's past; and to what is it all due? Not to the mere fact that antique objects have been dug out of the ground, but to their having been dug out scientifically.

But there is another point arising out of that first question "Why does anyone dig?" People sometimes put the accent in a different place, and ask "Why does anyone *dig*? Why do they have to use the spade to achieve these admirable results? How does it come about that things get buried and have to be dug up?"

Clearly, in the case of graves, which yield many of the archaeologist's treasures, the question does not arise, for the things were put underground deliberately and have remained there; but how do houses and cities sink below the earth's surface? They do not: the earth rises above them, and though people do not recognize the fact, it is happening all around them every day. Go no further than London. How many steps does one have to go down to enter the Temple Church? Yet it stood originally at ground level. The mosaic pavements of Roman Londinium lie twenty-five to thirty feet below the streets of the modern City. Wherever a place has been continuously occupied the same thing has happened. In old times municipal scavenging did not amount to much, the street was the natural receptacle for refuse and the street level gradually rose with accumulated filth; if it was re-paved the new cobbles were laid over the old dirt, at a higher level, and you stepped down into the houses on either side. When a house was pulled down and rebuilt the site would be partly filled in, and the new ground floor set at or above street level; the foundations of the older building would remain undisturbed below ground. The process would be repeated time after time so that when foundations are made for the huge buildings of to-day which go down nearly as far into the earth as they rise into the air, the excavating gangs cut through layer after layer of wall stumps and artificial filling of which each represents a stage in the city's growth. In the Near East the rate of rise is faster. The commonest building material is mud brick, and mud-brick walls have to be thick; when they collapse the amount of debris is very great, and fills the rooms to a considerable height, and as you cannot use mud bricks twice over, and the carting away of rubbish is expensive, the simplest course is to level the surface of the ruins and build on the top of them—which has the further advantage that it raises your new mud-brick building out of reach of the damp. In Syria and in Iraq every village stands on a mound of its own making, and the ruins of an ancient city may rise a hundred feet above the plain, the whole of that hundred feet being composed of superimposed remains of houses, each represented by the foot or so of standing wall which the fall of the upper part buried and protected from destruction.

I have not mentioned one way in which buildings may be buried, because it is so lamentably rare; that is by volcanic action. If the field archaeologist had his will, every ancient capital would have been overwhelmed by the ashes of a conveniently adjacent volcano. It is with a green jealousy that the worker on other sites visits Pompeii and sees the marvellous preservation of its buildings, the houses standing up to the second floor, the frescoes on the walls, and all the furniture and household objects still in their places as the owners left them when they fled from the disaster. Failing a volcano, the best thing that can happen to a city, archaeologically speaking, is that it should be sacked and very thoroughly burnt by an enemy. The owners are not in a position to carry anything away and the plunderers are only out after objects intrinsically valuable, the fire will destroy much, but by no means everything, and will bring down on the top of what does remain so much in the way of ashes and broken brickwork that the survivors, if there are any, will not trouble to dig down into the ruins; a burnt site is generally a site undisturbed. It is where cities have decayed slowly that least is to be found in their ruins.

Granted, then, that things do get buried in one or other of these ways, how, it may be asked, do you set to work to find them? Why do you dig just where you do?

Burial does not always mean obliteration, and there are generally some surface signs to guide the digger. In the Near East no one could possibly mistake the great mounds of "tells" which rose above the plain to mark the sites of ancient cities; very often, if the place was an important one, it can be identified from literary sources even before excavation begins; the difficulty is rather which point of attack to choose in so great an area. In Mesopotamia the highest mound will probably conceal the Ziggurat or staged tower attached to the chief temple; sometimes a low-lying patch will betray the position of the temple itself. Herodotus, visiting Egypt in the fifth century B.C., remarked that the temples there always lay in a hollow; the reason was that while the mud-brick houses of the town were shortlived and new buildings constructed over the ruins of the old quickly raised the ground level, the temples, built of stone and kept always in good repair, outlived many generations and remained at the same level throughout; on an Egyptian site, therefore, a square depression ringed about by mounds of crumbling grey brick gives the excavator a very obvious clue. Earthworks are enduring things, and the site, for instance, of a Roman camp in Britain can nearly always be traced by the low grass-clad lines of its ramparts, and the round barrows of the old British dead are still clear to see upon the Downs; but even where there is nothing upstanding, surface indications may not be lacking. In a dry summer the grass withers more quickly where the soil lies thin over the buried tops of stone walls, and I have seen the entire plan of a Roman villa spread out before me where no spade had ever dug; darker lines in a field of growing corn or, in the very early morning, a difference of tone given by the dew on the blades, will show where buildings run underground: nowadays air photographs bring to light masses of evidence invisible to one who stands upon the ground. An air photograph gives us the whole layout of the Roman village of Caistor, so that the excavator can confidently select the particular building he would like to dig, whereas, before, the site of Caistor was unknown; even more remarkable is it that an air photograph discovered Woodhenge, and showed on the plain surface of ploughed fields the concentric rings of dots where thousands of years ago wooden posts had been planted. From the ground such things are often quite invisible, or visible only at some lucky moment. At Wadi Halfa, in the northern Sudan, MacIver and I had dug a temple and part of the Egyptian town, but, search the desert as we might for two months, we had failed to find any trace of the cemetery which must have been attached to the place. One evening we climbed a little hill behind the house to watch the sunset over the Nile; we were grumbling at our ill luck when suddenly MacIver pointed to the plain at our feet; its whole surface was dotted with dark circles which, though we had tramped over it day after day, we had never seen. I ran down the hill and the circles vanished as I came close to them but, guided by MacIver from above, I made little piles of gravel here and there, one in the middle of each ring; and when we started digging there next morning our Arab workmen found under each pile the square, rock-cut shaft of a tomb. The original grave-diggers had heaped the splinters of stone round the mouth of the shaft, and when they filled it up again a certain amount remained over; 4000 years had produced a dead level of stone and gravel where the eye could distinguish no difference of arrangement or texture; but for the space of five minutes in the day the sun's rays, coming at a particular angle, brought out a darker tint in the stone which had been quarried from deeper underground—but, even so, the effect could be seen only from above, and perhaps from a single point.

The archaeologist, in fact, has to keep his eyes open for evidence of all sorts. At Carchemish, in North Syria, my old Greek foreman, Gregori, an experienced digger if ever there was one, and I, completely mystified our Turkish inspector. We told him we were going to excavate a cemetery, and as we had not previously found graves he was duly interested, and asked to be shown the spot. We took him outside the earth ramparts of the old city to a ploughed field by the river bank, lying fallow that year, and, pointing to the fragments of pottery which strewed the ground, explained that these constituted good evidence for the existence of a graveyard. Then Gregori and I, consulting together, started to make piles of stones marking the position of individual graves. This was too much for Fuad Beg, who protested that we were bluffing him; I betted that we should find a grave under every pile and no graves at all except where we had put a mark; he took the bet and lost it, and spent a month wondering why. It was really a simple case of deduction. The river bank was of hard gravel, the made soil overlying it very shallow, and disturbed to the depth of only about three inches by the feeble Arab plough; the field, being fallow, was covered with sparse growth, for the most part shallow-rooted, but with a mixture of sturdier weeds of a sort whose roots go deeply down; if one looked carefully it became manifest that these weeds were sometimes single, but often in clumps of four or five plants, but a clump never measured more than six feet across; at some time or another the gravel subsoil had been broken up, so that the plant roots could penetrate it, and it had been broken up in patches which would be just the right size for graves; the broken pottery on the surface represented either shallow burials or, more probably, offerings placed above the graves at ground level, and every deep-growing weed or group of weeds meant a grave-shaft. The deduction proved correct.

It was deduction of another sort that led to the discovery of Tutankhamen's tomb. The Valley of the Kings at Thebes contained the known graves of all the Pharaohs of the Eighteenth Dynasty except two: clearly it was the burial-ground of the dynasty, therefore *all* the kings of that dynasty ought to have been found there, and since they had not, the missing ones were still to seek, but to seek within the valley confines. For three years the late Lord Carnarvon worked in that part of the valley which was still unexplored, shifting the thousands of tons of limestone chips which filled the bottom of the ravine, scraping the cliff sides in search of a possible doorway, and it was only when the wearisome task was well-nigh done that the astonishing discovery was made: he and Howard Carter owed their success not to a stroke of good luck but to the patient following-out of a logical theory.

While the excavation is actually in progress the archaeologist's attention is necessarily devoted to each individual object in turn; as soon as it is over he has to consider his discoveries collectively, and what were museum exhibits become units in a series out of which history has to be made. We have been dealing with a cemetery which presumably contains a large number of graves; the cemetery must have been in use for a considerable length of time and therefore ought by its contents to illustrate the modifications of culture which took place during that time; very likely there were no dates known at all when the dig began, and the objects from the graves are the only material we possess for the history of a long period; how then is the archaeologist to use his material?

A visitor from Mars seeing a great collection of English domestic objects, costumes, etc., ranging in date from 1650 to 1900, but all mixed up together, could get a general idea of a moderately high level of civilization, but could not picture what the setting of life was like at any particular date; if the things were put in historical order our Martian, assuming that he were reasonably intelligent, could not only visualize each period but could trace the course of invention and evolution throughout three centuries. That is precisely what the archaeologist tries to do.

Where inscribed documents are found the history produced by digging may be extraordinarily detailed. At Meroë, in the Sudan, Dr Reisner excavated a number of pyramid tombs which had escaped the notice of less methodical diggers; inscriptions showed that they were the graves of Ethiopian kings and queens. Now for a short time, about the seventh century B.C., Egypt was ruled by Ethiopian Pharaohs; their names were recorded, but nothing was known as to how this conquering dynasty developed in its original southern home nor what happened to it after it was again driven out of Egypt, nor by what process there evolved out of it the Graecized royal house of Candace which ruled Ethiopia in the days of the deacon St Philip. Dr Reisner was able to put all his tombs in chronological order and to work out the genealogical tree of the entire family; as the result of a single excavation a complete chapter of ancient history could be written for the first time, and the growth of a civilization which at one time dominated Egypt could be traced in detail.

But supposing that there are no written records to define the order of our discoveries, what then? Then the archaeologist is thrown back on his own resources; he has to deduce the order from the facts which he has observed and it is on the fullness and accuracy of his notes that the value of his results will depend. Every point in which one grave differs from another may prove to be evidence for relative dating, and must be brought into the argument; because nothing is known nothing must be neglected.

Where the number of graves is large, and the objects from them are numerous, it will generally be possible to recognize with tolerable certainty an earlier and a later group. In some classes of objects there are sure to be signs of development of technique, of the gradual conventionalizing and degeneration of ornamental motives, of the evolution of vase types, and extreme instances of any such process may be taken as dating evidence for particular graves. Sometimes this modification in the contents of the graves may correspond with their position in the cemetery, and it will be clear that the latter expanded in a regular fashion, either along a line or outwards from a center, and then the plan of the cemetery will become the first basis of classification. Or the evidence may be more direct, as in the great cemetery at Ur, where very often the graves lie one directly below another in a series which may number half a dozen separate burials; obviously the lower grave must in every case be older than the upper, and wherever the series of superimposed graves is fairly long the lowest of all is likely to date fairly early in the period represented by the cemetery as a whole, and the topmost is likely to be reasonably late in the same period. This is the one certain fact on which all future argument must be based.

The archaeologist first analyses in tabular form the sum of his field notes; in parallel columns he will have the number of each grave, its depth, character, direction, and all its contents symbolized by type numbers—then he can proceed to make his comparisons. Taking first the score or so of graves which by their position at the bottom of a series he knows to be relatively early, he compares their contents, and will probably find that they have a good deal in common—that the same types of clay vessels and the same forms of weapons or tools appear in many of them. Then, taking the score or so of late graves, he may find again that there is a certain similarity between them, but that the pottery types of the early graves, and the metal forms, do not re-appear, or re-appear seldom, in the late graves, while the forms which characterize the latter are wanting in the early group. If he can establish that fact he is on fairly sure ground. Assuming that his two groups do represent approximately the beginning and the end of his cemetery period, he will go on to examine in the light of their contents the rest of his graves. Graves which contain only forms regarded as early will be added to the first group; those with some early types mixed up with others about which nothing is as yet known will provisionally be classed together as marking a step forward in time; those in which early types are outnumbered by unknown types will be attributed to the next phase of advance. Perhaps in this way a third of the graves may be placed in what is hoped to be a chronological grouping, and two-thirds will be left over as containing types still indeterminate. These provisional results must be checked. The graves of the group supposed to be the earliest but one, how do they lie in the ground? Does their depth in relation to the other graves justify our theory? Do the other contents, beads, gold ornaments, cylinder seals and so on agree with the evidence of the pottery and the bronze tools? If they do, we can assume that we are on the right track, and then the types of vessels and tools found in them but not in the graves of the earliest group of all can be taken as characteristic of their period, and can be used for classifying other graves in which they occur, but the earliest forms are not found at all. Gradually, fresh groups are formed at the expense of the undefined residue of graves left between the early and late groups, and with fresh evidence arising from each new classification, that undefined residue is in time reduced to nothing, and the whole cemetery is classified in a series of groups of graves which follow a really chronological order and illustrate a rational process of evolution.

Then begins a further correlation. On building sites we have found the remains of houses lying one above the other, each level producing its harvest of broken clay pots, copper utensils and what not. Comparing these with what our cemetery has given us, we may be able to connect various building strata with the sequence-periods of the graves; then for each phase we shall have something of the conditions of living as well as the habits of burial; if, in the building strata, there are ruins of temples, we can add elements of religious ritual and belief.

Just as, in the process of excavation, the archaeologist requires the help of the architect for the reconstruction of his buildings, and of the epigraphist for the reading of his inscriptions, if such be found, so, too, a measure of teamwork is necessary for dealing with the mass of material of all sorts which excavation provides for the reconstruction of social life. The graves, for instance, will have produced a number of human skulls and skeletons; the anthropologist will take charge of them, and from their physical characteristics determine the racial connections of the original people and perhaps trace the advent of new stocks, and the relative dating for that advent may coincide with the appearance of new fashions in weapons or pottery; the evidence of disease, arthritis, abscesses in the teeth and so on, will help to explain life conditions, and the setting of broken bones or marks of trepanning will illustrate the surgical knowledge of the period. Figures in stone or clay, drawings on pots or engraving on metal may give some idea of the looks and dress of the people; remains of cloth—sometimes only the impression preserved on metal of cloth whose substance has perished—will show their skill in weaving, and spindle-whorls, loom-weights, and combs will illustrate its process; the constant recurrence in the graves of a long pin lying near the shoulder and parallel with the bone of the upper arm will prove that the outer garment was an unshaped and unsewn shawl or cloak wrapped round the body under one arm and fastened by the pin over the other, a brooch under the chin will mean a shaped gown open at the neck, the remains of a belt will add to the picture which, with knowledge of materials and some idea of styles, we can begin to form. Preserved in their original order, bracelets, necklaces, and head-dresses reproduce rather than suggest the past. The elaborate head-dresses worn at Ur by the court dames of the period of the royal tombs are by now familiar—in the Sargonid age, about 2600 B.C., they have given place to simple ribbons of gold; had these been purchased in the market they would have told us nothing, but found in position they show how two long plaits with gold coiled about them were brought from behind the ears and fixed one above the other across the forehead.

The geologist will try to trace the sources from which were derived the raw materials, often imported from abroad, of the manufactured goods; foreign connections and trade routes become manifest. Etruscan graves in Italy, Crimean barrows, graves in Syria and Hungary show how the traders in Baltic amber pushed their business into the far South: the tools and weapons of the royal cemetery at Ur are of bronze, containing a certain percentage of nickel, and as the only ore known to contain nickel in that proportion comes from Oman, on the Persian gulf, we can safely assume that it was from Oman that the Sumerians of 3500 B.C. derived the metal for their foundries, while the lapis lazuli, which they employed so freely for ornaments, came from the Pamir mountains, NW. of India. Below the deposit of sand left by the Flood we found two beads of amazonite, a green stone for which the nearest known source is in the Nilghiri hills of Central India, or in the mountains beyond lake Baikal—and at once there is called up the astonishing picture of antediluvian man engaged in a commerce which sent its caravans across a thousand miles of mountain and desert from the Mesopotamian valley into the heart of India. Bones found in the middle-heaps of houses, or scattered on their floors, will tell the naturalist what breeds of domestic animals were kept, what wild animals were hunted and eaten; the dried contents of store-jars or pots of offerings will show what grains and what fruits were grown and used for food, while arrows of special types, fish-hooks and net-sinkers, hoes, plough-shares, sickles, and grindstones illustrate the manner in which the hunter and the farmer played their part. If written documents be forthcoming with which the epigraphist can deal, much more may be learnt of social organization and of positive chronology, but even without that, the comparison of the contents of different strata ought to bring out the main vicissitudes of a city's life, as well as the slower processes of development and decay. A single excavation is not likely to yield a complete or a continuous record, but by the time a number of sites have been dug the sum of the results worked out by the field archaeologist and his collaborators will be a genuine addition to history. To-day we can read, as our grandfathers could not, the story, vivid and circumstantial, of civilizations newly unearthed and of epochs in man's experience which until recently were literally "dark ages"; and realizing that of all this we have perhaps no contemporary written evidence, or virtually none, some may have been inclined to doubt its value, mistrusting the imagination which seems to base so much on a few potsherds. There must be imagination if life is to be breathed into the dry bones of a dead civilization, but imagination has not been allowed to run riot; the value of the "few potsherds" as documents for the building-up of history depends, as I have tried to show, on the scientific methods which the archaeologist employs in his work; accurate observation and faithful record are preliminary to any reconstruction.

The prime duty of the field archaeologist is to collect and set in order material with not all of which he can himself deal at first hand. In no case will the last word be with him; and just because that is so his publication of the material must be minutely detailed, so that from it others may draw not only corroboration of his views but fresh conclusions and more light. But no record can ever be exhaustive. As his work in the field goes on, the excavator is constantly subject to impressions too subjective and too intangible to be communicated, and out of these, by no exact logical process, there arise theories which he can state, can perhaps support, but cannot prove: their truth will depend ultimately on his own caliber, but, in any case, they have their value as summing up experiences which no student of his objects and his notes can ever share. It is true that he may not possess any literary gifts, and that, therefore, the formal presentation of results to the public may be better made by others; but it is the field archaeologist who, directly or indirectly, has opened up for the general reader new chapters in the history of civilized man; and by recovering from the earth such documented relics of the past as strike the imagination through the eye, he makes real and modern what otherwise might seem a far-off tale.

B. *The Human Machine*

•

The first truly scientific steps in our knowledge of the human body were taken in the sixteenth and seventeenth centuries, when Andreas Vesalius published his monumental Structure of the Human Body, *and William Harvey first demonstrated the circulation of the blood. The story of those great days, and of the fundamental discoveries on which rests the edifice of modern medicine, are told in* Vesalius and Harvey, *by one of the foremost modern teachers of physiology, Sir Michael Foster. The human attributes of the choleric little man who explained how the blood moved are illuminated in* The Only Contemporary Character Sketch of William Harvey *by John Aubrey.*

Since the days of Vesalius and Harvey an enormous body of knowledge, dealing in meticulous and microscopic detail with the functioning of the body's organs, has been amassed. Some of the greatest men in the history of science have contributed to this knowledge. Among them is Claude Bernard, the young provincial who came to Paris with an unacted play under his arm. It was fortunate for science that, in order to make a living, he abandoned the theater for medicine. A contemporary of Pasteur, he made contributions of comparable importance. In Examples of Experimental Physiological Investigation *Bernard describes his methods and some of his outstanding discoveries.*

One example of the exactness of our present understanding of bodily functions is contained in Margaret Shea Gilbert's Biography of the Unborn, *which describes the growth of the fetus from the entry of the male sperm into the female egg through the first nine months of our lives as individuals. From the beginning of human history the subject has been enveloped in mystery and fascination.*

Man, then, is a product of the basic processes of biology, of the laws of evolution, of the interaction of molecules according to chemical and physi-

cal law. This is the end product that Julian Huxley considers in Variations on a Theme by Darwin. *Although man is the most complex of living mechanisms, he must like all the others be studied as a product of his environment.*

SIR MICHAEL FOSTER

Vesalius and Harvey: The Founding of Modern Anatomy and Physiology

•

THE WHOLE STORY of the rise and growth of the art of healing is too vast to be gathered into one set of lectures, too varied to be treated of by one man alone. I will ask you to let me start with the middle of the sixteenth century, and indeed with the particular year 1543.

In this year 1543 the printing-press of J. Oporinus (or Herbst) in Basel gave to the world in a folio volume the *Fabrica Humani Corporis,* the *Structure of the Human Body,* by Andreas Vesalius. This marked an epoch in the history of Anatomy, and so of Physiology and of Medicine. Who was Andreas Vesalius, and why did his book mark an epoch?

Let me briefly answer the latter question first. In the times of the Greeks mankind had made a fair start in the quest of natural knowledge, both of things not alive and of things living; the search had been carried on into the second century of the Christian Era when Galen expounded the structure and the use of the parts of the body of man. As Galen passed away inquiry, that is to say inquiry into natural knowledge, stood still. For a thousand years or more the great Christian Church was fulfilling its high mission by the aid of authority; but authority, as with the growth of the Church it became more and more potent as an instrument of good, became at the same time more and more potent as a steriliser of original research in natural knowledge.

As spiritual truths were learned by the study of the revealed word, so anatomical and medical truths were to be sought for, not by looking directly into the body of man, not by observing and thinking over the phenomena of disease, but by studying what had been revealed in the writings of Hippocrates and Galen. As the Holy Scriptures were the Bible for all men, so the works of the Greek and Latin writers became the bible for the anatomist and the doctor. Truth and science came to mean simply that which was written, and inquiry became mere interpretation.

The "new birth" of the fifteenth and sixteenth centuries was in essence a revolt against authority as the guide in knowledge; and the work of Andreas Vesalius of which I am speaking marks an epoch, since by it the idol of authority in anatomical science was shattered to pieces never to be put together again. Vesalius described the structure of the human body such as he found it to be by actual examination, by appealing to dissection, by looking at things as they are. He dared not only to show how often Galen was wrong, but to insist that when Galen was right he was to be followed, not because he had said it, but because what he said was in accordance with what anyone who took the pains to inquire could assure himself to be the real state of things.

Who then was this Andreas Versalius?

He was born at Brussels at midnight as the last day of 1514 was passing into the first of 1515. His family, which had dwelt for several generations at Nymwegen and which originally bore the name of Witing, had produced many doctors and learned men, and his father was apothecary to Charles V. His mother, to judge by her maiden name, Isabella Crabbe, was probably of English extraction.

The young Vesalius (or Wesalius, for so it was sometimes spelt) was sent to school at Louvain and afterwards entered the University there, which then as later was of great renown. Though he diligently pursued the ordinary classical and rhetorical studies of the place, the bent of his mind early showed itself; while yet a boy he began to dissect such animals as he could lay his hands on. Such a boy could not do otherwise than study medicine, and in 1533, a lad of seventeen or eighteen, he went to Paris to sit at the feet of Sylvius, then rising into fame.

The ardent young Belgian was however no docile hearer, receiving open-mouthed whatever fell from the master. Sylvius' teaching was in the main the reading in public of Galen. From time to time however the body of a dog or at rarer intervals the corpse of some patient was brought into the lecture room, and barber servants dissected in a rough, clumsy way and exposed to the view of the student the structures which the learned doctor, who himself disdained such menial, loathsome work, bid them show. This did not satisfy Vesalius. At the third dissection at which he was present he, already well versed in the anatomy of the dog, irritated beyond control at the rude handling of the ignorant barbers, pushing them on one side, completed the dissection in the way he knew it ought to be done.

"My study of anatomy," says he, "would never have succeeded had I when working at medicine at Paris been willing that the viscera should be merely shewn to me and to my fellow-students at one or another public dissection by wholly unskilled barbers, and that in the most superficial way. I had to put my own hand to the business."

Besides listening to Sylvius, he was a pupil of Johannes Guinterius (Günther), a Swiss from Andernach, who also was teaching anatomy and surgery at Paris at the time, and with whom his relations seem to have been closer than with Sylvius.

Neither Sylvius, however, nor Guinterius, nor any one at the time was able to supply Vesalius with that for which he was obviously longing, the opportunity of dissecting thoroughly the human body. Complete dissection was then well-nigh impossible, the most that could be gained was the hurried examination of some parts of the body of a patient who had succumbed to disease. One part of the human body, the foundation of all other parts, the skeleton, could however be freely used for study. In those rude times burial was rough and incomplete, and in the cemeteries bones lay scattered about uncovered. In the burial-ground attached to the Church of the Innocents at Paris Vesalius spent many hours, studying the bones; and he also tells us how in another burial-ground, on what is now "Les Buttes Chaumont," he and a fellow-student nearly left their own bones, being on one occasion attacked and in great risk of being devoured by savage, hungry dogs who too had come there in search of bones. By such a rough, perilous study Vesalius laid the foundation of his great work, a full and exact knowledge of the human skeleton. He tells us how he and a fellow-student were wont to try their knowledge by a test which has been often used since, the recognition of the individual bones by touch alone, with the eyes shut.

After three years the wars drove him back from Paris to Louvain, where he continued to pursue his anatomical studies with unflagging zeal. Here as at Paris he was driven to use strange means to gain the material for his studies. Walking one day with a friend in the outskirts of the city and coming to the public gibbet, where "to the great convenience of the studious, the bodies of those condemned to death were exposed to public view," they came upon a corpse "which had proved such a sweet morsel to the birds that they had most thoroughly cleaned it, leaving only the bones and ligaments." With his friend's help he climbed up the gallows and attempted to carry off the skeleton, but in the hurry of such a theft in open daylight he only succeeded in getting part of it; accordingly that evening he got himself shut out of the city gates, secured in the quiet of night the rest of the skeleton, and returning home by a roundabout way and re-entering the city by a different gate, safely carried it in.

In 1537, after a year's stay at Louvain where, in the February of that year, he put forth his first juvenile effort, a translation of the ninth book of Rhazes, he migrated to Venice, the enlightened if despotic government of which was in all possible ways fostering the arts and sciences, and striving to develop in the dependent city of Padua a University which should worthily push on the new learning. It may be worth while to note, as an instance of how in the web of man's history threads of unlike kind are made to cross, that among the monks who had charge of the Hospital at Venice, at which Vesalius pursued his medical studies, was one who bore the name of Ignatius Loyola.

The brilliant talents of the young Belgian at once attracted the notice of the far-sighted rulers of Venice. He was in December of that same year, 1537, made Doctor of Medicine in their University of Padua, was immediately entrusted with the duty of conducting public dissections, and either then or very shortly afterwards, though he was but a lad of some one or two and twenty summers, was placed in a Chair of Surgery with care of Anatomy.

He at once began to teach anatomy in his own new way. Not to unskilled ignorant barbers would he entrust the task of laying bare before the students the secrets of the human frame; his own hand, and his own hand alone, was cunning enough to track out the pattern of structures which day by day were becoming more and more clear to him. Following venerated customs he began his academic labours by "reading" Galen, as others had done before him, using his dissections to illustrate what Galen had said. But time after time the body on the table said plainly something different from that which Galen had written.

He tried to do what others had done before him, he tried to believe Galen rather than his own eyes, but his eyes were too strong for him; and in the end he cast Galen and his writings to the winds and taught only what he himself had seen and what he could make his students see too.

Thus he brought into anatomy the new spirit of the time, and the men of the time, the young men of the time answered to the new voice. Students flocked to his lectures, his hearers amounted it is said to some five hundred, and an enlightened Senate recognized his worth by repeatedly raising his emoluments.

Such a mode of teaching laid a strain on the getting of the material for teaching. Vesalius was unwearied in his search for subjects to dissect. He begged all the doctors to allow him to examine the bodies of their fatal cases. He ingratiated himself with the judges, so that when a criminal was condemned to death they gave directions that the sentence should be carried out at such a time, and the execution should be conducted now in this manner, now in that as might best meet the needs of Vesalius' public dissections. Nor did he shrink apparently from robbing the grave, for he relates how, learning of the death and hurried burial of the concubine of a monk, he got possession of the body, and proceeded at once to remove the whole of the skin in order that the peccant holy man, who had got wind of the matter, might be unable to recognize his lost love. And he made dissections in Bologna as well as Padua.

Five years he thus spent in untiring labours at Padua. Five years he wrought, not weaving a web of fancied thought, but patiently disentangling the pattern of the texture of the human body, trusting to the words of no master, admitting nothing but that which he himself had seen; and at the end of the five years, in 1542, while he was as yet not 28 years of age, he was able to write the dedication to Charles V of a folio work, entitled the *Structure of the Human Body,* adorned with many plates and woodcuts, which appeared at Basel in the following year, 1543. He had in 1538 published, under the sanction of the Senate of Venice, *Anatomical Tables*, and in the same or succeeding year had brought forth an edition of Guinterius, a treatise on blood-letting, and an edition of Galen. There is a legend that the pictures in the great work were by the hand of Titian, but there seems no doubt that they, like the Tables, were done by one John Stephen Calcar, a countryman of Vesalius.

This book is the beginning not only of modern anatomy but of modern physiology.

We cannot, it is true, point to any great physiological discovery as Vesalius' own special handiwork, but in a sense he was the author of discoveries which were made after him. He set before himself a great task, that of placing the study of human anatomy on a sound basis, on the basis of direct, patient, exact observation. And he accomplished it. Galen had attempted the same thing before him; but the times were not then ripe for such a step. Authority laid its heavy hand on inquiry, and Galen's teaching instead of being an example and an encouragement for further research, was, as we have said, made into a bible, and interpretation was substituted for investigation. Vesalius, inspired by the spirit of the new learning, did his work in such a way as to impress upon his age the value not only of the results at which he arrived, but also and even more so, of the method by which he had gained them. He taught in such a way that his disciples, even when they thought him greater than Galen, never made a second Galen of him; they recognized that they were most truly following his teaching as a whole when they appealed to observation to show that in this or that particular point his teaching was wrong. After him backsliding became impossible; from the date of the issue of his work onward, anatomy pursued an unbroken, straightforward course, being made successively fuller and truer by the labours of those who came after.

Vesalius' great work is a work of anatomy, not of physiology. Though to almost every description of structure there are added observations on the use and functions of the structures described, and though at the end of the work there is a short special chapter on what we now call experimental physiology, the book is in the main a book of anatomy, the physiology is incidental, occasional, and indeed halting. Nor is the reason far to seek. Vesalius had a great and difficult task before him. He had to convince the world that the only true way to study the phenomena of the living body was, not to ask what Galen had said, but to see for oneself with one's own eyes how things really were. And not only was a sound and accurate knowledge of the facts of structure a necessary prelude to any sound conclusions concerning function, but also the former was the only safe vantage ground from which to fight against error. When he asserted that such a structure was not as Galen had described it but different, he could appeal to the direct visible proof laid bare by the scalpel. Even then he found it difficult to convince his hearers, so ready were men still to trust Galen rather than their own eyes. Much harder was the task when, in dealing with function, he had to leave the solid ground of visible fact, and to have recourse to arguments and reasoning.

Obviously his vigorous and active young mind was starting many inquiries of a purely physiological kind, and he was aware that much of the physiology which he had put into his book would not stand the test of future research. He knew more particularly that the chapter in that book in which he treated of the use of the heart and its parts was as he says "full of paradoxes." But he was no less aware that his bold attempt to expound the plain visible facts of anatomy, such as they appeared to one who had torn from his eyes the bandages of authority, was of itself enough to raise a storm of opposition; he feared to jeopardize his success in that great effort by taking upon himself further burdens.

Experience showed that in this he was right. Even while he was writing his book, timorous friends urged him not to publish it; its appearance they said would destroy his prospects in life. And in one sense it did. Towards the end of 1542 after the completion of his great task, although in August of that year he had been reappointed to the Chair of Surgery and Anatomy for three years, he, with the sanction of the Senate, left Padua for a while, his pupil Realdus Columbus being appointed his deputy. He made a short stay at Venice; he visited Basel either once or twice, chiefly it would seem to confer with his printers; but while in that city he prepared with his own hands from the body of an executed criminal a complete skeleton which is still religiously preserved there. He also probably made a hurried journey to the Netherlands. During his absence from Padua, after the appearance of his book the storm broke out. The great Sylvius and others thundered against him, reviling him in a free flow of adjectives. Coming back to Padua, after about a year's absence, he found opposition to his new views strong even there. The spirit shewn entered like iron into his soul. If the work on which he had laboured so long and which he felt to be so full of promise met with such a reception, why should he continue to labour? Why should he go on casting his pearls before swine? He had by him manuscripts of various kinds, the embodiment of observations and thoughts not included in the *Fabrica*. What they were we can only guess; what the world lost in their loss we shall never know. In a fit of passion he burnt them all, and the Emperor Charles V, offering him the post of Court Physician, he shook from his feet in 1544 the dust of the city in whose University he had done so much, and still a youth who had not yet attained the thirties, ended a career of science so gloriously begun.

Ended a career; for though in the years which followed he from time to time produced something, and in 1555 brought out a new edition of his *Fabrica*, differing chiefly from the first one, so far as the circulation of the blood is concerned, in its bolder enunciation of his doubts about the Galenic doctrines touching the heart, he made no further solid addition to the advancement of knowledge. Henceforward his life was that of a Court Physician much sought after and much esteemed, a life lucrative and honourable and in many ways useful, but not a life conducive to original inquiry and thought. The change was a great and a strange one. At Padua he had lived amid dissections; not content with the public dissections in the theatre, he took parts at least of corpses to his own lodgings and continued his labours there. No wonder that he makes in his *Fabrica* some biting remarks to the effect that he who espouses science must not marry a wife, he cannot be true to both. A year after his arrival at the Court he sealed his divorce from science by marrying a wife; no more dissections at home, no more dissections indeed at all, at most some few post-mortem examinations of patients whose lives his skill had failed to save.

When in 1556 Charles withdrew from the world and took refuge in the Cloister, Vesalius transferred to the son Philip II the services which he had paid to the father, and in 1559 returned with him to Spain.

Spain, as it then was, could be no home for a man of science. The hand of the Church was heavy on the land; the dagger of the Inquisition was stabbing at all mental life, and its torch was a sterilizing flame sweeping over all intellectual activity.

We cannot wonder that amid such surroundings the feelings that the past years had been years of a wasted life grew strong upon him, and that wistful memories of the earlier happy times gathered head. He was still in the prime of life, a man of some forty-five summers; many years of intellectual vigour were perhaps still before him. Was he to spend all these in marking time to the music of an Imperial Court?

Just at this time, in 1561, there came into his hands the anatomical observations of Falloppius (Gabrielo Falloppio), a man of whom I shall presently have to speak, who in 1551 had after a brief interval succeeded Vesalius in the Chair of Padua. This book came to the wearied and despondent Vesalius, banished to the intellectual desert of Madrid, as a living voice from a bright world outside. Putting everything else on one side, he gave himself, as he says, "wholly up to the instant greedy reading of the pages" which brought vividly back to him the delights of his youth. Calling back from the past the memory of things observed long ago, for new observations, as we have seen, were out of his power, he put together bit by bit some notes criticizing Falloppius' work, put them together hurriedly and rapidly, in order that Tiepolo, the Venetian ambassador, then at Madrid but about to return to Venice, might carry the manuscript with him. In that *Examen,* as he calls it, Vesalius says how the reading of Falloppius' notes had raised in him "a glad and joyful memory of that most delightful life which, teaching anatomy, I passed in Italy, the true nurse of intellects." He looks forward, he says, "to see the ornaments of our science continue to bud forth in the school from which I was while yet a youngster dragged away to the dull routine of medical practice and to the worries of continual journeys. I look forward to the accomplishment of that great work for which, to the best of my powers so far as my youth and my then judgment allowed, I laid foundations, such that I need not be ashamed of them."

And even more, he was nursing the idea that his present barren life might be exchanged for a more fruitful one. "I still," says he, "live in hope that at some time or other, by some good fortune I may once more be able to study that true bible, as we count it, of the human body and of the nature of man."

But it was not to be. In 1563 he suddenly determined to make a pilgrimage to Jerusalem. There are various legends as to the reasons which led him to this step. It is said that in making what was supposed to be a post-mortem examination on a noble man, or according to others a woman suffering from some obscure disease, it turned out that the body was still living, and that the Church insisted upon the pilgrimage as an expiation for an act deemed to be a sacrilege. The truer account is probably that told by the botanist Clusius, that Vesalius, ill in body, and we may add even more sick at heart, wearied of the Court, and harassed by the Church, seized an opportunity, and made the proposed pilgrimage an excuse for bringing to an end his then mode of life.

On his way to Jerusalem he stopped at Venice and renewed his intercourse with scientific friends. He there learnt that the manuscript on Falloppius had never reached that anatomist, who had somewhat suddenly died in 1562, but was still in Tiepolo's hands. His friends at once obtained it from Tiepolo, and it saw the light in the following May.

The Senate at Venice were just then at a loss for a fit successor to Falloppius, and it is possible that Vesalius during his stay in the city made known his willingness to desert the Court and to return to academic life; for it is said, though documentary evidence is lacking, that during his eastern journey he received an invitation to occupy his old Chair. Alas, on his way back in 1564 he was taken ill, or possibly a latent malady openly developed itself, he was put ashore on the island of Zante, and there he passed away.

The influence of Vesalius on the history of science may be regarded on the one hand in its general, on the other in its more special aspect.

Taking the general aspect first we may say that he founded modern anatomy. He insisted upon, and through his early unwearied labours by his conspicuous example he ensured the success of the new method of inquiry, the method of observation as against interpretation; he overthrew authority and raised up experience, he put the book of nature, the true book, in place of the book of Galen, and thus made free and open the paths of inquiry. Others before him, as we have said, Mundinus to wit and Carpi, had made like efforts, but theirs were partial and unsuccessful; Vesalius' efforts were great, complete, and successful. Upon the publication of the *Fabrica,* the pall of "authority" was once and for ever removed. Vesalius' results were impugned, and indeed were corrected by his compeers and his followers; but they were impugned and corrected by the method which he had introduced. Inquirers asserted that in this or that point Galen was right and Vesalius was wrong, but they no longer appealed to the authority of Galen as deciding the question, they appealed now to the actual things as the judge between the two, as the judge of Galen as of others. And even those who were Vesalius' most devoted disciples never made of him a second Galen; they never appealed to him as an authority, they were content to show on the actual body that what he had said was right.

Under a more special aspect he may be regarded as the founder of physiology as well as of anatomy in as much as he was the distinct forerunner of Harvey. For Harvey's great exposition of the circulation of the blood did, as we shall see, for physiology what Vesalius' *Fabrica* did for anatomy; it first rendered true progress possible. And Harvey's great work was the direct outcome of Vesalius' teaching.

When in 1542 after the completion of his great work Vesalius had leave to absent himself from Padua a young man, Matheus Realdus Columbus, a native of Cremona, was appointed as his deputy, and when in 1544 Vesalius finally left Padua, the Senate of Venice entrusted for two years the duty of reading the lectures on Surgery and Anatomy to the same Columbus. But Columbus did not remain Vesalius' successor even for the two years; in the next year, 1545, Cosimo de Medici appointed him as the first Professor of Anatomy in the newly renovated University of Pisa; and Vesalius' Chair was not adequately filled until 1551, when Gabrielus Falloppius was placed in it.

Falloppius, born in Modena in 1523, a favourite and a devoted pupil of Vesalius, an accomplished and travelled scholar, a careful and exact observer and describer, a faithful, modest, quiet man, has left his name in anatomy in the terms Falloppian canal and Falloppian tubes. We owe to him many valuable observations on the skeleton, especially on the skull, on the tympanum, on the muscles, and on the generative organs. But he made no large contribution to knowledge such as distinctly influenced the progress of physiology; and he left no mark on the doctrines of the circulation. I have already spoken of his anatomical observations as stirring up Vesalius in his later years to revived anatomical longings; in these

Falloppius says that if he had been able to advance any new truth, that was largely due to Vesalius "who so showed me the true path of inquiry that I was able to walk along it still farther than had been done before."

Born, in 1537, of humble parents, in the little Tuscan town or rather village bearing that name, Hieronymus Fabricius[1] studied under Falloppius at Padua, and, on the death of his master, in 1565, succeeded him in the Chair of Anatomy, holding it for 40 years, until 1619, when he died at the ripe old age of 82.

A distinguished surgeon and a learned anatomist, well acquainted with the anatomy not only of man but of other vertebrates, he was the author of many treatises, most of which had distinct physiological bearings and which contained many contributions to the advancement of knowledge. He was the first after Aristotle to describe the formation of the chick in the egg; he wrote well on locomotion, on the eye, on the ear, on the skin, on the larynx and on speech; but the one work which concerns the subject which we have in hand is that on the valves of the veins, the book *De venarum ostiolis*, "the little doors of the veins," which saw the light in 1574.

Johannus Baptista Cannanus, Professor at Ferrara, is said to have observed the valves long before, namely in 1547, and indeed to have told Vesalius of his observation; and even before that, these structures it is said were noticed by Sylvius. But they were not really laid hold of until Fabricius published his book. In that work he most carefully and accurately described their structure, position and distribution, illustrating his observations by fairly good figures. He moreover clearly recognized that the valves offered opposition to the flow of blood from the heart towards the periphery, and even gives the now well-known demonstration of their action on the living arm.

He says, *De venarum ostiolis:*

> Little doors of the veins is the name I give to certain very thin little membranes occurring on the inside of the veins, and distributed at intervals over the limbs, placed sometimes one by itself, and sometimes two together. They have their mouths directed towards the root of the veins (*i.e.* the heart), and in the other direction are closed. Viewed from the outside they present an appearance not unlike the swellings which are seen in the branches and stem of a plant. In my opinion they are formed by nature in order that they may to a certain extent delay the blood and so prevent the whole of it flowing at once like a flood either to the feet, or to the hands and fingers, and becoming collected there. For this would give rise to two evils; on the one hand the upper parts of the limbs would suffer from want of nourishment, and on the other the hands and feet would be troubled with a continual swelling. In order therefore that the blood should be everywhere distributed in a certain just measure and admirable proportion for maintaining the nourishment of the several parts, these valves of the veins were formed.

[1] Often spoken of, from the place of his birth, as *ab Aquapendente.*

But he wholly failed to recognize their true function. Still labouring under the influence of the old doctrines and bělieving that the use of the veins was that of carrying crude blood, blood not vivified by the vital spirits, from the heart to the tissues, he thought that he had fully explained the value of the veins, by pointing out that they opposed the flow from the heart to the tissues, not of all blood but only of an excess of blood; their purpose was to prevent the blood as it flowed along the veins from the heart being heaped up too much in one place. But he also thought that they were the means of furnishing temporary local reservoirs of blood; and he likens them to the devices by which in mills and elsewhere water is dammed up. He left for another, for a pupil of his, the opportunity of putting to its right use the discovery which he had made.

I need not take up time by entering largely into the details of the oft-told story of William Harvey's life.

Born at Folkestone, on the south coast of England, in April 1578, just four years after Fabricius had published his treatise on the valves of the veins, admitted to Gonville and Caius College, Cambridge, in 1593, taking his degree in Arts in 1597, he left England the following year to study medicine under the great master at Padua. There he spent the greater part of four years, years very nearly overlapping the period between the writing and the publication of Fabricius's treatise on Respiration, of which I have just spoken as being, in great measure, an exposition of the Galenic doctrine of the circulation. At the end of the period, in 1602, he received at Padua the degree of Doctor of Medicine, and on his return to England in the same year was incorporated into the Doctorate at Cambridge.

Setting up his abode in London, joining the Royal College of Physicians in 1604, and becoming Physician to St. Bartholomew's Hospital in 1609, he ventured in 1615 to develop, in his Lectures on Anatomy at the College of Physicians, the view which he was forming concerning the movements of the heart and of the blood. But his book, his *Exercitatio*, on that subject did not see the light until 1628.

"The little choleric man," as Aubrey calls him, attained fame among his fellows, and favour at Court. As Physician to King Charles I he accompanied that monarch on his unhappy wanderings, and every one knows the tale or legend of how at the battle of Edgehill, taking care of the Princes, he sat, on the outskirts of the fight under a hedge, reading a book. In 1646, after the events at Oxford, he retired into private life, publishing in 1651 his treatise, *De generatione animalium*, in which he followed up some of the researches of his Paduan master, and on June 3, 1667, he ended a life remarkable for its effects rather than for its events.

It is a fashion to speak of Harvey as "the immortal Discoverer of the Circulation"; but the real character of his work is put in a truer light when we say that he was the first to demonstrate the circulation of the blood.

His wonderful book, or rather tract, for it is little more, is one sustained and condensed argument, but an argument founded not on general principles and analogies but on the results of repeated "frequent appeals to vivisection" and ocular inspection. He makes good one position, and having done that advances on to another, and so marches victoriously from position to position until the whole truth is put clearly before the reader, and all that remains is to drive the truth home by further striking illustrations.

His first position is the true nature and purpose of the movements of the heart itself, that is, of the ventricles. When, in the beginning of the inquiry, he "first gave his mind to vivisections" he found the task of understanding the "motions and uses of the heart so truly arduous, so full of difficulties" that he began to think with Fracastorius (a Veronese doctor of the middle of the sixteenth century [1530] and more a poet than a man of science), "that the motion of the heart was only to be comprehended by God." But the patient and prolonged study of many hearts of many animals shewed him that "the motion of the heart consists in a certain universal tension, both of contraction in the line of its fibres, and constriction in every sense, that when the heart contracts it is emptied, that the motion which is in general regarded as the diastole of the heart is in truth its systole," that the active phase of the heart is not that which sucks blood in, but that which drives blood out. Caesalpinus alone of all Harvey's forerunners had in some way or other dimly seen this truth. Harvey saw it clearly and saw it in all its consequences. It is, he says, the pressure of the constriction, of the systole, which squeezes the blood into and along the arteries, it is this transmitted pressure which causes the pulses; the artery swells at this point or that along its course, not in order that it may suck blood into it, but because blood is driven into it, and that by the pressure of the constricting systole of the heart.

With this new light shining in upon him, he was led to a clear conception of the work of the auricles and the ventricles, with their respective valves. He saw how the vena cava, on the one side, and the vein-like artery, the pulmonary veins, on the other side, empty themselves into and fill the ventricles during the diastole, and how the ventricles in turn empty themselves during the systole into the artery-like vein, the pulmonary artery on the one side and the great artery or aorta on the other. And this at once led him to a truer conception of the pulmonary circulation than was ever grasped by Servetus or Columbus. On the old view, only *some* of the blood of the right ventricle passed through the septum into the left ventricle; the rest went back again to the tissues; and it was this "some" only which Servetus and Columbus believed to pass through not the septum but the lungs. Harvey saw that all the reasons for thinking that any of the contents of the ventricle so passed were equally valid for thinking that all passed, and that the latter view alone was consonant with the facts.

This new view, new in reality, though having so much resemblance to old ones that Harvey speaks of it as one "to which some, moved either by the authority of Galen or Columbus or the reasonings of others, will give their adhesion," led him at once to another conception which however "was so new, was of so novel and unheard of a character that in putting it forward he not only feared injury to himself from the envy of a few, but trembled lest he might have mankind at large for his enemies." This new view consisted simply in applying to the greater circulation the same conclusions as those at which he had arrived in regard to the lesser circulation.

It is important to note that to this new view he was guided by distinctly quantitative considerations. He argued in this way. At each beat of the heart a quantity of blood is transferred from the vena cava to the aorta. Even if we take a low estimate (he had made observations with a view to determining the exact amount but he leaves this aside for the present as unessential), say half an ounce, or three drachms, or only one drachm, and multiply this by the number of beats, say in half-an-hour, we shall find that the heart sends through the arteries to the tissues during that period as much blood as is contained in the whole body. It is obvious, therefore, that the blood which the heart sends along the arteries to the tissues cannot be supplied merely by that blood which exists in the veins as the result of the ingesta of food and drink; only a small part can be so accounted for; the greater part of that blood must be blood which has returned from the tissues to the veins; the blood in the tissues passes from the arteries to the veins, in some such way as in the lungs it passes from the veins (through the heart) to the arteries; the blood moves in a circle from the left side of the heart, through the arteries, the tissues and the veins to the right side of the heart, and from thence through the lungs to the left side of the heart.

This is what he says:

> I frequently and seriously bethought me, and long revolved in my mind, what might be the quantity of blood which was transmitted, in how short a time its passage might be effected, and the like; and not finding it possible that this could be supplied by the juices of the ingested aliment without the veins on the one hand becoming drained, and the arteries on the other hand becoming ruptured through the excessive charge of blood, unless the blood should somehow find its way from the arteries into the veins, and so return to the right side of the heart; I began to think whether there might not be *a motion, as it were, in a circle.* Now this I afterwards found to be true; and I finally saw that the blood, forced by the action of the left ventricle into the arteries, was distributed to the body at large, and its several parts, in the same manner as it is sent through the lungs, impelled by the right ventricle into the pulmonary artery, and that it then passed through the veins and along the vena cava, and so round to the left ventricle in the manner already indicated, which motion we may be allowed to call circular.

As the sun of this truly new idea rose in Harvey's mind, this new idea that the blood is thus for ever moving in a circle, the mists and clouds of many of the conceptions of old faded away and the features of the physiological landscape hitherto hidden came into view sharp and clear. This idea once grasped, fact after fact came forward to support and enforce it. It was now clear why the heart was emptied when the vena cava was tied, why it was filled to distension when the aorta was tied. It was now clear why a middling ligature which pressed only or chiefly on the veins made a limb swell turgid with blood, whereas a tight ligature which blocked the arteries made it bloodless and pale. It was now clear why the whole or nearly the whole of the blood of the body could be drained away by an opening made in a single vein. And now for the first time was clear the purpose of those valves in the veins, whose structure and position had been demonstrated doubtless to Harvey, by the very hands of their discoverer, his old master Fabricius, but "who did not rightly understand their use, and concerning which succeeding anatomists have not added anything to our knowledge."

Fabricius, as we have seen, had used the now well-worn experiment of pressing on the cutaneous veins of the bared arm to demonstrate the existence of the valves; but he had used it to demonstrate their existence only. Blinded by the conceptions of his time he could not see that the same experiment gave the lie to his explanation of the purpose of the valves, and demonstrated not only their existence, but also their real use. Harvey, with the light of his new idea, at once grasped the true meaning of the knotty bulgings.

These however were not the only phenomena which now for the first time received a reasonable explanation. Harvey was able to point to many other things, to various details of the structure and working of the heart, to various phenomena of the body at large both in health and in disease as intelligible on his new view, but incomprehensible on any other.

If we trust, as indeed we must do, Harvey's own account of the growth of this new idea in his own mind, we find that he was not led to it in a straight and direct way by Fabricius' discovery of the valves. It was not that the true action of these led to the true view of the motion of the blood, but that the true view of the motion of the blood led to the true understanding of their use. To that true view of the motion of blood he was led by a series of steps, each in turn based on observations made on the heart as seen in the living animal, or as he himself says "repeated vivisections," the great step of all being that one by which he satisfied himself that the quantity of blood driven out from the heart could not be supplied in any other way than by a return of the blood from the arterial endings in the body through the veins. As he himself says:

Since all things, both argument and ocular demonstration, show that the blood passes through the lungs and heart by the action of the ventricles, and is sent for distribution to all parts of the body, where it makes its way into the veins and pores of the flesh, and flows by the veins from the circumference on every side to the centre, from the lesser to the greater veins, and is by them finally discharged into the vena cava and right auricle of the heart, and this in such a quantity or in such a flux and reflux thither by the arteries, hither by the veins, as cannot possibly be supplied by the ingesta, and is much greater than can be required for mere purposes of nutrition; it is absolutely necessary to conclude that the blood in the animal's body is impelled in a circle, and is in a state of ceaseless motion; that this is the act or function which the heart performs by means of its pulse; and that it is the sole and only end of the motion and contraction of the heart.

The new theory of the circulation made for the first time possible true conceptions of the nutrition of the body, it cleared the way for the chemical appreciation of the uses of blood, it afforded a basis which had not existed before for an understanding of how the life of any part, its continued existence and its power to do what it has to do in the body, is carried on by the help of the blood. And in this perhaps, more than its being a true explanation of the special problem of the heart and the blood vessels, lies its vast importance.

JOHN AUBREY

The Only Contemporary Character Sketch of William Harvey

•

HE WAS WONT TO SAY that man was but a great mischievous baboon. He would say, that we Europaeans knew not how to order or governe our woemen, and that the Turkes were the only people used them wisely.

He was far from bigotry.

He had been physitian to the Lord Chancelor Bacon, whom he esteemed much for his witt and style, but would not allow him to be a great philosopher. "He writes philosophy like a Lord Chancelor," said he to me, speaking in derision; "I have cured him."

About 1649 he travelled again into Italy, Dr. George (now Sir George) Ent, then accompanying him.

At Oxford, he grew acquainted with Dr. Charles Scarborough, then a young physitian (since by King Charles II knighted), in whose conversation he much delighted; and whereas before, he marched up and downe with the army, he tooke him to him and made him ly in his chamber, and said to him, "Prithee leave off thy gunning, and stay here; I will bring thee into practice."

For 20 years before he dyed he tooke no manner of care about his worldly concernes, but his brother Eliab, who was a very wise and prudent menager, ordered all not only faithfully, but better then he could have donne himselfe.

He was, as all the rest of the brothers, very cholerique; and in his young days wore a dagger (as the fashion then was, nay I remember my old schoolemaster, old Mr. Latimer, at 70, wore a dudgeon, with a knife, and bodkin, as also my old grandfather Lyte, and alderman Whitson of Bristowe, which I suppose was the common fashion in their young days), but this Dr. would be to apt to draw out his dagger upon every slight occasion.

He was not tall; but of the lowest stature, round faced, olivaster complexion; little eie, round, very black, full of spirit; his haire was black as a raven, but quite white 20 yeares before he dyed.

I have heard him say, that after his booke of the Circulation of the Blood came-out, that he fell mightily in his practize, and that 'twas beleeved by the vulgar that he was crack-brained; and all the physitians were against his opinion, and envyed him; many wrote against him, as Dr. Primige, Paracisanus, etc. (vide Sir George Ent's booke). With much adoe at last, in about 20 or 30 yeares time, it was received in all the Universities in the world; and, as Mr. Hobbes sayes in his book "De Corpore," he is the only man, perhaps, that ever lived to see his owne doctrine established in his life time.

CLAUDE BERNARD

Examples of Experimental Physiological Investigation

•

IN SCIENTIFIC INVESTIGATIONS, various circumstances may serve as starting points for research; I will reduce all these varieties, however, to two chief types:

1. Where the starting point for experimental research is an observation;
2. Where the starting point for experimental research is an hypothesis or a theory.

I. *Where the Starting Point for Experimental Research Is an Observation*

Experimental ideas are often born by chance, with the help of some casual observation. Nothing is more common; and this is really the simplest way of beginning a piece of scientific work. We take a walk, so to speak, in the realm of science, and we pursue what happens to present itself to our eyes. Bacon compares scientific investigation with hunting; the observations that present themselves are the game. Keeping the same simile, we may add that, if the game presents itself when we are looking for it, it may also present itself when we are not looking for it, or when we are looking for game of another kind. I shall cite an example in which these two cases presented themselves in succession. At the same time I shall be careful to analyze every circumstance involved.

FIRST EXAMPLE. One day, rabbits from the market were brought into my laboratory. They were put on the table where they urinated, and I happened to observe that their urine was clear and acid. This fact struck me, because rabbits, which are herbivora, generally have turbid and alkaline urine; while on the other hand carnivora, as we know, have clear and acid urine. This observation of acidity in the rabbits' urine gave me an idea that these animals must be in the nutritional condition of carnivora. I assumed that they had probably not eaten for a long time, and that they had been transformed by fasting, into veritable carnivorous animals, living on their own blood. Nothing was easier than to verify this preconceived idea or hypothesis by experiment. I gave the rabbits grass to eat; and a few hours later, their urine became turbid and alkaline. I then subjected them to fasting and after twenty-four hours or thirty-six hours at most, their urine again became clear and strongly acid; then after eating grass, their urine became alkaline again, etc. I repeated this very simple experiment a great many times, and always with the same result. I then repeated it on a horse, an herbivorous animal which also has turbid and alkaline urine. I found that fasting, as in rabbits, produced prompt acidity of the urine, with such an increase in urea, that it spontaneously crystallizes at times in the cooled urine. As a result of my experiments, I thus reached the general proposition which then was still unknown, to wit, that all fasting animals feed on meat, so that herbivora then have urine like that of carnivora.

We are here dealing with a very simple, particular fact which allows us easily to follow the evolution of experimental reasoning. When we see a phenomenon which we are not in the habit of seeing, we must always ask ourselves what it is connected with, or putting it differently, what is its proximate cause; the answer or the idea, which presents itself to the mind, must then be submitted to experiment. When I saw the rabbits' acid urine, I instinctively asked myself what could be its cause. The experimental idea consisted in the connection, which my mind spontaneously made, between acidity of the rabbits' urine, and the state of fasting which I considered equivalent to a true flesh-eater's diet. The inductive reasoning which I implicitly went through was the following syllogism: the urine of carnivora is acid; now the rabbits before me have acid urine, therefore they are carnivora, i.e., fasting. This remained to be established by experiment.

But to prove that my fasting rabbits were really carnivorous, a counterproof was required. A carnivorous rabbit had to be experimentally produced by feeding it with meat, so as to see if its urine would then be clear, as it was during fasting. So I had rabbits fed on cold boiled beef (which they eat very nicely when they are given nothing else). My expectation was again verified, and, as long as the animal diet was continued, the rabbits kept their clear and acid urine.

To complete my experiment, I made an autopsy on my animals, to see if meat was digested in the same way in rabbits as in carnivora. I found, in fact, all the phenomena of an excellent digestion in their intestinal reactions, and I noted that all the chyliferous vessels were gorged with very abundant white, milky chyle, just as in carnivora. But *à propos* of these autopsies which confirmed my ideas on meat digestion in rabbits, lo and behold a fact presented itself which I had not remotely thought of, but which became, as we shall see, my starting point in a new piece of work.

SECOND EXAMPLE (Sequel to the last). In sacrificing the rabbits which I had fed on the meat, I happened to notice that the white and milky lymphatics were first visible in the small intestine at the lower part of the duodenum, about thirty centimeters below the pylorus. This fact caught my attention because in dogs they are first visible much higher in the duodenum just below the pylorus. On examining more closely, I noted that this peculiarity in rabbits coincided with the position of the pancreatic duct which was inserted very low and near the exact place where the lymphatics began to contain a chyle made white and milky by emulsion of fatty nutritive materials.

Chance observation of this fact evoked the idea which brought to birth the thought in my mind, that pancreatic juice might well cause the emulsion of fatty materials and consequently their absorption by the lymphatic vessels. Instinctively again, I made the following syllogism: the white chyle is due to emulsion of the fat; now in rabbits white chyle is formed at the level where pancreatic juice is poured into the intestine; therefore it is pancreatic juice that makes the emulsion of fat and forms the white chyle. This had to be decided by experiment.

In view of this preconceived idea I imagined and at once performed a suitable experiment to verify the truth or falsity of my suppositions. The experiment consisted in trying the properties of pancreatic juice directly on neutral fats. But pancreatic juice does not spontaneously flow outside of the body, like saliva, for instance, or urine; its secretory organ is, on the contrary, lodged deep in the abdominal cavity. I was therefore forced to use the method of experimentation to secure the pancreatic fluid from living animals in suitable physiological conditions and in sufficient quantity. Only then could I carry out my experiment, that is to say, control my preconceived idea; and the experiment proved that my idea was correct. In fact pancreatic juice obtained in suitable conditions from dogs, rabbits and various other animals, and mixed with oil or melted fat, always instantly emulsified, and later split these fatty bodies into fatty acids, glycerine, etc., etc., by means of a specific ferment.

I shall not follow these experiments further, having explained them at length in a special work. I wish here to show merely how an accidental first observation of the acidity of rabbits' urine suggested to me the idea of making experiments on them with carnivorous feeding, and how later, in continuing these experiments, I brought to light, without seeing it, another observation concerning the peculiar arrangement of the junction of the pancreatic duct in rabbits. This second observation gave me, in turn, the idea of experimenting on the behavior of pancreatic juice.

From the above examples we see how chance observation of a fact or phenomenon brings to birth, by anticipation, a preconceived idea or hypothesis about the probable cause of the phenomenon observed; how the preconceived idea begets reasoning which results in the experiment which verifies it; how, in one case, we had to have recourse to experimentation, i.e., to the use of more or less complicated operative processes, etc., to work out the verification. In the last example, experiment played a double rôle; it first judged and confirmed the provisions of the reasoning which it had begotten; but what is more, it produced a fresh observation. We may therefore call this observation an observation produced or begotten by experiment. This proves that, as we said, all the results of an experiment must be observed, both those connected with the preconceived idea and those without any relation to it. If we saw only facts connected with our preconceived idea, we should often cut ourselves off from making discoveries. For it often happens that an unsuccessful experiment may produce an excellent observation, as the following example will prove.

THIRD EXAMPLE. In 1857, I undertook a series of experiments on the elimination of substances in the urine, and this time the results of the experiment, unlike the previous examples, did not confirm my previsions or preconceived ideas. I had therefore made what we habitually call an unsuccessful experiment. But there are no unsuccessful experiments; for, when they do not serve the investigation for which they were devised, we must still profit by observation to find occasion for other experiments.

In investigating how the blood, leaving the kidney, eliminated substances that I had injected, I chanced to observe that the blood in the renal vein was crimson, while the blood in the neighboring veins was dark like ordinary venous blood. This unexpected peculiarity struck me, and I thus made observation of a fresh fact begotten by the experiment, but foreign to the experimental aim pursued at the moment. I therefore gave up my unverified original idea, and directed my attention to the singular coloring of the venous renal blood; and when I had noted it well and assured myself that there was no source of error in my observation, I naturally asked myself what could be its cause. As I examined the urine flowing through the urethra and reflected about it, it occurred to me that the red coloring of the venous blood might well be connected with the secreting or active state of the kidney. On this hypothesis, if the renal secretion was stopped, the venous blood should become dark: that is what happened; when the renal secretion was re-established, the venous blood should become crimson again; this I also succeeded in verifying whenever I excited the secretion of urine. I thus secured experimental proof that there is a connection between the secretion of urine and the coloring of blood in the renal vein.

But that is still by no means all. In the normal state, venous blood in the kidney is almost constantly crimson, because the urinary organ secretes almost continuously, though alternately for each kidney. Now I wished to know whether the crimson color is a general fact characteristic of the other glands, and in this way to get a clear-cut counterproof demonstrating that the phenomenon of secretion itself was what led to the alteration in the color of the venous blood. I reasoned thus: if, said I, secretion, as it seems to be, causes the crimson color of glandular venous blood, then, in such glandular organs as the salivary glands which secrete intermittently, the venous blood will change color intermittently and become dark, while the gland is at rest, and red during secretion. So I uncovered a dog's submaxillary gland, its ducts, its nerves and its vessels. In its normal state, this gland supplies an intermittent secretion which we can excite or stop at pleasure. Now while the gland was at rest, and nothing flowed through the salivary duct, I clearly noted that the venous blood was indeed dark, while, as soon as secretion appeared, the blood became crimson, to resume its dark color when the secretion stopped; and it remained dark as long as the intermission lasted, etc.

These last observations later became the starting point for new ideas which guided me in making investigations as to the chemical cause of the change in color of glandular blood during secretion. I shall not further describe these experiments which, moreover, I have published in detail. It is enough for me to prove that scientific investigations and experimental ideas may have their birth in almost involuntary chance observations which present themselves either spontaneously or in an experiment made with a different purpose.

Let me cite another case—one in which an experimenter produces an observation and voluntarily brings it to birth. This case is, so to speak, included in the preceding case; but it differs from it in this, that, instead of waiting for an observation to present itself by chance in fortuitous circumstances, we produce it by experiment. Returning to Bacon's comparison, we might say that an experimenter, in this instance, is like a hunter who, instead of waiting quietly for game, tries to make it rise, by beating up the locality where he assumes it is. We use this method whenever we have no preconceived idea in respect to a subject as to which previous observations are lacking. So we experiment to bring to birth observations which in turn may bring to birth ideas. This continually occurs in medicine when we wish to investigate the action of a poison or of some medicinal substance, or an animal's economy; we make experiments to see, and we then take our direction from what we have seen.

FOURTH EXAMPLE. In 1845, Monsieur Pelouze gave me a toxic substance, called *curare,* which had been brought to him from America. We then knew nothing about the physiological action of this substance. From old observations and from the interesting accounts of Alex. von Humboldt and of Roulin and Boussingault, we knew only that the preparation of this substance was complex and difficult, and that it very speedily kills an animal if introduced under the skin. But from the earlier observations, I could get no idea of the mechanism of death by curare; to get such an idea I had to make fresh observations as to the organic disturbances to which this poison might lead. I therefore made experiments to *see* things about which I had absolutely no preconceived idea. First, I put curare under the skin of a frog: it died after a few minutes; I opened it at once, and in this physiological autopsy I studied in succession what had become of the known physiological properties of its various tissues. I say physiological autopsy purposely, because no others are really instructive. The disappearance of physiological properties is what explains death, and not anatomical changes. Indeed, in the present state of science, we see physiological properties disappear in any number of cases without being able to show, by our present means of observation, any corresponding anatomical change; such, for example, is the case with curare. Meantime, we shall find examples, on the contrary, in which physiological properties persist, in spite of very marked anatomical changes with which the functions are by no means incompatible. Now in my frog poisoned with curare, the heart maintained its movements, the blood was apparently no more changed in physiological properties than the muscles, which kept their normal contractility. But while the nervous system had kept its normal anatomical appearance, the properties of the nerves had nevertheless completely disappeared. There were no movements, either voluntary or reflex, and when the motor nerves were stimulated directly, they no longer caused any contraction in the muscles. To learn whether there was anything accidental or mistaken in this first observation, I repeated it several times and verified it in various ways; for when we wish to reason experimentally, the first thing necessary is to be a good observer and to make quite certain that the starting point of our reasoning is not a mistake in observation. In mammals and in birds, I found the same phenomena as in frogs, and disappearance of the physiological properties of the motor nervous system became my constant fact. Starting from this well-established fact, I could then carry analysis of the phenomena further and determine the mechanism of death from curare. I still proceeded by reasonings analogous to those quoted in the above example, and, from idea to idea and experiment to experiment, I progressed to more and more definite facts. I finally reached this general proposition, that *curare causes death by destroying all the motor nerves, without affecting the sensory nerves.*

In cases where we make an experiment in which both preconceived idea and reasoning seem completely lacking, we yet necessarily reason by syllogism without knowing it. In the case of curare, I instinctively reasoned in the following way: no phenomenon is without a cause, and consequently no poisoning without a physiological lesion peculiar or proper to the poison used; now, thought I, curare must cause death by an activity special to itself and by acting on certain definite organic parts. So by poisoning an animal with curare and by examining the properties of its various tissues immediately after death, I can perhaps find and study the lesions peculiar to it.

The mind, then, is still active here, and an experiment in order to see is included, nevertheless, in our general definition of an experiment. In every enterprise, in fact, the mind is always reasoning, and, even when we seem to act without a motive, an instinctive logic still directs the mind. Only we are not aware of it, because we begin by reasoning before we know or say that we are reasoning, just as we begin by speaking before we observe that we are speaking, and just as we begin by seeing and hearing before we know what we see or what we hear.

FIFTH EXAMPLE. About 1846, I wished to make experiments on the cause of poisoning with carbon monoxide. I knew that this gas had been described as toxic, but I knew literally nothing about the mechanism of its poisoning; I therefore could not have a preconceived opinion. What, then, was to be done? I must bring to birth an idea by making a fact appear, i.e., make another experiment to see. In fact I poisoned a dog by making him breathe carbon monoxide and after death I at once opened his body. I looked at the state of the organs and fluids. What caught my attention at once was that its blood was scarlet in all the vessels, in the veins as well as the arteries, in the right heart as well as in the left. I repeated the experiment on rabbits, birds and frogs, and everywhere I found the same scarlet coloring of the blood. But I was diverted from continuing this investigation, and I kept this observation a long time unused except for quoting it in my course *à propos* of the coloring of blood.

In 1856, no one had carried the experimental question further, and in my course at the Collège de France on toxic and medicinal substances, I again took up the study of poisoning by carbon monoxide which I had begun in 1846. I found myself then in a confused situation, for at this time I already knew that poisoning with carbon monoxide makes the blood scarlet in the whole circulatory system. I had to make hypotheses, and establish a preconceived idea about my first observation, so as to go ahead. Now, reflecting on the fact of scarlet blood, I tried to interpret it by my earlier knowledge as to the cause of the color of blood. Whereupon all the following reflections presented themselves to my mind. The scarlet color, said I, is peculiar to arterial blood and connected with the presence of a large proportion of oxygen, while dark coloring belongs with absence of oxygen and presence of a larger proportion of carbonic acid; so the idea occurred to me that carbon monoxide, by keeping venous blood scarlet, might perhaps have prevented the oxygen from changing into carbonic acid in the capillaries. Yet it seemed hard to understand how that could be the cause of death. But still keeping on with my inner preconceived reasoning, I added: If that is true, blood taken from the veins of animals poisoned with carbon monoxide should be like arterial blood in containing oxygen; we must see if that is the fact.

Following this reasoning, based on interpretation of my observation, I tried an experiment to verify my hypothesis as to the persistence of oxygen in the venous blood. I passed a current of hydrogen through scarlet venous blood taken from an animal poisoned with carbon monoxide, but I could not liberate the oxygen as usual. I tried to do the same with arterial blood; I had no greater success. My preconceived idea was therefore false. But the impossibility of getting oxygen from the blood of a dog poisoned with carbon monoxide was a second observation which suggested a fresh hypothesis. What could have become of the oxygen in the blood? It had not changed into carbonic acid, because I had not set free large quantities of that gas in passing a current of hydrogen through the blood of the poisoned animals. Moreover, that hypothesis was contrary to the color of the blood. I exhausted myself in conjectures about how carbon monoxide could cause the oxygen to disappear from the blood; and as gases displace one another I naturally thought that the carbon monoxide might have displaced the oxygen and driven it out of the blood. To learn this, I decided to vary my experimentation by putting the blood in artificial conditions that would allow me to recover the displaced oxygen. So I studied the action of carbon monoxide on blood experimentally. For this purpose I took a certain amount of arterial blood from a healthy animal; I put this blood on the mercury in an inverted test tube containing carbon monoxide; I then shook the whole thing so as to poison the blood sheltered from contact with the outer air. Then, after an interval, I examined whether the air in the test tube in contact with the poisoned blood had been changed, and I noted that the air thus in contact with the blood had been remarkably enriched with oxygen, while the proportion of carbon monoxide was lessened. Repeated in the same conditions, these experiments taught me that what had occurred was an exchange, volume by volume, between the carbon monoxide and the oxygen of the blood. But the carbon monoxide, in displacing the oxygen that it had expelled from the blood, remained chemically combined in the blood and could no longer be displaced either by oxygen or by other gases. So that death came through death of the molecules of blood, or in other words by stopping their exercise of a physiological property essential to life.

This last example, which I have very briefly described, is complete; it shows from one end to the other, how we proceed with the experimental method and succeeded in learning the immediate cause of phenomena. To begin with I knew literally nothing about the mechanism of the phenomenon of poisoning with carbon monoxide. I undertook an experiment to see, i.e., to observe. I made a preliminary observation of a special change in the coloring of blood. I interpreted this observation, and I made an hypothesis which proved false. But the experiment provided me with a second observation about which I reasoned anew, using it as a starting point for making a new hypothesis as to the mechanism, by which the oxygen in the blood was removed. By building up hypotheses, one by one, about the facts as I observed them, I finally succeeded in showing that carbon monoxide replaces oxygen in a molecule of blood, by combining with the substance of the molecule. Experimental analysis, here, has reached its goal. This is one of the cases, rare in physiology, which I am happy to be able to quote. Here the immediate cause of the phenomenon of poisoning is found and is translated into a theory which accounts for all the facts and at the same time includes all the observations and experiments. Formulated as follows, the theory posits the main facts from which all the rest are deduced: Carbon monoxide combines more intimately than oxygen with the hemoglobin in a molecule of blood. It has quite recently been proved that carbon monoxide forms a definite combination with hemoglobin. So that the molecule of blood, as if petrified by the stability of the combination, loses its vital properties. Hence everything is logically deduced: because of its property of more intimate combination, carbon monoxide drives out of the blood the oxygen essential to life; the molecules of blood become inert, and the animal dies, with symptoms of hemorrhage, from true paralysis of the molecules.

But when a theory is sound and indeed shows the real and definite physico-chemical cause of phenomena, it not only includes the observed facts but predicts others and leads to rational applications that are logical consequences of the theory. Here again we meet this criterion. In fact, if carbon monoxide has the property of driving out oxygen by taking its place in combining with a molecule of blood, we should be able to use the gas to analyze the gases in blood, and especially for determining oxygen. From my experiments I deduced this application which has been generally adopted today. Applications of this property of carbon monoxide have been made in legal medicine for finding the coloring matter of blood; and from the physiological facts described above we may also already deduce results connected with hygiene, experimental pathology, and notably with the mechanism of certain forms of anemia.

As in every other case, all the deductions from the theory doubtless still require experimental verification; and logic does not suffice. But this is because the conditions in which carbon monoxide acts on the blood may present other complex circumstances and any number of details which the theory cannot yet predict. Otherwise, we could reach conclusions by logic alone, without any need of experimental verifications. Because of possible unforeseen and variable new elements in the conditions of a phenomenon, logic alone can in experimental science never suffice. Even when we have a theory that seems sound, it is never more than relatively sound, and it always includes a certain proportion of the unknown.

II. *When the Starting Point of Experimental Research Is an Hypothesis or a Theory*

In noting an observation we must never go beyond facts. But in making an experiment, it is different. I wish to show that hypotheses are indispensable, and that they are useful, therefore, precisely because they lead us outside of facts and carry science forward. The object of hypotheses is not only to make us try new experiments; they also often make us discover new facts which we should not have perceived without them. In the preceding examples, we saw that we can start from a particular fact and rise one by one to more general ideas, i.e., to a theory. But as we have just seen, we can also sometimes start with an hypothesis deduced from a theory. Though we are dealing in this case with reasoning logically deduced from a theory, we have an hypothesis that must still be verified by experiment. Indeed, theories are only an assembling of the earlier facts, on which our hypothesis rests, and cannot be used to demonstrate it experimentally. We said that, in this instance, we must not submit to the yoke of theories, and that keeping our mental independence is the best way to discover the truth. This is proved by the following examples.

FIRST EXAMPLE. In 1843, in one of my first pieces of work, I undertook to study what becomes of different alimentary substances in nutrition. I began with sugar, a definite substance that is easier than any other to recognize and follow in the bodily economy. With this in view, I injected solutions of cane sugar into the blood of animals, and I noted that even when injected in weak doses the sugar passed into the urine. I recognized later that, by changing or transforming sugar, the gastric juice made it capable of assimilation, i.e., of destruction in the blood.

Thereupon I wished to learn in what organ the nutritive sugar disappeared, and I conceived the hypothesis that sugar introduced into the blood through nutrition might be destroyed in the lungs or in the general capillaries. The theory, indeed, which then prevailed and which was naturally my proper starting point, assumed that the sugar present in animals came exclusively from foods, and that it was destroyed in animal organisms by the phenomena of combustion, i.e., of respiration. Thus sugar had gained the name of *respiratory nutriment*. But I was immediately led to see that the theory about the origin of sugar in animals, which served me as a starting point, was false. As a result of the experiments which I shall describe further on, I was not indeed led to find an organ for destroying sugar, but, on the contrary, I discovered an organ for making it, and I found that all animal blood contains sugar even when they do not eat it. So I noted a new fact, unforeseen in theory, which men had not noticed, doubtless because they were under the influence of contrary theories which they had too confidently accepted. I therefore abandoned my hypothesis on the spot, so as to pursue the unexpected result which has since become the fertile origin of a new path for investigation and a mine of discoveries that is not yet exhausted.

In these researches I followed the principles of the experimental method that we have established, i.e., that, in presence of a well-noted, new fact which contradicts a theory, instead of keeping the theory and abandoning the fact, I should keep and study the fact, and I hastened to give up the theory, thus conforming to the precept: "When we meet a fact which contradicts a prevailing theory, we must accept the fact and abandon the theory, even when the theory is supported by great names and generally accepted."

We must therefore distinguish, as we said, between principles and theories, and never believe absolutely in the latter. We had a theory here which assumed that the vegetable kingdom alone had the power of creating the individual compounds which the animal kingdom is supposed to destroy. According to this theory, established and supported by the most illustrious chemists of our day, animals were incapable of producing sugar in their organisms. If I had believed in this theory absolutely, I should have had to conclude that my experiment was vitiated by some inaccuracy; and less wary experimenters than I might have condemned it at once, and might not have tarried longer at an observation which could be theoretically suspected of including sources of error, since it showed sugar in the blood of animals on a diet that lacked starchy or sugary materials. But instead of being concerned about the theory, I concerned myself only with the fact whose reality I was trying to establish. By new experiments and by means of suitable counterproofs, I was thus led to confirm my first observation and to find that the liver is the organ in which animal sugar is formed in certain given circumstances, to spread later into the whole blood supply and into the tissues and fluids.

Animal glycogenesis which I thus discovered, i.e., the power of producing sugar, possessed by animals as well as vegetables, is now an acquired fact for science; but we have not yet fixed on a plausible theory accounting for the phenomenon. The fresh facts which I made known are the source of numerous studies and many varied theories in apparent contradiction with each other and with my own. When entering on new ground we must not be afraid to express even risky ideas so as to stimulate research in all directions. As Priestley put it, we must not remain inactive through false modesty based on fear of being mistaken. So I made more or less hypothetical theories of glycogenesis; after mine came others; my theories, like other men's, will live the allotted life of necessarily very partial and temporary theories at the opening of a new series of investigations; they will be replaced later by others, embodying a more advanced stage of the question, and so on. Theories are like a stairway; by climbing, science widens its horizon more and more, because theories embody and necessarily include proportionately more facts as they advance. Progress is achieved by exchanging our theories for new ones which go further than the old, until we find one based on a larger number of facts. In the case which now concerns us, the question is not one of condemning the old to the advantage of a more recent theory. What is important is having opened a new road; for well-observed facts, though brought to light by passing theories, will never die; they are the material on which alone the house of science will at last be built, when it has facts enough and has gone sufficiently deep into the analysis of phenomena to know their law or their causation.

To sum up, theories are only hypotheses, verified by more or less numerous facts. Those verified by the most facts are the best; but even then they are never final, never to be absolutely believed. We have seen in the preceding examples that if we had had complete confidence in the prevailing theory of the destruction of sugar in animals, and if we had only had its confirmation in view, we should probably not have found the road to the new facts which we met. It is true that an hypothesis based on a theory produced the experiment; but as soon as the results of the experiment appeared, theory and hypothesis had to disappear, for the experimental facts were now just an observation, to be made without any preconceived idea.

In sciences as complex and as little developed as physiology, the great principle is therefore to give little heed to hypotheses or theories and always to keep an eye alert to observe everything that appears in every aspect of an experiment. An apparently accidental and inexplicable circumstance may occasion the discovery of an important new fact, as we shall see in the continuation of the example just noted.

SECOND EXAMPLE (Sequel to the last). After finding, as I said above, that there is sugar in the livers of animals in their normal state, and with every sort of nutriment, I wished to learn the proportion of this substance and its variation in certain physiological and pathological states. So I began to estimate the sugar in the livers of animals placed in various physiologically defined circumstances. I always made two determinations of carbohydrate for the same liver tissue. But pressed for time one day, it happened that I could not make my two analyses at the same moment; I quickly made one determination just after the animal's death and postponed the other analysis till next day. But then I found much larger amounts of sugar than those which I got the night before with the same material. I noticed, on the other hand, that the proportion of sugar, which I had found just after the animal's death the night before, was much smaller than I had found in the experiments which I had announced as giving the normal proportion of liver sugar. I did not know how to account for this singular variation, got with the same liver and the same method of analysis. What was to be done? Should I consider two such discordant determinations as an unsuccessful experiment and take no account of them? Should I take the mean between these experiments? More than one experimenter might have chosen this expedient to get out of an awkward situation. But I disapprove of this kind of action for reasons which I have given elsewhere. I said, indeed, that we must never neglect anything in our observation of fact, and I consider it indispensable, never to admit the existence of an unproved source of error in an experiment and always to try to find a reason for the abnormal circumstances that we observe. Nothing is accidental, and what seems to us accident is only an unknown fact whose explanation may furnish the occasion for a more or less important discovery. So it proved in this case.

I wished, in fact, to learn the reason for my having found two such different values in the analysis of my rabbit's liver. After assuring myself that there was no mistake connected with the method of analysis, after noting that all parts of the liver were practically equally rich in sugar, there remained to be studied only the elapsed time between the animal's death and the time of my second determination. Without ascribing much importance to it, up to that time I had made my experiments a few hours after the animal's death; now for the first time I was in the situation of making one determination only a few minutes after death and postponing the other till next day, i.e., twenty-four hours later. In physiology, questions of time are always very important because organic matter passes through numerous and incessant changes. Some chemical change might therefore have taken place in the liver tissue. To make sure, I made a series of new experiments which dispelled every obscurity by showing me that liver tissue becomes more and more rich in sugar for some time after death. Thus we may have a very variable amount of sugar according to the moment when we make our examination. I was therefore led to correct my old determination and to discover the new fact that considerable amounts of sugar are produced in animals' livers after death. For instance, by forcibly injecting a current of cold water through the hepatic vessels and passing it through a liver that was still warm, just after an animal's death, I showed that the tissue was completely freed from the sugar which it contained; but next day or a few hours later, if we keep the washed liver at a mild temperature, we again find its tissue charged with a large amount of sugar produced after it was washed.

Once in possession of the first discovery that sugar is formed in animals after death as during life, I wished to carry my study of this singular phenomenon further; I was then led to find that sugar is produced in the liver with the help of an enzyme reacting on an amylaceous substance which I isolated and which I called *glycogenous matter*, so that I succeeded in proving in the most clear-cut way that sugar is formed in animals by a mechanism in every respect like the mechanism found in vegetables.

This second series of facts embodied results, which are also firmly acquired for science, and which have greatly advanced our knowledge of glycogenesis in animals. I have just very briefly told how these facts were discovered, and how they started with an experimental circumstance that was apparently inconsequential. I quote this case so as to prove that we must never neglect anything in experimental research, for every accident has a necessary cause. We must, therefore, never be too much absorbed by the thought we are pursuing, nor deceive ourselves about the value of our ideas or scientific theories; we must always keep our eyes open for every event, the mind doubting and independent, ready to study whatever presents itself and to let nothing go without seeking its reason. In a word, we must be in an intellectual attitude which seems paradoxical but which, in my opinion, expresses the true spirit of an investigator. We must have robust faith and not believe. Let me explain myself by saying that in science we must firmly believe in principles, but must question formulæ; on the one hand, indeed, we are sure that determinism exists, but we are never certain we have attained it. We must be immovable as to the principles of experimental science (determinism), but must not absolutely believe in theories. The aphorism which I just uttered is sustained by what we expounded elsewhere, to wit, that for experimental science principles are in our mind, while formulæ are external things. In practical matters, we are indeed forced to tolerate the belief that truth (at least temporary truth) is embodied in a theory or a formula. But in scientific experimental philosophy those who put their faith in formulæ and theories are wrong. All human science consists in seeking the true formula and true theory. We are always approaching it; but shall we ever find it completely? This is not the place to go into an explanation of philosophic ideas: let us return to our subject and pass on to a fresh experimental example.

THIRD EXAMPLE. About the year 1852, my studies led me to make experiments on the influence of the nervous system on the phenomena of nutrition and temperature regulation. It had been observed in many cases that complex paralyses with their seat in the mixed nerves are followed, now by a rise and again by a fall of temperature in the paralyzed parts. Now this is how I reasoned, in order to explain this fact, basing myself first on known observations and then on prevailing theories of the phenomena of nutrition and temperature regulation. Paralysis of the nerves, said I, should lead to cooling of the parts by slowing down the phenomena of combustion in the blood, since these phenomena are considered as the cause of animal heat. On the other hand, anatomists long ago noticed that the sympathetic nerves especially follow the arteries. So, thought I inductively, in a lesion of a mixed trunk of nerves, it must be the sympathetic nerves that produce the slowing down of chemical phenomena in capillary vessels, and their paralysis that then leads to cooling the parts. If my hypothesis is true, I went on, it can be verified by severing only the sympathetic, vascular nerves leading to a special part, and sparing the others. I should then find the part cooled by paralysis of the vascular nerves, without loss of either motion or sensation, since the ordinary motor and sensory nerves would still be intact. To carry out my experiment, I therefore sought a suitable experimental method that would allow me to sever only the vascular nerves and to spare the others. Here the choice of animals was important in solving the problem; for in certain animals, such as rabbits and horses, I found that the anatomical arrangement isolating the cervical sympathetic nerve made this solution possible.

Accordingly, I severed the cervical sympathetic nerve in the neck of a rabbit, to control my hypothesis and see what would happen in the way of change of temperature on the side of the head where this nerve branches out. On the basis of a prevailing theory and of earlier observation, I had been led, as we have just seen, to make the hypothesis that the temperature should be reduced. Now what happened was exactly the reverse. After severing the cervical sympathetic nerve about the middle of the neck, I immediately saw in the whole of the corresponding side of the rabbit's head a striking hyperactivity in the circulation, accompanied by increase of warmth. The result was therefore precisely the reverse of what my hypothesis, deduced from theory, had led me to expect; thereupon I did as I always do, that is to say, I at once abandoned theories and hypothesis, to observe and study the fact itself, so as to define the experimental conditions as precisely as possible. Today my experiments on the vascular and thermo-regulatory nerves have opened a new path for investigation and are the subject of numerous studies which, I hope, may some day yield really important results in physiology and pathology. This example, like the preceding ones, proves that in experiments we may meet with results different from what theories and hypothesis lead us to expect. But I wish to call more special attention to this third example, because it gives us an important lesson, to wit: without the original guiding hypothesis, the experimental fact which contradicted it would never have been perceived. Indeed, I was not the first experimenter to cut this part of the cervical sympathetic nerve in living animals. Pourfour du Petit performed the experiment at the beginning of the last century and discovered the nerve's action on the pupil, by starting from an anatomical hypothesis according to which this nerve was supposed to carry animal spirits to the eye. Many physiologists have since repeated the same operation, with the purpose of verifying or explaining the changes in the eye which Pourfour du Petit first described. But none of them noticed the local temperature phenomenon, of which I speak, or connected it with the severing of the cervical sympathetic nerve, though this phenomenon must necessarily have occurred under the very eyes of all who, before me, had cut this part of the sympathetic nerve. The hypothesis, as we see, had prepared my mind for seeing things in a certain direction, given by the hypothesis itself; and this is proved by the fact that, like the other experimenters, I myself had often divided the cervical sympathetic nerve to repeat Pourfour du Petit's experiment, without perceiving the fact of heat production which I later discovered when an hypothesis led me to make investigations in this direction. Here, therefore, the influence of the hypothesis could hardly be more evident; we had the fact under our eyes and did not see it because it conveyed nothing to our mind. However, it could hardly be simpler to perceive, and since I described it, every physiologist without exception has noted and verified it with the greatest ease.

To sum up, even mistaken hypotheses and theories are of use in leading to discoveries. This remark is true in all the sciences. The alchemists founded chemistry by pursuing chimerical problems and theories which are false. In physical science, which is more advanced than biology, we might still cite men of science who make great discoveries by relying on false theories. It seems, indeed, a necessary weakness of our mind to be able to reach truth only across a multitude of errors and obstacles.

What general conclusions shall physiologists draw from the above examples? They should conclude that in the present state of biological science accepted ideas and theories embody only limited and risky truths which are destined to perish. They should consequently have very little confidence in the ultimate value of theories, but should still make use of them as intellectual tools necessary to the evolution of science and suitable for the discovery of new facts. The art of discovering new phenomena and of noting them accurately should today be the special concern of all biologists. We must establish experimental criticism by creating rigorous methods of investigation and experimentation, which will enable us to define our observations unquestionably, and thus get rid of the errors of fact which are the source of errors in theory. A man who today attempted a generalization for biology as a whole would prove that he had no accurate feeling for the present state of the science. Today, the biological problem has hardly begun to be put; and, as stones must first be got together and cut, before we dream of erecting a monument, just so must the facts first be got together and prepared which are destined to create the science of living bodies. This rôle falls to experimentation; its method is fixed, but the phenomena to be analyzed are so complex that, for the moment, the true promoters of science are those who succeed in giving its methods of analysis a few principles of simplification or in introducing improvements in instruments of research. When there are enough quite clearly established facts, generalizations never keep us waiting. I am convinced that, in experimental sciences that are evolving, and especially in those as complex as biology, discovery of a new tool for observation or experiment is much more useful than any number of systematic or philosophic dissertations. Indeed, a new method or a new means of investigation increases our power and makes discoveries and researches possible which would not have been possible without its help. Thus researches as to the formation of sugar in animals could be made only after chemistry gave us reagents for recognizing sugar, which were much more sensitive than those we had before.

MARGARET SHEA GILBERT

Biography of the Unborn

•

First Month
Out of the Unknown

OUT OF THE UNKNOWN into the image of man—this is the miraculous change which occurs during the first month of human life. We grow from an egg so small as to be barely visible, to a young human embryo almost one fourth of an inch long, increasing 50 times in size and 8000 times in weight. We change from a small round egg cell into a creature with a head, a body and, it must be admitted, a tail; with a heart that beats and blood that circulates; with the beginnings of arms and legs, eyes and ears, stomach and brain. In fact, within the first 30 days of our life almost every organ that serves us during our allotted time (as well as some that disappear before birth) has started to form.

Shortly after fertilization the great activity which was stirred up in the egg by the entrance of the sperm leads to the division or "cleavage" of the egg into two cells, which in turn divide into four and will go on so dividing until the millions of cells of the human body have been formed.

In addition to this astounding growth and development we must also make our first struggle for food. For this purpose a special "feeding layer" —the trophoblast—forms on the outer edge of the little ball of cells, and "eats its way" into the tissues of the uterus. As these tissues are digested by the trophoblast, the uterus forms a protective wall—the placenta—which cooperates with the trophoblast in feeding the growing embryo. The maternal blood carries food, oxygen (the essential component of the air we breathe) and water to the placenta, where they are absorbed by the trophoblast and passed on to the embryo through the blood vessels in the umbilical cord. In return, the waste products of the embryo are brought to the placenta and transferred to the mother's blood, which carries them to her kidneys and lungs to be thrown out. In no case does the mother's blood actually circulate through the embryo—a prevalent but quite unfounded belief.

Meanwhile the new individual has been moving slowly along the path of changes which it is hoped will make a man of him. While the trophoblast has been creating a nest for the egg in the uterine wall, the inner cell-mass has changed from a solid ball of cells into a small hollow organ resembling a figure 8—that is, it contains two cavities separated in the middle by a double-layered plate called the embryonic disc which, *alone*, develops into a human being. The lower half of our hypothetical figure eight becomes a small empty vesicle, called the yolksac, which eventually (in the second month) is severed from the embryo. The upper half forms a water-sac (called the amnion) completely surrounding the embryo except at the thick umbilical cord. The embryo then floats in a water-jacket which acts as a shock-absorber, deadening any jolts or severe blows which may strike the mother's body.

Having now made sure of its safety, the truly embryonic part of the egg—the double-layered plate—can enter wholeheartedly into the business of becoming a human being. Oddly enough, it is his heart and his brain, in their simplest forms, which first develop.

Almost at once (by the age of 17 days at most) the first special cells whose exact future we can predict appear. They are young blood cells, occurring in groups called blood islands which soon fuse to form a single tube, the heart-tube, in the region that is to be the head end of the embryonic disc. This simple tube must undergo many changes before it becomes the typical human heart, but rather than wait for that distant day before starting work, it begins pulsating at once. First a slight twitch runs through the tube, then another, and soon the heart is rhythmically contracting and expanding, forcing the blood to circulate through the blood vessels in the embryonic disc. In must continue to beat until the end of life.

About the same time the nervous system also arises. In the embryonic disc a thickened oval plate forms, called the neural plate, the edges of which rise as ridges from the flat surface and roll together into a round tube exactly in the middle of what will be the embryo's back. The front end of this tube will later develop into the brain; the back part will become the spinal cord. Thus, in this fourth week of life, this simple tube represents the beginning of the nervous system—the dawn of the brain that is to be man's most precious possession.

The embryo now turns his attention to the food canal. The hungry man calls this structure his stomach, but the embryologist briefly and indelicately speaks of the gut. The flat embryonic disc becomes humped up in the middle into a long ridgelike pocket which has a blind recess at either end. Very shortly an opening breaks through from the foregut upon the under surface of the future head to form the primitive mouth, though a similar outlet at the hind end remains closed for some time.

Within 25 days after the simple egg was fertilized by the sperm, the embryo is a small creature about one tenth of an inch long with head and tail ends, a back and a belly. He has no arms or legs, and he lacks a face or neck, so his heart lies close against his brain. Within this unhuman exterior, however, he has started to form also his lungs, which first appear as a shallow groove in the floor of the foregut; his liver is arising as a thickening in the wall of the foregut just behind the heart; and he has entered on a long and devious path which will ultimately lead to the formation of his kidneys.

The development of the human kidneys presents a striking example of a phenomenon which might be called an "evolutionary hangover." Instead of forming at once the type of organ which he as a human will use, the embryo forms a type which a much simpler animal (say the fish) possesses. Then he scraps this "fish organ" and forms another which a higher animal such as the frog uses. Again the embryo scraps the organ and then, perhaps out of the fragments of these preceding structures, forms his own human organ. It is as if, every time a modern locomotive was built, the builder first made the oldest, simplest locomotive ever made, took this engine apart, and out of the old and some new parts built a later locomotive; and after several such trials finally built a modern locomotive, perhaps using some metal which had gone into the first. Scientists interpret this strange process common to the development of all higher animals as a hasty, sketchy repetition of the long process of evolution.

By the end of the month the embryo is about one fourth of an inch long, curled almost in a circle, with a short pointed tail below his belly, and small nubbins on the sides of his body—incipient arms and legs. On the sides of his short neck appear four clefts, comparable to the gill-slits of a fish—another "evolutionary hangover." Almost all the organs of the human body have begun to form. In the head the eyes have arisen as two small pouches thrust out from the young brain tube. The skin over the front of the head shows two sunken patches of thickened tissue which are the beginning of a nose. At a short distance behind each eye an ear has started to develop—not the external ear, but the sensitive tissue which will later enable the individual to hear. In 30 days the new human being has traveled the path from the mysteriously simple egg and sperm to the threshold of humanity.

Second Month
The Face of Man

From tadpole to man: so one might characterize the changes that occur during the second month of life. True, the embryo is not a tadpole, but it looks not unlike one. The tailed bulbous creature with its enormous drooping head, fish-like gill-slits, and formless stubs for arms and legs, bears little resemblance to a human form. By the end of the second month, however, the embryo has a recognizable human character, although it is during this period that the human tail reaches its greatest development. In this month the embryo increases sixfold in length (to almost an inch and a half) and approximately 500 times in weight. Bones and muscles, developing between the skin and the internal organs, round out the contours of the body.

But the developing face and neck are the main features that give a human appearance, however grotesque. The mouth, now bounded by upper and lower jaws, is gradually reduced in size as the fused material forms cheeks. The nasal-sacs gradually move closer together until they form a broad nose. The eyes, which at first lie on the sides of the head, are shifted around to the front. During the last week of the month eyelids develop which shortly afterwards close down.

The forehead is prominent and bulging, giving the embryo a very brainy appearance. In fact, the embryo is truly brainy in the sense that the brain forms by far the largest part of the head. It will take the face many years to overcome this early dominance of the brain and to reach the relative size the face has in the adult.

The limbs similarly pass through a surprising series of changes. The limb "buds" elongate, and the free end of the limb becomes flattened into a paddlelike ridge which forms the finger-plate or toe-plate. Soon five parallel ridges separated by shallow grooves appear within each plate; the grooves are gradually cut through, thus setting off five distinct fingers and toes. At the same time, transverse constrictions form within each limb to mark off elbow and wrist, knee and ankle.

The human tail reaches its greatest development during the fifth week, and the muscles which move the tail in lower animals are present. But from this time it regresses, and only in abnormal cases is it present in the newborn infant. Along with the muscles develop the bones. In most instances of bone development a pattern of the bone is first formed in cartilage, a softer translucent material, and later a hard bony substance is laid down in and around the cartilage model. As a sculptor first fashions his work in clay and then, when he knows that his design is adequate, casts the statue in bronze, so the developing embryo seems to plan out its skeleton in cartilage and then cast it in bone. This process continues through every month of life before birth, and throughout childhood and adolescence. Not until maturity is the skeleton finally cast.

Perhaps the most interesting feature of the second month of life is the development of the sexual organs. At the beginning of the month there is no way of telling the sex of the embryo except by identifying the sex chromosomes. By the end of the month the sex is clearly evident in the internal sex organs and is usually indicated externally. The most surprising aspect of sexual development is that the first-formed organs are identical in the two sexes. Even milk glands start to develop in both sexes near the end of the second month. Nature seems to lay down in each individual all the sexual organs of the race, then by emphasizing certain of these organs and allowing the remainder to degenerate, transforms the indifferent embryo into male or female.

Is each human being, then, fundamentally bisexual with the organs and functions of the apparent sex determined at fertilization holding in abeyance the undeveloped characters of the opposite sex? Laboratory experiments with sex-reversal in lower animals suggest that there may be various degrees of sexual development, even in mankind, and that between the typical male and female there may occur various degrees of intersexuality.

So the second month of life closes with the stamp of human likeness clearly imprinted on the embryo. During the remaining seven months the young human being is called a fetus, and the chief changes will be growth and detailed development.

Third Month
Emergence of Sex

Now the future "lords of all they survey" assert their ascendency over the timid female, for the male child during the third month plunges into the business of sexual development, while the female dallies nearer the neutral ground of sexual indifference. Or if sexual differences are overlooked, the third month could be marked the "tooth month," for early in this period buds for all 20 of the temporary teeth of childhood are laid down, and the sockets for these teeth arise in the hardening jaw bones.

Although six months must pass before the first cry of the infant will be heard, the vocal cords whose vibrations produce such cries now appear, at present as ineffective as a broken violin string. Only during the first six months after birth do they take on the form of effective human vocal cords. It must be remembered that during the period of life within the uterus no air passes through the larynx into the lungs. The fetus lives in a watery world where breathing would merely flood the lungs with amniotic fluid, and the vocal cords remain thick, soft and lax.

The digestive system of the three-months-old fetus begins to show signs of activity. The cells lining the stomach have started to secrete mucus—the fluid which acts as a lubricant in the passage of food through the digestive organs. The liver starts pouring bile into the intestine. The kidneys likewise start functioning, secreting urine which gradually seeps out of the fetal bladder into the amniotic fluid, although most of the waste products of the fetus's body will still be passed through the placenta into the mother's blood.

Overlying the internal organs are the bones and muscles which, with their steady development, determine the form, contours, and strength of the fetal body. In the face, the developing jaw bones, the cheek bones, and even the nasal bones that form the bridge of the nose, begin to give human contours and modeling to the small, wizened fetal face. Centers of bone formation have appeared in the cartilages of the hands and feet, but the wrists and ankles are still supported only by cartilage.

No longer is there any question about whether or not the fetus is a living, individual member of mankind. Not only have several of the internal organs taken on their permanent functions, but the well-developed muscles now produce spontaneous movements of the arms, legs and shoulders, and even of the fingers.

Fourth Month
The Quickening

Death throws its shadow over man before he is born, for the stream of life flows most swiftly through the embryo and young fetus, and then inexorably slows down, even within the uterus. The period of greatest growth occurs during the third and fourth fetal months, when the fetus grows approximately six to eight inches in length, reaching almost one half its height at birth. Thereafter the rate of growth decreases steadily.

However, the young fetus is not a miniature man, but a gnomelike creature whose head is too large, trunk too broad, and legs too short. At two months the head forms almost one half of the body; from the third to fifth months it is one third, at birth one fourth, and in the adult about one tenth the body height.

Nevertheless, the four-month fetus is not an unhandsome creature. With his head held more or less erect, and his back reasonably straight, he bears a real resemblance to a normal infant. The face is wide but well modeled, with widely spaced eyes. The hands and feet are well formed. The fingers and toes are rather broad, and are usually flexed. At the tip of each finger and toe patterned whorls of skin ridges appear—the basis of future fingerprints and toeprints. As might be expected, the pattern of these skin ridges is characteristically different for each fetus; at four months each human being is marked for life with an individual, unchangeable stamp of identity.

The skin of the body is in general dark red and quite wrinkled at this time; the redness indicates that the skin is so thin that the blood coursing through the underlying vessels determines its color. Very little fat is stored in the fetus's body before the sixth month, and the skin remains loose and wrinkled until underlain by fat.

Now the still, silent march of the fetus along the road from conception to birth becomes enlivened and quickened. The fetus stirs, stretches, and vigorously thrusts out arms and legs. The first movements to be perceived by the mother may seem to her like the fluttering of wings, but before long his blows against the uterine wall inform her in unmistakable terms that life is beating at the door of the womb. For this is the time of the "quickening in the womb" of folklore.

Fifth Month
Hair, Nails and Skin

Man is an enigma; indivisible and yet complex; he is composed of hundreds of separate parts that are constantly dying and being renewed, yet he retains a mysterious "individuality." The human being may be compared to a cooperative society whose members band together for mutual support and protection, presenting a common front to the external world, and sharing equally in the privileges and responsibilities of their internal world. Division of labor, specialization, and the exchange of produce are just as important in the society of cells and organs as in the society of men. The digestive organs convert the materials taken in as food into the components of living cells. The circulating fluids of the body form an extensive transportation system. Nerves are the cables of the communications system while the brain is the central exchange. The potent endocrine glands determine the speed and constancy of many activities. Overlying all of the body's specialized systems is the skin—the protector, conservator, and inquirer of the society of organs.

Now that the internal organs are well laid down, the skin and the structures derived from it hasten to attain their final form. The surface of the skin becomes covered with tough, dried and dead cells which form a protective barrier between the environment and the soft tissues of the body. Even as in life after birth, the outer dead cells are being constantly sloughed off and replaced from below by the continually growing skin. Sweat glands are formed, and sebaceous glands, which secrete oil at the base of each hair. During the fifth month these glands pour out a fatty secretion which, becoming mixed with the dead cells sloughed off from the skin, forms a cheesy paste covering the entire body. This material, called the *vernix caseosa*, is thought to serve the fetus as a protective cloak from the surrounding amniotic fluid, which by this time contains waste products which might erode the still tender skin.

Derivatives of the skin likewise undergo marked development. Fine hair is generally present all over the scalp at this time. Nails appear on the fingers and toes. In the developing tooth germs of the "milk teeth," the pearly enamel cap and the underlying bonelike dentine are formed.

But the most striking feature of the month's development is the straightening of the body axis. Early in the second month the embryo forms almost a closed circle, with its tail not far from its head. At three months the head has been raised considerably and the back forms a shallow curve. At five months the head is erectly balanced on the newly formed neck, and the back is still less curved. At birth the head is perfectly erect and the back is almost unbelievably straight. In fact, it is more nearly straight than it will ever be again, for as soon as the child learns to sit and walk, secondary curvatures appear in the spinal column as aids in body balance.

The five-month fetus is a lean creature, with wrinkled skin, about a foot long and weighing about one pound. If born (or, strictly speaking, aborted) it may live for a few minutes, take a few breaths, and perhaps cry. But it soon gives up the struggle and dies. Although able to move its arms and legs actively, it seems to be unable to maintain the complex movements necessary for continued breathing.

Sixth Month
Eyes That Open on Darkness

Now the expectant parents of the six-months-old human fetus may become overwhelmingly curious about the sex of their off-spring, especially when they realize that the sex is readily perceived in the fetus. Yet to the external world no sign is given.

During the sixth month the eyelids, fused shut since the third month, reopen. Completely formed eyes are disclosed which, during the seventh month, become responsive to light. Eyelashes and eyebrows usually develop in the sixth or seventh month.

Within the mouth, taste buds are present all over the surface of the tongue, and on the roof and walls of the mouth and throat, being relatively more numerous than in the infant or adult. It seems odd that the fetus, with no occasion for tasting, should be more plentifully equipped, and some biologists believe that this phenomenon is but another evidence of the recurrence of evolutionary stages in development, since in many lower animals taste organs are more widely and generously distributed than they are in man.

The six-month fetus, if born, will breathe, cry, squirm, and perhaps live for several hours, but the chances of such a premature child surviving are extremely slight unless it is protected in an incubator. The vitality, the strength to live, is a very weak flame, easily snuffed out by the first adverse contact with the external world.

Seventh Month
The Dormant Brain

Now the waiting fetus crosses the unknown ground lying between dependence and independence. For although he normally spends two more months within the sure haven of the uterus, he is nonetheless capable of independent life. If circumstances require it and the conditions of birth are favorable, the seven-month fetus is frequently able to survive premature birth.

One of the prime causes of the failure of younger fetuses to survive birth is believed to be the inadequate development of the nervous system, especially of those parts concerned in maintaining constant rhythmic breathing movements, in carrying out the sequence of muscular contractions involved in swallowing, and in the intricate mechanism for maintaining body temperature.

The human nervous system consists of a complex network of nerves connecting all the organs of the body with the brain and spinal cord, the centralized "clearinghouse" for all the nervous impulses brought in from the sense organs and sent out to the muscles. By the third month of life special regions and structures have developed within the brain: the cerebellum, an expanded part of the brain that receives fibers coming mostly from the ear; and two large saclike outpocketings, the cerebral hemispheres, which are the most distinctive feature of man's brain. They are destined to become the most complex and elaborately developed structures known in the nervous system of any animal. They are alleged by some to be the prime factor in man's dominance over other animals.

At seven months these hemispheres cover almost all the brain, and some vague, undefined change in the minute nerve cells and fibers accomplishes their maturation. Henceforth the nervous system of the fetus is capable of successful functioning.

The seven-month fetus is a red-skinned, wrinkled, old-looking child about 16 inches long and weighing approximately three pounds. If born he will cry, breathe, and swallow. He is, however, very susceptible to infection and needs extra protection from the shocks which this new life in the external world administers to his delicate body. He is sensitive to a light touch on the palm. He probably perceives the difference between light and dark. Best of all—he has a chance to survive.

Eighth and Ninth Months
Beauty That Is Skin-Deep

Now the young human being, ready for birth, with all his essential organs well formed and able to function, spends two more months putting the finishing touches on his anatomy, and improving his rather questionable beauty. Fat is formed rapidly all over his body, smoothing out the wrinkled, flabby skin and rounding out his contours. The dull red color of the skin fades gradually to a flesh-pink shade. The fetus loses the wizened, old-man look and attains the more acceptable lineaments of a human infant.

Pigmentation of the skin is usually very slight, so that even the offspring of colored races are relatively light-skinned at birth. Even the iris of the eye is affected; at birth the eyes of most infants are a blue-gray shade (which means that very little pigment is present) and it is usually impossible to foretell their future color.

The fetus is by no means a quiet, passive creature, saving all his activity until after birth. He thrashes out with arms and legs, and may even change his position within the somewhat crowded quarters of the uterus. He seems to show alternate periods of activity and quiescence, as if perhaps he slept a bit and then took a little exercise.

Exodus

Just what specific event initiates the birth sequence remains unknown. For some weeks or even months previous to birth, slow, rhythmic muscular contractions, similar to those which cause labor pains, occur in a mild fashion in the uterus. Why the uterus, after withstanding this long period of futile contractions, is suddenly thrown into the powerful, effective muscular movements which within a few hours expel the long-tolerated fetus remains the final mystery of our prenatal life. It is quite probable that the birth changes occur as a complex reaction of the mother's entire body, especially those potent endocrine glands which may pour into the blood stream chemicals that stimulate immediate and powerful contractions of the uterine muscle.

There is nothing sacrosanct about the proverbial "nine months and ten days" as the duration of pregnancy; but 10 per cent of the fetuses are born on the 280th day after the onset of the last true menstrual period and approximately 75 per cent are born within two weeks of that day.

As soon as the infant is born, he usually gasps, fills his lungs with air and utters his first bleating cry, either under the influence of the shock which this outer world gives to his unaccustomed body or from some stimulus administered by the attending doctor. The infant is still, however, connected through the umbilical cord with the placenta lodged within the uterine wall. Their usefulness ended, the placenta and umbilical cord are cut off from the infant. The stump soon degenerates, but its scar, the defect in the abdominal wall caused by the attachment of the cord to the fetus, remains throughout life as the navel—a permanent reminder of our once parasitic mode of living.

The newborn infant is by no means a finished and perfect human being. Several immediate adjustments are required by the change from intra-uterine to independent life. The lungs at birth are relatively small, compact masses of seemingly dense tissue. The first few breaths expand them until they fill all the available space in the chest cavity, and as the numerous small air sacs are filled with air, the lungs become light and spongy in texture. But it is not yet a complete human lung, for new air sacs are formed throughout early childhood, and even those formed before birth do not function perfectly until several days of regular breathing have passed.

The heart, which is approximately the size of the infant's closed fist, gradually beats more slowly, approaching the normal rate of the human heart. Shortly after birth the material which has been accumulating in the intestine during the last six months of fetal life is passed off. One peculiarity of the newborn infant is that the intestine and its contents are completely sterile; the elaborate and extensive bacterial population present in the intestine of all human beings appears only after birth.

Neither tear glands nor salivary glands are completely developed at birth; the newborn infant cries without tears, and his saliva does not acquire its full starch-digesting capacity until near weaning time. The eyes, although sensitive to light, have not yet acquired the power of focusing on one point so that the newborn infant may be temporarily cross-eyed.

Thus the first nine months of life are completed. The manifold changes occurring during this period form the first personal history of each member of the race. It is the one phase of life which we all have in common; it is essentially the same for all men.

JULIAN HUXLEY

Variations on a Theme by Darwin

•

During the present century we have heard so much of the revolutionary discoveries of modern physics that we are apt to forget how great has been the change in the outlook due to biology. Yet in some respects this has been the more important. For it is affecting the way we think and act in our everyday existence. Without the discoveries and ideas of Darwin and the other great pioneers in the biological field, from Mendel to Freud, we should all be different from what we are. The discoveries of physics and chemistry have given us an enormous control over lifeless matter and have provided us with a host of new machines and conveniences, and this certainly has reacted on our general attitude. They have also provided us with a new outlook on the universe at large: our ideas about time and space, matter and creation, and our own position in the general scheme of things, are very different from the ideas of our grandfathers.

Biology is beginning to provide us with control over living matter—new drugs, new methods for fighting disease, new kinds of animals and plants. It is helping us also to a new intellectual outlook, in which man is seen not as a finished being, single lord of creation, but as one among millions of the products of an evolution that is still in progress. But it is doing something more. It is actually making us different in our natures and our biological behaviour. I will take but three examples.

The application of the discoveries of medicine and physiology is making us healthier: and a healthy man behaves and thinks differently from one who is not so healthy. Then the discoveries of modern psychology have been altering our mental and emotional life, and our system of education: taken in the mass, the young people now growing up feel differently, and will therefore act differently, about such vital matters as sex and marriage, about jealousy, about freedom of expression, about the relation between parents and children. And as a third example, as a race we are changing our reproductive habits: the idea and the practice of deliberate birth-control has led to fewer children. People living in a country of small families and a stationary or decreasing population will in many respects *be* different from people in a country of large families and an increasing population.

This change has not been due to any very radical new discoveries made during the present century. It has been due chiefly to discoveries which were first made in the previous century, and are at last beginning to exert a wide effect. These older discoveries fall under two chief heads. One is Evolution—the discovery that all living things, including ourselves, are the product of a slow process of development which has been brought about by natural forces, just as surely as has to-day's weather or last month's high tides. The other is the sum of an enormous number of separate discoveries which we may call physiological, and which boil down to this: that all living things, again including ourselves, work according to regular laws, in just the same way as do non-living things, except that living things are much more complicated. The old idea of "vital-force" has been driven back and back until there is hardly any process of life where it can still find any foothold. Looked at objectively and scientifically, a man is an exceedingly complex piece of chemical machinery. This does not mean that he cannot quite legitimately be looked at from other points of view—subjectively, for instance; what it means is that so far as it goes, this scientific point of view is true, and not the point of view which ascribed human activities to the working of a vital force quite different from the forces at work in matter which was not alive.

Imagine a group of scientists from another planet, creatures with quite a different nature from ours, who had been dispassionately studying the curious objects called human beings for a number of years. They would not be concerned about what we men felt we were or what we would like to be, but only about getting an objective view of what we actually were and why we were what we were. It is that sort of picture which I want to draw for you. Our Martian scientists would have to consider us from three main viewpoints if they were to understand much about us. First they would have to understand our physical construction, and what meaning it had in relation to the world around and the work we have to do in it. Secondly, they would have to pay attention to our development and our history. And thirdly, they would have to study the construction and working of our minds. Any one of these three aspects by itself would give a very incomplete picture of us.

An ordinary human being is a lump of matter weighing between 50 and 100 kilograms. This living matter is the same matter of which the rest of the earth, the sun, and even the most distant stars and nebulae are made. Some elements which make up a large proportion of living matter, like hydrogen and especially carbon, are rare in the not-living parts of the earth; and others which are abundant in the earth are, like iron, present only in traces in living creatures, or altogether absent, like aluminum or silicon. None the less, it is the same matter. The chief difference between living and non-living matter is the complication of living matter. Its elements are built up into molecules much bigger and more elaborate than any other known, often containing more than a thousand atoms each. And of course, living matter has the property of self-reproduction; when supplied with the right materials and in the right conditions, it can build up matter which is not living into its own complicated patterns.

Life, in fact, from the "public" standpoint, which Professor Levy has stressed as being the only possible standpoint for science, is simply the name for the various distinctive properties of a particular group of very complex chemical compounds. The most important of these properties are, first, feeding, assimilation, growth, and reproduction, which are all aspects of the one quality of self-reproduction; next, the capacity for reacting to a number of kinds of changes in the world outside—to stimuli, such as light, heat, pressure, and chemical change; then the capacity for liberating energy in response to these stimuli, so as to react back again upon the outer world—whether by moving about, by constructing things, by discharging chemical products, or by generating light or heat; and finally the property of variation. Self-reproduction is not always precisely accurate, and the new substance is a little different from the parent substance which produced it.

The existence of self-reproduction on the one hand and variation on the other automatically leads to what Darwin called "natural selection." This is a sifting process, by which the different new variations are tested out against the conditions of their existence, and in which some succeed better than others in surviving and in leaving descendants. This blind process slowly but inevitably causes living matter to change—in other words, it leads to evolution. There may be other agencies at work in guiding the course of evolution; but it seems certain that natural selection is the most important.

The results it produces are roughly as follows. It *adapts* any particular stream of living matter more or less completely to the conditions in which it lives. As there are innumerable different sets of conditions to which life can be adapted, this has led to an increasing diversity of life, a splitting of living matter into an increasing number of separate streams. The final tiny streams we call species; there are perhaps a million of them now in existence. This adaptation is progressive; any one stream of life is forced to grow gradually better and better adapted to some particular condition of life. We can often see this in the fossil records of past life. For instance, the early ancestors of lions and horses about 50 million years ago were not very unlike, but with the passage of time one line grew better adapted to grass-eating and running away from enemies. And finally natural selection leads to general progress; there is a gradual rising of the highest level attained by life. The most advanced animals are those which have changed their way of life and adapted themselves to new conditions, thus taking advantages of biological territory hitherto unoccupied. The most obvious example of this was the invasion of the land. Originally all living things were confined to life in water, and it was not for hundreds of millions of years after the first origin of life that plants and animals managed to colonize dry land.

But progress can also consist in taking better advantage of existing conditions: for instance, the mammal's biological inventions, of warm blood and of nourishing the unborn young within the mother's body, put them at an advantage over other inhabitants of the land; and the increase in size of brain which is man's chief characteristic has enabled him to control and exploit his environment in a new and more effective way, from which his pre-human ancestors were debarred.

It follows from this that all animals and plants that are at all highly developed have a long and chequered history behind them, and that their present can often not be properly understood without an understanding of their past. For instance, the tiny hairs all over our own bodies are a reminder of the fact that we are descended from furry creatures, and have no significance except as a survival.

Let us now try to get some picture of man in the light of these ideas. The continuous stream of life that we call the human race is broken up into separate bits which we call individuals. This is true of all higher animals, but is not necessary: it is a convenience. Living matter has to deal with two sets of activities: one concerns its immediate relations with the world outside it, the other concerns its future perpetuation. What we call an individual is an arrangement permitting a stream of living matter to deal more effectively with its environment. After a time it is discarded and dies. But within itself it contains a reserve of potentially immortal substance, which it can hand on to future generations, to produce new individuals like itself. Thus from one aspect the individual is only the casket of the continuing race; but from another the achievements of the race depend on the construction of its separate individuals.

The human individual is large as animal individuals go. Size is an advantage if life is not to be at the mercy of small changes in the outer world: for instance, a man the size of a beetle could not manage to keep his temperature constant. Size also goes with long life: and a man who only lived as long as a fly could not learn much. But there is a limit to size; a land animal much bigger than an elephant is not, mechanically speaking, a practical proposition. Man is in that range of size, from 100 lb. to a ton, which seems to give the best combination of strength, and mobility. It may be surprising to realize that man's size and mechanical construction are related to the size of the earth which he inhabits; but so it is. The force of gravity on Jupiter is so much greater than on our own planet, that if we lived there our skeletons would have to be much stronger to support the much increased weight which we would then possess, and animals in general would be more stocky; and conversely, if the earth were only the size of the moon, we could manage with far less expenditures of material in the form of bone and sinew, and should be spindly creatures.

Our general construction is determined by the fact that we are made of living matter, must accordingly be constantly passing a stream of fresh matter and energy through ourselves if we are to live, and must as constantly be guarding ourselves against danger if we are not to die. About 5 per cent of ourselves consists of a tube with attached chemical factories, for taking in raw materials in the shape of food, and converting it into the form in which it can be absorbed into our real interior. About 2 per cent consists in arrangements—windpipe and lungs—for getting oxygen into our system in order to burn the food materials and liberate energy. About 10 per cent consists of an arrangement for distributing materials all over the body—the blood and lymph, the tubes which hold them and the pump which drives them. Much less than 5 per cent is devoted to dealing with waste materials produced when living substance breaks down in the process of producing energy to keep our machinery going—the kidneys and bladder and, in part, the lungs and skin. Over 40 per cent is machinery for moving us about—our muscles; and nearly 20 per cent is needed to support us and to give the mechanical leverage for our movements—our skeleton and sinews. A relatively tiny fraction is set apart for giving us information about the outer world—our sense organs. And there is about 3 per cent to deal with the difficult business of adjusting our behaviour to what is happening around us. This is the task of the ductless glands, the nerves, the spinal cord and the brain; our conscious feeling and thinking is done by a small part of the brain. Less than 1 per cent of our bodies is set aside for reproducing the race. The remainder of our body is concerned with special functions like protection, carried out by the skin (which is about 7 per cent of our bulk) and some of the white blood corpuscles; or temperature regulations, carried out by the sweat glands. And nearly 10 per cent of a normal man consists of reserve food stores in the shape of fat.

Other streams of living matter have developed quite other arrangements in relation to their special environment. Some have parts of themselves set aside for manufacturing electricity, like the electric eel, or light, like the firefly. Some, like certain termites, are adapted to live exclusively on wood; others, like cows, exclusively on vegetables. Some, like boa-constrictors, only need to eat every few months; others, like parasitic worms, need only breathe a few hours a day; others, like some desert gazelles, need no water to drink. Many cave animals have no eyes; tapeworms have no mouths or stomachs; and so on and so forth. And all these peculiarities, including those of our own construction, are related to the kind of surroundings in which the animal lives.

This relativity of our nature is perhaps most clearly seen in regard to our senses. The ordinary man is accustomed to think of the information given by his senses as absolute. So it is—for him; but not in the view of our Martian scientist. To start with, the particular senses we possess are not shared by many other creatures. Outside backboned animals, for instance, very few creatures can hear at all; a few insects and perhaps a few crustacea probably exhaust the list. Even fewer animals can see colours; apparently the world as seen even by most mammals is a black and white world, not a coloured world. And the majority of animals do not even see at all in the sense of being given a detailed picture of the world around. Either they merely distinguish light from darkness, or at best can get a blurred image of big moving objects. On the other hand, we are much worse off than many other creatures—dogs, for instance, or some moths—in regard to smell. Our sense of smell is to a dog's what an eye capable of just distinguishing big moving objects is to our own eye.

But from another aspect, the relativity of our senses is even more fundamental. Our senses serve to give us information about changes outside our bodies. Well, what kind of changes are going on in the outside world? There are ordinary mechanical changes: matter can press against us, whether in the form of a gentle breeze or a blow from a poker. There are the special mechanical changes due to vibrations passing through the air or water around us—these are what we hear. There are changes in temperature—hot and cold. There are chemical changes—the kind of matter with which we are in contact alters, as when the air contains poison gas, or our mouth contains lemonade. There are electrical changes, as when a current is sent through a wire we happen to be touching.

And there are all the changes depending on what used to be called vibrations in the ether. The most familiar of these are light-waves; but they range from the extremely short waves that give cosmic rays and X-rays, down through ultra-violet to visible light, on to waves of radiant heat, and so on to the very long Hertzian waves which are used in wireless. All these are the same kind of thing, but differ in wave-length.

Now of all these happenings, we are only aware of what appears to be a very arbitrary selection. Mechanical changes we are aware of through our sense of touch. Air-vibrations we hear; but not all of them—the small wave-lengths are pitched too high for our ears, though some of them can be heard by other creatures, such as dogs and bats. We have a heat sense and a cold sense, and two kinds of chemical senses for different sorts of chemical changes—taste and smell. But we possess no special electrical sense—we have no way of telling whether a live rail is carrying a current or not unless we actually touch it, and then what we feel is merely pain.

The oddest facts, however, concern light and kindred vibrations. We have no sense organs for perceiving X-rays, although they may be pouring into us and doing grave damage. We do not perceive ultra-violet light, though some insects, like bees, can see it. And we have no sense organs for Hertzian waves, though we make machines—wireless receivers—to catch them. Out of all this immense range of vibrations, the only ones of which we are aware through our senses are radiant heat and light. The waves of radiant heat we perceive through the effect which they have on our temperature sense organs; and the light-waves we see. But what we see is only a single octave of the light waves, as opposed to ten or eleven octaves of sound-waves which we can hear.

This curious state of affairs begins to be comprehensible when we remember that our sense organs have been evolved in relation to the world in which our ancestors lived. In nature, there are large-scale electrical discharges such as lightning, and they act so capriciously and violently that to be able to detect them would be no advantage. The same is true of X-rays. The amount of them knocking about under normal conditions is so small that there is no point in having sense organs to tell us about them. Wireless waves, on the other hand, are of such huge wave-lengths that they go right through living matter without affecting it. Even if they were present in nature, there would be no obvious way of developing a sense organ to perceive them.

As regards light, there seem to be two reasons why our eyes are limited to seeing only a single octave of the waves. One is that of the ether vibrations raying upon the earth's surface from the sun and outer space, the greatest amount is centered in this region of the spectrum; the intensity of light of higher or lower wave-lengths is much less, and would only suffice to give us a dim sensation. Our greatest capacity for seeing is closely adjusted to the amount of light to be seen. The other is more subtle, and has to do with the properties of light of different wavelengths. Ultra-violet light is of so short a wave-length that much of it gets scattered as it passes through the air, instead of progressing forward in straight lines. Hence a photograph which uses only the ultraviolet rays is blurred and shows no details of the distance. A photograph taken by infra-red light, on the other hand, while it shows the distant landscape very well, over-emphasizes the contrast between light and shade in the foreground. Leaves and grass reflect all the infra-red, and so look white, while the shadows are inky-black, with no gradations. The result looks like a snowscape. An eye which could see the ultra-violet octave would see the world as in a fog; and one which could see only the infra-red octave would find it impossible to pick out lurking enemies in the jet-black shadows. The particular range of light to which our eyes are attuned gives the best-graded contrast.

Then of course there is the pleasant or unpleasant quality of a sensation; and this, too, is in general related to our way of life. I will take one example. Both lead acetate and sugar taste sweet; the former is a poison, but very rare in nature; the latter is a useful food, and common in nature. Accordingly most of us find a sweet taste pleasant. But if lead acetate were as common in nature as sugar, and sugar as rare as lead acetate, it is safe to prophesy that we should find sweetness a most horrible taste, because we should only survive if we spat out anything which tasted sweet.

Now let us turn to another feature of man's life which would probably seem exceedingly queer to a scientist from another planet—sex. We are so used to the fact that our race is divided up into two quite different kinds of individuals, male and female, and that our existence largely circles round this fact, that we rarely pause to think about it. But there is no inherent reason why this should be so. Some kinds of animals consist only of females; some, like ants, have neuters in addition to the two sexes; some plants are altogether sexless.

As a matter of fact, the state of affairs as regards human sex is due to a long and curious sequence of causes. The fundamental fact of sex has nothing to do with reproduction; it is the union of two living cells into one. The actual origin of this remains mysterious. Once it had originated, however, it proved of biological value, by conferring greater variability on the race, and so greater elasticity in meeting changed conditions. That is why sex is so nearly universal. Later, it was a matter of biological convenience that reproduction in higher animals became indissolubly tied up with sex. Once this had happened, the force of natural selection in all its intensity became focused on the sex instinct, because in the long run those strains which reproduce themselves abundantly will live on, while those which do not do so will gradually be supplanted.

A wholly different biological invention, the retention of the young within the mother's body for protection led to the two sexes becoming much more different in construction and instincts than would otherwise have been the case. The instinctive choice of a more pleasing as against a less pleasing mate—what Darwin called sexual selection—led to the evolution of all kinds of beautiful or striking qualities which in a sexless race would never have developed. The most obvious of such characters are seen in the gorgeous plumage of many birds; but sexual selection has undoubtedly modelled us human beings in many details—the curves of our bodies, the colours of lips, eyes, cheeks, the hair of our heads, and the quality of our voices.

Then we should not forget that almost all other mammals and all birds are, even when adult, fully sexed only for a part of the year: after the breeding season they relapse into a more or less neuter state. How radically different human life would be if we too behaved thus! But man has continued an evolutionary trend begun for some unknown reason among the monkeys, and remains continuously sexed all the year round. Hunger and love are the two primal urges of man: but by what a strange series of biological steps has love attained its position!

We could go on enumerating facts about the relativity of man's physical construction; but time is short, and I must say a word about his mind. For that too has developed in relation to the conditions of our life, present and past. Many philosophers and theologians have been astonished at the strength of the feeling which prompts most men and women to cling to life, to feel that life is worth living, even in the most wretched circumstances. But to the biologist there is nothing surprising in this. Those men (and animals) who have the urge to go on living strongly developed will automatically survive and breed in greater numbers than those in whom it is weak. Nature's pessimists automatically eliminate themselves, and their pessimistic tendencies, from the race. A race without a strong will to live could no more hold its own than one without a strong sexual urge.

Then again man's highest impulses would not exist if it were not for two simple biological facts—that his offspring are born helpless and must be protected and tended for years if they are to grow up, and that he is a gregarious animal. These facts make it biologically necessary for him to have well-developed altruistic instincts, which may and often do come into conflict with his egoistic instincts, but are in point of fact responsible for half of his attitude towards life. Neither a solitary creature like a cat or a hawk, nor a creature with no biological responsibility towards its young, like a lizard or a fish, could possibly have developed such strong altruistic instincts as are found in man.

Other instincts appear to be equally relative. Everyone who has any acquaintance with wild birds and animals knows how much different species differ in temperament. Most kinds of mice are endowed with a great deal of fear and very little ferocity; while the reverse is true of various carnivores like tigers or Tasmanian devils. It would appear that the amounts of fear and anger in man's emotional make-up are greater than his needs as a civilized being, and are survivals from an earlier period of his racial history. In the dawn of man's evolution from apes, a liberal dose of fear was undoubtedly needed if he was to be preserved from foolhardiness in a world peopled by wild beasts and hostile tribes, and an equally liberal dose of anger, the emotion underlying pugnacity, if he was to triumph over danger when it came. But now they are on the whole a source of weakness and maladjustment.

It is often said that you cannot change human nature. But that is only true in the short-range view. In the long run, human nature could as readily be changed as feline nature has actually been changed in the domestic cat, where man's selection has produced an amiable animal out of a fierce ancestral spit-fire of a creature. If, for instance, civilization should develop in such a way that mild and placid people tended to have larger families than those of high-strung or violent temperament, in a few centuries human nature would alter in the direction of mildness.

Pavlov has shown how even dogs can be made to have nervous breakdowns by artificially generating in their minds conflicting urges to two virtually exclusive kinds of action; and we all know that the same thing, on a higher level of complexity, happens in human beings. But a nervous breakdown puts an organism out of action for the practical affairs of life, quite as effectively as does an ordinary infectious disease. And just as against physical germ-diseases we have evolved a protection in the shape of the immunity reactions of our blood, so we have evolved oblivion as protection against the mental diseases arising out of conflict. For, generally speaking, what happens is that we forget one of the two conflicting ideas or motives. We do this either by giving the inconvenient idea an extra kick into the limbo of the forgotten, which psychologists call suppression, or else, when it refuses to go so simply, by forcibly keeping it under in the sub-conscious, which is styled repression. For details about suppression and repression and their often curious and sometimes disastrous results I must ask you to refer to any modern book on psychology. All I want to point out here is that a special mental machinery has been evolved for putting inconvenient ideas out of consciousness, and that the contents and construction of our minds are different in consequence.

But I have said enough, I hope, to give you some idea of what is implied by calling man a relative being. It implies that he has no real meaning apart from the world which he inhabits. Perhaps this is not quite accurate. The mere fact that man, a portion of the general stuff of which the universe is made, can think and feel, aspire and plan, is itself full of meaning, but the precise way in which man is made, his physical construction, the kinds of feelings he has, the way he thinks, the things he thinks about, everything which gives his existence form and precision—all this can only be properly understood in relation to his environment. For he and his environment make one interlocking whole.

The great advances in scientific understanding and practical control often begin when people begin asking questions about things which up till then they have merely taken for granted. If humanity is to be brought under its own conscious control, it must cease taking itself for granted, and, even though the process may often be humiliating, begin to examine itself in a completely detached and scientific spirit.

C. *The Conquest of Disease*

•

Infested by parasites, surrounded by bacteria, a prey to viruses, the human machine fills us with a persisting amazement. How can it function so well so much of the time? Much of the answer lies in man's understanding of his own bodily enemies. The study goes back to the most primitive of medicine men. But as Clendening shows in Hippocrates, the Greek—the End of Magic, *there came a moment when the order of nature and not the whim of the gods was recognized as causing disease. Perhaps Hippocrates was not the first to make this initial discovery; perhaps his* Oath, *which still hangs in doctors' offices, was the work of a school or schools. However, the man or men responsible for this achievement were the intellectual fathers of the investigators of today.*

We have space for only a few of the highlights of the subsequent story. Jenner's classic paper An Inquiry into the Causes and Effects of the Variolae Vaccinae *shows the steps in his discovery of the methods of inoculation against smallpox. This discovery carried his name to the ends of the earth. An equally great investigator was Louis Pasteur. Vallery-Radot's story of his life is one of the greatest biographies in any field. The blend of scientific insight and human emotion contained in* Louis Pasteur and the Conquest of Rabies *makes it an indispensable contribution.*

The story is brought up to date with three selections on modern medicine. Sir Alexander Fleming has won world-wide recognition for his discovery of penicillin, still the most effective and the most widely used of the antibiotics. In Chemotherapy *he discusses for the layman not only his own discovery but also the sulfonamides, streptomycin and others. The search for new "magic bullets" goes on at a feverish pace. One reason is that, astonishingly enough, some of them, such as aureomycin, when fed to animals like pigs and chicks, result in phenomenal increases in growth.*

Our second selection on modern medicine discusses advances in The Battle Against Cancer. *This paper by Walter Goodman, which emphasizes the virus theory of the disease, has been checked for accuracy by the staff of the American Cancer Society and the National Cancer Institute.*

With the advent of the atomic bomb and increasingly effective methods for its delivery, the specter of the destruction of the human race confronts us. We think immediately of the results of blast and fire such as those which occurred at Hiroshima. But, as Warren Weaver points out in Radiations and the Genetic Threat, *there is a more insidious effect of equal or even greater long-range significance.*

LOGAN CLENDENING

Hippocrates, the Greek—the End of Magic

•

Philiscus, who lived by the wall in Athens, lay sick of a fever. The year, according to our reckoning, was 410 B.C. The Battle of Marathon had been fought eighty years before. Athens was still the greatest city in the world—great in the sunset of its golden age.

The members of Philiscus' family were uneasy about him, for the malady had not progressed favourably.

They sat sadly on the doorstep awaiting the report of his wife, who had gone in to help him.

She appeared with an unhappy frown on her brow.

"He doth not know me," she explained. "And he hath not slept. He hath passed water that is black."

"Ah! I have seen that," exclaimed her father. "It is a bad omen." His voice sank to a whisper. "I tell you it is the hounds of Hekate that rend him."

Another elder shook his head.

"It was a sudden affliction that seized him—it came from Pan, or, mayhap, one of the arrows of Apollo," he averred.

"What physicians have treated him?" inquired this sage, after an interval of silence.

"Im-Ram, the Egyptian, came by two days ago and gave him an emetic of white hellebore. But he was no better."

The elder looked stolidly ahead at this. He did not approve of Egyptians or Egyptian remedies. He wanted to placate the angry Apollo.

"Then there was the Babylonian, Mother," the son of Philiscus reminded her.

"What did he do?" inquired the elder.

"He sacrificed a goat and made divination by the liver."

"Ah! and what did that show?" asked the elder, somewhat more approvingly.

"He laid the liver out and explained it to me carefully," said the son, eagerly. "There was the lobus dexter and the lobus sinister—and they were inequal."

"The omens were not clear," sighed the wife.

"And the *vesica fellea,*" continued the lad—"the gall-bladder—it was full of stones."

"How many?" demanded the old man.

"There were three large ones and many small ones."

"Three?" the elder shook his head, dubiously—"that is grave. One element is missing. There should be four."

"Water, perhaps," suggested the wife, "he cries, when he cries sensibly at all, always for water."

"Fire, air, earth, and water," repeated the old man, sententiously. "The elements of Pythagoras, the Samian. If one is taken away by the demons or the hounds of Hekate, it must be replaced. Now here the sick man is hot and dry—fire is in the ascendancy. Water is cold and moist—just the opposite. It is water he must have." And he nodded his head emphatically, pleased with his own reasoning.

"I give him water morning, noon, and night, every hour," answered the wife, distractedly.

"If we could take him to a temple of Aesculapius," suggested the father-in-law, "and let the priests treat him."

The wife shook her head. "He is too sick to move," she said, "and out of his senses—we could not leave him alone."

"I went to a temple once when I was a young man—for this eye," the old man said, reminiscently. "It was a very good temple, and a very good treatment, to my way of thought. My eye got better soon after; whereas before, it had been painful and running like a sore."

"What temple did you go to?" the other old man inquired.

"At Epidaurus—naturally," the narrator replied. "I remember it very well. The priests of the temple made me cleanse myself first. There was a bath of salt water, too, as well as the clear water which they made me enter. Then I purified my soul with prayer. And then the oblation."

"What was your oblation?"

"I was too poor to offer a sheep or a cock, so I offered a popana—a small cake dipped in oil. The priests sell it to you. Then I starved four days and was allowed to enter the sanctuary for the incubation sleep."

"What was that?" asked the boy.

"Inside the temple—you slept. There was the great image of Aesculapius at the high altar. It was an awe-inspiring sight. The representation of the flesh was of ivory, and the rest was of gold enamelled in colours."

By this time he had acquired the attention of his audience, and he launched into his narrative.

"Sufferers were all over the floor of the temple. Each of us had his pallet. The night came down and we composed ourselves to sleep. And whether it was a dream or not I cannot tell, but it seemed to me the god himself came down from the altar and walked among us. He had two great yellow snakes and a dog. He stopped a moment at my pallet and leaned over me. One of the snakes licked my eye. The god put some ointment in it. And the next day I found a box of ointment at my side. I took it away with me. And soon my eye was well, and I placed a votive tablet in the temple."

The youth laughed incredulously.

"Ay!" the elder reproved, "in this age of doubt you fall away from the old things, but I tell you they are good—those temple rites. I know of things wrought there that would outdo your modern treatments. While I was being cured, Proklos, the philosopher, himself, was also there: he was afflicted with a rheum of his knee—very painful: and he covered it with a cloth. The night he slept in the temple, a sacred sparrow plucked the cloth away, and the pain left with it. His knee was as good as ever."

This account of success seemed to impress his audience.

"Yes, you doubt!" he continued. "You doubt the old ways, and you doubt the old gods. I heard of a new drama of Aristophanes—what is the name of it—*Plutus*—played in the theatre of Dionysus—and what does it amount to? Making fun of a poor sick person who goes to the temple for help—that's what. It jests at the priests—says they steal the offerings of food brought by the patients and eat the food themselves and give it to the sacred serpents—and all this—" the old man's voice rose excitedly—"all this played out in the theatre of Dionysus—and the priests do not interfere. Why, in my time—"

A wild cry from the delirious patient interrupted the discourse. The wife hurried in to attend her patient.

The boy crept to his grandfather's feet and said: "Grandfather, can we not fetch the physician Hippocrates to counsel about father?"

"Hippocrates? Yes, I have heard of this healer," assented his grandfather. "He hath a good name and is highly esteemed."

"He is in Athens now," declared the youth.

"He was the son of a temple priest—I know, I have heard," said the other old man. "His father was a priest in the Temple of Aesculapius at Cos. Ah, that is a wonderful temple for healing! If we cannot take Philiscus to a temple, the next best thing were to have a priest of the Asclepiadæ come to him."

The wife returned to their circle again, gravely troubled.

"He is worse even than before," she answered their interrogatory glances.

"Think you we could get Hippocrates, the physician, to see him?" asked her father.

A look of hope came to her face.

"I have heard of that Hippocrates," she answered. "Is he not the one who treated the Clazomenian who was lodged by the wall of Phynichides?"

"Yes, that was he—now that you recall it to me."

"The Clazomenian was cured."

"He was indeed—and his case is much like that of Philiscus. He had a pain in the neck and head, and fever. And he could not sleep, and besides, like Philiscus, he became delirious. He was sick for many days, but this Hippocrates came to see him every day and wrote down on his tablets the condition of the patient every day."

"What was his treatment?" asked the elder.

"That I cannot recall, but he prescribed diets and baths, that I know."

"Mayhap he leaves the patient and propitiates a god," suggested the elder.

"Mayhap, but his method was good in the case of the Clazomenian."

"Let us send for him quickly—quickly," cried the wife. "Who will help us?"

"I will run through the streets, Mother, and bring him back," said the boy, eagerly starting off.

The watchers waited impatiently; time seemed to them to pass slowly, but in reality in a short while the boy returned, leading a radiant stranger, the physician Hippocrates. He was between forty-five and fifty years of age—tall, erect, godlike in presence and calmness.

Three young men accompanied him, disciples learning the art. They were his sons Thessalus and Dracon, and one named Dexippus.

He entered the home of Philiscus gravely, greeted the wife and the two older men with a smile, and then walked quickly to the bed where the patient lay.

He put his hand on the sick man's forehead.

"Have you any pain?" he asked.

The patient stared at him vacantly, his lips trembling in a muttering delirium, and then he suddenly started as if to rise from his bed.

The younger physicians restrained him.

"Has he been delirious long?" Hippocrates asked the wife.

"Since yesterday evening," she answered.

"How long has he been sick?"

"This is the third day. He went to the market-place to discuss some matter and stood in the sun, and something he said must have incurred the anger of the god."

Hippocrates lifted his hand to stop her.

"Tell the story just as it happened, without bringing in the gods," he said, somewhat severely.

The woman looked at him with some fear. Then seeing a reassuring smile from the physician, she continued:

"He came home and took to his bed. He sweated and was very uneasy. Yesterday, the second day, he was worse in all these points."

"Did he have a stool?" asked Hippocrates.

"Yes, late in the evening—a proper stool from a small clyster."

"Write that down, Dexippus," commanded the master—and "Go on," he said to the wife.

"Today he has been much worse. He has been very hot. He trembles; he sweats and is always thirsty. He hath been delirious on all subjects. He has passed black water."

"Oh! When was that?" asked the physician, suddenly alert.

"This afternoon."

"Let me see some of it."

A slave boy was summoned and brought an earthen vessel with some of the sick man's urine in it.

"Notice, my sons," said the physician to his disciples. "The black water again. We have seen it often this season. And always the prognostic is unfavourable."

Anything more to tell us?" he asked, turning again to the wife.

"That is all, I think—O mighty physician, invoke the gods to drive this devil from my husband."

"Your husband hath no devil—he hath a disease. We will do our best. More we cannot promise. Pray to the gods, but pray for piety and good works. Do not ask them for things they cannot grant."

He left some instructions about the patient's diet and recommended limewater to drink. He instructed the slave in bathing his master by sponging him with cloth, unless he chilled. He left a draught of medicine to be given for delirium.

"I will return tomorrow and observe the patient," he announced to the family. "Let us see now, Dexippus, if you have that description right." He took the scroll from his pupil and read it.

"Shall we sacrifice to any of the gods?" asked the elder, tremulously.

"I do not practise by the gods," answered Hippocrates. "I try to discover the nature of the disease and to follow that. To read nature—believe me, friend, it is better than relying on the gods."

When he came the next day, the patient was unimproved. The physician noted all points about his condition and ordered Dexippus to write them down—which he did.

On the fifth day, however, the patient was worse. There was something very peculiar about the breathing.

Hippocrates motioned for the members of Philiscus' family to leave the room. Then "What think you of that breathing?" the physician asked his pupils.

"It is passing strange," answered Thessalus.

"How would you describe it?" demanded his father.

The young man watched the patient for a few minutes and then said: "Sometimes it is very rapid and deep—then it becomes shallower."

"Then what?"

"Then it stops altogether for a moment, and then he begins again like a person recollecting himself."

"That is good," approved the master. "Write that down, Dexippus—a splendid description—'like a person recollecting himself.'—Good. There is no better description I ever heard. See, there it is—'like a person recollecting himself.' Have you ever seen it before?"

"Yes—the Thessalonian had something like it."

"Quite true. And what was the outcome of his case?"

"He died."

"So he did. Do you remember anyone else who had it?"

"Was not that woman we saw in the little house yonder breathing in this way?" inquired Dracon, diffidently.

"She was indeed. Do you not all remember? Exactly the same. And what was the outcome of her case?"

"She, too, died," answered Dracon.

"That is the rule," said Hippocrates. "I have never seen one recover. So it will be here. I am sorry, for the wife loves her husband, but the rules of nature are immutable. Feel the spleen, Thessalus."

"It is large and round," answered Thessalus, after placing his hands on the abdomen.

"Extremities altogether cold," dictated Hippocrates, for his notes. "The paroxysms on the even days. Sweats cold throughout. So—"

The physician gave the family such comfort as he could, but his prognosis was fulfilled and Philiscus died that night.

The method of Hippocrates the Greek was to ignore all of the gods. Disease, he preached, was a part of the order of nature, and to conquer it, to understand it, one must study it as one does any other natural event. Many useful facts about treatment and diagnosis and the classification of disease were gathered together before Hippocrates. But with him the doctrine that disease is a natural event and follows natural laws comes out clear and strong.

That is why Hippocrates is called the Father of Medicine. Yet how long it took men to learn the simple thing he taught! How many hundreds of years elapsed between the medicine man and Hippocrates is a matter for the conjecture of anthropologists. Certainly not less than fifty thousand. But from his time to ours his influence extends in a clear stream, never changing in the great essential doctrine that disease is a part of nature.

The case of Philiscus is a good subject of study in order to analyse the elements which Hippocrates contributed to human thought.

Here we see him in the midst of his regular daily life, expounding by precept and example those principles. I have tried to show how far ahead of his time he was—how the older men in the scene harked back to the superstitions and to the ways of the gods of their youth—to Babylonian liver prognostication, to the idea of Apollo as the dealer of death, to the influence of Pan, and to the hounds of Hekate.

How scornful the Hippocratic writings are about the last: "But terrors which happen during the night, and fevers and delirium, and jumpings out of bed, and frightful apparitions and fleeing away—all these hold to be the plots of Hekate!"

The case of "Philiscus who lived by the wall" is actual enough. It is the first of those many little case histories found in Hippocrates—that earliest collection of clinical cases recorded from the standpoint of science.

The case is described simply day by day. There is no embroidery—simply the symptoms as they appeared and the outcome of the case.

The name of the patient, the address (by the wall), the circumstances, are all set down. The picture of the sick man tossing through the hot Athenian night comes to us across two thousand years, stabbing us like a personal anxiety.

The peculiarity of breathing which Hippocrates noted—"as of a person recollecting himself"—is now known as Cheyne-Stokes respiration. It is a common symptom of approaching death and is due to exhaustion or lack of oxygenation of the respiratory centre.

Hippocrates as a historical figure, aside from his writings, is very vague. In this he corresponds to Homer in the epic literature of Greece. It is doubtful if there was any single personality known as Hippocrates. The Hippocratic writings are probably the work of many men, the crystallization of the thought of a school.

Tradition dates his birth at 460 B.C. Plato mentions him as if he were a living man known to him. He is said to have travelled widely, teaching as he went. He died at Larissa, it was said, at the age of a hundred and ten.

It is difficult for anyone who reads the Hippocratic writings to escape the conviction that the best of them were the product of a single mind—their unity of thought, their clarity, their radiance, preclude any other idea.

The best are the "Aphorisms"—short, descriptive, clinical facts. The most famous, of course, is the first: "Life is short and art is long."

But "Persons who are naturally very fat are apt to die earlier than those who are slender" might have a place in a modern life-insurance actuary's summary of his studies.

"Consumption most commonly occurs between the ages of eighteen and thirty-five."

"From a spitting of blood there is a spitting of pus" shows that Hippocrates has watched people with tuberculosis of the lungs have as the initial symptom a hemorrhage and then begin ordinary expectoration.

"Eunuchs do not take the gout nor become bald."

"If a dropsical patient be seized with hiccup, the case is hopeless."

"Anxiety, yawning, and rigour—wine drunk with equal proportions of water removes these complaints."

Into the domain of treatment also he tried to bring some order.

His diets, for instance, as Dr. Singer points out, were to be prescribed according to certain sensible rules. First the age of the patient was to be considered—"Old persons use less nutriment than young." Then the season—"In winter abundant nourishment is wholesome; in summer a more frugal diet." The physical state of the patient—"Lean persons should take little food, but this little should be fat; fat persons, on the other hand, should take much food, but it should be lean." Digestibility of the food—"White meat is more digestible than dark."

The typical Greek myth has always seemed to my mind that of Prometheus. He stole the fire of the gods from heaven and brought it down to earth for man's use. The Greeks constantly did that. The Mediterranean basin was hag-ridden and god-ridden until they appeared. They took the drama—a service to the god—and they wrenched it away from the god and subdued it to the services of man. They made it not a service in a temple, but a story to charm the mind; they filled it with music and dancing and song for their fellow-men's entertainment. So with Hippocrates. He took once and for ever "the art"—the art of healing. He wrested it from the gods and made it man's.

With Hippocrates—with all the Greeks—we first find people of our own kind. We come out, as Osler says—"out of the murky night of the East, heavy with phantoms, into the bright daylight of the West." Here are men speaking our words, following our devotions, thinking our thoughts, pursuing objects which seems to us worth gaining and to us understandable.

EDWARD JENNER

An Inquiry into the Causes and Effects of the Variolae Vaccinae, Known by the Name of the Cow-Pox

•

THE DEVIATION of man from the state in which he was originally placed by nature seems to have proved to him a prolific source of diseases. From the love of splendour, from the indulgence of luxury, and from his fondness for amusement he has familiarized himself with a great number of animals, which may not originally have been intended for his associates.

The wolf, disarmed of ferocity, is now pillowed in the lady's lap. The cat, the little tiger of our island, whose natural home is the forest, is equally domesticated and caressed. The cow, the hog, the sheep, and the horse, are all, for a variety of purposes, brought under his care and dominion.

There is a disease to which the horse, from his state of domestication, is frequently subject. The farriers call it the grease. It is an inflammation and swelling in the heel, from which issues matter possessing properties of a very peculiar kind, which seems capable of generating a disease in the human body (after it has undergone the modification which I shall presently speak of), which bears so strong a resemblance to the smallpox that I think it highly probable it may be the source of the disease.

In this dairy country a great number of cows are kept, and the office of milking is performed indiscriminately by men and maid servants. One of the former having been appointed to apply dressings to the heels of a horse affected with the grease, and not paying due attention to cleanliness, incautiously bears his part in milking the cows, with some particles of the infectious matter adhering to his fingers. When this is the case, it commonly happens that a disease is communicated to the cows, and from the cows to dairy maids, which spreads through the farm until the most of the cattle and domestics feel its unpleasant consequences. This disease has obtained the name of the cow-pox. It appears on the nipples of the cows in the form of irregular pustules. At their first appearance they are commonly of a palish blue, or rather of a colour somewhat approaching to livid, and are surrounded by an erysipelatous inflammation. These pustules, unless a timely remedy be applied, frequently degenerate into phagedenic ulcers, which prove extremely troublesome. The animals become indisposed, and the secretion of milk is much lessened. Inflammed spots now begin to appear on different parts of the hands of the domestics employed in milking, and sometimes on the wrists, which quickly run on to suppuration, first assuming the appearance of the small vesications produced by a burn. Most commonly they appear about the joints of the fingers and at their extremities; but whatever parts are affected, if the situation will admit, these superficial suppurations put on a circular form, with their edges more elevated than their centre, and of a colour distantly approaching to blue. Absorption takes place, and tumours appear in each axilla. The system becomes affected—the pulse is quickened; and shiverings, succeeded by heat, with general lassitude and pains about the loins and limbs, with vomiting, come on. The head is painful, and the patient is now and then even affected with delirium. These symptoms, varying in their degrees of violence, generally continue from one day to three or four, leaving ulcerated sores about the hands, which, from the sensibility of the parts, are very troublesome, and commonly heal slowly, frequently becoming phagedenic, like those from whence they sprung. The lips, nostrils, eyelids, and other parts of the body are sometimes affected with sores; but these evidently arise from their being heedlessly rubbed or scratched with the patient's infected fingers. No eruptions on the skin have followed the decline of the feverish symptoms in any instance that has come to my inspection, one only excepted, and in this case a very few appeared on the arms: they were minute, of a vivid red colour, and soon died away without advancing to maturation; so that I cannot determine whether they had any connection with the preceding symptoms.

Thus the disease makes its progress from the horse to the nipple of the cow, and from the cow to the human subject.

Morbid matter of various kinds, when absorbed into the system, may produce effects in some degree similar; but what renders the cow-pox virus so extremely singular is that the person who has been thus affected is forever after secure from the infection of the smallpox; neither exposure to the variolous effluvia, nor the insertion of the matter into the skin, producing this distemper.

In support of so extraordinary a fact, I shall lay before my reader a number of instances.

Case I. Joseph Merret, now as under gardener to the Earl of Berkeley, lived as a servant with a farmer near this place in the year 1770, and occasionally assisted in milking his master's cows. Several horses belonging to the farm began to have sore heels, which Merret frequently attended. The cows soon became affected with the cow-pox, and soon after several sores appeared on his hands. Swellings and stiffness in each axilla followed, and he was so much indisposed for several days as to be incapable of pursuing his ordinary employment. Previously to the appearance of the distemper among the cows there was no fresh cow brought into the farm, nor any servant employed who was affected with the cow-pox.

In April, 1795, a general inoculation taking place here, Merret was inoculated with his family; so that a period of twenty-five years had elapsed from his having the cow-pox to this time. However, though the variolous matter was repeatedly inserted into his arm, I found it impracticable to infect him with it; an efflorescence only, taking on an erysipelatous look about the centre, appearing on the skin near the punctured parts. During the whole time that his family had the smallpox, one of whom had it very full, he remained in the house with them, but received no injury from exposure to the contagion.

It is necessary to observe that the utmost care was taken to ascertain, with the most scrupulous precision, that no one whose case is here adduced had gone through the smallpox previous to these attempts to produce that disease.

Had these experiments been conducted in a large city, or in a populous neighborhood, some doubts might have been entertained; but here, where population is thin, and where such an event as a person's having had the smallpox is always faithfully recorded, no risk of inaccuracy in this particular can arise.

Case II. Sarah Portlock, of this place, was infected with the cow-pox when a servant at a farmer's in the neighborhood, twenty-seven years ago.

In the year 1792, conceiving herself, from this circumstance, secure from the infection of the smallpox, she nursed one of her own children who had accidentally caught the disease, but no indisposition ensued. During the time she remained in the infected room, variolous matter was inserted into both her arms, but without any further effect than in the preceding case.

Case XVII. The more accurately to observe the progress of the infection I selected a healthy boy, about eight years old, for the purpose of inoculating for the cow-pox. The matter was taken from a sore on the hand of a dairymaid, who was infected by her master's cows, and it was inserted on the 14th day of May, 1796, into the arm of the boy by means of two superficial incisions, barely penetrating the cutis, each about an inch long.

On the seventh day he complained of uneasiness in the axilla and on the ninth he became a little chilly, lost his appetite, and had a slight headache. During the whole of this day he was perceptibly indisposed, and spent the night with some degree of restlessness, but on the day following he was perfectly well.

The appearance of the incisions in their progress to a state of maturation were much the same as when produced in a similar manner by variolous matter. The difference which I perceived was in the state of the limpid fluid arising from the action of the virus, which assumed rather a darker hue, and in that of the efflorescence spreading round the incisions, which had more of an erysipelatous look than we commonly perceive when variolous matter has been made use of in the same manner; but the whole died away (leaving on the inoculated parts scabs and subsequent eschars) without giving me or my patient the least trouble.

In order to ascertain whether the boy, after feeling so slight an affection of the system from the cow-pox virus, was secure from the contagion of the smallpox, he was inoculated the 1st of July following with variolous matter, immediately taken from a pustule. Several slight punctures and incisions were made on both his arms, and the matter was carefully inserted, but no disease followed. The same appearances were observable on the arms as we commonly see when a patient has had variolous matter applied, after having either the cow-pox or smallpox. Several months afterwards he was again inoculated with variolous matter, but no sensible effect was produced on the constitution.

After the many fruitless attempts to give the smallpox to those who had had the cow-pox, it did not appear necessary, nor was it convenient to me, to inoculate the whole of those who had been the subjects of these late trials; yet I thought it right to see the effects of variolous matter on some of them, particularly William Summers, the first of these patients who had been infected with matter taken from the cow. He was, therefore, inoculated from a fresh pustule; but, as in the preceding cases, the system did not feel the effects of it in the smallest degree. I had an opportunity also of having this boy and William Pead inoculated by my nephew, Mr. Henry Jenner, whose report to me is as follows: "I have inoculated Pead and Barge, two of the boys whom you lately infected with the cow-pox. On the second day the incisions were inflamed and there was a pale inflammatory stain around them. On the third day these appearances were still increasing and their arms itched considerably. On the fourth day the inflammation was evidently subsiding, and on the sixth day it was scarcely perceptible. No symptoms of indisposition followed.

"To convince myself that the variolous matter made use of was in a perfect state I at the same time inoculated a patient with some of it who never had gone through the cow-pox, and it produced the smallpox in the usual regular manner."

These experiments afforded me much satisfaction; they proved that the matter, in passing from one human subject to another, through five gradations, lost none of its original properties, J. Barge being the fifth who received the infection successively from William Summers, the boy to whom it was communicated from the cow.

Although I presume it may not be necessary to produce further testimony in support of my assertion "that the cow-pox protects the human constitution from the infection of the smallpox," yet it affords me considerable satisfaction to say that Lord Somerville, the President of the Board of Agriculture, to whom this paper was shown by Sir Joseph Banks, has found upon inquiry that the statements were confirmed by the concurring testimony of Mr. Dolland, a surgeon, who resides in a dairy country remote from this, in which these observations were made.

RENÉ VALLERY-RADOT

Louis Pasteur and the Conquest of Rabies

•

AMIDST the various researches undertaken in his laboratory, one study was placed by Pasteur above every other, one mystery constantly haunted his mind—that of hydrophobia. When he was received at the Académie Française, Renan, hoping to prove himself a prophet for once, said to him: "Humanity will owe to you deliverance from a horrible disease and also from a sad anomaly: I mean the distrust which we cannot help mingling with the caresses of the animal in whom we see most of nature's smiling benevolence."

The two first mad dogs brought into the laboratory were given to Pasteur, in 1880, by M. Bourrel, an old army veterinary surgeon who had long been trying to find a remedy for hydrophobia. He had invented a preventive measure which consisted in filing down the teeth of dogs, so that they should not bite into the skin; in 1874, he had written that vivisection threw no light on that disease, the laws of which were "impenetrable to science until now." It now occurred to him that, perhaps, the investigators in the laboratory of the Ecole Normale might be more successful than he had been in his kennels in the Rue Fontaine-au-Roi.

One of the two dogs he sent was suffering from what is called *dumb madness:* his jaw hung, half opened and paralyzed, his tongue was covered with foam, and his eyes full of wistful anguish; the other made ferocious darts at anything held out to him, with a rabid fury in his bloodshot eyes, and, in the hallucinations of his delirium, gave vent to haunting, despairing howls.

Much confusion prevailed at that time regarding this disease, its seat, its causes, and its remedy. Three things seemed positive: firstly, that the rabic virus was contained in the saliva of the mad animals; secondly, that it was communicated through bites; and thirdly, that the period of incubation might vary from a few days to several months. Clinical observation was reduced to complete impotence; perhaps experiments might throw some light on the subject.

One day, Pasteur having wished to collect a little saliva from the jaws of a rabid dog, so as to obtain it directly, two of Bourrel's assistants undertook to drag a mad bulldog, foaming at the mouth, from its cage; they seized it by means of a lasso, and stretched it on a table. These two men, thus associated with Pasteur in the same danger, with the same calm heroism, held the struggling, ferocious animal down with their powerful hands, whilst the scientist drew, by means of a glass tube held between his lips, a few drops of the deadly saliva.

But the same uncertainty followed the inoculation of the saliva; the incubation was so slow that weeks and months often elapsed whilst the result of an experiment was being anxiously awaited. Evidently the saliva was not a sure agent for experiments, and if more knowledge was to be obtained, some other means had to be found of obtaining it.

Magendie and Renault had both tried experimenting with rabic blood, but with no results, and Paul Bert had been equally unsuccessful. Pasteur tried in his turn, but also in vain. "We must try other experiments," he said, with his usual indefatigable perseverance.

As the number of cases observed became larger, he felt a growing conviction that hydrophobia has its seat in the nervous system, and particularly in the medulla oblongata. "The propagation of the virus in a rabid dog's nervous system can almost be observed in its every stage," writes M. Roux, Pasteur's daily associate in these researches, which he afterwards made the subject of his thesis. "The anguish and fury due to the excitation of the grey cortex of the brain are followed by an alteration of the voice and a difficulty in deglutition. The medulla oblongata and the nerves starting from it are attacked in their turn; finally, the spinal cord itself becomes invaded and paralysis closes the scene."

As long as the virus has not reached the nervous centres, it may sojourn for weeks or months in some point of the body; this explains the slowness of certain incubations, and the fortunate escapes after some bites from rabid dogs. The *a priori* supposition that the virus attacks the nervous centres went very far back; it had served as a basis to a theory enunciated by Dr. Duboué (of Pau), who had, however, not supported it by any experiments. On the contrary, when M. Galtier, a professor at the Lyons Veterinary School, had attempted experiments in that direction, he had to inform the Academy of Medicine, in January, 1881, that he had only ascertained the existence of virus in rabid dogs in the lingual glands and in the buccopharyngeal mucous membrane. "More than ten times, and always unsuccessfully, have I inoculated the product obtained by pressure of the cerebral substances of the cerebellum or of the medulla oblongata of rabid dogs."

Pasteur was about to prove that it was possible to succeed by operating in a special manner, according to a rigorous technique, unknown in other laboratories. When the post-mortem examination of a mad dog had revealed no characteristic lesion, the brain was uncovered, and the surface of the medulla oblongata scalded wtih a glass stick, so as to destroy any external dust or dirt. Then, with a long tube previously put through a flame, a particle of the substance was drawn and deposited in a glass just taken from a stove heated up to 200° C., and mixed with a little water or sterilized broth by means of a glass agitator, also previously put through a flame. The syringe used for inoculation on the rabbit or dog (lying ready on the operating board) had been purified in boiling water.

Most of the animals who received this inoculation under the skin succumbed to hydrophobia; that virulent matter was therefore more successful than the saliva, which was a great result obtained.

"The seat of the rabic virus," wrote Pasteur, "is therefore not in the saliva only: the brain contains it in a degree of virulence at least equal to that of the saliva of rabid animals." But, to Pasteur's eyes, this was but a preliminary step on the long road which stretched before him; it was necessary that all the inoculated animals should contact hydrophobia, and the period of incubation had to be shortened.

It was then that it occurred to Pasteur to inoculate the rabic virus directly on the surface of a dog's brain. He thought that, by placing the virus from the beginning in its true medium, hydrophobia would more surely supervene and the incubation might be shorter. The experiment was attempted: a dog under chloroform was fixed to the operating board, and a small, round portion of the cranium removed by means of a trephine (a surgical instrument somewhat similar to a fret-saw); the tough fibrous membrane called the dura-mater, being thus exposed, was then injected with a small quantity of the prepared virus, which lay in readiness in a Pravaz syringe. The wound was washed with carbolic and the skin stitched together, the whole thing lasting but a few minutes. The dog, on returning to consciousness, seemed quite the same as usual. But, after fourteen days, hydrophobia appeared: rabid fury, characteristic howls, the tearing up and devouring of his bed, delirious hallucination, and finally, paralysis and death.

A method was therefore found by which rabies was contracted surely and swiftly. Trephinings were again performed on chloroformed animals —Pasteur had a great horror of useless sufferings, and always insisted on anæsthesia. In every case, characteristic hydrophobia occurred after inoculation on the brain. The main lines of this complicated question were beginning to be traceable; but other obstacles were in the way. Pasteur could not apply the method he had hitherto used, *i.e.* to isolate, and then to cultivate in an artificial medium, the microbe of hydrophobia, for he failed in detecting this microbe. Yet its existence admitted of no doubt; perhaps it was beyond the limits of human sight. "Since this unknown being is living," thought Pasteur, "we must cultivate it; failing an artificial medium, let us try the brain of living rabbits; it would indeed be an experimental feat!"

As soon as a trephined and inoculated rabbit died paralyzed, a little of his rabic medulla was inoculated to another; each inoculation succeeded another, and the time of incubation became shorter and shorter, until, after a hundred uninterrupted inoculations, it came to be reduced to seven days. But the virus, having reached this degree, the virulence of which was found to be greater than that of the virus of dogs made rabid by an accidental bite, now became fixed; Pasteur had mastered it. He could now predict the exact time when death should occur in each of the inoculated animals; his predictions were verified with surprising accuracy.

Pasteur was not yet satisfied with the immense progress marked by infallible inoculation and the shortened incubation; he now wished to decrease the degrees of virulence—when the attenuation of the virus was once conquered, it might be hoped that dogs could be made refractory to rabies. Pasteur abstracted a fragment of the medulla from a rabbit which had just died of rabies after an inoculation of the fixed virus; this fragment was suspended by a thread in a sterilized phial, the air in which was kept dry by some pieces of caustic potash lying at the bottom of the vessel and which was closed by a cotton-wool plug to prevent the entrance of atmospheric dusts. The temperature of the room where this desiccation took place was maintained at 23° C. As the medulla gradually became dry, its virulence decreased, until, at the end of fourteen days, it had become absolutely extinguished. This now inactive medulla was crushed and mixed with pure water, and injected under the skin of some dogs. The next day they were inoculated with medulla which had been desiccating for thirteen days, and so on, using increased virulence until the medulla was used of a rabbit dead the same day. These dogs might now be bitten by rabid dogs given them as companions for a few minutes, or submitted to the intracranial inoculations of the deadly virus: they resisted both.

Having at last obtained this refractory condition, Pasteur was anxious that his results should be verified by a Commission. The Minister of Public Instruction acceded to this desire, and a Commission was constituted in May, 1884, composed of Messrs. Béclard, Dean of the Faculty of Medicine, Paul Bert, Bouley, Villemin, Vulpian, and Tisserand, Director of the Agricultural Office. The Commission immediately set to work; a rabid dog having succumbed at Alfort on June I, its carcase was brought to the laboratory of the Ecole Normale, and a fragment of the medulla oblongata was mixed with some sterilized broth. Two dogs, declared by Pasteur to be refractory to rabies, were trephined, and a few drops of the liquid injected into their brains; two other dogs and two rabbits received inoculations at the same time, with the same liquid and in precisely the same manner.

Bouley was taking notes for a report to be presented to the Minister:

"M. Pasteur tells us that, considering the nature of the rabic virus used, the rabbits and the two new dogs will develop rabies within twelve or fifteen days, and that the two refractory dogs will not develop it at all, however long they may be detained under observation."

On May 29, Mme. Pasteur wrote to her children:

"The Commission on rabies met to-day and elected M. Bouley as chairman. Nothing is settled as to commencing experiments. Your father is absorbed in his thoughts, talks little, sleeps little, rises at dawn, and, in one word, continues the life I began with him this day thirty-five years ago."

On June 3, Bourrel sent word that he had a rabid dog in the kennels of the Rue Fontaine-au-Roi; a refractory dog and a new dog were immediately submitted to numerous bites; the latter was violently bitten on the head in several places. The rabid dog, still living the next day and still able to bite, was given two more dogs, one of which was refractory; this dog, and the refractory dog bitten on the 3rd, were allowed to receive the first bites, the Commission having thought that perhaps the saliva might then be more abundant and more dangerous.

On June 6, the rabid dog having died, the Commission proceeded to inoculate the medulla of the animal into six more dogs, by means of trephining. Three of those dogs were refractory; the three others were fresh from the kennels; there were also two rabbits.

On the 10th, Bourrel telegraphed the arrival of another rabid dog, and the same operations were gone through.

"This rabid, furious dog," wrote Pasteur to his son-in-law,

> had spent the night lying on his master's bed; his appearance had been suspicious for a day or two. On the morning of the 10th, his voice became rabietic, and his master, who had heard the bark of a rabid dog twenty years ago, was seized with terror, and brought the dog to M. Bourrel, who found that he was indeed in the biting stage of rabies. Fortunately a lingering fidelity had prevented him from attacking his master.
>
> This morning the rabic condition is beginning to appear on one of the new dogs trephined on June 1, at the same time as two refractory dogs. Let us hope that the other new dog will also develop it and that the two refractory ones will resist.

At the same time that the Commission examined this dog which developed rabies within the exact time indicated by Pasteur, the two rabbits on whom inoculation had been performed at the same time were found to present the first symptoms of rabic paralysis. "This paralysis," noted Bouley, "is revealed by great weakness of the limbs, particularly of the hind quarters; the least shock knocks them over and they experience great difficulty in getting up again." The second new dog on whom inoculation had been performed on June 1 was now also rabid; the refractory dogs were in perfect health.

Bouley's report was sent to the Minister of Public Instruction at the beginning of August. "We submit to you to-day," he wrote, "this report on the first series of experiments that we have just witnessed, in order that M. Pasteur may refer to it in the paper which he proposes to read at the Copenhagen International Scientific Congress on these magnificent results, which devolve so much credit on French Science and which give it a fresh claim to the world's gratitude."

The Commission wished that a large kennel yard might be built, in order that the duration of immunity in protected dogs might be timed, and that other great problem solved, viz., whether it would be possible, through the inoculation of attenuated virus, to defy the virus from bites.

By the Minister's request, the Commission investigated the Meudon woods in search of a favourable site; an excellent place was found in the lower part of the Park, away from dwelling houses, easy to enclose and presumably in no one's way. But, when the inhabitants of Meudon heard of this project, they protested vehemently, evidently terrified at the thought of rabid dogs, however securely bound, in their peaceful neighbourhood. Another piece of ground was then suggested to Pasteur, near St. Cloud, in the Park of Villeneuve l'Etang. Originally a State domain, this property had been put up for sale, but had found no buyer, not being suitable for parcelling out in small lots; the Bill was withdrawn which allowed of its sale and the greater part of the domain was devoted by the Ministry to Pasteur's and his assistants' experiments on the prophylaxis of contagious diseases.

Pasteur pondered on the means of extinguishing hydrophobia or of merely diminishing its frequency. Could dogs be vaccinated? There are 100,000 dogs in Paris, about 2,500,000 more in the provinces: vaccination necessitates several preventive inoculations; innumerable kennels would have to be built for the purpose, to say nothing of the expense of keeping the dogs and of providing a trained staff capable of performing the difficult and dangerous operations. And, as M. Nocard truly remarked, where were rabbits to be found in sufficient number for the vaccine emulsions?

Optional vaccination did not seem more practicable; it could only be worked on a very restricted scale and was therefore of very little use in a general way.

The main question was the possibility of preventing hydrophobia from occurring in a human being, previously bitten by a rabid dog.

The successful opposition of the inhabitants of Meudon had inspired those of St. Cloud, Ville d'Avray, Vaucresson, Marnes, and Garches with the idea of resisting in their turn the installation of Pasteur's kennels at

Villeneuve l'Etang. People spoke of public danger, of children exposed to meet ferocious rabid dogs wandering loose about the park, of popular Sundays spoilt, picnickers disturbed, etc., etc.

Little by little, in spite of the opposition which burst out now and again, calm was again re-established. French good sense and appreciation of great things got the better of the struggle; in January, 1885, Pasteur was able to go to Villeneuve l'Etang to superintend the arrangements. The old stables were turned into an immense kennel, paved with asphalt. A wide passage went from one end to the other, on each side of which accommodation for sixty dogs was arranged behind a double barrier of wire netting.

The subject of hydrophobia goes back to the remotest antiquity; one of Homer's warriors calls Hector a mad dog. The supposed allusions to it to be found in Hippocrates are of the vaguest, but Aristotle is quite explicit when speaking of canine rabies and of its transmission from one animal to the other through bites. He gives expression, however, to the singular opinion that man is not subject to it. More than three hundred years later we come to Celsus, who describes this disease, unknown or unnoticed until then. "The patient," said Celsus, "is tortured at the same time by thirst and by an invincible repulsion towards water." He counselled cauterization of the wound with a red-hot iron and also with various caustics and corrosives.

Pliny the Elder, a worthy precursor of village quacks, recommended the livers of mad dogs as a cure; it was not a successful one. Galen, who opposed this, had a no less singular recipe, a compound of cray-fish eyes. Later, the shrine of St. Hubert in Belgium was credited with miraculous cures; this superstition is still extant.

Sea bathing, unknown in France until the reign of Louis XIV, became a fashionable cure for hydrophobia, Dieppe sands being supposed to offer wonderful curing properties.

In 1780 a prize was offered for the best method of treating hydrophobia, and won by a pamphlet entitled *Dissertation sur la Rage,* written by a surgeon-major of the name of Le Roux.

This very sensible treatise concluded by recommending cauterization, now long forgotten, instead of the various quack remedies which had so long been in vogue, and the use of butter of antimony.

Le Roux did not allude in his paper to certain tenacious and cruel prejudices, which had caused several hydrophobic persons, or persons merely suspected of hydrophobia, to be killed like wild beasts, shot, poisoned, strangled, or suffocated.

It was supposed in some places that hydrophobia could be transmitted through the mere contact of the saliva or even by the breath of the victims; people who had been bitten were in terror of what might be done to them. A girl, bitten by a mad dog and taken to the Hôtel Dieu Hospital on May 8, 1780, begged that she might not be suffocated!

Those dreadful occurrences must have been only too frequent, for, in 1810, a philosopher asked the Government to enact a Bill in the following terms: "It is forbidden, under pain of death, to strangle, suffocate, bleed to death, or in any other way murder individuals suffering from rabies, hydrophobia, or any disease causing fits, convulsions, furious and dangerous madness; all necessary precautions against them being taken by families or public authorities."

In 1819, newspapers related the death of an unfortunate hydrophobe, smothered between two mattresses; it was said à propos of this murder that "it is the doctor's duty to repeat that this disease cannot be transmitted from man to man, and that there is therefore no danger in nursing hydrophobia patients." Though old and fantastic remedies were still in vogue in remote country places, cauterization was the most frequently employed; if the wounds were somewhat deep, it was recommended to use long, sharp and pointed needles, and to push them well in, even if the wound was on the face.

One of Pasteur's childish recollections (it happened in October, 1831) was the impression of terror produced throughout the Jura by the advent of a rabid wolf who went biting men and beasts on his way. Pasteur had seen an Arboisian of the name of Nicole being cauterized with a red-hot iron at the smithy near his father's house. The persons who had been bitten on the hands and head succumbed to hydrophobia, some of them amidst horrible sufferings; there were eight victims in the immediate neighbourhood. Nicole was saved. For years the whole region remained in dread of that mad wolf.

As to the origin of rabies, it remained unknown and was erroneously attributed to divers causes. Spontaneity was still believed in. Bouley himself did not absolutely reject the idea of it, for he said in 1870: "In the immense majority of cases, this disease proceeds from contagion; out of 1,000 rabid dogs, 999 at least owe their condition to inoculation by a bite."

Pasteur was anxious to uproot this fallacy, as also another very serious error, vigorously opposed by Bouley, by M. Nocard, and by another veterinary surgeon in a *Manual on Rabies*, published in 1882, and still as tenacious as most prejudices, viz., that the word hydrophobia is synonymous with rabies. The rabid dog is *not* hydrophobe, he does *not* abhor water. The word is applicable to rabid human beings, but is false concerning rabid dogs.

Many people in the country, constantly seeing Pasteur's name associated with the word rabies, fancied that he was a consulting veterinary surgeon, and pestered him with letters full of questions. What was to be done to a dog whose manner seemed strange, though there was no evidence of a suspicious bite? Should he be shot? "No," answered Pasteur, "shut him up securely, and he will soon die if he is really mad." Some dog owners hesitated to destroy a dog manifestly bitten by a mad dog. "It is such a good dog!" "The law is absolute," answered Pasteur; "every dog bitten by a mad dog must be destroyed at once." And it irritated him that village mayors should close their eyes to the non-observance of the law, and thus contribute to a recrudescence of rabies.

Pasteur wasted his precious time answering all those letters. On March 28, 1885, he wrote to his friend Jules Vercel—

> Alas! we shall not be able to go to Arbois for Easter; I shall be busy for some time settling down, or rather settling my dogs down at Villeneuve l'Etang. I also have some new experiments on rabies on hand which will take some months. I am demonstrating this year that dogs can be vaccinated, or made refractory to rabies *after* they have been bitten by mad dogs.
>
> I have not yet dared to treat human beings after bites from rabid dogs; but the time is not far off, and I am much inclined to begin by myself—inoculating myself with rabies, and then arresting the consequences; for I am beginning to feel very sure of my results.

In May, everything at Villeneuve l'Etang was ready for the reception of sixty dogs. Fifty of them, already made refractory to bites or rabic inoculation, were successively accommodated in the immense kennel, where each had his cell and his experiment number. They had been made refractory by being inoculated with fragments of medulla, which had hung for a fortnight in a phial, and of which the virulence was extinguished, after which further inoculations had been made, gradually increasing in virulence until the highest degree of it had again been reached.

All those dogs, which were to be periodically taken back to Paris for inoculations or bite tests, in order to see what was the duration of the immunity conferred, were stray dogs picked up by the police. They were of various breeds, and showed every variety of character, some of them gentle and affectionate, others vicious and growling, some confiding, some shrinking, as if the recollection of chloroform and the laboratory was disagreeable to them. They showed some natural impatience of their enforced captivity, only interrupted by a short daily run. One of them, however, was promoted to the post of house-dog, and loosened every night; he excited much envy among his congeners. The dogs were very well cared for by a retired *gendarme*, an excellent man of the name of Pernin.

A lover of animals might have drawn an interesting contrast between the fate of those laboratory dogs, living and dying for the good of humanity, and that of the dogs buried in the neighbouring dogs' cemetery at Bagatelle, founded by Sir Richard Wallace, the great English philanthropist. Here lay toy dogs, lap dogs, drawing-room dogs, cherished and coddled during their useless lives, and luxuriously buried after their useless deaths, while the dead bodies of the others went to the knacker's yard.

Rabbit hutches and guinea-pig cages leaned against the dogs' palace. Pasteur, having seen to the comfort of his animals, now thought of himself; it was frequently necessary that he should come to spend two or three days at Villeneuve l'Etang. The official architect thought of repairing part of the little palace of Villeneuve, which was in a very bad state of decay. But Pasteur preferred to have some rooms near the stables put into repair, which had formerly been used for non-commissioned officers of the Cent Gardes; there was less to do to them, and the position was convenient. The roof, windows, and doors were renovated, and some cheap paper hung on the walls inside. "This is certainly not luxurious!" exclaimed an astonished millionaire, who came to see Pasteur one day on his way to his own splendid villa at Marly.

On May 29 Pasteur wrote to his son—

> I thought I should have done with rabies by the end of April; I must postpone my hopes till the end of July. Yet I have not remained stationary; but, in these difficult studies, one is far from the goal as long as the last word, the last decisive proof is not acquired. What I aspire to is the possibility of treating a man after a bite with no fear of accidents.
>
> I have never had so many subjects of experiments on hand—sixty dogs at Villeneuve l'Etang, forty at Rollin, ten at Frégis', fifteen at Bourrel's, and I deplore having no more kennels at my disposal.
>
> What do you say of the Rue Pasteur in the large city of Lille? The news has given me very great pleasure.

What Pasteur briefly called "Rollin" in this letter was the former *Lycée Rollin*, the old buildings of which had been transformed into outhouses for his laboratory. Large cages had been set up in the old courtyard, and the place was like a farm, with its population of hens, rabbits, and guinea-pigs.

Two series of experiments were being carried out on those 125 dogs. The first consisted in making dogs refractory to rabies by preventive inoculations; the second in preventing the onset of rabies in dogs bitten or subjected to inoculation.

On Monday, July 6, Pasteur saw a little Alsatian boy, Joseph Meister, enter his laboratory, accompanied by his mother. He was only nine years old, and had been bitten two days before by a mad dog at Meissengott, near Schlestadt.

The child, going alone to school by a little by-road, had been attacked by a furious dog and thrown to the ground. Too small to defend himself, he had only thought of covering his face with his hands. A bricklayer, seeing the scene from a distance, arrived, and succeeded in beating the dog off with an iron bar; he picked up the boy, covered with blood and saliva. The dog went back to his master, Théodore Vone, a grocer at Meissengott, whom he bit on the arm. Vone seized a gun and shot the animal, whose stomach was found to be full of hay, straw, pieces of wood, etc. When little Meister's parents heard all these details they went, full of anxiety, to consult Dr. Weber, at Villé, that same evening. After cauterizing the wounds with carbolic, Dr. Weber advised Mme. Meister to start for Paris, where she could relate the facts to one who was not a physician, but who would be the best judge of what could be done in such a serious case. Théodore Vone, anxious on his own and on the child's account, decided to come also.

Pasteur reassured him; his clothes had wiped off the dog's saliva, and his shirt-sleeve was intact. He might safely go back to Alsace, and he promptly did so.

Pasteur's emotion was great at the sight of the fourteen wounds of the little boy, who suffered so much that he could hardly walk. What should he do for this child? could he risk the preventive treatment which had been constantly successful on his dogs? Pasteur was divided between his hopes and his scruples, painful in their acuteness. Before deciding on a course of action, he made arrangements for the comfort of this poor woman and her child, alone in Paris, and gave them an appointment for 5 o'clock, after the Institute meeting. He did not wish to attempt anything without having seen Vulpian and talked it over with him. Since the Rabies Commission had been constituted, Pasteur had formed a growing esteem for the great judgment of Vulpian, who, in his lectures on the general and comparative physiology of the nervous system, had already mentioned the profit to human clinics to be drawn from experimenting on animals.

His was a most prudent mind, always seeing all the aspects of a problem. The man was worthy of the scientist: he was absolutely straightforward, and of a discreet and active kindness. He was passionately fond of work, and had recourse to it when smitten by a deep sorrow.

Vulpian expressed the opinion that Pasteur's experiments on dogs were sufficiently conclusive to authorize him to foresee the same success in human pathology. Why not try this treatment? added the professor, usually so reserved. Was there any other efficacious treatment against hydrophobia? If at least the cauterizations had been made with a red-hot iron! but what was the good of carbolic acid twelve hours after the accident. If the almost certain danger which threatened the boy were weighed against the chances of snatching him from death, Pasteur would see that it was more than a right, that it was a duty to apply antirabic inoculation to little Meister.

This was also the opinion of Dr. Grancher, whom Pasteur consulted. M. Grancher worked at the laboratory; he and Dr. Straus might claim to be the two first French physicians who took up the study of bacteriology; these novel studies fascinated him, and he was drawn to Pasteur by the deepest admiration and by a strong affection, which Pasteur thoroughly reciprocated.

Vulpian and M. Grancher examined little Meister in the evening, and, seeing the number of bites, some of which, on one hand especially, were very deep, they decided on performing the first inoculation immediately;

the substance chosen was fourteen days old and had quite lost its virulence: it was to be followed by further inoculations gradually increasing in strength.

It was a very slight operation, a mere injection into the side (by means of a Pravaz syringe) of a few drops of a liquid prepared with some fragments of medulla oblongata. The child, who cried very much before the operation, soon dried his tears when he found the slight prick was all that he had to undergo.

Pasteur had had a bedroom comfortably arranged for the mother and child in the old Rollin College, and the little boy was very happy amidst the various animals—chickens, rabbits, white mice, guinea-pigs, etc.; he begged and easily obtained of Pasteur the life of several of the youngest of them.

"All is going well," Pasteur wrote to his son-in-law on July 11:

> the child sleeps well, has a good appetite, and the inoculated matter is absorbed into the system from one day to another without leaving a trace. It is true that I have not yet come to the test inoculations, which will take place on Tuesday, Wednesday and Thursday. If the lad keeps well during the three following weeks, I think the experiment will be safe to succeed. I shall send the child and his mother back to Meissengott (near Schlestadt) in any case on August 1, giving these good people detailed instruction as to the observations they are to record for me. I shall make no statement before the end of the vacation.

But, as the inoculations were becoming more virulent, Pasteur became a prey to anxiety: "My dear children," wrote Mme. Pasteur, "your father has had another bad night; he is dreading the last inoculations on the child. And yet there can be no drawing back now! The boy continues in perfect health."

Renewed hopes were expressed in the following letter from Pasteur—

> My dear René, I think great things are coming to pass. Joseph Meister has just left the laboratory. The three last inoculations have left some pink marks under the skin, gradually widening and not at all tender. There is some action, which is becoming more intense as we approach the final inoculation, which will take place on Thursday, July 16. The lad is very well this morning, and has slept well, though slightly restless; he has a good appetite and no feverishness. He had a slight hysterical attack yesterday.

The letter ended with an affectionate invitation. "Perhaps one of the great medical facts of the century is going to take place; you would regret not having seen it!"

Pasteur was going through a succession of hopes, fears, anguish, and an ardent yearning to snatch little Meister from death; he could no longer work. At nights, feverish visions came to him of this child whom he had seen playing in the garden, suffocating in the mad struggles of hydrophobia, like the dying child he had seen at the Hôpital Trousseau in 1880.

Vainly his experimental genius assured him that the virus of that most terrible of diseases was about to be vanquished, that humanity was about to be delivered from this dread horror—his human tenderness was stronger than all, his accustomed ready sympathy for the sufferings and anxieties of others was for the nonce centered in "the dear lad."

The treatment lasted ten days; Meister was inoculated twelve times. The virulence of the medulla used was tested by trephinings on rabbits, and proved to be gradually stronger. Pasteur even inoculated on July 16, at 11 A.M., some medulla only one day old, bound to give hydrophobia to rabbits after only seven days' incubation; it was the surest test of the immunity and preservation due to the treatment.

Cured from his wounds, delighted with all he saw, gaily running about as if he had been in his own Alsatian farm, little Meister, whose blue eyes now showed neither fear nor shyness, merrily received the last inoculation; in the evening, after claiming a kiss from "Dear Monsieur Pasteur," as he called him, he went to bed and slept peacefully. Pasteur spent a terrible night of insomnia; in those slow dark hours of night when all vision is distorted, Pasteur, losing sight of the accumulation of experiments which guaranteed his success, imagined that the little boy would die.

The treatment being now completed, Pasteur left little Meister to the care of Dr. Grancher (the lad was not to return to Alsace until July 27) and consented to take a few days' rest. He spent them with his daughter in a quiet, almost deserted country place in Burgundy, but without however finding much restfulness in the beautiful peaceful scenery; he lived in constant expectation of Dr. Grancher's daily telegram or letter containing news of Joseph Meister.

By the time he went to the Jura, Pasteur's fears had almost disappeared. He wrote from Arbois to his son August 3, 1885: "Very good news last night of the bitten lad. I am looking forward with great hopes to the time when I can draw a conclusion. It will be thirty-one days to-morrow since he was bitten."

On his return to Paris, Pasteur found himself obliged to hasten the organization of a "service" for the preventive treatment of hydrophobia after a bite. The Mayor of Villers-Farlay, in the Jura, wrote to him that, on October 14, a shepherd had been cruelly bitten by a rabid dog.

Six little shepherd boys were watching over their sheep in a meadow; suddenly they saw a large dog passing along the road, with hanging, foaming jaws.

"A mad dog!" they exclaimed. The dog, seeing the children, left the road and charged them; they ran away shrieking, but the eldest of them, J. B. Jupille, fourteen years of age, bravely turned back in order to protect the flight of his comrades. Armed with his whip, he confronted the infuriated animal, who flew at him and seized his left hand. Jupille, wrestling with the dog, succeeded in kneeling on him, and forcing his jaws open in order to disengage his left hand; in so doing, his right hand was seriously bitten in its turn; finally, having been able to get hold of the animal by the neck, Jupille called to his little brother to pick up his whip which had fallen during the struggle, and securely fastened the dog's jaws with the lash. He then took his wooden *sabot,* with which he battered the dog's head, after which, in order to be sure that it could do no further harm, he dragged the body down to a little stream in the meadow, and held the head under water for several minutes. Death being now certain, and all danger removed from his comrades, Jupille returned to Villers-Farlay.

Whilst the boy's wounds were being bandaged, the dog's carcase was fetched, and a necropsy took place the next day. The two veterinary surgeons who examined the body had not the slightest hesitation in declaring that the dog was rabid.

The Mayor of Villers-Farlay, who had been to see Pasteur during the summer, wrote to tell him that this lad would die a victim of his own courage unless the new treatment intervened. The answer came immediately: Pasteur declared that, after five years' study, he had succeeded in making dogs refractory to rabies, even six or eight days after being bitten; that, he had only once yet applied his method to a human being, but that once with success, in the case of little Meister, and that, if Jupille's family consented, the boy might be sent to him. "I shall keep him near me in a room of my laboratory; he will be watched and need not go to bed; he will merely receive a daily prick, not more painful than a pin-prick."

The family, on hearing this letter, came to an immediate decision; but, between the day when he was bitten and Jupille's arrival in Paris, six whole days had elapsed, whilst in Meister's case there had only been two and a half!

Yet, however great were Pasteur's fears for the life of this tall lad, who seemed quite surprised when congratulated on his courageous conduct, they were not what they had been in the first instance—he felt much greater confidence.

A few days later, on October 26, Pasteur in a statement at the Academy of Sciences described the treatment followed for Meister. Three months and three days had passed, and the child remained perfectly well. Then he spoke of his new attempt. Vulpian rose—

"The Academy will not be surprised," he said, "if, as a member of the Medical and Surgical Section, I ask to be allowed to express the feelings of admiration inspired in me by M. Pasteur's statement. I feel certain that those feelings will be shared by the whole of the medical profession.

"Hydrophobia, that dread disease against which all therapeutic measures had hitherto failed, has at last found a remedy. M. Pasteur, who has been preceded by no one in this path, has been led by a series of investigations unceasingly carried on for several years, to create a method of treatment, by means of which the development of hydrophobia can *infallibly* be prevented in a patient recently bitten by a rabid dog. I say infallibly, because, after what I have seen in M. Pasteur's laboratory, I do not doubt the constant success of this treatment when it is put into full practice a few days only after a rabic bite."

Bouley, then chairman of the Academy, rose to speak in his turn—

"We are entitled to say that the date of the present meeting will remain for ever memorable in the history of medicine, and glorious for French science; for it is that of one of the greatest steps ever accomplished in the medical order of things—a progress realized by the discovery of an efficacious means of preventive treatment for a disease, the incurable nature of which was a legacy handed down by one century to another. From this day, humanity is armed with a means of fighting the fatal disease of hydrophobia and of preventing its onset. It is to M. Pasteur that we owe this, and we could not feel too much admiration or too much gratitude for the efforts on his part which have led to such a magnificent result."

As soon as Pasteur's paper was published, people bitten by rabid dogs began to arrive from all sides to the laboratory. The "service" of hydrophobia became the chief business of the day. Every morning was spent by Eugène Viala in preparing the fragments of marrow used for inoculations: in a little room permanently kept at a temperature of 20° to 23° C., stood rows of sterilized flasks, their tubular openings closed by plugs of cotton wool. Each flask contained a rabic marrow, hanging from the stopper by a thread and gradually drying up by the action of some fragments of caustic potash lying at the bottom of the flask. Viala cut those marrows into small pieces by means of scissors previously put through a flame, and placed them in small sterilized glasses; he then added a few drops of veal broth and pounded the mixture with a glass rod. The vaccinal liquid was now ready; each glass was covered with a paper cover, and bore the date of the medulla used, the earliest of which was fourteen days old. For each patient under the treatment from a certain date, there was a whole series of little glasses. The date and circumstances of the bites and the veterinary surgeon's certificate were entered in a register, and the patients were divided into series according to the degree of virulence which was to be inoculated on each day of the period of treatment.

Pasteur took a personal interest in each of his patients, helping those who were poor and illiterate to find suitable lodgings in the great capital. Children especially inspired him with a loving solicitude. But his pity was mingled with terror, when, on November 9, a little girl of ten was brought to him who had been severely bitten on the head by a mountain dog, on October 3, thirty-seven days before! The wound was still suppurating. He said to himself, "This is a hopeless case: hydrophobia is no doubt about to appear immediately; it is much too late for the preventive treatment to have the least chance of success. Should I not, in the scientific interest of the method, refuse to treat this child? If the issue is fatal, all those who have already been treated will be frightened, and many bitten persons, discouraged from coming to the laboratory, may succumb to the disease!" These thoughts rapidly crossed Pasteur's mind. But he found himself unable to resist his compassion for the father and mother, begging him to try and save their child.

After the treatment was over, Louise Pelletier had returned to school, when fits of breathlessness appeared, soon followed by convulsive spasms; she could swallow nothing. Pasteur hastened to her side when these symptoms began, and new inoculations were attempted. On December 2, there was a respite of a few hours, moments of calm which inspired Pasteur with the vain hope that she might yet be saved. This delusion was a short-lived one. Pasteur spent the day by little Louise's bedside, in her parents' rooms in the Rue Dauphine. He could not tear himself away; she herself, full of affection for him, gasped out a desire that he should not go away, that he should stay with her! She felt for his hand between two spasms. Pasteur shared the grief of the father and mother. When all hope had to be abandoned: "I did so wish I could have saved your little one!" he said. And as he came down the staircase, he burst into tears.

He was obliged, a few days later, to preside at the reception of Joseph Bertrand at the Académie Française; his sad feelings little in harmony with the occasion. He read in a mournful and troubled voice the speech he had prepared during his peaceful and happy holidays at Arbois. Henry Houssaye, reporting on this ceremony in the *Journal des Débats*, wrote, "M. Pasteur ended his speech amidst a torrent of applause, he received a veritable ovation. He seemed unaccountably moved. How can M. Pasteur, who has received every mark of admiration, every supreme honour, whose name is consecrated by universal renown, still be touched by anything save the discoveries of his powerful genius?" People did not realize that Pasteur's thoughts were far away from himself and from his brilliant discovery. He was thinking of the child he had been unable to snatch from the jaws of death; his mind was not with the living, but with the dead.

A telegram from New York having announced that four children, bitten by rabid dogs, were starting for Paris, many adversaries who had heard of Louise Pelletier's death were saying triumphantly that, if those children's parents had known of her fate, they would have spared them so long and useless a journey.

The four little Americans belonged to workmen's families and were sent to Paris by means of a public subscription opened in the columns of the *New York Herald;* they were accompanied by a doctor and by the mother of the youngest of them, a boy only five years old. After the first inoculation, this little boy, astonished at the insignificant prick, could not help saying, "Is this all we have come such a long journey for?" The children were received with enthusiasm on their return to New York, and were asked "many questions about the great man who had taken such care of them."

A letter dated from that time (January 14, 1886) shows that Pasteur yet found time for kindness, in the midst of his world-famed occupations.

My dear Jupille, I have received your letters, and I am much pleased with the news you give me of your health. Mme. Pasteur thanks you for remembering her. She, and every one at the laboratory, join with me in wishing that you may keep well and improve as much as possible in reading, writing and arithmetic. Your writing is already much better than it was, but you should take some pains with your spelling. Where do you go to school? Who teaches you? Do you work at home as much as you might? You know that Joseph Meister, who was first to be vaccinated, often writes to me; well, I think he is improving more quickly than you are, though he is only ten years old. So, mind you take pains, do not waste your time with other boys, and listen to the advice of your teachers, and of your father and mother. Remember me to M. Perrot, the Mayor of Villers-Farlay. Perhaps, without him, you would have become ill, and to be ill of hydrophobia means inevitable death; therefore you owe him much gratitude. Good-bye. Keep well.

Pasteur's solicitude did not confine itself to his two first patients, Joseph Meister and the fearless Jupille, but was extended to all those who had come under his care; his kindness was like a living flame. The very little ones who then only saw in him a "kind gentleman" bending over them understood later in life, when recalling the sweet smile lighting up his serious face, that Science, thus understood, unites moral with intellectual grandeur.

ALEXANDER FLEMING

Chemotherapy

•

SCIENTIFIC CHEMOTHERAPY dates from Ehrlich and scientific chemotherapy of a bacterial disease from Ehrlich's Salvarsan, which in 1910 revolutionized the treatment of syphilis. The story of Salvarsan has often been told, and I need not go further into it except to say that it was the first real success in the chemotherapeutic treatment of a bacterial disease. Ehrlich originally aimed at "Therapia magna sterilisans," which can be explained as a blitz sufficient to destroy at once all the infecting microbes. This idea was not quite realized, and now the treatment of syphilis with arsenical preparations is a long-drawn-out affair. But it was extraordinarily successful treatment, and stimulated work on further chemotherapeutic drugs. While they had success in some parasitic diseases the ordinary bacteria which infect us were still unaffected.

It was in 1932 that a sulfonamide of the dye chrysoidine was prepared, and in 1935 Domagk showed that this compound (Prontosil) had a curative action on mice infected with streptococci. It was only in 1936, however, that its extraordinary clinical action in streptococcal septicaemia in man was brought out. Thus just ten years ago and twenty-six years after Ehrlich had made history by producing Salvarsan, the medical world woke up to find another drug which controlled a bacterial disease. Not a venereal disease this time, but a common septic infection which unfortunately not infrequently supervened in one of the necessary events of life—childbirth.

Before the announcement of the merits of the drug Prontosil, the industrialists concerned had perfected their preparations and patents. Fortunately for the world, however, Téfouel and his colleagues in Paris soon showed that Prontosil acted by being broken up in the body with the liberation of sulfanilamide, and this simple drug, on which there were no patents, would do all that Prontosil could do. Sulfanilamide affected streptococcal, gonococcal and meningococcal infections as well as *B. coli* infections in the urinary tract, but it was too weak to deal with infections due to organisms like pneumococci and staphylococci.

Two years later Ewins produced sulfapyridine—another drug of the same series—and Whitby showed that this was powerful enough to deal with pneumococcal infections. This again created a great stir, for pneumonia is a condition which may come to every home.

The hunt was now on and chemists everywhere were preparing new sulfonamides—sulfathiazole appeared, which was still more powerful on streptococci and pneumococci than its predecessors, and which could clinically affect generalized staphylococcal infections.

Since then we have had sulfadiazine, sulfamerazine, sulfamethazine and others. But of these we need not go into detail, so much has already been written about them. Meantime there had appeared other sulfonamide compounds, such as sulfaguanidine, which were not absorbed from the alimentary tract, and these were used for the treatment of intestinal infections like dysentery.

The sulfonamides were very convenient for practice, in that they could be taken by the mouth. The drug was absorbed into the blood, where it appeared in concentration more than was necessary to inhibit the growth of sensitive bacteria. From the blood it could pass with ease into the spinal fluid, so it was eminently suited for the treatment of cerebrospinal infections. The sulfonamides were excreted in high concentration in the urine, so that although they were unable to control generalized infections with coliform bacilli they rapidly eliminated similar infections of the urinary tract. In contrast to the older antiseptics they had practically no toxic action on the leucocytes. There were disadvantages in that they were not without toxicity to the patient.

Soon after the sulfonamides came into practice, also, it was discovered that some strains of what were generally sensitive microbes were resistant to their action. The result of widespread treatment was that the sensitive strains were largely displaced by insensitive strains. This was especially noticeable in gonococcal infections, and after a few years something like half of the gonococcal infections were sulfonamide insensitive.

This could be due to one of two things; the sensitive organisms might have been eliminated by treatment with the drug, while the insensitive ones persisted and were passed on from one individual to another; or that by insufficient treatment with the drug a sensitive microbe might have acquired a resistance or "fastness" to the drug.

It is not difficult in the laboratory to make sensitive bacteria resistant to the sulfonamides, but this is not peculiar to the sulfonamides. There is probably no chemotherapeutic drug to which in suitable circumstances the bacteria cannot react by in some way acquiring "fastness."

In the first year of the war the sulfonamides had the field of chemotherapy of septic infections to themselves, but there were always the drawbacks I have mentioned. Later another type of sulfonamide, "Marfanil," was introduced in Germany which for systemic administration had relatively little potency, but which was not inhibited by pus or the usual sulfonamide inhibitors. This was largely used in Germany throughout the war, but there is no doubt that their methods of dealing with sepsis were far behind ours.

The sulfonamides did not directly kill the organisms—they stopped their growth, and the natural protective mechanisms of the body had to complete their destruction. This explained why in some cases of rather long-continued streptococcal septicaemia sulfanilamide failed to save the patient, although the *Streptococcus* was fully sensitive to the drug; the protective mechanism of the body—the opsonic power and phagocytes—had become worn out and failed.

Fildes introduced a most attractive theory of the action of chemotherapeutic drugs. It was that these drugs had a chemical structure so similar to an "essential metabolite" of the sensitive organism that it deluded the organism into the belief that it was the essential metabolite. The organism therefore took it up, and then its receptors became filled with the drug so that it was unable to take up the essential metabolite which was necessary for its growth. Thus it was prevented from growing and died or was an easy prey for the body cells. This theory had been supported by many experimental facts and may give a most profitable guide to future advances in chemotherapy.

But another completely different type of chemotherapeutic drug appeared, namely, penicillin. This actually was described years before the sulfonamides appeared, but it was only concentrated sufficiently for practical chemotherapeutic use in 1940.

The story of penicillin has often been told in the last few years. How, in 1928, a mold spore contaminating one of my culture plates at St. Mary's Hospital produced an effect which called for investigation; how I found that this mold—a *Penicillium*—made in its growth a diffusible and very selective antibacterial agent which I christened Penicillin; how this substance, unlike the older antiseptics, killed the bacteria but it was nontoxic to animals or to human leucocytes; how I failed to concentrate this substance from lack of sufficient chemical assistance, so that it was only ten years afterwards, when chemotherapy of septic infections was a predominant thought in the physician's mind, that Florey and his colleagues at Oxford embarked on a study of antibiotic substances, and succeeded in concentrating penicillin and showing its wonderful therapeutic properties; how this happened at a critical stage of the war, and how they took their information to America and induced the authorities there to produce penicillin on a large scale; how the Americans improved methods of production so that on D-day there was enough penicillin for every wounded man who needed it, and how this result was obtained by the closest cooperation between governments, industrialists, scientists and workmen on both sides of the Atlantic without thought of patents or other restrictive measures. Everyone had a near relative in the fighting line and there was the urge to help him, so progress and production went on at an unprecedented pace.

Penicillin is the most powerful chemotherapeutic drug yet introduced. Even when it is diluted 80,000,000 times it will still inhibit the growth of *Staphylococcus*. This is a formidable dilution, but the figure conveys little except a series of many naughts. Suppose we translate it into something concrete. If a drop of water is diluted 80,000,000 times it would fill over 6,000 whisky bottles.

We have already seen that all the older antiseptics were more toxic to leucocytes than to bacteria. The sulfonamides were much more toxic to bacteria than to leucocytes, but they had some poisonous action on the whole human organism. Here in penicillin we had a substance extremely toxic to some bacteria but almost completely nontoxic to man. And it not only stopped the growth of the bacteria, it killed them, so it was effective even if the natural protective mechanism of the body was deficient. It was effective, too, in pus and in the presence of other substances which inhibited sulfonamide activity.

Penicillin has proved itself in war casualties and in a great variety of the ordinary civil illnesses, but it is specific, and there are many common infections on which it has no effect. Perhaps the most striking results have been in venereal disease. Gonococcal infections are eradicated with a single injection and syphilis in most cases by a treatment of under ten days. Subacute bacterial endocarditis, too, was a disease which until recently was almost invariably fatal. Now with penicillin treatment there are something like 70 per cent recoveries.

But I am not giving you a discourse on penicillin. Suffice it to say that it has made medicine and surgery easier in many directions, and in the near future its merits will be proved in veterinary medicine and possibly in horticulture.

The spectacular success of penicillin has stimulated the most intensive research into other antibiotics in the hope of finding something as good or even better.

Gramicidin and Tyrothricin

But even before penicillin was publicized another antibiotic had been introduced by Dubos in 1939. This was a substance made by the *Bacillus brevis*, which had a very powerful inhibitory action on the Gram-positive bacteria. This substance was originally named gramicidin, but later the name was changed to tyrothricin, when it was found to be a mixture of two antibiotic substances—true gramicidin and tyrocidin. Gramicidin has proved to be a very useful local application to infected areas. It has an inhibitory power on bacteria far in excess of its antileucocytic power, but unfortunately it is toxic when injected, so that it cannot be used for systemic treatment. If penicillin had not appeared it is likely that gramicidin or tyrothricin would have been much more extensively used, but penicillin, which is quite nontoxic, can be used either locally or systemically for almost every condition which would be benefited by gramicidin.

Streptomycin

Waksman in 1943 described this antibiotic, which is produced by *Streptomyces griseus.* This substance has very little toxicity and has a powerful action on many of the Gram-negative organisms. It has been used in tularaemia, undulant fever, typhoid fever, and *B. coli* infections, but the greatest interest has been in its action on the tubercle bacillus. *In vitro* it has a very powerful inhibitory action on this bacillus, and in guinea pigs it has been shown to have a definite curative action. In man, however, the clinical results have not been entirely successful, but in streptomycin we have a chemical which does have *in vivo* a definite action on the tubercle bacillus and which is relatively nontoxic. This is a great advance and may lead to startling results. One possible drawback may be that bacilli appear to acquire rapidly a fastness to streptomycin, much more rapidly than they do to penicillin or even the sulfonamides.

Many other antibiotics have been described in the last five years. Most of them are too toxic for use, but there are some which so far have promise in preliminary experiments. Whether they are going to be valuable chemotherapeutic agents belongs to the future.

Tomorrow

Let us now consider the future. There are now certain definite lines on which research is proceeding in antibacterial chemotherapy.

Fildes's theory of the action of chemotherapeutic drugs has already led to certain results—not sufficiently powerful to have made wonderful advances in practical therapeutics—but the work goes on, and from it at any time some new antibacterial chemical combination may emerge. All this is dependent on further fundamental research on the essential metabolites necessary for the growth of different bacteria.

Bacteriologists and mycologists are, by more or less established methods, investigating all sorts of molds and bacteria to see if they produce antibiotic substances. The chemist concentrates or purifies the active substance, and then the experimental pathologist tests the concentrate for activity and toxicity. There are teams of workers who are thus investigating every bacillus and every mold in the collections which exist in various countries. This is useful team work and may lead to something of practical importance, but it is reminiscent of the momentous German researches lacking in inspiration but which by sheer mass of labor bear some fruit.

It seems likely that in the next few years a combination of antibiotics with different antibacterial spectra will furnish a "cribrum therapeuticum" from which fewer and fewer infecting bacteria will escape.

Then the work on antibiotics has led to the discovery of many new chemical combinations possessing antibacterial power. Most of the antibiotics have certain disadvantages—many of them are too toxic—but it may not be beyond the powers of the organic chemists to alter the formula in such a way that the antibiotic power is retained, but the toxic power reduced to such an extent that these substances can be used therapeutically.

As to chemotherapeutic research in general, I would like to conclude with a quotation from Mervyn Gordon: "No research is ever quite complete. It is the glory of a good bit of work that it opens the way for something still better, and this rapidly leads to its own eclipse. The object of research is the advancement, not of the investigator, but of knowledge."

WALTER GOODMAN

The Battle Against Cancer

•

NEW EXCITEMENT is running through the ranks of cancer researchers, generated by mounting evidence that human cancer may be caused by viruses.

Until now, doctors have been limited to fighting cancer after it has been diagnosed—with surgery, radiation, drugs. Their accomplishments have been remarkable, but not remarkable enough. Of the Americans who will die this year, one in six will die of cancer.

Today our scientists face the challenge not only of treating cancer but of preventing it altogether. In laboratories around the world teams of researchers representing disciplines old and new are closing in on the long-hidden nature of the disease. They may be on their way to fulfilling the prophecy that H. G. Wells made years ago:

> The disease of cancer will be banished from life by calm, unhurrying, persistent men and women, working with every shiver of feeling controlled and suppressed, in hospitals and laboratories, and the motive that will conquer cancer will not be pity nor horror; it will be curiosity to know how and why.

These seekers after the how and the why are civilization's troops in this war, and many of them are converging just now on that strange bit of life (if it is, in fact, alive), the virus.

Dr. Wendell M. Stanley, director of the University of California Virus Laboratory and one of America's most distinguished virologists, has been preaching the virus hypothesis to his colleagues, with results that we can measure in dollars and cents: Not many years ago only a few hundred thousand dollars were available for virus-cancer studies; today the figure is upwards of $10 million.

If Dr. Stanley's hypothesis is correct, if human cancer-causing viruses can be identified, then we are on the way to a cancer vaccine. Virus hunters are already looking toward the time, perhaps only a few years away, when they will have a vaccine that will eliminate for once and all the incurable disease of leukemia. In addition to being a momentous achievement in itself, this could be a harbinger of farther-reaching discoveries.

"The leukemia vaccine," says one specialist, "will probably be the first significant advance in cancer prevention." The great hope is that, just as today's children are protected against crippling polio, tomorrow's children will be protected against murderous cancer.

The accumulating evidence that cancer viruses exist is impressive, dramatic, and slightly eerie. To begin with the eerie, in February, 1961, Sister Mary Viva, principal of the St. John Brebeuf School in Niles, Illinois, wrote a letter to the executive director of the American Cancer Society's Illinois Division, to report "an unusual number of deaths of children from leukemia." On investigation it turned out that over a three-year period, between the fall of 1957 and the summer of 1960, eight children in her small close-in suburb of Chicago had come down with leukemia, a form of cancer which strikes severely at the young.

Eight cases in three years may not seem like many to the untrained observer, but experts knew that it was more than four times the number of leukemia cases that could have been expected for the town of Niles (population then 19,000) in that period—enough to alert epidemic specialists from the U.S. Public Health Service's Communicable Disease Center in Atlanta. They were not prepared for what they found. Seven of the eight children, they discovered, either attended or had brothers or sisters attending the same parochial school; the eighth child had close friends there. Leukemia in Niles was behaving suspiciously like an infectious disease.

That is not the same as saying that leukemia is contagious. There is no evidence that leukemia, or any other kind of cancer, can be transmitted from individual to individual—if there were, leukemia patients would be quarantined. Even so, leukemia may be caused by a microorganism or a virus. Tetanus, for example, is such a disease. It is not passed directly from one person to another, but it threatens entire communities when the microorganism that causes it gets abroad.

As "clusters" of leukemia have been reported from other places—from Philadelphia to Cheyenne, from Bergen County, New Jersey, to western Idaho, and from other lands as well—suspicion of infection has mounted. Because Americans move around so much, it is difficult to get accurate reports on where leukemia victims lived when the disease first struck. The largest cluster located occurred in a small, isolated area out West. Between 1950 and 1961 there were at least 24 cases of leukemia, 16 of them in children, in the twin cities of Clarkston, Washington, and Lewiston, Idaho. Within Lewiston itself, the cases were bunched in a single suburb of fewer than 10,000 people.

The numbers of cases are small, and occasional clusters of this sort may be only statistical coincidences. But experts tend to take them seriously, as pointers to how the disease works. (Happily for the peace of mind of the inhabitants of Niles, Clarkston and Lewiston, their leukemia "epidemics" seem to have abated.)

The most imposing effort to find out where cancer settles in a community, and why—though not directly related to the virus hunt—took place in Washington County, Maryland. For five years, from 1957 to 1962, a team of scientists from the National Cancer Institute kept this entire county—selected because it had a stable population—under a microscope. In search of environmental factors in cancer, they studied the air, the water, the soil, the plants, the houses people lived in, the pets they owned. They probed deeply into the medical histories of 15,000 families. And here again they found that the incidence of cancer does vary from place to place, although they could uncover no reason for the variation except sheer chance.

For the laboratory scientist the circumstantial evidence of cancer clusters is tantalizing.

Convinced though he may be that viruses are at work, he has not been able to catch the culprits. A human cancer virus has yet to be discovered. However, many different viruses have been pinned down as causes of cancer in animals—and these are highly suggestive findings.

It is more than 50 years now since Dr. Peyton Rous, working at the Rockefeller Institute demonstrated that a virus caused cancer in chickens. He was met with cool skepticism, "But, my dear fellow," one prominent authority is supposed to have said, "don't you see, it can't be cancer, because you have found the cause." The story may be apocryphal, but decades passed before a concerted effort was made to pursue this momentous discovery. Virus studies were actually banned from many cancer-research institutions. "Not so long ago," one scientist told me, "suggesting that cancer might be caused by viruses was like inviting somebody to spit in your eye."

As late as 1956, Nobel Laureate Wendell Stanley would still complain: "I find it very difficult to understand why so many investigators have continued to have such a firm blind spot with respect to the virus causation of cancer. . . . I continue to be amazed at the willingness of so many investigators to accept viruses as (causative) agents for animal cancers and their unwillingness to consider them . . . in cancers of man."

But interest soared in the late '50's. By 1959, pioneer Peyton Rous was able to say: "The study of viruses, and of the tumor viruses in special, has never been so rewarding and exciting as now. All is in flux."

Viruses have been found to cause cancer not only in chickens but in rabbits, frogs, mice, hamsters, deer, bear and monkeys, as well as in some varieties of plant life. One versatile virus brings on 20 different kinds of tumors in mice; significantly, it can cross species, to afflict rats and hamsters too.

Finding a virus that works on different species of animals is significant because we are ultimately concerned with men, not with mice. Experimenting on animals is difficult enough; experimenting directly on humans is virtually impossible. People do not roll up their sleeves eagerly for an injection of a virus suspected of causing a fatal disease. So our scientists are restricted in the main to working with lesser animals, or with human-tissue culture in the artificial setting of the laboratory.

Still, they have made provocative discoveries. They have found recently, for example, that two common viruses known to cause a mild respiratory disease in man cause tumors in hamsters. Working from the other direction, they have found that a monkey virus produces bizarre changes in human cells.

Particles similar to the viruses that cause mouse leukemia have been discovered in human leukemia patients. At New York's Sloan-Kettering Institute a virus was isolated from human cancers which has never been found either in animals or in human beings without cancer. Suspicious findings, but not proof positive.

Even though viruses have been caught loitering around the scene of human cancer crimes, investigators are not yet able to build a foolproof case against them. And even if a specific virus is someday soon incriminated in a specific kind of cancer—say, leukemia—that will not mean that every case of cancer is virus-produced.

To understand why it is proving so difficult to trap a virus while it actually does its alleged dirty work on human cells, we have to know something about the nature of the virus itself, whose *modus operandi* is still shrouded in mystery.

There are hundreds of different viruses; some we know are involved with specific diseases such as smallpox or polio; others we know nothing about except that they exist. Some look like spheres, some like rods, some like tadpoles. Each virus is an incredibly tiny particle, as small as one millionth of an inch across, far smaller than any bacteria. It is difficult to imagine anything that tiny, and impossible to see it without the aid of an electron microscope, which magnifies up to 200,000 times.

Each virus consists of nucleic acid in a jacket of protein, which makes it chemically similar to the body's genes. The 1960's have seen a revolutionary breakthrough in our knowledge of how the nucleic acid in our cells passes along hereditary information from parents to children. It has produced a whole new scientific discipline—molecular biology—bringing together biology, biochemistry, physical chemistry, virology and genetics.

Revelations about the Genetic Code have come fast and fascinating, opening up the possibility that man may one day control heredity and direct the future evolution of the race. That possibility remains, for the time being, in the realm of science fiction. But the prospect that the secrets of cancer will be bared along with the secrets of normal cell life is not so farfetched.

We know that viruses, like our normal genes, can wreak permanent changes on the body cells which they enter. Until it gets into a cell a virus has no life of its own; one researcher describes the virus as a "naked wandering gene in search of a cell." The big difference between an authentic gene and a virus is the difference between an educator and a con man; the gene passes along information, the virus passes along misinformation. When it finds a hospitable cell, the virus may set it off on the wild growth that we call cancer. And having committed this crime, to the great annoyance of our molecular biologists, the virus sometimes seems to shed its protein jacket and disappear into the berserk cell's own nucleus, which also consists of nucleic acid. The cells keep multiplying, forming more and more abnormal cells, but the real culprit has vanished, like a hit-and-run driver.

Unlike bacteria, which work in clumsy ways, sprouting inflammations and secreting poisons, viruses operate with exasperating subtlety. We don't know precisely how they get into cells in the first place, and we can only guess at what happens to them afterward. University of Pennsylvania investigators have been able to develop malignant properties in human tissue by infecting it with a monkey virus, only to be left in some cases without a trace of the virus itself. The investigators put the virus in the tissue, but where is it?

One of the scientists engaged in the search to implicate viruses in human cancer is known to all America. He is Dr. Albert B. Sabin, developer of the live polio vaccine. Dr. Sabin announced in May that he and his co-workers at the Children's Hospital Research Foundation in Cincinnati had isolated a virus from the thymus gland of an 18-year-old youth suffering from chest cancer. It appears to be related to a class of viruses that causes tumors in some animals, and it may have a causal relationship to certain kinds of human cancer. "It would be too good to be true," says Dr. Sabin, "but it could be true." However, he is quick to add, "I'm always prepared to fall flat on my face."

No one knows enough to predict just what, if anything, a given virus will do to a given human cell under given circumstances. Still considerable evidence indicates that viruses do not begin their destructive work on the human body until a physical or chemical irritant gives them their chance.

"Viruses alone will not do the trick," says one very experienced microbiologist flatly. A person may have smoked too many cigarettes for too many years, or breathed in too much polluted air or had too many X-rays. It may be, some researchers believe, a constitutional susceptibility or hormonal maladjustment or simply the aging process itself, or even emotional stress, that opens the body to infection. Some investigators are convinced that cancer viruses are present in all of us, possibly from childhood, just lying in wait for an opportunity to strike.

None of this would make the workings of the alleged cancer viruses unique. Dr. Robert J. Huebner, of the National Institute of Allergy and Infectious Diseases, reminds us that the tuberculosis bacillus is most dangerous to a person who has a hereditary disposition to the disease, or who lives in an unhealthy environment. Other people can harbor the TB bug for years and never come down with TB. But merely living in a slum will not give anybody tuberculosis unless the bacillus is present.

To get back to our subject, this could mean that if a lung-cancer virus is someday found and successfully attacked before it has a chance to do the attacking, then heavy cigarette smokers will be able to breathe, or rather inhale, more easily. (Smokers had best keep in mind, though, that this is still highly speculative. Viruses have not yet been connected with lung cancer; cigarettes have been.)

Beyond the present all-out search for human cancer viruses is the prospect that, if they are found, microbiologists will be able to grow their own viruses and so produce cancer vaccine. As vaccine developer Sabin cautions, it is still early to talk about such a vaccine; but researchers cannot keep their speculations from turning in that direction. They know that an antileukemia-virus vaccine has proved successful in mice. They know that vaccines, by using a mild form of the virus they are combating, help make antibodies man's natural defense against disease. And they are encouraged to find that this process appears to be feasible in the case of human cancer.

A vicious cycle seems to be at work on the cancer patient. If his defenses are not strong enough to fight the cancer cell when it first appears, it can gain a foothold. As the disease becomes more severe, the body is further weakened and the cancer cells are able to grow and spread more and more rapidly. How to help our own defense forces stave off the invader is a prime challenge to cancer fighters.

Why doesn't the body's "immunological defense" come into play more potently against cancer? This puzzle is being pondered by some of our most distinguished researchers, including Dr. Jonas Salk, who now heads his own Institute for Biological Studies in La Jolla, California.

The reason, surmises Dr. Salk, may be fairly simple: the cancer cell is so similar to a normal cell that the body has trouble recognizing it as an interloper. It is as though a foreign-born traitor were playing his role as an ordinary American so well that the FBI could not spot him among our millions of loyal citizens.

But the body is not always fooled. The rate of leukemia for Japanese who received the heaviest radiation in the A-bomb explosions reached the unprecedented figure of one in every 100 persons. But 99 persons out of the 100 *did not* come down with leukemia. Scientists would love to know why.

They do have some clues. In his intensive studies of the Niles "cluster" families, blood specialist Dr. Steven O. Schwartz, director of the Department of Hematology of Chicago's Hektoen Institute for Medical Research, found leukemia antibodies in about a third of the members of the victims' families. Mothers of the victims and the brothers and sisters closest to them turned out to be most likely to have such antibodies.

To Dr. Schwartz, the evidence indicates that exposure to a leukemia virus has different consequences for different individuals. Some people develop no antibodies, develop no immunity and are stricken with leukemia. Others develop antibodies, develop immunity and recover completely. Perhaps early exposure to a weak strain of the virus results in immunity; perhaps there are hereditary factors involved. So far the experts can do no more than make educated guesses.

But summing up what we have learned about leukemia in just a few years, Dr. Schwartz, a man who thoroughly enjoys a good story, a good dinner and a good cigar, comes to a hopeful conclusion: "In view of the evidence that leukemia is virus-caused, and that man is able to produce antibodies, it is logical to foresee the time when leukemia will be either prevented or its treatment radically altered.

"As soon as someone has the luck to grow a leukemia virus—which should happen within a few years—there should be no delay in developing a vaccine."

In anticipation of the day when a human cancer virus will be identified, the National Cancer Institute is already working out techniques for growing viruses in the large quantities that will be needed to develop protective vaccines.

The anticancer investment of the past decades has yielded rich dividends in the prevention, detection and treatment of the disease, and the search for viruses should be thought of as one chapter, a major one, in the bigger story. Large-scale studies, such as those into the cigarette-lung-cancer link, have alerted millions of people to the hazards of some common activities.

We are also learning to detect certain kinds of cancer in their very early stages—and, obviously, the earlier a disease is identified, the better the chance of fighting it. Scores of thousands of women owe their lives to the "pap" smear for the early diagnosis of uterine cervical cancer. According to the American Cancer Society, 1,200,000 Americans are alive today who have been "cured" of cancer—that is, they have survived for five years or more after treatment with no evidence of the disease—and new diagnostic techniques are continually being developed.

While surgeons and radiation specialists have been refining their methods of treatment with encouraging results, biochemists have begun an epic search for effective anticancer drugs, including those which will work against viruses. Hundreds of thousands of chemicals have been tested, and some 30 drugs are now being used clinically, with noteworthy success against the scourge of leukemia in particular.

But biochemists have yet to find a drug that will act against cancer cells without harming healthy cells, or building up resistance in the body.

The fight against cancer is not over, not by any means, and millions of men, women and children will suffer and die before it is over.

"Anyone who 'understands' cancer is not well-informed," Dr. Albert Szent-Györgyi cautioned me at lunch one day, after I had been expounding on the cancer-virus hypothesis.

Dr. Szent-Györgyi, winner of a 1937 Nobel Prize for discovering ascorbic acid and isolating Vitamin C, is skeptical of the theory that viruses cause human cancer; indeed, he is skeptical of theories in general.

"Scientists, like other people," he points out, "build up vested interests in their theories. They are influenced by fame, honor, vanity, money, career and many other things." His advice to advocates who are carried away by their enthusiasm is, "Smile when you say that."

Yet theories are necessary, he concedes, to give direction to a scientist's work, and he attributes his own "small discoveries" to his efforts to prove theories that invariably turned out to be all wrong.

The seekers after a human cancer virus may never quite reach their destination, but the discoveries they make en route may be lifesavers.

After all the properly cautious reservations have been made, the fact remains that never has there been such hope, such a sense among the nation's scientists of being on a fruitful track. They are on the threshold of one of medicine's great mysteries. The challenge that lures them on is not only to cure but to *understand*, to get to the heart of the enigma known as cancer.

For the rest of us, their final success will mean that the most fearsome disease of our time will be forever blocked from beginning its deadly work.

WARREN WEAVER

Radiations and the Genetic Threat

•

The Plot

WE ALL REMEMBER that Dickens used to start some of his novels with two or three chapters which appeared to be wholly unconnected. Then the relationship would gradually and dramatically come to light. Since our common purpose here is clarity rather than suspense, I will tell you at the outset that our plot will, in a way, be similar. Chapter I will be devoted to the villain of the piece—radiations. Chapter II will deal with the innocent victim—the genes. Chapter III will describe the crime—mutations. And Chapter IV will then give the verdict of society, will indicate, at least in modest part, what we ought to do about this.

The Villain—Radiations

What is radiation? It is energy on the move, energy being transmitted from one location in space to another. But this remark requires an immediate modification. For not all energy on the move is radiation. A thrown baseball or the moving stream of water from a hose—these involve moving energy, but these are purely mechanical effects, rather than radiation. So we must be more accurate and say that radiation is electromagnetic energy being transmitted. I am not speaking here of electricity in wires, but rather of electromagnetic waves—radio or television waves—moving freely through space, or as we very inaccurately sometimes say, "through the air." (This is really a bad phrase, for insofar as the air plays any role at all, it tends to stop such waves rather than transmit them. A century ago scientists used to say that these waves moved through the "aether," but they just invented this word to diminish their worry over the fact that they didn't understand what was happening. We don't either, but we have abandoned the verbal tranquilizer.)

The most familiar instance of an electromagnetic wave is light, the visible light which affects our eyes. And fortunately this familiar instance is a completely typical instance, as we will see in just a moment.

Any wave disturbance can be partly characterized by its wave length. In the case of a water wave, this wave length is simply the distance from one crest to the next adjacent crest. One can also speak of the frequency of a wave motion, this being simply the number of waves which in each second pass a fixed point. Granting a fixed speed for the waves (and this is the case with electromagnetic waves), the longer the wave length, the fewer of them pass a given point. Quantitatively, if you double the wave length, you halve the frequency. It is often useful to speak both of wave length, which is familiar to us from the water-wave case, and of frequency, which is familiar from the case of sound waves. When one speaks of an "octave" on the piano, the frequency of the upper "do" is just twice the frequency of the lower "do"; so using the sound analogy one can speak of two electromagnetic radiations being an "octave" apart, one of them having a frequency twice that of the other.

In these terms (mixing the sound and light cases) we can think of a great "light-piano." Suppose its keyboard covers about seventy octaves. In the center of it, rather less than one octave wide, is the visible light that affects our eyes. To the right stretch out octave after octave of progressively higher frequencies (smaller and smaller wave lengths). First come about eleven octaves of so-called ultraviolet, the light that is bluer than blue. Then come four octaves of the still shorter wave length X-rays, six octaves of so-called gamma rays, and finally, shortest of all in wave length and highest of all in frequency, some sixteen octaves of very high energy gamma rays derived from cosmic rays.

To the left of the central octave of visible light we find about fifteen octaves of infrared light, then about six octaves of radar waves,[1] and finally some twelve octaves of ordinary radio waves.

That is to say, these physical entities with differing names—radio waves, infrared, visible light, ultraviolet light, X-rays, gamma rays, and cosmic rays—are now all known to be electromagnetic radiations which are alike except that they differ in wave length and hence in frequency. They range from the deep base of the radio waves at the left end of our radiation piano, up to the ultra-high tenor of the cosmic rays at the right end. The only part of this whole spectrum of which we are immediately conscious is the less than one octave of visible light in the middle.

Speaking broadly, the very long wave length radiations tend to flow around any obstacle they meet, while the very short wave lengths tend to penetrate right into an obstacle. Since we are concerned here with radiations which are able to penetrate into our bodies, we see that we are

[1] That is to say, waves of wave lengths which are very short when considered as radio waves, although not in the least short from the point of view of the entire electromagnetic spectrum.

dealing with the tenor half of the radiation piano, namely, with X-rays, the still more penetrating gamma rays and the most of all penetrating cosmic rays.

At this point it will be useful to take note of the most common source of radiations. Most ordinary atoms are stable—their insides stay put. But some atoms are inherently unstable. Their insides have a tendency to readjust into a new pattern of arrangement. It is not possible to predict, for one particular atom, when this readjustment will occur. These unstable atoms—they are often called radioactive—are like alarm clocks wound up and set for unknown times. Eventually the alarm goes off, and the inside of that atom readjusts into a more stable arrangement. When that readjustment occurs, the atom sends out a burst of radiant energy, and this process is in fact the commonest origin of radiations.

Although one cannot at all say when one given radioactive atom will pop off, one can give a useful description of the time behavior of a lot of similar radioactive atoms. One does this by specifying the interval of time within which half of the atoms will pop off. This interval is called the half-life. Starting out with a large number of unpopped atoms of half-life equal to, just as an example, one day, half would be popped by the end of the first day, half of the remaining half of unpopped atoms would then pop in the second day (or a total of three fourths of the original number), again half of the remaining quarter of unpopped atoms in the third day (or a total of seven eighths of the original number), and so on. Some radioactive atoms have half-lives of minutes, hours or days. There are some so transient that they have half-lives as short as a millionth of a second, and others so nearly stable that they have half-lives of thousands or even millions of years.

To return now to the general subject of this chapter, namely the radiations which are capable of penetrating our bodies, part are furnished by nature, and part are caused by man. The part furnished by nature is often referred to as the background radiation, this implying an inevitable and omnipresent base to which is added whatever man causes. Of this background radiation a certain amount comes up from the radioactive material in the rocks and soil. In the top layer of depth one foot there exists on the average, per square mile of earth surface, two grams of radium, eight tons of uranium, and twelve tons of thorium; so clearly the earth under our feet is an important and inescapable source of radiation. On the other hand, part of the background radiation comes down from the sky. This part is due to cosmic rays, so very penetrating that they can pass through ten centimeters of lead, and so universally present that as you read these lines some two to three hundred bursts are passing through your body each second. These two contributions, up from the soil and down from the sky, add together to form the background radiation.

In addition to this background that has been flooding man throughout the centuries are the radiations which man has recently learned how to produce. In this latter category there are two main kinds. First, there are the X-rays so widely used in medicine for both diagnostic and therapeutic purposes. Second, there are the rays (from our point of view the gamma rays are the important ones) produced in nuclear experiments, in atomic weapons testing and in nuclear power plants. In addition to these two main kinds there are various other radiations, usually of minor importance or affecting fewer persons, such as are produced by luminous dials, encountered in certain industries, experienced by certain miners, etc.

The advent of atomic weapons has drawn attention to the possible dangers to man from all sorts of radiations. In the explosion of an atom bomb, in fact, there are produced about 175 kinds of what the physicist calls isotopes—abnormal variants of the ordinary elements. Some of these isotopes from an atom bomb are stable, and the unstable radioactive ones have half-lives which vary from a few seconds to 1,000,000,000 years. About 1,000 pounds of radioactive material are produced per ten-megaton shot.

In the explosion of an atom bomb there are three discernible stages. The first, involving truly awful destruction from immediate radiation, heat, and blast, lasts for the order of one second and extends out, one supposes, to a ten-to-twenty-five-mile radius. The second stage involves a radiation dose due to immediate fallout, which is directly dangerous or even lethal to those receiving it. This lasts for a few days, and extends over an area which presumably may be 10,000 or perhaps in special circumstances even 100,000 square miles. The third stage is that of the eventual "fallout." The finely dispersed radioactive material is carried high into the atmosphere, drifts with the winds, settles down at various rates, or is brought down with rain. This stage lasts over months or years, and extends over the entire planet. This third stage is what is usually meant by "fallout," and the word will be used here in this sense.

So now we have met the villain. It consists of blobs of penetrating energy produced by the popping off of wound-up atoms. Some of these atoms nature herself winds up, but many of these the modern physicist has learned how to wind.

The Innocent Victim—The Genes

Every cell of a person's body contains a great collection, passed down from the parents, the parents' parents, and so on back, of diverse hereditary units called genes. These genes singly and in combination control our inherited characteristics.

These genes exist in every cell of the body. But from the point of view of heredity the ordinary "body cells," which make up the body as a whole, are not comparably as important as the "germ cells" which exist in the reproductive organs, and which play the essential roles in the production of children.

Ordinarily genes are passed on unchanged to children, grandchildren and more remote descendants, but occasionally they do change. They are changed by certain agents, notably by heat, by some chemicals and by radiation. It is at this point that we begin to see the villain plotting against the innocent victim.

The Crime—Mutations

When a gene becomes thus permanently altered we say it mutates. The gene is then duplicated in its altered form in each subsequent cell division. If the mutant gene is in an ordinary body cell, it is merely passed along to other body cells. The mutant gene, under these circumstances, is not passed on to progeny, and the effect of the mutant gene is limited to the person in whom the mutation occurred.

However, it cannot safely be assumed that this body-cell effect is a negligible one on the person in whom the mutation occurred. For various kinds of cellular abnormalities are known to be perpetuated within an individual through body-cell divisions; so these effects are genetic in the broad sense. In fact, although the quantitative relations are not yet clear, it is nevertheless clear that certain malignancies such as leukemia, and certain other cellular abnormalities, can be induced by ionizing radiations. There is also some evidence that effects of this sort measurably reduce the life expectancy of the individual receiving the radiation.

But to return to a consideration of the risks which are passed on to progeny: the mutant gene may exist in a sperm or an egg cell as a result of a mutation having occurred either in that cell or at some earlier cell stage. In this case, a child resulting from this sperm or egg will inherit the mutant gene.

We are now in a position to indicate why it is that radiations, such as X-rays or gamma rays, can be so serious from the genetic point of view. For although the genes, as described above, normally remain unchanged as they multiply and are passed on from generation to generation, they do very rarely change, or mutate; and radiation, as we have already mentioned, can give rise to such changes or mutations in the genes. Mutation ordinarily affects each gene independently; and once changed, an altered gene then persists from generation to generation in its new or mutant form.

Moreover, the mutant genes, in the vast majority of cases, and in all the species so far studied, lead to some kind of harmful effect. In extreme cases the harmful effect is death itself, or loss of the ability to produce offspring, or some other serious abnormality. What in a way is of even greater ultimate importance, since they affect so many more persons, are those cases that involve much smaller handicaps, which might tend to shorten life, increase disease, reduce fertility, or to be otherwise detrimental.

In assessing the harm done to a population by deleterious genes, it is clear that society would ordinarily consider the death of an early embryo to be of much less consequence than that of a child or young adult. Similarly a mutation that decreases the life expectancy by a few months is clearly less to be feared than one that in addition causes its bearer severe pain, unhappiness, or illness throughout his life. Perhaps most obviously tangible are the instances, even though they be relatively uncommon, in which a child is born with some tragic handicap of genetic origin.

A discussion of genetic damage necessarily involves, on the one hand, certain tangible and imminent dangers, certain tragedies which might occur to our own children or grandchildren; and on the other hand certain more remote trouble that may be experienced by very large numbers of persons in the far distant future.

This is not a suitable occasion on which to go into details. But due to well understood genetic principles it is possible to state some important conclusions concerning the danger which is inherent in a radiation-mutated gene. First of all, the change produced by mutation is practically always a change for the worse. Second, the amount of mutation varies directly with the amount of radiation. Third, there is no minimum amount of radiation which is genetically safe—all radiation is genetically bad. A little radiation is a little bad, and a lot is a lot bad. Fourth, once exposed to some radiation, this never "wears off": that is to say, the genetically important number of mutations depends on the total dose that one accumulates from his own conception up to the time of conception of his last child. Fifth, the radiation that is important genetically is only that which reaches the gonads—that is to say, the male testicles and the female ovaries. Sixth, what counts from the point of view of society as a whole is the total number of mutated genes. Thus a small radiation dose to a large number of persons is, socio-genetically speaking, equivalent to a large dose to a few.

The resulting damage may, in a small proportion of cases, appear promptly in one's children or grandchildren, or it may be hidden for many generations; but it is usually not completely hidden and almost always imposes some small handicap on all generations. Moreover these small handicaps accumulate, and the mutated gene eventually eliminates itself through disaster—the disaster of a person whose life span is so shortened or whose fertility is so impaired that no progeny is possible, and this particular genetic line dies out.

The Verdict of Society

Crime, we ordinarily say, does not pay. One's natural inclination, knowing that any and all radiation is genetically bad, is to say, "Let's just eliminate radiation." But we couldn't do that if we really wanted to, and we wouldn't dare if we could.

We couldn't if we wanted to because of the background radiation which comes up from the soil and rocks, and down from the sky. This background is such as to give each person, on the average in the United States, a reproductive lifetime dose (say, over thirty years, from conception to the birth of the "average" child) of about 3 roentgens.[2]

As a practical matter, moreover, it would be virtually impossible to eliminate man-made radiation also; but this is the part that we wouldn't really dare eliminate even if we could. To consider one type of man-made radiation: at the present time a person in the United States receives a reproductive lifetime dose of about 4.5 roentgens from diagnostic and therapeutic medical X-rays. Any of this that could be avoided without interfering with really necessary medical procedures should, of course, be eliminated. But obviously this involves careful and technical judgments in deciding, in each instance, which is the more acceptable risk—the genetic risk or the medical risk which would result in not using the X-rays.

In addition to the substantial doses from background and from medical X-rays, there is the dose—up to the present time small—due to the radioactive fallout from atomic tests. And in the future we certainly face the possibility of significant doses from nuclear power installations.

The reproductive lifetime fallout dose has recently been estimated (assuming no increase in number or size of weapons tested) to be about 0.1 roentgen. There is a considerable uncertainty, and fluctuation from place to place, in this figure; and it may be a fifth as large as stated or, on the other hand, may perhaps be five times as large.

When we think of the genetic risk from any of these sources of radiation we should always, of course, think in terms of comparing two risks—the risk from the radiation, and the risk we would incur if we eliminated the radiation. We all have to compare risks every day, even though we usually do not do so explicitly, but rather in a vague and unformulated way which invokes experience.

If a person must go from New York to San Francisco, he could look up the traffic statistics, and could thus compare two actual numbers representing the traffic deaths per mile of automobile or of air travel. But in the case before us there are reasons which make any attempted comparison of risk very difficult indeed.

We wish to compare Risk A, the genetic risk from radiation, with Risk B, the medical, economic, political and military risk which might result

[2] A roentgen is the common unit in which radiation dose is measured. You get a gonad dose of about 0.005 roentgen from a dental X-ray, from 0.1 to 1.0 roentgen from a pelvic X-ray, and up to 2 roentgens in a fluoroscopic examination of the pelvis. In the original report of the Genetics Committee of the National Academy of Sciences the background dose was estimated at 4.3 roentgens and the average dose from diagnostic and therapeutic medical X-rays was estimated at about 3 roentgens. A large amount of additional data has now been analyzed, and the result has been to decrease the estimate of the background dose to 3 roentgens, and to increase the estimate of the medical X-ray dose to about 4.5 roentgens. The estimate for the sum of the two thus remains nearly unchanged.

from decreasing X-rays, from handicapping the development of nuclear power, and from weakening our position of world leadership and our capacity to defend ourselves. Our difficulties result from the very basic facts that we do not as yet know enough about human radiation genetics to give precise and quantitative estimates of the radiation Risk A; that we certainly cannot give any accurate estimates of the medical, economic, political or military Risk B; and that even if we could describe both Risk A and Risk B, there would be the final and baffling difficulty that these two risks are inherently unlike and hence essentially incomparable.

To speak only of Risk A, we must remember that our knowledge of genetics is very largely derived from experiments with lower forms of life —fruit flies, corn, mice. Large-scale and controlled genetic experiments with human beings are obviously out of the question. Genetics, moreover, is an inherently complicated and subtle subject, almost no quantitative facts concerning radiation harm are known with high precision, and the great but necessary leap from mice to men is one which unavoidably introduces uncertainties. But do not make the mistake of concluding, from these discouraging comments, that the geneticists have little to offer in the way of knowledge and advice. On the contrary, and despite uncertainties about details and exact values, geneticists are in firm agreement on practically all of the really basic points.

Many of you will, at this point, want to make a protest, or will at least want to voice your confusion. If the basic genetic facts are indeed firm and agreed, how can different informed persons, all of whom are clearly intelligent and socially sensitive, appear to hold such diverse views? Take the much publicized case of fallout. How can some be worried about this as a serious menace, while others even refer to it as "harmless" or "negligible"?

A report of the National Academy of Sciences suggested that we ought to plan our medical and nuclear affairs so that an average U. S. citizen would receive a reproductive lifetime dose, from man-made radiation, of not more than 10 roentgens. Such an amount would probably not double our present genetic load, in the sense of doubling the long-established and long-tolerated rate of natural mutations from background radiation and from other agents (heat and chemicals) which cause radiations. Thus, the report concluded, perhaps 10 roentgens of man-made radiation will not result in an unreasonable burden to society. Well, if 10 roentgens thus gets enthroned as reasonable,[3] isn't the 0.1 roentgen from fallout negligible?

It is certainly not surprising that some persons, deeply and properly concerned over the military and political importance of nuclear weapons, answer that question in the affirmative. But the geneticists, if I interpret them correctly, answer in the negative. How can this be?

I think that this is to be explained, and that some of the differences in

[3] Reasonable, mind you; not harmless.

emphasis among the geneticists are also to be explained, in terms of two paradoxes, one numerical, the other temporal.

The numerical paradox is the one which applies most directly to the fallout problem. The paradox arises by virtue of the fact that some persons are impressed by relative figures, some by absolute amounts.

A fallout dose of 0.1 roentgens, is for example, only 1/100 of the 10 roentgens set as "reasonable" by the NAS report. It is only about 1/500 of the dose that would presumably be required to double the natural rate of mutation. If I am already running a certain risk (and after all surviving in spite of it) then ultimately to add to this risk by only one part in 500 doses, when put in this relative way, seems pretty negligible.

But look at the question in other terms. At the present time, roughly 4.5 per cent of the babies born in the United States have serious defects (congenital malformation, mental defects, epilepsy, cutaneous and skeletal defects, visual and aural defects, etc.). Of these, it seems likely that about half are genetic in origin. Let us roughly assume that 2 per cent of the babies born have defects of mutational origin.

Now all the persons now alive in the world—all the persons who at this moment face this problem—will have, before they are all dead, something of the order of 15,000,000,000 babies (1.5×10^9). The immediate genetic risk to this vast set of babies—the world's next set of persons—may well be increased, due to fallout, only by one part in 5,000 (2×10^{-4}).[4] The increase in risk is very small, but the increase applies to a vast number of persons. The estimated result, in fact, is 6,000 additional handicapped babies. Now what impresses you as more significant: that 6,000 is a good many babies to subject to serious handicap, or that 1/5000 is a very small fraction and correspondingly is a very small relative addition to the 30,000,000 babies that, without fallout, will have serious genetic handicaps? And remember that this calculation underestimates the total radiation effect in two ways: it speaks of first-generation damage (whereas there will be increasing amounts of damage in later generations), and it speaks only of gross abnormalities (which, if we accept the evidence from lower forms of life, constitute only a small part of the total genetic damage).

In connection with these remarks about the numerical paradox there is one aspect of the problem of genetic risk which probably deserves some explanation. Increasing some types of risk to individuals by one part in 5,000 might just result in each individual person experiencing a small amount of additional harm—each one might, in an average year, say, have a certain type of physical distress for a total of 5,001 minutes rather than for 5,000 minutes, as previously. But genetic harm does not work that way. Mutations differ a great deal in their seriousness, of course. But for a given

[4] This factor is the product of 1/500 (a reasonable ratio of fallout to the doubling dose) by 1/10 (a reasonable estimate of the fraction of total damage which would be expressed in the first generation).

mutation, it either occurs or it does not occur. Thus when a lot of persons are subjected to a low radiation dose, almost all of them experience no harm whatsoever, but in the case of a few persons mutation will occur. When a mutation occurs it occurs, so to speak, completely. The result for the person in whom it occurs is just the same as though the mutation had been caused by a larger dose of radiation. In other words, a small dose actually affects a small proportion of the exposed population, and a larger dose affects a larger proportion; but those individuals who are affected experience the same result in the two cases.

The temporal paradox is also a difficult one. What impresses you as more important—a relatively little tangible and tragic suffering encountered promptly, by your own children and grandchildren, say—or a great deal more of rather vague and remote suffering to be encountered by the next fifty or one hundred generations?

Some sincere and intelligent persons, including some geneticists, think it is difficult enough to try to play short-range God, without attempting, or worrying about, the problems of a long-range God. These persons, moreover, have a substantial comfort in their confidence that man's intellect will succeed in finding ways out of the long-range difficulties, so that we are justified in trying to deal only with the next few generations.

On the other hand, there are equally sincere and intelligent persons, again including some geneticists, who think that we may well have no greater responsibility than that of protecting the genetic heritage of the future; and that that responsibility does not in the least excuse our committing genetic crimes simply on the grounds that they will not be found out for a long time.

I hope that the statement of these two paradoxes may help you interpret certain statements which might otherwise confuse you and—worst of all—might lead you to think that this situation is so mixed up that the best thing to do is disregard it. Whatever we do, we must not disregard this problem. A massive discontinuity was introduced into life by the discovery of nuclear fission. We have to learn to live with it, for the alternative is that we do not live.

D. *Man's Mind*

•

Only in comparatively recent times has psychology, the study of man's mind, attained scientific respectability. Its history has involved more than its fair share of pseudo-science and quackery. In Psychology, Science and Man *George A. Miller defines the subject and explores its background.*

One of the most rewarding areas of psychological research has been the study of brain physiology. The brain is an organ of incredible complexity. As John Pfeiffer points out in Introducing the Brain, *"It may take years to explore an area [on the cortex] no bigger than a postage stamp." Yet only through such exploration can we hope to arrive at an understanding of its functions.*

The treatment of mental abnormality has ranged from the trepanning of skulls, first practiced by primitive man, to the unspeakable cruelties of the "insane asylums" which existed until recent times, to the theories of the subconscious of Freud and the psychoanalysts. A recent clinical development has been the use of drugs in the treatment of mental disorders. The results have in many instances been astonishing. The number of cases in mental hospitals has been sharply reduced. Robert S. de Ropp describes some of these drugs and their effects in Sick Minds, New Medicines.

The mystery of the processes of thought has not yet been solved. It assumes its most dramatic form in the inspiration of genius. Out of his long study of the creative process, John Livingston Lowes investigates its workings in Imagination Creatrix. *This whole* Treasury *has been an example of what Lowes discusses. We follow it in the flashes of inspiration of Copernicus, Newton and Darwin; in the laboratory of Madame Curie; in the garden of Gregor Mendel; in the musings of James Hutton as he gazes at the seemingly eternal hills.*

GEORGE A. MILLER

Psychology, Science and Man

•

SEVERAL YEARS AGO a professor who teaches psychology at a large university had to ask his assistant, a young man of great intelligence but little experience, to take over the introductory psychology course for a short time. The assistant was challenged by the opportunity and planned an ambitious series of lectures. But he made a mistake. He decided to open with a short definition of his subject. When the professor got back to his classroom two weeks later he found his conscientious assistant still struggling to define psychology.

An alternative approach is to assume at the very outset that everybody knows, more or less, what psychology is all about. "Psychology," said William James in the first sentence of his classic text, "is the science of mental life." Although this definition no longer means what it did when James wrote *The Principles of Psychology* in 1890, it is relatively familiar and mercifully short. We can use it to launch our discussion of psychology without prolonged introductions.

Psychology is the science of mental life. The key words here are *science* and *mental.*

Our conception of what a science of mental life should be has changed considerably since James's time. In 1890 mental life seemed to be a well-defined thing. No one doubted that a mind was there waiting for scientists to study it. But today, after seventy years of trying to study it scientifically, we are less certain. No longer is it self-evident what a psychologist means when he says that he studies mental life. The modern mind seems to be concealed from view, a mental iceberg floating nine-tenths hidden in a vague, unconscious sea; even its owner can do little more than guess which way it will drift next. At the time James wrote, scientific psychology was very young and the mental life that psychologists had been able to study was largely limited to the *conscious* mental life of *sane, well-educated, adult, Western European, human* beings. Today, every one of the restrictions implied by those adjectives has been removed. As the science of mental life developed, its base broadened to include children, animals, preliterate peoples, the mentally retarded, the insane. It is not obvious that all these newcomers share anything we could call a mental life, in the sense understood during the nineteenth century.

At the time James wrote, his claim that psychology was a science was little more than an expression of hope and enthusiasm. In 1890 scientific psychology was still a possible future development. A few men had begun to ask what they might do to make this branch of philosophy more empirical in its methods and conceptions. A few small laboratories had been founded, a few methods of measurement had been adopted, a few preliminary results had been reported. Whipped together with physiology, philosophy, and great common sense in the delightful Jamesian prose, the result was engaging and full of promise, but still a good deal less than a true science of the mind.

Scientific methods, however, are notoriously successful. Since James wrote his *Principles* there has been a remarkable growth in both the quantity and the quality of scientific research on psychological problems. Today when we say that psychology is a science we support the claim with several impressive accomplishments. Indeed, the rapid development of this young science has disrupted the daily pattern of our lives in scores of ways.

Scientific accomplishments usually affect us on at least two levels. On the one hand, scientific knowledge provides a foundation for technological advances, for the solution of practical problems that arise in the daily affairs of ordinary people. In this aspect, science is something that we exploit, just as we would exploit a natural resource. Many people think that this is all there is to science; they are confused by distinctions between scientists and engineers, between science and technology. But in its essence, science is something more than a useful art. Science has understanding, as well as control, as one of its aims. Thus science affects us at a deeper level by changing the way we understand the world we live in. Scientific advances mold our vision of reality, our fundamental and often unspoken set of assumptions about how the world *really* runs and what people are *really* like. Such effects of science are less tangible than the technological ones, but it is perilous to assume they are less important.

Like all sciences, psychology has influenced our lives on both levels. It has given us technical tricks and it has changed our conception of human nature.

When new fields of scientific activity first take form they begin, almost necessarily, with things and ideas that are part of the common experience of all men. During this early period of growth the science is widely intelligible, and the discoveries it makes can be understood, argued, resisted, supported, or ridiculed by millions of people. At a later stage the science may become more precise, may achieve deeper insight or soar to greater heights of intellectual virtuosity, but it will never again have quite the same impact on the average man's view of himself and the world around him. At this later stage the science may be supported for the technical miracles it mysteriously provides, but it is no longer a living reality to more than a handful of specialists. As its technological impact on society tends to increase, its impact on the common understanding often fades into the background.

Psychology is currently passing through its initial stage. It is still intelligible to most people. It is not unusual to hear a layman say, "I'm something of a psychologist myself, and I think . . ." What he thinks is often subtle and interesting and would not embarrass a more professional practitioner. In order to stay alive among our fellow men, we must all be psychologists. Of course, survival requires us to be mathematicians, physicists, chemists, and biologists, too, but there the distance has grown too great; no layman claims brotherhood without a prolonged initiation ritual conducted at some accredited university. It was not always so. There was a time when Everyman was a physicist, when Shakespeare would interrupt a play to argue against the heliocentric theory of the universe, just as a modern playwright may digress to illustrate or to oppose some new psychological theory today. It is in this initial stage of development that a science is most visible, most controversial, and most likely to change our vision of reality.

In spite of psychology's youth, however, the little knowledge it has painfully gained has fed a thousand different human needs. In some quarters demand has so far outrun supply that many psychologists fear their science has been oversold to an overwilling public. Yet even when we try to be conservative in our appraisal, it is plain that the new psychotechnology has already changed the way we live.

Consider our public schools. Everyone in the United States has felt the influence of modern psychology through its effects on our educational system. Indeed, there was a time when our schools seemed little more than a vast laboratory to test the psychological theories of John Dewey. The modern teacher has tried to use psychology—he has thumbed many a textbook ragged in search of the psychological principles underlying good teaching. Frequently the answer he seeks is not to be found, and the educator's pressing responsibilities to the young force him to extrapolate far beyond the established facts of scientific psychology; he hopes his guesses will be more intelligent if he tries to use psychology. Psychologists have tried to find answers for him. They have carefully explored a variety of conditions that affect how quickly a child can learn. They have painstakingly charted the stages of mental and social development. They have developed better techniques for measuring the progress of the child and the effectiveness of the teaching. They have provided counselling and guidance services outside the classroom. And they have given teachers that indispensable tool, the intelligence test. Yet all this is far too little, for the teacher's needs are great and vitally important.

The mention of intelligence tests is a reminder of another area of psychotechnology, the mental testing business. It is a sizable business.

Mental tests, like the airplane, are part of our heritage from World War I. Before that time the tests were given individually to school children, and they tested nothing but intelligence. During the war, however, psychologists in the U. S. Army developed a pencil-and-paper test of adult intelligence that could be given to thousands of draftees—the famous Army Alpha Test—and so the large-scale testing procedures became firmly established in the public's consciousness. After that the testers began to branch out. They began to test aptitudes, to classify interests, to evaluate achievements. Now they can pigeonhole your personality, assess your emotional stability, your masculinity, your imagination, executive potential, chances of marital bliss, conformity to an employer's stereotype, or ability to operate a turret lathe. Whatever you plan to do, there seems to be a psychological test you should take first. Citizens who resent the many hours spent answering pointless questions are apparently in the minority, since enterprising newspapers and magazines have found that they can boost sales by providing daily or weekly questionnaires for their readers to answer. The flood of tests that has poured out across the nation has included many frauds—tests that are poorly conceived, confusingly phrased, completely unstandardized, and never validated. Psychologists have maintained reasonable professional standards among themselves, but it is not always easy to restrain the amateurs—it is as if everyone who bought a knife became a surgeon. Yet, in spite of these problems, the mental testing movement in the United States has managed to perform a needed service for both the individual and the community.

Once the Army saw how useful psychologists could be in the assessment of men, it began to discover other problems of a similar nature. Soon the psychologist became a familiar member of the military team. For example, during World War II much highly technical military equipment was developed that had never existed before. In the developmental stages it often seemed that no one less gifted than Superman would be able to operate the equipment. The task of making the equipment fit the man was tackled by psychologists, who were able to contribute their knowledge of what a human eye could see or a human ear could hear, how far and how fast a human hand could move, how much interference and distraction a human mind could overcome. Psychologists can help to design trainers and simulators, to plan training programs, to select men who are likely to succeed in each type of job. Moreover, in addition to man-machine problems, the military services have a vast range of psychological problems in the area known as mental health, where psychologists work together with psychiatrists to maintain morale and to heal the mentally wounded. A military branch is a small society unto itself—each application of psychology in our larger society has its parallel in this more limited world of warriors.

One large and active sector of psychotechnology goes under the trade name of industrial psychology. Many of industry's concerns are similar to the Army's—how to select men who will be successful at different types of jobs, how to train workers to do their jobs better. Industrial psychologists have worked on the problem of fatigue: how should intervals of rest and work be alternated to give the greatest output with the least fatigue for the worker? The discovery that an employer often got less for his money from a laborer who worked a ten-hour day than from the one who worked an eight-hour day helped to change management's attitude toward many of labor's demands. Questions of fatigue lead quickly into questions of morale; industrial psychologists have worried mightily over this important factor. And morale, in turn, leads into questions of emotional adjustment. Clinical psychologists and psychiatrists have found their niche in the industrial scene, with a consequent reduction, so it is claimed, in illness, absenteeism, and accidents. Even the executives have succumbed to the psychologist's charms, and many a firm's management has been overhauled on the recommendation of a psychological consultant. There are people who feel that if the traditionally hardheaded American businessman is enough convinced of the usefulness of psychology to spend his own dollars on it, then there must be something to it after all!

A possible reason why some businessmen are willing to tolerate a psychologist underfoot is that they may have made a good profit by following his advice about advertising and selling the company's product. The psychologist has been keenly interested in techniques of persuasion, and his discoveries have colored our advertising, propaganda, politics, and entertainment as these are distributed broadside through our mass media of communication. And by probing around in the consumer's unconscious, a psychologist may turn up some useful information for the advertising agency. Just how far one can go in shaping the public mind with a television screen is debatable. But it is apparent that there are both good and bad ways to advertise; psychologists can often help distinguish between them in advance.

Business is not the only place where careful attention is paid to surveys of public opinion. Government agencies have used polls for years to guide our public policies; politicians are particularly sensitive to fluctuations in their popularity with the voters. And feedback from the grass roots is just one of several ways that social psychology is involved in the processes of government. For example, in 1954 psychological evidence was used in the United States Supreme Court decisions against racial segregation in the public schools, where it was held that separate but equal facilities for both races were impossible because the psychological consequences of segregation were too great a handicap for the minority group. The Court's decision rested as much on a point of psychology as on a point of law.

This recital could be extended for several pages. Psychological dogma influences the way we discipline our children, manage our businesses, and run our marriages. Studies of abnormal behavior modify our conception and treatment of mental illness, incompetence, perversion, criminality, and delinquency. The priest and the rabbi agree in their use of psychological techniques to guide their flocks to salvation. Novels, plays, and movies now feature psychological themes as one of their standard formulas. Psychological drugs have already changed the situation in our mental hospitals, and more are yet to come. Wherever people are involved, psychology can be useful—and that is almost everywhere. Whether we like it or not, the practical application of psychology to our daily affairs is already in an advanced stage.

It must be admitted, however, that not every application of psychology is firmly grounded in scientific evidence. Those who apply psychology to the dynamic processes of an evolving society often jump to conclusions that make their laboratory colleagues tremble and turn pale. But when decisions must be made here and now, they must be made in the light of the evidence at hand, no matter how fragmentary and inconclusive that may be. In the past the same decisions had to be made with even less help; today the man who must take the responsibility can at least console himself that he tried to be intelligent, that his guess was informed by whatever evidence existed. The sun will not stand still while he discovers and verifies every fact he needs to know. He works by guess and hunch and intuitive feel, searching always for what will work, for what will meet the present needs. By a shrewd mixture of intelligence, science, and salesmanship, the applied psychologists have given us better answers to hundreds of practical questions. And they will improve those answers just as fast as our growth in basic, scientific psychology permits.

But, if those are some of the practical consequences of scientific psychology, what are some of the impractical ones? What subtler influences has psychology had upon our contemporary attitudes toward life and the universe? Those subtler effects are not easily converted into 8 per cent investments, yet there is a sense in which they are more deeply significant than any merely technological advances.

Scientific psychology educates public psychology. It informs and enriches the picture of man that we all share and that guides so much of our daily conduct. It modifies the public image that is taken for granted in our literature, in our schools, in our theatres, in art and music, in religion and government. It has been said that if human nature ever changes, it is because we learn to see ourselves in a new way. Our feeling for right and wrong, our sense of what is comic and what is tragic, our judgment of what will perish and what will survive are shaped and reshaped by our silently assumed psychology.

Consider, for example, the shadow that our implicit psychology casts on our conception of power, of how human behavior is controlled, of how man is governed. In every age the standards by which laws are written and enforced, goals are set, promises are kept or broken, actions are judged and rewards are given derive from a loose consensus about human nature, about the gap between what is humanly desirable and what is humanly possible. Change man's image of himself and you send a jar reverberating through the foundations of his society. Those who sit in positions of power are particularly sensitive to tremors in the structure that supports them. They will not let man move from the center of the universe or evolve from a monkey without protest. And their protest can be passionate and merciless.

The extent to which the political system of a country can affect the kind of psychology carried on there is eloquent testimony to the investment that our rulers have in our public image of human nature. Psychologists in the United States during World War II were appalled to see their colleagues in Germany twist psychology to support the Nazi's fantastic claims of racial superiority. The history of Russian psychology also illustrates this danger. The leaders of the revolution were slow at first to recognize the importance of psychology; but by 1923 it was clear that Russian psychology, if it wanted to survive, would have to base its theories on materialistic philosophy. For a brief period, therefore, the official image of Soviet man was that of a physiological robot. When a government decides to impose its preconceived views, science, which is never easy, can become virtually impossible.

Our concern here, however, is not with direct interactions between psychology and government, but with the indirect influence psychology can exert by modifying slowly the opinions that every man holds of himself and his neighbors. What are these influences? A citizen should find it in his own interest to learn which way he is being pushed. Where does scientific psychology seem to lead? What image of man is the psychologist trying to promote? Unfortunately—or, perhaps, fortunately—no simple answer will suffice, for there are many psychologists and many different images.

There is a general scientific ethos shared by most psychologists. They expect to base their image of man on empirical knowledge, not upon political dogma or traditional opinion or divine revelation or esthetic appeal. Once this much is said, however, it is difficult to continue until we know which psychologist we are talking about. There are many ways to be scientific, there are many different psychological problems to be studied, and there are innumerable ways to fit our scraps of evidence together into an image of man.

Up until a century ago psychology was a branch of philosophy; the great thinkers somehow knew intuitively what was true and spent their days inventing clever arguments designed to prove it. Then, beginning with Fechner and Wundt and some of their contemporaries, they began to buttress arguments with observations and experiments: at that point the shift into scientific modes of thinking began. But it was still a philosophical kind of psychology, concerned primarily with the source and nature of man's conscious knowledge.

In the background, however, a tremendous development was taking place in the biological sciences. At the first sign of trouble with the introspective analysis of mental life, therefore, the philosophical preoccupation with Man as Knower was swept away and replaced with the newer vision of Man as Animal. The new focus was not knowledge, but adaptation, not thought, but behavior. The mental life that psychology now began to study was not something to be experienced, but something inferred from action.

Eventually, however, problems inherent in a purely behavioral conception of psychology also began to appear. So the vision of man was once more revised and extended, this time emphasizing Man as Social Animal—buffeted as much by the strange whims of his fellow men as by the stern demands of physiology. Developments in the social sciences—in anthropology and sociology—enabled psychologists to recognize the extent to which all mental life is conditioned by cultural traditions, by personal participation in the social process. The adaptation that man struggled to achieve was now seen to be largely a social adaptation. The knowledge he accumulated was seen to be largely symbolic knowledge, encoded in whatever language his culture provided. And this concern with socially significant symbols led back once more to a renewed concern with Man as Knower, but now in a vastly expanded context of new methods and new theories.

That is about where psychology stands today—partly social science, partly biological science, and still partly psychology.

Where is it going in the future?

We can only guess. It is unlikely that we will see any more revolutions that completely redefine what we mean by mental life. Probably we will see increasing specialization as our factual information continues to grow in depth and detail. The dream of a single philosophical principal that explains everything it touches seems to be fading before the realization that man is vastly curious and complicated, and that we need a lot more information about him before we can formulate and test even the simplest psychological laws. Perhaps a whole set of psychological sciences will eventually emerge, although where the divisions between them should be is not yet clear.

How psychology develops in the future will depend to a large and increasing extent upon what it can contribute to our lives, both individually and collectively. As science in recent years has become more and more an instrument of national policy we have tended more and more to support those scientific enterprises that are relevant to our social, economic, political situation. There are today enormous problems facing us—facing all mankind—where psychological knowledge would be invaluable: education, race prejudice, mental health, old age, population control, international cooperation, and many others. These problems do not themselves pose scientific questions, of course; asking the right questions will always be just as difficult in psychology as it is everywhere in science. But if it is possible for scientific psychology to contribute to the solution of practical problems such as these, its future will be bright indeed.

On the basis of the record so far, there is some reason to be optimistic.

JOHN PFEIFFER

Introducing the Brain

•

THE HUMAN BRAIN is three pounds of "messy substance shut in a dark warm place"—a pinkish-gray mass, moist and rubbery to the touch, about the size of a softball. Shock-absorbing fluid cushions it against bumps, sharp blows and other impacts. It is wrapped in three membranes, including an extra-tough outer envelope, and sets snugly in a crate of bone.

The brain is perched like a flower on the top of a slender stalk which in a six-foot man is not quite a yard long. The top three inches of the stalk, a thick white cable of nerve fibers known as the brainstem, lies entirely within the skull and is partly buried by the bulging halves or hemispheres of the brain. The rest of the long stalk, the spinal cord, is a direct continuation of the cable outside the skull. It runs down through holes in the vertebrae of the spine and ends at the small of the back.

Many branches extend from the central stalk, like the roads that feed traffic in and out of a superhighway. From the right side of the spinal cord thirty-one nerves pass through special windows between the vertebrae to the right side of the body. The same number of nerves pass by similar routes to the left side of the body. Besides the spinal nerves, there are a dozen pairs of cranial nerves which arise from the brainstem in the skull. Thus, eighty-six nerves connect the brain and the rest of the body. Through their finest fibers they reach into the remotest places, and into every nook and cranny from the roots of hairs and teeth to the tips of the toes. This is the general structure of the nervous system.

The nervous system is made up of a large number of cells with long extensions or "tendrils." They come in assorted shapes—ovals, pyramids, bulbs, irregular blobs. The biggest have main bodies about two thousandths of an inch in diameter, and networks of fibers which may extend from a fraction of an inch to several feet. Under the microscope a single brain cell with its fibers may resemble the crown of a tree. Growing out from each branch are smaller branches, and from each of them comes a succession of smaller and smaller offshoots down to the most delicate twig. The brain contains some thirteen billion such cells, five times more than the total number of people in the world.

Some twenty-three hundred years ago Hippocrates discussed this organ in terms which still make sense today:

> And men ought to know that from nothing else but from the brain comes joys, delights, laughter and sports, and sorrows, griefs, despondency and lamentations. And by the brain in a special manner we acquire wisdom and knowledge, and see and hear, and know what are foul and what are fair, what are bad and what are good, what are sweet and what unsavory. . . . By the brain we distinguish objects of relish and disrelish; and the same things do not always please us. And by the same organ we become mad and delirious, and fears and terrors assail us, some by night and some by day.

Controls and Adjustments

What is the brain for? Judging by what we know today, it is the great organ of adjustment. It plays the basic biological role of keeping us adjusted to unpredictable events in the outside world, of preserving our identities in an environment of swift and ceaseless chemical change.

Parts of you are continually dying and being born again. Some three million of your red blood cells die every second. Or to look at it another way, three million red cells are born every second, because the body continuously calls up fresh reserves to keep the total count the same.

Your brain keeps you alive by balancing the processes of birth and decay. These basic reactions have top priority. Everything else either helps in carrying them out, or else waits its turn. We pay a high price when the balance of any vital process is upset. For example, sugar is one of the body's energy-providing substances and we must have just the right amount, no more and no less. You are walking a biological tightrope between coma and convulsion, the possible results of relatively slight changes in blood sugar levels.

But the brain usually receives advance notice of impending trouble. It receives a steady flow of information about current sugar levels, and makes adjustments as effectively as a pilot guiding an airplane through a storm. If there is too much sugar, the excess is burned up and excreted. If there is too little, the liver is instructed to release the proper amount of reserve sugar. Notice what such control implies. The brain must "know" the desired sugar level, about a sixtieth of an ounce for every pint of blood, on the average. It must go by similar standards in regulating breathing (you probably inhale and exhale seventeen to twenty times a minute) and heart-beat rates (about seventy times a minute), and in holding body temperature at 98.6 degrees Fahrenheit.

The brain must also be in constant communication with all parts of the body. Indeed, it turns out to be the headquarters of the most elaborate communications network ever devised. Its activities are the result of the combined and patterned activities of billions of nerve cells. A nerve cell is a living wire which produces and conducts rapid electrical impulses. It keeps itself "loaded" and ready for action with the aid of a built-in battery which runs on an oxygen-sugar mixture and recharges automatically. It fires—that is, emits up to several hundred impulses a second—when triggering impulses reach it from sense organs or from other nerve cells.

These outside signals enter the body of the cell through special receiving fibers which are usually short, fine and highly branched. The cell also has a transmitting or "exit" fiber which, as a rule, is relatively long and thick. Its receiving fibers make delicate contact with the transmitting fiber of fifty other nerve cells on the average; some cells make contact with more than a thousand others. Thus signals are relayed from cell to cell as they pass through the nervous system. The rate at which a signal travels depends on the diameter of the fiber conducting it, the thickest fibers being high-speed express routes.

The slenderest fibers, about 1/25,000 of an inch in diameter, have speed limits of a foot a second or two thirds of a mile an hour. But in large-gauge fibers, which measure about ten times thicker, nerve impulses flash along at speeds up to 150 yards a second, a respectable 300 miles an hour. Thick fast fibers generally connect remote parts of the nervous system; thin slow fibers connect neighboring regions. Thus, if a cell communicates with several other cells at varying distances, the messages all tend to arrive at about the same time. This means that widely scattered parts of the nervous system can be stimulated, inhibited or alerted at once—a distinct advantage in coordinating complex behavior.

The brain uses this network to adjust us to the outside world. Generally speaking its operation can be divided into three parts: (1) it receives input in the form of messages from the sense organs; (2) it organizes the input on the basis of past experience, current events and future plans; and (3) it selects and produces an appropriate output, an action or series of actions. In the following sections we will discuss input and output, leaving the matter of what happens in between for later discussions.

Streams of Sensation

The brain keeps in constant touch with the flow of events. It is stirred up by lights, sounds, odors and other disturbances in the environment. Each sensation produces electrical impulses in nerves leading to the brain, "shocks" which stream into higher nerve centers and cause cell after cell to fire in a series of chain reactions. The nervous system includes one-way "sensory" channels, fibers carrying nothing but messages from sense organs.

The sense organs most remote from your brain are those located in your toes. Fibers originating in these outlying stations carry messages concerning heat, cold, muscle tension, touch, pain. They are joined by more and more fibers from your foot, leg, knee and thigh.

By the time the collected fibers reach the lower part of the spinal cord they form a thick cable. The cable continues to thicken as it climbs and is joined by millions of fibers from other organs of the body on the way up to the brain. This is the great ascending, sensory part of the nervous system. It subjects the brain to constant proddings. Although its lines are less busy during sleep, even then it is occupied with various duties—keeping your heart and lungs going, dreaming, and listening with a somewhat reduced vigilance. The brain relaxes but as long as it is alive, it finds no rest.

The brain's informers are sense organs, sentinels located at strategic points throughout the body. Imbedded in the skin are some 3,000,000 to 4,000,000 structures sensitive to pain, 500,000 touch or pressure detectors, more than 200,000 temperature detectors. These tiny organs—plus the ears, eyes, nose and tongue—are some of your windows to the outside world. Reports about the state of things inside your body come from other built-in sense organs which give rise to sensations of muscular tension, hunger, thirst, nausea. The number of senses is not known exactly. It is certainly more than five, and probably somewhere around twenty.

Each sense has its own fibers which carry its messages exclusively. But the fibers run together in the great ascending cable of sensory messages, and they are sorted and separated into smaller bundles in the brain. Each bundle ends in a different region of the cortex, the brain's outer layer or bark. This gray sheet of cells is the highest center of the nervous system, in two ways. It occupies top position, overspreading the brain like the dome of a cathedral, and it carries out some of the most advanced mental processes.

If the folded and crumpled cortex were spread out flat, it would cover more than two square feet (an area almost as large as the front page of your newspaper). Part of this area is reserved for maps of a special sort. For example, when a person listens to music nerve impulses representing the notes are flashed to a horizontal strip of cortex at the side of his head. Now imagine that the strip is exposed as he listens. Fine wires are placed at many points along the surface and attached to electronic recording equipment. Electrical signals representing different notes arrive at the strip and set up currents in the wires located there. These currents are detected and charted.

Roughly similar experiments have been performed on surgical patients and laboratory animals. The findings reveal that the strip at the side of your head is a kind of natural keyboard. The highest notes you can hear come in at the back end of the strip. The lowest notes come in at the front end. In between, the entire range of notes is represented by a sequence of precisely placed points. The octaves are marked off at regular intervals of about a tenth of an inch from the back to the front of the strip, that is, from high to low notes. Actually, of course, there are two strips—one on each side of the head, for the right and left ears.

Every note you hear has its sites on these strips and sends electrical pulses there. And, conversely, if the site were stimulated artificially—say, by touching them with electrical probes—you would hear that note clearly, even though there was complete silence in the outside world. Theoretically, this offers the possibility of a new kind of subjective music played without instruments of any sort. You simply stimulate the proper points on the "hearing maps" of a person's cortex in the proper order. This doesn't mean that you would have to touch the surface of his brain with a probe. The trick might be done by remote control, with radio waves. In any case, by playing on the cerebral piano you could entertain him with classical music or perhaps a modern symphony composed especially for direct high-fidelity transmission to the cortex.

There is also an area which maps general skin sensations in fine detail. When you stub your right big toe or use it to test the water before you go in swimming, signals travel to the upper end of a strip of cortex running down the left side of your head. Sensations in your left big toe send signals to a corresponding part of a similar strip of the right cortex. Touch your ankles and messages flash to areas just below the toe areas on the strips. Leg signals arrive just below the ankle area and so on down the sides of the cortex, from hips to fingers and from eyes to throat. Together the two strips form a map of the entire skin surface. Again, as in the case of sound, stimulating a point on this brain map produces a definite sensation—usually, a numbness or tingling—at a point on the skin.

The brain has other sensory maps. On the cortex at the back of the head are visual maps, screens made up of a mosaic of nerve cells. Every pattern you see around you, every tree and building and face, produces patterns on these screens as various cells in the mosaic fire. Other sensory fibers lead to the smell areas of the cortex, which are buried deep down in the walls of the chasm between the cerebral hemispheres. Each sense thus has its map on the cortex, its exclusive zone in the highest center of the nervous system. In this way, the brain sorts the information upon which its activities are based.

In nerve messages, as in dot-dash telegraph codes, patterns of pulses stand for the items of information being sent. But the interpretation of nerve signals depends first of all on the place they arrive at. No matter how accurately senses have been coded, no matter how meaningful the signals are, they will be misinterpreted if they arrive at the wrong place. A happy-birthday telegram means just that, even if it should happen to reach the wrong person. But a slip-up in the nervous system is something else again.

Supposing you were listening to fast music—say, the Benny Goodman version of "Sing, Sing, Sing"—and the nerve signals somehow got switched to the wrong line, arriving at the visual areas of the cortex instead of the hearing areas. You'd "see" the music as a mad rush of flashing lights, moving forms, vivid colors. Such mix-ups actually occur, and may result from "crosstalk" between nerve fibers. Crosstalk is familiar to repair men of your local telephone cable, electricity leaks away and you may find yourself listening in on someone else's conversation.

Similar leaks in the nervous system may account for many peculiar sensory disorders. Current escaping from a touch fiber to a nearby sound fiber, for example, might make you hear crashing noises when you bumped your elbow. Somehow certain drugs increase crosstalk among sensory fibers, and nerve injuries may produce the same effect. There is no reason to doubt that a certain amount of crosstalk takes place in the normal nervous system, the nerve signals traveling through neighboring fibers interact in some way. We do not yet know the significance of this effect. But new evidence indicates that crosstalk between fibers of the right and left eyes have something to do with the mechanism whereby we see objects as three-dimensional solids.

Nerves, Muscles, Action

For all its maps, however, the cortex is a good deal more than an atlas. Most of its cells belong to unmapped association areas, where different kinds of sensory information are brought together and related. Fresh sensory evidence from many maps is pooled and compared with remembered evidence. We feel emotions and conceive abstract ideas. As already mentioned, later chapters will present some theories about what happens in the association areas and how they are involved in surgery for mental illness. The following paragraphs deal with the net effects of these processes in directing certain forms of everyday behavior.

In tracing fibers upward from toes to brain, we have been considering only that part of the nerve network concerned with meaningful messages—sensory information arriving from the body and the outside world. It is also the origin, the point of departure, for outgoing messages addressed to muscles throughout the body. Since the messages cause us to move about, the fibers that carry them are called "motor" fibers and they make up the "motor" part of the nervous system. In other words, the brain serves as headquarters for descending motor as well as ascending sensory messages.

Sensory signals flow steadily into the brain; motor signals flow out. Sooner or later—and usually sooner—we go into action. All the centers of the brain exist to help us act more efficiently. Whatever the brain does, whatever problems it must solve, its decisions and conclusions and orders become impulses in the great system of descending motor nerves. These nerves carry messages from the cortex and brainstem and make contact with large motor nerve cells in the spinal cord, completing the last relays of the descending pathways. The pathways lead to muscle tissue. Each motor cell has a signal-transmitting extension which divides into tiny branches, each branch ending on a separate muscle fiber. The average motor nerve cell controls more than a hundred muscle fibers.

Action involves a shift from brain to brawn; from one type of remarkable tissue to another. We rarely put our muscles to the test, except perhaps in the heat of athletic competition or during emergencies. Several summers ago part of the grandstand collapsed during a baseball game in a small Pennsylvania town. A ticket-taker rushed to the scene, saw a young boy pinned under the wreckage, and lifted a large beam so that the child's body could be pulled to safety. The ticket-taker, an average-sized man in his late fifties, had lifted a total weight of more than five hundred pounds. Professional strong men have trained themselves to lift one thousand to fourteen hundred pounds, the weight of two concert-type grand pianos.

Muscles are composed of fibers that pull together in teams. The biceps in your arm contain 600,000 fibers too slender to be seen with the naked eye. Each fiber is a kind of cable made up of thousands of still smaller strands called fibrils. Magnified enormously under the electron microscope, the structure of each fibril appears as a series of fine dark bands, one above the other like the rungs of a ladder. Further study reveals that each fibril, in turn, consists of interlocked filaments—long-chain molecules less than two-millionths of an inch wide.

These ultimate units of muscle are designed to do work, and a single fiber can lift a thousand times its own weight. We speak of muscles of steel, but the comparison fails to do justice to the real thing. Actually, your muscles are more like jelly. They are made up of protoplasm, a slushy semifluid which is three-quarters water. When you are relaxing, fibers and fibrils and filaments form a soft limp mass. The instant you start working, however, they contract and are transformed into a thick, tough elastic substance. Moreover, they can change from jelly to gluey plastic and back again hundreds of times a minute.

The entire nervous system participates, to some degree or other, in the control and coordination of these living fibers. They are always prepared for action. Motor nerve cells in the spinal cord discharge and charge five to ten times a second. This is their idling rate. Enough impulses stream to the muscles to keep them in proper states of tension. They are ready to stretch further or relax at an instant's notice from higher centers. As a matter of fact, certain muscles are continually relaxing and stretching. The brain is continually adjusting and readjusting the tensions of many muscles so that you maintain your posture and balance.

Simply standing up represents an acrobatic feat which is no less remarkable because it is performed automatically. Everyone naturally sways a bit in an upright position, and a failure in balance-controlling centers of the brain would send you sprawling. There is one powerful muscle which, if uncontrolled, would snap your leg back at the knee pressing your calf hard against your thigh. Another muscle would keep your leg stiff as a ramrod. The brain receives messages specifying the tensions of more than two hundred pairs of opposing muscles, every one of which must be properly adjusted to keep you standing.

Things become more complicated during a walk over uneven ground—and even more complicated when you dive from a high board, lower a sail in a storm or ride a surfboard. Every action, however simple, is made up of many individual muscle contractions and large-scale movements. These movements must follow one another at just the right time in just the right order. The brain does the timing. It coordinates all sequences of movements so that we move smoothly and not in a series of jerks. When it comes to pursuing the activities of everyday life, we are thus reasonably sure of ourselves and our positions in the world.

Action involves other centers besides those which smooth our movements and control posture and balance. When you decide to move your foot, the decision is a nerve signal that comes from a particular region of the cortex known as the motor area. This strip of tissue is another map, and runs parallel and next to the strip for general sensation. Different parts of the body have their special sections on the map. The most active, not the biggest, parts rate the largest sections. If regular maps were similarly designed, they would look quite different from the familiar variety. In New York City, for example, Central Park is many times larger than Times Square. But busy Times Square would be a huge section on an "activity" map, while the less hectic park would occupy a much smaller area.

On the motor-cortex map the trunk of your body and shoulders requires relatively little territory. But talking is such a frequent activity that your tongue and lips are considerably bigger than your back, as far as their areas on the motor strip are concerned. Your hands have the largest territory of all. As the most active parts of the entire body they need extra brain space, and provide a striking example of the close interworkings of nerve and muscle. To adjust the positions of the fingers and hands in space—which also means adjusting wrist, arm and shoulder—the brain controls thirty different joints and more than fifty muscles, for each hand.

Exploring the Cortex

The cortex, with its maps and association areas, is one of the most intensively studied tissues in biology. I have a 630-page book which is devoted entirely to a few square inches of the center, the motor strip. No single scientist or group of scientists can understand the whole cortex, and it may take years to explore an area no bigger than a postage stamp. Specializing in research has come in for a great deal of criticism recently, because it produces strange technical jargons and people who find communicating with one another increasingly difficult. But there is no other way to study things as complex as the cortex. The task is an endless one. No scientist expects that the brain of man, or of any other animal sufficiently advanced to be interesting, will ever be fully charted. Attempts have been made to estimate how many possible pathways exist in a cortex containing only a million nerve cells. The number is meaninglessly large. Merely writing it down would fill several volumes—and the human cortex contains ten billion nerve cells, about twice as many as the cortex of the highest ape.

The Old Brain

Since our behavior is so complicated, it is perhaps natural to identify human qualities with the most complicated part of the brain. But there are indications that the cortex, like intelligence itself, may have been overrated. So far we have focused on the "top man" of the cerebral totem pole, and have ignored the chain of sub-brains underneath. The brainstem and the nerve centers associated with it make up the so-called old brain which plays a major role in regulating blood pressure, breathing and many other automatic functions. It is also involved in raw emotions and all primitive drives.

The Brain in Action

What is the relationship between the old brain and the new brain? According to one theory, they are in continual conflict as raw emotion and intellect jockey for control of the body. That seems to be part of the story, but it's far too simple for a full explanation. Emotion and intellect cannot be located and divided quite so easily. The brain functions as a unit. The forces of evolution formed it through the ages. It developed as a servant of the muscles, a way of increasing the chances of survival in a deadly game of hide-and-seek.

In earlier times the game was played by sluggish or awkward creatures. But the interplay of environment and altered genes produced new species with more finesse; hunter and hunted became more clever. Special nerve centers evolved which inhibited gross movements and triggered the precisely timed contractions of small bundles of muscle. The brain was changed from a crude mass-action machine to an instrument capable of marvelously subtle and varied controls. Together with these changes came another which was, if anything, more radical and spectacular—it became possible for higher animals to delay the satisfaction of their wants, to look and wait and think before they leaped.

Something important had happened in the organization of the brain and nervous system. The simplest possible reflex consists of two cells only. A sensory cell carries nerve impulses from a sense organ to the muscles it controls. This two-link chain is capable of fast reactions, the delay being at a minimum because only one cell-to-cell relay is involved. But it is a rigid arrangement. A sensory impulse produces the same reaction as invariably as pushing a button rings an electric bell. Touch a sea anemone, and it collapses instantly like a pricked balloon—and it will collapse in the same way every time you touch it.

The situation changes, however, if an extra cell is placed between the sensory and motor cells. There is a longer nerve chain and a somewhat greater delay. But now the reaction is less predictable; behavior can be more versatile. The added unit is known as an internuncial cell (from "internuncio," an envoy or diplomatic representative of the Pope). It intervenes between sensory stimulus and muscle response, and may introduce a measure of finesse. It may alter routine reflexes on the basis of information it happens to receive from other parts of the nervous system.

You have many such three-link chains in your spinal cord. Usually, your hand jerks away when you touch a hot plate—but not invariably. If you're passing the plate to a guest and it happens to be part of your favorite china set, you may hold on until you put it down. Such restraint would be impossible without spinal internuncial cells. In higher centers like the cortex many, many such cells are placed between incoming sensory and outgoing motor fibers. We take account of a great deal of information from many sources. We analyze, compare, figure out. An abundant supply of internuncial units in the cortex—small cells with short transmitting fibers—provides channels and sufficient delays for all this mulling over.

Consequently, a nerve cell in the brain seldom fires as the result of a signal from only one other cell. Its decision to fire or not to fire is usually based on information from several sources. Usually it must receive all the signals within an extremely brief period, about two tenths of a millisecond (a millisecond, a thousandth of a second, is the unit used to clock many nerve reactions). The cell is so designed that the effects of separate electrical "pushes" must be added together before it discharges.

What you do depends on summations of signals throughout the brain's nerve network. Before eating, you may have to be hungry *and* finish reading that report *and* telephone Mr. Jones. Even when a sufficient number of properly timed signals arrive at a nerve cell, however, it still may not go into action. It delays for about half a millisecond before firing. If during this period the cell receives a "don't fire" signal, it will remain inactive despite the fact that it has previously received its quota of "go ahead and fire" signals. Although your morning's work is done—the report is read, you have telephoned Mr. Jones and you are hungrier than ever—you may wait for the boss to leave first before you go out for luncheon.

We do not know why some electrical impulses prevent a cell from firing, while others act as triggers. But inhibition, which has come to imply something bad and abnormal in human behavior, plays a necessary part in the normal workings of the brain and nervous system. Both inhibiting impulses and internuncial cells permit us to postpone our actions. If a nerve cell's half-millisecond delay period passes without inhibiting signals, it fires. After sending its signal, the cell is "dead" for nearly half a millisecond as it "reloads" or recharges for another burst of impulses.

Evolution has lengthened the time span "between the emotion and the response . . . the desire and the spasm." In us, as in lower animals, the act of consummation is largely a matter of mass automatic movements and the satisfying of basic desires. It is primitive. Once we have food in front of us, reflexes take over. Part of us, the part controlled by the old brain, becomes a robot.

Getting at the food, however, is something else again. For this we need the cortex. A tiger must wait for a while after it sights its victim. Its mouth may water and it may tremble in anticipation. But it does not pounce. Many of its muscles go into action; many more are inhibited. Of the sense impressions coming into its brain, only a few are used and the rest are ignored. The cortex is operating with a terrible efficiency. The tiger remembers the features of the piece of jungle before it, and the habits and escape tactics of former preys. It stalks, perhaps for hours.

Our time scale is different; we can wait years or generations for the attainment of our goals. But in the hushed interval preceding the final leap of a tiger are contained all the tensions that in us produce guided missiles and social planning.

Our abstract ideas are probably conceived in the cortex. It has a great deal to say about what we do and how we do it. But it is mainly a top-level consultant. It does not make final decisions. It helps set up broad policies in close consultation with the old brain. And if the old brain is busy with more important matters of day-to-day operations, the cortex waits its turn. The cortex is a bit of the professor, slightly on the academic side. Left to itself, it would speculate endlessly and have little to do with the real world.

Most of the time the old brain nags the cortex into useful activity. It says: "Hurry! Here's my problem—analyze it and report back within a minute!" The cortex replies: "Now, that's an extremely interesting problem. It may have some implications you haven't explored thoroughly. I must look into them. It reminds me of . . ." "Hurry—within a minute!" interrupts the old brain.

The old brain drives the cortex. Cut some of the fibers running between them, and the cortex idles. Surgeons do this in frontal lobotomy, an operation used for the treatment of certain severe mental disorders. The scalpel severs nerve pathways carrying messages from lower centers to the front parts of the cortex, and produces significant changes in personality. People no longer feel as strongly about getting ahead as they used to. They still seem to know the rules and strategies of their respective games. But they are perfectly content to live without improving their skills or competing with others. The old brain represents whatever it is that makes us desire and care.

Brains of the Future

The human brain has come a long way since the sea anemone, and it is still evolving. Fossil remains show that the brains of horses, whales, deer and other mammals became larger during the course of their evolution. Most investigators expect a similar growth record for man. Adults with swollen brain-cases holding perhaps as much as an extra pound or so of gray matter will look a good deal more like children, because their heads will be much larger in proportion to their bodies. Learning will take longer. A monkey runs when it is two weeks old, while a child requires eighteen to twenty-four months. Professor Haldane predicts that the man of the future "will not speak until he is five years of age and will continue to learn until he is forty."

The most easily educated areas of the brain, the ones whose detailed structure is least determined by heredity, will grow more than any other parts of the nervous system. These areas allow for the greatest variety of activities and ideas, and have expanded most rapidly in the past. Furthermore, certain nerve centers concerned with instinctive, automatic behavior are shrinking in size. As far as we can tell from the changing anatomy of the human brain, the men and women of the distant future will be even more difficult to regiment than they are now. The chances are exceedingly slim that we will freeze into the sort of system evolved by the so-called social insects. Ant brains have about 250 nerve cells, bee brains about 900. We have 13,000,000,000. *Brave New World, 1984* and other novels that picture future races of semi-zombies may make good reading; they are very poor science.

The prospect of bigger brains has led to another dismal prediction. According to one theory, the nervous system is a kind of creeping ivy on the human frame, a parasite that has been growing fat on the rest of the body. It already uses up about 25 per cent of the oxygen you inhale (about twice as much as the oxygen requirements of ape-brain), and may be interfering with the workings of organs much more important as far as survival is concerned. The high incidence of stomach ulcers, high blood pressure and other diseases that flare up during emotional stress is cited to support the notion that our enlarging brains may kill us.

It could happen. But the argument sounds too much like the old tale about the saber-toothed tiger, which is supposed to have become extinct because its teeth grew so long that it could not bite effectively. Actually, it survived for nearly forty million years and when it vanished during the last Ice Age, it could still bite effectively. The Irish Elk whose antlers grew too large for his head and the oyster with shells that curled so much it couldn't open are also evolutionary myths. There is no evidence that any creature has died out because a part of its body was overgrown. If we become extinct, the odds are that the size of our brains will have nothing to do with it.

Three possibilities are open to Homo sapiens. He may be replaced by another species, disappearing altogether or continuing as a "poor relation"—an animal which almost made it but not quite, like the chimpanzee or gorilla. What sort of animal could replace us? It will not be a strange breed of microbe or insect—these organisms have had their day, and evolution never gives a species a second chance. It might be an advanced type of ape or one of the living prosimians, like the large-eyed lemur. But we have no clues.

If an observer had been asked a similar question in the days of the dinosaurs, his guess would almost certainly have been wrong. No one could have foreseen that the species of the future would evolve from the ratlike creatures which quivered with fear as monster reptiles thundered by. The problem is just as difficult today—assuming, of course, that the species exists. If it's to be a new species, guessing is even more difficult. One scientist points out that "there are no truly aerial . . . organisms living and reproducing in air . . . as seaweeds and fishes do in water."

Another possibility is that we may wipe ourselves out. Anyone acquainted with recent work on self-guiding missiles designed to carry atomic warheads cannot keep the prospect out of his mind. About the only positive thing that can be said along such lines is that no other species has been able to accomplish the feat. Nature has always done its own scrapping of species.

Finally, and this has never happened either, we may be the animal that survives—and keeps on evolving. Most long-lived species that are with us today represent evolutionary dead-ends. Some of them found their niches, settled down and have reproduced practically unchanged for as much as 400,000,000 years. The tendency is strong in even the most highly advanced animals. Give a chimpanzee, a few chimpanzees, a peaceful stretch of jungle and plenty of bananas, and it will live happily for the rest of its life. Give a man an environment correspondingly idyllic, say a Garden of Eden, and he will get into trouble. Getting into trouble is our genius and glory as a species.

Of all animals, we are the only ones with brain sufficiently complex to keep us in a constant state of "maladjustment." We are always trying to go to places where we have never been before—the New World, darkest Africa, the North Pole. Our drive toward the planets and the stars is in the grand old tradition. The first creatures that came out of the seas and flapped about experimentally on a primeval shore were great pioneers. But they stopped there, and left further advances to other species.

Our Christopher Columbuses and Daniel Boones keep coming. We keep inventing new devices, new desires for the devices to satisfy and new ways of arousing those desires. We design radar, radio telescopes and other instruments which extend our range of vision by millions of light-years. And we organize the search for new problems into a full-fledged profession. Science with its appalling output of more than a million technical papers a year is a sophisticated method of discovering things too complex for our brains to analyze. Then we build electronic calculating machines, accessory brains, to help us.

Evolution seems to have put an extra energy supply somewhere in the old brain, cortex and their interconnecting fibers. It has been called everything from keeping up with the Joneses (which, of course, means getting ahead of them) to divine discontent. It may be enough to make us the first exception to the iron-clad rule that nature either "freezes" her species, or else discards them entirely.

ROBERT S. DE ROPP

Sick Minds, New Medicines

•

Mental illness takes many forms. A *psychosis* draws a veil between its victim and the outer world, clouds the mind with hallucinations which make purposive action difficult or impossible. The psychotic individual can play no active part in life. He must, for his own good and that of others, be separated from his normal surroundings and cared for in a special hospital until he re-establishes contact with reality.

A *neurotic*, on the other hand, does not lose contact with reality. He can continue his work and deal with most of the situations that confront him. Nonetheless he is sick emotionally and mentally and his sickness colors his waking and possibly also his sleeping hours. Because of it he can never really enjoy his existence. His neurosis hovers over him like the mythological harpy and whatever choice morsel life offers him in the way of pleasure it swoops upon and carries off. It distorts his every feeling and colors his every impression, poisoning with suspicion, fear, guilt, apprehension, envy, or malice the very fountainhead of his existence. The psychiatrists spend much time delving into the subconscious of such a one to discover the old griefs, traumas, repressions, complexes which set this poison flowing. The chemist prefers to leave the complexes alone and to pin his faith on the dictum, "All is chemical." He believes that the sufferings of these hapless neurotics have a chemical basis, that there can be neither guilt, anxiety, depression, nor agitation without some sort of chemical unbalance within the body.

Where should we seek the basis for such unbalance? If we consider the mental and emotional life of man we see that it changes its tone from day to day and from hour to hour. Today he is elated, tomorrow depressed; in the morning an optimist, in the afternoon a pessimist; a lover after lunch, a misanthrope before it. And on what do these ceaseless variations of mood depend? They depend on an endless sequence of minor changes in the outpourings of those glands whose blended secretions make up the chords of man's inner symphony. From pituitary and adrenals, from thyroids and gonads flows the stuff of which man's feelings are created, partially regulated by processes in the brain which, like a conductor struggling through a difficult symphony, does not always produce a very distinguished performance. Neurosis and psychosis alike must be the result of a breakdown in glandular harmony: too much adrenalin here, too little thyroxine there, a shade too much testosterone or too little progesterone, a shortage of ACTH, an insufficiency of cortisone, too little serotonin or perhaps too much. Why should we enmesh ourselves in a tangle of complexes when the root of all evil lies in chemical disharmony? Let us take as our motto the dictum of R. W. Gerard: "There can be no twisted thought without a twisted molecule."

So, from this standpoint, to use a slightly different analogy, we can depict the ever changing moods of man as a more or less continuous spectrum composed of many colors. From hour to hour man's ego, that which he feels to be himself, moves to and fro across this spectrum under the influence of inward and outward events. At one end of the spectrum lies the infrared of melancholia or depression. At the opposite end lies the ultraviolet of mania or extreme agitation. A normal, balanced man remains for the most part in the middle region of the spectrum and strays into the extreme regions only rarely. If he does enter those regions he can, without too much difficulty, remove himself from them. The dark or the frenzied mood passes. The needed chemical adjustments are carried out. Harmony is restored, the inward symphony trips along smoothly again, *allegro ma non troppo*.

In the mentally sick individual, however, this healthy chemical adjustment does not take place. Such a one may become permanently stuck at one end or the other of the psychological spectrum. If stuck at one end he is said to be suffering from depression or melancholia; if stuck at the other he is said to be suffering from agitation or mania. Quite commonly such a sick individual fluctuates between the two extremes in a condition known as a manic-depressive psychosis. Now like a god he strides on the clouds above Olympus, feeling himself to be capable of anything and everything; a few hours later, falling with a crash from the heights, he creeps through the glooms of the infernal regions, feeling lower than a worm. There is, in this case, an obvious effort on the part of the ruling chemical mechanism to correct the unbalance which has arisen among the lesser hormones. The correction, however, is always overdone, so that the mood of such an unfortunate swings from one extreme to the other and his personal symphony fluctuates between a frenzied *presto agitato* and an almost unendurably dreary *largo*.

To treat conditions such as these the physician will seek a remedy among two very different classes of drugs. The patient at one end of the spectrum, overactive, agitated, tense, nerves "frayed" with anxiety, requires a medicament that will soothe and tranquilize. But the patient at the opposite end of the spectrum, whose load of depression is so heavy that he can scarcely raise his head, whose life is an empty, meaningless, valueless void, and whose pale apathetic face gazes indifferently alike at the prizes and penalties offered by this life, obviously needs a very different sort of drug, one which, by opening the dampers that regulate our inner fires, will restore that healthy glow now almost stifled in a cloud of poisonous smoke.

We will consider first the tranquilizing agents, the "ataraxics." What are these drugs whose action is so special that we have to borrow a new Greek word to describe it? The first of the ataraxics is not new at all. It is an extremely ancient remedy and has been used for at least twenty-five hundred years in India by practitioners of a system of medicine known as the Ayur-Veda. This drug, known in India by the name *sarpaganda*, is the powdered root of a small bush belonging to the family Apocynaceae, the Latin name of which is *Rauwolfia serpentina*. In English the plant is commonly referred to as snake root, a practice inviting errors, for this name is also applied to several entirely different drug plants (e.g., *Eryngium aquaticum, Asarum canadensis, Polygala senega*). Confusion can be avoided if one simply refers to the plant as Rauwolfia, a name bestowed upon it by Plumier in honor of Dr. Leonhard Rauwolf, a sixteenth-century German physician who had traveled widely in India collecting medicinal plants.

It is curious indeed that a remedy so ancient and one on which so much excellent research had been carried out by several Indian scientists should have been ignored by Western researchers until the year 1947. This situation resulted, in part at least, from the rather contemptuous attitude which certain chemists and pharmocologists in the West have developed toward both folk remedies and drugs of plant origin, regarding native medicines as the by-products of various old wives' tales and forgetting that we owe some of our most valued drugs (digitalis, ephedrine, and quinine, to name only a few) to just such "old wives' tales."

That the secret of Rauwolfia's potent action was finally brought to light was due to the curiosity of an eminent biochemist, Sir Robert Robinson, and the enterprise of Dr. Emil Schlittler of the Swiss pharmaceutical firm of Ciba, at Basle. Sir Robert was interested in an alkaloid of Rauwolfia called adjmaline and persuaded Dr. Schlittler to prepare this substance from the ground roots of *Rauwolfia serpentina*. After the adjmaline had been crystallized there remained large amounts of muddy, unattractive residue which Schlittler, with that thrift which is the hallmark of every good chemist, refused to discard until he had further explored its make-up. His exploration of this muddy resinous residue proved profitable beyond his wildest dreams, for the pharmacologists to whom he sent this material discovered, on testing it in animals, indications of that curious tranquilizing effect for which the drug has now become justly famous. Spurred on by this report, Dr. Schlittler set out to isolate the chemical substance responsible for this activity.

In September of 1952, just five years after Sir Robert Robinson had presented his request for some adjmaline, three Ciba scientists, Schlittler, Muller, and Bein, finally published an account of their labors. The few grams of shining white crystals they had obtained from the muddy resinous extract of Rauwolfia represented the fruit of a prodigious amount of work. Every crystal was equivalent in activity to more than ten thousand times its weight in the crude drug. "We have long intended," wrote Schlittler and his colleagues, "to isolate the sedative substance of crude Rauwolfia extracts. This hypnotic principle had been examined earlier by Indian authors, but they did not get any further than the crude 'oleoresin fractions.' Starting from these fractions, we have now been able to isolate the carrier of the sedative effect in pure crystalline form." To this crystalline substance they gave the name reserpine.

A few months later Dr. Bein published a second report which revealed that reserpine, besides producing sedation, also lowered the blood pressure of the experimental animals. The drug reduced blood pressure slowly and safely, taking a fairly long period to attain its maximum effect. As high blood pressure is a particularly common ailment in America it is not surprising to find that one enterprising American physician, Dr. Robert W. Wilkins of Boston University, had already given the crude Indian drug a trial. Pure reserpine was not available to him. It had not at that time been isolated. Instead he used tablets of the crude drug imported from India with which he treated more than fifty patients suffering from high blood pressure.

By 1952, Wilkins and his colleagues were able to report progress:

> We have confirmed the clinical reports from India on the mildly hypotensive [blood-pressure lowering] effect of this drug. It has a type of sedative action that we have not observed before. Unlike barbiturates or other standard sedatives, it does not produce grogginess, stupor or lack of coordination. The patients appear to be relaxed, quiet and tranquil.

One of the doctors at a later scientific meeting supplied this statement: "It makes them feel as if they simply don't have a worry in the world."

It was this observation, that the drug not only lowered blood pressure but also relaxed the tensions and anxieties by which high blood pressure is often accompanied, that aroused the interest of psychiatrists. Here, they reflected, might be the drug for which they had so long been seeking. Until the discovery of Rauwolfia no drug available to psychiatrists would really tranquilize the agitated, anxious, restless patients who so often came to them seeking help. The bromides were short-acting and apt to be toxic.

The barbiturates made the patients too sleepy to carry on with their work; chloral and paraldehyde suffered from the same drawbacks.

As soon as the drug did become available a flood of scientific publications poured from the presses; indeed so great was the interest that for a time one rarely opened a medical journal without finding within it at least one article on Rauwolfia.

Reserpine is an extraordinary drug in more ways than one and its mode of action is hard to understand. It acts slowly and takes several weeks to exert its full effects, and these effects when they come follow a definite pattern. Dr. Nathan S. Kline, who has used reserpine extensively on mental patients in Rockland State Hospital, New York, summarizes his findings as follows: When reserpine is given by mouth, very little response is noted for several days. This suggests that the drug is transformed in some way in the body and that the substance which really produces the effect may not be reserpine itself but some product of reserpine. When the effects do begin to be seen they follow a very definite sequence. First comes the *sedative phase*. Patients behave more normally. They become less excited, assaultive, and agitated, appetite improves, and they begin to gain weight. Then, at the end of the first week, the patient enters the *turbulent phase*. During this phase the mental state seems suddenly to worsen. Delusions and hallucinations increase. Patients complain of a sense of strangeness; they do not feel like themselves, do not know what they are going to do next, have no control over their impulses. A physician who does not expect such manifestations may be alarmed at these symptoms and discontinue the use of the drug. Medication, however, should not be reduced until the patient has been able to get "over the hump." The *turbulent phase* may last for two or three weeks or may pass in a few hours. In some patients it was not observed at all. Finally, if all goes well, the patient enters the *integrative phase*, becomes quieter, more cooperative, friendly, and more interested in his environment. Delusions and hallucinations become less marked. This is followed by recognition on the patient's part that he has actually been ill.

The statistics offered by Dr. Kline are impressive. In a series of 150 chronically disturbed psychotics who had failed to improve when treated with electroshock or insulin, 84 per cent showed improvement with reserpine, and 21 per cent of these patients maintained their improvement after medication had been discontinued. Electroconvulsion treatment was largely abandoned. Dr. L. E. Hollister and his colleagues report from California that reserpine produced significant improvement in 98 out of 127 chronic schizophrenics. Drs. Tasher and Chermak (Illinois) report excellent results in 221 chronically ill schizophrenics. The drug has been used with success in the treatment of emotionally disturbed children, in the treatment of skin diseases in which nervous factors were involved, in headache of the tension and migraine type.

Needless to say, this chorus of praise contains a few discordant notes. Dr. J. C. Muller and his co-workers, in an article in the *American Journal of Medicine,* declare that the tranquilizing action of reserpine may on occasion go too far and lead to a depression. High doses of the drug produce definite side effects which may be troublesome. The nose may become stuffy and the patient may experience drowsiness and dizziness. The drug, of course, lowers blood pressure and this effect may have to be watched rather carefully. It is definitely not a medicament to be taken without medical supervision, but the side effects it produces are of minor importance compared with the tremendous benefits it can confer.

The second of the new ataraxics has a history entirely different from that of reserpine. Here there was no romantic background of ancient folk medicine. The remedy originated in the chemical laboratory and its full title, 3-dimethyl-amino-propyl-2-chlorphenothiazine hydrochloride, is awe-inspiring to anyone but a chemist. The Rhone-Poulenc Special Laboratories in France, which developed this valuable drug, gave it the name chlorpromazine, by which it is now generally described. To ensure the greatest possible confusion, however, various trade names were also given to this substance. In the United States it is met with as "Thorazine," in Britain and Canada it goes under the name of "Largactil." It has also been called "R.P.4569" and "Megaphen."[1]

Chlorpromazine, like reserpine, rose to fame with rocketlike velocity. In 1953 it was almost unheard of, in 1955 it was known to every physician in the country and reports on its use were eagerly studied, especially by those responsible for the care of the mentally sick. Dr. Douglas Goldman of Cincinnati published one of the first reports on large-scale use of this medicament in a mental hospital. So encouraging were the effects that, in the words of his colleague, Dr. Fabing, he took a new lease on life.

> The reduction in assaults, the lessened use of restraint, the increased granting of privileges to locked ward patients, the lessened need for repeated electroshock treatment to control explosive behavior, and the beginnings of an improved discharge rate of patients from the hospital all stem from the use of this drug in his hands and parallel the kind of improved state of affairs which Kline reports with reserpine at Rockland.
>
> Goldman likes to tell the story about Willie. Willie was a dishevelled, mute, untidy schizophrenic who had to be spoon fed and who managed to tear off just about all the clothes anyone tried to put on him. Willie received an eight weeks' trial with chlorpromazine but at the end of that time Goldman was not greatly impressed with his improvement. He announced that he was going to withdraw Willie's drug, whereupon an orderly raised a clamor, pleading for its continu-

[1] Reserpine may also be met with under various trade names, such as "Serpasil" (Ciba). "Raunormine" (Renick) is not reserpine but a closely related alkaloid from *Rauwolfia canescens.*

ance, insisting that Willie was much better. He said, "Wait a minute. I'll prove it to you. I'll get Johnny." In a moment he returned with Willie's identical twin. "See, they were both alike two months ago," he said. They stood side by side. Johnny's hair fell in his face, he was soiled, his pants were torn, and he was barefoot. Willie was fully clothed, barbered, shaved, clean and wore shoes. The difference was obvious. Instead of taking a patient off chlorpromazine he put another on.

Goldman also pointed out that chlorpromazine, when used with barbiturates, so greatly enhanced the effectiveness of these drugs that excited patients could be sedated with doses of a barbiturate which would barely have produced somnolence if given by itself.

In the *Journal of the American Medical Association* Dr. Robert Gatskie reported enthusiastically on the value of chlorpromazine in the treatment of emotionally maladjusted children. Such children, rejected by their parents on account of their aggressive, violent, and destructive behavior, were housed in a cottage-type treatment center, 150 of all ages ranging from four to sixteen years. Nine of these children were treated with chlorpromazine and within a week all showed improved behavior. They became calm, cooperative, and more communicative. Their social behavior improved and they became more amenable to cottage supervision. Last but not least, they established rapport with the therapist.

On occasion chlorpromazine exerts an influence that may quite justifiably be called miraculous. An example is given by Dr. L. H. Margolis and coworkers in their paper "Psychopharmacology." The patient on whom the drug was tried was the despair of psychologists, a thirty-four-year-old paranoid schizophrenic who had been treated with insulin coma and electroconvulsions, despite which his condition had remained unchanged. His brain was filled with delusions of grandeur and of persecution and his whole life was spent amid a collection of systematized delusions. As neither insulin nor electroconvulsion had helped him lobotomy was recommended, but his wife refused to consent to the operation. Finally in August, 1954, chlorpromazine was recommended "as a desperation measure in a hopeless case." If it failed lobotomy and/or return to the state hospital were planned.

Soon after treatment with chlorpromazine was started the night staff began to report a subtle change in the patient's attitude. On the fiftieth day of treatment he began to emerge from his world of delusions. By the fifty-seventh day he ceased to show any evidence of mental derangement. He developed an interest in the world of reality, broadened his interests and soon began to lay plans for his future. For the first time since he had been committed to the state hospital he was allowed to go home, where his wife was so impressed by his improvement that she began at once to make plans for his discharge and return to normal life. This patient was fortunate indeed. Only by his wife's refusal of her consent was he saved from a mutilating operation which, while it might have freed him from some of his delusions, would have left him with an irreparably injured brain. Chlorpromazine accomplished all that might have been done by the surgeon's knife *without* doing any damage to those precious lobes on the integrity of which the highest aspects of the personality depend. Some workers have referred to the action of chlorpromazine as "chemical lobotomy." It produces some of the good effects of the operation without the mutilation.

Passing to the opposite end of the psychological spectrum, the gloomy infrared of depression and melancholia, we must now consider remedies for these conditions. Melancholia, to use the time-honored name whose origin goes back to the days of Hippocrates, when the condition was thought to be due to an overproduction of black bile, is a much commoner condition than is generally realized.

"What potions have I drunk of Siren tears, distilled from limbecs foul as hell within," writes Shakespeare who, to judge by certain passages in *Hamlet,* was personally familiar with every aspect of melancholia. The chemist must now try to reach that "foul alembic" and analyze its products, a task which is likely to tax his skill to the utmost. What shall he seek, where shall he seek it? Is melancholia also the result of an error in metabolism which leads to the production of a poison similar to the hypothetical "M" substance in schizophrenia? If so where shall we look for the poison? In blood, in urine, in lymph, in spinal fluid? But perhaps no poison is involved. Perhaps melancholia results simply from an imbalance of those potent hormones on whose quantitative relationships depend the inner harmonies of man's emotional life.

Since we cannot uncover the causes of melancholia our quest for agents that will cure this condition has to be on a strictly trial-and-error basis. In the old days a good deal of reliance was placed on various weird concoctions known collectively as "nerve tonics."

Far more specific in their action on the melancholy humor are various drugs belonging to the amphetamine group whose best-known member is amphetamine itself, more familiar to the public under its trade name of "Benzedrine." "Benzedrine" acts directly on the central nervous system. It stimulates, cheers, elevates, and enlivens. Its relative, Dexedrine, is even more active in this respect. Considerable studies have been carried out on this drug, especially on its use under war conditions to combat the fatigue that results from prolonged strain or effort. Reifenstein and his coworkers reported progress of a depressed individual under the influence of this drug. He was, at the outset, "hopeless in his outlook and lacking ambition." An hour after taking the drug he became talkative and three hours later was feeling much better. On the fourth day he remarked that he was very happy, cheerful, and alert. On the sixth he was jovial, by the tenth he was singing and appeared to have reached approximately his normal state. The improvement, however, was not maintained and by the fourteenth day the depression began to return.

The same authors obtained favorable results in a case of catatonic schizophrenia in which the prevailing symptoms were dullness, listlessness, inactivity, and passivity.

It appears from this that these "analeptic" drugs, as they are called, do have some value in the treatment of depressed states, especially the condition called narcolepsy, in which the patient keeps falling asleep at inappropriate moments. But both "Benzedrine" and its close relative "Dexedrine" are apt to produce annoying side effects. For one thing they reduce appetite, so much so that "Dexedrine" is incorporated into several varieties of reducing pills. For another they tend to overstimulate the nervous system in such a way that sleep becomes difficult or impossible. On this account their use in the treatment of depression has been limited.

More recently pipradrol, a close relative of "Frenquel," produced by its makers (Wm. S. Merrell Co.) under the name of "Meratran," has gained some fame as an "anti-melancholic." Dr. Howard Fabing found it a valuable drug, for it does not, as do the amphetamines, seriously reduce appetite or interfere with natural sleep. "On occasion the response of patients with reactive depression is sudden and dramatic, in that they note an elevation of mood and a quickening of their retarded psychomotor state within two hours after ingesting the first tablet."

That curious nervous disorder "narcolepsy," which may in some ways be allied to melancholia, also responds to medication with "Meratran." The disorder can be well illustrated by one of Dr. Fabing's cases. A housewife began at the age of twenty to have attacks of narcoleptic sleep. She slept in the car, she slept in the cinema, she slept through the sermon, she slept while watching television. To keep awake she had to keep on the move, but even when on the move she sometimes became lost in a cataleptic trancelike state. So profound were these states that she would often cook a whole meal in a trance. Regular doses of "Meratran" relieved her of these symptoms but if she stopped taking the drug the symptoms recurred within forty-eight hours.

"Meratran" appears to be useful only in states of pure depression. Where depression is mixed with anxiety it tends to make the anxiety more severe. As Fabing puts it, "The manic patient becomes more excited, the deluded patient becomes more actively paranoid, an obsessive patient becomes more obsessive, an anxious patient becomes more anxious, and an agitated patient becomes more agitated." In short, to quote from Dr. W. Begg's article in the *British Medical Journal,* "The chief drawback to this drug's therapeutic usefulness is its tendency to exacerbate pre-existing anxiety."

On the whole it must be admitted that the ideal drug for the treatment of melancholia seems not yet to have been discovered. It is a hard problem for the pharmacologist, for one cannot produce melancholia in experimental animals. What we really need is a naturally melancholic guinea pig, the depth of whose gloom can be measured by some means or other. We could then try, by chemical agents, to restore its *joie de vivre*. But until someone devises such a beast the quest for the perfect anti-melancholic will be almost as problematical as was the hunting of the Snark.

We can now consider the impact which some of these newly discovered drugs have made on that complex, costly, and prolonged procedure loosely referred to as psychoanalysis. Behind the analyst, says Freud, stands the man with a syringe. Shall we now put all our faith in the syringe and forget about the analyst?

It is still far too early to answer this question. We can say, however, that the task of the therapist may be eased if he makes intelligent use of some of the chemical agents now available. Psychoanalysis is a procedure frequently rendered impossible by the inner fears of the patients which persistently "block" those very memories and damaging experiences from rising to the surface from the depths of the subconscious. In this way the health-giving cleansing or "catharsis" is prevented. It is this blocking which can to some extent be overcome by the judicious use of drugs, particularly such barbiturates as thiopental ("Pentathol"), which, for reasons known only to journalists, has been frequently referred to in the press as "truth serum." The drug is injected intravenously and, as it begins to take its effect, the tense, anxious, uncommunicative patient becomes more or less unguarded, receptive, friendly, and expansive. Unfortunately, this communicative stage lasts for a rather short time. As the action of the barbiturate continues the patient becomes increasingly drowsy and is apt to fall asleep on the analyst's couch, which makes the procedure highly unprofitable for the patient.

Attempting to overcome this difficulty, Drs. Rothman and Seward combined the soporific "Pentathol" with a stimulant of the "Benzedrine" type called methamphetamine, administering both drugs intravenously. This procedure decreased the excessive tension of the patient and at the same time promoted a state of alertness, spontaneity, and well-being. The isolation of the patient was broken down, a gate opened in the wall of fear and tension with which he had formerly been surrounded, wide enough for the analyst to squeeze through and establish contact. Sixteen patients formerly unanalyzable were thus rendered accessible for the first time.

Another chemical key that has been used to unlock the closed rooms of the mind is LSD. It seems strange that this potent drug, whose effect so closely resembles the symptoms of schizophrenia, should prove of value in the psychotherapy of neuroses. Dr. R. A. Sandison and his colleagues have tried it, however, and report favorably on its effects when used under proper conditions:

> Our clinical impressions have convinced us that LSD, *when used as an adjunct to skilled psychotherapy,* is of the greatest value in the obsessional and anxiety groups accompanied by mental tension. *We cannot emphasize too strongly, however, that the drug does not fall into the group of "physical" treatments and that it should be used only by experienced psychotherapists and their assistants.*

Before we leave the subject of the sick minds mention should be made of that much-misunderstood ailment, epilepsy. This illness, so dramatic in its manifestations, has occupied the attention of physicians from the earliest times. In the days of Hippocrates it was known as the "Sacred Disease," a concept which aroused the scorn of the Father of Medicine who, with his usual common sense, rejected the idea "that the body of man can be polluted by a god." He wrote a treatise on epilepsy and announced, with an insight surprising for the times, that "its origin, like that of other diseases, lies in heredity." By the Jews, however, it was regarded as a form of demonic possession as may be seen from the well-known passage in the Gospels: "And lo, a spirit taketh him, and he suddenly crieth out, and it teareth him that he foameth again, and bruising him hardly departeth from him."

It is in connection with epilepsy that that wonderful instrument, the electroencephalograph, has given us so much information. All the outward symptoms of epilepsy are the direct results of an electrical storm in the brain. The storm begins with violent electrical discharges from a small group of neurones. This violence, like panic in a densely packed crowd, spreads to the other neurones until in a few seconds the whole great mass of the cortex is discharging in unison. These massive discharges, registered by the pen of the electroencephalograph, are so distinctive that the veriest amateur can spot them. Each kind of epilepsy shows it own kind of disturbed brain wave. The three-per-second "dome and spike" of "petit mal" are entirely different from the eight-per-second spikes of "grand mal" which, in turn, differ from the slow waves of the psychomotor seizure. Oddly enough these abnormal brain waves may occur in people who have never had an epileptic fit. They have, however, a tendency to the disease and, if exposed to certain stimuli, such as a flickering light flashing at a critical rate per second, may develop the outward symptoms of epilepsy.

Few ailments have yielded more dramatically to the combined attack of the modern chemist and pharmacologist than has epilepsy. Research on the disease was made possible by the discovery that convulsions typical of epilepsy could be induced in cats by passing an electric current through their heads. Here was a tool which could be used for the mass screening of chemical substances for anti-convulsive activity. It was seized upon by Merritt and Putnam of Parke, Davis, who tested seven hundred chemicals for their ability to prevent such artificially induced fits, and emerged triumphantly with diphenylhydantoin ("Dilantin"). "Dilantin" differs from the bromides and such barbiturates as phenobarbital, both of which have been used in the treatment of epilepsy, in not rendering the patient drowsy. It appears to act by preventing the spread through the brain of that "electrical storm" of which the convulsions and unconsciousness are the outward and visible signs. "Dilantin" is effective against "grand mal" and psychomotor epilepsy. Complete relief from seizures is generally experienced by 60 to 65 per cent of patients suffering from "grand mal" and in 20 per cent the number and severity of convulsions are reduced. For those afflicted with "petit mal" another drug, trimethadoine ("Tridione"), is available. It will generally keep the patient completely free from seizures. In addition to these two agents several other anti-convulsants are on the market. Their names are legion: "Mesantoin," "Mysolin," "Miltonin," "Hibicon," "Diamox," "Paradione," "Phenurone," "Gemonil," "Peganone." If one proves ineffectual the physician can always try another. In this particular disease he has a remarkably wide choice of remedies.

Unfortunately for the epileptic, public education has not kept pace with these triumphs of the pharmacologist. The old horror which, in the past, was associated with epilepsy is still far too prevalent today and the epileptic suffers more from public ostracism than he does from his illness. In actual fact even "grand mal" epilepsy need not interfere too seriously with the life of the individual who suffers from it. Epileptics are frequently perfectly normal intellectually. They may be outstanding. Dostoevsky and Julius Caesar both suffered from the disease. With modern medication it is generally possible to prevent the development of convulsions. Even when they cannot be prevented there is no reason why one who suffers them should be treated as a leper. His ailment actually, except in extreme cases, is no more serious than migraine or dysmenorrhea. Obviously one prone to epileptic seizures should not work with dangerous machinery or drive a car, but otherwise there is no reason why he should not perform a useful function in society. If he happens to develop a fit it is merely necessary to loosen his clothing, prevent him from hurting himself, and put a gag in his mouth to stop him from biting his tongue. Horror and disgust will not help him and are not called for. There are about 1,000,000 epileptics in the United States, the majority of whom can perform a useful function if society will let them and abandon its rather medieval attitude toward this disease.

Editors' Note: It is impossible to tell what new drugs will develop from current research or how valuable they will be. In a note appended to this article, the author has written as follows. "An immense amount of research has been devoted to discovering other substances like iproniazid ('Marsilid') which Dr. Nathan

Kline defined as a psychic energizer. Substances in this group have one thing in common. They inhibit the action of an enzyme called monoamide oxidase which plays a role in the inactivation of adrenalin. Substances having this action are often quite active anti-depressants and a number of new ones have now appeared on the market under such trade names as 'Catron,' 'Nardil,' and 'Niamid.' 'Deaner,' another new anti-depressant, belongs in a somewhat different chemical category."

JOHN LIVINGSTON LOWES

Imagination Creatrix

•

I

Every great imaginative conception is a vortex into which everything under the sun may be swept. "All other men's worlds," wrote Coleridge once, "are the poet's chaos." In that regard "The Ancient Mariner" is one with the noble army of imaginative masterpieces of all time. Oral traditions—homely, fantastic, barbaric, disconnected—which had ebbed and flowed across the planet in its unlettered days, were gathered up into that marvel of constructive genius, the plot of the *Odyssey,* and out of "a tissue of old *märchen*" was fashioned a unity palpable as flesh and blood and universal as the sea itself. Well-nigh all the encyclopedic erudition of the Middle Ages was forged and welded, in the white heat of an indomitable will, into the steel-knot structure of the *Divine Comedy.* There are not in the world, I suppose, more appalling masses of raw fact than would stare us in the face could we once, through some supersubtle chemistry, resolve that superb, organic unity into its primal elements. It so happens that for the last twenty-odd years I have been more or less occupied with Chaucer. I have tracked him, as I have trailed Coleridge, into almost every section of eight floors of a great library. It is a perpetual adventure among uncharted Ophirs and Golcondas to read after him—or Coleridge. And every conceivable sort of thing which Chaucer knew went into his alembic. It went in *x*—a waif of travel-lore from the mysterious Orient, a curious bit of primitive psychiatry, a racy morsel from Jerome against Jovinian, alchemy, astrology, medicine, geomancy, physiognomy, Heaven only knows what not, all vivid with the relish of the reading—it went in stark fact, "nude and crude," and it came out pure Chaucer. The results are as different from "The Ancient Mariner" as an English post-road from spectre-haunted seas. But the basic operations which produced them (and on this point I may venture to speak from first-hand knowledge) are essentially the same.

As for the years of "industrious and select reading, steady observation, insight into all seemly and generous arts and affairs" which were distilled into the magnificent romance of the thunder-scarred yet dauntless Rebel, voyaging through Chaos and old Night to shatter Cosmos pendent from the battlements of living sapphire like a star—as for those serried hosts of facts caught up into the cosmic sweep of Milton's grandly poised design, it were bootless to attempt to sum up in a sentence here the opulence which countless tomes of learned comment have been unable to exhaust. And what (in apostolic phrase) shall I more say? For the time would fail me to tell of the *Æneid,* and the *Orlando Furioso,* and the *Faërie Queene,* and *Don Juan,* and even *Endymion,* let alone the cloud of other witnesses. The notion that the creative imagination, especially in its highest exercise, has little or nothing to do with facts is one of the *pseudodoxia epidemica* which die hard.

For the imagination never operates in a vacuum. Its stuff is always fact of some order, somehow experienced; its product is that fact transmuted. I am not forgetting that facts may swamp imagination, and remain unassimilated and untransformed. And I know, too, that this sometimes happens even with the masters. For some of the greatest poets, partly by virtue of their very greatness, have had, like Faust, two natures struggling within them. They have possessed at once the instincts of the scholar and the instincts of the artist, and it is precisely with regard to facts that these instincts perilously clash. Even Dante and Milton and Goethe sometimes clog their powerful streams with the accumulations of the scholar who shared bed and board with the poet in their mortal frames. "The Professor still lurks in your anatomy"—*Dir steckt der Doktor noch im Leib*—says Mephistopheles to Faust. But when, as in "The Ancient Mariner," the stuff that Professors and Doctors are made of has been distilled into quintessential poetry, then the passing miracle of creation has been performed.

II

But "creation," like "creative," is one of those hypnotic words which are prone to cast a spell upon the understanding and dissolve our thinking into haze. And out of this nebulous state of the intellect springs a strange but widely prevalent idea. The shaping spirit of imagination sits aloof, like God as he is commonly conceived, creating in some thaumaturgic fashion out of nothing its visionary world. That and that only is deemed to be "originality"—that, and not the imperial moulding of old matter into imperishably new forms. The ways of creation are wrapt in mystery; we may only marvel, and bow the head.

Now it is true beyond possible gainsaying that the operations which we call creative leave us in the end confronting mystery. But that is the fated terminus of all our quests. And it is chiefly through a deep-rooted reluctance to retrace, so far as they are legible, the footsteps of the creative faculty that the power is often thought of as abnormal, or at best a splendid aberration. I know full well that this reluctance springs, with most of us, from the staunch conviction that to follow the evolution of a thing of beauty is to shatter its integrity and irretrievably to mar its charm. But there are those of us who cherish the invincible belief that the glory of poetry will gain, not lose, through a recognition of the fact that the imagination works its wonders through the exercise, in the main, of normal and intelligible powers. To establish that, without blinking the ultimate mystery of genius, is to bring the workings of the shaping spirit in the sphere of art within the circle of the great moulding forces through which, in science and affairs and poetry alike, there emerges from chaotic multiplicity a unified and ordered world.

Creative genius, in plainer terms, works through processes which are common to our kind, but these processes are superlatively enhanced. The subliminal agencies are endowed with an extraordinary potency; the faculty which conceives and executes operates with sovereign power; and the two blend in untrammelled interplay. There is always in genius, I imagine, the element which Goethe, who knew whereof he spoke, was wont to designate as "the Dæmonic." But in genius of the highest order that sudden, incalculable, and puissant energy which pours up from the hidden depths is controlled by a will which serves a vision—the vision which sees in chaos the potentiality of Form.

III

"The imagination," said Coleridge once, recalling a noble phrase from Jeremy Taylor's *Via Pacis*, "*. . . sees all things in one*." It sees the Free Life—the endless flux of the unfathomed sea of facts and images—but it sees also the controlling Form. And when it acts on what it sees, through the long patience of the will the flux itself is transformed and fixed in the clarity of a realized design. For there enter into imaginative creation three factors which reciprocally interplay: the Well, and the Vision, and the Will. Without the Vision, the chaos of elements remains a chaos, and the Form sleeps forever in the vast chambers of unborn designs. Yet in *that* chaos only could creative Vision ever see *this* Form. Nor without the cooperant Will, obedient to the Vision, may the pattern perceived in the huddle attain objective reality. Yet manifold though the ways of the creative faculty may be, the upshot is one: from the empire of chaos a new tract of cosmos has been retrieved; a nebula has been compacted—it may be!—into a star.

Yet no more than the lesser are these larger factors of the creative process—the storing of the Well, the Vision, and the concurrent operation of the Will—the monopoly of poetry. Through their conjunction the imagination in the field of science, for example, is slowly drawing the immense confusion of phenomena within the unfolding conception of an ordered universe. And its operations are essentially the same. For years, through intense and unremitting observation, Darwin had been accumulating masses of facts which pointed to a momentous conclusion. But they pointed through a maze of baffling inconsistencies. Then all at once the flash of vision came. "I can remember," he tells us in that precious fragment of an autobiography—"I can remember the very spot in the road, whilst in my carriage, when to my joy the solution occurred to me." And then, and only then, with the infinite toil of exposition, was slowly framed from the obdurate facts the great statement of the theory of evolution. The leap of the imagination, in a garden at Woolsthorpe on a day in 1665, from the fall of an apple to an architectonic conception cosmic in its scope and grandeur is one of the dramatic moments in the history of human thought. But in that pregnant moment there flashed together the profound and daring observations and conjectures of a long period of years; and upon the instant of illumination followed other years of rigorous and protracted labour, before the *Principia* appeared. Once more there was the long, slow storing of the Well; once more the flash of amazing vision through a fortuitous suggestion; once more the exacting task of translating the vision into actuality. And those are essentially the stages which Poincaré observed and graphically recorded in his "Mathematical Discovery." And that chapter reads like an exposition of the creative processes through which "The Ancient Mariner" came to be. With the inevitable and obvious differences we are not here concerned. But it is of the utmost moment to more than poetry that instead of regarding the imagination as a bright but ineffectual faculty with which in some esoteric fashion poets and their kind are specially endowed, we recognize the essential oneness of its function and its ways with all the creative endeavours through which human brains, with dogged persistence, strive to discover and realize order in a chaotic world.

For the Road to Xanadu is the road of the human spirit, and the imagination voyaging through chaos and reducing it to clarity and order is the symbol of all the quests which lend glory to our dust. And the goal of the shaping spirit which hovers in the *poet's* brain is the clarity and order of pure beauty. Nothing is alien to its transforming touch. "Far or forgot to (it) is near; Shadow and sunlight are the same." Things fantastic as the dicing of spectres on skeleton-barks, and ugly as the slimy spawn of rotting seas, and strange as a star astray within the moon's bright tip, blend in its vision into patterns of new-created beauty, *herrlich, wie am ersten Tag*. Yet the pieces that compose the pattern are not new. In the world of the shaping spirit, save for its patterns, there is nothing new that was not old. For the work of the creators is the mastery and transmutation and reordering into shapes of beauty of the given universe within us and without us. The shapes thus wrought are not that universe; they are "carved with figures strange and sweet, All made out of the carver's brain." Yet in that brain the elements and shattered fragments of the figures already lie, and what the carver-creator sees, implicit in the fragments, is the unique and lovely Form.